ANALOG ELECTRONICS:
Devices, Circuits, and Techniques

GERALD E. WILLIAMS
ITT Technical Institute

West Publishing Company

Minneapolis/St. Paul ■ New York ■ Los Angeles ■ San Francisco

To Patty, my wife and
my life's greatest blessing
for 38 years (so far)

Production: The Book Company
Cover Photograph: Doug Armand/Tony Stone Worldwide
Composition: Omegatype Typography, Inc.
Design: Wendy LaChance/By Design
Art: Gerald E. Williams
Production, Prepress, Printing, and Binding by West Publishing Company

WEST'S COMMITMENT TO THE ENVIRONMENT
In 1906, West Publishing Company began recycling materials left over from the production of books. This began a tradition of efficient and responsible use of resources. Today, up to 95 percent of our legal books and 70 percent of our college and school texts are printed on recycled, acid-free stock. West also recycles nearly 22 million pounds of scrap paper annually—the equivalent of 181,717 trees. Since the 1960s, West has devised ways to capture and recycle waste inks, solvents, oils, and vapors created in the printing process. We also recycle plastics of all kinds, wood, glass, corrugated cardboard, and batteries, and have eliminated the use of Styrofoam book packaging. We at West are proud of the longevity and the scope of our commitment to the environment. Printing and binding by West Publishing Company

 TEXT IS PRINTED ON 10% POST CONSUMER RECYCLED PAPER  PRINTED WITH SOY INK

British Library Cataloguing-in-Publication Data. A catalogue record for this book is available from the British Library.

Copyright © 1996   West Publishing Company
                    610 Opperman Drive
                    P.O. Box 64526
                    St. Paul, MN 55164-0526

Printed in the United States of America

03 02 01 00 99 98 97 96   8 7 6 5 4 3 2 1 0

Library of Congress Cataloging-in-Publication Data

Williams, Gerald Earl,   1931–
    Analog electronics : devices, circuits, and techniques / Gerald E.
    Williams.
        p.   cm.
    Includes index.
    ISBN 0-314-04553-8
    1. Analog electronic systems.   2. Electronic circuits.   I. Title.
TK7867.W53   1996
621.3815--dc20   94-49430
                                                                   CIP

Preface

This book is designed for the practical technician, someone who must understand how to solve real-world problems. The text is intended for students at community colleges and public and private technical schools who plan to use their knowledge in the workplace. It is written in a near-conversational style and written to communicate ideas in a clear and concise way.

The book is highly structured as a learning continuum, moving from the simple to the more complex, with careful attention to smooth bridges from each concept to the next more complex related concept. There are over 1,000 carefully-designed drawings to ensure that each concept is fully understood.

Nearly all circuit discussions are linked together by the common theme of feedback, without which most analog circuits could not exist. Most modern texts are primarily a collection of circuits with independent discussions of each, as though they had little or nothing in common. This text pulls usually fragmented ideas and circuit concepts together using feedback, a property common to nearly all analog circuits.

Organization

There are a number of specific learning tools that have been tested and revised in the classroom over many years.

- Each chapter begins with a section called **In Brief,** where each of the most important concepts of the chapter are presented in a nutshell, using text, drawings, and associated equations. This section serves as a condensed version of the chapter, to be used as both a preview and a review.

- Also at the beginning of the chapter, specific **Performance Objectives** are provided to help guide the reader.

- Each new technical term is highlighted and briefly defined the first time it is used in the text, and a **Glossary** provides a more complete definition of each new term.

- Most chapters contain a practical real-world **Troubleshooting** section that uses non-invasive, live tests to locate defective components. Parts need not be removed from the circuit board unless and until the student is certain the part is defective. The tests are based on the latest industry troubleshooting practices and are fast and certain.

- A verbal **Summary** is frequently provided at the end of a chapter to help student retention.

- A unique section, the **Digital Connection,** provides some insights into cases where analog and digital circuits are combined into hybrid circuits. This section also provides a brief discussion of common digital circuits that are parts of modern analog integrated circuits. If the student has already taken a digital course, these can be used as a review and can help the student understand how the digital circuit fits into the analog integrated circuit. The digital circuit explanations are presented from the perspective of *how they work as part of an otherwise analog integrated circuit*. A few unusual digital applications, not normally found in digital books, are also included.

- The subject of basic transistor analysis has been streamlined by using two universal analysis flow charts, the **Worksheets,** which provide a consistent format for transistor circuit analysis. These are introduced in Chapter 5. They make it easy to learn the analysis process, and emphasize the common elements in transistor circuits, while pointing out those specific minor circuit differences, which yield quite different circuit characteristics for the various configurations. The analysis examples and practice problems in Chapter 5 serve to reinforce the transistor theory presented in the previous chapters. The flow chart approach is also helpful if you want to use one of the computer circuit analysis or design programs such as Workbench or SPICE, or if you wish to write an analysis program. The worksheets are the result of many years of classroom testing and they have proved to facilitate learning and retention while using less course time and textbook space. They are an efficient learning tool.

- The mathematics level has been kept to the lowest level consistent with proper understanding of the material. Special **Boxes** contain more advanced mathematical analysis of important concepts. Those students with strong mathematical backgrounds can use the special boxes to augment their learning. The math boxes are not essential, but can be used to supplement and provide extra depth for students who can profit from them.

Greater efficiency in learning basic transistor circuits based on the worksheet method allows far more book space and course time to be devoted to the many modern analog integrated circuits currently used in industry. This text goes well beyond the usual operational amplifier, 555 timer, etc., to include a wide variety of modern integrated circuits. The text provides an unusually comprehensive coverage of comparators, which form the basic element in many integrated circuit timers, voltage controlled oscillators, sound samplers, analog memory systems, modulators, mixers, multipliers, *FM* demodulators, analog-to-digital converters, and a number of other circuits covered in the text.

The Information Highway

The information highway is already a reality, and it is important for technicians to know about the latest modulators, demodulators, analog memories and so on that are an essential part of the technology. Nature is analog and all known transmission media are analog, even though they may carry digital information. Much of the information highway is, of course, digital, but once the information is placed on the highway, the digital information must travel an analog highway.

There was a time, not very long ago, when modulation and demodulation and associated techniques were of interest to only the people in the communications industry, but now these techniques and concepts have become an essential part of both digital and analog communications systems. Once you move beyond the desktop computer to a modem, fiber optic line, radio or other transmission medium, new side-bands appear and analog activities come into play whether you want them to or not. Even the process of sampling an analog waveform for conversion into digital data is a form of amplitude modulation which produces sidebands. These sidebands influence sampling rates and other factors.

This text provides the essential ideas and modern techniques required to use the information highway. Older modulators and other older communication techniques have been omitted, because they remain of interest only to a small group of communications specialists.

A modest coverage of transmission lines is also included. Transmission-line theory is essential to any kind of data transmission, and also has other applications to the study of analog electronics. For example, the termination resistor pack at the end of a computer disk drive cable may serve as a pull-up resistor, but must be there to deal with the analog properties of the cable.

This text has profited greatly from my consulting practice, which has kept me well informed about what circuits and devices are actually being used in industry. My specialty is new product development, so my industry knowledge is state-of-the-art. Twenty-five years of teaching has allowed me to test and develop the processes and pedagogy in the text, and to adjust to changing student reading levels and mathematical abilities.

This text may be a little different from what you have been used to, but you will find that it is a powerful learning tool, and easy to adjust to.

Learning Aids

The LAB MANUAL, Experiments in Analog Electronics, written by Wayne M. Hope, provides:

- 26 experiments, two for each text chapter
- 8-fi ˇ 11" shrink-wrapped three-hole punched looseleaf format (the student provides the binder) saves money and allows easy hand-in and return of exercises

- summarized data sheets can be handed in while student keeps original measurement data
- simplified component requirements with device specifications included
- 20 of the 26 labs were developed for easy use with Electronics Workbench™ or similar simulation programs as well as traditional hands-on lab stations
- An Instructor's Guide, containing answers to all data sheets and questions is available

The STUDENT STUDY GUIDE, written by the author of this text, provides:

- a quick review of critical ideas
- a self-test for each critical review topic
- a troubleshooting reference guide
- blank transistor circuit analysis worksheets for use in working the problems in the text.

In addition to its usefulness in coursework, the Study Guide will prove to be a valuable reference source that can be used in the workplace. The INSTRUCTOR'S MANUAL, also written by the textbook author, includes:

- course outline
- answers to the end-of-chapter problems, with comments
- an introduction to powerful spreadsheet programming, for those instructors who prefer to have students develop their own analysis programs
- instructions on making quick and inexpensive custom circuit boards, using a photocopier and common overhead transparency film
- a foolproof audio amplifier/power supply laboratory starter project, with circuit board layout and assembly drawings
- blank transistor circuit analysis worksheets, which can be copied for student examinations and other uses

The TEST BANK, written by Richard Burchell, provides 200 questions, organized by chapter.

COLOR OVERHEAD TRANSPARENCIES are available for 50 text illustrations. Critical parts of In Brief sections are the core of this teaching aid, and provide a framework for organizing lectures. Other drawings are included because they are difficult or time-consuming to draw. In addition, 75 MASTERS in black and white are available free of cost to adopters on request.

Acknowledgments

I would like to thank the following people, who all contributed so much to this book:

- my wife, Patty Williams; her contributions, if listed individually, would take up an entire page

- my son Geoffrey Williams, who created the fractal scene in Chapter 1
- my son Kelly Williams, for his advice on the artwork
- Richard Burchell, who created the Test Bank and for his excellent technical accuracy checking
- Phillip Denham, for helping me make more time to work on this project
- Wayne M. Hope, for developing an outstanding laboratory manual
- Ron Johnson, for his excellent photography and for contributing the interest boxes
- Habib Karaky, for his lab work and circuit testing
- Donald Lillard, for developing and testing the Norton op-amp function generator in Chapter 10
- Alan W. Lowe, who taught me most of what I know about publishing
- Sean Twigg, for his excellent computer work, circuit testing, and work on the glossary.

My thanks to a small army of graphic artists, all of whom went the extra mile:

- Earl J. Averill, with special thanks for developing drawing specifications
- Tom Beale
- Randy Lee Steehler
- Derrel West, with special thanks for developing the standard primitives for the drawings

I also wish to thank the reviewers, from whom I shamelessly stole a multitude of good ideas. Each and every member of this group provided ideas and constructive criticism that I used to make the book better. They could not have been nicer about deadlines. I am in their debt.

- Melvin E. Duvall, Sacramento City College
- John E. Ephraim, East Tennessee State University
- Douglas F. Fuller, Humber College
- John Hamilton, Spartan School of Aeronautics
- Stephen Harsany, Mt. San Antonio College
- David Hata, Portland Community College
- Bruce Koller, Diablo Valley College
- Clay Laster, San Antonio College
- T. Wayne Lee, Northern Alberta Institute of Technology
- Bruce Sargent, Middlesex Community College
- Gerald Schickman, Miami-Dade Community College
- James J. Schreiber, DeVry Institute of Technology—Phoenix

The efforts of the West Editorial group are greatly appreciated:

- Christopher Conty, probably the best editor in the business. Without his vision and support this book would be much less than it is.
- Elizabeth Riedel, who made it all work, kept the loose ends tied up, and served as damage control expert when things went wrong.
- Tamborah Moore, my production editor, who managed the production details and paid the bills.
- The members of the West Publishing staff who worked on the book, but whom I didn't get to know personally

The Book Company saw the book through its design and production. Thanks to:

- George Calmenson, who put all the pieces together and kept them from falling through the cracks
- Wendy LaChance, who designed the book and was the art editor

Brief Contents

Contents

CHAPTER 5 Practical Transistor Circuit Analysis 227

CHAPTER 1

Analog Electronics: The Quiet Revolution

Objectives

Upon completion of this chapter, you should be able to:

1 Identify the only pure waveform.

2 Explain how waveforms are created by adding sine waves.

3 Name each of the archetype waveforms.

4 Describe how waveshaping is usually accomplished.

5 Explain attack and decay envelopes.

6 Explain the difference between **ac** and **dc** waveforms.

7 Understand what happens to a **dc** component when signals are coupled by transformers and capacitors.

8 Explain the problem of unwanted waveforms.

9 Describe the composition of an analog rectangular waveform.

10 Define the term *resolution*.

11 Identify the characteristics of a sound that allow you to identify the musical instrument you are hearing.

12 Explain how a 20-MHz square-wave signal can interfere with the 88- to 108-MHz FM band.

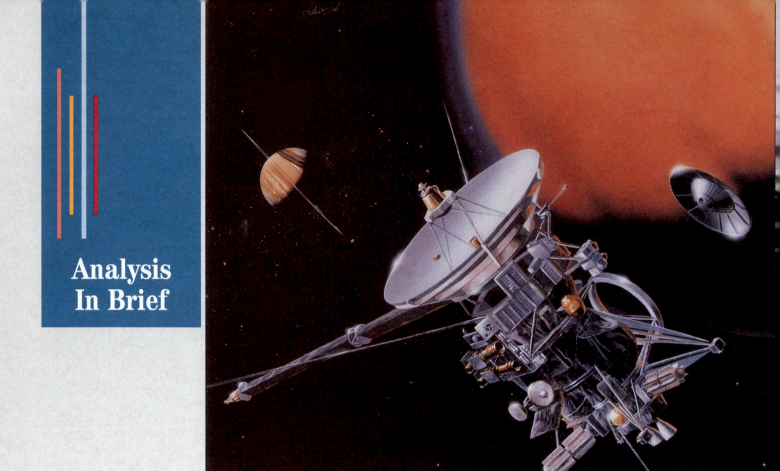

Saturn orbiter and Titan probe (Source: NASA)

Analog Electronics in Brief

ANALOG/DIGITAL PERSPECTIVE

The information superhighway already exists. There are hundreds of satellites in space and a worldwide telephone network. Databases for every conceivable subject are rapidly being created. Fiber-optic cable networks are rapidly making thousands of new communications channels available. The information superhighway will carry all kinds of information. The data transmitted may be either analog or digital, but the superhighway itself is—and must always be—natural and analog. Analog electronics must play a vital part.

The Analog Perception

The public perception of analog electronics is that it is old-fashioned, outdated, and soon to be replaced by digital technology. Analog electronics is just something stodgy old professors make electronics students study because that is what they learned when they were young.

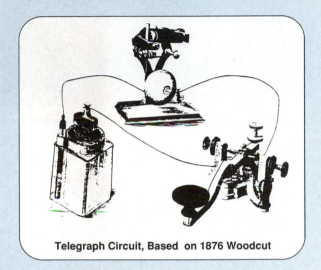

Telegraph Circuit, Based on 1876 Woodcut

THE TRUTH ABOUT ANALOG TECHNOLOGY

Take away analog electronics, and you would still have personal computers and other stand-alone computers. But there would be no radio, no television, no cellular phones, no satellite communications, no computer bulletin boards, no Fax machines, and no future in the form of an information superhighway.

Even "digital" answering machines, with no moving parts, actually use an analog storage device to record and play back messages. Digital circuits are used to select and activate the analog memory cells. Analog technology is a natural technology because all of nature is analog. Whenever electronics must interface with the natural world, analog electronics must be part of the mix.

The Digital Perception

The public perception of digital electronics is one of power-packed computers that will soon put analog electronics out of work. Digital is the future of electronics. Digital technology can do anything!

Using a modern personal computer (Photo: Ron Johnson)

THE TRUTH ABOUT DIGITAL TECHNOLOGY

There is no question that digital technology has revolutionized nearly everything. The digital computer has changed the way the world works. It is also true that digital technology has some limitations, and some tasks can only be performed by analog circuits.

Research efforts to make faster and better digital circuits have also given us the processing technology to make highly reliable—and formerly impossibly complex or impossibly expensive—analog integrated circuits. Without digital memory research, the analog memory used in many non-mechanical voice and video recorders would never have been developed. Analog technology is finally becoming a full partner with digital systems.

Analog electronics is all around us as part of many so-called "digital" systems. The term *analog* doesn't sell widgets; the term *digital* does.

3

Earth from space (Photo: NASA)

The Analog Perception

THE WORLD IS ANALOG; NATURE IS STRICTLY ANALOG

Not only is nature analog, but it tends to be more complex than our technology can handle. Linear things are easy to work with; even simple machines can handle linear actions. Unfortunately, nature seems to despise anything simple or linear. Electronics started out as analog, but we couldn't make our electronics do simple things like simple arithmetic.

Analog electronics worked fine as long as we stuck to natural events and electronic signals derived from them. But our analog machines were not suitable for business purposes—counting money, doing arithmetic, keeping the time of day, and so on.

Check your analog watch. If the second-hand jumps from second to second, it has been digitized. Your fingers (digits) are the original digital computer! (Photo: Ron Johnson)

4

Digital electronics in action (Photo: Ron Johnson)

The Digital Perception

DIGITAL ELECTRONICS, THE MATHEMATICAL ENGINE THAT DRIVES THE BUSINESS WORLD

It didn't take humankind long to discover the difficulty of making machines in the face of nature's complexity. Seventeenth-century clockmakers went digital in order to get around nature's complexity.

Digital clocks in the seventeenth century? Is that a typesetting error? No, the seventeenth century is correct. Early clockmakers tried a lot of power sources to run their clocks, but they found that one of the best was a wound-up spiral spring.

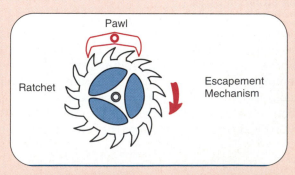

The escapement mechanism is a mechanical analog-to-digital converter. Analog spring tension is converted into 1-second jumps.

The Analog Perception

Whenever we wish to interact with the natural world, we must use nature's analog language and analog electronics. Nature does not speak digital! Radio waves, sound waves, light, and even electronic signals sent down a cable become analog events.

The electronics used to impress digital information on the communications channel must be analog. You can't log on to the internet or download a program from your favorite computer bulletin board without a modem (modulator/demodulator). Modulators and demodulators are analog circuits, because they must interact with the real world.

"Mars Observer calling planet Earth, over . . . " (Photo: NASA)

The Digital Perception

THE ESCAPEMENT MECHANISM

There was one small problem: the spring was non-linear, so the clock ran fast when newly wound and slow when run down. The escapement mechanism was added to advance the clock in increments of one tooth (or one second) for each swing of the pendulum, no matter how tight the spring was wound. The clock was now, at least in part, digital. Timekeeping had gone digital; and more important, timekeeping had become a linear and simple mechanical process. We can make reliable reproducible clocks, so everyone can give the same answer to the question, "What time is it?"

By adding the escapement, clockmakers reduced the resolution of the clock from an infinite number of possible speeds, depending on spring tension, to a constant value of exactly 1 second per second. The spring's nonlinear behavior no longer mattered, because the system could no longer resolve differences in spring tension into differences in speed.

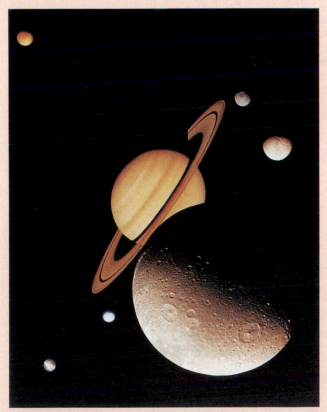

"Received the pictures, Observer. Well done from Earth." (Photo: NASA)

The Analog Perception

The electronics involved in all of our communications channels is analog, and somehow it must convert all forms of data into natural phenomena, like light beams, radio waves, signals on a cable, or voltages to make a loudspeaker produce music for our ears.

Casting stones into the water in digital order creates an analog wave carrying the digital information.

NATURE IS NONLINEAR AND COMPLEX

In electronics, we plot voltage against time to make a graph called a *waveform*. Waveforms are usually complex and almost always nonlinear. Interestingly, the capacitor charge/discharge curve (waveform) is the same as the graph of the winding and unwinding force of a spiral spring. Both capacitors and springs are used in timing devices, and the nonlinearity is a problem in both cases. We will see how to deal with the linearity problem later on.

Complex analog waveform. (Photo: Ron Johnson)

The Digital Perception

DIGITAL COMPUTING

The invention of digital computing did the same thing for the business world that the escapement mechanism did for clocks.

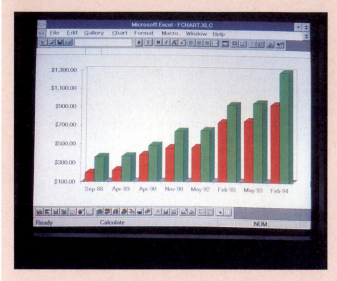

Computer screen doing business. (Photo: Ron Johnson)

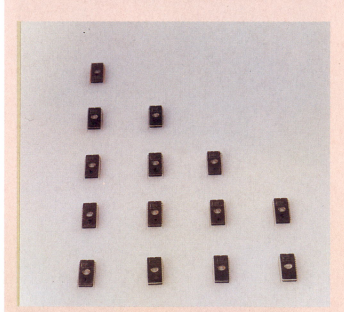

Stones represent data in digital form. (Photo: Ron Johnson)

The Analog Perception

First we had analog computers, which were fine for controlling temperatures and pressures in an oil refinery but (like the spring) were unsuitable for the human-simplified money system.

Analog electronics is not a good tool for organizing and managing the vast databases that provide the foundation for the information superhighway. Interactive hypertext organization will allow us to find any information we need, anywhere in the world in seconds. Only digital technology is suitable for this monumental task.

ANALOG WAVEFORMS ARE NONLINEAR AND COMPLEX

Because analog waveforms are voltage variations that mimic environmental changes in pressure, temperature, and so on, they are as complex as nature itself. In fact, the waveforms are often derived directly from nature, with the help of a transducer. As a result, analog waveforms are infinitely complex; but the human mind learns to interpret these graphs of natural events quite easily. In fact, we make graphs of all sorts of things, because graphs are easy to understand.

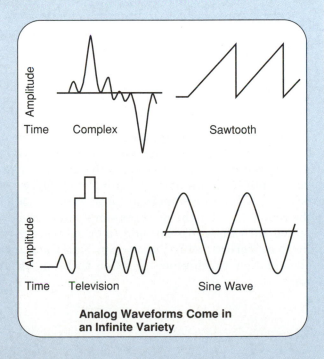

Analog Waveforms Come in an Infinite Variety

The Digital Perception

HUMANS LIKE THINGS LINEAR AND SIMPLE

We divide the dollar into 100 equal parts. We divide every day into 24 equal hours. We split day and night (A.M. and P.M.) into two equal parts, even though the ratio of light hours to dark hours varies throughout the year. We simplify nature's dates and times to make life less complex. Every four years, we adjust the calendar to put it back into synch with nature (more or less).

THE DIGITAL WAVEFORM IS SIMPLE AND LINEAR

The analog spring in early clocks had an infinite number of power levels available as it unwound. We wanted to retain the power, but we wanted it in manageable bursts worth 1 second each.

The purpose of digitizing anything is to reduce an infinite range of natural possibilities to a limited and well-defined range of useful values, rejecting all values that lie outside the defined system. We then have complete control (total power) over those values within our system—decimal numbers or dates during the year, for example. Hence, we don't have to be concerned with scheduling activities for the 25.54325th of June, despite its being a legitimate analog value (converted into digital for our convenience).

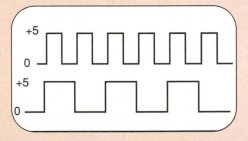

A digital waveform can only be square or rectangular. It can have only one of only two amplitude values: +5 volts or 0 volts. A value of +5 volts represents a binary one (on). A value of 0 volts (off) represents a binary zero. The digital frequency can vary.

The Analog Perception

THE CONCEPT OF SAMPLING

Whenever we need to convert analog information from a transducer into digital (numerical) form, we must break the waveform down into slices or sample voltages along the waveform.

Each sample is then converted into a digital numerical value, which is stored as a particular analog or digital voltage charge on a capacitor memory cell. The dual-trace oscilloscope samples the two waveforms. You are really seeing only half of the original waveforms; the missing part of each waveform is recreated by analog filters.

The Digital Perception

Because digital technology only has two valid digits and is represented electronically by the two voltages 0 and +5 volts (or the new low-power standard, 0 and +3 volts). If we plot the voltage changes from 0 to +5 V (against time) in a digital circuit, we get a square or rectangular waveform. This is the only valid waveform in digital technology.

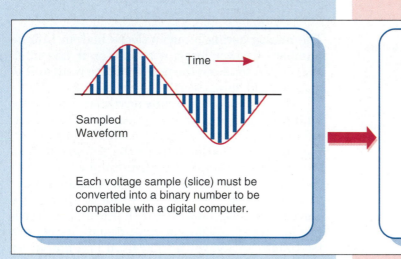

Time →

Sampled Waveform

Each voltage sample (slice) must be converted into a binary number to be compatible with a digital computer.

```
1000110011010101101000100101110
0001110010110011110101101011100
1100010010101110001010010100001
0101010111011111001100110101110
1110101000111101000111111010101
1101011001010111001101010111011
1100110101011101101000100100011
```

The peak voltage of each slice can be converted into a binary number representing the amplitude of the slice. (the binary values shown are not intended to be correct; they are for example only.)

DIGITAL COMPUTERS ARE MATHEMATICAL ENGINES

When using digital computer circuits to analyze or control natural phenomena, such as music or visual information, we must convert the analog waveforms into their numerical equivalents. Once we have everything in numerical form, we can modify the sounds or alter the pictures simply by changing numerical values. We can add to or subtract from these numerical values, or we can plug them into an equation to enhance or alter them in other ways.

SAMPLING AND THE TELEPHONE

Our nation's telephone network has used sampling techniques for many years to carry two or more conversations simultaneously on a single phone line. Fiber-optic cables combined with new modulation and data compression techniques permit a virtually unlimited amount of information to travel the information highway.

SOUND LOSS

Contrary to popular belief, some information is always lost when we convert sound (for example) into digital form. The advantage of digital recording is that we can make sure that noise, tape hiss, and other annoyances are also lost. The loss of a little sound is a small price to pay for the crystal-clear sound you get in the absence of noise and tape hiss.

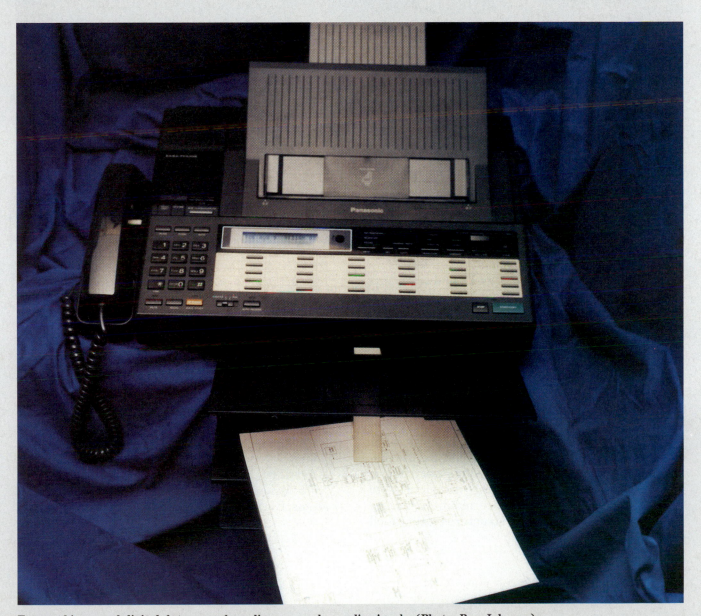

Fax machines send digital data over phone lines as analog audio signals. (Photo: Ron Johnson)

Fractal scene (Source: computer generated by Geoffrey Williams)

DATA COMPRESSION

The idea that information can be compressed into a smaller disk space or a lower bandwidth and then restored to its original condition is not new. The Joint Photographic Experts Group (JPEG) has established compression standards and mathematical techniques for compressing video information. After compression, a picture may take up only $\frac{1}{2}$ to $\frac{1}{3}$ the disk space of uncompressed data. A 2- or 3-to-1 compression ratio is useful, but not spectacular.

Benoit Mandlebrot, an IBM scientist, discovered that a hidden geometry exists in what appears to be random patterns in nature. He discovered that a tiny fraction of the image of a mountain might carry the geometric instructions to produce an image of the entire mountain range. These hidden geometrical structures are called *fractals*. And like the genes in a cell, which carry instructions for constructing an entire organism, fractals carry instructions for constructing large areas of a visual scene.

This suggests that, instead of storing the entire scene on disk, as we do now, we could store a few fractals and recreate the scene from those instructions. Compression ratios of 20,000 to 1 are theoretically possible using fractals. Ratios of 200 to 1 are common now. Fractals are samples of the real world, so new analog circuits may help reduce the compression time involved, which can be many hours. Decompression of fractal compressed scenes is very fast, however. Fractals were once called the "cold fusion" of electronics. As unbelievable as it seems, fractals are real and functional.

An ordinary super-highway
(Photo: Steve Schneider/
Frozen Images)

AN INVENTORY OF THE DEVICES AND CIRCUITS WE WILL STUDY

Diodes

Nearly all electronic devices are made of silicon, an element of a type called a *semiconductor*. Silicon tends to be very temperature-sensitive, so the first problem in building silicon devices is to make silicon conductivity less temperature-sensitive and more responsive to voltage variations. Adding special doping elements to silicon produces two kinds of electrically conductive silicon: p-type and n-type. A device called a *junction diode* is created from a tiny bit of silicon that is half n-type and half p-type. Diodes are one-way electronic valves that control the direction of current flow. Some diodes are designed to be sensitive to light.

Thyristors

A four-layer diode with an added control element creates a family of power diodes that can be turned on by a small control signal. Silicon Controlled rectifiers, triacs, and diacs are thyristor family members.

Bipolar Junction Transistors

Adding a third silicon layer forms either an n–p–n or p–n–p device called a *bipolar junction transistor* or *BJT*. The bipolar is the primary amplifying device in both analog and digital circuits. BJTs are current-controlled amplifier devices.

Field-effect Transistors

Voltage-controlled amplifiers, called *field-effect transistors,* can also be fabricated from silicon.

Integrated Circuits

Many transistors, diodes, resistors, and so on can be manufactured on a tiny piece of silicon called a chip. Metal film wiring patterns can be vapor-deposited on top of the chip to interconnect the transistors and diodes to form complete microcircuits called *integrated circuits*. A few individual transistors are still used, but integrated circuits are much more common.

Light-emitting Diodes

Light-emitting diodes (LEDs) and semiconductor lasers are made of gallium arsenide compounds, instead of silicon. LEDs produce light when current flows through them. LEDs do not replace the workhorse silicon diodes.

Color-enhanced silicon crystal
magnified 250 times
(Source: Photo micrograph by Gerald Williams)

Diodes

1. Rectifier circuits convert alternating current (**ac**) power line voltages into pulsating direct current (**dc**).

2. Most electronic devices require pure direct current.

3. Diode snubbers protect circuits from collapsing field voltages when semiconductors control electromagnetic devices.

4. Diode reference voltage sources take advantage of the constant 0.6 V diode junction voltage.

5. Special Zener diodes provide reference voltages other than 0.6 V.

6. Voltage variable capacitors select TV channels and tune resonant circuits.

Anode ◀ Cathode

Diode Symbol

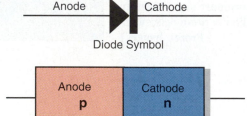

Diode Structure

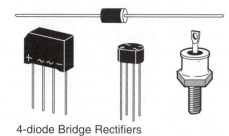

4-diode Bridge Rectifiers
Diode Case Styles

Thyristors
Gate Controlled, 4-Layer Diodes

1. Duty cycle control, using thyristors, can control high voltage, current and power levels.

2. Lamp dimmers and motor speed controllers are typical Thyristor applications.

3. Thyristors are also used in industrial applications, such as spot welder control.

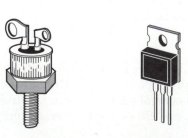

Thyristor Case Styles

Transistors are Basic Amplifying Devices

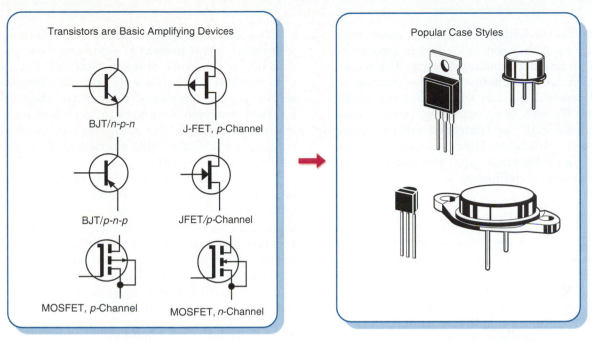

BJT/*n-p-n*

J-FET, *p*-Channel

BJT/*p-n-p*

JFET/*p*-Channel

MOSFET, *p*-Channel

MOSFET, *n*-Channel

Popular Case Styles

Many transistors, diodes and resistors on a silicon microchip are "wired" to form complete functional circuits called Integrated Circuits. ICs are modern electronic building blocks.

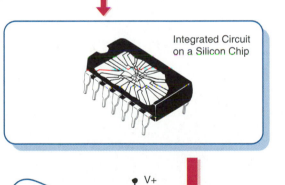

Integrated Circuit on a Silicon Chip

Analog Intergrated Circuit Inventory

1. Operational Amplifiers
2. Comparators
3. Timer Circuits
4. Voltage Controlled Oscillators
5. Phase Locked Loops
6. Analog Multipliers and Balanced Modulators
7. AM and FM Modulators and Demodulators for radio, wire, cable and Fiber-Optic cable
8. Radio Mixers and Frequency Synthesizers
9. Audio and Video Analog Storage recording and playback
10. Power control and electronic Voltage regulators
11. Analog-to-Digital and Digital-to-Analog Converters
12. Audio Power Amplifiers

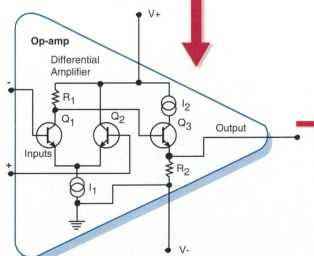

Op-amp

V+

Differential Amplifier

R_1

I_2

Q_1 Q_2 Q_3

Output

Inputs

R_2

I_1

V-

Common Integrated Circuit Packages

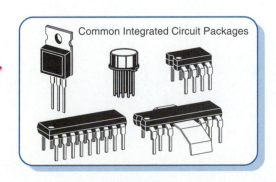

13

Electronics begins with some kind of amplifying device—anything that controls a larger amount of power with a smaller amount of power. The power amplification factor is defined as the amount of controlled power divided by the amount of controlling power. We call the controlling power *input power,* and we call the controlled power *output power.* The amplification factor A_P (where the A represents amplification and the subscript $_P$ stands for power) is defined as

$$A_P = \frac{\text{Power output}}{\text{Power input}}$$

We generally call the amplification factor *gain*. In some cases we are only interested in the voltage amplification (A_V) or the current amplification (A_i). Power is a voltage–current product.

Before transistors were invented, the vacuum tube was the primary amplifying device. Vacuum tubes were large, power-hungry, heat-generating monsters; but they were not temperature-sensitive, and they could be built to very accurate specifications.

The invention of the transistor revolutionized electronics, but it presented engineers with a very difficult and serious problem that did not exist with vacuum tubes. Not only were the new transistors highly temperature-sensitive, they could not hold reasonable manufacturing tolerances. As it turned out, there was really no way to prevent temperature changes from altering critical specifications of the transistor.

For a while, it appeared that transistors would be unfit for use in analog circuits and would forever be restricted to use in a digital on/off mode. But then, since we couldn't do anything about the temperature and manufacturing problems, we adopted nature's trick, called *negative feedback*.

Negative feedback is a self-regulating, automatically self-adjusting system. It is easy and cheap to use but a bit difficult to understand. Negative feedback will be a continuing theme throughout the rest of the text.

1.1   Preliminary Concepts

The term *analog* conjures up images of old-fashioned watches with big hands and little hands, of vacuum tubes, and of other archaic technology. The word *digital* conjures up visions of computers, and other state-of-the-art wonders. However, a quiet revolution is occurring in analog technology, mostly ignored by the popular press and by digital enthusiasts. Analog technology has once again become an integral part of electronics.

This book is about analog electronics—the electronic processing of data in a form that emulates (as nearly as possible) the way nature works. Nature is so complex that our best attempts always fall short. We have gone through a period in which many technologists would have abandoned analog electronics completely if it had been possible to do so. But it is impossible to avoid resorting to analog electronics, and indeed analog electronics often do the job better and far more cheaply than digital circuits.

One decade ago, digital electronics was seen as the ultimate, and analog electronics was expected to fade away altogether. For a number of reasons, this didn't happen and can't happen.

Analog and *Digital* Briefly Defined

Analog and digital technology are defined in relation to the way information is processed in each. In **analog technology,** natural phenomena such as temperatures or sound pressures are converted into electrical voltages that exactly follow those temperature or sound-pressure variations. The variations often produce very complex waveforms on an oscilloscope or chart recorder.

Digital technology is strictly mathematical and uses a special two-digit number system, called the **binary system.** Digital circuits are a natural for counting and all kinds of arithmetic operations. Although digital circuits can only do basic arithmetic, complex mathematical operations can be reduced to a very large number of simple arithmetic operations that digital circuits can calculate. Because they can complete millions of simple operations per second, digital computers can perform complex mathematical operations in a very short time.

When digital circuits must process real-world analog data, the data first have to be converted into numerical values or numerical codes. **Analog-to-digital electronic circuits** automatically convert varying voltages from analog sources or sensors into numerically equivalent values that digital circuits can process.

In analog circuits, complex waveforms are amplified or otherwise processed just as they come from the microphone or other transducer. The waveforms in analog circuits can vary from the only pure waveform—the sine wave—to highly complex waveforms that require expert interpretation.

The vast majority of analog circuits interact directly with the natural environment, using transducers (like the microphone) to convert changes in environmental conditions into their electrical analogs. Other transducers (like the loudspeaker) are used to reconvert the electrical analog signals into changes in the environment.

For example, a singer sings into a microphone. The microphone converts sound pressure variations into an electrical analog. An amplifier increases the power level to fill a large theater, and perhaps adds a little electronic echo to the singer's voice. A loudspeaker (transducer) converts the electrical analog into real sound. The audience hears a louder voice plus a little added echo.

It is possible to perform analog mathematical operations. For example, 5 volts (representing the digit 5) can be added to 10 volts (representing the number 10) to get 15 volts (representing the number 15). Analog computers have been used in bomb sights, petroleum refineries, and so on for many years. Some of these systems still exist. In many ways, they performed quite satisfactorily when direct interaction with the environment was required.

However, nature deals with the mathematics of infinity and seems to dislike anything linear or limited. Analog computers are not practical—and perhaps are impossible—to use for ordinary numerical and money-based calculations. The effort is somewhat like using a shotgun to kill a fly. One of the pellets might hit the fly, but it is hard to accomplish this every time. Trying to use an infinitely complex analog computer signal to do simple arithmetic and to balance a checkbook obviously falls into the shotgun-versus-fly category. Analog computers are simply the wrong tool for business.

Analog electronics uses infinite variations in an infinite number of different waveforms. The science of analog electronics focuses on complex waveforms and their interpretation. Analog electronics is useful for direct interactions with nature. It is not useful for business calculations.

Digital Computers

Digital computers were designed to crunch numbers for business. They are basically programmable digital calculators: everything they do involves manipulating numbers. Within the machine text, pictures and everything else must be encoded into numerical form and manipulated as numerical values. The computer can only do simple arithmetic, perform basic logic operations, and move binary numbers around.

The real power of a digital computer rests in its ability to make a logical decision based on the outcome of a calculation. For example, if two values are equal, the program can jump to a new subprogram and perform a different set of operations. If the two resulting values are not alike, the computer can continue executing the step-by-step procedure it was working on. The computer's remarkable speed and its ability to make logical decisions account for its amazing accomplishments.

The digital computer was intended to be a closed system that communicated with the outside world via a keyboard, a monitor screen, and a printer. It was not originally intended to interact with the natural world of temperatures, sound pressures, heat, and light.

Only a few years ago, it was possible to divide integrated circuits into analog and digital types just as easily. At the circuit level, however, it is no longer as easy to distinguish between the two in some of the newer integrated circuits, because they are a mix of both technologies.

Newer forms of analog technology and analog/digital hybrid systems have brought analog circuits back into the industry as a major player. Some modern integrated circuit chips can't honestly be classified as either digital or analog because they are about half-analog and half-digital. The manufacturer may classify them as digital or as analog depending on their intended function, but not based on the kind of circuits on the chip.

Modern analog technology incorporates the best of both analog and digital circuits, and we will see many examples of that marriage in this book. For several years, analog electronics has been an orphan. The most exciting analog integrated circuit was the operational amplifier, a holdover from analog computers; and it was extremely useful, but pretty boring.

Ironically, research and development of digital microchips have provided techniques for inexpensively producing analog circuits of previously impossible complexity and reliability. Advances in digital technology have been responsible for a rebirth of analog systems just in time to meet the challenges of the planned fiber-optic information superhighway (or whatever it may ultimately be called).

Electronics technology has matured to such an extent that we now recognize that some tasks are better done by digital circuits and others by analog circuits. In many cases, we find circuits and systems that are distinct mixes of the older analog methods and the newer digital methods, with both methods using the latest in microchip-processing technology.

We have even come full circle in some areas and are abandoning certain digital electronics in favor of an analog approach. There have also been some interesting technology exchanges between analog and digital technologies. The best example is what has happened in computer memories.

1.2 Analog Memory—A New Tool, with Thanks to Digital Research

Modern computer **RAM** memory is called **dynamic memory;** it uses an analog storage device, called a **capacitor,** as a digital memory element (no doubt, you have heard of the analog capacitor before). In digital computers there are only two digits, 0 and 1, represented by +5 volts and 0 volts. A memory capacitor charged to +5 volts stores a binary 1. A discharged capacitor represents a 0.

A great deal of research has gone into finding ways to pack millions of silicon memory storage capacitors, along with memory cell selection circuits, onto a tiny microchip. Thirty or forty million memory cell capacitors are contained in a typical modern personal computer (PC).

Recording Sounds—The Digital Approach

If we want to store sounds in a digital computer's memory capacitors, we must first use an **analog-to-digital converter** circuit to convert the analog voltage signals from a microphone into digital binary numerical values to input to a computer I/O port. Once the sounds have been converted into binary numerical values, we can store the numerical values as charged and discharged memory capacitor cells.

When we are ready to play back the stored sounds, we must first use a **digital-to-analog converter** circuit to convert the numerical values (stored in the memory cells as +5 volt charges and discharged capacitors) back into analog signals, which we can then amplify and send to a speaker.

This digital sound storage process is common, but it requires a lot of memory, an analog-to-digital converter, and a digital-to-analog converter. Good sound resulting from this approach is common, but it is a very costly way to do the job. Good sound quality requires at least a 16-bit data word, and sound of compact disk quality usually takes a 32-bit computer. Good sound quality is very demanding in digital form.

If a computer is involved, the digital approach permits some highly sophisticated sound modification methods. One example of such modification involved using old recorded speech from a deceased celebrity to computer-create a voice-over product endorsement. In this case, the celebrity's family objected in court and the ad was scrapped, but the potential for both good and evil here is clearly enormous.

The Bright Analog Idea

At some point, someone realized that a capacitor is an analog device that can be charged to just about any voltage, not just to +5 volts. Why not adapt this computer memory technology to pack large numbers of capacitors on a microchip in order to store analog voltages just as they come from the microphone?

First, this would save money by eliminating the analog-to-digital converter and the digital-to-analog converter. And second, because the data would be stored in analog form, less memory would be required for a given sound recording. In fact, you can record roughly 40 to 50 times as much sound in analog form as you can in the same number of digital memory cells. There is no way to compare relative storage capacity between the two systems directly, because data compression schemes exist in both systems. However, the analog system is generally simpler and less expensive for simple sound storage or for video recording.

The introduction of **analog storage devices** has resulted in telephone answering machines with no moving parts, and even greeting cards in which you can record a personal message. A new generation of sound and video products is being designed around this new analog storage technology. Often, these products are characterized as *digital* this or that, because the word *digital* has the glitz and because there is some digital circuitry involved. However, the central technology at work here is analog. People who work with other kinds of nonnumerical data, such as television signals, are also finding analog storage devices a better answer than strictly digital ones—and a good way to eliminate the mechanical problems that plague video recording equipment.

We will study analog storage systems in detail in Chapter 12.

1.3   The Communications Challenge

The amount of information we send over wires, cables, television channels, and radio channels is already too much for existing transmission channels to handle. The information superhighway will multiply the amount of information that must be transmitted, many times over. The communications challenge is to find ways to allow each cable or radio wave to carry many channels of information at the same time. Fiber optics holds the greatest potential for multichannel transmission over a single fiber. As yet, however, our electronics technology isn't sophisticated enough to take full advantage of the fiber-optic system. We must also find ways to make existing wires, cables, and so on carry more information channels than they do at present.

Analog Multiplier

Communications used to be a special analog field, chiefly of interest to broadcast engineers. Now, however, digital data—along with speech, music, and pictures—are being transmitted over a variety of telephone, radio, satellite, fiber-optic, and cable links. Like it or not, the modulation/demodulation process is essentially analog because it slips into the natural world of complex analog waveforms.

We must perform a kind of analog encoding called **modulation** on any kind of information before it becomes suitable for transmission over any distance via radio, fiber-optic cable, or virtually any other transmission medium. There is no way to move data from point A to point B without prodding nature into action and turning the enterprise into an analog process. Modulation is an analog activity even if the data themselves are in digital form. After the information has been modulated and transmitted to a remote location, it must then be demodulated to recover the original analog or digital information when it arrives at its destination.

The **analog multiplier** is an analog circuit that can modulate and demodulate nearly any kind of data—analog or digital—for transmission (or reception) over nearly any medium, from radio to fiber-optic cable. The analog multiplier can handle frequency and amplitude modulation, including the many clever digital encoding schemes.

The **analog multiplier** circuit is not new; it was invented for the analog computer. But like most of the circuits originally designed for analog computers, the analog multiplier was extremely complex and expensive and consequently was never even considered for use in commercial broadcast transmitters. We have had the perfect modulator/demodulator circuit for many years, but we couldn't afford to use it.

Again, research and experience in digital integrated circuits has enabled us to build integrated-circuit versions of the analog multiplier today that are silicon cheap and far superior to the very expensive analog multipliers available during the analog computer era.

The integrated-circuit version of the analog multiplier is a good example of an integrated circuit that is part digital and part analog. The device serves an analog function, but part of its internal circuitry is standard digital. The analog multiplier is actually sold under two or three different names, depending on the intended function.

The term *analog multiplier* implies that the circuit is used in analog computation to perform multiplication, and this is indeed the most precise and most expensive version. Less precise versions of the same basic circuit are available under the names *modulator / demodulator* and *mixer.*

The **modem** (modulator/demodulator) that connects your computer to the telephone line is a mix of complex analog and digital circuitry. And fiber-optic cable will demand enormous sophistication in both analog and digital electronics if we are to realize its full information-carrying potential.

Analog electronics is rapidly changing its status from unloved orphan to full partner with digital electronics.

Interfaces

Since it has become a part of the general language, the word *interfacing* has lost much of its scientific meaning; in this text, we will use its scientific meaning. An **interface** is the place where two distinctly different elements join together. Where the gas of the sky meets the liquid of a lake is an interface. The term *interface* also implies that the two elements on either side of the interface treat some signal differently. In the case of an air/water interface, a light beam (the signal) is bent at the air/water interface because the wave travels at different speeds in air and in water. Objects under water appear displaced and distorted when viewed from the air side of the interface. If we use an optical interface device, we can correct the distortion.

Suppose that we place a special lens partly in the water and partly in the air. If the lens is ground properly, the view from the air side of the interface will no longer be distorted. Actually, the lens introduces a new air-to-glass and a new glass-to-water interface and makes the necessary corrections within the lens.

It is technically incorrect to talk about interfacing with another person who speaks the same language, since the signal (speech) passes between the two people in the same medium. In terms of speech, the two people are not distinctly different, so there is no interface. When two people speak different languages, they need an interface person—an interpreter—to transfer the signal data correctly between the two.

In electronics, several interface problems exist that usually require some kind of interface device. We have already noted that an interface device called an analog-to-digital converter is required when electronic data must be transmitted from an analog to a digital system; and that a digital-to-analog converter interface device is required when digital data must be transmitted to an analog device or circuit. But electronic systems often require other kinds of interface devices, as well—for example, a level converter, which converts one voltage, current, or power level into another. Thus, if a transistor or integrated circuit operating at a low-voltage **dc** were required to control the 120-volt lights in your home, you would need a level converter. Because analog systems use a virtually infinite variety of signal waveforms and voltage levels, it is often hard to distinguish between ordinary analog circuits and interface circuits. Digital circuits use a single standard waveform and a standard voltage level, so interfaces between them are easy to identify. For analog circuits, the best we can do is to apply a fairly useful rule of thumb: an analog interface exists if one device or circuit delivers signal information to a second device or circuit, and the two circuits use different power supply voltages or different kinds of power supply voltage.

Contrasting Analog and Digital Technologies

For the first fifty years of the electronic era, all electronic circuits were analog. Digital technology (as we know it) was not possible until the invention of integrated circuits. Digital techniques owe their popularity to the fact that human-made machines have never quite measured up to the complexities of the natural world, so we have found ways to build less complex machines to meet our needs.

The first electronic computers were analog machines based on electronic analogs of other natural systems. For example, an electronic resonant circuit can be made to behave exactly like the mechanical resonant circuit of a plucked guitar string. The flight of a missile can be duplicated by using electronic voltages, inductors, and capacitors as analogs of gravitational force, mass, inertia, and so on. Real-world electronic components were used as analogs of their corresponding real-world counterparts in mechanical, thermodynamic, acoustical, and other physical systems.

Analog systems must deal with an infinite number of values along a continuous line of values. Their usefulness generally depends on how accurately we can—or need to—measure a given quantity. Digital systems are counting systems in which values are measured by counting the number of predefined units.

Digital electronics is easier to work with and more certain than analog electronics, but analog electronics permits an infinite resolution of values with less circuitry. Digital circuits trade infinite resolution of values for a comparatively coarse resolution based on a definite and repeatable set of values.

The decision about whether to use analog or digital circuits depends on which technology is better suited for the particular job. Both technologies have strengths and weaknesses, and neither is perfect for all applications.

In this book, we will examine a number of electronic devices that are definitely not digital but are not truly analog either. Overall, however, these devices and circuits are more nearly analog than digital, because they use complex waveforms of the kind normally associated with analog systems.

Digital waveforms are standard square or rectangular waveforms with an amplitude of +5 volts (or the recent +3-volt standard). All other waveforms fall into the category of analog signals and must be processed by analog circuits and devices. The kind of waveform to be processed is a useful basis for separating the two technologies. If it doesn't use standard digital waveforms, it must be analog. There are a couple of small exceptions to this rule—aren't there always?—but it is generally a good rule, nonetheless.

1.4   Analog Waveforms

The only pure waveform in the electronics world is the **sine wave.** All other waveforms are made up of combinations of sine waves of differing frequencies, phases, and amplitudes. In a composite waveform, one sine wave is the master or **fundamental waveform,** which determines the frequency of the composite waveform.

FIGURE 1-1   **Square Waves**

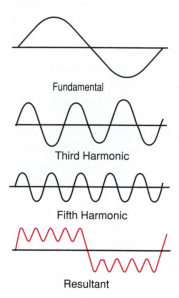

a. True Square Wave

Fundamental

Third Harmonic

Fifth Harmonic

Resultant

b. Adding Sine Waves to Form a Square Wave

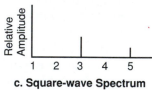

c. Square-wave Spectrum Diagram

To develop a square, triangular, sawtooth, or other waveform, we must add complementary sine waves, called **harmonics,** to the fundamental waveform. Harmonics are always integral frequency multiples of the fundamental frequency. Harmonics of a 1-kHz fundamental would be 2 kHz, 3 kHz, 4 kHz, 5 kHz, and so on. Nonsinusoidal (anything but a sine wave) waveforms can be created by adding the proper mix of harmonics to the fundamental, but this approach is practical only in limited cases. The problem is that you have to use a lot of individual sine waves to produce a true square, triangular, sawtooth, or other waveshape. A perfect square wave, for example, requires a fundamental sine wave and an infinite number of odd harmonics.

Figure 1-1 shows how a fundamental and a combination of odd harmonics can be used to form a rough approximation of a square wave. The more odd harmonics added (at their proper amplitudes) the more nearly square the waveform becomes; a truly square wave would be composed of an infinite number of odd harmonics. The spectrum diagram in Figure 1-1(c) is a bar graph showing the relative amplitude of each harmonic. Special oscilloscopes are available that can display **spectrum diagrams,** and some home stereos have light-bar displays that produce a simplified spectrum diagram bar graph. Graphic equalizers use special filters to cut or boost specific frequencies along the spectrum line, altering waveforms at will.

The waveforms produced by musical instruments tend to be quite complex, but even the most complex of these waveforms can be broken down into fundamental and harmonic sine waves.

Standard or Archetype Waveforms

Although analog waveforms come in an infinite variety of shapes and amplitudes, they can be categorized into four (more or less) basic waveshapes (or some variation or combination thereof). Figure 1-2 shows the ideal form of

FIGURE 1-2   **Archetype Waveforms**

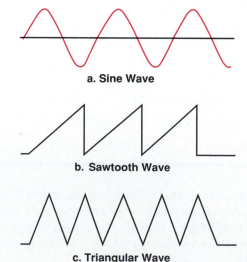

a. Sine Wave

b. Sawtooth Wave

c. Triangular Wave

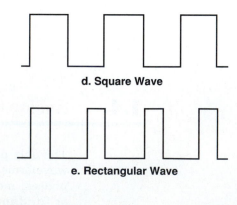

d. Square Wave

e. Rectangular Wave

FIGURE 1-3   Triangular Waves

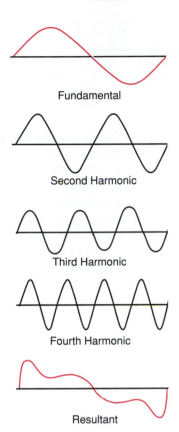

Fundamental

Second Harmonic

Third Harmonic

Fourth Harmonic

Resultant

a. Adding Sine Waves to Form a Triangular Wave

b. Results of Adding an Infinite Number of Even and Odd Harmonics

each of the four archetype waveforms. Figure 1-2(a) depicts a sine wave, the raw material from which all other waveforms are assembled. Figure 1-2(b) shows a **sawtooth waveform,** whose important features are a gradual ramp up and a sudden drop to the baseline. The waveform in Figure 1-2(c) is called a **triangular waveform;** it is similar to the sawtooth except that it features a gradual ramp up, followed by a gradual ramp down instead of a sudden drop. Figure 1-2(d) presents a **square waveform**—a special case of the **rectangular waveform** (Figure 1-2(e)). The square wave spends the same amount of time in both its maximum and its minimum amplitude conditions. Although it is a special case of the rectangular wave, the square wave seems to be the most common type of rectangular wave.

Figure 1-3 illustrates how a triangular wave is formed from a combination of even and odd harmonics. A perfectly triangular waveform would be composed of an infinite number of even and odd harmonics. Analog waveforms tend to be complex in real systems. Figure 1-4 shows some typical examples of complex analog waveforms.

Waveshaping

In general, so many harmonics are required to synthesize a composite waveform that the technique of adding individual harmonics is seldom used. Excellent approximations of true square, triangular, sawtooth, and other common waveforms can be obtained by using simple methods and circuits. For example, you can convert a sine wave into a square wave by cutting off the top of an oversized sine wave near its base, as shown in Figure 1-5. The example

FIGURE 1-4   **Examples of Complex Waveforms**

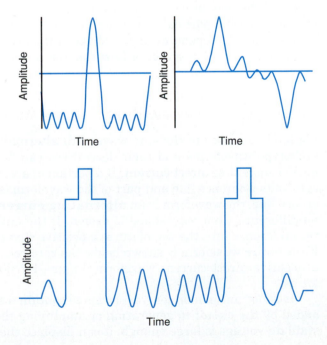

FIGURE 1-5   Shaping a Sine Wave into a Square Wave by Clipping

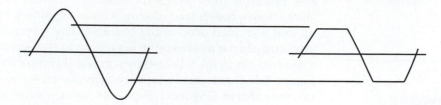

shown in Figure 1-5 is a rather poor approximation, because the sine wave has been clipped so far above the baseline. In practice, an amplifier can be used to clip the waveform as close to the baseline as is required and still produce a square wave with the required amplitude. All of the archetype waveforms in Figure 1-2 can be formed by using similar gross waveshaping techniques.

If we create the waveform through simple waveshaping methods, the spectrum analyzer will still find all of the appropriate harmonic sine waves contained in the reshaped waveform. When we reshape a waveform, the harmonics are added to it as part of the shaping process—or perhaps adding the proper harmonics reshapes the waveform. But this is a philosophical question. The important thing is that every waveform, whatever its source, always contains the proper set of harmonics for its particular shape.

> All of the proper harmonics exist, no matter how the waveform is created.

Attack and Decay Envelopes

Waveforms do not necessarily exist as a continuous stream of waves of unchanging amplitude. When a guitar string is plucked, it sounds when it is released, vibrates, and gradually dies out. Each musical instrument has a distinctive sound, resulting from a combination of the instrument's harmonic content and the way in which the sound rises to peak volume and decays out. A waveform with a particular harmonic content may sound like an organ note, a string, or a bell, depending on the shape of the attack and decay envelope. The envelope is simply a graph of how the amplitude of the waveform rises and falls (see Figure 1-6).

Alternating Current and Direct Current Waveforms

We tend to associate the sine wave with alternating current, but any of the archetype waveforms and their derivatives can be produced by either alternating current or direct current. If a portion of a waveform is shown below the zero volts reference line and part of the waveform is shown above the zero reference line, the waveform is an **alternating current (ac)** waveform. It makes no difference what waveshape is drawn. If the entire waveform is above the zero reference line, the waveform is a positive **direct current (dc)** waveform. If the entire waveform is shown below the zero reference line, the waveform is a negative direct current waveform. Figure 1-7 illustrates **ac** and **dc** waveforms.

It is not uncommon for waveforms to contain a steady-state **dc** component added by the circuit in generating or amplifying the waveform. If the steady-state **dc** voltage is large enough, it can displace the entire waveform above or

FIGURE 1-6 **Attack and Decay Envelopes**

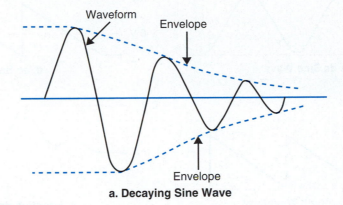

a. Decaying Sine Wave

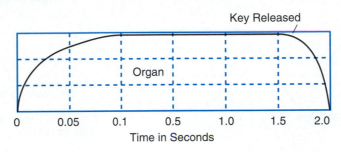

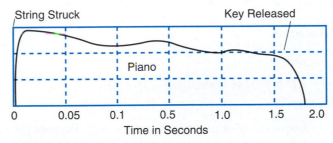

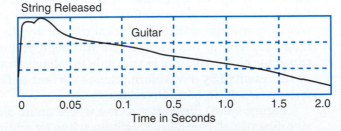

b. Typical Musical Instrument Envelopes

below the zero reference line. If the entire waveform is on one side of the zero line, the current flows in the same direction for all points on the waveform graph.

To qualify as alternating current, some part of the waveform must cause the current in the circuit to reverse direction. The change of current direction is shown by having the waveform cross the zero reference line in the waveform drawing.

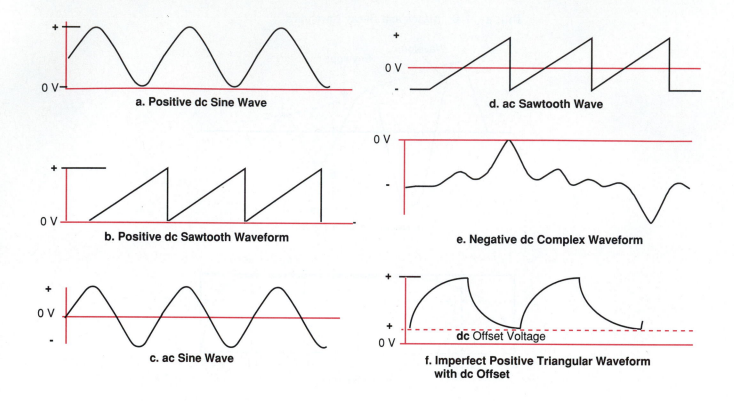

a. Positive dc Sine Wave

b. Positive dc Sawtooth Waveform

c. ac Sine Wave

d. ac Sawtooth Wave

e. Negative dc Complex Waveform

f. Imperfect Positive Triangular Waveform
with dc Offset

FIGURE 1-7   ac and dc Waveforms

Effect of a Coupling Capacitor or Transformer on a dc Waveform

Capacitors and transformers are frequently used as coupling elements between two analog circuits. Whenever such a coupling capacitor or transformer is used, a **dc** waveform is transformed into an **ac** waveform by the charge/discharge action of the capacitor or by the expanding and contracting magnetic field of the transformer. Although you often hear people say that a capacitor blocks **dc,** this is true only for steady-state **dc.** A varying **dc** voltage, as represented by a waveform, causes current to flow in the capacitor or transformer circuit, in both directions.

Figure 1-8(a) illustrates a case where an **ac** sine wave is shifted to a point significantly above the zero reference line by adding a steady state **dc** voltage (the battery). The added **dc** voltage converts the **ac** sine wave produced by the generator into a positive **dc** sine wave with a measurable **dc** component, often called a **dc** offset voltage. When the circuit is first turned on, the capacitor quickly charges to the battery voltage. During the initial charging period, current flows through the output resistor, resulting in a voltage drop across the output resistor.

This output voltage exists only until the capacitor charges to the battery voltage. Once the capacitor is fully charged, the battery can no longer cause

FIGURE 1-8 **Capacitor and Transformer Coupling**

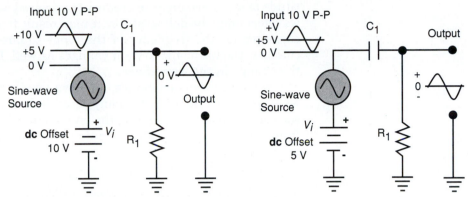

a. Capacitor Coupling—Converting a dc Waveform into an ac Waveform by Removing the dc Component

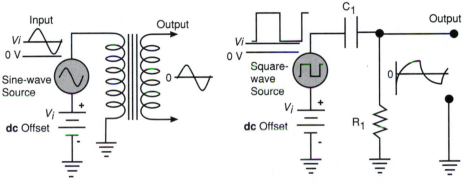

b. Transformer Coupling—Converting a dc Waveform into an ac Waveform by Removing the dc Component

c. Waveshape Alteration Due to Capacitor That Is Too Small

any current flow, and no voltage drop persists across the output resistor. The battery thenceforth has no effect on the output voltage, which drops to zero.

As the capacitor charge follows the sine wave voltage, current flows through the output resistor, producing a zero crossing (**ac**) sine wave–shaped output voltage. The output voltage across the output resistor has the same shape as the input sine wave, but (as you learned in your **ac** theory course) it is shifted in phase by approximately 90 degrees.

Figure 1-8(b) shows a sine waveform with an added **dc** component (the battery). The battery causes a brief transfer of energy to the transformer secondary when the battery is first connected, as the primary magnetic field builds. After this initial burst, the battery induces no further energy into the secondary. The transformer output voltage varies only when the triangular waveform changes from one voltage value to another. This, too, is basic **ac** theory. The **dc** component is removed, and the output is an **ac** triangular waveform.

The circuit in Figure 1-8(c) illustrates one way in which a waveform can easily be converted into a different waveform without first generating a multitude of new harmonics to produce the desired waveform. The alteration of the waveform may be deliberate, or it may result from the selection of too small a capacitor for the frequency of the input square wave. When the waveshape changes, the harmonic content changes as well. It doesn't matter what caused the change in the waveshape.

In the previous examples, we have used a **dc** voltage (battery) to offset waveforms to a **dc** level. In many cases, a **dc** waveform is due to some **dc** offset voltage that is present in the circuit. In other cases, however, a **dc** waveform is originally generated as a **dc** waveform, and no extra **dc** offset voltage is present.

Analog Waveform Summary

Analog circuits require us to deal with a wide variety of waveshapes: sine, square, rectangular, sawtooth, triangular, and many less common waveshapes. Analog waveforms come in a wide variety of amplitudes (voltage levels); and in many cases, their amplitudes vary with time.

In analog systems, all of the information being processed is carried by the waveform. Thus, the waveform is the information or data. Everything happens in **real time**—natural time, our time—and the waveform you observe represents what is happening at that instant. Because analog waveforms represent all of the information in an analog system, any unplanned alteration of the waveshape distorts the information by adding unintended harmonic content to the waveform. Electrical noise can also get into the system and alter the waveshapes. Using negative feedback and careful wiring practices can minimize distortion and noise problems.

Digital Waveforms

In digital systems, the waveform is (almost) always square or rectangular, with a standardized amplitude of 5 volts. (Some of the newer integrated circuits use a 3-volt square wave to reduce the power consumption in laptop and notebook computers.) A square or rectangular waveform can be viewed as representing a voltage source switched on for a given time period, and then switched off. Because digital circuits are composed entirely of on/off high-speed electronic switches, this is an appropriate digital view. This implies that we can create a square wave signal, with its infinite number of odd harmonics, simply by switching some **dc** voltage on and off at regular intervals; and that is, indeed, the most popular method used to generate a square wave.

Digital systems do not ordinarily operate in real time. The timing is controlled by a master digital waveform generator, appropriately called the *clock*. Nearly all operations of the digital system are governed by the heartbeat of the clock. The digital clock typically ticks off tens of millions of computer "seconds" for each real-world second.

The on/off nature of digital systems demands that thousands of these on/off operations be involved in performing even relatively simple arithmetic, so the system must operate much faster than real time. But even on this light-

FIGURE 1-9   **Typical Digital Waveforms**

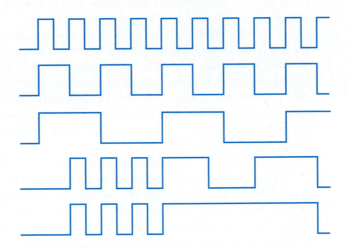

ning-paced time scale, computers are often hard-pressed to keep up with complex real-world events.

Computer speed has risen rapidly over the years, mostly because people have demanded color moving images and increasingly high-resolution pictures on the computer monitor screen. High-quality video is the most demanding kind of computer operation and the main source of the demand for ever-increasing computer speed. In many cases, digital color computer monitors have been replaced by analog monitors in an effort to get more natural color without further straining the computer's capabilities.

Multimedia uses demand that the computer be capable of producing both high-quality color pictures and high-quality sounds. We are accustomed to virtually infinite resolution in both natural pictures and natural sounds. The computer is hard-pressed to provide a resolution high enough and fast enough to appear natural to our very keen analog senses.

Digital clocks bear some relationship to real time, in that a specific number of computer "seconds" occur in one real-time second. This relationship allows digital systems to keep track of the real-world time and date. Figure 1-9 illustrates some typical digital waveforms. Notice that they all are based on a common time scale, because they all fall under the timing control of a master clock. All of the waveforms in Figure 1-9 have an amplitude, at any given instant, of either zero volts or the standard level of +5 volts.

Unwanted Analog Elements in a Waveform

Square waves contain an infinite number of odd harmonic sine waves. These harmonic sine waves actually exist, and each harmonic contains energy just like any other sine wave.

If you generate a square wave with a fundamental frequency of 20 MHz, there will be a fifth harmonic sine wave at 100 MHz. This 100-MHz signal contains a fair amount of energy—less than the fundamental or the third harmonic, but still a considerable amount. And in spite of efforts to counteract the problem, the 100-MHz signal can occasionally cause annoying interference with the 88- to 108-MHz FM radio band. The key message here is that harmonics are real and contain energy.

The little metal covers that close the unused openings for expansion slots on the back of your computer are there to prevent leakage of these higher harmonics into space. The FCC compliance label on all digital equipment certifies that leakage of higher harmonic energies is low enough not to interfere with other electronic equipment. But occasionally the higher harmonic leakage can cause problems, notwithstanding FCC certification to the contrary. At 100 MHz, two traces on a circuit board running close together form a significantly large capacitor. Even a straight piece of wire (or circuit board trace) carrying a current shows some inductive behavior at 100 MHz.

This distributed inductance and capacitance can combine to form a resonant circuit. In any circuit of reasonable complexity, quite a few of these distributed resonant circuits may exist. If (as is likely) the circuit includes a distributed resonant circuit with a resonant frequency near 100 MHz, its circulating resonant current can be provided by the energy from digital square wave's fifth harmonic energy.

As the square wave rises from zero volts to its maximum, a strong fifth harmonic causes the resonant circuit to ring, and the ringing energy adds to the top of the square wave to produce a distorted waveform, as shown in Figure 1-10.

The resonant circuit rings again at the base of the waveform, causing distortion there as well. Most digital circuits can ignore a certain amount of this analog distortion, but problems can result if the ringing voltage is too high.

FIGURE 1-10　Square Wave with Distortion and Sine-wave Ringing

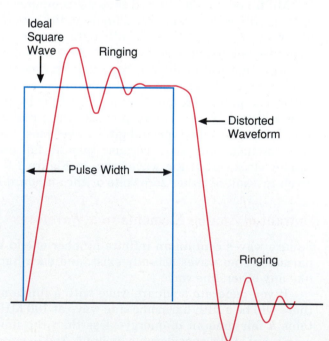

This kind of analog distortion is frequently observed on an oscilloscope, when you test real amplifiers, regulated power supplies, and some digital systems.

1.5 The Digital Connection

Because some of the integrated circuits discussed in this text include digital elements and binary numbers, you should know how to count from 0 to 15 in binary. Counting to 15 in binary is quite easy, and you won't have to count higher than that for the purposes of this text. This section of text offers a brief explanation of how binary numbers work and how to count in binary.

Digital Representation of Values

In analog systems, we represent various quantities by the voltage amplitude of the waveform. For example, we can represent a temperature of 10° by a voltage of 10 V, 20° by 20 V, and so on. Or we can scale the voltage—for example, letting 1 V equal 10° and 2 V equal 20°.

In digital systems, all waveforms have a fixed amplitude of +5 volts. How then, can we represent different values of temperature or other real-world quantities? Let's compare the analog and digital approches to the following simple problem: construct a lamp dimmer control to provide a range of control from 0 brightness (off) to a full 15-watt brightness.

Figure 1-11(a) shows a simple analog method for accomplishing this result. In the analog circuit, we simply vary the brightness of the 15-watt lamp by introducing more or less resistance in series with the lamp. The circuit is simple and can be adjusted over the range of 0 to 15 watts in increments of a very small fraction of a watt. Theoretically, the resolution is infinite, in which case the lamp brightness could be varied in infinitely small fractions of a watt.

FIGURE 1-11   **Electronic Representation of Values**

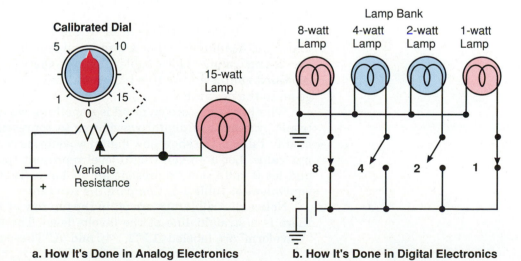

a. How It's Done in Analog Electronics   **b. How It's Done in Digital Electronics**

In practice, the variable resistor does have some limits to its resolution, but very small increments are possible.

The analog circuit in Figure 1-11(a) has two disadvantages. First, power is wasted in the series resistance of the variable resistor. Second, like most analog devices, the incandescent lamp is not a linear device. Consequently, the marks on the dial scale will not be evenly spaced, and accurate adjustments may be difficult where the marks on the dial are crowded.

In the digital version in Figure 1-11(b), we use a bank of four lamps, each of which is either fully on or totally dark. Variations in brightness are accomplished by using combinations of fully-on lamps. For example, the dimmest light available is the 1-watt lamp by itself. The second brightness level uses only the 2-watt lamp, and the third level uses the 2-watt and the 1-watt lamp at the same time. If we activate the lamps in a binary sequence (see Table 1-1) we can vary the brightness from 0 to 15 watts in 1-watt increments.

The resolution of the analog version is virtually infinite, while the resolution of the digital version is 1 watt. We can increase the resolution of the digital version by adding more properly scaled lamps. The digital version provides very precise control of the total brightness, but only at exactly 1-watt increments. Achieving resolutions of fractions of a watt requires adding more switches and lamps.

To get a $1/2$-watt resolution, for example, we would have to add one more lamp, rated at $1/2$ watt, and one more switch to the circuit in Figure 1-11(b). Digital circuits always have finite and limited resolutions, but because modern integrated circuits can put thousands of electronic switches on a silicon chip that is less than $1/4$-inch square, any required practical resolution is possible. In the digital version, lamps are either completely on or totally off, and there is no power-wasting series resistor.

Table 1-1 shows all of the possible on/off conditions for the switches (and lamps) depicted in Figure 1-11(b). Notice that switch 1 is connected to the 1-watt lamp, switch 2 is connected to the 2-watt lamp, and so on. The lowest light level, 1 watt, requires that only the 1-watt lamp be turned on (all other lamps are off). Similarly, 2 watts requires that the 2-watt lamp be turned on, with all others off. The third state (3 watts) demands that two lamps be turned on at the same time. The combination of the 2-watt lamp plus the 1-watt lamp produces 3 watts.

Follow Table 1-1 through all of the possible lamp conditions. The data under the heading "C" offer a numerical summary of the conditions shown in the "Lamp" column of the table. Compare these numerical data with the lamp conditions for each row. It is important that you see how this works before we go on to the next table.

Whether lamps are on or off is something we can observe, but digital electronic systems use digital waveforms to represent values. Suppose that we modify Table 1-1 to show how digital waveforms could represent each light output value. Let a +5-volt (high) level represent the equivalent of an on lamp, and let 0 volts (low) represent an off lamp. Table 1-2 shows the waveform equivalent of Table 1-1. Compare the two tables.

Notice that all lamps are off in the top row of Table 1-1, while in Table 1-2 there is a straight line at the 0-volts level. The columns under the heading "Waveform" are labeled "1," "2," "4," and "8." These are the same numerical val-

TABLE 1-1
Binary Lamp Conditions

A	B	LAMPS				C
Decimal Number Watts	Binary Number Watts	8 W ▼	4 W ▼	2 W ▼	1 W ▼	Total Watts ▼
0	0 0 0 0	●	●	●	●	0 + 0 + 0 + 0 = 0
1	0 0 0 1	●	●	●	○	0 + 0 + 0 + 1 = 1
2	0 0 1 0	●	●	○	●	0 + 0 + 2 + 0 = 2
3	0 0 1 1	●	●	○	○	0 + 0 + 2 + 1 = 3
4	0 1 0 0	●	○	●	●	0 + 4 + 0 + 0 = 4
5	0 1 0 1	●	○	●	○	0 + 4 + 0 + 1 = 5
6	0 1 1 0	●	○	○	●	0 + 4 + 2 + 0 = 6
7	0 1 1 1	●	○	○	○	0 + 4 + 2 + 1 = 7
8	1 0 0 0	○	●	●	●	8 + 0 + 0 + 0 = 8
9	1 0 0 1	○	●	●	○	8 + 0 + 0 + 1 = 9
10	1 0 1 0	○	●	○	●	8 + 0 + 2 + 0 = 10
11	1 0 1 1	○	●	○	○	8 + 0 + 2 + 1 = 11
12	1 1 0 0	○	○	●	●	8 + 4 + 0 + 0 = 12
13	1 1 0 1	○	○	●	○	8 + 4 + 0 + 1 = 13
14	1 1 1 0	○	○	○	●	8 + 4 + 2 + 0 = 14
15	1 1 1 1	○	○	○	○	8 + 4 + 2 + 1 = 15

○ = Lamp On   ● = Lamp Off

ues as those under the "Lamp" heading in Table 1-1. The 1, 2, 4, and 8 headings are the generic heading values for any table of binary values; and the values represented can stand for watts, dollars, or any other quantity we might ordinarily assign a numerical value to.

Examine row 5 in Table 1-1 (5 under "Total Watts"). Notice that the 4-watt and 1-watt lamps are lit. Now, look at row 5 in Table 1-2. The columns headed "C" are identical in the two tables. Notice that the waveform rises to +5 volts in the "4" and "1" columns, but is 0 volts under the other column headings. This correlates exactly with the conditions of the 4-watt and 1-watt lamps in Table 1-1. If you inspect the two tables row by row, you will find an exact correlation between "Lamp" and "Waveform."

Notice that the numbers in column "A" and the right-most line of numbers in column "C" of Table 1-2 are identical. They are also identical to the "Total Watts" column under the heading "C" in Table 1-1. These are ordinary decimal

TABLE 1-2

Digital Values Represented by Waveforms

A	B	Waveform				C
Decimal Number (Watts)	Binary Number (Watts)	8	4	2	1	Explanation 1 = +5 V 0 = 0 V
0	0 0 0 0					0 + 0 + 0 + 0 = 0
1	0 0 0 1					0 + 0 + 0 + 1 = 1
2	0 0 1 0					0 + 0 + 2 + 0 = 2
3	0 0 1 1					0 + 0 + 2 + 1 = 3
4	0 1 0 0					0 + 4 + 0 + 0 = 4
5	0 1 0 1					0 + 4 + 0 + 1 = 5
6	0 1 1 0					0 + 4 + 2 + 0 = 6
7	0 1 1 1					0 + 4 + 2 + 1 = 7
8	1 0 0 0					8 + 0 + 0 + 0 = 8
9	1 0 0 1					8 + 0 + 0 + 1 = 9
10	1 0 1 0					8 + 0 + 2 + 0 = 10
11	1 0 1 1					8 + 0 + 2 + 1 = 11
12	1 1 0 0					8 + 4 + 0 + 0 = 12
13	1 1 0 1					8 + 4 + 0 + 1 = 13
14	1 1 1 0					8 + 4 + 2 + 0 = 14
15	1 1 1 1					8 + 4 + 2 + 1 = 15

values. The numbers in column "B" of Table 1-1 are the equivalent values for the decimal numbers in column "A" written in the binary number system.

Table 1-2 shows the standard binary counting table as it is usually presented. With a little practice, you will find that counting to 15 in binary is easy.

SUMMARY

1 The sine wave is the only pure waveform. All other waveforms are composed of a mixture of sine waves of different frequencies and amplitudes.

2 All waveforms other than the sine wave contain harmonics, which are sine waves with frequencies 2, 3, 4, 5, and so on times greater than the basic or fundamental sine wave frequency.

3 The square wave is composed of a fundamental sine wave and an infinite number of odd harmonics (1, 3, 5, . . .).

4 The archetype or standard waveforms are the sine wave, the square or rectangular wave, the triangular wave, and the sawtooth waveform.

5 Attack and decay envelopes define how a sound builds up and decays in volume. These envelopes greatly influence how a musical instrument sounds.

6 When a signal is coupled through a transformer or capacitor, the waveform is converted into an **ac** waveform.

7 Analog signals come in an infinite variety of waveforms and amplitudes. Digital waveforms are always square or rectangular, with an amplitude of +5 volts.

8 Because all waveforms, except the sine wave, contain energy at many frequencies (harmonics), stray resonant circuits can respond to harmonic signals. Stray capacitance and stray inductance are always present and excitable at some resonant frequency. When stray resonant circuits produce unwanted signals, the phenomenon is called *ringing*.

Q U E S T I O N S   A N D   P R O B L E M S

1.1 What is the only pure waveform?

1.2 Which waveform contains no harmonics?

1.3 Define the term *harmonic*.

1.4 List the archetype (standard) waveforms.

1.5 Make a sketch of each of the archetype waveforms.

1.6 Which of the archetype waveforms contains only odd harmonics?

1.7 Explain attack and decay envelopes.

1.8 Explain the term *ringing*.

1.9 Why can a 200-kHz square wave sometimes cause interference in the 88- to 108-MHz FM radio band?

1.10 Describe the effect of clipping off the top of a sine wave. In what way are the harmonics affected?

1.11 Write the numbers 0 through 15 in binary.

1.12 Explain the concept of *resolution*.

1.13 Compare the levels of resolution in digital and analog systems.

1.14 What kind of transducers are available for direct conversion of environmental temperatures into digital data?

1.15 What is a fractal? (Please see the Analysis In Brief Section)

1.16 What can be done with fractals? (Please see the Analysis In Brief Section)

1.17 Why is an analog computer unsuitable for business accounting purposes?

1.18 What device is used to store information (memory) in both analog and digital systems?

1.19 What are the advantages of using an analog storage system instead of a digital memory system in an answering machine?

1.20 Nature is:
   a. analog.
   b. digital.
   c. some of each.

1.21 Modulating a signal with digital data always produces:
   a. an analog signal carrying digital data.
   b. an analog signal carrying analog data.
   c. a digital signal only.
   d. an analog signal only.

1.22 A radio wave is:
   a. always analog.
   b. always digital.
   c. either digital or analog.

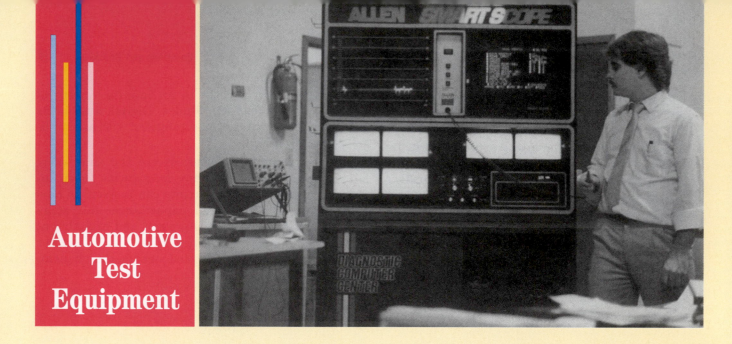

Automotive Test Equipment

Until just a few years ago, the electrical system of an automobile engine was quite simple: a battery, mechanical distributor, and points (switch contacts); a coil (step-up transformer); and spark plugs. Now, most vehicles use solid-state electronics, including several microcomputers and various sensors, to monitor and control the operation of the engine, transmission, and other systems.

For the most part, the increased use of electronics in the vehicle has not been the province of the average electronics technician. Automotive technicians are responsible for troubleshooting and repairing automotive electrical systems. The area that *has* affected electronics technicians relates to the test equipment that auto techs use to analyze and tune engines.

In the same way that an oscilloscope can be used to display waveforms from a transistor amplifier or a motor control circuit, specialized automotive scopes are used to display the ignition and charging waveforms from a vehicle. Automotive scopes use special cable harnesses and sensors that connect to several points on the engine. They monitor parameters such as the battery voltage and current, the ignition coil primary and secondary waveforms, the "number one" spark-plug signal, the engine vacuum, and the coolant temperature.

Some engine analysis systems include a separate gas analyzer that samples and analyzes the components of the exhaust gas. Oxygen, carbon monoxide, carbon dioxide, and hydrocarbon concentrations indicate whether pollution emission standards are being met.

The latest computerized analyzers are really sophisticated data-acquisition systems that perform a complete engine analysis in just a few minutes. Usually they prompt the automotive technician through a series of tests. Data specific to the model of engine being tested are recalled from computer memory and compared to test data. Diagnostic software analyzes this comparison, and the computer prints out a list of recommended service procedures.

Of course, any kind of equipment can fail, and that's where the electronics technician comes in. Working as a manufacturer's service representative or as an independent contractor, the technician provides regular maintenance and emergency service to the equipment owner. Typically, this kind of service involves traveling to the customer's location, where the technician uses an "engine simulator" to test and calibrate the analyzer. The simulator is an electronic signal source that produces waveforms similar to those produced in the engine. To calibrate the exhaust analyzer, the technician injects special calibration gases into the system and adjusts the electronics.

Repairing these units may involve something as simple as replacing an engine harness cable or as complex as troubleshooting a printed circuit board. Circuitry ranges from analog amplifiers

Automotive Ignition System

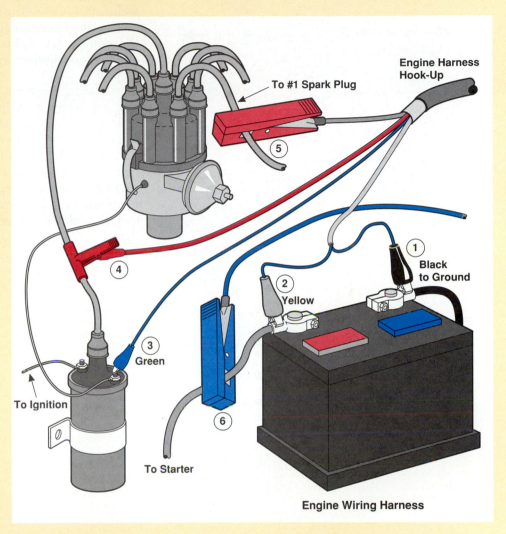

Automotive Ignition System

Engine Harness Hook-Up

To #1 Spark Plug

5

4

3 Green

To Ignition

To Starter

6

2 Yellow

1 Black to Ground

Engine Wiring Harness

that use transistors, FETs, and op-amps to switching circuits that implement SCRs and TRIACs.

The scope section of the analyzer operates similarly to a conventional oscilloscope, but it has specialized triggering and deflection circuitry. Computerized systems use typical microcontroller chips, but the signal-processing circuitry is designed specifically for accessing automotive signals. Gas analyzers have complex plumbing arrangements for filtering and removing water vapor from the vehicle's exhaust. They also use infrared optical and electrochemical sensors to detect and measure the gases.

Automotive test equipment is very reliable, but it undergoes continuous use and is subjected to a harsh environment in the automotive service shop. Consequently, it requires regular service. To perform this kind of service, the electronics technicians must thoroughly understand the field of electronics in general, as well as the specific elements of how an automobile engine operates. They must also possess good mechanical aptitude and problem-solving abilities.

As with any job, the employee who is versatile and can improvise has the advantage. This is especially true with automotive test equipment. In

Ignition System Waveform

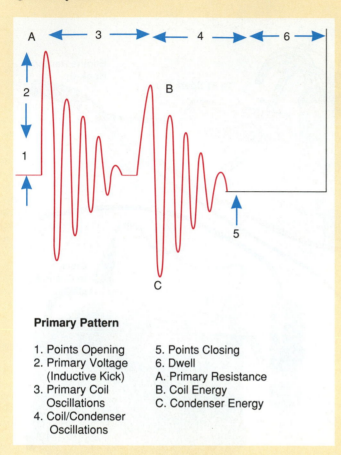

Primary Pattern

1. Points Opening
2. Primary Voltage (Inductive Kick)
3. Primary Coil Oscillations
4. Coil/Condenser Oscillations
5. Points Closing
6. Dwell
A. Primary Resistance
B. Coil Energy
C. Condenser Energy

addition to having service responsibilities, the technician is sometimes called upon to train automotive shop personnel in the operation and use of the equipment. Sales and customer service are often responsibilities of the field service technician as well.

Growing concern over environmental issues has led many governments to mandate periodic testing and certification of automobiles to ensure their conformation to emissions-control standards. Add to this the fact that automobiles are becoming increasingly high-tech, and it seems likely that every automotive shop will soon need to have an analyzer. This, in turn, creates a demand for electronics technicians who can perform service on them.

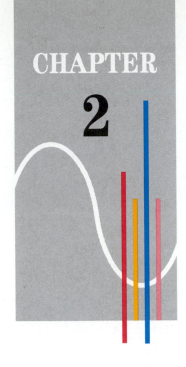

CHAPTER 2

Semiconductor Diodes

O B J E C T I V E S

Upon completion of this chapter, you should be able to:

1 Give the details (with numbers) of the influence of temperature on silicon conduction.

2 Explain (with sketches) how doping makes temperature less of a problem in silicon devices.

3 Define the characteristics of *n*-type and *p*-type silicon.

4 Explain how electrons and holes are distributed in a newly formed *p–n* junction.

5 Describe the forward- and reverse-bias conditions in a junction diode.

6 Explain Zener and avalanche operating modes in a junction diode.

7 List the factors that effect junction switching time.

8 Explain the difference between static and dynamic junction resistances.

9 Describe an optical isolator, and explain its function.

10 Define all of the important diode specifications, and identify the symbol for each.

11 Explain the advantages and the dangers of connecting diodes in series and in parallel.

12 Identify the rules for replacing defective diodes.

13 Explain the operation of Zener- and avalanche-mode devices, and list the important characteristics of avalanche-mode devices.

14 Explain the operating principles of the varactor diode.

15 Explain the theory underlying operation of light-emitting diodes (LEDs).

16 Explain the operating principle of the semiconductor laser.

17 State the meaning of the acronym *laser*.

18 Calculate the value of the current-limiting resistor for a LED at any supply voltage.

19 Explain how a diode functions as a one-way electrical valve.

20 Identify the standard junction voltage for any silicon diode.

Analysis In Brief

Diodes in Brief

FORWARD-BIASED DIODES
(Current Flows Through the Diode)

1　The anode must be positive with respect to the cathode.

2　There is always 0.6 to 0.7 volts across a working silicon diode junction. The value of 0.6–0.7 V is a constant, independent of diode type, size, or current.

3　Since the junction voltage is a constant 0.6–0.7 V even when the junction current changes, the junction resistance must change with the current. The junction resistance goes down as the junction current increases. The junction resistance is defined by the equation:

$$R_j = 25/I_j$$

where R_j is the junction resistance (in ohms), 25 is a constant 25 millivolts, and I_j is the junction current (in milliamps).

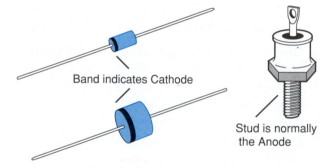

Band indicates Cathode

Stud is normally the Anode

Common Diode Case Styles

Anode ⟶　　　⟵ Cathode

Diode Symbol

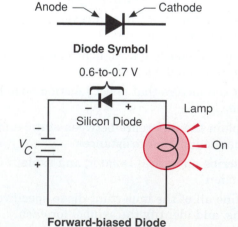

0.6-to-0.7 V

Silicon Diode

Lamp

On

V_C

Forward-biased Diode

4 The most important forward-biased diode specification is the maximum continuous forward current, $I_{F(max)}$.

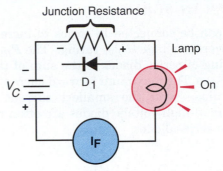

Forward-biased Diode

REVERSE-BIASED DIODE
(No Current Flows Through the Diode)

1 The anode is negative with respect to the cathode.

2 The diode is an open circuit. The reverse-bias voltage creates a wide depletion zone, which contains no free carriers and is an effective insulator.

3 A reverse-biased diode is a *capacitor*. The depletion zone is a *dielectric* with a dielectric constant of about 12. The carrier-rich anode and cathode areas serve as the plates. Increasing the reverse-bias voltage widens the depletion zone, lowering the capacitance value. The diode can be used as a voltage-variable capacitor.

4 The key reverse-biased diode specification is the maximum peak reverse voltage, $P_{RV(max)}$. The peak reverse voltage is often called the *peak inverse voltage, $P_{IV(max)}$*. The maximum reverse **dc** voltage, V_{RDC}, is also used. If this specification is exceeded, the diode goes into Zener/avalanche conduction.

Diode Symbol

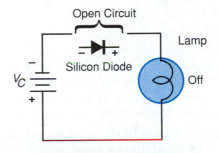

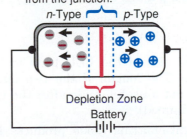

Reverse-biased Diodes

A higher reverse-bias voltage produces a wide depletion zone—equivalent to a thick capacitor dielectric and a lower capacitance value.

DIODES IN SERIES

Diodes can be connected in series to increase the peak reverse voltage (P_{RV}) rating. The P_{RV} rating of a string of series diodes is the sum of the individual P_{RV}s. The maximum forward current is that of the diode with the smallest current rating. Commercial high-voltage diodes are often made of multiple series diodes.

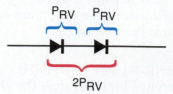

Diodes in Series Increase the P_{RV} Value

DIODES IN PARALLEL

It is not normally a good idea to connect diodes in parallel. One diode will turn on first, clamping the voltage across the other diodes at a voltage too low to turn them on. The first diode to turn on will conduct all the demanded current ("current hogging"). The other diodes will conduct no current and might as well not be there.

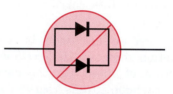

Current Hogging Makes Diodes in Parallel Impractical

Note: The relative diameter of a diode's leads is a better guide to its current rating than is body style or size. The leads carry the heat, so their diameter is related to the heat produced. This does not apply to diodes intended to be mounted on a heat sink.

REPLACING DIODES

1   The replacement diode must have a forward current capability as great as or greater than the diode it replaces. It is okay if the replacement has a much higher current rating.

2   The replacement diode must have a maximum P_{IV} (P_{RV}) capability as great as or greater than the diode it replaces. It is okay if the replacement has a much higher P_{IV} (P_{RV}) rating.

LIGHT-EMITTING DIODES

1   The flat side of the LED body indicates the cathode. The longer of the two leads is the anode (not a useful guide after the leads have been trimmed).

2   The light-emitting diode produces light only when it is forward biased.

3   The LED should not be connected in the reverse-bias direction or to **ac** if this can be avoided. The typical LED peak reverse voltage is only about 5 volts. A silicon diode in series with a LED can be used to extend the P_{RV}.

4   The LED junction voltage varies from about 1.2 to 2 volts.

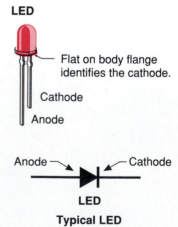

LED

Flat on body flange identifies the cathode.

Cathode

Anode

Anode → ← Cathode

LED

Typical LED

2.1   Introduction

The semiconductor diode is the simplest practical semiconductor device, but it is also one of the most useful. Some of the circuits we will study rely on the fact that the diode acts as a one-way electronic valve, allowing current to flow through it in only one direction.

Other circuits rely on the diode's constant voltage drop in the face of varying current flow. Still other circuits take advantage of a reverse-bias condition known as the *Zener,* or *avalanche mode*. The junction diode is also sensitive to heat and light, so special versions of it can be used to sense or measure light and heat values.

Special diodes, called *Varactor diodes,* behave as voltage-variable capacitors and are used to provide automatic electronic tuning and channel selection for radio, television, and communications equipment. Diodes made of gallium arsenide compounds emit light when current flows through them. Special versions of this light-emitting diode (LED) become semiconductor lasers, for use in compact-disk players and various other laser-based electronics systems.

Constructing multilayer diodes results in the almost magical transformation of diodes into transistors and a family of power control devices called *thyristors*. In this chapter, we will study basic diode theory, the way diodes work, and the special characteristics that make them so useful. In the next chapter, we will see how diodes are used in a variety of circuit applications.

2.2   Conductivity in Semiconductors

The physics of semiconductor materials is a highly complex subject; and no matter how complex you try to make your explanation, it is bound to present only partly the truth. Scientific investigation of the subject is a bit like peeling an onion: each level of discovery reveals a deeper layer with new questions to be answered and new lessons to be learned. Often, the most recently uncovered level indicates that some of what was learned at the previous level was not quite true. But if we must be liars by omission, let us try to adopt the half-truths that have the most practical value. What we will try to do is to examine the shape and form of the concepts that will be essential later on, as well as sketching in a few necessary details. We will stick with the ideas that best promote a practical understanding of real devices.

Two aspects of semiconductor physics are vitally important: the nature of conduction in semiconductors, and the effect of heat on that conduction. Our preoccupation with temperature may at times seem to be excessive; but heat has tremendous influence on the operating characteristics of semiconductor devices.

Conductors, Insulators, and Semiconductors

Materials can be categorized into three more or less distinct classes according to their electrical conductivity:

1 **Conductors,** which are mostly metals.

2 **Insulators,** which include many kinds of materials, from wood to complex organic material to plastics to glass (which is largely silicon and oxygen).

3 **Semiconductors,** which are really insulators driven into a state of conduction by heat energy. At temperatures near absolute zero (−273° C), semiconductors would be excellent insulators. But the higher the temperature, the better conductors semiconductor materials become. Within the range of temperature in which human life can exist, the conductivity of semiconductors ranges from poor to fair. And yet the picture is not quite complete, because the change in conductivity in relation to temperature is not simple and linear. In fact, an increase in temperature of only 10 Celsius degrees will increase silicon's conductivity by a factor of three. This means that, with a given applied voltage, the 10-degree rise in temperature triples the circuit current.

If we take a block of silicon and connect it in the circuit shown in Figure 2-1, we discover that the current in the circuit increases as the temperature of the block increases. In the case of Figure 2-1, increasing the temperature from 20°C (part a) to 40°C (part b) results in nine times the original 1mA current (assuming a times 3 figure), or 9 mA.

We should also expect the current to increase if the voltage is increased—and it does. But increasing the voltage from 10 to 20 volts (Figure 2-1(c)) increases the current (linearly) to only 2 mA, obeying Ohm's law.

Increasing the voltage increases the current at a linear rate, while increasing the temperature increases the current (without changing the voltage) at an exponential rate (even if we assume the times 2 value). This gives us a problem, because the temperature (over which we have no control) exerts far more influence on the current than does the voltage (which we ordinarily use to control circuit current).

There are two ways to make the current more voltage-dependent than temperature-dependent.

First, we can make free electrons available in large quantities by adding certain impurities in controlled amounts to the pure semiconductor material. These impurity-produced electrons are not temperature-dependent but they are voltage-dependent, and their conductivity is thus under our control. The impurity-produced electrons simply overwhelm the temperature-produced electrons by their greater numbers. The device manufacturer takes care of this part; we, the users of the device, are not responsible for it.

Second, we can take care to produce a good circuit design. In analog (often linear) circuits, this nearly always involves using negative feedback. Proper use of negative feedback is often our responsibility, as device users, and negative feedback will be a major theme of this text.

Structure of Semiconductor Crystals

Silicon has been the mainstay of the semiconductor industry for almost as long as there has been a semiconductor industry. Compounds such as gallium

FIGURE 2-1 Effects of Doubling the Voltage and of Doubling the Temperature on the Current in a Block of Silicon

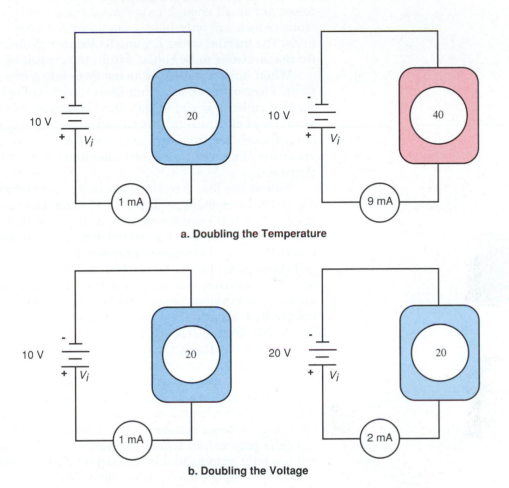

a. Doubling the Temperature

b. Doubling the Voltage

arsenide and silicon carbide have been used, but they have not yet presented any real threat to supplant silicon. Germanium was used early on, but gave way to silicon as soon as silicon manufacturing and processing techniques were perfected. Silicon is a tetravalent element, which simply means that four electrons lie in the atom's outermost orbit. These outlander electrons are the only ones that can leave the atom. We shall find that electrons behave differently when there are many atoms in close proximity.

In the case of metallic elements, the valence electrons retain nearly complete freedom of movement throughout the structure at all reasonable temperatures. This high degree of freedom of the valence electrons makes metals good conductors. Insulators, on the other hand, tend to form relatively strong electrostatic bonds that prevent valence electrons from leaving specific boundaries within the solid.

Aggregates of silicon atoms form a crystalline structure in which the electrostatic bonding forces that tend to restrict electron movement within specific boundaries are only moderately strong. These forces in semiconductors are sufficiently great that large external electric fields are needed to force valence electrons beyond their normal prescribed boundaries. At the same time, these forces are small enough to permit thermal energy to force some valence electrons (which are only lightly bound to the atom) beyond the normal boundaries. The inertial forces are small (about 0.05 electron volt), but it is not usual for the electrons to be bound within the crystal by them alone.

When large numbers of atoms form into a crystal, they tend to form additional electrostatic bonds that keep the electrons belonging to a specific community, called a *grain,* within the boundaries of that atomic community. The behavior of electrons in this situation is very complex. The model we will use here, though far from complete, has the distinct advantage of being easy to visualize. Moreover, this more understandable model is quite adequate for our purposes.

Silicon has four free electrons in its outer orbit. When many atoms of silicon exist in crystalline form, an electron-sharing phenomenon occurs. The result of all this complex sharing behavior is that, in effect, the formerly free valence electrons become bound within their particular grain boundaries inside the crystal structure. Figure 2-2 presents a stylized diagram of this so-called **covalent bond.** An electron around an individual silicon atom requires only 0.05 electron volt to free it from the atom. In a crystal, multitudes of atoms form covalent bonds, and these bind the electron to a pair of atoms so tightly that about 0.7 eV (more than ten times as much energy) is needed to break the electron free. In the covalent bonds illustrated in Figure 2-2, only the electrons in the outer ring (the valence electrons) can participate in current flow; we will concentrate on them.

Covalent bonds in semiconductors are so weak that many are broken at any temperature above absolute zero. The higher the temperature, the more covalent bonds are broken. Whenever a covalent bond is ruptured by thermal energy, an electron becomes free to wander and to be moved by an electric field (if one is present). In a pure silicon crystal, electrons freed by thermal energy are the only ones available for current flow; it would take impractically high voltages to rip electrons from their covalent moorings. When the electron leaves its home grain in the crystal, it leaves a **dangling bond** behind.

FIGURE 2-2   Covalent Bonds

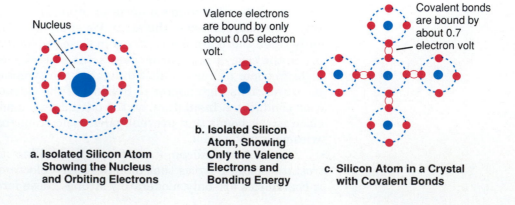

Nucleus

a. Isolated Silicon Atom Showing the Nucleus and Orbiting Electrons

Valence electrons are bound by only about 0.05 electron volt.

b. Isolated Silicon Atom, Showing Only the Valence Electrons and Bonding Energy

Covalent bonds are bound by about 0.7 electron volt

c. Silicon Atom in a Crystal with Covalent Bonds

FIGURE 2-3 **Electron/Hole Pairs Created by Heat Energy**

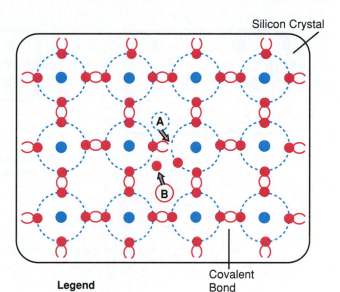

Legend
A-Dangling Bond (Hole)
B-Free Electron

This dangling bond has a definite attraction for any electron passing close by and may capture it if it strays too near. The dangling bond is called a **hole** in semiconductor jargon; and because of its attraction for electrons, it is conveniently treated as though it were a positively charged body. It is not really a body, of course. It is just a force field. Pretending that it is an object allows us to visualize complex semiconductor actions.

Once we agree to treat the hole as a positively charged particle, we must accord it all the rights and privileges of a duly authorized, charged body. Strange as it may seem, one of these privileges is mobility: the hole must have the freedom to move through the structure, just as the electron does.

This concept of hole mobility is, of course, a fiction, but it is a very convenient bit of fiction. Although we cannot really visualize holes moving under the influence of an electric field, we can easily visualize changing concentrations of holes.

For a simple sketch demonstrating this concept, see Figure 2-3, in which B is an electron whose covalent bond has been broken and has gained its freedom. The dangling bond A is the hole.

Electron and Hole Distribution

When no electric field is involved, free electrons are pretty evenly distributed throughout the semiconductor block (see Figure 2-4). Electrons are also being captured by dangling bonds (holes), and each capture annihilates an electron–hole pair. This is because, when an electron is captured by a hole, the hole is filled and no longer exists, and the captured electron is no longer a free electron. Electron–hole pairs are constantly being created and destroyed in about equal numbers. Consequently, a kind of equilibrium is established.

FIGURE 2-4   Distribution of Electrons and Holes in a Block of Silicon

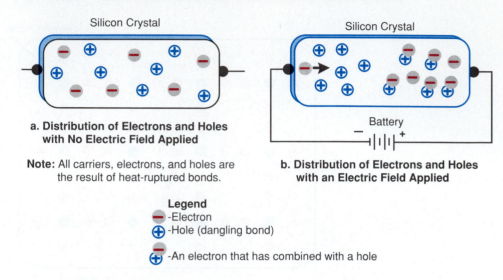

a. Distribution of Electrons and Holes with No Electric Field Applied

Note: All carriers, electrons, and holes are the result of heat-ruptured bonds.

b. Distribution of Electrons and Holes with an Electric Field Applied

Legend
⊖ -Electron
⊕ -Hole (dangling bond)
⊖⊕ -An electron that has combined with a hole

This uniform distribution is upset when an electric potential is applied across the block. Electrons rush to the positive end of the block as soon as heat energy frees them. Electron–hole pairs are still being formed at both ends of the block (and in the middle, too) at about the same rate, but the electric field produces a heavy concentration of electrons at the positive end and a shortage of electrons at the negative end.

Meanwhile, the holes that are being created at the negative end of the block are not capturing electrons, since almost none are available to capture, most of them having fled to the positive end of the block. Thus, we have an excess of unfilled holes at the negative end of the block.

Meanwhile, the positive end of the block possesses such an abundance of electrons that holes are filled almost as soon as thermal energy creates them. We now have a block of semiconductor material with a high concentration of electrons at one end and a high concentration of holes at the other.

n-Type Semiconductors

If a small quantity of some impurity (actually a different pure element) with five valence electrons is introduced into the crystal, we gain one free electron. Figure 2-5 presents a sketch of how this comes about.

The atom in the center of Figure 2-5 is called a **donor** because it donates a free electron to the cause. In addition to having five valence electrons, donor atoms must fit into the crystal structure comfortably. This restricts potential donor atoms to a few members of the ninety-odd natural elements. Two of the most commonly used donor atoms are arsenic and phosphorus.

On average, there may be 1 impurity atom to each 100,000 to 1 million atoms of silicon. If this relatively small quantity of donor atoms hardly seems worthwhile, it is because your perspective is shaped by the macroscopic world in which we live—a world where $1 million is a lot of money, and where $1 profit for $1 million invested would be a poor proposition indeed. In the realm

FIGURE 2-5   *n*-Type Silicon

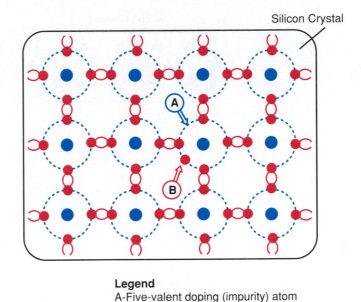

Legend
A-Five-valent doping (impurity) atom
B-Free electron–not bound by a covalent bond

of the atom, however, the perspective is different. For example, there are roughly 10^{22} (that is, 10,000,000,000,000,000,000,000) atoms in a cube of silicon that measures 1 centimeter (about 0.4 inch) on a side.

At the rate of 1 donor atom per 1 million silicon atoms, our 1-cubic-centimeter block would receive 10^{16} (10,000,000,000,000,000) free electrons from the donors. In addition to the donated electrons, another 10^9 electrons (approximately) are freed as a result of heat-rupturing of bonds between silicon atoms.

The great majority of the electrons available to participate in conduction are derived from the donors and are free at all temperatures above absolute zero (–273°C). In fact, at room temperature we have a system with roughly 1 million times more donor electrons than electrons freed due to heat-broken bonds. Donor electrons constitute the vast majority, and their number is essentially constant at all usual temperatures. This is because they are not part of a covalent bond.

So far, all we have accomplished by adding donor atoms—a process called **doping**—is to make a less temperature-sensitive semiconductor with an abundance of electrons. Because the electrons carry a negative charge, a semiconductor doped with donor atoms is called **n-type** (**n** for negative). The *n*-type semiconductor (if we ignore the few heat-generated electron–hole pairs) has only free electrons.

At temperatures near absolute zero, no heat, no heat-ruptured bonds, and no heat-produced electron–hole pairs are generated. In the real world, heat always produces some electron–hole pairs. But the heat-generated electrons simply join the donor electrons and become members of the team.

The holes constitute a small minority of the active particles (remember, we have agreed to treat holes—which are really immobile dangling bonds—as if

FIGURE 2-6   *n*-Type Silicon Crystal

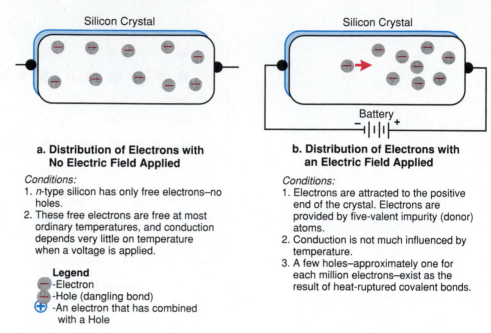

a. Distribution of Electrons with No Electric Field Applied

Conditions:
1. *n*-type silicon has only free electrons–no holes.
2. These free electrons are free at most ordinary temperatures, and conduction depends very little on temperature when a voltage is applied.

Legend
⊖ -Electron
⊖ -Hole (dangling bond)
⊕ -An electron that has combined with a Hole

b. Distribution of Electrons with an Electric Field Applied

Conditions:
1. Electrons are attracted to the positive end of the crystal. Electrons are provided by five-valent impurity (donor) atoms.
2. Conduction is not much influenced by temperature.
3. A few holes–approximately one for each million electrons–exist as the result of heat-ruptured covalent bonds.

they were positively charged particles) and always moves in a direction exactly opposite that taken by the majority electrons. Figure 2-6 illustrates electron and hole distributions in an *n*-type silicon crystal.

p-Type Semiconductors

In order to make practical semiconductor devices, we need another type of semiconductor to complement the *n*-type. This complementary type is called the **p-type;** it has only free holes (at absolute zero) and no free electrons.

Like *n*-type semiconductors, *p*-type semiconductors are made by doping pure semiconductor crystals. In this case, however, the doping impurities consist of atoms that have only three valence electrons. The trivalent doping atoms are called **acceptors,** because each one contributes a dangling bond (or hole) to the structure. These holes can accept electrons supplied by a battery (or other power source) from outside the block. Figure 2-7 presents a simple sketch of this process.

In doping the semiconductor crystal with trivalent atoms, such as those of the element aluminum, we create a crystal that has, at absolute zero, no free electrons but an abundance of holes (dangling bonds). The dangling bonds can accept electrons, which must be supplied by an external battery or other source of electrons if current is to flow through a *p*-type crystal.

If we connect a battery across a *p*-type crystal, as indicated in Figure 2-8, electrons jump off the negative electrode of the battery into nearby holes in the crystal, because they are attracted by the positive field. But the positive end of the block doesn't just have to attract the electrons; it also contains the positive

FIGURE 2-7 *p*-Type Silicon Crystal

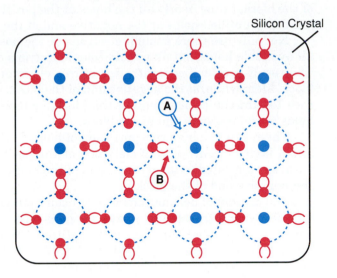

Silicon Crystal

Legend
A-Three-valent doping (impurity) atom
B-Dangling bond (hole)

FIGURE 2-8 Electron/Hole Distribution in *p*-Type Silicon

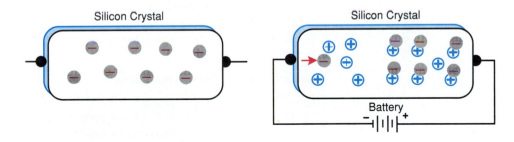

Silicon Crystal   Silicon Crystal

Battery

Conditions:
1. p-type silicon has only holes—no electrons (at absolute zero).
2. Holes are provided by three-valent impurity (acceptor) atoms.
3. Some electrons (1/million holes) exist, owing to heat-ruptured bonds.

Legend
⬤ -Electron
⊕ -Hole (dangling bond)
⬤⊕ -An Electron that has Combined with a Hole

a. Distribution of Holes with No Electric Field Applied

Conditions:
1. All the electrons involved in conduction must be supplied from outside the crystal, in this case by the battery.
2. Electrons injected into the crystal move freely in and out of available holes.
3. Conductivity is not a function of temperature, except for the small current resulting from electron-hole pairs produced by heat-ruptured bonds.

b. Distribution of Holes with an Electric Field Applied

field of the battery. The net result is that electrons race toward the positive end of the block, filling nearly all the holes at the positive end of the block but leaving many unfilled holes at the negative end of the block.

Not only are there enough electrons at the positive end of the block to fill all available holes, but many additional electrons can find no hole to fall into. The result is a high concentration of free electrons at the positive end of the block. Meanwhile, at the negative end of the block, the newly injected electrons race toward the positive end of the block, bouncing lightly in and out of the holes at the negative end of the block.

The result is a high concentration of holes at the negative end of the block and a high concentration of electrons at the positive end. It is as though the electrons migrated to the positive end of the block, while the holes migrated to the negative end of the block.

The concept of hole movement is crucial to the following discussion of junction diodes and transistors. Again, in the *p*-type semiconductor, electron–hole pairs are generated through heat-ruptured bonds. The holes generated by thermal activity join the *p* (positive) carriers (holes), which in this situation are the majority carriers. The heat-generated electrons join the injected electrons.

In the *p*-type semiconductor, electrons are called *minority carriers,* because all electrons either originate outside or result from thermally broken bonds. There are no free electrons at absolute zero in an electrically isolated crystal of *p*-type semiconductor. In practice, of course, it is rare to have such extraordinary control that no *p*-type impurities exist in an *n*-type crystal, or vice versa. The reality of mixed impurities may bend our imaginations a bit; but when one kind of impurity is sufficiently dominant, the opposing impurity may be ignored.

Table 2-1 shows the most common doping elements.

TABLE 2-1

Common Doping Elements for Silicon

p-Type	*n*-Type
Aluminum (A1)	Phosphorus (P)
Boron (B)	Arsenic (As)
Gallium (Ga)	Antimony (Sb)
Indium (In)	

2.3   Junction Diodes

There are dozens of manufacturing methods for accomplishing the doping process, but most of them start by diffusing donor and acceptor impurities into a block of carefully refined pure silicon. The process is much like diffusing butter into your warm breakfast toast.

The diffusion is accomplished at temperatures near 1000°C. Silicon is much like glass in that it can be almost (but not quite) liquid, and can be worked as a solid at very high temperatures. Watch a glass worker at the county fair sometime.

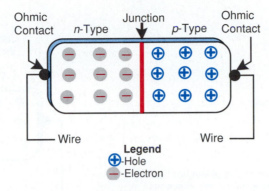

FIGURE 2-9 Semiconductor Junction Diode

To make a **diode,** the manufacturer introduces more p-type (acceptor) impurities into one end and more n-type (donor) impurities into the other end of the block. Within the block, a more or less abrupt transition from p-type to n-type silicon occurs. This transition is called a **junction** (see Figure 2-9).

The junction diode is essentially a one-way electrical valve—a **rectifier,** in electronics jargon. Our primary interest, at this stage of the game, is in the junction properties rather than in the device properties; but the device properties (with the exception of some power considerations) are the junction properties. Therefore, any distinction between device characteristics and junction characteristics is not very significant.

Depletion Zone

Figure 2-9 features a vertical line between the n-type area and the p-type area. The line seems to indicate some sort of barrier between the two regions; and indeed, there must be such a barrier. If there were not, what would prevent all the free electrons in the n-type region from gradually diffusing into the holes in the p-type region, leaving neither free electrons nor holes?

What is the nature and the origin of this barrier? In the beginning, when the doped crystal was being formed and cooled, electrons began to diffuse into holes, forming what is variously called a **transition zone,** a **depletion zone,** or a **barrier zone,** containing neither free electrons nor unoccupied holes. The zone thus acts as an insulator, because it has no available electrical carriers. But the zone gets only so wide; the zone-building process is self-arresting and stops upon completion of a very narrow transition (depletion) zone.

Figure 2-10 indicates that, as the transition zone grows wider, electrons must cross a widening area where carriers (electrons and holes) are few and far between. The amount of recombination decreases exponentially as the distance from the junction increases. The exponential gradient is shown by the **barrier-voltage** or **junction-voltage curve** in Figure 2-10.

Both the n and the p regions are electrically neutral in the beginning; but as the crystal is cooled, electrons diffuse across the junction and fall into holes on the other side. The n side loses some of its free electrons to holes on the p side, leaving the n side slightly positive. This makes the area next to the junction on the p side slightly negative, because of the electrons that have come over from the n side.

FIGURE 2-10 **Transition (Depletion) Zone**

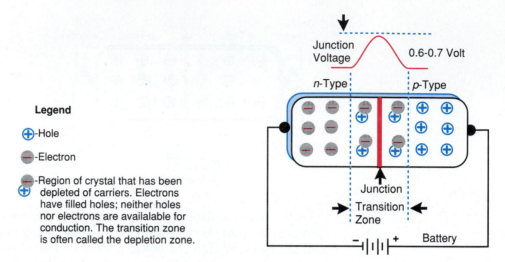

Legend

⊕ -Hole

⬤ -Electron

⬤ -Region of crystal that has been
⊕ depleted of carriers. Electrons
  have filled holes; neither holes
  nor electrons are availalable for
  conduction. The transition zone
  is often called the depletion zone.

In silicon, a barrier voltage equal to approximately 0.6 to 0.7 volt is formed, and further movement across the junction is prohibited until an external voltage greater than this junction voltage is applied across the device. The value of the junction voltage is a function of the semiconductor material used. All silicon devices have a junction voltage of 0.6 to 0.7 volts, give or take a few hundred millivolts, depending on the temperature (+2.5 mV/C°).

We cannot measure the diode junction voltage with a voltmeter, unless we put the diode (junction) into a circuit. In a circuit, we can verify its existence by applying a voltage across the device (with a series current-limiting resistor) starting at 0 volts, increasing the voltage to a little more than 0.6 V, and plotting current against voltage on a graph.

We will always be able to measure this 0.6 to 0.7-volt junction voltage in real working circuits, when the diode is in the conducting mode. That fact will prove very helpful when we begin troubleshooting circuits.

The important thing to remember about the transition zone junction voltage is that it establishes a minimum threshold voltage. Current cannot flow through the junction until a voltage of at least 0.6 to 0.7 V is applied across the junction. This critical junction voltage is sometimes (but not often) called the *hook voltage* or *potential hill*. The term *potential hill* refers to an energy diagram commonly used to describe transition zone formation. *Junction voltage* is today's common term.

Reverse-biased Junction Diode

The reverse-bias condition is the *off,* or nonconducting, condition. Figure 2-11(a) shows the reverse-bias battery polarity and the hole and electron distributions in a reverse-biased junction diode.

As the diagram in Figure 2-11(a) shows, electrons are attracted to the positive field (provided by the battery) at the left end of the drawing. Meanwhile, the holes in the *p* end of the block are attracted to the negative pole of the battery, which is connected to the right end of the drawing. The carriers—both

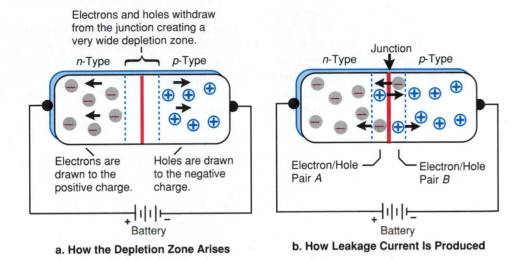

FIGURE 2-11 Reverse-bias Junction and Depletion Zone

Electrons and holes withdraw from the junction creating a very wide depletion zone.

n-Type   *p*-Type

Electrons are drawn to the positive charge.

Holes are drawn to the negative charge.

+ Battery −

a. How the Depletion Zone Arises

Junction

n-Type   *p*-Type

Electron/Hole Pair *A*

Electron/Hole Pair *B*

+ Battery −

b. How Leakage Current Is Produced

holes and electrons—withdraw from the junction, leaving a zone that is almost completely depleted of carriers. This area is known as the *depletion zone.*

There is no real difference between the depletion zone and the transition zone. The transition zone is a small permanent depletion zone that exists under any conditions. The depletion zone is a wider version of the transition zone, produced by applying an external reverse-bias voltage. The higher the reverse-bias voltage, the wider the depletion zone becomes. The depletion zone is essentially an insulator; it has no available carriers. No current flows through the junction area in the reverse-bias condition, except for a small leakage current.

Reverse-bias Leakage Current

So far, the description of reverse-bias conditions has been valid only for the donor- and acceptor-supplied carriers. It is time now to examine the behavior of the heat-generated electron-hole pairs. Remember that an electron that has been freed by heat energy leaves a hole behind. Heat energy always generates an electron-hole pair. This is true in both the *p* and the *n* ends of the diode crystal. Figure 2-11(b) shows a reverse-biased diode with electron-hole pairs being thermally generated within the depletion zone.

The arrows in the drawing in Figure 2-11(b) show the direction of carrier movement under the influence of the field produced by the battery. If you examine electron-hole pair *A,* you will see that the electron moves away from the junction to join the rest of the electrons in the *n* group, while the hole is repelled toward the junction.

In electron-hole pair *B,* the hole is attracted away from the junction to join the rest of the holes in the *p* group. The electron, on the other hand, is repelled toward the junction, where it combines with the hole from pair *A.*

This upsets the balance on each side of the crystal; the negative pole of the battery delivers an electron to the *p* end of the crystal, and the positive pole draws one from the *n* side of the crystal. One-half of the electron-hole pair *A* is forward-biased, and these carriers (the electron on the right and the hole on the left (in Figure 2-11(b)) constitute a small current flow that is called **leakage current.**

The other half of each pair is reverse-biased and does not cause any current flow. This leakage current is the only current in the reverse-bias direction; its magnitude is normally from 1/10,000 to 1/10,000,000 that of the forward-bias current. The amount of leakage current increases exponentially with temperature, but increasing the reverse-bias voltage has very little effect on the reverse-bias leakage current.

Only a few free electrons are available to become current, and even a small voltage will move all that are available. If there are no more free electrons to move, an increase in voltage can't increase the current.

In silicon, leakage current triples with each rise in temperature of 10 Celsius degrees.

Although leakage currents are present in the forward-bias condition as well, they are always measured in the reverse-bias condition; and the magnitude of the forward-bias leakage is assumed, for good theoretical reasons, to be equal to the reverse-bias leakage (at any given temperature).

The number of thermally generated electron-hole pairs is a function of temperature, not bias condition. In the forward-bias condition, all electrons in the n region and all the holes in the p region participate in forward conduction, regardless of whether they originated from doping atoms or were thermally generated. Thus, the leakage current is simply added to the forward conduction current, making it virtually impossible to measure the separate components. Forward-bias leakage is largely irrelevant anyway.

In a junction diode, leakage currents are generally of little importance; but the transistor has the ability to amplify, and leakage currents amplified by a factor of 100 or more can be a problem.

Junction Capacitance

A **capacitor** consists of two conductive plates separated by an insulating material called the **dielectric.** Although its physical dimensions are small, a reverse-biased diode meets all the requirements to become a full-fledged capacitor. It has two areas that are rich in free electrons and available holes (conductors, by definition), separated by the depletion zone, which has no available carriers and is therefore (by definition) an insulator. The dielectric constant of a silicon depletion zone is approximately 12.

The amount of capacitance in a junction diode is sufficiently small that it becomes a factor only at quite high frequencies. Special junction diodes are manufactured in such a way as to exhibit relatively high capacitances (on the order of 10 pF, more or less). These diodes are called **Varactor diodes** (or sometimes *Vari-cap diodes*). Varactor diodes are often used as voltage-variable capacitors in oscillator tank circuits and other applications such as for television channel selection. We will examine some applications later in the text.

The capacitance of a reverse-biased junction is varied by varying the bias voltage. Increasing the voltage increases the width of the depletion zone (dielectric), decreasing the capacitance. Making the dielectric thicker makes the capacitance smaller.

This relatively small capacitance in a diode junction can become the equivalent of a fairly large capacitor in a transistor, because capacitive effects are often amplified along with the signal and some other intrinsic parameters.

Forward-biased Junction Diode

The **forward-bias** condition is the conducting condition for a junction diode. Figure 2-12 shows the battery polarity and the distribution of electrons and holes in the forward-bias connection. Perhaps the best way to explain the action that takes place is to list the sequence of events involved:

1   The negative field at the *n* end of the device repels the free electrons toward the junction (the voltage must be higher than the junction (transition) zone voltage, 0.7 V, or the process ends at the transition zone).

2   The positive field at the *p* end of the block repels the holes toward the junction. The arrows within the block show the direction of carrier movements within the block.

3   If the battery voltage is higher than the junction voltage (approximately 0.7 V, for silicon), electrons are forced through the transition zone, across the junction, and into the *p* region. An electron that crosses the junction into the *p* region is promptly captured by a hole.

4   The electron has left the *n* region, leaving it one electron short. The *n* crystal, which was electrically neutral, now becomes one unit positive, since it now has one more proton than it has electrons. This positive charge draws an electron from the negative terminal of the battery. One electron has left the battery.

5   Meanwhile, back at the *p* side of the block, our wandering electron has dropped into a hole. Once the hole is filled, it ceases to be a hole. For practical purposes, one hole has been lost and an electron has been gained by the *p* crystal.

FIGURE 2-12   **Forward-biased Junction Diode**

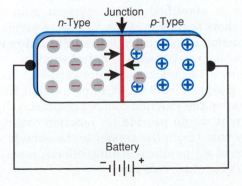

6 The *p* part of the crystal now has an excess electron and consequently a 1-unit negative charge. The positive terminal of the battery draws an electron out of the right-hand end of the block to restore the charge balance in the *p* end of the crystal (see Figure 2-12). One electron has now left the negative battery terminal, and one electron has returned to the positive battery terminal.

7 If you can visualize some 10^{14} to 10^{19} electrons crossing the junction and jumping into holes in the *p* region near the junction, while a like number of electrons leave the negative terminal of the battery and a like number are returned to the positive pole of the battery, you have a rough picture of a forward-biased diode in action. Forward bias is the conducting condition for a junction diode.

Diode Conduction Curve

The diode conduction curve plots diode voltage against diode current for both forward and reverse-bias conditions. The forward-bias part of the curve shows the sudden transition at 0.6 to 0.7 V, from almost no current to a very high current as the forward voltage increases. The reverse-bias part of the curve illustrates a similar transition at the Zener knee.

Forward-bias Conduction

Now let's take a quantitative look at the nature of forward-conduction current and leakage current. Nature being what it is, we could hardly expect the current-voltage relationship to be simple. The best way to describe these relationships is to draw a graph like the one shown in Figure 2-13.

If you examine the upper right-hand quadrant, you will notice that almost no current flows until the junction voltage is reached. The small current that does flow is mostly due to thermally generated carriers within the depletion zone. As soon as the junction voltage is reached, a small increase in forward-bias voltage produces a large increase in current. But the curve is not quite linear, and that means that the junction resistance changes in a nonlinear fashion as the voltage across the junction changes.

If our graph had numerical values and we calculated junction resistance $(R = V/I)$ at several points along the graph, we would find that the junction resistance decreases as the voltage increases. Again this change in junction resistance is only occasionally important in a junction diode; but bipolar transistors involve forward-biased junctions, and in a transistor the junction resistance value is amplified by the current gain of the transistor. A careful analysis of short segments of the V/I curve yields an important empirical formula for junction resistance called the **Shockley relationship.** We will examine the Shockley relationship in detail shortly.

As Figure 2-13 indicates, the forward-bias conduction curve rises steeply and shows no sign of leveling off, implying that there is no limit to the current. If we could keep the junction cool, we could run the current up quite high; and to some extent we do provide for junction cooling in the form of heat sinks, extra-heavy wires from the crystal to the outside world, and conductive cases. But in spite of all practical cooling efforts, junction temperature determines the upper limits of current flow.

WARNING
The principal failure mechanism in a forward-biased junction is excessive junction temperature. Excessive junction temperature may occur before it is noticeable on the outside of the diode. A diode must never be operated without some series resistance. Normally, the load resistance serves to limit the current so it won't go off the chart in Figure 2-13. A shorted load spells "dead diode."

HINT
You can always replace an ordinary diode with one of a much higher forward-current rating. It may even be cheaper to do so. CAUTION: Higher-current diodes may have leads too large for the circuit board holes.

FIGURE 2-13   Diode Conduction Characteristics

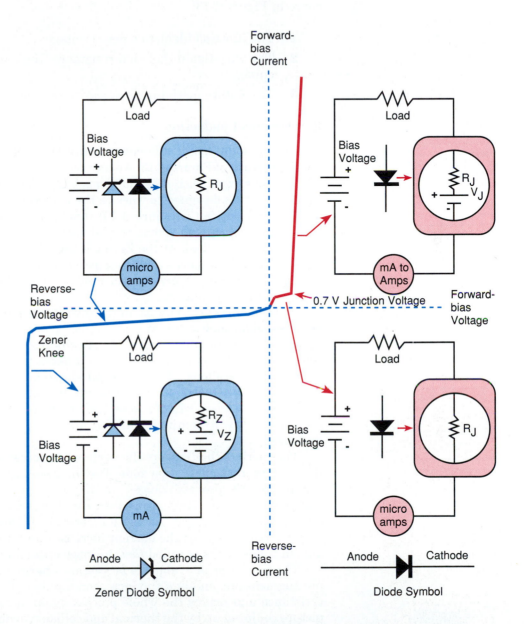

To avoid excessively high junction temperatures, we must follow the manufacturer's instructions with regard to maximum voltage, current, and mounting arrangements. The manufacturer has probably burned out a fair number of diodes in determining these ratings, and it would be foolish for us to repeat those efforts. It is wise, however, to assume that those tests were conducted in the laboratory, under conditions that we are not likely to encounter. We should play it safe and operate diodes only up to 80 percent or so of the manufacturer's maximum ratings. How conservative we can permit ourselves to be is often determined by cost factors.

There are three principal points of interest on the forward-bias conduction curve in Figure 2-13:

1 The junction (depletion zone) voltage.
2 The variation in junction resistance (the nonlinearity and slope of the curve).
3 The failure mechanism: excessive junction temperature.

Reverse-bias Conduction

Now, let's take a look at the **reverse-bias** part of the curve in the lower left-hand quadrant of Figure 2-13. We have encountered a new symbol in Figure 2-13: a diode symbol with diagonal arms added. This is the symbol for a **Zener diode,** which is designed to work in the reverse-bias mode.

You will notice (in Figure 2-13) that increasing the reverse-bias voltage increases the reverse current only slightly until a critical voltage called the *Zener point* is reached. At the **Zener knee** of the curve, the current suddenly rises from microamperes to milliamperes or amperes. There are actually two principal mechanisms involved in production of this current: field emission, and ionization by collision.

As we increase the voltage in the reverse-bias direction, the depletion zone becomes wider and wider. At some critical point, the field is large enough and the path a carrier can travel without colliding with an atom is long enough that an electron can achieve the speed it needs to knock electrons out of atoms with which it collides. The dislodged electrons, in turn, pick up speed until they collide with other atoms, liberating still more electrons. Figure 2-14 depicts this process, which is often called an **avalanche** because, unless it is controlled, it will continue building like an avalanche until the diode is destroyed.

The carriers that initiate the avalanche process are heat-generated minority carriers in the depletion zone. The increase in current at the beginning of the avalanche process causes an increase in temperature, because any increase in current through a given resistance causes a larger voltage drop and an increase in heat. As the temperature increases, more thermally produced carriers appear; the current increases and produces more heat; and on and on it goes. This is called *thermal regeneration* or *thermal runaway*.

The two kinds of avalanche processes—thermal and collision—do not simply augment one another; the increased number of carriers lowers the system resistance and makes the whole process dynamic and complex. However, the picture conjured up by the thermal and collision avalanche mechanisms is adequate for our purposes. We need not complicate it with further details and possible confusion.

Field emission was hypothesized by Zener as the sole generative mechanism; and at lower voltages, it is. But, at higher voltages, the avalanche mechanism becomes dominant. Field emission is simply the concept that an atom's electrons can be torn out of orbit if the field voltage is high enough.

Members of the thyristor family—TRIACs and SCRs—are designed to operate in a controlled avalanche mode. Controlled avalanche and Zener operation modes are permissible as long as the junction temperature is not permitted to go too high. This is normally accomplished by restricting the current

FIGURE 2-14 Avalanche Effect

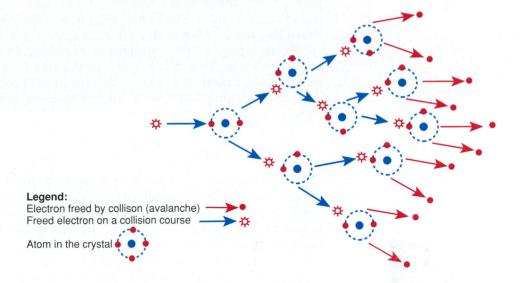

Legend:
Electron freed by collison (avalanche)
Freed electron on a collision course

Atom in the crystal

through the diode by placing a limiting resistance in the circuit and by removing the heat that is generated.

Most of the diodes that operate in this part of the curve are specially designed for such service. Zener and avalanche diodes have their Zener knee placed at commonly useful voltages—typically from 1 to 100 volts. Ordinary diodes typically have the Zener knee placed at 200 to 1000 volts. Members of the thyristor family of avalanche devices are specifically constructed for optimum characteristics in avalanche mode.

This same limited avalanche operation was once one of the chief failure mechanisms in Zener and avalanche devices, and in a transistor's reverse-biased junction. There is a tendency for the avalanche breakdown to occur first in a limited area, forming a sort of tunnel of avalanche current and (more often than not) causing a fatal hot spot. This thermally tunneling failure mechanism is no longer a serious problem except when voltages and currents are too close to the manufacturer's absolute maximum ratings, and the junction temperature rises.

Junction Switching Time

One specification sometimes given for certain junction devices is **recovery time.** It takes a finite amount of time to switch a junction from a forward- to a reverse-bias condition—that is, from a conducting to a nonconducting state. (The reverse is also true.) This specification is important to computer-destined devices, to high-power devices that tend to be slow, and to others whose functions involve rapid turn-on or turn-off.

When a junction is forward-biased, the depletion zone is saturated with carriers because of the large number of recombinations (holes and electrons) taking place at the junction. If the bias is suddenly removed, conduction does not cease instantly, because the nonconducting (reverse-bias) state cannot be obtained until all carriers (holes and electrons) have been cleared out of the depletion zone. This short period of current flow after the bias has dropped to zero is called the *storage phenomenon*. The time required to clear the carriers out of the transition (depletion) zone and restore a proper depletion zone is called the **storage time.** You will read more about this later.

There is also a delay involved in switching from a reverse-bias to a forward-bias state. This delay is due to the fact that the junction capacitance soaks up the initial forward-bias current in the process of taking on a charge. After the charging is complete, current becomes available for normal conduction.

Resistance of a Junction Diode

The resistance of a forward-biased junction diode consists of two principal components: the bulk resistance and the junction resistance. The **bulk resistance** is simply the ohmic resistance of the p and n blocks of silicon; and with the exception of some variation due to temperature changes, this resistance is fairly constant. Except under conditions involving very high current, the bulk resistance is usually too low to be of any concern.

Figure 2-15 shows forward- and reverse-bias connections and the diode symbol. The arrow shape of the diode symbol indicates the direction of forward-bias conventional current flow. If you are used to thinking in terms of electron current, it is better to view the diode symbol as being just a symbol with a triangle-shaped part: forget any arrow or directional ideas. This can help you avoid some confusion. If you view the flat (line) of the symbol as being wrapped around one end (as a band) of a real diode body, it will help you to remember which end is the cathode.

Notice that the actual forward current in the circuit (in Figure 2-15) is slightly reduced because the constant 0.7-volt drop across the diode junction must be subtracted from the 10-volt battery voltage. The 0.7-volt drop across the diode between part (a) and part (b) of the figure reduces the calculated current slightly. This may or may not be important in a particular circuit. Here, we have

$$10\text{ V} - 0.7\text{ V} = 9.3\text{ V}$$

and

$$I = V/R = 9.3\text{ V}/10\text{ }\Omega = 0.93\text{ A}$$

An additional reduction in circuit current occurs because of the diode junction resistance, which we have not yet considered. The **junction resistance** is the effective resistance of the depletion zone. It depends on operating current and, to a small extent, on temperature. An empirical equation derived by Shockley gives the junction resistance (the Shockley relationship) as

$$R_J \cong 25\text{ mV}/I\text{ mA}$$

where R_J is the junction resistance (in ohms), 25 is a constant (in millivolts), I is the current through the junction (in milliamps), and $\cong$ means "approximately equal to." The millivolts may be implied, with only the 25 shown in the numerator.

FIGURE 2-15 **Forward- and Reverse-biased Diode Connections**

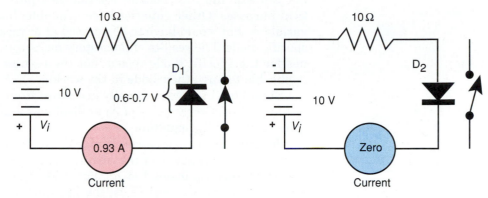

a. Forward-bias Connection **b. Reverse-bias Connection**

Example 2.1

Finding the Resistance of a Forward-biased Diode

Calculate the junction resistance for the forward-biased diode in Figure 2-15.

Solution

$$R_J = 25 \text{ mV}/I_j = 25 \text{ mV}/930 \text{ mA} = 0.027 \ \Omega$$

Obviously, the junction resistance is quite small at nearly 1 amp (ampere) of current and would have an insignificant effect on the current in Figure 2-15. An ampere is quite a large current in modern electronic circuits, however, so let's try another example.

Example 2.2

Finding the Voltage Drop Across a Forward-biased Diode

Calculate the junction resistance for the forward-biased diode in Figure 2-15, using a current of 0.5 mA. Then find the voltage drop across the junction.

$$R_J = 25 \text{ mV}/0.5 \text{ mA} = 50 \ \Omega$$

The voltage drop across the junction, resulting from the 0.5-mA current through it, is only 0.025 volts:

$$V = IR \quad \text{(Ohm's law)}$$
$$= 0.0005 \text{ A} \times 50 \ \Omega = 0.025 \text{ V}$$

CRITICAL FACT

Don't forget: there is always a 0.6 to 0.7-volt drop across a forward-bias silicon junction, regardless of the current or the physical size of the diode.

Students trained in the sciences are inclined to include all small-order effects, and to carry values out to as many decimal places as possible. But if you feel inclined to add the 0.025-volt junction resistance voltage drop to the 0.7-volt drop for greater accuracy, forget it.

The 0.025-volt figure is considerably smaller than the probable error in assuming that 0.6 to 0.7 volt is the precise value of the junction voltage. The value of 0.025 volt is also probably inaccurate, because the equation used to calculate it (the Shockley relationship) is a statistically derived empirical equation and can only act as a useful approximation.

There is no other useful junction resistance equation for the technologist, but you will find the Shockley relationship quite adequate for nearly all practical purposes. Other equations are available, but they all require values for variables that would have to be guessed at or measured in the laboratory on a specific diode. The results of laboratory measurements could yield precise values for the specific diode tested, but the results could not be assumed to be applicable to any other diode in the world.

Shockley's little gem is really just a restatement of Ohm's law, but he gave us a useful statistical value of 25 millivolts as a mathematical constant for the voltage part of the equation.

Junction resistance is rarely significant in most diode circuits, but it can become very significant in transistor circuits. In transistors, we call the junction resistance R'_e instead of R_J, but aside from the symbols, R_J and R'_e are identical. In transistor circuits, the junction resistance value is often multiplied by the gain of the transistor amplifier. Amplified resistance is another fascinating aspect of negative feedback.

Dynamic and Static Resistances

A semiconductor junction is a complex system, and part of its complexity is that it has two distinct resistances in both the forward- and reverse-bias directions. In one sense, these two resistances depend on how the resistance is measured; but in a broader sense, the two resistances reflect different reactions to different operating conditions. **Static resistance** is easy to understand because it involves the simplest form of Ohm's law:

$$R = V/I$$

Dynamic resistance is a little more difficult to understand because it must be stated in terms of changes in current caused by changes in voltage. In mathematical form:

$$R_D = \Delta V/\Delta I$$

> **NOTE**
> Δ [delta] means "the change in . . ."

where R_D is the dynamic resistance (in ohms), ΔV (delta V) is the change in voltage (in volts), and ΔI (delta I) is the change in current (in amperes).

In the case of forward-biased diodes, we know that the junction resistance depends on the junction current, as expressed in the Shockley relationship. If we know the minimum and maximum currents in a particular application, we can calculate the dynamic forward-bias junction resistance, because the Shockley relationship tells us that the voltage part of the resistance equation is a constant 25 millivolts.

Dynamic resistance is a trivial parameter in forward-biased junctions. But in a reverse-biased junction (such as the collector-base junction in a transistor), it is quite a different story. The static resistance of a reverse-biased junction is simply

$$R = V/I$$

where V is the voltage impressed across the junction, and I is the current through the junction (which in a simple diode is only the leakage current). In this case, the concept of reverse-bias static resistance is not useful, because leakage current is a function of temperature and has almost nothing to do with the applied voltage. The equation is a proper expression of Ohm's law, but it is not useful.

The dynamic resistance is a measure of how changes in voltage affect changes in current. If you look again at the reverse-bias part of the diode conduction curve (Figure 2-13), you will notice that the reverse leakage current increases only very slightly as the reverse-bias voltage increases. In a more or less typical case, we would expect no more than a 10% increase in current for a 100% increase in the applied voltage.

Example 2.3

Calculating the Static and Dynamic Resistance of a Reverse-biased Diode

Suppose that we have a diode connected in the reverse-bias direction and that, with 10 volts applied, we measure a current of 10 μA (microamps). Compute the static resistance of the junction.

Solution

$$R_S = V/I$$
$$= 10 \text{ V}/10 \text{ μA}$$
$$= 1 \text{ M}\Omega \text{ (megohm)}$$

Now, assume that a doubling of the voltage produces a current increase of 10%; and compute the dynamic resistance of the reverse-bias diode when the voltage is increased from 10 V to 20 V:

1   10% of 10 μA is equal to 1 μA.

2   ΔV is equal to 10 V (the change from 10 V to 20 V).

3   ΔI is equal to 1 μA.

$$R_D = \Delta V/\Delta I$$
$$= 10 \text{ V}/1 \text{ μA}$$
$$= 10 \text{ M}\Omega$$

Of course, this is just a made-up example, but the results obtained from it are fairly typical. The absolute values will vary over a wide range, but the implication here that the dynamic resistance is always greater than the static resistance in a junction diode in the reverse-bias state is valid. The static resistance of a reverse-biased diode is controlled by temperature. The dynamic resistance is mostly independent of temperature.

This is only a description of the dynamic resistance of a reverse-biased junction, not an explanation. The explanation will have to be a bit oversimplified, but here it is:

1   We would expect an increase in reverse-bias voltage to scavenge available heat-generated electrons more efficiently from the depletion zone, resulting in an increase in current.

2   However, working against the increased voltage is the fact that simultaneously, the depletion zone is widened by the increased potential.

3   The longer path through the high-resistance depletion zone that the carriers must travel very nearly balances out the effect of the increased voltage applied across the reverse-biased junction.

4   The result is that the current does not increase as much as Ohm's law (static resistance) would lead us to expect for a given increase in voltage. The static and dynamic resistances of an ordinary resistor are the same, but a reverse-biased diode is not at all like an ordinary resistor.

The dynamic resistance property is not particularly important in most ordinary diode applications; at least, we don't have to be concerned with it. But, bipolar transistors use reverse-biased junctions; and the concept of dynamic reverse-bias junction resistance is absolutely essential to understanding and applying transistors. So don't write it off. It will be very important later.

Junction Diode Ratings and Specifications

Diodes come in a wide variety of types, shapes, and sizes. Each special type has some special ratings, but certain ratings are common to nearly all diode devices. We discuss these next.

Important Diode Ratings

Five diode ratings are especially important and merit individual attention.

Peak Reverse Voltage (P_{RV}) or Peak Inverse Voltage (P_{IV})   The **peak inverse voltage (P_{IV})** or **peak reverse voltage (P_{RV})** defines the maximum momentary reverse-bias voltage that can be applied before the diode slips into Zener or avalanche conduction. This reverse voltage must not be exceeded in devices not designed for avalanche-mode operation. For devices that are designed for Zener- or avalanche-mode operation, the peak reverse bias voltage is variously called *reverse breakover voltage, reverse breakdown voltage, Zener voltage,* and or *avalanche voltage.*

Maximum Reverse dc Voltage (V_{Rdc})   The **maximum reverse dc voltage (V_{Rdc})** is theoretically the same as the peak reverse voltage, but it is often listed as having a slightly lower voltage. Specifying a slightly lower voltage gives some leeway for internal heating (caused by leakage current flow), which may (given a little time) shift the Zener knee slightly. Some manufacturers use the same values for both P_{RV} (or P_{IV}) and V_{Rdc} because the P_{RV} rating is generally quite conservative. If the manufacturer specifies a P_{RV} or V_{Rdc} of 200 volts, for example, this signifies that 100% of the diodes of that type number will have the Zener knee located at 200 volts or higher. For perhaps 85% of those diodes, the Zener knee will be located somewhere between 300 and 400 volts; but the manufacturer only guarantees a P_{RV} of 200 volts.

Maximum Forward dc Current ($I_{F(max)}$) **Maximum forward dc current ($I_{F(max)}$)** is the maximum allowable continuous direct current flow in the forward-bias direction. The $I_{F(max)}$ contains a temperature specification. The

$I_{F(max)}$ decreases as the temperature increases and derating curves or data are generally included in the manufacturer's data sheet. In the case of power diodes, where special cooling considerations are involved, the manufacturer generally provides some information about required heat-sink size and so on.

Forward Voltage (V_F) **Forward voltage (V_F)** is the voltage drop across the diode in the forward-conducting mode. This voltage drop includes the 0.6- to 0.7-V junction potential plus any bulk resistance voltage drop. Both current and temperature are usually included in this specification. Typical values are very close to the junction voltage of 0.6 to 0.7 volt, but they may be as high as 1.2 volts for diodes conducting very high currents or having unusually high internal bulk resistances. The junction voltage itself decreases at the rate of about 2.5 millivolts per Celsius degree of increase in temperature.

Reverse-bias Leakage Current (I_R) The **reverse-bias leakage current** is not always listed in the manufacturer's data sheet. In many diode applications, it is insignificant. When the values are given, they are totally temperature-dependent. The leakage current triples for each 10 Celsius degrees of increase in temperature in all silicon devices.

Manufacturers may use different symbols, but they will list the parameters we have discussed. They will also list other parameters, which may be of interest to design engineers. Figure 2-16 illustrates some typical diode case styles, mounting arrangements, and heat sinks.

Diodes Connected in Series

Connecting diodes in series increases their peak inverse (reverse) voltage rating. For example, a 200-P_{IV} diode and a 400-P_{IV} diode connected in series yield a peak inverse voltage rating of 600 volts. It is more common, in practice, to place identical diodes in series, because of other considerations. High-voltage diodes often consist of a package of several to many diodes connected in series to get the required peak inverse voltage. High-voltage power supply diodes in color television receivers, for example, must have peak reverse voltages of 20 to 30 kV, and they may consist of 25 to 50 series diode cells in a package.

When diodes are connected in series, there is no improvement in the maximum forward current $I_{F(max)}$. The maximum forward current is equal to that of the diode in the series string that has the lowest $I_{F(max)}$ rating. If all diodes in the string are identical, the maximum forward current of the string of series diodes is the same as that of any one of the individual diodes.

Reverse-bias leakage current in a string of diodes is approximately equal to the leakage current of the diode with the lowest leakage current. The leakage current is limited by the number of electron-hole pairs formed by thermal rupture of covalent bonds that are available to form leakage current. As a practical matter, the leakage current in a string of identical diodes is about the same as the leakage current in any individual diode at a particular temperature. Anyway, normal leakage current in diodes is usually unimportant. Figure 2-17 illustrates series diode characteristics.

FIGURE 2-16 **Diode Case Styles and Mounting Arrangements**

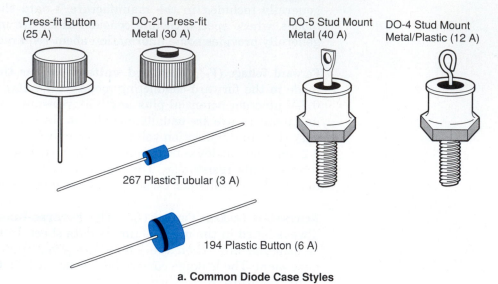

Press-fit Button
(25 A)

DO-21 Press-fit
Metal (30 A)

DO-5 Stud Mount
Metal (40 A)

DO-4 Stud Mount
Metal/Plastic (12 A)

267 PlasticTubular (3 A)

194 Plastic Button (6 A)

a. Common Diode Case Styles

Cathode — Anode

b. Diode Symbol

c. Bridge Rectifier Packages

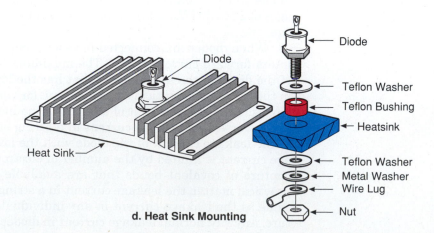

Diode

Diode

Teflon Washer

Teflon Bushing

Heatsink

Heat Sink

Teflon Washer

Metal Washer

Wire Lug

Nut

d. Heat Sink Mounting

FIGURE 2-17   Connecting Diodes in Series

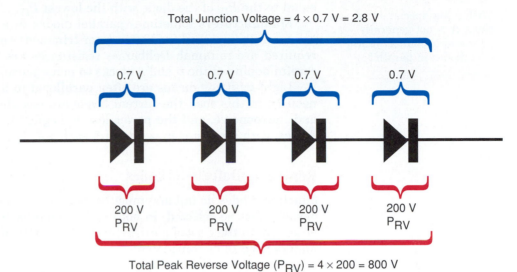

Total Junction Voltage = 4 × 0.7 V = 2.8 V

0.7 V   0.7 V   0.7 V   0.7 V

200 V   200 V   200 V   200 V
P_RV   P_RV   P_RV   P_RV

Total Peak Reverse Voltage (P_RV) = 4 × 200 = 800 V

Diodes Connected in Parallel

As a general rule, connecting diodes in parallel is not a good idea. The total forward-current capability can theoretically be increased by paralleling diodes, but a phenomenon known as **current hogging** complicates the situation.

When diodes are connected in parallel and power is applied, one of the diodes nearly always starts conducting before the others. The first diode to start conducting pulls the voltage across the other diodes down to its junction voltage.

Because the first diode to start conducting has a slightly lower junction voltage than do the rest of the diodes, the forward voltage across the remaining diodes is never high enough to turn them on. It only takes a few millivolts to make the difference. The diode that takes off first acts as if it thinks it is alone, and it tries to conduct all the current demanded by the system.

This is generally catastrophic for the conducting diode, which usually shorts. And if you replace that diode, the laws of chance predict that one of the other diodes will conduct first, hog the current, and short. You can find yourself playing a frustrating game of musical diodes if you don't realize what is going on. In each case, only one diode will turn on, and it will have to do all of the work by itself.

It is possible to parallel diodes successfully if each diode is equipped with a series resistor. The price that must be paid in this case is the power lost as heat in each resistor. For low and medium current levels, individual diodes with adequate current ratings are available at reasonable cost, and are probably cheaper than combinations of power resistors and several diodes.

At higher power levels (50 Amps and above), diodes get more expensive, but resistors also get more expensive, and they produce a great deal of heat and wasted power. The reverse-bias leakage current also increases when diodes are connected in parallel. Each diode adds its individual leakage current to the system. The peak inverse voltage in a group of parallel diodes is equal to the P_{IV} of the diode with the lowest P_{IV}.

Manufacturers sometimes parallel diodes in a package, but those diodes are carefully matched (and often laser-trimmed), so that the series resistances required are minimal. Deliberate resistances are sometimes added through careful doping of the p and n blocks to make paralleling them practical.

Light-emitting diodes are often paralleled to light several LEDs simultaneously. In this case, the current levels are low, the series resistors are small and inexpensive, and the power loss is negligible. Paralleling light-emitting diodes, with a resistor in series with each one, is a common practice.

Replacing Defective Diodes

Junction diodes do fail and must be replaced, and diodes for an experimental design must be selected. In both cases, some basic rules of thumb should be observed. In the case of garden-variety silicon diodes, only two electrical parameters need to be considered:

1   The peak inverse (reverse) voltage (P_{IV} or P_{RV}).
2   The maximum forward-current rating (I_F).

Always select a replacement diode where P_{IV} is as high as or higher than the P_{IV} of the diode being replaced. A 50-volt P_{IV} diode can successfully be replaced with a 200-, a 400-, or even a 600-volt P_{IV} unit. The P_{IV} of the replacement diode could be even higher, but the diode's cost and physical dimensions might make that impractical.

Always select a replacement diode whose maximum forward current (I_F) is equal to or greater than the I_F of the diode being replaced. A 500-mA diode may be replaced with a 1-amp, a 5-amp, or an even higher-current diode. The major problem with higher-current diodes is that they tend to get larger and take on different case styles, so they may require different mounting arrangements from that of the diode being replaced. Higher-current diodes may or may not be more expensive than a matching replacement.

For example, if you are replacing a 500-mA, 50-P_{IV} diode, your best choice would probably be a 1-amp, 200-P_{IV} diode or a 1-amp, 600-P_{IV} device. These two types of 1-amp diodes are so popular that they are generally cheaper and more readily available than less popular lower-current types.

Selecting a diode with a higher P_{IV} and I_F rating tends to extend the maximum instantaneous peak currents and voltages and other parameters in a direction that provides a greater safety factor. Higher-current diodes tend to have higher leakage currents that could theoretically pose problems, but you probably could not find the space to install such a diode in an existing circuit.

There are of course exceptions to these general rules of thumb, but the rules will answer for about 99% of all generally encountered replacement problems. Physical mounting problems generally prove to be the most vexing aspect of selecting a replacement diode. Many smaller diodes have large diameter leads to provide conduction cooling; and as a result, the leads in a higher-current diode may not fit the holes in a circuit board. Higher-power devices may have stud-mounted or press-fit case styles. It is important to find the proper case style to ensure adequate cooling.

2.4 Special Diodes

There are a number of special diodes, some of which are made of silicon and simply take advantage of special properties of silicon diodes. Other special diodes are made of other materials.

Fixed 0.6-volt Regulators

Silicon diodes can be used as voltage-regulating diodes. The forward-based silicon junction has a reliable 0.6- to 0.7-V junction voltage and can be used as a fixed-voltage regulator. The junction voltage is constant over a wide range of currents.

Varactor Diodes (Voltage-variable Capacitors)

Silicon diodes, in the reverse-bias mode, are small capacitors consisting of a *p*-type conducting area, an *n*-type conducting area, and a depletion zone that serves as the insulating dielectric. The depletion zone has no free electrical carriers and thus acts as an insulator. Varying the applied reverse-bias voltage varies the thickness of the depletion-zone dielectric, thus varying the capacitance. The manufacturer can control the capacitance range during the manufacturing process. Specially designed diodes that are intended to be used as voltage-controlled capacitors are called *varactor diodes*.

Zener- and Avalanche-mode Devices

Zener (avalanche) diodes are available that start to conduct at reverse-bias voltages of between about 2 volts and 200 volts. Most of the electronics literature identifies diodes that conduct with reverse-bias voltages up to about 4 volts (where very little avalanche behavior is involved) as true Zeners. At voltages of between 4 and 6 volts, both mechanisms—Zener and avalanche—are believed to be more or less equally involved. For Zener voltages in excess of 6 volts, the avalanche mechanism dominates.

The transition from Zener to avalanche mode is so smooth that no easily measurable difference in the diode's action occurs as the transition is made. For practical purposes, the diode behaves the same regardless of whether Zener or avalanche mode predominates. The particular operating mode is thus chiefly of theoretical interest.

The term *avalanche diode* is not often used. Zener diodes are normally rated in terms of the Zener voltage (V_Z)—the voltage at which reverse-bias Zener conduction begins—and the maximum power dissipation (P_Z), in watts.

Zener Diodes

The Zener diode is a silicon diode primarily intended for use as a voltage regulator. The reverse-bias part of the curve in Figure 2-13 indicates that the voltage drop across the Zener diode, when it is operating on the Zener part of the curve, is essentially constant over a wide range of currents.

The voltage drop across the Zener diode is nearly (but not completely) independent of the current through the diode. Typically, a 100% change in current through a Zener diode will cause the voltage drop across it to vary by only 1% to 5%.

The most common application for Zener diodes is as a shunt (parallel) voltage regulator. A regulator has two basic functions: holding the output (load) voltage constant in spite of input voltage variations; and keeping the output (load) voltage constant in spite of load current variations. Details of Zener diodes are best discussed in connection with the circuits in which they are used, and we will do that in Chapter 3.

Schottky Barrier Diodes

Conventional silicon junction diodes require a finite (and sometimes excessively long) time to switch from reverse bias to forward bias and from forward bias to reverse bias. When a silicon junction diode must switch from reverse to forward bias, the junction capacitance must first be charged. This takes time. When a silicon junction diode must switch from forward conduction to the reverse-bias condition, carriers must be swept out of the depletion zone. This storage time also slows down the diode switching speed.

The **Schottky barrier** (or *hot carrier*) **diode** has been developed for use in circuits where high-speed transitions from one bias state to the other are required. Applications include use in switching power supplies that operate at frequencies of 20 kHz or more, microwave mixer and detector circuits up to frequencies of about 20 GHz, and high-speed digital integrated circuits (where they serve as clamping diodes).

Schottky barrier diodes are metal–silicon junction devices. Unlike p(silicon)–n(silicon) junction devices, they have very little junction capacitance and virtually no storage time problems. The characteristics of a Schottky diode vary greatly, depending on the metal used for the metal–silicon junction. Platinum, tungsten, chrome, and molybdenum are commonly used junction metals.

The principal disadvantages of Schottky devices are their large reverse-bias leakage current compared to silicon p–n junction devices (particularly at higher temperatures) and their relatively low breakdown voltages (typically, 50 to 100 volts or so). Schottky devices are available with up to 100-amp forward current ratings, and that upper-end value could increase in the future. Schottky devices are currently not recommended for temperatures of 100°C or higher. The high reverse-bias leakage current yields a relatively poor forward-to-reverse current ratio.

Theory of Operation

Figure 2-18 shows the typical construction of a Schottky diode schematically. Unlike p–n junctions, where both minority and majority carriers are involved, the metal–silicon junction uses only majority carriers (electrons). The diode action—one-way current flow—depends on the metal's being virtually saturated with totally free electrons. In contrast, the n-type silicon diode consists of more tightly bound electrons distributed comparatively thinly throughout the silicon block.

FIGURE 2-18   Schottky Barrier Diode

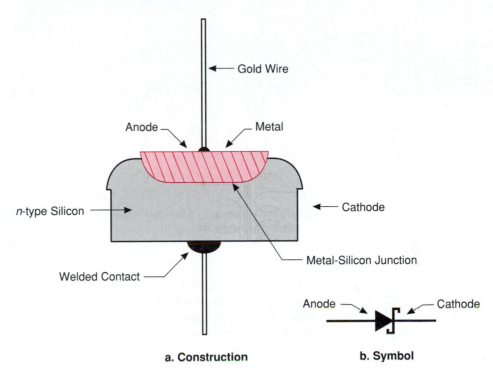

a. Construction   **b. Symbol**

When the metal anode is positive with respect to the silicon cathode, a kind of electron vacuum is created near the junction. The electrons that are moving most rapidly toward the junction are drawn into the metal near the junction. These rapidly moving electrons have higher-than-average energy content and so are called *hot carriers*. This is why the Schottky diode is sometimes called a *hot carrier* diode.

In the reverse-bias direction, the higher-energy electrons in the n-type silicon tend to form a barrier that repels electrons in the metal away from the junction, inhibiting current flow in the reverse-bias direction.

As a general rule, silicon p–n junction devices are superior to Schottky devices, except in one respect: the vastly greater switching speed of Schottky devices.

Light-emitting Diodes and Lasers

Light-emitting diodes (LEDs) are currently one of the most common indicator lights used in electronic equipment. The light-emitting diode is a specially constructed gallium arsenide or gallium arsenide phosphide junction diode.

The diode junction **laser** is essentially a light-emitting diode with a built-in optical resonant circuit. The unmodified light-emitting diode emits light of several different colors (frequencies) and random phases. The optical resonant circuit in the laser diode causes the light output to be produced at a single frequency, with all output signals in phase. The laser's single frequency, in-phase,

FIGURE 2-19   Coherent and Incoherent Radiation

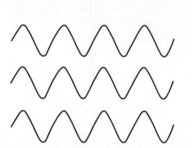

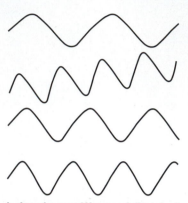

a. Coherent Radiation: A Single Frequency
   with Waves in Phase

b. Incoherent Waves: A Band of
   Frequencies and a Variety of
   Phases

output beam is called a **coherent beam.** The light-emitting diode's multifrequency, random-phase output is called an **incoherent beam.** Figure 2-19 illustrates the difference between coherent and incoherent waves.

All light sources, including the common incandescent lamp, the fluorescent lamp, and the laser, operate on the same physical principle. Energy introduced into the device excites electrons from their stable orbital state to a semistable excited state. If the electrons were not disturbed further, they would remain in the semistable excited state; but there is always some stray energy around (usually thermal energy), and this eventually disturbs the excited electrons and causes them to fall to the lower energy, stable or ground, state.

The electrons must absorb enough energy to advance to the excited state, and they must give that energy up when they fall to the ground state. Under certain circumstances the energy given up by the falling electrons takes the form of electromagnetic energy in the visible part of the spectrum.

An Analogy

Suppose that we drive a fi-inch diameter rod into the ground so that the top of the rod stands 1 foot above the ground. Imagine that a 6-inch-diameter, 50-pound cannon ball is sitting on the ground next to the rod. The cannon ball is in its stable (ground) state. We can poke it, hit it, and otherwise disturb it, but it will not fall.

Now, suppose that we lift the 50-pound cannon ball up to the top of the rod and carefully balance it on top of the rod. You will be well aware that you have expended some energy in lifting the 50-pound ball 1 foot off the ground. But energy can neither be created nor destroyed, so the energy you expended in lifting the cannon ball must have been transferred somewhere.

The energy you expended in lifting the cannon ball is stored in the ball (as potential energy) while it is balanced on top of the rod. Nothing happens unless something disturbs the cannon ball and upsets its critical balance. A strong gust of wind or a light tap may be enough to cause the ball to fall off the rod and hit the ground. When the cannon ball hits the ground, it transfers the energy it was storing to the ground.

A shock wave will then radiate outward from the ball's impact, much like the wave that you create when you toss a pebble into a calm lake, or perhaps like the electromagnetic energy radiating out from a light source.

In the analogy, the energy from your muscles was transferred to the ball, stored for a time, and finally transferred from the ball to the ground in the form of a mechanical radiant wave. The energy went from your body system to the ball, and then—unless you were careless enough to have your foot under the ball when it fell—the energy was transferred a second time into a totally different system: into a shock wave. The mechanism in all light-emitting devices is similar to that described in the previous analogy.

Some energy source (often, an electric current) pumps the electrons up to a higher semistable quantum level. In most light sources, the electrons are disturbed by thermal energy, causing them to fall back to the ground state, where the energy is given up as electromagnetic energy (waves) in the visual spectrum. Our eyes and brain interpret this electromagnetic energy as light.

In all light-emitting devices except the laser, the light emission is called *spontaneous* because the electrons are disturbed and caused to fall by chance, owing to random thermal or other energy.

In a laser (*l*ight *a*mplification by *s*timulated *e*mission of *r*adiation), a portion of the output light is fed back into the device so that electrons are knocked down in synchronism with the light wavefront. This results in an output of lightwaves that are all of the same frequency (color) and in phase (coherent).

In virtually all materials, the electrons present can be pumped up to one of several quantum levels (excited states). Each individual quantum level of excitation produces a specific electromagnetic frequency when an electron falls to ground state. The excitation quantum levels must be quite high before any electromagnetic energy can leave the material and radiate into space.

The lowest frequencies at which electromagnetic waves normally leave the system at ambient temperatures are in the range of 380 THz in the infrared part of the spectrum. Light-emitting devices involve very energetic electrons. High-microwave radio frequencies have been produced but only at temperatures low enough to eliminate most thermal energy and its consequent spontaneous emission.

Light-emitting diodes are spontaneous emission devices that produce a band of output frequencies and various phases. Junction-diode lasers use optically tuned feedback to synchronize the output phase, producing a single-frequency output light wave. The junction diode laser uses an optical resonant circuit, analogous to an inductor–capacitor resonant circuit or a quartz crystal electromechanical resonant circuit. The optically tuned (resonant) circuit effectively tunes out all frequencies (colors) but the one to which the laser is tuned.

Most light-emitting junction diodes (and lasers) are made from gallium arsenide or gallium arsenide phosphide to produce a light output in the visible red or invisible infrared part of the spectrum.

Some LEDs produce a yellow or green output. LEDs are constructed so that the area about the junction is open to allow the light to escape the device. Light-emitting diodes produce light only in the forward-bias condition. The forward-bias voltage pumps the electrons up to an excited state. When a moving electron combines with a hole at the junction, a packet of light is emitted.

Fɪɢᴜʀᴇ 2-20 Light-emitting Diodes (LEDs)

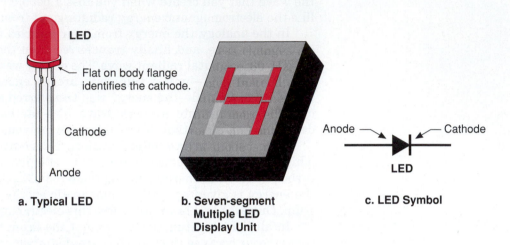

a. Typical LED

b. Seven-segment Multiple LED Display Unit

c. LED Symbol

Practical LED Considerations

The light-emitting diode behaves electrically like any other junction diode. The general diode curve shown in Figure 2-13 is valid for LEDs. The junction voltage in a light-emitting diode is typically 1.2 to 1.4 volts for a red LED and a bit higher for a green or yellow LED. The junction voltage is different because the diodes are made of gallium arsenide instead of silicon.

LEDs are always operated in forward-bias mode. Reverse-bias avalanche operation is not permitted. The P_{IV} of a typical LED is only about 5 or 6 volts. A current-limiting resistor is always used in series with a LED—except in a few special cases where the current is limited in some other way.

LEDs are available in a variety of case styles and in special multiple-LED arrays. Figure 2-20 illustrates some of the more common LED devices.

Current-limiting Resistors (Dummy Load)

Like any other type of diode, LEDs will self-destruct if precautions to limit the current are not taken. In rectifier circuits and so on, the load itself limits the current, so we don't have to do anything extra. But the LED is intended to function only as a light source; there is no load involved.

We must always add a dummy-load resistor in series with a light-emitting diode to limit the current through it to a safe value. It only takes a millisecond or so to kill a LED if you forget to include the dummy-load resistor. Common LEDs require operating currents of from 10 to 20 mA. Typically, 12 to 15 mA is a comfortable range. A fairly large current variation must occur to produce a noticeable change in brightness level. Figure 2-21 shows how to connect the dummy-load resistor, and what you must do if the LED is to be connected to an **ac** source. The extra diode is essential because the P_{RV} (P_{IV}) of the typical LED is only about 5 volts. The maximum reverse-bias voltage is then the sum of the two maximum reverse-bias voltages. The diode has a much higher maximum reverse-bias voltage than the LED.

FIGURE 2-21   **Light-emitting Diodes—Practical Considerations**

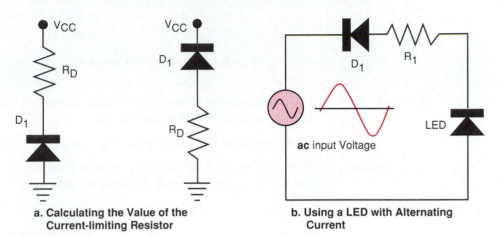

a. Calculating the Value of the Current-limiting Resistor

b. Using a LED with Alternating Current

Example 2.4

Calculating the Value of a LED Current-limiting Resistor

Calculate the value of an appropriate dummy-load current-limiting resistor (R_D) for a LED circuit with the following specifications (include the LED junction voltage drop):

1   LED operating current (I_{LED}) range: 10 mA (minimum) to 20 mA (maximum).
2   LED junction voltage drop (V_{LED}): 1.2 volts, independent of operating current.
3   Power supply voltage (V_{CC}): 12 volts **dc.**

Solution

Assumptions

Let's select a mid-range operating current for the LED of 12 mA. It is almost never a good idea to operate an electronic device at a level close to its maximum or minimum rating. Operating the device too close to its maximum shortens the life of the device, and devices almost always display some odd behavior when they are operated near their minimum rating.

Calculations

1   Now that we have selected the desired current, we need to find the actual working voltage. We have a power supply voltage of 12 volts, but the LED will take its 1.2 volts off the top, so we will have

$$V_{Eff} = V_{CC} - V_{LED}$$
$$= 12\text{ V} - 1.2\text{ V}$$
$$= 10.8\text{ V}$$

2   We now have a current of 12 mA and a voltage of 10.8 volts, so we can use Ohm's law to find the required resistor value:

$$R_D = V_{Eff}/I_{LED}$$
$$= 10.8 \text{ V}/0.012 \text{ A}$$
$$= 900 \text{ }\Omega$$

We would probably use a 1 kΩ resistor.

Example 2.5

Calculating the Current-limiting Resistor Value Ignoring the LED Voltage Drop

What happens if we ignore the LED junction voltage drop in Example 2.4?

Solution

If we ignore the LED junction voltage drop and go directly to Ohm's law, using the full V_{CC} power supply voltage, we get

$$R_D = V_{CC}/I_{LED}$$
$$= 12 \text{ V}/0.012 \text{ A}$$
$$= 1000 \text{ }\Omega \quad (1 \text{ k}\Omega)$$

If we don't subtract the junction voltage, the result is a lower LED current. This is the equivalent of a built-in safety factor and probably will not make much difference in our perceived brightness of the LED.

Light-sensitive (Photo) Diodes

In ordinary junction diodes, the depletion zone is shielded from light energy by its structure and (generally) by an opaque case. If light is shined on the junction area while the diode is reverse-biased, the light energy will create electron-hole pairs.

In most diodes, the reverse-bias leakage current is almost entirely a function of temperature. But if the diode is designed to absorb light energy in the depletion zone efficiently (as light-sensitive (photo) diodes are), the reverse-bias leakage current will largely be a function of the intensity (and color) of the light. Greater light energy will create a proportionally larger number of electron-hole pairs. The reverse-bias leakage current will increase almost linearly with an increase in the amount of light striking the depletion zone.

Thermal energy still produces electron-hole pairs, but light energy will dominate over a normal range of temperatures. In addition, a number of simple circuit design techniques can be used to compensate for or cancel the temperature-induced part of the leakage current.

Optical Isolator (Opto-isolator)

The **optical isolator (opto-isolator),** in its simplest form, uses a LED that has its beam directed at a photo diode and is sealed in a light-tight package. If the intensity of the LED is varied in some fashion, the diode leakage current will vary in the same fashion.

The advantage of such an arrangement is that the input signal is coupled to the output signal without there being any electrical connection between the input and output. Thus, the circuit driving the LED is totally independent, electrically, of the circuit controlled by the photo diode. The opto-isolator can

FIGURE 2-22 Light-sensitive Receivers for Optical Isolators

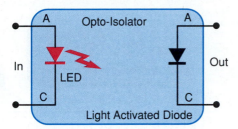
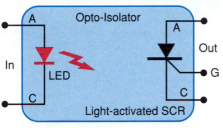

a. Light-sensitive Diode Receiver

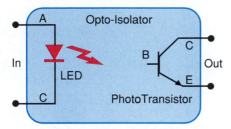

b. Light-sensitive Transistor Receiver

c. Light-sensitive Silicon-controlled
Rectifier Receiver

d. Dual In-line Package
(Number of Pins Varies from 6 for
Single to 16 for Multiple Isolators)

be used to solve the sticky problem of coupling the output signal from a low-voltage, low-power, integrated circuit to a high-voltage or high-power device that must operate at voltages that would normally destroy the integrated circuit. Figure 2-22 illustrates a typical diode optical isolator (coupler). We will examine other varieties of opto-isolators in a later chapter.

Troubleshooting Diodes

Junction diodes and LEDs are frequent objects of testing. In this section, we briefly discuss how to proceed with such testings.

Testing Junction Diodes

Figure 2-23 shows a quick test procedure for testing junction diodes. Most digital multimeters have a special low-ohm setting for testing diodes. The range-selector dial often shows a diode symbol. The special range is often required because the ohms circuitry in the meter may not deliver a high enough voltage (or current) to turn the diode on in the forward-bias direction.

The diode should be tested when connected in each direction, as shown in the drawing. The low-ohm (or diode test) range is fine for a forward-bias test, but serious leakage problems may not show up except on the higher range. Use the highest available ohms range for the reverse-bias test. Use the special lowest-ohms (diode test) range for the forward-bias test.

FIGURE 2-23 Testing Junction Diodes

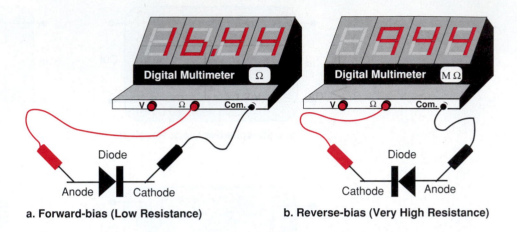

a. Forward-bias (Low Resistance)

b. Reverse-bias (Very High Resistance)

FIGURE 2-24 Testing LEDs

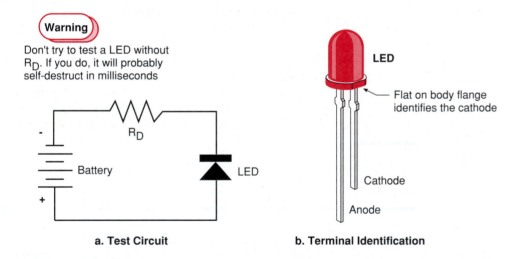

a. Test Circuit

b. Terminal Identification

Because of the nonlinearity of the diode's operating curve, absolute resistance values vary over a wide range.

Testing LEDs

The best way to test a light-emitting diode is to make it emit light. The diode test ranges on a multimeter may or may not test a LED because of the 1.2- to 2-volt junction turn-on voltage. Use the circuit shown in Figure 2-24, and calculate a resistance value for about 10 to 15 mA through the LED. Typical R_D values are 330 ohms for 5 volts and 1 kilohm for 12 volts.

QUESTIONS AND PROBLEMS

2.1 Define the *transition (depletion) zone* of a *p–n* junction.

2.2 What is the junction voltage or transition voltage in a *p–n* junction?

2.3 What factor determines the value of the junction voltage in a junction diode?
  a. How it is manufactured.
  b. The doping level.
  c. The element or compound from which it is made.
  d. The temperature at which the junction is formed.

2.4 How does a semiconductor junction prevent current flow in the reverse-bias state?
  a. Depletion zone carriers are absent.
  b. The doping level on the anode side is very high compared to the doping level in the cathode block.
  c. The doping level on the anode side is very low compared to the doping level in the cathode block.
  d. The forward-bias junction resistance prevents current flow.

2.5 What is the depletion zone?
  a. A zone in the *p*-type block that has a high resistance.
  b. A zone in the *n*-type block that has a high resistance.
  c. A zone in both the *n*- and the *p*-type blocks that has a high resistance.
  d. A zone on either side of the junction that has very few available carriers.

2.6 What happens to thermally generated electron-hole pairs when the junction diode is reverse-biased?
  a. They are trapped at the junction.
  b. Both electrons and holes move toward the *p*-block.
  c. Electrons move toward the *n*-block, and holes move toward the *p*-block.
  d. Both electrons and holes move toward the *n*-block.

2.7 How does conduction take place in the forward-bias condition?
  a. Electrons move toward the *n*-block, and holes move toward the *p*-block.

  b. Both electrons and holes move toward the *p*-block.
  c. Electrons move toward the *p*-block, and holes move toward the *n*-block.
  d. Both electrons and holes move toward the *n*-block.

2.8 What happens to thermally generated electron-hole pairs when the junction is forward biased?
  a. They are trapped at the junction.
  b. Both electrons and holes move toward the *p*-block.
  c. Electrons move toward the *p*-block, and holes move toward the *n*-block.
  d. Both electrons and holes move toward the *n*-block.

2.9 How does temperature affect the amount of reverse current in an operating junction diode circuit?
  a. The current (in silicon devices) triples for each 10 Celsius degrees' increase in temperature.
  b. The temperature has virtually no effect on the current.
  c. The current (in silicon devices) doubles for each 10 Celsius degrees' increase in temperature.
  d. The current decreases linearly with a temperature increase.
  e. Either (a) or (c)—the author is uncertain.

2.10 How does temperature affect the amount of forward current in an operating junction diode circuit?
  a. The current (in silicon devices) increases exponentially.
  b. The temperature has virtually no effect on the current.
  c. The current (in silicon devices) depends only on the applied voltage.
  d. The current decreases linearly with a temperature increase.

2.11 How does voltage affect the amount of current through a reverse-biased junction diode circuit?
  a. The voltage has virtually no effect on the current.
  b. The current decreases linearly with a voltage increase.

c. The current increases linearly with a voltage increase.

d. The current increases exponentially with a voltage increase.

2.12 How does voltage affect the amount of current through a forward-biased junction diode circuit?

a. The voltage has virtually no effect on the current.

b. The current decreases linearly with a voltage increase.

c. The current increases linearly with a voltage increase.

d. The current increases exponentially with a voltage increase.

2.13 Define *junction capacitance.*

2.14 What is *carrier storage,* and what is *storage time*?

2.15 Define *recovery time* in relation to a diode that is switched from a forward-bias to a reverse-bias condition.

2.16 How does reverse bias influence the junction capacitance?

a. Increasing the reverse bias voltage increases the capacitance.

b. Increasing the reverse bias voltage decreases the capacitance.

c. Increasing the reverse bias voltage has no effect on the capacitance.

2.17 State the Shockley relationship, and explain its meaning.

2.18 What is the Zener region, and what is the Zener knee in a junction diode?

2.19 What is an avalanche condition?

2.20 What external component can be used to control avalanche current?

2.21 List three of the most important electrical ratings of a junction diode.

2.22 What is the chief failure mechanism in a forward-biased diode?

2.23 What is the chief failure mechanism in a reverse-biased diode?

2.24 How is forward-bias resistance normally computed?

a. As dynamic resistance.

b. As static resistance.

2.25 How is reverse-bias resistance normally computed?

a. As dynamic resistance.

b. As static resistance.

2.26 What is the difference between static resistance and dynamic resistance?

2.27 Define the diode rating P_{IV}.

2.28 Define the diode rating P_{RV}.

2.29 Define the diode rating $I_{F(max)}$.

2.30 Define the diode rating V_{Rdc}.

2.31 Define the diode rating V_F.

2.32 At what rate does the junction voltage of a silicon diode vary as the temperature increases?

a. +2.5 mV per Celsius degree.

b. –2.5 mV per Celsius degree.

2.33 What is the approximate junction voltage of a silicon diode?

2.34 What is the approximate junction voltage of a red LED?

2.35 Four silicon diodes are connected in series. The diodes have the following specifications:

$$V_F = 0.7 \text{ V at 1 Amp}$$

$$V_{Rdc} = 600 \text{ V}$$

$$I_{F\,(max)} = 5 \text{ Amps}$$

a. What is the total voltage drop across the four diodes when they are forward-biased?

b. What is the total V_{Rdc} of the four diodes?

c. What is the total junction voltage of the four diodes?

d. If the leakage current of the four diodes is 0.001 mA at 25°C, will the total leakage current be higher or lower at 35°C?

2.36 Name and describe the phenomenon that generally makes connecting diodes in parallel a poor idea.

2.37 What components must be added to a group of diodes connected in parallel to make diode paralleling work?

2.38 What are the disadvantages of adding the required components mentioned in Question 2.37?

2.39 You have discovered a faulty power diode in a circuit. It is rated $I_{F(max)}$ of 2 A and a $V_{Rdc} = 200$ V. The only diode at hand has $I_{F(max)} = 6$ A and $V_{Rdc} = 600$ V. Would this diode probably serve (electrically) as a satisfactory replacement?

2.40 In Question 2.39, are there problems other than electrical ones that you might have to consider? If so, what are they?

2.41 What is the only advantage of the Schottky diode over a silicon diode?

2.42 What characteristics of the Schottky diode give it the advantage referred to in Question 2.41?

2.43 In what ways is the Schottky diode inferior to the silicon junction diode?

2.44 What are the differences between a LED and a laser diode?

2.45 Define *coherent* and *incoherent waves*.

2.46 A LED produces light output only when the junction is:
  a. Forward-biased.
  b. Reverse-biased.

2.47 LEDs are made of what semiconductor materials?

2.48 How do light-sensitive diodes differ physically from ordinary junction diodes?

2.49 Photo diodes operate in:
  a. Forward-bias mode.
  b. Reverse-bias mode.

2.50 What is the purpose of an opto-isolator?

2.51 Give an example of a set of practical circumstances under which an opto-isolator could be used to advantage.

Given the circuit in Figure 2-25, answer the following questions.

FIGURE 2-25   Circuit for Problems 2.52 Through 2.56

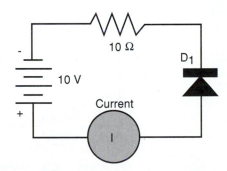

2.52 The diode is:
  a. Reverse-biased.
  b. Forward-biased.

2.53 What is the current flowing in the circuit?

2.54 What is the voltage drop across the diode?

2.55 What is the voltage drop across the resistor?

2.56 What is the diode junction resistance?

Given the circuit in Figure 2-26, answer the following questions.

FIGURE 2-26   Circuit for Problems 2.57 Through 2.61

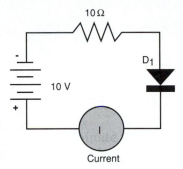

2.57 The diode is:
  a. Reverse-biased.
  b. Forward-biased.

2.58 What is the current flowing in the circuit?

2.59 What is the voltage drop across the diode?

2.60 What is the voltage drop across the resistor?

2.61 What is the diode junction resistance?

FIGURE 2-27 Graphical Summary of Diode Conduction Characteristics

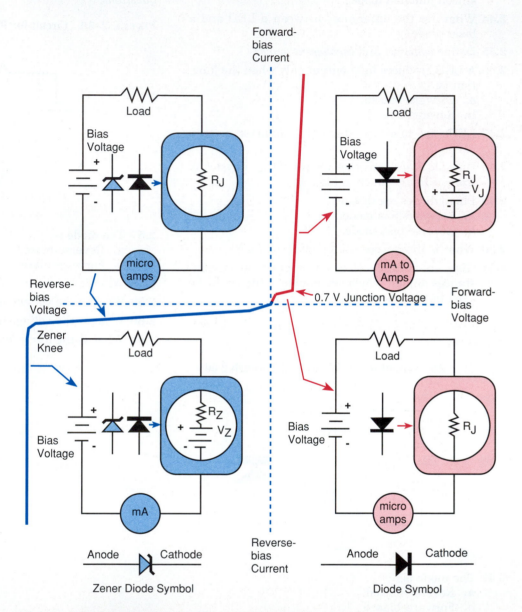

Critical Ideas About the Diode Conduction Curve

1. The junction voltage is a minimum voltage required to turn the junction on. When the diode is conducting, you can measure 0.6 to 0.7 V across the junction.

2. Both forward bias and Zener (avalanche) conduction are potentially destructive. Some resistance must be used to limit the current to a safe value.

3. The junction voltage is always 0.6 to 0.7 V in the forward conduction condition. The manufacturer can't change the value. The manufacturer can adjust the Zener voltage.

It's a "Smart" World After All

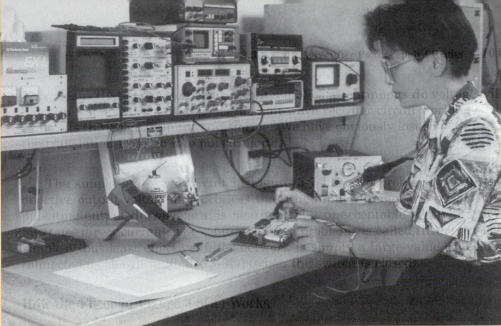

When computers became common in the business world, one school of thought predicted the total elimination of paper. In the "paperless office," all documentation, memos, and files would be stored on disk, tape, and other electronic media. There would be no further need for paper, filing cabinets, or any of the other accessories associated with hard copy. The notion sounded plausible; but we know today that computers have, if anything, increased the flow of paper.

The same idea of physical obsolescence has sometimes been proposed in relation to electronics technicians. The trend in electronics manufacturing over the last several years has been to build systems so that repairs can be performed simply by replacing easily removable modules. Even more recently, some electronic systems have been designed to perform self-diagnostic checks and indicate which part or module is faulty. Some equipment manufacturers have moved toward a "throw-away" approach. Rather than providing field service or technical support, these companies price their equipment such that it is cheaper to replace a module than to repair it. The old module is simply discarded.

At the same time, components and board designs have become more reliable. Pretesting components, burn-in cycles, and quality-control procedures in the manufacturing plants ensure that very few boards fail due to faulty components.

These factors have led some observers to believe that the role of the electronics technician is diminishing and that eventually, if an electronic device fails at all the equipment user will be able to replace it without requiring technical assistance. However, a closer look at history—as in the case of the "paperless office"—suggests a different outcome.

The role of an electronics technician is not simply to know a great deal about electronic components and circuits, although that is important. The best technicians use their knowledge—electronic or otherwise—to solve problems. This requires skills such as knowing how to test, how to solder, and how to repair. But the most important ability of all is the capacity to perceive the nature of a problem and to pursue it to its source. The next step is to rectify the problem expeditiously.

The ability to solve problems will never become obsolete. Vacuum-tube circuits have become all but nonexistent in the last few years. But did the technicians who had become expert on tube circuits become obsolete? No, they acquired new knowledge. They learned semiconductor theory, transferring their troubleshooting skills to the analysis and repair of transistor and diode circuitry. The same applies to operational amplifiers, microprocessor circuits, and then new technologies just now becoming available. Technicians who learn the important skill of problem solving will adapt to emerging technology and keep apprised of the latest developments.

Troubleshooting skills are also transferable from the component level upward through the module level and on to the system level. For example, the area of data communications—which often deals with communicating between systems—is currently in need of skilled technicians. The complexity and the variety of the different systems that must be interfaced require people who can grasp technical difficulties and find specific solutions. The technician working in this area must be knowledgeable about transistors, op-amps, special communications circuits, and the standards and protocols used by various systems. More important, the technician must be able to accumulate the necessary knowledge, tie it together, and then use it to find answers to specific problems.

The latest buzzword in the technical area is the term *smart*. We have "smart" cars, "smart" appliances, and "smart" telephones. Computers use "smart" modems. In the industrial area, we have "smart" transmitters, "smart" controllers, and even "smart" valves. The "smart" concept is based on the use of microcontrollers, which can adapt the equipment to the application. For example, some "smart" modems sense the type of signals and equipment they are communicating with and automatically adapt themselves to the correct communication standard and protocol.

Regardless of how "smart" equipment may become, however, there will still be a need for technicians who can choose, install, configure, and solve the problems that inevitably will occur. Comprehensive knowledge of available options—components and circuits, hardware and software, standards and systems, will facilitate those solutions. But the "smart" technician with the necessary problem-solving skills is the individual who will be assured of a long-term niche and ultimate success in this area of technology.

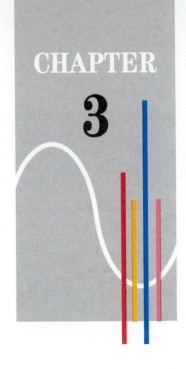

3

Diode Circuits

OBJECTIVES

Upon completion of this chapter, you should be able to:

1 List and define the basic diode circuit properties (parameters).

2 Use the proper abbreviations and acronyms for each diode parameter.

3 Describe the operation of diode snubbers, and explain why they are important when inductive loads are used with semiconductor devices.

4 Draw the schematic diagram of a basic half-wave rectifier circuit, including a step-down transformer and a load resistor.

5 Explain concisely how the half-wave rectifier works.

6 Make a list of the half-wave circuit's limitations.

7 Describe the operation and purpose of basic power-supply filter networks.

8 Identify the differences between half-wave and full-wave rectifiers.

9 Account for the fact that the full-wave circuit is more common than the half-wave rectifier circuit.

10 Draw a schematic diagram for a full-wave center-tapped transformer rectifier circuit.

11 Explain concisely how the full-wave center-tapped transformer rectifier circuit operates.

12 Draw a schematic diagram for a full-wave bridge rectifier circuit.

13 Explain concisely how the full-wave bridge rectifier circuit operates.

14 Define *ripple voltage*.

15 Explain how ripple voltage is measured.

16 List the most common methods of reducing ripple to an acceptable value.

17 Draw the schematic diagram of a Zener diode regulator circuit.

18 List the advantages of using linear integrated circuit regulators in modern power-supply circuits.

19 Explain how the full-wave voltage doubler works.

20 Explain how voltage triplers and voltage quadruplers can be developed from the basic concept involved in the voltage-doubler circuit.

Analysis In Brief

Diode Circuits in Brief

USING THE ONE-WAY VALVE ACTION

Diode circuits take advantage of two common properties of diodes. The first of these is the fact that current can flow in only one direction through the diode. This property is most often used to convert a sine wave from the **ac** power line into pulsating **dc** that can then be smoothed into reasonably pure **dc** to operate electronic circuits.

The one-way valve action can also be used to protect delicate transistors and integrated circuits from accidental reverse polarity connections. When a magnetic field collapses, the polarity of the voltage produced is reversed and could damage electronic components. But a diode can be set up to turn on and bypass the magnetic-field energy under these conditions.

USING THE CONSTANT VOLTAGE DROP

The second common property of diodes is their constant voltage drop of 0.6 to 0.7 volts, regardless of the current through the diode. This property permits the diode to operate as a voltage regulator or series voltage drop that is independent of the circuit current.

Power supply uses a diode to convert ac power into pulsating dc.

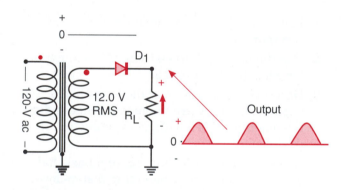

Diode's one-way conduction serves as a protective snubber.

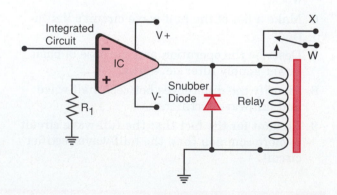

88

Series diodes provide a constant voltage drop.

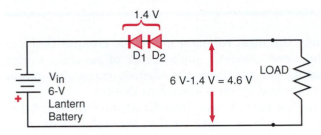

Shunt diodes form a simple voltage regulator or reference source.

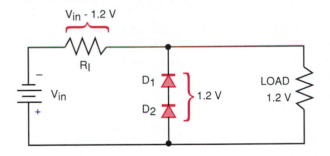

FILTERING AND RIPPLE

A diode can be used to convert power-line **ac** into pulsating **dc.** But pulsating **dc** is not suitable for electronic power supplies. A large capacitor can be used to store and deliver power when the line voltage heads down.

THE DIODE BRIDGE

Four diodes in a full-wave bridge can reroute the current on the negative part of the cycle and deliver it to the load as a positive pulse. Power from the line is then delivered to the load on both half-cycles of the **ac** waveform. Smaller capacitors can be used to filter the pulses. Variations that are not filtered out are called *ripple*.

ELECTRONIC REGULATORS

Nearly all modern power supplies use the bridge circuit and electronic regulators. Many electronic circuits demand accurate voltages in spite of changing power-line voltages, load-current demands, and so on. Negative-feedback electronic regulators can provide precise voltages at very little expense. Negative feedback also cancels any remaining ripple.

Filtering the Full-wave Pulse

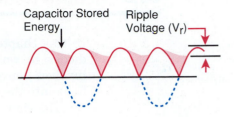

Typical Diode Bridge Circuit with a Capacitor Filter

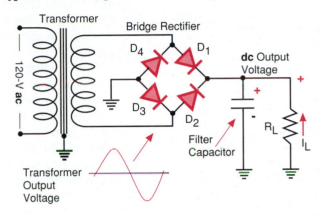

Modern Multiple-voltage Power Supply

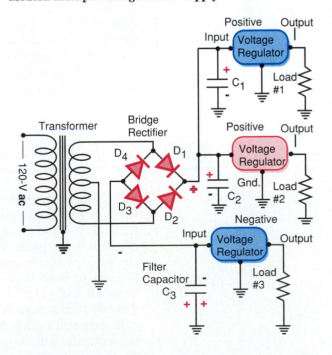

3.1 Introduction

In this chapter, we will study the most common and basic circuits that put diodes to work. One of the most common applications of junction diodes involves converting alternating current from the electric company's power lines into the direct current required by most electronic devices.

The conversion of alternating current (**ac**) into direct current (**dc**) takes advantage of the fact that a diode allows current to flow through it in only one direction. These rectifier circuits are essential to our being able to power electronic devices from the common power line. We will also examine basic filtering and regulating techniques, which serve to filter out sine-wave power-line variations to get the nearly pure direct current most electronic systems require.

Most modern electronic systems are not very tolerant of variations in the **dc** voltage provided by line-operated power supplies. Electronic integrated-circuit voltage regulators are by far the most popular device used in modern power supplies. These regulators are actually complex circuits on a chip, but we use them as though they were a simple three-terminal device.

We will briefly examine the all-important integrated circuit in this chapter, since you will be using IC-regulated power supplies in your laboratory work from the very start. We will examine the internal circuitry in a later chapter, when you have learned enough about transistors and other relevant components to understand how they work. We can use these important ICs very nicely without knowing exactly what is inside. Linear regulator ICs also provide a great opportunity to introduce the concepts and power of negative feedback.

In this chapter, we will examine linear-type integrated circuit regulators. A second class of electronic regulators, called *switching regulators,* are much too specialized and complex to introduce in this chapter. Switching-type power supplies are not usually repaired anyway, because any component failure tends to initiate a kind of chain reaction that destroys most of the rest of the parts. We will study them simply as another type of feedback circuit later in the text.

3.2 Simple Diode Circuits

The diode is an electronic one-way valve. Current can flow through the diode when the anode is positive with respect to the cathode. Current cannot flow through it when the anode is negative with respect to the cathode. Figure 3-1 illustrates the point. The circuit in Figure 3-1(a) shows the diode connected in the forward-biased condition, with current flowing and the lamp lit. The anode is positive with respect to the cathode, even though the lamp is connected between the positive battery pole and the diode. The cathode is connected directly to the negative battery terminal—the most negative point in the circuit. Any other place in the circuit must, therefore, be positive with respect to the cathode. This is pretty obvious in a simple circuit; but in more complex diode circuits, it may be far less obvious. Looking for the most positive or most

FIGURE 3-1 Simple Diode Circuits

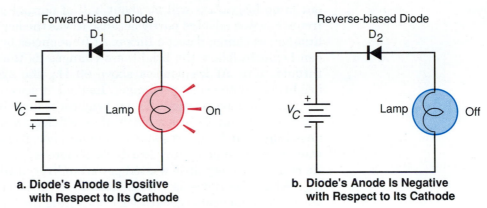

a. Diode's Anode Is Positive
   with Respect to Its Cathode

b. Diode's Anode Is Negative
   with Respect to Its Cathode

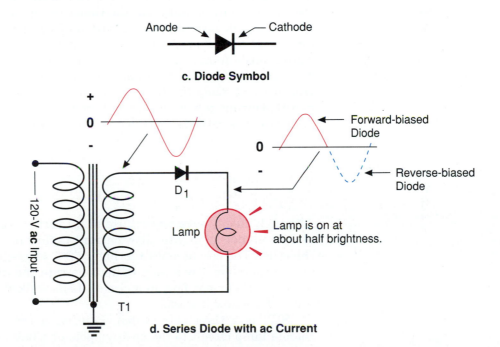

c. Diode Symbol

d. Series Diode with ac Current

negative point in the circuit, to serve as a reference point, is often a good trick for tracing more complex circuits.

The circuit in Figure 3-1(b) connects the diode in the reverse-bias condition. No current flows in the circuit, and the lamp is not lit. In this case, the cathode is connected to the most positive part of the circuit; and the anode must, therefore, be negative with respect to the cathode. The diode symbol is repeated in Figure 3-1(c), just in case you have forgotten it.

The circuit in Figure 3-1(d) shows the diode connected to a step-down transformer, delivering an alternating current to the circuit. During the half-cycle in which the diode anode is positive with respect to the cathode, current flows and the lamp lights. Because your eye has a slow response time, and because the lamp filament does not cool and heat instantaneously, the apparent lamp brightness will be about half of normal brightness. The 60-Hz line frequency was adopted partly because it was the lowest frequency at which the human eye cannot detect a flicker. At frequencies below 60 Hz, the human eye can begin to follow the brightness changes as the sine-wave voltage passes through zero. At frequencies above 60 Hz, the eye cannot see the varying brightness as the voltage changes. Instead, it sees an average brightness.

You can make a practical dual-brightness lamp by switching a diode in and out of the circuit. This is a much more efficient way of controlling the brightness than switching a resistor in and out, because the resistor is a big power waster and heat generator. The diode is more efficient because it offers very little resistance, very little wasted power and very little heat when it is conducting. When the diode is reverse-biased (off), no current flows, no power is wasted, and no heat is generated. Controlling the lamp brightness with a diode is called **duty-cycle** power control. In the case of the simple diode controller, the lamp is only on duty half the time (50% duty cycle), or it is on for the full cycle. The four-layer-diode thyristor family provides much more sophisticated duty-cycle control for power circuits. Thyristor devices can vary the duty cycle continuously from near 0% to 100%, allowing for a full range of brightnesses. The common home light-dimming controls use thyristors and duty-cycle control. We will study the thyristor family later on; now let's go back to Figure 3-1(d). During the alternate half-cycle, the diode anode is negative with respect to the cathode, the diode is reverse-biased, and the lamp is dark.

How to Control Two Lamps
Independently with Only Two Wires

Although diodes have several useful properties, their most important characteristic is their ability to allow current to flow in one direction and block it in the other. The circuit shown in Figure 3-2 represents a somewhat uncommon application, but it will help to demonstrate the diode's ability to control the direction of current flow in a circuit. Let's walk through the circuit in Figure 3-2 and see how it works.

SW_1 is a three-pole switch. Position A connects the two diodes in the remote lamp circuit to the positive pole of a battery. The anode of diode D_1 is positive with respect to its cathode. Diode D_1 is forward-biased, and current flows through D_1 and through the red lamp that is connected in series with it. The red lamp is on. Diode D_2 is reverse-biased because its anode is negative with respect to its cathode. No current can flow through diode D_2 or through the green lamp that is connected in series with it. The green lamp is off. Switch (SW_1) position B is not connected to any power source, so both lamps are off in position B.

Position C on the switch (SW_1) connects the negative pole of the battery to the two diodes. In this position, D_1 has a negative voltage on the anode and therefore is reverse-biased. No current flows through D_1 or through the red lamp. The red lamp is off. Meanwhile, diode D_2 has a positive anode and a negative cathode, so it is forward-biased. Current flows through D_2 and through the green lamp. The green lamp is on.

FIGURE 3-2   Diodes Allowing Independent Control of Two Lamps, with Only Two Wires

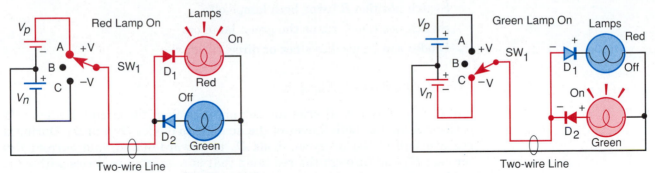

a. Version with Direct Current, Parallel Lamps, and Series Diodes

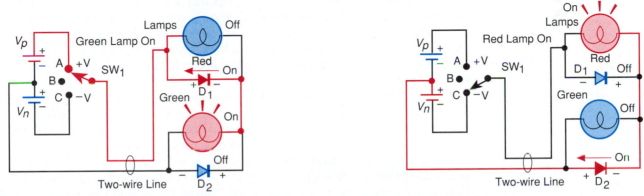

b. Version with Direct Current, Series Lamps, and Parallel Diodes

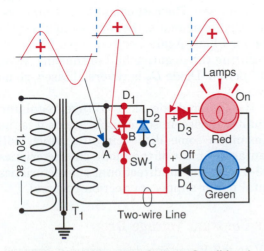

c. Version with Alternating Current, Condition 1

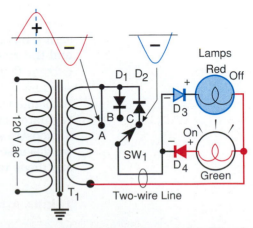

d. Version with Alternating Current, Condition 2

Summary

Switch position A turns the red lamp on.

Switch position B turns both lamps off.

Switch position C turns the green lamp on.

Diodes can be used to steer or direct a signal.

Alternating-current Version

Figure 3-2c shows an alternating-current version of the circuit. Switch (SW_1) position A applies both halves of the **ac** cycle to diodes D_3 and D_4. During the positive half of the **ac** cycle, diode D_3 is forward-biased, and current flows through D_3 and through the red lamp that is connected in series with it. The red lamp is on.

When the **ac** voltage goes through zero and into the negative half-cycle, diode D_3 turns off, and the red lamp goes out. During the negative half of the **ac** cycle, diode D_4 becomes forward-biased, and current flows through the green lamp. Thus the two lamps are alternately turned on and off, but our eyes see them as both being on at about half-brightness. The red lamp is on during the positive half of the **ac** cycle, and the green lamp is on during the negative half of the **ac** cycle.

With switch SW_1 in position B, diode D_1 is forward-biased and so is diode D_3. Since the two diodes are in series, the minimum turn-on voltage is no longer 0.6 volts, but instead roughly twice that (about 1.2 volts). The positive half of the **ac** waveform must rise from 0 to 1.2 volts before the two diodes will turn on. Other than that, the two diodes in series behave much as a single diode does. Current flows through diodes D_1 and D_3, and through the red lamp that is connected in series with them. Diode D_4 is reverse-biased, and the green lamp is off. The red lamp stays on for the duration of the positive half of the **ac** sine wave.

With switch SW_1 in position C, diodes D_2 and D_4 are forward-biased during the negative half-cycle of the **ac** sine wave. The cathode of D_2 is connected to the most negative point in the circuit; consequently, the anode must be positive with respect to the cathode-forward bias. Again, we have two diodes in series, but in this case they conduct during the negative half of the sine wave, and the green lamp is on during that period. Diode D_3 is reverse-biased, so no current flows through the red lamp.

In this chapter, we will study the Zener diode as a voltage regulator. Although the Zener diode is a very limited device, in some applications it is the regulator of choice. We will also examine some diode circuits, which will demonstrate how we can take advantage of most of the diode characteristics we discussed in Chapter 2. Figure 3-3 uses the unidirectional characteristics of a diode to protect against accidental power-supply connection reversal.

Taking Advantage of the Diode's Constant Voltage Drop

In some circuits, the constant 0.6-V to 0.7-V voltage drop across a forward-biased diode junction can serve a number of uses, including use as a diagnostic tool. Every forward-biased silicon junction diode must have a junction

FIGURE 3-3   **How Diodes Can Prevent Damage to an Electronic Unit If Power Supply Connections Are Accidentally Reversed**

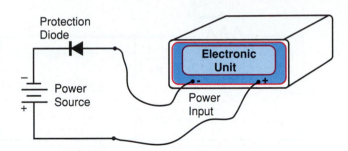

voltage drop of 0.6 to 0.7 volts; if it does not, you can be certain that the diode is not working. It doesn't matter how big or small the diode is, or what its operating current is. This diagnostic tool also applies to transistors, because one of the transistor junctions is normally forward-biased. There aren't many useful constants in electronics, but this is a valuable one.

Example 3.1

Applications Example

Suppose that we have a very valuable antique (made around 1500 B.C.) precision-made glockenspoggen. As everyone knows, the glockenspoggen is an imaginary device and is very critical. It must have an operating voltage of 5 volts, plus or minus 10%. The acceptable voltage range is therefore from 4.5 to 5.5 volts. We wish to operate the device from a 6-volt battery, but 6 volts will destroy the glockenspoggen. How do we proceed?

Solution

If the glockenspoggen always demanded a specific constant current, we could consider using a simple voltage-dropping resistor. But the glockenspoggen demands a wide range of currents, depending on constantly changing conditions. Fortunately, because a diode's voltage drop is independent of current demand, we can install two diodes in series with the 6-volt battery to do the job.

As Figure 3-4 illustrates, if we subtract 1.4 volts (two diode junction voltages) from the 6 volts supplied by the battery, we get 4.6 volts, which is within the 4.5- to 5.5-V acceptable voltage range.

FIGURE 3-4   **Taking Advantage of the Fact that a Diode's Voltage Drop Is a Constant 0.6 to 0.7 Volts, Regardless of Its Operating Current**

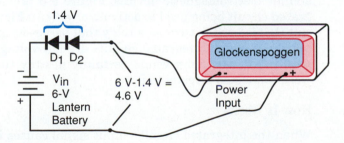

FIGURE 3-5   Diodes Acting as a Shunt Regulator for a 1.2-Volt Reference

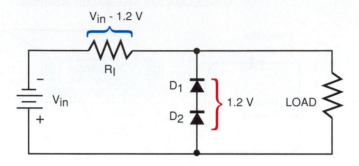

Using Diode Junctions as Voltage Regulators

Figure 3-5 shows a pair of diodes connected as a 1.2-volt shunt (parallel) regulator. This scheme is commonly used for on-the-chip reference voltages for a variety of integrated circuits. 1.2 volts is the most commonly used value because it is high enough to forward-biased other junctions. More junctions could be placed in series for higher reference voltages, but this is not often done, because temperature problems are additive and could become significant. A voltage of 1.2 volts is adequate for most reference functions.

The junction voltage of a forward-biased diode junction increases at the rate of about 2.5 mV for each Celsius degree of increase in temperature.

Diode Inductive Snubber or Free-wheeling Diode

Flashback

An inductor stores energy in its magnetic field. When the magnetic field collapses, it induces a (self-induced) voltage across the coil. Thus, the energy stored in the magnetic field returns to the coil's circuit when the field collapses. The self-induced voltage has a polarity opposite that of the original applied voltage, and the voltage can be much higher than the original applied voltage.

Such (potentially) high collapsing field voltages can spell disaster for transistors, integrated circuits, and other semiconductor devices. The collapsing field voltage can cause tunneling-avalanche destruction of reverse-biased junctions in transistors, and it can punch holes in the gate insulator in field-effect transistors.

Electronic circuits are often called on to control relays, solenoid valves, and similar electromagnetic devices. Figure 3-6 provides an example of an integrated circuit being used to activate a relay. In Figure 3-6(a), an integrated circuit (IC) is used to turn on a relay that serves as an interface device to control a 120-volt lamp. Integrated circuits are low-voltage devices, and the 120 volts required by the lamp would certainly destroy the IC if the 120-volt supply were connected directly to it.

How It Works

When the integrated circuit's input signal causes its output to produce a voltage of +5 volts, current flows through the relay coil. The coil becomes an elec-

FIGURE 3-6 Diode Snubber to Protect Integrated Circuits and Transistors from Inductive Load Hazards

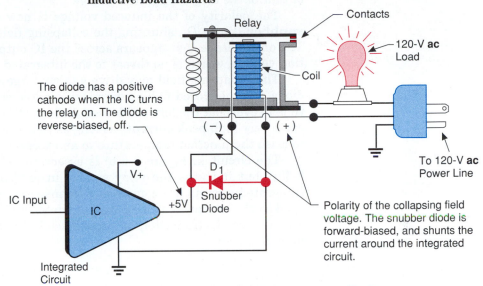

The diode has a positive cathode when the IC turns the relay on. The diode is reverse-biased, off.

Relay

Contacts

120-V **ac** Load

Coil

To 120-V **ac** Power Line

V+

IC Input

IC

+5V

Snubber Diode

D_1

(−)   (+)

Polarity of the collapsing field voltage. The snubber diode is forward-biased, and shunts the current around the integrated circuit.

Integrated Circuit

a. Circuit with Relay Shown in Pictorial Form

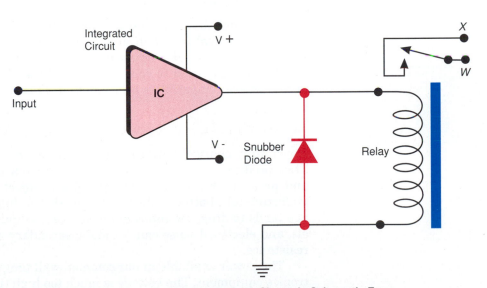

Integrated Circuit

V +

X

Input

IC

W

V −

Snubber Diode

Relay

b. Circuit with Relay Shown in Schematic Form

tromagnet that pulls the relay armature bar down, closing the contacts and turning on the lamp. Notice that the integrated circuit's +5-volt output voltage (which closes the relay) reverse-biases the diode (D_1). While the integrated circuit drives current through the coil, the diode is an open circuit and has no significant effect on anything. When the integrated circuit's input signal causes its output circuit to turn off (and the output goes to 0 volts), the IC no longer drives current through the relay coil. When the drive current stops, the magnetic field collapses, inducing a high voltage whose polarity is the opposite of that of the original applied voltage.

The polarity of the induced voltage is now suitable to forward-bias the **snubber diode** D_1, shunting the collapsing field current through the diode. The only voltage that appears across the IC output is the 0.7-volt diode junction voltage, which is no threat to the integrated circuit.

The collapsing field may drive a fairly large current through the snubber diode, but the stored field energy is dissipated so quickly that it doesn't have time to overheat the forward-biased diode junction. The junction diode can tolerate very high peak currents as long as those currents don't last long enough to raise the junction temperature to an excessive level.

The circuit in Figure 3-6(b) is essentially the same circuit as in Figure 3-6(a), but it shows the relay symbol instead of a pictorial representation. Integrated circuits are composed of transistors, and transistors can be damaged by the collapsing field voltage. A snubber must be used whenever an integrated circuit, transistor, or other semiconductor device must drive an inductive load.

3.3   Diode-rectifier Power-supply Circuits

Most electronic devices require a low-voltage **dc** power source. Batteries are not very practical to use in most situations. The **ac** power-line voltage is **ac** and is too high for most applications. In the following sections, we will examine the methods used to convert household **ac** outlet power into low **dc** voltages that are compatible with solid-state electronic circuits.

Introduction to Line-operated Power Supplies

Fresh, fully charged batteries are perfect for powering electronic equipment. They produce a constant output voltage, have a very low internal resistance, and produce almost no electrical noise to inject into the associated circuit. Unfortunately, batteries do not stay fresh and fully charged. The output voltage tends to drop, the internal resistance gradually rises to unacceptable levels, and electrical noise can become a secondary result of the rise in internal resistance.

The power available in our common wall plugs is destructive to most electronics equipment. The voltage is much too high (in most cases), and alternating current is death to most semiconductor devices. A power-line-operated power supply must modify the power-line voltage in several ways to convert it into a constant direct-current voltage with very little noise or other undesired voltage variations. In this process, the diode is a key player, because connecting a diode between an **ac** power source and a load (electronic circuit) automatically converts alternating current into direct current. The diode allows current to flow through the circuit in only one direction. A current that flows through a circuit in only one direction is direct current, by definition.

Converting **ac** into **dc** is only the first part of the problem. Because a sine wave presents continuously changing voltage, we must find a way to smooth out those changes into a steady, unchanging **dc** voltage. If, for example, we wish to convert a 120-volt **ac** voltage into a 12-volt **dc** power supply, we must ultimately obtain a direct current that is always within a few millivolts of the desired 12 volts.

A transformer is normally used to step the voltage down to a safe level of 24 volts or less. The transformer also isolates the rest of the circuit from the power lines, for our personal safety. There is no other practical way to get the power-line voltage down to an acceptable voltage and level of safety.

Perhaps the single advantage of switching power supplies is that they allow us to use much cheaper and more compact transformers. We won't cover switching supplies until Chapter 13; they are much less reliable than the circuits we will deal with in this chapter.

The final problem involved in attempting to eliminate the natural **ac** sine-wave voltage variations and hold a constant output voltage is solved by using energy-storing capacitors and devices called **voltage regulators.** Let's examine the various power-supply techniques, starting with the simplest version: the half-wave rectifier power supply.

Half-wave Rectifier

The **half-wave rectifier circuit** is the simplest but least effective of common rectifier circuits. Figure 3-7 shows the basic half-wave rectifier circuit. The voltage across the load in Figure 3-7 consists of one-half of the input **ac** waveform. The output waveform is called a **half-wave dc pulse** waveform.

Let's look at the work that can be done by the power delivered to the load. The diode has simply opened the circuit and turned off the power during one-half of the cycle. The other half of the sine wave delivers power to the load, but the delivered power has anything but a constant voltage. The average value of the **dc** output voltage for the half-wave circuit is given by the following equation:

$$V_{\textbf{dc}} = V_P/\pi = \text{Peak Voltage}/3.14$$

Example 3.2

Calculating the Average Voltage of a Half-wave Pulse

Calculate the average voltage of the half-wave pulse in Figure 3-7.

Solution

We will proceed in three steps:

1   Assume that the output voltage of the transformer in Figure 3-7 is 12.0 volts RMS.

2   The peak voltage is

$$1.4 \times 12.0 \text{ V} = 16.8 \text{ V}$$

3   Now, we can calculate the average **dc** voltage as

$$V_{\textbf{dc}} = V_P/3.14 = 16.8/3.14 = 5.35 \text{ V}$$

FIGURE 3-7 **The Half-wave Rectifier, Power Supply Circuit**

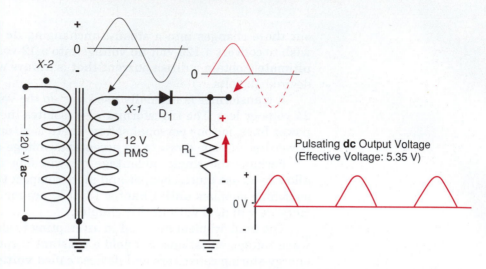

Pulsating **dc** Output Voltage
(Effective Voltage: 5.35 V)

a. Overview

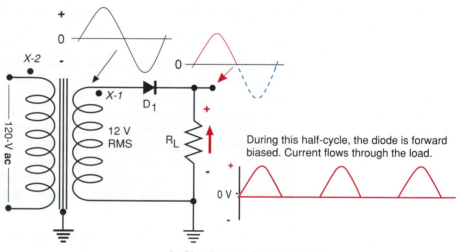

During this half-cycle, the diode is forward biased. Current flows through the load.

b. Circuit when Current Flows

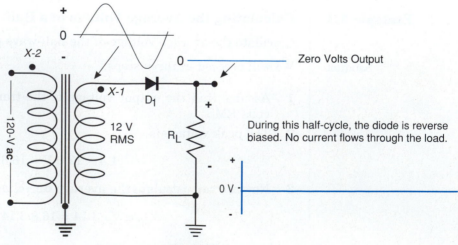

Zero Volts Output

During this half-cycle, the diode is reverse biased. No current flows through the load.

c. Circuit when No Current Flows

This means that a 5.35-volt battery that supplies a continuous **dc** voltage will have the same average voltage as the half-wave rectifier circuit working with 12.0 volts (RMS) **ac** from the transformer. We have obviously lost something in the translation from **ac** into pulsating **dc.**

The simple half-wave circuit in Figure 3-7 has two problems. First, the effective output voltage is much less than the input voltage; and second, the pulsating output **dc** waveform is nearly always unacceptable to circuits that must use direct current. The solution to both problems is called **filtering.** The simplest (but least effective) filter network is simply a large-value capacitor, connected as shown in Figure 3-8(a) (when the switch is closed).

How the Filter in Figure 3-8(a) Works

1   During the half of the **ac** cycle when the diode is turned on, the transformer delivers current to the load; at the same time, the transformer delivers current to charge the capacitor (C) to the peak voltage.

2   On the alternate half-cycle, the diode turns off and the transformer delivers no current to anything. (The capacitor was charged to the peak voltage during the previous half-cycle.)

3   During this second half-cycle, no power is available from the transformer, and the capacitor begins to discharge through the load. Thus, the capacitor operates the load, using the energy it stored during the previous half-cycle. If the capacitor is large enough, it will supply nearly full-load current during most of the diode's reverse-biased (off) half-cycle.

So the transformer supplies load current during the first half of the **ac** input cycle, and the stored energy in the capacitor supplies load current during the other half of the cycle. The shaded area in Figure 3-8(a), labeled "Filtered **dc** Waveform," represents the energy supplied to the load by the capacitor during the diode's off time.

Pi Filter

A much more efficient filter arrangement is called the **pi (π) filter resistor-capacitor (R-C) network.** The basic pi filter is shown in Figure 3-8(b). The circuit uses two capacitors and a resistor instead of a single much larger capacitor. The pi configuration is useful only for relatively low-load currents, but it provides better filtering at a lower cost than one gigantic capacitor. The first capacitor (C_1) serves as a prefilter. The resistor and the second capacitor form a voltage divider for pulsating variations (called **ripple**), and they also provide an R-C time delay that produces a better distribution of the discharge current over time. The resistance is selected to be several times the capacitive reactance of the second capacitor (C_2) at the ripple frequency.

The ripple frequency in a half-wave circuit is the same as the **ac** power-line frequency—in most cases, 60 Hz. The frequency of the ripple is the same as the line frequency because one ripple pulse occurs for each power-line cycle.

FIGURE 3-8   Basic Power Supply Filter Circuits

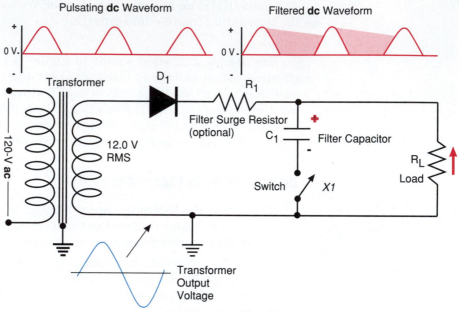

a. Half-wave Rectifier with Simple Filter

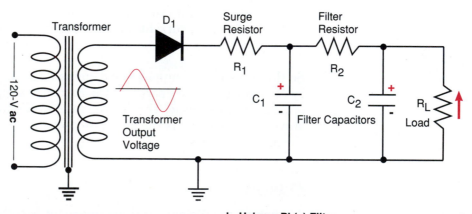

b. Using a Pi (π) Filter

Several voltage-divider filter sections can be cascaded to reduce the ripple further, but each section increases series resistance, limiting the available **dc** current for a heavy load. A large-value resistance provides the best ripple reduction, but a low-value resistance is required to prevent excessive **dc** voltage drop. These conflicting requirements make the R-C filter suitable only for loads that require relatively small currents. In general, half-wave rectifiers with several stages of R-C (pi) filtering are not very popular.

Full-wave Rectifier Circuits

The problem with trying to filter the half-wave circuit is that the filter capacitors must deliver continuous current to the load for a relatively long period of time (one-half the time required for an **ac** cycle) without discharging to a substantially lower voltage. This implies a need for a large amount of stored energy, which in turn demands a large-value capacitor; and large-value capacitors tend to be physically large and expensive. **Full-wave rectifier circuits** are arranged so that the transformer (power line) delivers current to the load, in the proper polarity, during both halves of the alternating-current cycle.

Figure 3-9 compares half-wave- and full-wave-circuit pulsating **dc** waveforms and identifies the relative amount of energy the capacitors must store in each case. In Figure 3-9, the shaded areas are proportional to the energy that must be stored in capacitors to provide power to the load during the period when the rectifier diode (or diodes) disconnects power delivery from the transformer. It is quite obvious in Figure 3-9 that the full-wave circuit demands far less capacitor energy storage for a given load current and ripple content than does the half-wave circuit. Of course, there is a price to pay for that full-wave pulse, but it is a small one. Extra diodes are required for both common full-wave circuits, but diodes are inexpensive compared to the cost of large-value capacitors. Diodes also tend to be far less prone to failure than the common electrolytic capacitors.

Comparison of Half-wave and Full-wave Circuits

In the center-tapped transformer version of the full-wave rectifier, a tapped transformer secondary with twice the total output voltage is required. Such a transformer is rather common. Extra windings and taps add little to the transformer's cost. For most practical situations, transformers are priced by the pound. The frame, the laminated core, and the wire in the primary winding constitute 80 to 90% of the cost of the transformer. A little extra copper wire or a tap in the secondary adds very little to the cost. The cost of extra diodes and/or the minimal extra transformer cost is a small price to pay for the much greater efficiency of a full-wave circuit and for the much smaller and less expensive filter capacitors required.

An excellent way to gauge the efficiency of a full-wave circuit as opposed to a half-wave circuit is to compare the average **dc** voltage outputs of the two.

FIGURE 3-9 **Relative Amounts of Stored Energy Needed to Supply Power to the Load Between Power Pulses**

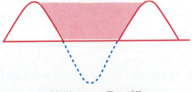

a. Half-wave Rectifier

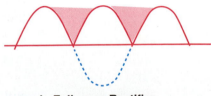

b. Full-wave Rectifier

Example 3.3

Solution

Comparing Full-wave and Half-wave Efficiencies

Compare the efficiencies of full-wave and half-wave unfiltered rectifier circuits.

We will examine the two kinds of circuits separately and then compare them.

Half-wave Circuit

1   In our previous discussion of the half-wave circuit, we used the following equation for average dc output voltage:

$$V_{dc} = \text{Peak Voltage}/\pi = V_P/3.14$$

2   With a 12-V (RMS) transformer we get peak voltage:

$$V_P = 1.4 \times 12 = 16.8$$

or about 17 V.

3   Therefore,

$$V_{dc} = 17/3.14 = 5.4 \text{ V} \qquad \text{(average)}$$

4   Let's rewrite the equation to make the eventual comparison a bit more obvious:

$$V_{dc} = 1/3.14 \times V_P = 0.318 \times V_P$$

5   If we translate 0.318 into a percentage, we get a voltage efficiency of 31.8% or roughly 32%.

Full-wave Circuit

1   Because the full-wave circuit produces two pulses for each **ac** power-line cycle, we multiply by 2:

$$2 \times 0.318 = 0.636 \times V_P$$

2   So, for the full-wave circuit, we get

$$0.636 \times V_P = 63.6\%$$

3   This 0.636 (63.6%) factor is the same as the average **dc** (equivalent) value of sine-wave alternating current and is close to the 0.707 (70.7%) RMS figure so commonly used to describe the work done by alternating current.

In terms of average **dc** values, then, the half-wave circuit is 31.8% efficient and the full-wave circuit is 63.6% efficient. In terms of RMS equivalent values, the half-wave circuit is about 50% efficient, whereas the full-wave circuit is about 70% efficient.

Now let's look again at the example we used earlier with the half-wave circuit, this time comparing the results with those of a full-wave circuit.

Example 3.4

Comparing Full-wave and Half-wave Voltages

Compare the half-wave circuit voltage to the full-wave circuit voltage.

Solution

1 Assume a transformer output voltage of 12 V RMS.

2 The peak voltage is $1.4 \times 12 = 16.8$ V.

3 The average **dc** output voltage for the half-wave circuit is

$$V_{dc} = 0.318 \, V_P = 0.318 \times 16.8 = 5.34 \text{ V}$$

4 The average **dc** output voltage of the full-wave rectifier with the same transformer output voltage is

$$V_{dc} = 0.636 \, V_P = 0.636 \times 16.8 = 10.68 \text{ V}$$

It thus appears that the half-wave circuit doesn't have much going for it other than its basic simplicity, and that simplicity is almost always offset by the extra filtering components required to produce a useful half-wave circuit.

Full-wave Center-tapped Transformer Circuit

The circuit in Figure 3-10 requires two diodes and a **center-tapped transformer.** Each half of the secondary winding must be capable of producing the desired output voltage and current, because only one-half the winding is on (and in operation) at any given time. The diodes serve as switches that alternately turn on one winding half and then the other. The diodes should have a peak inverse voltage rating of $2 \times V_{ac}$ (peak) or greater. Only one of the diodes is in the reverse-bias condition at any given time, so nearly the entire transformer voltage ($2 \times V_{ac}P$) appears across the reverse-biased diode. The diodes should have a forward-current rating equal to or greater than the maximum expected load current.

How the Circuit Works

Examine Figure 3-10 as we discuss the circuit. The center tap of the transformer is used as the common reference point and is generally the ground connection for the load.

With reference to the center tap, points A and B in Figure 3-10 are 180° out of phase. Thus, when point A is positive with respect to the center tap, point B is negative. During the first 180° of the **ac** cycle, diode A is forward-biased and diode B is reverse-biased. Current flows through the forward-biased diode (A) and through the load as shown in Figure 3-10(a).

The waveform across the load for the first 180° of sine-wave transformer voltage is indicated in Figure 3-10(c). During the 180° to 360° portion of the transformer sine wave, diode A is reverse-biased and turned off; at the same time, diode B is forward-biased (turned on), and current flows through diode B and the load as shown in Figure 3-10(c). Figure 3-10(e) shows the waveform across the load for the 180° to 360° portion of the transformer sine wave.

The waveform in Figure 3-10(d) is the combination of the 0° to 180° and the 180° to 360° load voltages. This waveform is called a **full-wave dc pulse.**

Figure 3-10 Full-wave Rectifier Power Supply Circuit, Using a Center-tapped Transformer

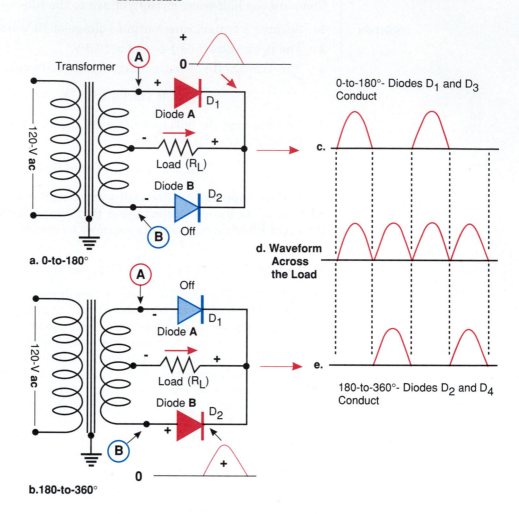

Filtering is still required for the full-wave circuit, but it is far easier and cheaper to perform than is filtering for a half-wave pulse.

Full-wave Bridge Circuit

The **full-wave bridge** is the most widely used rectifier circuit in current technology. The center-tapped transformer full-wave circuit was the most popular for many years, when diodes were comparatively expensive. Now diodes are so cheap that it usually makes more sense to use the four diodes required by the full-wave bridge than to use two diodes and a center-tapped transformer. In recent years, the full-wave bridge has become even more economically attractive because manufacturers package the four diodes wired internally in the bridge configuration. These bridge rectifier packages cost little more than a pair of diodes do; and because they can be installed on a printed circuit board as a single component, they cut assembly costs as well.

FIGURE 3-11 Full-wave Rectifier Power Supply Circuit Using a Diode Bridge

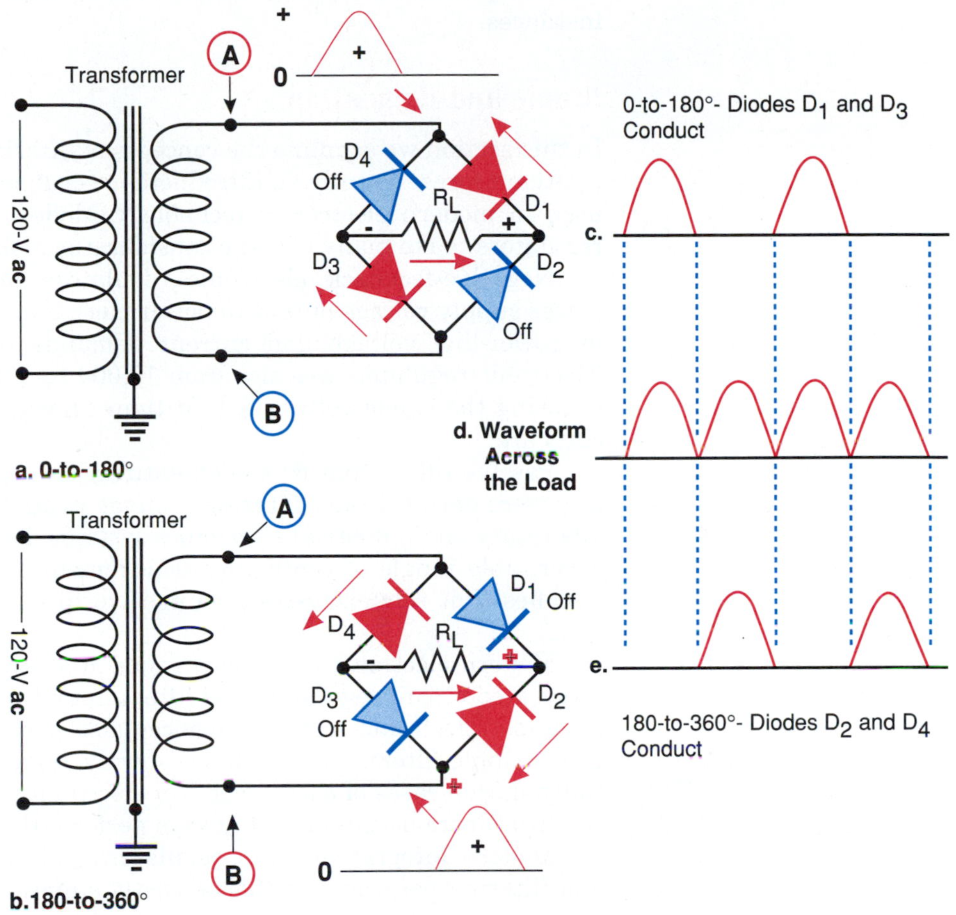

a. 0-to-180°

b.180-to-360°

0-to-180°- Diodes D_1 and D_3 Conduct

c.

d. Waveform Across the Load

e.

180-to-360°- Diodes D_2 and D_4 Conduct

How It Works

During the 0° to 180° part of the sine wave (from the transformer), point A in Figure 3-11 is positive and point B is negative. Diodes D_1 and D_3 are in series with the load, and both are forward-biased. Current flows through diodes D_1 and D_3 and through the load. Diodes D_2 and D_4 are reverse-biased and are effectively open circuits. Figure 3-11(c) shows the voltage waveform across the load for this half of the cycle.

When the transformer output sine wave passes through 180°, the polarities of points A and B reverse, with point A becoming negative while point B goes positive. Diodes D_2 and D_4 are now forward-biased, and current flows through them and through the load as illustrated in Figure 3-11(b) and (e). Figure 3-11(d) shows the waveform voltage across the load for one complete alternating-current (transformer) cycle.

Each diode in the bridge must be capable of carrying the full load current, and the P_{IV} (P_{RV}) rating must be equal to the peak output voltage (1.4 ×

RMS) of the transformer secondary. In this era of mostly low-voltage circuits and inexpensive, fairly high-voltage diodes, this may not be important in most instances.

Ripple and Regulation

In this section we examine the concepts of filtering and regulation. Our investigation represents only an introduction, because brute-force filtering is rarely used in modern electronics technology. Modern integrated-circuit electronic regulators make many of these considerations obsolete.

Negative-feedback electronic regulators maintain a constant desired power supply voltage output to power electronic circuits, in spite of variations in power-line voltage and current demands placed on the power supply. Electronic regulators are also from 10,000 to 100,000 times more effective in reducing the ripple voltage output than conventional R-C (passive) filter circuits are.

Passive filters that depended entirely on capacitors, resistors, and inductors were once the only filtering systems available, except for very expensive laboratory and industrial electronics systems. Passive filter systems introduce a veritable jungle of conflicting requirements, and truly well-filtered power supplies that rely exclusively on passive filters present many difficult problems.

Modern electronic technology relies almost entirely on negative-feedback integrated-circuit regulators and filter circuits, or discrete electronic regulator circuits. Electronic regulator devices also perform remarkably well as electronic ripple filters. As we will see shortly, regulation and filtering are essentially special cases of each other. An electronic device that can perform one of the two functions can almost always perform the other function equally well.

Modern integrated-circuit regulator/ripple filters can provide regulation and filtering performance that is vastly superior to that of passive filters for pennies instead of dollars, ounces instead of pounds, and cubic inches instead of cubic feet. In this chapter we will study some basic applications of electronic integrated-circuit regulators. Later in the text, we will examine the internal circuitry of these regulators to see how they actually perform their magic. Some simple passive filtering techniques must be used even by these high technology IC devices. And for reasons peculiar to that kind of circuit, switching power supplies take even more advantage of passive filters than do the simple power-supply systems covered in this chapter. We will also examine these complex switching power supplies later in the text. For now, let's look at some passive filter basics.

Ripple Defined

Both ripple and regulation involve power-supply output voltage variations that result from varying input voltages or varying load currents. *Ripple* is a variation in output voltage that occurs regularly at the ripple frequency (usually 60 or 120 Hz). It is caused by the pulsating voltages, full- or half-wave, produced by any rectifier circuit with an **ac** input. A filter circuit of some sort is used to reduce those ripple voltage variations to some reasonable level that the electronic regulator can manage initially.

A filter capacitor can be charged almost to the peak voltage of the half-wave or full-wave pulse; and if the capacitor is large enough, it can deliver current to the load without its voltage dropping very far below the fully charged voltage value. But the capacitor voltage must drop somewhat if any current at all is demanded by the load.

The difference between the voltage to which the capacitor is charged by the rectified pulse and its charge voltage just prior to delivery of the next charging pulse is termed the **peak-to-peak ripple voltage.**

If the load demands more current, the capacitor will discharge to a lower voltage between charging pulses and the peak-to-peak ripple voltage will increase. The greater the current demand, the more ripple you can expect from a particular filter.

Figure 3-12 shows the waveform and ripple component for a simple capacitor filter. If you look only at the shape the top envelope of the waveform which follows, you will notice that the ripple wave shape is not a sine wave. However, it has become common practice to pretend that the ripple wave shape is a sine wave and to discuss its RMS value. Figure 3-12(c) shows an alternative way to describe ripple that may help make the concept a little easier to understand. The approach adopted in Figure 3-12(c) is often used in commercial specification sheets.

There are good reasons for pretending that the ripple waveform is a sine-wave. Some kind of average value is generally more useful than a peak-to-peak value, but the equation and measurements for a mathematically precise average value for this odd wave shape are ponderous and difficult. And once having measured parameters and solved the equation, we would find that the numerical value differed only slightly from the value we would have gotten by pretending that the waveform was a sine wave and finding the RMS value

FIGURE 3-12   **Ripple Voltage**

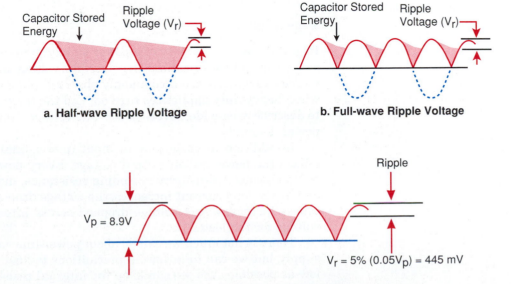

a. Half-wave Ripple Voltage

b. Full-wave Ripple Voltage

c. Ripple Calculations

(0.707 × Peak) for it. Since we would like to do away with ripple altogether, all we really need is some standard reference to allow us to compare the ripple among power supplies or in a given power supply under various conditions.

Actually, there are several common ways to discuss ripple. The ripple may be specified in peak-to-peak millivolts, or in millivolts RMS, or by a figure of merit called the **ripple factor.** The ripple factor is often converted into a percentage of the **dc** power supply voltage.

For the most part, a technician is not much concerned with ripple except when its value is high enough to indicate that it presents a problem that must be fixed. We will study that and other power-supply problems in Section 3.7 of this chapter. Sometimes you must read the manufacturer's specifications to find out whether a certain ripple value is excessive for that particular unit. With this in mind, let's consider the following numerical definition of ripple factor:

$$\% \text{ Ripple} = (\text{Ripple Voltage [RMS]}/\textbf{dc Output Voltage}) \times 100$$
$$\%R = (V_{r\,(\text{RMS})}/V_{\textbf{dc}}) \times 100$$

where $\%R$ is the ripple percentage, V_r is the RMS ripple voltage, and $V_{\textbf{dc}}$ is the **dc** output voltage of the power supply.

Example 3.5

Calculating the Ripple Factor

Find the ripple factor in a power supply that has a $V_{\textbf{dc}}$ output of 5 volts and an RMS ripple voltage of 5 millivolts.

Solution

Plugging the values given above into our equation, we get

$$\%R = (V_{r(\text{RMS})}/V_{\textbf{dc}}) \times 100$$
$$= (0.005\text{ V}/5\text{ V}) \times 100 = 0.1\%$$

Regulation

Regulation is the ability of a power supply to maintain a constant **dc** output voltage in spite of variations in power-line voltage and load current. Power-line voltage variations are a commonly observed phenomenon. Lights dim briefly when heavy-duty appliances turn on, and the term *brown-out* has been coined to describe area-wide, severe power-line voltage drops resulting from excessive power demands.

In addition to variations in input (power-line) voltage, the load current affects the power-supply output voltage. Every power supply has some internal resistance (transformer winding resistance, diode resistance, and so on); and as the load current increases, the voltage drop across these internal resistances increases. Any voltage dropped across internal resistance leaves less voltage for the load.

There is not much we can do about power-line variations in a simple power supply, but we can take special precautions to keep the internal resistance as low as possible. And we can keep the internal resistance low if we are willing to trade high ripple values for good regulation. When dealing with simple resistor-capacitor filter networks, we must always trade one for the other. This conflicting demand for both low ripple values and good regulation makes electronic regulators critically important. With electronic regulation, the trade-offs are not necessary.

A Zener diode or more complex integrated-circuit regulator can hold the output voltage constant in spite of line-voltage variations. Electronic regulators respond so rapidly that ripple variations are effectively regulated out of the system. The added costs of a well-regulated, low-ripple power supply consist of a very inexpensive regulator device and a 20 to 100% increase in the transformer output voltage. Nearly all modern power supplies use some form of electronic regulator. In a power supply that does not use electronic regulation, the cost of the very large filter capacitors needed is much greater.

Voltage Regulation Figure of Merit

To determine the relative regulation of different power supplies or the regulating ability of a given power supply, we need some standard of comparison. This standard, called the **regulation figure of merit,** is defined by the following equation:

% Regulation = [(No-load Voltage − Full-load Voltage)/Full-load Voltage] × 100

$$\%V_{Reg} = [(V_{NL} - V_{FL})/V_{FL}] \times 100$$

Example 3.6

Calculating the Regulation Figure of Merit

A certain power-supply circuit has a no-load **dc** output voltage of 10 volts, but that **dc** output voltage drops to 9.5 volts under full load. Find the regulation percentage (figure of merit).

Solution

We proceed with the regulation percentage equation, as follows:

$$\%V_{reg} = [(V_{NL} - V_{FL})/V_{FL}] \times 100$$
$$= [(10 - 9.5)/9.5)] \times 100$$
$$= 5.26\%$$

Selecting a Filter Capacitor Value

The shortcut procedure developed in the following example (Example 3.7) has been adopted by most power-supply designers. This procedure is mostly included here to illustrate how much larger a filter capacitor would have to be if we were not using an electronic regulator. Even at that, the 375-millivolt ripple in the example is at least 10 times greater than we would expect from an IC regulator with a (typical) filter capacitor value of only 1000 microfarads.

The simple filter circuit for the example is shown in Figure 3-13. The shortcut calculation applies only to this simple filter. A pi filter would yield better ripple reduction, but at the expense of poorer regulation. Although the shortcut procedure yields a small error, the result is more than adequate. The error is considerably smaller than the expected tolerance error in off-the-shelf electrolytic capacitors, which are often allowed a tolerance of −0%, +50% to +100%.

FIGURE 3-13   Filter Capacitor Calculation

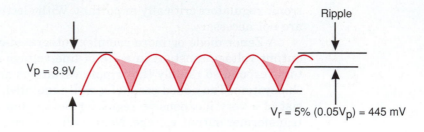

a. Filtered dc Waveform

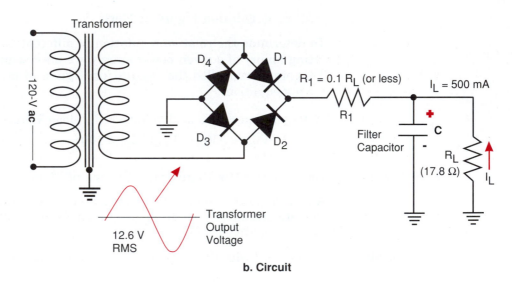

b. Circuit

Example 3.7 | **Calculating Filter Capacitor Value**

Calculate the filter capacitor value for nonelectronic filtering.

Solution | Figure 3-13(a) defines the values needed for calculating the filter capacitor value. The relevant equation is

$$C = 0.0083/\%V_r\ (P–P) \times R_L$$

where C is the minimum capacitor value required for the desired ripple voltage, and 0.0083 is the time period for one ripple cycle:

$$T = 1/\text{Frequency}$$
$$= 1/120\ \text{Hz} = 0.0083$$

The power supply uses a full-wave rectifier with a ripple frequency of 120 Hz. The equivalent value of R_L is calculated by Ohm's law, using the peak voltage V_P and the current that the power supply must deliver to the load:

$$R_L = V_P/I\ (\text{load})$$

Example 3.8

Filter Capacitor Calculations with Numbers

Calculate the value required for a passive capacitor filter.

Solution

Refer to Figure 3-13, and assume the following conditions:

1 The RMS output voltage of the transformer is 12.6 V. The peak output voltage is 17.8 V.

2 The junction voltage of the two diodes in series in the diode bridge circuit is 0.7 V × 2 = 1.4 V. The diode voltage drops reduce the actual peak voltage to the load: 17.8 V − 1.4 V = 16.4 V, so the actual peak voltage across the load (V_P) is 16.4 V.

3 The load current demanded of the supply is 500 mA.

Now, we can calculate the equivalent value of R_L:

$$R_L = V_P/I \text{ (load)}$$
$$= 16.4 \text{ V}/0.5 \text{ A} = 32.8 \text{ } \Omega$$

4 We want to limit the ripple to 5%.

Now, we can calculate the value of the capacitor:

$$C = 0.0083/0.05 \times 32.8 \text{ } \Omega = 5{,}061 \text{ } \mu\text{F}$$

or realistically, 5,000 microfarads.

If a larger capacitor were used, the only difference would be a little less ripple.

3.4   Integrated-circuit Regulators

Integrated-circuit regulators are a classic example of the use of **negative feedback** in its most obvious form. Negative feedback is the single most important concept in the field of analog electronics. Without it, there would be no analog IC technology. We will encounter negative feedback in many—if not most—of the circuits in the balance of this text. The integrated-circuit voltage regulator offers a splendid opportunity to introduce negative feedback, as well as to explain the operation of a most useful device. Negative feedback is not mysterious, but it certainly is magic. The voltage regulator gives us a chance to study it in its most primitive form, which is a good thing because it will seem less obvious and more tricky in later circuits. The influence of feedback on such circuits is enormous, to the point of determining nearly all of their characteristics.

What Is Negative Feedback?

Negative feedback is an automatic error corrector. Negative feedback always involves holding some quantity as constant as possible, in spite of any influences that might ordinarily change the value of that quantity. The value of the

controlled quantity is continuously corrected for any slight change that occurs as a result of outside influences. Although the controlled quantity may change slightly, it is always corrected back to the desired value before it can deviate significantly from that value.

Take, for example, a thermostatically controlled household heating and cooling system. Temperature is the quantity you wish to hold constant at some comfortable level. Suppose that it is a cold day, and you have set the thermostat to 75°F. If the temperature in the house drops a little below 75°F, the heater magically turns on. As soon as the temperature rises to 75°F again, with similar magic, the heater turns off. This system illustrates a typical practical application of negative feedback.

All negative feedback systems must have the following five well-defined elements:

1   A controlled quantity—temperature, output voltage, motor speed, or the like.

2   A reference quantity—reference battery, reference diode, or the like.

3   An error detector—a device that compares the controlled quantity value to the reference quantity value, and then produces an output signal that is proportional to the difference between them. The output signal is called the error signal. No error signal is produced when the reference quantity and controlled quantity values are equal.

4   An amplifier to boost the output signal from the error detector. This output signal is proportional to the difference between the reference and controlled quantities. If that difference is slight, the unamplified error signal will be too small to activate the error corrector. A small error yields a small error signal.

5   An error corrector adjusts the controlled quantity value to correct the detected error, until it again equals the value of the reference quantity. When the adjustment has been made, the error detector produces a zero error output signal. No further correction will be made until a new error signal is produced.

How the Negative-feedback Voltage Regulator Works

The block diagram in Figure 3-14 is a typical negative-feedback voltage-regulator block diagram. This block diagram fits both integrated-circuit regulators and regulators composed of discrete parts, and it includes all of the elements of a negative-feedback control system.

Controlled Quantity

In Figure 3-14, the load voltage is the controlled quantity. In this case, we have a 12-volt regulator, so the value of the controlled quantity is 12 volts. The only job the regulator circuit has is to hold the voltage across the load (regulator output) at a constant 12 volts, no matter what may happen.

An important idea in negative-feedback control systems is that the cause of the change in the controlled quantity is irrelevant to the system's functioning. If a small difference between the controlled and reference values occurs,

WARNING

Beware of used electrolytic capacitors. An electrolytic capacitor with 50 volts printed on it will become a 25-volt unit if it is operated over a period of time in a 25-volt circuit. The dielectric is an electrically formed oxide and will be re-formed if used in a lower-voltage circuit. Nearly all capacitors used for power-supply filter service are polarized. Installing a polarized capacitor backward (reverse polarity) or applying **ac** to it can result in a small explosion. Most electrolytic capacitors are clearly marked.

You can restore electrolytic capacitors to their original voltage rating by connecting them to a variable voltage power supply and very gradually increasing the voltage up to the capacitor's original rated voltage. This process takes a lot of time—sometimes hours or days or weeks—and is not often attempted.

A capacitor of any kind should not generate heat. A warm or hot capacitor should be replaced if any of the heat is being generated internally.

FIGURE 3-14

Use of Negative Feedback by an Integrated Circuit Electronic Regulator

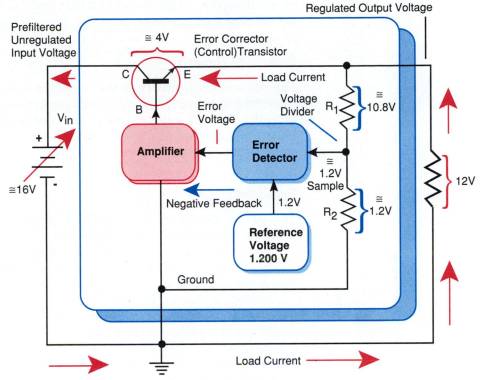

a. Block Diagram and Circuit

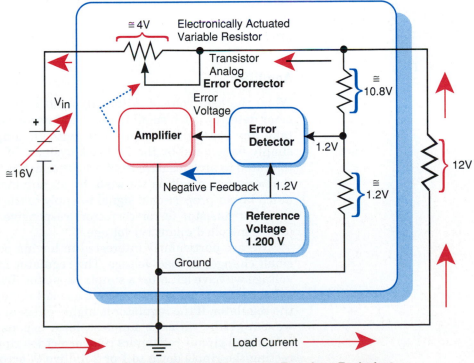

b. Variable Resistor Shown as a Transistor Equivalent

an error voltage is produced, and that error is then corrected. It doesn't matter what caused the error.

Reference Quantity

The **reference quantity** in the regulator is a 1.2-volt constant voltage, often consisting of two diode junction voltages in series (0.6 volts each). More sophisticated reference voltage devices are also used, but we must save those for later. For this example, assume that the reference voltage is provided by two series diode junction voltages.

Error Detector

The **error detector** is normally a difference (differential) amplifier, the details of which we will cover in a later chapter. For now, we will concentrate on what it does rather than on how it works. The error detector compares the reference voltage with the load voltage and produces an output error voltage that is proportional to the difference between the two.

A small problem arises here. We are trying to compare the controlled quantity of 12.00 volts with a reference voltage of only 1.2 volts. The error detector would see this difference as a very large error. We have two options: increase the reference voltage to 12.00 volts, or use a 10-to-1 voltage divider to divide the load voltage (the controlled quantity) down to 1.2 volts. Because 12-volt reference devices are not readily available, especially on an integrated-circuit chip, we have used the voltage divider approach. Now, with 12.00 volts across the load, we have a 1.2-volt input to the error detector. As a result, the error detector is now comparing 1.2 volts to 1.2 volts (when the load voltage is correct). If the load voltage changes, the output of the voltage divider will change proportionally, yielding an error voltage output from the error detector.

Amplifier

When the load voltage drifts only a little, the error voltage produced by the error detector is too small to cause the **error corrector** (control transistor) to take any corrective action. Thus, without amplification, small errors would simply be ignored by the feedback system. Of course, some degree of error is tolerated in any negative feedback system; but the amplifier allows us to make that error as small as we wish, by amplifying the error signal to produce a much larger proportional signal to apply to the error corrector. This makes the control transistor (error corrector) responsive to very slight changes in the load (controlled quantity) voltage.

We are particularly interested in having our voltage regulator correct for small changes in load voltage. The regulator must eliminate whatever ripple voltage we have left after a simple capacitor filter has attempted to smooth the voltage. Ripple voltage represents a small regular variation in input voltage to the regulator. If the regulator is highly efficient at correcting for ripple voltage, we can get by with the simplest, least expensive prefilter possible. Modern integrated circuit regulators can correct for ripple voltages of 1 or 2 volts, correcting the ripple down to 1 or 2 millivolts across the load.

Error-corrector (Control Transistor)

The load-current circuit in Figure 3-14 includes a 16-volt battery with an arrow through it, symbolizing the output voltage of a simple rectifier circuit with a simple filter. This simple power supply is subject to line-voltage variation, internal-resistance voltage drops that vary with current, and ripple voltage variations. The regulator's job is to eliminate all of these voltage variations and to maintain a constant voltage across the load.

If you examine the load-current loop of Figure 3-14, you will notice two series voltage drops: one of 12.00 volts across the load, and one of 4 volts across the error-corrector control transistor. If the unregulated voltage (V_{in}) is actually 16 volts, 12 volts will appear across the load and 4 volts will appear across the control transistor.

We have not yet studied transistors, and now we need to take a little mystery out of the error-corrector transistor. To visualize what is going on, think of the error-corrector control transistor as a special variable resistor that can be controlled by the output signal from the error amplifier. Figure 3-14 offers an analogous version of the error-corrector control transistor that may help you visualize the situation a little better. After you have studied transistors, this will all be obvious; but at this point you may have to take a few details on faith.

Because negative feedback is the glue that holds analog electronics together, it is important to start thinking about feedback in general terms as soon as you can. You will encounter many specific applications of feedback as we go through the text.

What Happens When the Unregulated Input Voltage Changes

Suppose that the unregulated voltage rises to 17 volts. Assume that the circuit has not yet had time to adjust for the increase. The voltage across the load rises to 13 volts, while the control transistor's voltage drop remains at 4 volts. The 13 volts across the load is divided by 10 in the voltage divider. This places 1.3 volts (voltage sample) on one input of the error detector. The other input of the error detector remains the 1.2-volt reference voltage.

Next, 1.3 volts from the voltage divider is subtracted (in the error detector) from the 1.2-volt reference voltage. The error voltage at the output of the error detector is −0.1 volt. The amplifier then increases the error voltage level to −1.0 volt.

When the −1.0-volt amplifier output voltage is applied to the base of the error-corrector control transistor, the voltage across load decreases by 1.0 volt, bringing the load voltage back to the desired level of 12.00 volts. The control transistor accomplishes the 1-volt reduction in the load voltage by increasing its own voltage drop by 1 volt. The voltage drop across the control transistor is now 5 volts. With the new 17-volt regulator input voltage, we now have a voltage drop of 5 volts across the control transistor and 12 volts across the load, for a total of 17 volts.

If the input voltage drops, instead of increasing, the resulting error voltage is positive instead of negative. The error voltage then causes the control transistor to decrease its own voltage and increase the load voltage.

Because the regulator feedback circuit only compares and corrects voltages, the load resistance and the current it demands do not affect the operation of the regulator. The regulator output voltage is completely independent of the actual load value, as long as the regulator's maximum current rating is not exceeded.

Because ripple voltage consists of a regular periodic rise and fall of the regulator input voltage, the regulator feedback circuit treats ripple voltage as it would any other change in input voltage. The regulator corrects for the variations in ripple voltage, virtually eliminating the ripple at the load.

Comparison

Back in the subsection on regulation, we examined the task of selecting a passive filter capacitor value for a power supply, without electronic regulation. In that example, we came up with a value of 5,000 microfarads for a ripple voltage of 375 mV, with 500 mA delivered to the load. Suppose that we connect a bridge-rectifier circuit like the one shown in Figure 3-13, to a commercial integrated-circuit voltage regulator; and suppose further that we replace the 5,000-microfarad filter capacitor with a 1000-microfarad capacitor.

The typical ripple voltage output of the regulator will be something less than 10 millivolts. By adding an IC regulator to the simple rectifier/filter circuit shown in Figure 3-8, we have obtained a constant, dependable power-supply voltage with a ripple voltage of less than 10 mV (compared to one of 375 mV).

In dollar terms, we have replaced a $5.00 capacitor with a $1.00 capacitor; the regulator adds about $1.00 to the replacement system's cost. We have traded up from a very mediocre power supply to an excellent one, and we have saved $4.00 in the process. Our excellent power supply also weighs less and takes up less space.

It's not hard to understand why electronic negative-feedback regulators are so popular. In the next section we will look at some common IC regulators and their circuits. In a later chapter, after you learn more about transistor circuits, we will look at the electronic circuits inside the regulators.

3.5   Commercial Integrated-circuit Regulators

Now let's look at some real integrated-circuit regulators. Actually, quite a variety of IC regulators are available; but in this chapter, we will concentrate on a few of the most common types. It is customary to defer this material on commercial regulators until later, when you can follow a discussion of their internal circuitry. By Chapter 13, you will have that knowledge and we will study the internal circuitry.

This text operates on the assumption that you need to know the following material as soon as possible, so that you can use the devices intelligently in your laboratory work. Every lab experiment you perform requires power, and most of the power supplies you use will be based on the commercial IC regulators that are about to be presented.

N O T E
In decibel notation, an attenuation is often indicated by the use of a negative sign: –80 dB, for example. Because the word *attenuation* coupled with a minus sign is a double negative, other people leave out the negative sign. You can expect to find both conventions in the literature.

In some colleges (including the author's), lab instructors give students the assignment of building a regulated power supply as a first construction project. It is a simple starter experience, and it encourages students to tinker at home because it gives them a power source with which to work.

Ripple Reduction

The regulators we will examine here have manufacturer-specified ripple rejection factors ranging from about 74 dB to about 80 dB. The 74 dB figure translates into a voltage attenuation of 5000; simply stated, any ripple voltage present at the input of the regulator will be reduced by a factor of 5000. Similarly, 80 dB translates into a voltage reduction ratio of 10,000.

Example 3.9

Calculating Ripple Reduction

A certain regulator has a ripple rejection ratio of 80 dB; and 1 volt of ripple is applied to the regulator input. What is the regulator output ripple voltage?

Solution

Our answer will be based on the following equation for ripple:

$$\text{Ripple Out} = \text{Ripple In/Ripple Rejection Ratio}$$

Since 80 dB = 10,000, we know that

$$\text{Ripple Out} = 1\ \text{V}/10{,}000 = 0.0001\ \text{V} = 0.1\ \text{mV}$$

We may not always obtain this much ripple reduction in practice, but it is more than we usually need in practice. The ripple reduction provided by commercial integrated-circuit regulators is nearly always adequate.

Drop-out Voltage

As we saw in the negative-feedback circuit block diagram (Figure 3-14), some voltage drop across the error-corrector control transistor must occur to enable the circuit to make corrections. The control transistor and its associated circuitry must have some minimum voltage across them or the transistor will simply quit working—as will the rest of the regulator circuit along with it. This minimum keep-alive voltage usually ranges from 1.2 to 2.0 volts. A few special low-dropout-voltage regulators require only 0.3 to 0.6 volts to keep them alive. It is inadvisable to work at a voltage level that is close to the dropout voltage. If for any reason the unregulated input voltage drops below the sum of the regulator output voltage plus the dropout voltage, the regulator will not work until the voltage goes up again.

G U I D E L I N E
In practice, the unregulated input voltage should always be several volts higher than the regulated output voltage.

Output Voltage Accuracy

Regulators have an output voltage that is normally within plus or minus 5% (or 10%) of the manufacturer's specified voltage; 5-volt regulators are intended for digital service, and are within plus or minus 5% of the specified value.

Regulator Protection Circuits

Nearly all modern IC regulators have two distinct protection circuits. One of them is a fast-acting overcurrent shut-down circuit. If the regulator's maximum current rating is exceeded, this circuit kicks in, reducing the regulator output voltage sufficiently to prevent damaging values of current flow. This overcurrent circuit usually prevents short-circuit damage.

A built-in temperature-sensing circuit shuts down the regulator if its internal temperature gets too high. An excessively high internal temperature can result from restricted airflow or from an inadequate heat sink.

Other Important IC Regulator Data

Figures 3-15 through 3-18 and Tables 3-1 through 3-4 provide the following data about some of the most popular IC regulator types:

1 Case styles, standard case designations, and associated current ratings.

2 IC terminal connections for input, output, ground, and adjust (for adjustable regulators).

3 Popular regulator type numbers, together with their available voltage, current rating, and case types.

FIGURE 3-15 **Positive Fixed-voltage Regulators**

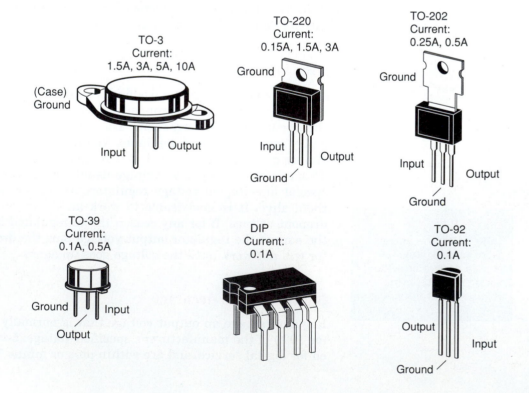

FIGURE 3-16

Negative Fixed-voltage Regulators

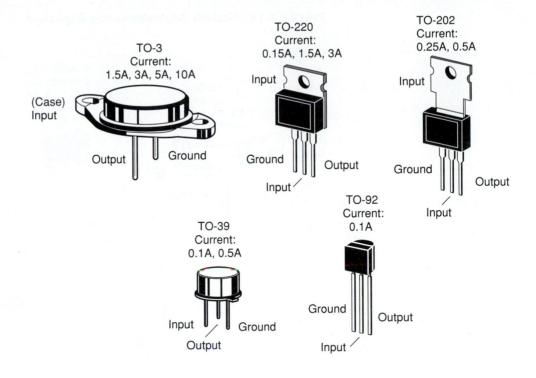

FIGURE 3-17

Positive Adjustable-voltage Regulators

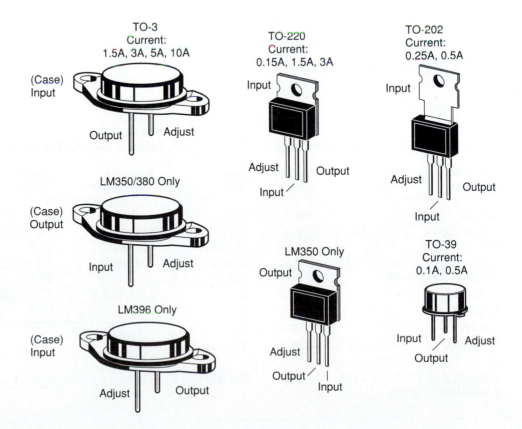

FIGURE 3-18   Negative Adjustable-voltage Regulators

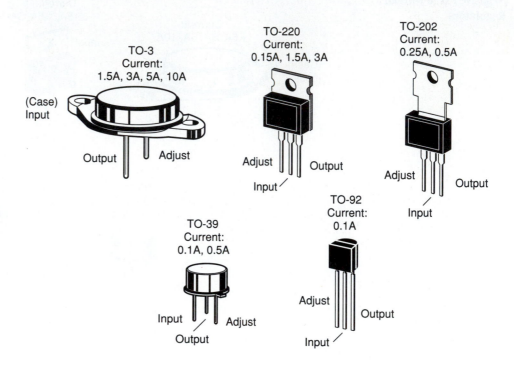

TABLE 3-1
Common Positive Fixed-voltage Regulators

Regulator Type Number	Voltages Available	Maximum Current	Case Type
LM323	+5.0 V	3 A	TO-3
LM340XX LM78XX	+5.0 V, +12 V, +15 V	1.5 A	TO-3 TO-220
LM340LAXX 78LXX	+5.0 V, +12 V, +15 V	0.1 A	TO-39 TO-92
LM2931	+5.0 V	0.15 A	TO-92
LM2930	+5.0 V, +8.0 V	0.15 A	TO-220
LM330	+5.0 V	0.15 A	TO-220
LM342XX	+5.0 V, +12 V, +15 V	0.25 A	TO-202
LM341XX LM78MXX	+5.0 V, +12 V, +15 V	0.5 A	TO-202

TABLE 3-2
Common Negative Fixed-voltage Regulators

Regulator Type Number	Voltages Available	Maximum Current	Case Type
LM345	−5.0 V, −5.2 V	3 A	TO-3
LM320XX LM79XX	−5.0 V, −12 V, −15 V	1.5 A	TO-3 TO-220
LM320ML	−5.0 V, −12 V, −15 V	0.25 A	TO-202
LM320LXX	−5.0 V, −12 V, −15 V	0.1 A	TO-39 TO-92

TABLE 3-3
Common Positive Adjustable-voltage Regulators

Regulator Type Number	Voltage Range	Maximum Current	Case Type
LM396	+1.2 V to +15 V	10 A	TO-3
LM317	+1.2 V to +37 V	1.5 A	TO-3 TO-220
LM350	+1.2 V to +33 V	3 A	TO-3 TO-220
LM338	+1.2 V to +32 V	5 A	TO-3
LM317HV	+1.2 V to +57 V	1.5 A	TO-3 TO-220
LM317M	+1.2 V to +37 V	0.5 A	TO-202 TO-39

TABLE 3-4
Common Negative Adjustable-voltage Regulators

Regulator Type Number	Voltage Range	Maximum Current	Case Type
LM337L	−1.2 V to −37 V	0.1 A	TO-92
LM337	−1.2 V to −37 V	1.5 A	TO-3 TO-220
LM337HV	−1.2 V to −47 V	1.5 A	TO-3 TO-220
LM337M	−1.2 V to −37 V	0.5 A	TO-202 TO-39

Some Typical Regulated Power-supply Circuits

The circuit diagram for a positive voltage-regulated power supply is shown in Figure 3-19. The circuit consists of a step-down transformer, a standard diode bridge full-wave rectifier, a simple capacitor filter, and an integrated circuit regulator. The 2.2-ohm resistor (R_1) is optional. The purpose of R_1 is to limit capacitor in-rush current when the power is first turned on. Because most diodes can handle very large short-term overcurrents, the resistor is often left out.

The circuit shown in Figure 3-20—a negative-voltage version of the circuit in Figure 3-19—differs only in the negative IC voltage regulator and in the fact that the input of the regulator is connected to the negative side of the bridge rectifier.

FIGURE 3-19   Typical Positive-regulated Power Supply

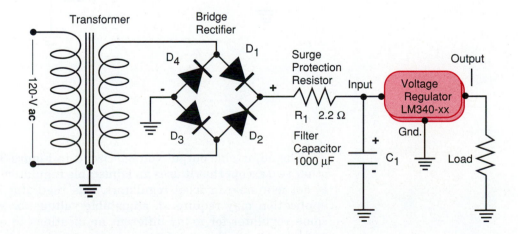

FIGURE 3-20

Typical Negative-regulated Power Supply

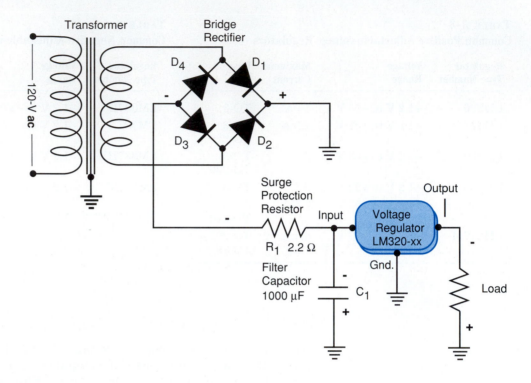

FIGURE 3-21

Typical Variable-voltage-regulated Power Supply

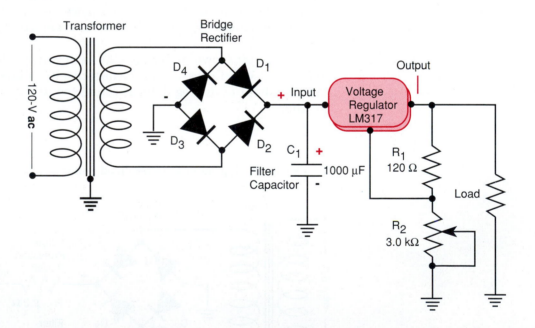

The adjustable output voltage circuit in Figure 3-21 closely resembles the other two except that it uses an adjustable regulator. If you need a voltage that is not available in fixed regulators, this regulator solves the problem. Your application may require an adjustable voltage, or you may want to use the same regulator for many different applications to obtain quantity discounts and to make stocking easier.

FIGURE 3-22 Typical Multiple-output Regulated Power Supply

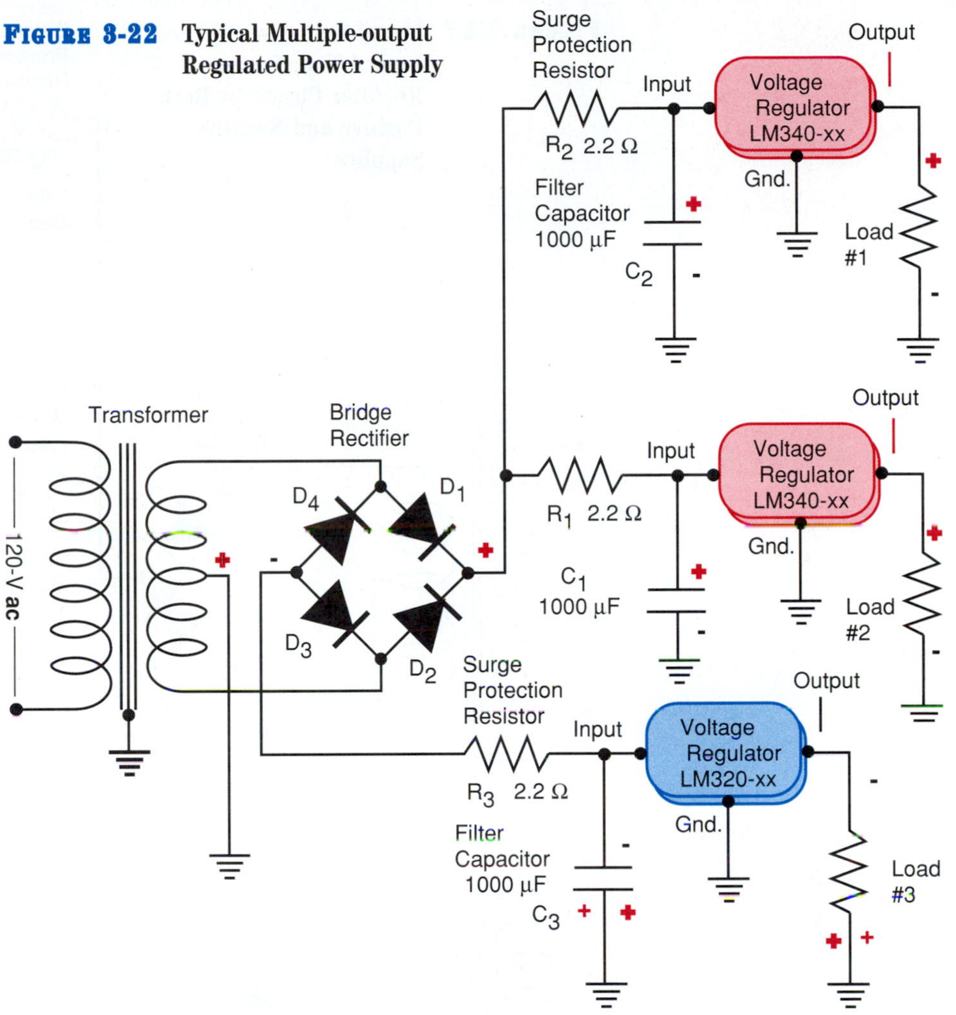

The multiple-output power supply in Figure 3-22 differs in one significant way from the previous circuits. The previous circuits have all used full-wave diode bridge rectifiers; but the rectifier circuit in Figure 3-22, although it appears to be a standard diode bridge, is actually two full-wave center-tapped transformer rectifier circuits. One of the center-tapped transformer rectifier circuits is used for the positive regulators, and the other is used for the negative regulator.

Figure 3-23 breaks the two circuits down, so you can better see what is actually going on. A standard full-wave diode bridge package could be used even though the two halves of the bridge are used independently.

Zener Diode Voltage Regulator

The Zener diode is the simplest voltage-regulating device. It requires only a single series current-limiting resistor to form a complete regulator circuit like

Figure 3-23 Use of Full-wave Center-tapped Transformer Rectifier Circuit by Both Positive and Negative Supplies

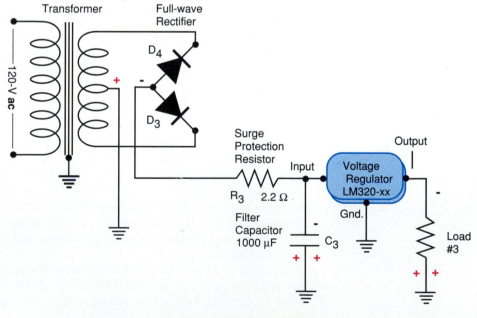

a. Positive Half of the Circuit

b. Negative Half of the Circuit

the one shown in Figure 3-24(b). The curve in Figure 3-24(a) shows the key operating points on the Zener part of the diode curve. I_{ZT} is the Zener test current used by the manufacturer. The test current is often selected at approximately the lowest useful operating current, designated I_{ZK}. Several tests are made at the test current; but equally important, it is a key to the minimum operating current. The second important value is I_{ZM}, the maximum allowable Zener current. The third value is the Zener breakover voltage, V_Z. The breakover voltage is also the regulated output voltage for the circuit in Figure 3-24. The manufacturer may not provide a specification for the maximum Zener current (I_{ZM}), and you may have to calculate the value, given the Zener breakover voltage and the wattage rating of the diode:

$$\text{Maximum Current} = \text{Watts/Breakover Voltage}$$

At all supply voltages and load currents, the Zener diode must be biased into its operating region between the test current (I_{ZT}) or the specified minimum value and the maximum current (I_{ZM}). If either the input voltage or the load current varies by more than about 30%, however, or if load currents are greater than about 50 mA, a Zener diode is probably not the best choice as the voltage regulator. Integrated-circuit regulators are more effective and, in many cases, are cheaper than the Zener diode in those instances.

A device called a *variable Zener diode* permits the Zener voltage to be set to any desired value within its range. The variable Zener is not really a Zener diode at all, but it serves the same function.

Zener Diode Resistance

Figure 3-24(c) presents a Zener diode–equivalent circuit, showing the Zener's internal resistance. Zener resistance is not constant; it decreases as the operating current increases. The resistance generated at the manufacturer's test current is usually a worst-case value and can be used for most practical purposes. The amount of resistance variation is not very great over the Zener's operating range. When you have a choice, try to select the Zener diode with the lowest resistance.

The Zener diode selected should have a maximum current rating that is higher than the anticipated maximum load current. This will allow the diode to survive if the load is disconnected. Table 3-5 provides some technical data for common Zener diodes.

Calculating the Current-limiting Resistance

The formula for calculating the current-limiting resistance is as follows:

Current-limiting Resistance = (Peak Voltage – Zener Voltage)/
(Load Current + Minimum Zener Current)

How the Circuit in Figure 3–24 Works

The input voltage to the circuit must always be somewhat higher than the Zener voltage, to allow for the voltage drop across the series resistor (R_1). The

FIGURE 3-24 Zener Diode Voltage Regulator

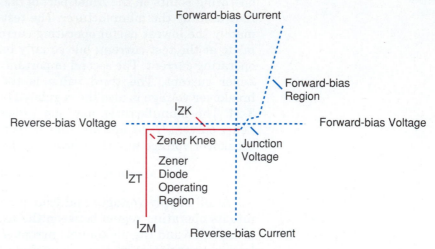

a. Zener Diode Conduction Curve

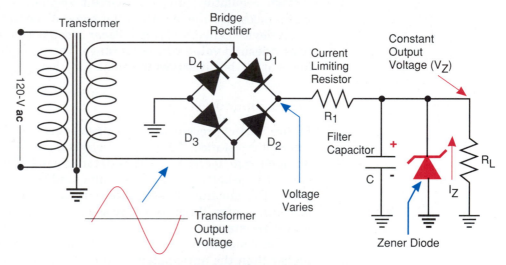

b. Zener Diode Voltage Regulator Circuit

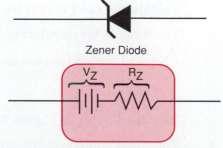

c. Zener Diode Equivalent Circuit

TABLE 3-5

Summary of Common *Zener* Diode Specifications

Voltage Values Commonly Available in 0.5-Watt, 1-Watt, 1.5-Watt, and 5-Watt Ratings (Ratings in Volts at *Zener* Breakdown)
1.8, 2.0, 2.2, 2.5, 2.7, 2.8, 3.0, 3.3, 3.6, 3.9, 4.3, 4.7, 5.1, 5.6, 6.0, 6.2, 6.8, 7.5, 8.2, 8.7, 9.1, 10.0, 11, 12, 13, 14, 15, 16, 17, 18, 19, 20, 22, 24, 25, 27, 28, 30, 33, 36, 39, 43, 47, 51, 56, 60, 62, 68, 75, 82, 87, 91, 100, 110
Voltage Values Normally Available Only in 1-Watt, 1.5-Watt, and 5-Watt Sizes
120, 130, 150, 160, 170, 175, 180, 200

resistor and the Zener diode form a series circuit. The Zener has a constant voltage drop across it, but the voltage drop across the resistor varies with variations in input voltage and with load current variations.

If the input voltage increases, the current through the Zener diode increases, too, causing a greater voltage drop across the resistor. The voltage drop across the Zener plus the resistor's voltage drop must add up to the input voltage (as in any series circuit). But the voltage drop across the Zener is essentially constant, so any input voltage increase must be absorbed as an increased voltage drop across the resistor. If the input voltage decreases, the Zener current decreases. The voltage drop across the Zener remains the same (V_Z), but the voltage drop across the resistor decreases proportionally. If the load demands more current, the voltage drop across the resistor increases and the current through the Zener decreases; but the voltage across the Zener and the voltage across its load remain constant. The Zener simply operates on a different part of its curve. A decrease in current demand by the load increases the Zener current, working it farther down on its curve, but still the voltage across the Zener and the load voltage remain essentially constant.

Multiple-voltage Zener-regulated Power Supplies

On many occasions in electronic systems, several different voltages are required. Figure 3-25 shows a full-wave rectifier with a simple capacitor filter and two Zener diodes. The two Zeners are in series and provide a regulated 12 volts across the series pair. The two Zener voltages are additive (10 V + 2 V = 12 V). The second voltage output is taken across only one of the Zener diodes (the 10-volt Zener) and provides a regulated voltage of 10 volts. Several Zeners can be connected in series in this fashion to provide several different voltages. The fact that the Zener voltages are additive also allows us to obtain voltages for which no standard Zener diode is available. Ordinary forward-biased diodes are sometimes placed in series with Zener diodes to extend the Zener voltage in 0.6-volt increments.

3.6   Other Diode Circuits

The following diode circuits are not often used in new designs, but you won't always be repairing the very latest equipment. The diodes in this section are

FIGURE 3-25   **Full-wave Rectifier, Filter, and Two Zener-regulated Output Voltages**

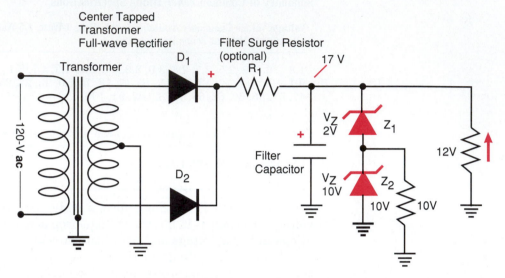

used to alter a waveshape by clipping or altering the **dc** reference level of the waveform. Waveshapes are usually altered by using a diode to clip off part or parts of the input waveform. Voltage stored in a capacitor is used in conjunction with one or more diodes to shift the waveform's **dc** reference level. The circuits are all variations of the simplest diode rectifier that clips off half of an **ac** waveform.

Diode Clippers and Clampers

The **clipper** is a circuit that clips off a portion of a waveform. Junction diodes and Zener diodes can be used in simple circuits to perform clipping operations. Because a clipper circuit limits the amplitude to which a waveform can rise, the clipper circuit is also called a **limiter.** A clamper circuit adds a **dc** component to a waveform. The result is a shift in the zero reference voltage level. Some circuits will not work properly unless some **dc** level accompanies the waveform involved.

In other circuits, the **dc** component is blocked by some device (for example, a capacitor or a transformer) and must be restored. This is often the case in television video amplifiers, where the clamper circuit is generally given the special title of **dc restorer.** Both clipping and clamping functions can also be performed by transistor and integrated-circuit amplifiers, but the simplicity of the diode versions often make them the most attractive choice.

Parts (a), (b), (c), and (d): Zener Diode Shunt Clipper

The circuit in Figure 3-26(a) clips the positive half of the input waveform at approximately 7 volts, and removes all but about 0.7 volts from the negative half of the input waveform.

How It Works

The following steps are involved in the clipping process:

1   As the sine wave progresses from 0° to 45°, the voltage seen by the Zener diode is insufficient to initiate Zener diode conduction. The Zener diode is effectively an open circuit, as illustrated in part (b). Over the portion of the sine wave from 0° to 45°, the equivalent circuit is a simple resistive voltage divider, and the voltage appearing across R_L is a slightly attenuated version of the input voltage. R_L is deliberately made large with respect to R_1 to minimize the attenuation.

2   At the 45° point on the input signal waveform, the voltage reaches 7 volts and the Zener diode starts conducting. The circuit now becomes an ordinary Zener regulator circuit, equivalent to the circuit of part (c).

3   Between 45° and 135°, the voltage of the input signal is greater than the 7-volt Zener turn-on voltage, and the surplus signal voltage is dropped across R_1. The output voltage is a constant 7 volts.

4   At very slightly more than 135° of the input signal's progression, the input signal voltage drops to less than 7 volts and the Zener diode turns off. The equivalent circuit is again the one shown in part (b), where the Zener is effectively an open circuit.

5   As the input signal passes through 180°, the Zener becomes an ordinary forward-biased junction diode as soon as the signal voltage reaches –0.7 volt. A junction diode in forward-bias mode has a voltage-current characteristic very similar to the Zener diode's voltage-current curve and behaves very much like a 0.7-volt Zener diode. The equivalent circuit for this condition is shown in part (a). Notice that the output waveform in Figure 3-26(a) goes 0.7 volt below the 0-volt reference line, reflecting the 0.7-V regulating action of a forward-biased diode.

Part (e): Zener Diode Clipper

The circuit in Figure 3-26(e) is the same circuit as the one in part (a), except that the Zener diode has been reversed. The theory of operation is the same as that for the circuit in part (a).

Part (f): Clippers and Nonsinusoidal Waveforms

Clippers do their job regardless of the shape of the input waveform. Whatever the shape of the waveform, the circuit will clip (limit) when the waveform voltage reaches the clipping level. Figure 3-26(f) illustrates the response of the circuit in part (a) to a sawtooth waveform. The clipper circuit only responds to specific voltage levels. The circuit's function remains the same, whatever the input waveform.

Parts (g) and (h): Diode/Zener Diode Clippers

Figure 3-26(g) shows an ordinary diode and a Zener diode in series in a shunt (parallel) clipper circuit, on the positive half of the cycle. The Zener is reverse-biased to begin with, and the current through the load follows the input waveform. The waveform across the load is the same as the input waveform. As soon as the positive-going input waveform reaches the Zener voltage, the Zener conducts while holding a constant voltage across the load. The conventional diode

FIGURE 3-26   Zener Diode Shunt (*Parallel*) Clipper Circuits

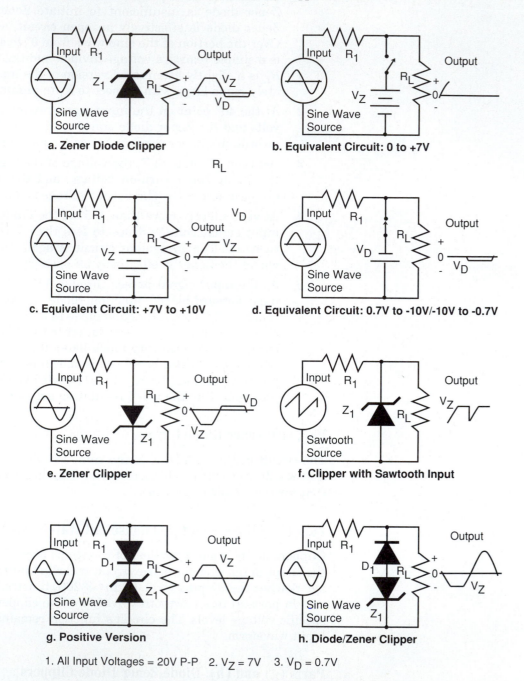

a. Zener Diode Clipper

b. Equivalent Circuit: 0 to +7V

c. Equivalent Circuit: +7V to +10V

d. Equivalent Circuit: 0.7V to -10V/-10V to -0.7V

e. Zener Clipper

f. Clipper with Sawtooth Input

g. Positive Version

h. Diode/Zener Clipper

1. All Input Voltages = 20V P-P   2. V_Z = 7V   3. V_D = 0.7V

is forward-biased, so both diodes are conducting. The voltage across the load is the sum of the Zener voltage and the 0.7-V junction voltage of the conventional diode.

When the input waveform has passed its peak and is on its way toward zero, it crosses the Zener voltage and the Zener turns off. The voltage across the load again follows the input voltage down to zero.

When the input voltage heads in the negative direction, the Zener diode is ready to behave the way a conventional diode would. It would ordinarily turn on at about –0.7 V, but in this case the conventional diode is reverse-biased, preventing current from flowing through the series diodes. As a result, the voltage across the load follows the input voltage for the entire negative half of the cycle.

Parallel (Shunt) Diode Clippers

Figure 3-27(a) through (f) shows a variety of diode clippers.

Parts (a) and (b): Basic Shunt Diode Clippers

The circuit in part (a) of Figure 3-27 is the most basic clipper circuit. During the positive half of the cycle, diode D_1 is turned on. The voltage drop across D_1 and across the load is the junction voltage (0.7 V). The rest of the input voltage is dropped across R_1. During the negative half of the cycle, the diode is turned off, and the voltage across the load follows the input voltage.

The circuit in part (b) is the same as the one in part (a), except that the diode is reversed and the negative half of the cycle is clipped.

Parts (c) and (d): Biased Diode Clipper

The circuits in Figure 3-27(c), (d), (e), and (f) use a conventional diode with a bias voltage to get several variations of clipping action. In the circuit in part (c), the cathode of the diode has a positive bias. If the diode is to conduct, the anode must have a positive voltage that is higher than the bias voltage.

At the beginning of the positive half of the input cycle, the anode of the diode is less positive than the cathode (the diode is off). The voltage across the load follows the input voltage.

At some point in the input voltage's positive-going rise, the anode becomes more positive than the cathode of the diode. The diode then turns on and holds the voltage across the load at the bias voltage. As the waveform moves down toward zero, it crosses the bias voltage and the diode turns off. The voltage across the load again follows the input voltage. On the negative half of the cycle, the diode is reverse-biased (an open circuit) and the voltage across the load follows the input voltage for the entire negative half of the input cycle.

The circuit in part (d) is a negative peak clipper. It works the same way as the one in part (c). The diode and bias polarity have been reversed to clip the negative instead of the positive peak.

Parts (e) and (f): Biased Baseline Clippers

The circuit in Figure 3-27(f) starts the positive-rising half of the input cycle with the diode biased on. The voltage across the load is equal to the bias voltage V. As the input voltage rises to a voltage that is more positive than the bias voltage, the diode is reverse-biased (open circuit). The voltage across the load then begins to follow the input voltage.

When the input voltage heads back down toward zero, it eventually crosses the bias voltage again. The diode then becomes forward-biased and holds the load equal to the bias voltage. The diode stays forward-biased for the rest of the positive half and for all of the negative half of the input cycle.

FIGURE 3-27   Shunt (Parallel) Diode Clippers

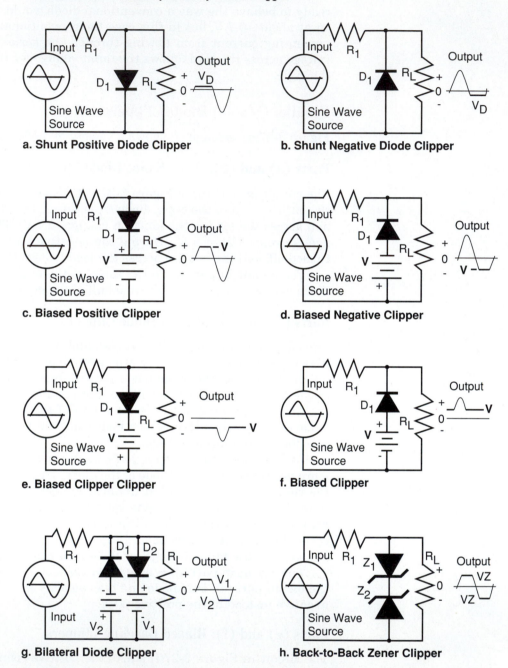

a. Shunt Positive Diode Clipper

b. Shunt Negative Diode Clipper

c. Biased Positive Clipper

d. Biased Negative Clipper

e. Biased Clipper Clipper

f. Biased Clipper

g. Bilateral Diode Clipper

h. Back-to-Back Zener Clipper

The circuit in part (e) is the negative peak-clipping version of the circuit in part (f).

Part (g): Bilateral Clipper

The circuit in part (g) is a combination of the circuits in parts (e) and (f). It is really two independent clippers—one for each half of the input signal waveform.

FIGURE 3-28 Series Diode Clippers

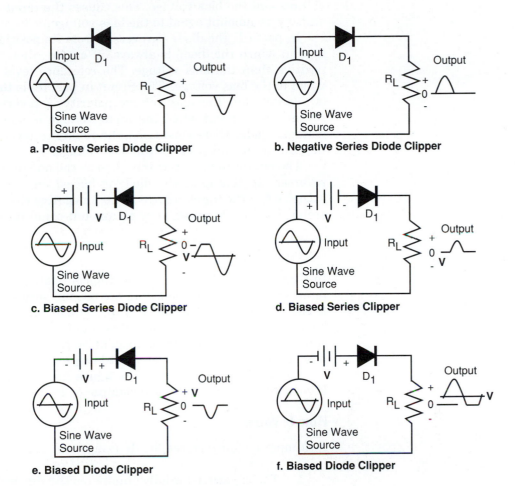

a. Positive Series Diode Clipper

b. Negative Series Diode Clipper

c. Biased Series Diode Clipper

d. Biased Series Clipper

e. Biased Diode Clipper

f. Biased Diode Clipper

Part (h): Symmetrical Zener Diode Clipper

The Zener diode clipper in Figure 3-26(h) uses two Zener diodes back-to-back to produce a symmetrically clipped sine wave. The circuit in part (f) works in the same fashion as the single Zener circuit in part (e), except that the positive and negative peaks of the input signal are limited to (clipped at) +7 volts and −7 volts.

Series Diode Clippers

The **series diode clipper** circuits in Figure 3-28 are load-current-dependent. The voltage across the load follows the input voltage only when the series diode is forward-biased (on). The circuits in parts (a) and (b) are the now-familiar half-wave rectifier circuit, with the diode of part (a) reversed in part (b).

Notice that the circuits in Figure 3-28, parts (c), (d), (e), and (f), shift the reference level. In part (c), the diode is biased on all the time—except when the input voltage is more positive than the reference voltage. Meanwhile the diode conducting the voltage across the load is the sum of the instantaneous input voltage and the bias voltage. This causes the input voltage's 0-V reference to shift by an amount equal to the bias voltage.

In part (d), the diode is reversed from its position in part (c), producing a case in which the diode is always off except when the input voltage is more positive than the bias voltage. The reference level is shifted by an amount equal to the bias voltage. The circuit in part (e) is the same as the one in part (c), except that the bias voltage polarity is reversed. The diode is held in reverse bias except when the input voltage is more negative than the bias voltage. Under those conditions, the reference level shifts in a positive direction by an amount equal to the bias voltage.

The circuit in part (f) is the same as the one in part (e), but with the diode reversed. In this case, the diode is biased on (forward-biased) at all times except when the input voltage is more negative than the bias voltage. The reference level shifts negative by an amount equal to the bias voltage.

Clamper Circuits

In Chapter 1, we noted that a waveform is always converted into a 0-V centered waveform when capacitors are used in series with the waveform source. If the waveform had included a **dc** voltage component, the capacitor would have removed it. In many cases this is unimportant, but in some cases the circuit demands that the **dc** component be restored. The circuit used to accomplish this is called a *dc restorer* or **clamper.**

Figure 3-29(a) is a flashback figure illustrating the **dc** component loss. Figure 3-29(b) shows a typical clamper or **dc** restorer circuit.

How It Works

A clamper circuit restores the **dc** component to a circuit as follows:

1 The capacitor initially eliminates the **dc** component.
2 The diode serves as a half-wave rectifier, charging the capacitor up to the peak value of the incoming sine wave.

FIGURE 3-29   **DC Reference Level Yielded by a Diode Clamper**

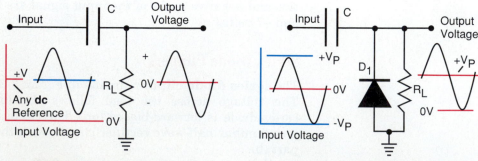

a. Zero Reference Level from Capacitor Coupling

b. Clamper (dc Restorer) Circuit

3   The output consists of a composite of the input sine wave and the stored charge on the capacitor.

4   The output now consists of a sine wave with a **dc** offset. The **dc** component has been restored.

Half-wave Voltage Doubler/Peak-to-Peak Detector

If we add an additional half-wave rectifier to the clamper circuit, as in Figure 3-30, we get a circuit that can produce an output **dc** voltage that has twice the peak value of the input sine wave. The underlying theory involves charging the two capacitors in parallel and discharging them in series. The circuit is not very common, because it is useful only for small current values. As is always the case, some ripple occurs in the output.

Most Common Full-wave Doubler Circuit

The circuit shown in Figure 3-31 still charges capacitors in parallel, and discharges them in series, but it has a ripple frequency of 120 Hz, which makes filtering it easier, yielding a cleaner **dc.**

Full-wave Voltage Doubler and Voltage Multiplier

The circuit in Figure 3-32 is a more sophisticated voltage multiplier, which provides output voltages of double, triple, or quadruple the peak value of the input sine-wave voltage. All or part of this circuit may be used, but the low current output remains a serious limitation.

You may have to study the circuit carefully to recognize it, but the diodes are all arranged to charge the capacitors in parallel, while allowing the loads

FIGURE 3-30   Half-wave Voltage Doubler/Peak-to-Peak Detector

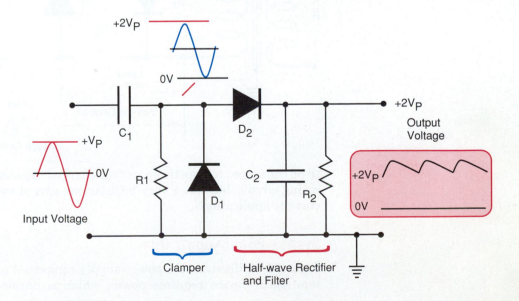

FIGURE 3-31   **Full-wave Voltage Doubler**

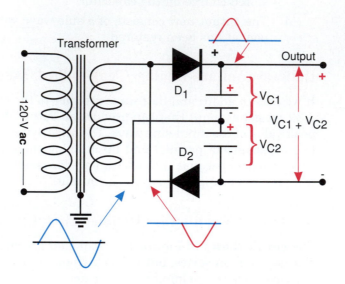

FIGURE 3-32   **Voltage Multiplier**

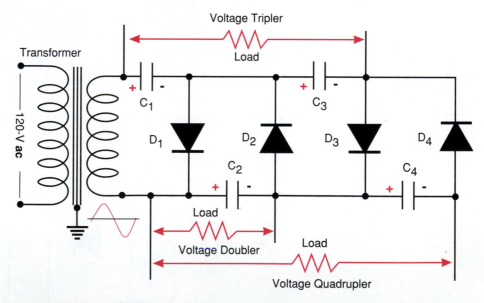

to be connected to two, three, or four charged capacitors in series. It is an interesting circuit, but not a very popular one except in certain high-voltage, low-current applications.

System Power-supply Bus

Figure 3-33 illustrates a power supply connected to several circuits in a system. Each device requires power, which is normally provided by a central

FIGURE 3-33 System Power Supply Bus

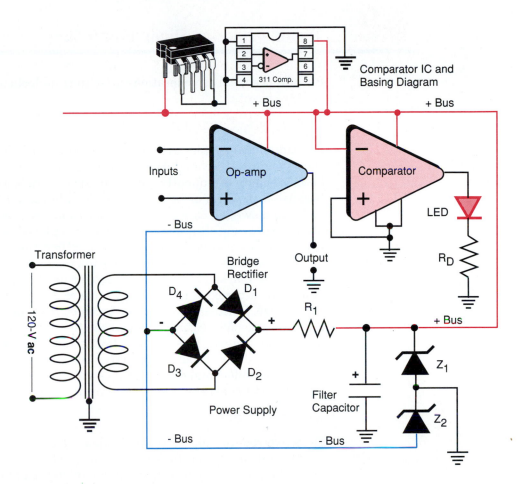

power supply. Along with the central power supply, local Zener or electronic IC regulators may be included to meet special requirements for specific circuits.

Operational amplifiers normally require one power supply voltage that is positive with respect to ground, and one voltage that is negative with respect to ground. Ground is normally the common reference for all voltages in the system.

In systems involving both analog and digital circuits, there may be an analog ground and a digital ground. The two grounds are normally returned to the same physical ground point at the power supply, but they often have one special wire (or trace) leading from the power supply to the analog circuits and another dedicated to the digital part of the system. Digital and analog currents cause small digital and analog voltage drops across the ground power wires. Mixing analog and digital voltage drops in the ground wire can cause mysterious and hard-to-find problems. Keeping the small analog and digital ground voltage drops separate helps avoid problems. Don't try any short-cuts here.

Troubleshooting Diode Power-supply Circuits

Several practical considerations arise in troubleshooting and repairing power-supply circuits.

Electrolytic Capacitors

Electrolytic capacitors are the components most likely to fail. In such cases, they generally become shorted or leaky (a resistive short); rarely do they become open. Frequently you can spot a defective electrolytic capacitor because it is warm or hot to the touch. A bulging rubber plug in the positive end and a swollen case are other indications. A capacitor should never generate internal heat. If it does, it should be replaced. A shorted electrolytic filter capacitor can also cause one or more rectifier diodes to fail.

Installing used electrolytic capacitors can cause problems. If, for example, a capacitor marked as a 25-volt capacitor has been operating in a 5-volt power supply for a period of time, it may have re-formed toward the existing operating voltage. In time, the capacitor becomes more nearly a 5-volt capacitor than a 25-volt capacitor (which it is marked to be). If you later install it in a circuit at a higher voltage, it is likely to fail in a short time. On the other hand, you can nearly always replace a defective filter capacitor with one that has a higher voltage and/or higher capacitance value. The space available to install the new one will generally prevent you from going too far with this rule.

Rectifier Diodes

Defective rectifier diodes typically short. They seldom open except as a result of mechanical damage, broken leads, or the like. If a diode shorts, it may apply **ac** to the electrolytic filter capacitor, causing it to fail. If you encounter a bad electrolytic, you must check to ensure that a shorted diode did not cause it to fail. Otherwise, the shorted diode will destroy the new electrolytic filter capacitor, too. Generally, diodes are very reliable; so if you find a bad one, look for some other problem that may have caused it to fail. A shorted capacitor or shorted load circuit may be the culprit.

Replacement diodes may have much higher current and/or voltage ratings than the diodes being replaced. A practical guideline for selecting a replacement diode is that, if there is room to install it, its current rating will not be too high.

The heavy leads on power diodes serve to carry heat away from the diode. Small signal diodes have much smaller-diameter leads because they are not intended to produce any heat.

Transformers

Transformers tend to be very reliable components, but they can fail. Failure is usually caused by overheating, which in turn usually occurs when a circuit or component demands more current than the transformer can comfortably provide. It is difficult to be exact about how hot a transformer can safely run. A rule-of-thumb says that you must be able to grip the transformer with your bare hand for 10 or more seconds. This is a pretty imprecise rule, though.

FIGURE 3-34 Power-supply Troubleshooting Guide

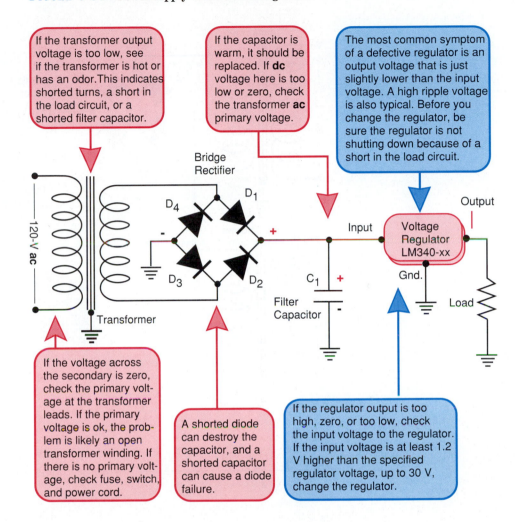

If you disconnect the load from the secondary of the transformer and the transformer fails the hand test, it probably has shorted turns and should be replaced. Transformers also emit a distinctive odor when they are truly overheated. Your nose can be an excellent diagnostic tool in this regard.

Figure 3-34 offers a handy power-supply troubleshooting guide.

Live-circuit Diode Bridge Testing

Figure 3-35 shows a method for testing diode bridge circuits without disconnecting any parts. This noninvasive test involves comparing the ripple waveform with a sine-wave reference voltage derived from a test transformer. The extra transformer is needed for personal safety and because the diode bridge has no point that is common to the **ac** line ground. A full-wave ripple waveform tells you that all diodes are okay. A bad diode yields a half-wave ripple.

FIGURE 3-35 Live Circuit Diode-bridge Testing

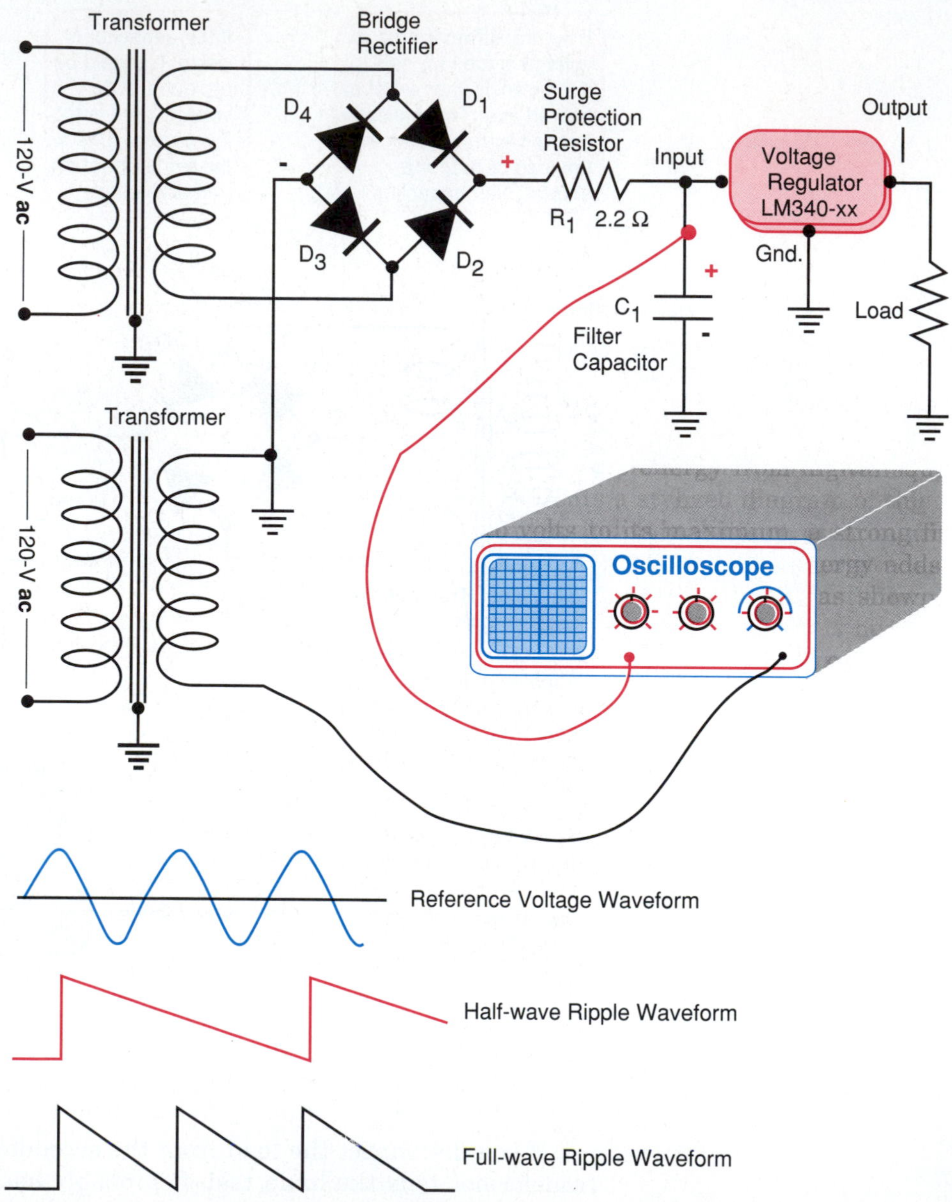

It is almost impossible to tell the difference, however, without the reference sine wave from the extra transformer (see Figure 3-35).

Testing Voltage Regulators

If the voltage regulator is in a live circuit, follow the troubleshooting guide in Figure 3-34. If the regulator is not in a circuit, you can test it by using the procedure in Figure 3-36. Don't remove the regulator from a board to test it until

FIGURE 3-36 **Testing IC Voltage Regulators**

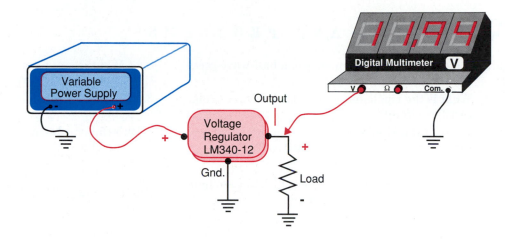

you have used the power-supply troubleshooting guide and are reasonably sure that the regulator is at fault. You can test regulators from the parts supply or junk box by using the procedure in Figure 3-36.

Power-supply Problem Casebook Example

The power-supply problem casebook example (Figure 3-37) demonstrates how the power-supply troubleshooting guide was used to solve a real power-supply problem. The problem illustrated in the figure is a very common one.

FIGURE 3-37 **Power-supply Problem Casebook Example**

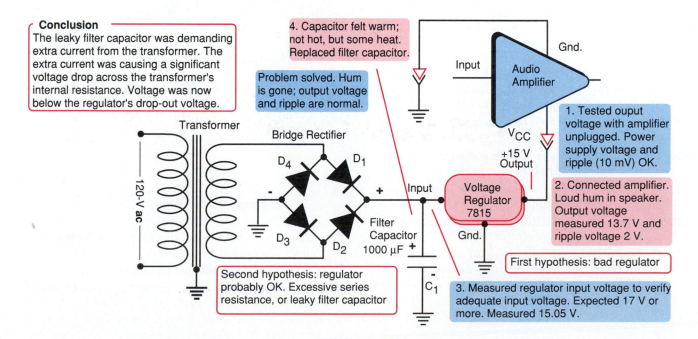

Conclusion
The leaky filter capacitor was demanding extra current from the transformer. The extra current was causing a significant voltage drop across the transformer's internal resistance. Voltage was now below the regulator's drop-out voltage.

4. Capacitor felt warm; not hot, but some heat. Replaced filter capacitor.

Problem solved. Hum is gone; output voltage and ripple are normal.

Input

Audio Amplifier

Gnd.

1. Tested ouput voltage with amplifier unplugged. Power supply voltage and ripple (10 mV) OK.

V_{CC}
+15 V Output

2. Connected amplifier. Loud hum in speaker. Output voltage measured 13.7 V and ripple voltage 2 V.

Transformer

Bridge Rectifier

D_4 D_1

D_3 D_2

120-V ac

Input

Voltage Regulator 7815

Gnd.

First hypothesis: bad regulator

Filter Capacitor 1000 µF

Second hypothesis: regulator probably OK. Excessive series resistance, or leaky filter capacitor

C_1

3. Measured regulator input voltage to verify adequate input voltage. Expected 17 V or more. Measured 15.05 V.

QUESTIONS AND PROBLEMS

3.1 What is the ripple frequency of a half-wave rectifier circuit?

3.2 What is the ripple frequency of a full-wave bridge rectifier circuit?

3.3 What is the ripple frequency of a full-wave center-tapped transformer rectifier circuit?

3.4 Define *ripple*.

3.5 Compare the output voltage efficiency of half-wave and full-wave rectifier circuits.

3.6 The cathode of the LED shown in Figure 3-38 is identified by which labeled part?

FIGURE 3-38 **Drawing for Problem 3.6**

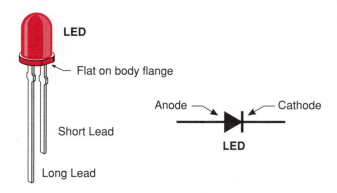

3.7 What is the function of a rectifier circuit?

3.8 What is the purpose of the filter capacitor in a line-operated power-supply circuit?

3.9 Why are line-operated power-supply circuits necessary?

3.10 What is the purpose of using diodes in a line-operated power supply?

3.11 What does the filter capacitor in a line-operated power supply accomplish?

3.12 Why is it easier to filter a full-wave rectifier circuit than a half-wave circuit?

3.13 Given an input **ac** voltage of 40 volts, calculate the effective (average) voltage delivered to the load for:
  a. a half-wave rectifier.
  b. a full-wave rectifier.

3.14 What is a pi filter? Why is it superior to a simple capacitor filter? Why is it called a pi filter?

3.15 Draw the schematic diagram for a two-stage pi filter.

3.16 Define *regulation* as the term applies to a power supply.

3.17 What is the purpose of a clipper circuit?

FIGURE 3-39 **Clipper Circuits for Problem 3.18**

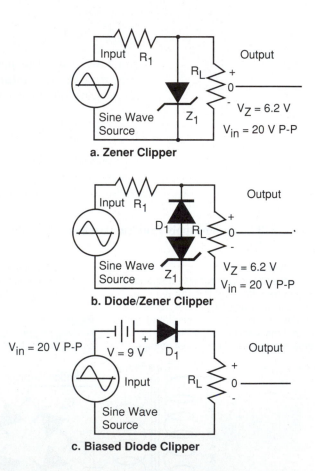

a. Zener Clipper

b. Diode/Zener Clipper

c. Biased Diode Clipper

3.18 Given the clipper circuits in Figure 3-39(a)–(c), draw the output waveforms.

3.19 What is the purpose of a clamper circuit?

3.20 Given the clamper circuit in Figure 3-40, draw the output waveform.

FIGURE 3-40 Clamper Circuit for Problem 3.20

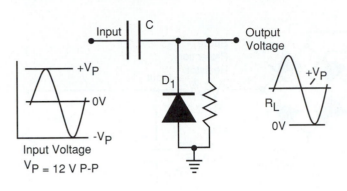

FIGURE 3-41 Circuit for Problems 3.21 Through 3.23

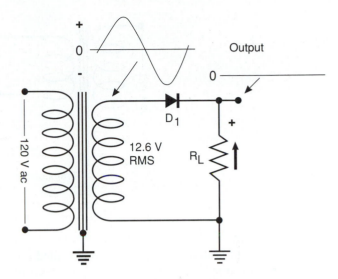

FIGURE 3-42 Schematic Diagram for Problems 3.24 Through 3.29

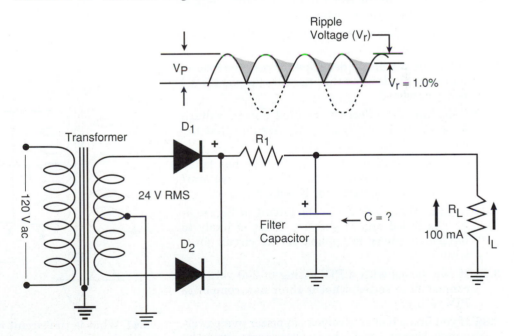

3.21 Given the schematic diagram in Figure 3-41, identify the circuit.

3.22 Draw the waveform developed across the load in Figure 3-41.

3.23 Calculate the peak rectified output voltage for the circuit in Figure 3-41.

3.24 Given the schematic diagram in Figure 3-42, identify the circuit.

3.25 Draw the waveform developed across the load in Figure 3-42. (Don't include the filter capacitor.)

3.26 Calculate the effective rectified **dc** output voltage for the circuit in Figure 3-42 after the filter capacitor (*C*) has been connected.

FIGURE **3-43**   **Circuit for Problems 3.30 Through 3.34**

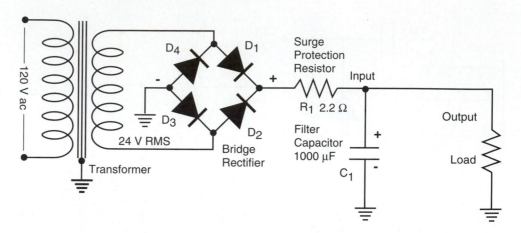

3.27 What would probably happen if you installed the filter capacitor (C) with its polarity reversed in the circuit in Figure 3-42?

3.28 If diode D_2 in Figure 3-42 fails and is shorted, what is likely to happen?

3.29 What happens in the circuit in Figure 3-42 if diode D_1 fails and is open?

3.30 Given the schematic diagram in Figure 3-43, identify the circuit.

3.31 Draw the waveform developed across the load in Figure 3-43. Assume that the filter capacitor (C) is disconnected.

3.32 Calculate the effective rectified output voltage for the circuit in Figure 3-43. Assume that the capacitor (C) is disconnected.

3.33 Assume that diode D_3 in Figure 3-43 has failed and is open. What is the output waveform? Assume that the capacitor is not connected.

3.34 Assume that diode D_3 in the circuit in Figure 3-43 has failed and is shorted. What is likely to happen if power is applied to the circuit for a while?

3.35 If two diodes with a PIV rating of 200 volts are connected in series, what is their new composite PIV rating?

3.36 If you find a bad electrolytic capacitor in a power-supply circuit, what other part or parts should be replaced?

3.37 If an electrolytic capacitor is warm to the touch, what action should you take?

3.38 Why is a full-wave bridge rectifier circuit generally used instead of a full-wave center-tapped transformer circuit?

3.39 Why is a full-wave rectifier circuit nearly always used instead of a half-wave rectifier circuit, considering that the latter requires only one diode instead of four diodes?

3.40 Compare the cost of a full-wave bridge power-supply circuit and a half-wave power-supply circuit for a 12-volt, 1-amp power supply.

FIGURE **3-44**   **Circuit for Problem 3.41**

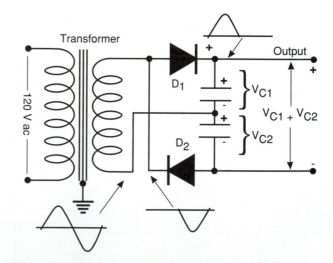

3.41 What is the circuit in Figure 3-44 called?

3.42 What is the purpose of a clamper circuit?

3.43 List as many reasons as you can why integrated-circuit voltage regulators (7805, etc., class) are used in nearly all modern power supplies. A switching regulator circuit is also fairly common, but coverage of it must be saved for a later chapter.

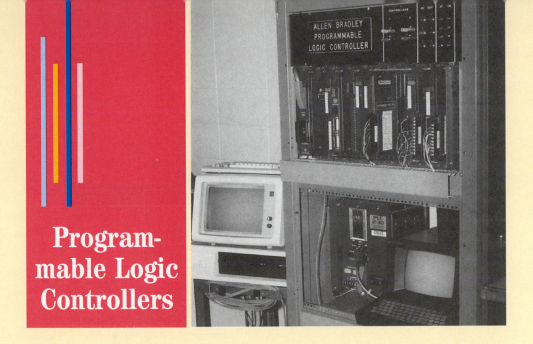

Program-mable Logic Controllers

Computers are popping up everywhere today—in homes, in offices, and in commercial and industrial settings. Automobile manufacturers, power-generating plants, petroleum and petro-chemical production facilities, textile manufacturers, and the pulp and paper industry are examples of facilities that use industrial electronics and computer systems.

Regardless of the application, computers must connect to the real world if they are to be useful. Real-world applications include turning electrical equipment on and off at specified times or when the appropriate conditions exist. More sophisticated systems require measurement and control of analog signals reflecting data from temperature, pressure, flow, and other continuously varying parameters.

Programmable logic controllers (PLCs) are computers specifically designed for these kinds of real-world applications, especially in industrial and commercial environments. PLCs are rugged, modular computers with specialized input/output modules that connect to external electrical equipment. A PLC implements software programs that control the I/O modules; these programs are usually downloaded into the PLC from a portable computer or a special hand-held programming device.

In a typical industrial application, a PLC may receive inputs from discrete devices such as hand-operated switches; temperature, pressure, and flow switches; and external relay contacts. Some systems also accept analog parameters, including low-level sensor signals and standardized voltage and current signals. Through various kinds of output modules, the PLC may control the start/stop functions of dozens of electric motors, solenoid valves, lights, relays, and other types of electrical equipment. Analog output modules control valves, servo motors, and other devices that require continuously variable signals. In larger systems, several PLCs may be connected in a network to permit information sharing between subsystems.

The electronic circuitry employed in these I/O modules varies depending on the application. For example, discrete (on/off) input modules are designed to sense the presence or absence of a voltage at their terminals. Depending on the application, the voltage may range from a few volts to more than 100 volts **ac** or **dc.** Each individual module is designed to accept a specified voltage and to adjust that signal to a standard level acceptable to the computer circuitry inside the PLC. In the process, **ac** voltages must be converted to **dc,** noise must be filtered out, and (usually) the external signal must be isolated from the internal circuitry through the use of optical couplers (see Figure A).

Analog input modules also perform signal conditioning. Analog-to-digital converters scale and convert the varying input signal into digital data signals that the PLC's computer can accept.

Output modules also perform conversions. The low-level computer signals must be converted into

Figure A: Typical Input Circuit

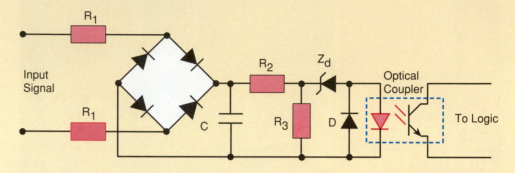

Figure B: Typical ac Output Circuit

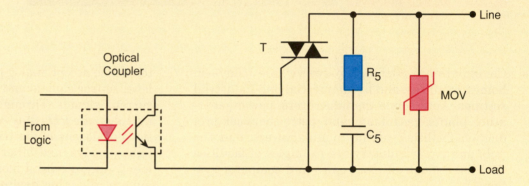

voltages usable in the field—often **ac** voltages such as 120 V **ac** or even 600 V **ac.** When a discrete output module must switch **ac** voltages and be capable of driving higher currents, TRIAC circuits (like the one shown in Figure B) are often used.

Before PLCs were available, electrical systems were hard-wired. All of the switches, indicators, relays, motors, and so on were connected according to a schematic called a *ladder diagram.* The PLCs job is to simulate the connections and operation of the ladder diagram. Most PLCs use a programming language that allows ladder-type diagrams to be programmed on a computer screen or on a handheld programming unit. We call this a *ladder logic program* (see Figure C). By examining the states of the inputs and how they have been connected in the ladder logic program, the PLC's computer can decide when to turn on and off the outputs that control devices such as motors and solenoids.

Programmable logic controllers (and the larger area of industrial electronics) offer interesting opportunities for the electronics technician. While much of the electrical wiring and some of the programming of PLCs is done by electricians and engineers, electronic technicians often find themselves involved in troubleshooting and repairing this kind of equipment. Troubleshooting can proceed at several levels: locating and repairing or replacing faulty modules, debugging ladder logic programs, and troubleshooting system and communications problems.

In some organizations, electricians or instrumentation maintenance personnel troubleshoot PLCs at the system level. When faulty modules are identified and removed for service, the electronics technician's job is to repair the circuitry in the shop. A suitable test jig and some schematics—coupled with a knowledge of electronic circuits and some test equipment—are the basic tools needed to isolate faulty components. The technician then replaces and retests the module before approving it for use.

Figure C: Ladder Schematic Diagram

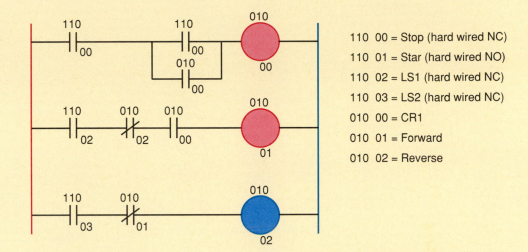

110 00 = Stop (hard wired NC)

110 01 = Star (hard wired NO)

110 02 = LS1 (hard wired NC)

110 03 = LS2 (hard wired NC)

010 00 = CR1

010 01 = Forward

010 02 = Reverse

Sometimes electronic technicians become involved in debugging the ladder logic diagrams. The same kind of troubleshooting skills used in tracking down faulty components are invaluable in troubleshooting ladder logic programs. Just as a technician might isolate specific sections of circuitry, the debugger can isolate and test sections of ladder logic programs, using a programming terminal. The same troubleshooting concepts apply in solving communications and networking problems between PLCs.

The industrial environment presents unique problems for electronic equipment and for electronics technicians. Industrial plants are sometimes located in remote areas, and the technician may be required to work in less than ideal environments. Such conditions, however, are commonly offset by good pay and benefits, considerable opportunity for advancement, and the challenge of developing practical troubleshooting skills.

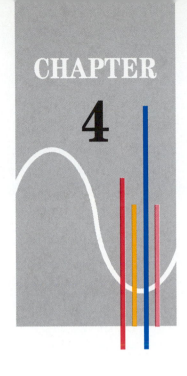

CHAPTER 4

Introduction to Transistors

OBJECTIVES

Upon completion of this chapter, you should be able to:

1 Understand the importance of negative feedback in transistor circuits.

2 Explain that the collector–base junction in bipolar transistors is normally reverse-biased, while the base-emitter junction is normally forward-biased.

3 Understand the concept of beta, and identify the factors that can cause it to vary.

4 Define the beta stability factor, and explain how beta is stabilized.

5 Describe the relationship between beta and circuit current gain.

6 Explain what causes the base–emitter junction voltage to vary, and tell how those variations are controlled in a transistor circuit.

7 Understand why beta independent circuits result when negative feedback is used, and recognize why that allows for beta independent circuit design and analysis of transistor circuit parameters.

8 Explain current-mode negative feedback in a transistor circuit.

9 Identify the effects of current-mode feedback on current gain, voltage gain, and input resistance in a transistor circuit.

10 State the equation for voltage gain, and describe how to use it.

11 Distinguish between the working load and a dummy load.

12 Understand the purpose of establishing a quiescent collector current.

13 Interpret the forward-current transfer (beta) curve.

14 Understand the purpose of bias, and specify (step by step) the two most common bias methods.

15 Calculate the quiescent collector current, given appropriate circuit values in an emitter-bias circuit.

16 Calculate the quiescent collector current, given appropriate circuit values in a base-bias circuit.

17 Understand the concept of swamping, and discuss how it is used.

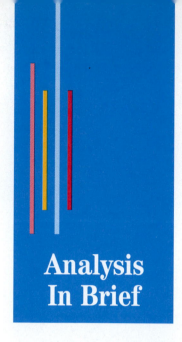

Analysis In Brief

Transistors in Brief

JUNCTION VOLTAGES

The forward-biased junction must always be 0.6 to 0.7 V for any transistor or any circuit. If this junction voltage is wrong, you have found a problem.

The V_{CE} (Voltage between the collector and the emitter) includes the base-emitter junction (which is always 0.6–0.7 V) and collector–base junction (which can range from 1.0 to many volts). The V_{CE} must be from an absolute minimum of 1.5 V. The normal minimum V_{CE} is 3 V. If the V_{CE} is 0.8 V or less, the transistor is saturated ($V_{CE(sat)}$), and the base no longer has control over the collector current.

CURRENTS

The base current controls the collector current. (The base–emitter junction must be forward-biased.)

1   The collector current is a function of base current and beta (β). $I_C = \beta \times I_B$

2   Collector voltage does not control the collector current. Increasing the collector voltage widens the depletion zone, increasing the junction's dynamic resistance. When the voltage increases, the resistance increases, so the current stays the same.

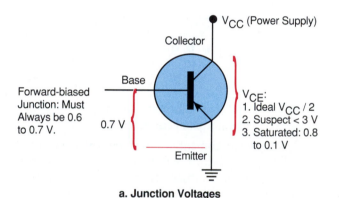

V_{CC} (Power Supply)

Collector

Base

Forward-biased Junction: Must Always be 0.6 to 0.7 V.

0.7 V

V_{CE}:
1. Ideal V_{CC} / 2
2. Suspect < 3 V
3. Saturated: 0.8 to 0.1 V

Emitter

a. Junction Voltages

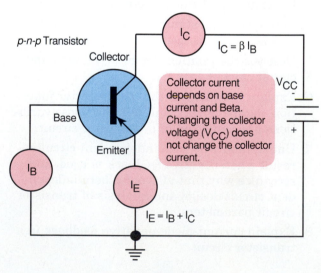

p-n-p Transistor

Collector

Base

Emitter

I_C

$I_C = \beta\, I_B$

Collector current depends on base current and Beta. Changing the collector voltage (V_{CC}) does not change the collector current.

V_{CC}

I_B

I_E

$I_E = I_B + I_C$

b. Currents

Emitter current consists of collector current plus base current. Since the collector current is always much larger than the base current, the common assumption is that emitter current equals collector current.

JUNCTION RESISTANCES

Base-to-emitter junction: Must always be forward-biased. Junction resistance decreases as junction current increases. The equation for junction resistance is

$R'_e = 25/Q_{IC}$ (the Shockley relationship).

Collector-to-base junction: Must be reverse-biased. Dynamic resistance is from 0.5 to 2 megohms. This value is high enough to make collector current nearly independent of collector voltage. The value is so high, it is not usually included as part of a circuit output resistance calculation.

CIRCUIT RESISTANCES

Input resistance is equal to the value of any resistance in series with the emitter, multiplied by beta (β). Normally, emitter resistance includes the junction resistance ($R'_e = 25/Q_{IC}$) plus the current-mode feedback resistor (R_f). If the bypass capacitor is not connected, the input resistance and impedance values are the same. If the bypass capacitor is connected, the equation for the input impedance is $Z_{in} = β × R'_e$.

Note: Any added base-bias resistors must be included as resistors-in-parallel with the basic input impedance.

Output Resistance (Impedance): Consists of the dummy load (R_L) effectively in parallel with the working load (R_{LW}). The collector–base junction's dynamic resistance is usually ignored.

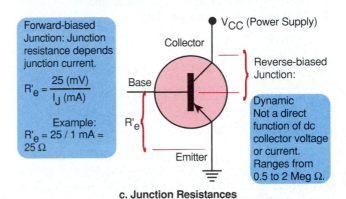

Forward-biased Junction: Junction resistance depends junction current.

$$R'_e = \frac{25\ (mV)}{I_J\ (mA)}$$

Example:
$R'_e = 25\ /\ 1\ mA = 25\ Ω$

Reverse-biased Junction:

Dynamic Not a direct function of dc collector voltage or current. Ranges from 0.5 to 2 Meg Ω.

c. Junction Resistances

Input Impedance:
1. For signal frequencies: Z_{in} is β x R_f, with C_1 disconnected.
2. For signal frequencies: Z_{in} is β x R'_e, with C_1 connected.

Output Impedance is: R_L in parallel with R_{LW}. Collector-base junction resistance is high enough to ignore in practice.

d. Output Impedance

VOLTAGE GAIN

1 Voltage gain is defined as:
Voltage Gain = Output Voltage/Input Voltage

2 Circuit voltage gain is controlled by the effective emitter impedance and the effective collector impedance. Beta is not a factor.
Voltage gain (unbypassed):

$$A_V = R_{Lac}/(R_f + R'_e)$$

Voltage gain (bypassed):

$$A_V = R_{Lac}/R'_e$$

Output Impedance

3 In most cases, the signal output impedance (R_{Lac}) consists of a dummy load (R_L) in parallel with a working load (R_{LW}). If inductors and capacitors are used, however, the total impedance at the frequency of interest must be calculated.

Emitter Impedance

4 The emitter impedance is normally the sum of the feedback resistance (R_f) and the junction resistance (R'_e).

5 The bypass capacitor, if used, is selected to have a reactance of near 0 Ω at the lowest frequency of interest. It effectively shorts out the feedback resistor for signal frequencies, leaving only the junction resistance to serve as the emitter resistance.

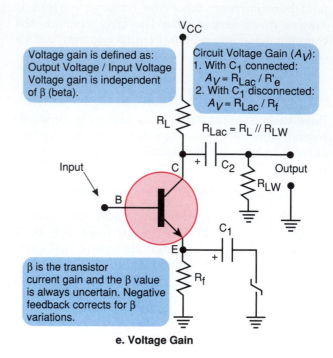

Voltage gain is defined as:
Output Voltage / Input Voltage
Voltage gain is independent of β (beta).

Circuit Voltage Gain (A_V):
1. With C_1 connected:
   $A_V = R_{Lac} / R'_e$
2. With C_1 disconnected:
   $A_V = R_{Lac} / R_f$

$R_{Lac} = R_L // R_{LW}$

β is the transistor current gain and the β value is always uncertain. Negative feedback corrects for β variations.

e. Voltage Gain

4.1   Introduction

So far, we have been dallying on the outskirts of electronics proper, but in this chapter we investigate the transistor and enter the heart of that realm.

Amplifiers and Amplification

Technically at least, electronics begins with the introduction of amplifiers. Until we employ amplification in a circuit, the circuit presents an example of electricity, not electronics. Everybody has some concept of what an amplifier does, but that concept is usually incomplete and not quite correct.

An amplifier doesn't actually make a small signal larger. The small electrical signal from a microphone is not somehow boosted or made larger so that it can drive a loudspeaker to fill an auditorium. In fact, none of the original electrical signal ever reaches the speaker.

What actually happens is that a source of considerable power is varied in step with the microphone's electrical signal to form a more powerful replica of the original signal variations. The power for the louder sound comes from a power supply, and a device called an *amplifier* is used to control the available power. An amplifier is thus a power controller, in which a small signal power controls the rest of the amplifier to produce an exact but larger copy of the original input signal, drawing the necessary power for control from a special power source.

Distortion

If the enlarged replica of the original signal is a true and accurate copy, the amplifier is termed *undistorted*. If the amplifier produces an output signal that is not a true and accurate copy of the input signal, the amplifier is characterized as *distorted*. A true copy can be determined by comparing the input signal waveform to the output signal waveform, using an oscilloscope.

The **amplitudes** of the two waveforms can be different, because increasing the amplitude is the purpose of an amplifier. The input and output waveshapes, however, must be identical. Any difference between the input waveshape and the output waveshape constitutes distortion. Any change in the output waveshape indicates that some new harmonic frequencies have been added by the amplifier circuitry. In an audio amplifier, the extra frequencies represent sounds added by the amplifier circuit that were not part of the original input signal sounds. Such added signal information is always considered to be distortion and is usually (but not always) undesirable.

Linear and Nonlinear Amplifiers

Before we go any further, let's clear up a little confusion in terminology. Old hands in the field of electronics sometimes get a little sloppy in their use of certain terms. For experienced technicians, this doesn't pose a problem; but it can be confusing to a newcomer trying to learn the business. The terms *analog* and *linear* are often used interchangeably, but the underlying concepts differ.

Linear amplifier circuits are actually a subset of analog amplifier circuits, because not all analog circuits are linear. Even manufacturers' data manuals are often called *linear device manuals,* although they should be titled *analog device manuals.* The manuals contain data about many analog devices that are not linear devices. A few definitions are in order here:

1   A **linear amplifier** is an amplifier in which the output waveshape is identical to the input waveshape. (A tiny bit of unavoidable distortion may be present, but theoretically there is no distortion of any kind.)

2   In a **nonlinear amplifier,** the output waveshape is very different from the input waveshape. We often alter the waveshape intentionally in analog circuits. They can be called waveshaping or wave-reshaping circuits, and they are analog circuits.

3   **Nonlinear switching amplifiers** are a special case. In switching amplifiers, the output waveshape is always a rectangular or square wave. The input waveform can be any waveshape, and the output waveform can have any desired amplitude (peak voltage), but the output waveshape must always be square or rectangular, representing a voltage that is either fully on or fully off. A perfect square wave is often more difficult to produce than a perfect sine wave, because of unwanted (but ever-present) distributed capacitance and inductance.

4   **Digital amplifiers** are a special case of the nonlinear switching amplifier. Not only are digital waveshapes always square or rectangular, but they also (always) have a peak voltage of +5 volts and a minimum voltage of 0 volts (ground). The +5-V/Ground square-wave standard was the only digital pulse standard for many years. More recently, a second digital standard of +3 volts and ground has emerged. The new standard was introduced to prolong battery life in portable computers and other digital systems.

Waveshaping Amplifiers

An amplifier designed for near zero distortion is called a *linear amplifier.* If we try to get a larger output voltage than the power supply can provide, we get a particular kind of distortion called *flat-top distortion* or *clipping.* When an amplifier produces clipping, it ceases to function as a linear amplifier; and if it is used for music, it will sound something like a torn loudspeaker cone.

On the other hand, such an overdriven amplifier can be used to convert a sine-waveshape input into a good approximation of a square-wave output. Digital amplifiers are nearly always overdriven on purpose, to produce a digital type pulse or switching action.

Nonlinear amplifiers are often used to convert one waveshape into another. The amplifier circuit is still analog, but it is not a linear amplifier circuit.

We will see examples of both linear and nonlinear amplifier applications in this text. We will start with linear applications, using the **Bipolar Junction Transistor (BJT).** The BJT is by far the most common amplifying device in both discrete- and integrated-circuit technology. Dozens to hundreds of BJTs may be integrated onto a silicon chip. A micro-miniature printed circuit is used to interconnect individual BJTs to form complete functional building blocks.

A second kind of amplifying transistor is the **Field-Effect Transistor (FET).** There are several kinds of FETs—actually, a whole family of them. FETs are much more common in digital technology than in analog technology. Since the primary subject of this book is analog technology, we will mostly be concerned with BJTs and their circuits; but we will also look at some special digital and analog/digital hybrids. We will study FETs at some length in Chapter 7, and will find FETs in a few analog applications in later chapters.

Basic Amplifier Classifications

All amplifiers are power amplifiers, but because power consists of voltage times current, they are also voltage amplifiers and current amplifiers. We can emphasize the aspect of power amplification, voltage amplification, or current amplification by the way in which we connect resistors, capacitors, and so on to the transistor.

We classify amplifiers as voltage amplifiers, current amplifiers, or power amplifiers, according to what kind of amplification we want; and we can design a circuit to optimize the particular kind of amplification we are interested in. Of course, the amplifier may also amplify the other two quantities (to a lesser extent), but they are of no interest and cause no problems. If we design a voltage amplifier, we don't much care if it also amplifies the current a bit, as long as we get the voltage amplification we want.

Amplification/Gain

We term the amount of relevant amplification provided by an amplifier **voltage gain, current gain,** or **power gain.** Let's examine a summary of the symbols and relationships for the three kinds of amplification:

1 Voltage Gain = Signal Output Voltage/Signal Input Voltage

$$A_V = V_O/V_{in}$$

where A_V represents voltage amplification.

2 Current Gain = Signal Output Current/Signal Input Current

$$A_i = I_O/I_{in}$$

where A_i represents current amplification.

3 Power Gain = Signal Output Power/Signal Input Power

$$A_P = P_O/P_{in}$$

where A_P represents power amplification.

For example, if an amplifier has an input signal voltage of 1 volt and an output signal voltage of 10 volts, the amplifier has a voltage gain of 10. Gain is a simple ratio, with no attached units (such as volts). The voltage gain is constant and independent of actual signal voltage values, as long as we don't attempt to get more output voltage than the power supply can provide.

Comparing BJTs and FETs

Bipolar junction transistors (BJTs) use a small input current to control a larger output current, whereas field-effect transistors (FETs) control a larger output current with a smaller input voltage. Both are amplifiers, but the bipolar device is a current amplifier and the field-effect device is called a **transconductance** amplifier. The devices are quite different, even though the circuits used in association with the two devices appear to be nearly identical. In practice, the same component in a circuit may serve completely different functions in the two technologies.

In bipolar circuits, large amounts of negative feedback must be used to render the circuits only slightly dependent on individual device properties. Bipolar characteristics are too variable to permit the transistor to be used in analog circuits without considerable negative feedback. Negative feedback provides a system that is self-adjusting and automatically compensates or corrects for large variations in transistor parameters. Although negative feedback is demanded by bipolar transistors, using it is by no means a liability. Negative feedback allows us to custom-design a circuit for nearly any desired characteristics, easily. We are not bound by the limitations and characteristics of the transistor itself.

Although FETs are often superior to BJTs when used as switching amplifiers, they have limited analog circuit potential—mostly because of their comparatively low level of voltage, current, or power gain.

In field-effect transistors, transistor characteristics tend to define analog circuit limitations to a much greater extent than they do in bipolar devices. Negative feedback is important in field-effect devices, but such devices don't usually have enough excess gain to allow for very much of it; and negative feedback must be used to stabilize the operating point of both kinds of transistor.

Both transistors have input capacitance that gives rise to a type of negative feedback called the **Miller effect.** The Miller effect significantly affects the operation of higher-frequency circuits.

Bipolar transistor circuits tend to have relatively low input resistances, whereas field-effect transistor circuits tend to have very high input resistances. There are other differences, too, as well as similarities at the applications level. Either device can be installed into circuits that allow both kinds of transistor to operate as voltage, current, or power amplifiers.

As Figure 4-1 illustrates, both kinds of transistor can be used in three different circuit configurations:

Bipolar Junction Transistors

1   Common emitter
2   Common collector
3   Common base

FIGURE 4-1 Three Possible Circuit Configurations for Bipolar (BJT) and for Field-effect (FET) Transistors

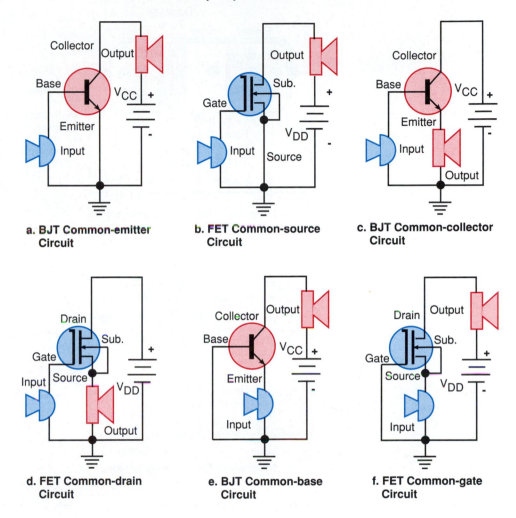

a. BJT Common-emitter Circuit

b. FET Common-source Circuit

c. BJT Common-collector Circuit

d. FET Common-drain Circuit

e. BJT Common-base Circuit

f. FET Common-gate Circuit

Field-Effect Transistors

1 Common source
2 Common drain
3 Common gate

Structure of the Bipolar Junction Transistor

The bipolar junction transistor consists of two silicon junctions in a continuous crystal. The configuration may be *p–n–p* or *n–p–n* (see Figure 4-2). A welded contact is made to each crystal area, yielding a three-terminal device. One end block is called the **emitter;** the middle block is called the **base;** and the other end is called the **collector.** The middle section is always called the base. Theoretically, either end could be called the emitter or the collector. In prac-

FIGURE 4-2 Structure of the Bipolar Transistor

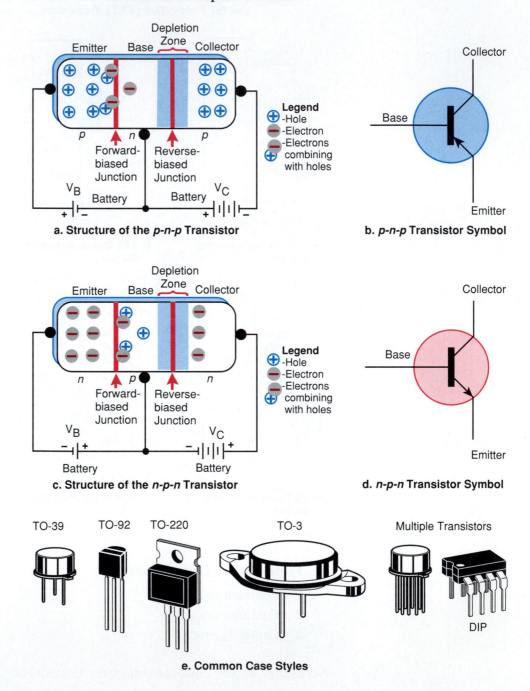

a. Structure of the *p-n-p* Transistor

b. *p-n-p* Transistor Symbol

c. Structure of the *n-p-n* Transistor

d. *n-p-n* Transistor Symbol

e. Common Case Styles

tice, the manufacturer designates one end the *collector* and the other end the *emitter*.

If you take a real transistor and exchange the emitter and collector leads, you still have a functional transistor. It will not conform to the specifications published by the manufacturer, but it will still be a transistor. The critical differences between correct and incorrect connections are due to the fact that the manufacturer has doped one end block more heavily than the other. The manufacturer has also made certain that the block designated the *collector* has been designed to promote good heat transfer between it and the case.

The heat-conducting case is exposed to the open air or mounted on a heat radiator (heat sink). Convection air currents moving around the case or heat radiator carry the heat away into the atmosphere. The area designated *emitter* is less able to get rid of internally generated heat.

The base region of the crystal is very thin—0.8 microns or less. This thin base region is essential to normal transistor operation, as we shall see shortly. A length of 25 microns is approximately equal to 0.001 inch. In the following discussion, two underlying points are crucial. First, the base region is very thin and lightly doped. As a result, the base has very few dangling bonds (holes in an *n–p–n* transistor) in comparison to the vast number of free electrons available in both the emitter and the collector regions. Second, the collector–base junction is reverse-based.

How the Transistor Amplifies Current

Refer to the sketch in Figure 4-3 as you follow the stages of the operation point by point.

1   Assume that the base–emitter bias voltage is, at first, less than the junction voltage (the 0.6- to 0.7-V depletion zone potential) of the base–emitter junction. The base–emitter junction is therefore intrinsically reverse-biased by 0.6 to 0.7 V, and there is no base–emitter current.

2   The collector–base junction is reverse-biased by (let's say) 10 volts. If the junction is reverse-biased, no collector-to-base current flows. In addition, no emitter-to-collector current flows because both junctions are reverse-biased.

3   So far, the transistor has been in the off (nonconducting) state. Now, suppose that we bring the emitter-to-base voltage up above the 0.6- to 0.7-V junction voltage, forward-biasing the junction and initiating base–emitter circuit conduction. A multitude of electrons rushes through the emitter block toward the junction. The electrons reach the depletion zone, but there (in the base region) holes are very scarce.

4   Electrons that do find holes in the base region combine with them and generate base current in the fashion of any forward-biased junction. For each 100 or so electrons drawn into the base region (at any given instant), the holes in the base can capture only a fraction (1% or less) of them. The rest of the electrons cannot go back to the emitter, because the negative field pushes them away from the emitter. The majority of electrons, which don't become base–emitter current, accumulate near the collector–base junction. The electrons are attracted by the 10-volt positive field provided by the collector voltage supply. They are then drawn into the collector, forming an emitter-to-collector current loop.

FIGURE 4-3 How Transistor Current Amplification Works

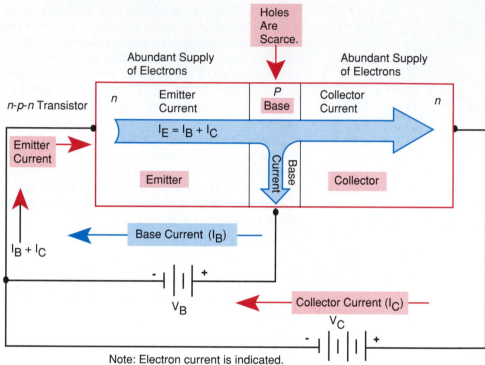

Note: Electron current is indicated.

Thus, electrons leaving the emitter block take two paths. A small percentage combine with the available holes in the base to form a forward-bias base–emitter current, while the great majority of the electrons move through the collector block to form a collector–emitter current.

The large infusion of electrons into the collector–base depletion zone effectively cancels the effect of the collector–emitter reverse-bias voltage. The electrons in the collector–base deletion zone flow easily through the collector region to become external collector-to-emitter current.

This relatively large emitter–collector current does not begin to flow until emitter–base current begins to flow. The large emitter–collector current depends on the flow of the much smaller emitter–base current. If the base–emitter voltage is increased slightly, more electrons are drawn into the base region—far more than the base can use. The result is a small increase in base current and a much larger (but proportional) increase in collector current. This current gain is called *beta* (ß).

Beta (ß), the Forward-current Transfer Ratio

The **forward current transfer ratio, beta,** is a measure of how base current controls collector current in a transistor. Figure 4-4 presents a graphical representation of beta. Figure 4-5 shows the circuit used to plot the beta curve, which represents collector current vs. base current (I_C vs. I_B). Refer to these two drawings for the following discussion. Beta is a simple unitless ratio. It is often called $\mathbf{h_{fe}}$ (in **ac** settings) or $\mathbf{H_{FE}}$ (in **dc** settings) to reflect the manufac-

FIGURE 4-4   Forward Current Transfer Curve
(*Beta Transfer Curve*)

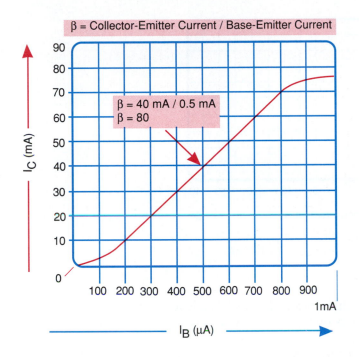

FIGURE 4-5   Circuit for Plotting the Beta
Transfer Curve

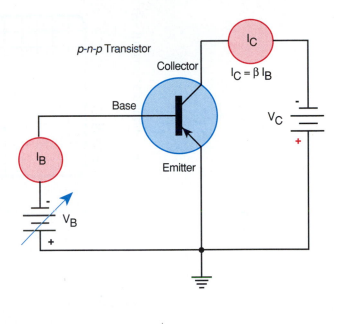

turer's adoption of a measurement system, called the **hybrid parameters,** that is widely used in specification sheets. When we use ß, we seldom distinguish between **ac** β and **dc** ß. Whichever notation is used, beta (or h_{fe} or H_{FE}) is defined as:

$$\text{Beta} = \text{Collector-circuit current/Base-circuit current}$$
$$\text{ß} = I_C/I_B$$

For example, if a base-circuit current of 1 microamp yields a collector-circuit current of 100 microamps, the transistor has a beta of 100. If we rearrange the equation to solve for collector current, we get:

$$\text{Collector Current} = \text{Beta} \times \text{Base Current}$$
$$I_C = \text{ß} \times I_B$$

The manufacturer's literature specifies a beta or H_{FE} value for a particular type number, but this beta value must not be taken too literally. To begin with, the beta (H_{FE}) specification is a statistical value; and even at that, it is only valid for the particular temperature and operating current at which the measurements were made.

If you buy a 2nxxx transistor that the manufacturer says has a beta value of 100, you will find that its actual beta values range from 50 to 200 (or so), depending on temperature and operating current.

Don't let these wide variations in the transistor's most important parameter worry you. Negative feedback can compensate for such unusually large

FIGURE 4-6   **Beta vs. Collector Current Graph**

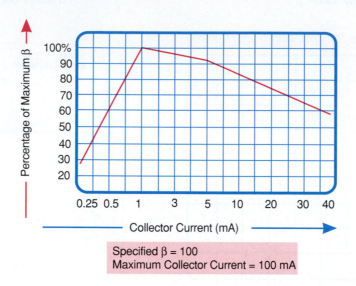

Specified β = 100
Maximum Collector Current = 100 mA

variations. High beta values are important because proper use of negative feedback normally depends on the availability of considerable surplus current gain (a high beta value).

So, in spite of the fact that beta numbers are always a bit suspect, they are useful. We will see shortly how to deal with normal variations in beta values. There is also an **ac** beta, but because there is very little difference (except at high frequencies) between the two, and because of inherent uncertainties in beta values, the distinction is seldom made (except in the manufacturer's literature, which does not assume the use of negative feedback). We will use beta instead of the device parameter H_{FE}, because we will be relying on negative feedback and taking a lot of liberties with beta values.

It is sometimes useful to know how beta varies with the operating current. Figure 4-6 is a generic graph of beta vs. collector current. Sometimes a transistor must be operated at an unusually low or high current, and we need to have some idea of how beta might be affected by such operation.

Current Gain in a Real Common Emitter Circuit

Beta or h_{fe} (or H_{FE}) defines the current gain of the transistor, as measured under very specific conditions. These special measurement conditions do not represent the conditions the transistor will face in a real circuit. Once a transistor is installed in a real circuit that includes feedback components, bias resistors, and so on, nearly all of its fundamental parameters (such as beta) become relatively unimportant.

What our measuring instruments see in a circuit is a set of circuit parameters quite different from the transistor's internal parameters. The overall circuit parameters are almost totally governed by external component values, primarily those involved in the negative-feedback loop.

FIGURE 4-7   Current Gain in a Real Circuit

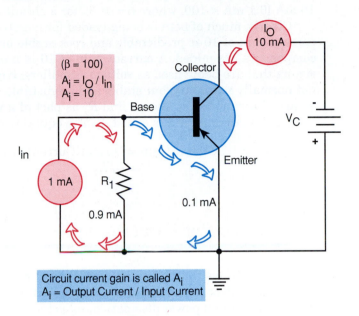

Beta, the current gain of the transistor, is probably the single most important device parameter in bipolar transistors, but the current gain of a real circuit is nearly always much lower than the transistor's beta value.

Because most transistor circuits emphasize either voltage gain or power gain, it is seldom necessary to have high current gains. We must use negative feedback if we want a stable and predictable circuit, however, and feedback demands that we trade some gain (current gain or voltage gain) for the many benefits of negative feedback. We therefore need high beta numbers so that we have plenty of excess current gain to trade for effective feedback.

All practical linear (analog) circuits use some kind of negative feedback, and nearly all of them include a resistor from base to ground as part of the feedback circuit. The base-to-ground resistor (part of a feedback voltage or current divider) may also be part of a bias circuit, but its value is determined by feedback needs. Bias requirements must be worked around the feedback-determined value of this resistor.

Figure 4-7 shows this resistor connected from base to ground. Let's examine how it reduces the circuit's current gain. Assume that a value for R_1 has been selected that is one-tenth the value of the transistor's base-to-ground (input) resistance. If the total signal input current is 1 mA, the current will divide between the base-to-ground resistance and R_1. Because the value of R_1 is one-tenth the base-to-ground resistance, 0.9 mA will flow through R_1, and 0.1 mA will flow into the transistor's base circuit.

The current flowing through R_1 never enters the transistor and so does not get amplified. The 0.1 mA that enters the transistor's base gets amplified by the factor beta.

Without resistor R_1, assuming a beta of 100, a 1-mA input current produces a 100-mA collector (output current), for a current gain of 100 (equal to ß). In the case where R_1 is added, only one-tenth of the input current gets amplified, so a 1-mA input current yields a collector (output) current of only 10 mA [0.1 mA × 100, where ß = 100], for a circuit current gain of 10.

Clearly, much of beta is being traded for negative feedback, and the circuit current gain of 10 is predictable and rock stable once the rest of the feedback components are added. A current gain of 10 is more than adequate for transistors that are being used as voltage amplifiers. High circuit current gains are not normally necessary, but stable and predictable current gain is always necessary. You can almost always find R_1 as part of a working circuit.

Circuit current gain is called A_i. The equation for it is as follows:

$$\text{Circuit Gain} = \text{Output Current/Input Current}$$
$$A_i = I_O/I_{in}$$

4.2   Basic Transistor Parameters

In the subsections that follow, we will consider six basic transistor parameters: maximum collector current, quiescent collector current, maximum collector voltage, maximum power dissipation, collector–emitter saturation voltage, and base–emitter junction resistance. Then we will turn to some more general considerations of working with these parameters.

Maximum Continuous Collector Current ($I_{C\,(max)}$)

The **maximum continuous collector current** $I_{C\,(max)}$ is an absolute maximum rating. It is temperature-dependent, but it is not altered by negative feedback. The value specified must be derated (reduced) if the environmental temperature is higher than the manufacturer's rated temperature.

The maximum collector-current rating is normally accompanied by a collector voltage value, because power ($P = V \times I$) dissipation is a measure of heat production. Excessive junction temperature is the primary cause of transistor failure. The amount of accumulated junction heat is a function of the outside temperature and the internally generated heat.

Internally generated heat is a function of power dissipation, which is a voltage-times-current function ($P_D = V \times I$). In a properly designed circuit, the transistor's continuously operating collector current almost never exceeds 40% of the maximum collector-current rating. The normal operating current is more typically between 1% and 20% of maximum.

In power transistors, an operating current of 40% of the rated maximum is typical; but there is no reason to exceed that 40%, and there are many reasons not to. In some cases, the manufacturer leaves this value out, and we must calculate it from power dissipation ratings if we want to know it.

Quiescent Collector Current (QI_C)

The **quiescent collector current** is steady-state, continuously operating current that is established by a fixed base–emitter bias current. The base–emitter bias current causes a beta-times-larger continuous collector current to flow. This is the primary controlled quantity and must be held at a constant value by using negative feedback.

The quiescent or **Q-point** collector current is the practical value of I_C. It ranges from about 1% to a maximum of 40% of $I_{C\ (max)}$. The exact value of the quiescent collector current is set by the circuit designer and is dictated by the current required by the working load that the transistor must drive. Quiescent current is a resting current when no varying input signal is present.

Maximum Collector Voltage ($V_{CE\ (max)}$)

For practical reasons, the **maximum collector voltage** is generally measured from collector to emitter. This voltage is an absolute maximum and is not subject to feedback control. In normal operation, the transistor's base–emitter junction is forward-biased, so it will have a voltage of 0.7 V across it (see Figure 4-8). The power-supply voltage is usually designated as V_{CC}, V_C, or V+; and V_{CC} is divided between the base–emitter and collector–base junctions. The forward-biased base–emitter junction absorbs 0.7 V, and the rest of the supply voltage (V_{CC}) may appear across the reverse-biased collector–base junction.

$V_{CE(max)}$ is within 0.7 V of the maximum allowable value of the power-supply voltage V_{CC}. We can thus assume that $V_{CE\ (max)}$ is the maximum supply voltage (V_{CC}), with a 0.7-V safety factor. Exceeding the maximum collector voltage will drive the reverse-biased collector junction over the Zener knee and into avalanche, possibly destroying the junction.

FIGURE 4-8 **Collector-side Voltage Distribution**

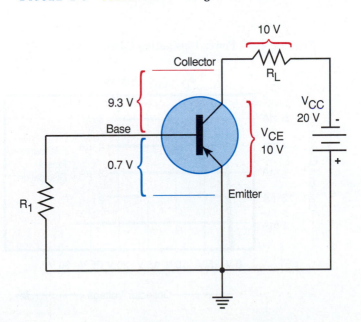

Maximum Power Dissipation ($P_{D\ (max)}$)

Maximum power dissipation is a measure of the heat developed as current flows through the internal resistances of a transistor. Power dissipation is defined as:

$$P_{D\ (max)} = V_{CE} \times I_C$$

If we want to find the maximum collector current, we can rearrange the equation to solve for $I_{C\ (max)}$ for a specific collector voltage (V_{CE}).

$$I_{C(max)} = P_{D(max)}/V_{CE}$$

And for nearly all practical circuits, we can write the equation as:

$$I_{C(max)} = P_{D(max)}/V_{CC}$$

where V_{CC} is the power-supply voltage.

Manufacturers frequently provide a curve like the one illustrated in Figure 4-9. This power dissipation curve shows the safe operating conditions for a device. The 40%-of-maximum-collector-current limit (rule) generally keeps users out of trouble automatically, except in connection with some power applications, where referring to the power dissipation curve may be necessary.

The maximum power-dissipation rating assumes a specific air temperature and that air flow around the transistor is not restricted. It also assumes that the manufacturer's heat sink recommendations (when applicable) are followed. If the outside temperature is higher than the temperature at which the maximum power dissipation is specified, the manufacturer will provide a derating factor when applicable. Derating usually applies only to power transistors.

FIGURE 4-9   Power Dissipation Curve

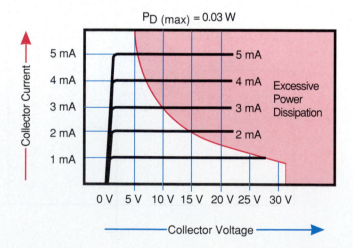

Collector–Emitter Saturation Voltage ($V_{CE\ (sat)}$)

Collector–emitter saturation voltage applies to switching transistor circuits. It represents a case where an input signal has raised the base voltage to a value 0.7 V (or more) above the collector voltage, causing the (normally) reverse-biased collector–base junction to switch into forward bias. The situation is illustrated in Figure 4-10.

The circuit in Figure 4-10 is an inverting amplifier. As the base voltage rises, it causes an increase in base current and a beta-times-larger increase in collector current. An increase in collector current causes a larger voltage drop across the load resistor (R_L) and a consequent decrease in the collector voltage.

As the base voltage increases, the collector voltage (V_{CE}) decreases proportionally. As the base voltage rises and the collector voltage falls, a crossover point is reached where the base voltage is 0.6 to 0.7 V higher than the collector voltage. At this crossover point, the previously reverse-biased collector–base junction suddenly switches to a forward-bias condition. This condition is called **saturation.** At this point, both junctions are forward-biased and the collector–base junction voltage (V_{CE}) drops to a value of from 0.2 to 0.8 volt.

We would expect a value of twice 0.6 V for two forward-biased junctions in series, but the two junctions are part of the same crystal and share the very thin base block. This condition is more complex than one that would be produced by simply connecting two diodes in series, and it results in a voltage across the two junctions that is lower than expected. The collector–emitter voltage in the saturated condition is called $V_{CE(sat)}$. The transistor produces a small voltage drop and a consequently low power dissipation in the saturated condition. The combination of low voltage drop and low power dissipation makes the transistor a very good electronic switch. $V_{CE(sat)}$ can also serve as a useful diagnostic condition in circuits that were not intended to act as switches.

Base–Emitter Junction Resistance (R_b)

The **base–emitter junction resistance** is made up of a combination of the ordinary resistance of the base and emitter blocks and the current-dependent junction resistance. We saw how to calculate the junction resistance by using the Shockley relationship when we studied diodes. The ordinary resistance of the emitter and base blocks is called **bulk resistance (R_b)**; and although it is not large, it is larger than you might expect. A phenomenon called **base spreading resistance** makes the base resistance larger than it ordinarily would be.

The base region in a transistor is a very thin sheet, with comparatively few available holes (electrons in a p–n–p transistor). Electrons entering the base tend to wander along the surface for a considerable distance before finding a hole and becoming a part of the base–emitter current. The electron must travel a much longer than usual resistive path, which increases the total effective bulk resistance.

We can also use the Shockley relationship to calculate the resistance of the base–emitter junction in a working transistor, but there is one slight compli-

FIGURE 4-10   Collector-to-Emitter Voltage

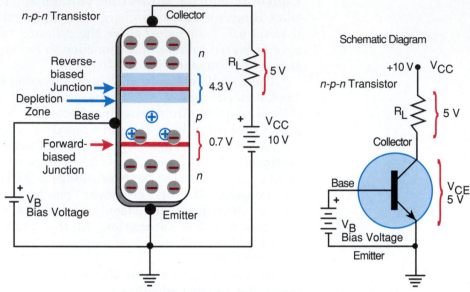

a. Normal Voltage Distribution in a Properly Biased Transistor

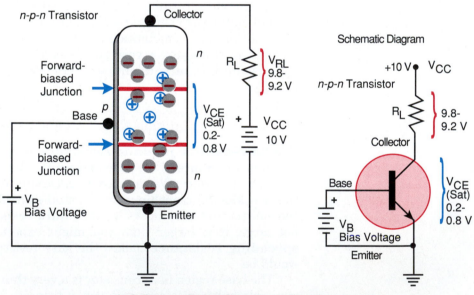

b. Collector-to-Emitter Saturation Voltage [$V_{CE (Sat)}$]

FIGURE 4-11   Base–Emitter Junction Resistance

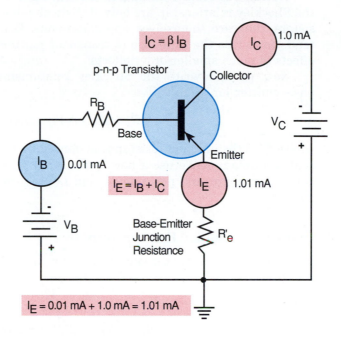

cation here. As indicated in Figure 4-11, if a base–emitter junction current is flowing, then beta times as much collector current is flowing through that same junction. The current through the base–emitter junction is the sum of the base and collector currents. For example, if the base–emitter current is 0.01 mA, and if the transistor has a beta of 100, then 1 mA of collector current is also flowing through the base–emitter junction. The total base–emitter junction current is thus:

$$I_E = I_B + I_C$$
$$= 0.01 \text{ mA} + 1.0 \text{ mA} = 1.01 \text{ mA}$$

Because the base–emitter current is typically only 1% (or less) of the total, the base current component can be ignored without significant error. Again, this goes a bit against the scientific grain, since it may appear sloppy to throw out the base current and call the emitter and collector currents equal.

Approximations in Context

The transistor circuit business involves so many quantities that we cannot know exactly that we generally obtain better total accuracy by ignoring small-order effects than by taking them (or as many of them as we possibly can) into account.

Statistically, there is a bias toward small-order effects canceling one another out if they are left alone. But if we insist on coming up with (usually by guessing at) numerical values for all of the small-order effects, we risk skewing the statistics by introducing a consistent (noncanceling) error, and we end up with less accuracy than if we had ignored them altogether.

We call our results an approximation, but remember that both beta and the Shockley relationship are only statistical values; thus, for individual transistors, they are themselves approximations. Remember, too, that feedback tends to reduce the influence of transistor parameters, reducing small-order effects to much smaller-order effects.

Now we can write the Shockley relationship (approximation) for the base–emitter junction:

$$R'_e = 25 \ (\text{mV})/I_C \ (\text{mA})$$

where R'_e is the junction resistance in ohms, and I_C is the collector-emitter current in milliamps. We use a new symbol (R'_e) to indicate that we are dealing with a transistor and not a diode; and I_C is normally the quiescent collector current (QI_C).

Example 4.1	**Using the Shockley Relationship**

Suppose that the collector current is 1.0 mA. Find the base–emitter junction resistance.

Solution   We calculate R'_e as follows:

$$R'_e = 25/I_C = 25/1.0 = 25 \ \Omega$$

The Curious Case of Higher-than-Expected Input Resistance

The input resistance to the common-emitter circuit is equal to beta times the junction resistance: not just the value of R'_e, but beta times that value. Let's examine how this unexpected multiplication of R'_e comes about. But before we look at this particular case, examine the mystery box in Figure 4-12. What is in the box? The obvious answer is a 10-ohm resistor; but there is another possibility, as shown in Figure 4-13. The 5-volt battery in the box there is series-opposing, and cancels 5 of the 10 volts provided by the battery, yielding an effective input voltage of only 5 volts. The reality is, however, that the effective input resistance of the mystery box is 10 ohms, because Ohm's Law demands a total series circuit resistance of 20 ohms, and the visible resistor is only a 10-ohm resistor. Something in the box must produce the effect of a 10-ohm resistance.

This kind of opposing-voltage current-impeding situation is often called **impedance** to distinguish it from ordinary resistance. It would be nice if we had a special, universally accepted word to describe current reduction owing to an opposing voltage, but we don't. So far you have associated the term *impedance* with reactance and **ac** circuits. Reactance is similar in that it also impedes current as a result of an opposing voltage stored in the capacitor or in an inductor's magnetic field. But reactance is frequency-dependent, whereas the kind of impedance involved in the mystery box is not. Still, for want of a special word, we will use the term *impedance*. The conditions in the mystery box can apply equally well to an opposing **ac** voltage, as illustrated in the sec-

FIGURE 4-12 Mystery Box

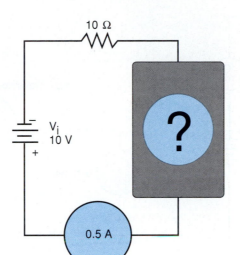

FIGURE 4-13 Mystery Box Revealed

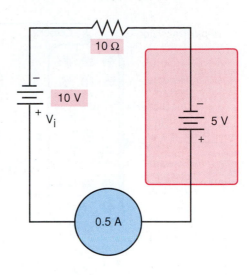

FIGURE 4-14 Not-So-Mysterious ac Box

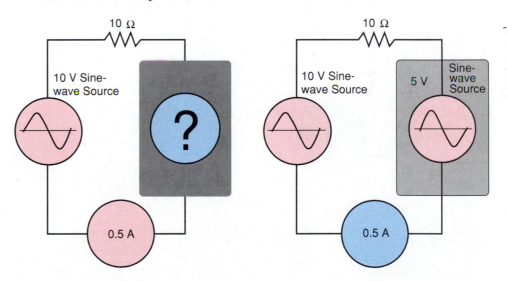

ond—and not so mysterious—box in Figure 4-14. Again, the box has an input impedance of 10 ohms.

Now that we have established the basic concept, let's see how it applies to the input resistance (impedance) of a common-emitter transistor circuit (see Figure 4-15).

The transistor has a beta value of 100; so with a base current of 0.01 mA, we get a collector current of 1 mA. Notice in Figure 4-15(a) that two currents flow through the junction: the base current and the (larger) collector current.

FIGURE 4-15 Transistor Circuit Input Resistance (Impedance)

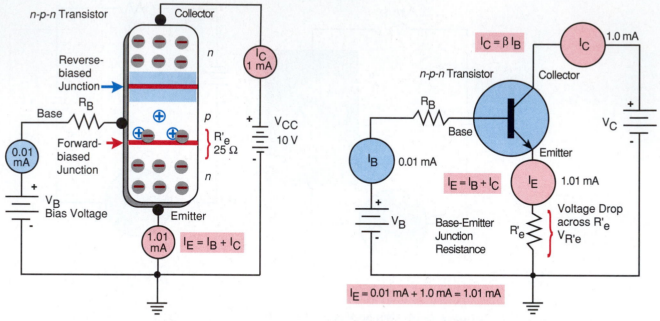

a. Block Diagram

b. Schematic Equivalent of the Block Diagram

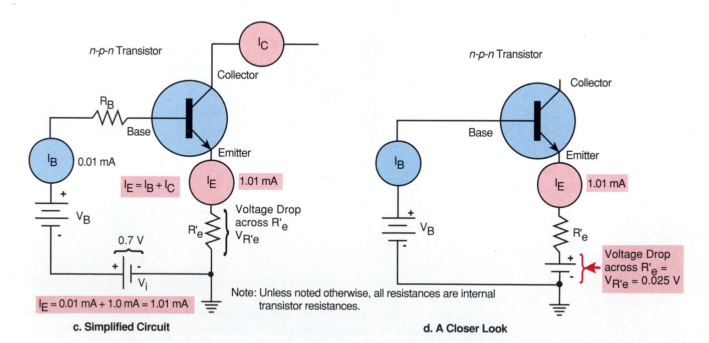

c. Simplified Circuit

d. A Closer Look

Since the collector current is 100 times larger than the base current ($\beta = 100$), we will ignore the base current in calculating the junction resistance. The junction resistance is approximately equal to

$$R'_e = 25 \text{ mV/Collector Current (in mA)}$$

Now let's calculate the voltage drop across the junction resistance (see Figure 4-15(c)):

$$V_{R'_e} = I_C \times R'_e = 0.001 \times 25 = 0.025 \text{ V}$$

Figure 4-15(c) shows the equivalent circuit, with the voltage drop across the junction resistance shown as a voltage source. The second battery of 0.7 V is just enough to forward-bias the junction so that we don't have to include those two factors in our example. Now, look at the equivalent circuit in Figure 4-15(d). What input voltage would be required for a current of 0.01 mA, if the junction resistance voltage drop were not there, and if the input resistance were the expected 25 ohms? We calculate this as

$$V_{in} = I_{base} \times 25 \ \Omega = 0.00001 \text{ A} \times 25 = 0.00025 \text{ V}$$

But because of the opposing voltage drop across the junction resistance of 0.025 V, we must increase the input voltage to get the same current:

$$0.00025 \text{ V} + 0.025 \text{ V} = 0.02525 \text{ V}$$

If a voltage of 0.02525 is needed to generate an input current of 0.01 mA, what is the effective input resistance? To determine this, we use the following equation:

$$R_{in} = V_{in}/I = 0.02525 \text{ V}/0.00001 \text{ A} = 2525 \ \Omega$$

The value of 2525 ohms is approximately equal to beta times the junction resistance:

$$R_{in} = \beta \times R'_e = 100 \times 25 = 2500 \ \Omega$$

Let's write a complete **Kirchhoff's loop equation** to prove that the assumed equation is valid. Refer to Figure 4-16 as we go through the loop.

Example 4.2

Analyzing the Base-emitter Circuit

Prove that the correct formula for effective input resistance is

$$R_{in} = R_B + R'_e + \beta R'_e$$

Solution

We start by writing the Kirchhoff's loop equation:

$$V_B = I_B R_B + (I_C + I_B)R'_e$$

Because $I_C = \beta \times I_B$, we can write

$$V_B = I_B R_B + (I_B + \beta I_B)R'_e$$

FIGURE 4-16   Circuit for the Input Resistance
(Impedance) Loop Analysis (Example 4-1)

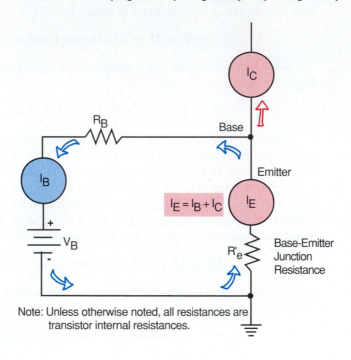

Note: Unless otherwise noted, all resistances are
transistor internal resistances.

Clearing the parentheses, we have

$$V_B = I_B R_B + I_B R_e' + ßI_B R_e'$$

Dividing through by I_B, we get

$$V_B/I_B = R_B + R_e' + ßR_e'$$

Finally, since $V_B/I_B = R_{in}$, we obtain

$$R_{in} = R_B + R_e' + ßR_e'$$

Example 4.3   **Adding Values to the Base-emitter Circuit Analysis**

Let's try some numbers:

ß = 100
$R_e' = 25 \ \Omega$
$R_B = 5 \ \Omega$

These are fairly typical real values. What is the effective input resistance?

Solution   Doing the arithmetic, we get:

$$R_{in} = 5 + 25 + (100 \times 25) = 2530 \ \Omega$$

Because the bulk resistance is comparatively small—particularly when
some external emitter resistance is added—and because beta times R_e' is large

compared to R'_e, the input resistance (impedance) equation is often written as:

$$R_{in} = \text{ß}R'_e \qquad \text{or} \qquad Z_{in} = \text{ß}R'_e$$

(Z_{in} is often used in place of R_{in}.)

Perhaps you feared that discussion of the often-mentioned negative feedback would be saved for the final chapter. Not so. We have just tripped over our first example of negative feedback in transistors, and it had a drastic effect on the input resistance of the circuit. Let's look at what happened here and see if it fits one of the common definitions of negative feedback. Later, we will study general feedback equations in some depth. The feedback equations are quick and simple, and they can be used for any circuit in which feedback is involved.

The feedback we just encountered is accidental in the sense that we didn't deliberately introduce it; it was just there. There is not enough incidental feedback (provided by R'_e) in the example for it to be of much value, but we can easily add as much external resistance in series with the emitter as we need. The external resistor is functionally equivalent to a very high junction resistance. We can't do much about the junction resistance itself, since we are stuck with whatever it happens to be at a particular collector-current value, but we can add any value of external feedback resistor we need to get the desired amount of negative feedback. The applications for each circuit configuration are roughly the same, regardless of which kind of transistor is used.

The bipolar transistor was once the undisputed king of integrated-circuit technology. Field-effect devices have become very important in the latest digital ICs, but bipolar devices are still the mainstay of analog integrated circuits.

Bipolar Junction Transistor (BJT) Parameters

Transistor technology has produced a small jungle of electrical and mathematical models to describe how transistors behave, but these models can all be classed into three categories:

1   Models that describe complex electrical activities going on inside the device.
2   Models used by manufacturers to define the characteristics of the devices they sell.
3   Models that define the characteristics of complete, working, transistor circuits.

Because negative feedback tends to make the first two models relatively unimportant in practical circuits, we will spend most of our time studying the third kind of model.

In any workable analog transistor circuit, the circuit's behavior is dictated mostly by feedback components and not by the manufacturer's specifications. This implies that, once a circuit is properly designed, negative feedback—and not individual transistor characteristics—dictates the circuit's specifications. If the transistor's characteristics are made irrelevant by negative feedback, any bipolar junction transistor can be installed in a properly designed circuit (as a replacement) and it will work just fine.

The existence of thousands of transistor types makes it hard to believe that, if just any transistor can be installed in an existing circuit, the circuit will work. If that is so, why are there so many types? Actually, the answer is that it just happened. Each manufacturer assigned company numbers to its devices, and no coordination was maintained or developed among the many manufacturers. In the early years, process control was very poor: a batch of transistors would be sorted according to their characteristics, and each group in the sort would get a different number. Because it is virtually impossible to exercise precise control over transistor characteristics, there are still no serious standards, and type numbers continue to proliferate.

Of course, it is only partly true that any bipolar transistor will work in a given circuit, but (because of negative feedback) a dozen or so well-chosen devices can replace several thousand different type numbers. As a matter of fact, several manufacturers make a series composed of a handful of replacement transistors that can replace a whole cross-reference book of several thousand transistor types.

The idea that external circuit parameters can dominate the system, with the amplifying device (transistor) playing a subservient role, is not odd or mysterious. The influence of negative feedback is essential to account for the wide variations in manufacturing tolerances, transistor temperature dependence, and the fact that transistor parameters vary with operating currents. Once sufficient negative feedback has been designed into the circuit, the circuit characteristics become largely independent of the transistor parameters. Even complex multitransistor operational amplifiers depend on feedback.

Transistors exhibit unusually large manufacturing tolerances and an uncomfortable degree of temperature sensitivity. Obviously, however, in light of the advanced state of the technology, these difficulties have been overcome—although not by tightening manufacturing tolerances or making transistors less temperature-sensitive. These problems have been solved through the use of negative feedback and other circuit design procedures that render overall circuit parameters almost completely independent of the included transistor's internal parameters.

Current manufacturing technology does not permit very close tolerances for internal transistor parameters; and if it did, little would be gained by it. Large variations in these important parameters produced by moderate temperature variations pose a problem every bit as troublesome as broad manufacturing tolerances. Even if manufacturers could hold closer tolerances, we would still have to cope with temperature sensitivity; it is an inherent property of semiconductors.

But as it happens, the negative-feedback techniques used to deal with temperature problems also serve to accommodate large manufacturing tolerances. By the time the circuit designer has compensated for the inherent temperature problems, very large manufacturing tolerances have also been compensated for. As a result, the designer does not press the manufacturer for closer tolerances, and the manufacturer has no incentive to establish them.

In addition to tolerance and temperature variations, some important transistor parameters vary with the operating current, which is selected according to the needs of the designer. But these variations also yield to the same negative feedback techniques that allow for large tolerance and temperature variations.

The use of negative feedback—an essential requirement for any linear transistor circuit—makes both circuit design and circuit analysis for troubleshooting purposes comparatively easy. In this chapter we will examine bias and bias stability, the keys to reliable transistor circuit operation.

The equations we will use are fairly simple. They are only valid, however, if adequate negative feedback is provided. Circuit analysis procedures must assume that the designer provided adequate feedback as an integral part of the circuit design. That is always the case if the circuit has once worked properly.

Symbols and Terminology

The following list identifies some of the more common symbols and terminology used in work with transistors:

$\approx$ – Approximately equal to. (Example: $A \approx B$ means A is approximately equal to B.)

Δ – A change or variation in some quantity. (Example: ΔV means a change (or variation) in the voltage V.)

// or || – In parallel with. (Example: R_1 // R_2 means resistor R_1 is in parallel with resistor R_2, or $(R_1 \times R_2)/(R_1 + R_2)$.)

typ – Typically, or typical value.

4.3   Output Resistance of a Common Emitter Circuit

The output resistance of the transistor itself has two different values: one is for static voltages, and the other is a dynamic resistance that exists only when changing voltages are involved. A third (and usually more important) output resistance exists for the circuit as a whole. It is totally controlled by the value of the load being driven by the transistor.

Dynamic Output Resistance

In normal transistor operation, the collector–base junction is always reverse-biased. There are only two varieties of carriers that can become collector current in a reverse-bias junction: carriers (holes or electrons) injected into the collector–base depletion zone by the emitter, which are controlled entirely by the base current; and heat-generated carriers in the collector–base junction depletion zone. Nearly all of the injected carriers are harvested by the collector battery, even at very low collector voltages.

If we increase the collector voltage, we might reasonably expect a proportional increase in collector current. But that doesn't happen. Has Ohm's law let us down? In this case, a 100% increase in collector voltage only yields a 1% to 5% or so increase in collector current. This is not a violation of Ohm's law, however, because a reverse-bias junction is not a simple resistance.

In a reverse-bias junction, you might think that increasing the collector voltage should pull more electrons from the emitter, through the base-collector junction, and into the collector block. But the emitter only injects carriers in proportion to the base–emitter current. Because the injected carriers are nearly all captured even by very low collector voltages, increasing the collector voltage only picks up the strays that didn't get caught before.

This minor gleaning of missed carriers would be more significant if they did not reside in a depletion zone. When we increase the collector voltage, we also widen the collector–base depletion zone, so the carriers must travel longer distances to escape the zone. Any energy gained from the increase in collector voltage is largely nullified by the additional energy expended in dragging carriers over longer distances through the widened depletion zone.

Temperature-generated carriers originate in the depletion zone, and they, too, must travel greater distances when higher collector voltages widen the depletion zone. This behavior, which does not seem to conform to the simple terms of Ohm's law, is called the **dynamic output resistance** of the junction. The result of this dynamic resistance behavior is that large changes in collector voltage produce very small changes in collector current.

The collector current is nearly completely independent of the collector voltage. And if the collector current is independent of the collector voltage, the only things that determine how much collector current flows are the base–emitter current and the value of ß.

If an increase in collector voltage caused no increase in collector current, we would have the dynamic equivalent of an open circuit: an infinite resistance. The reverse-bias collector–base junction does not have an infinite resistance, because collector voltage does have a slight effect on collector current, but the collector–base junction has a dynamic resistance of about 0.5 to 2 megohms.

Dynamic Output Resistance Equation

The equation for dynamic output resistance (R_O) is

$$R_O = \Delta V_C / \Delta I_C$$

where Δ represents a change in the value of some quantity.

The dynamic resistance is a transistor parameter, and it varies from transistor to transistor. Let's look at some made-up (but typical) numbers.

Example 4.4

Calculating Collector-base Junction Dynamic Resistance

Given the circuit shown in Figure 4-17, find the dynamic output resistance of a transistor when a change in collector voltage from 10 volts to 20 volts results in a change in collector current from 1.0 mA to 1.01 mA.

Solution

Our solution consists of the following steps:

1   1.01 mA – 1.0 mA = 0.01 mA   [a change of 1%]
2    $R_O = \Delta V_C / \Delta I_C = 10 \text{ V}/0.00001 \text{ A} = 1 \text{ Meg } \Omega$

As long as we hold the base–emitter current constant, we obtain a beta-times-larger collector current, and the collector current does not change, even with fairly large changes in collector voltage.

FIGURE 4-17 Independence of Collector Current from Collector Voltage

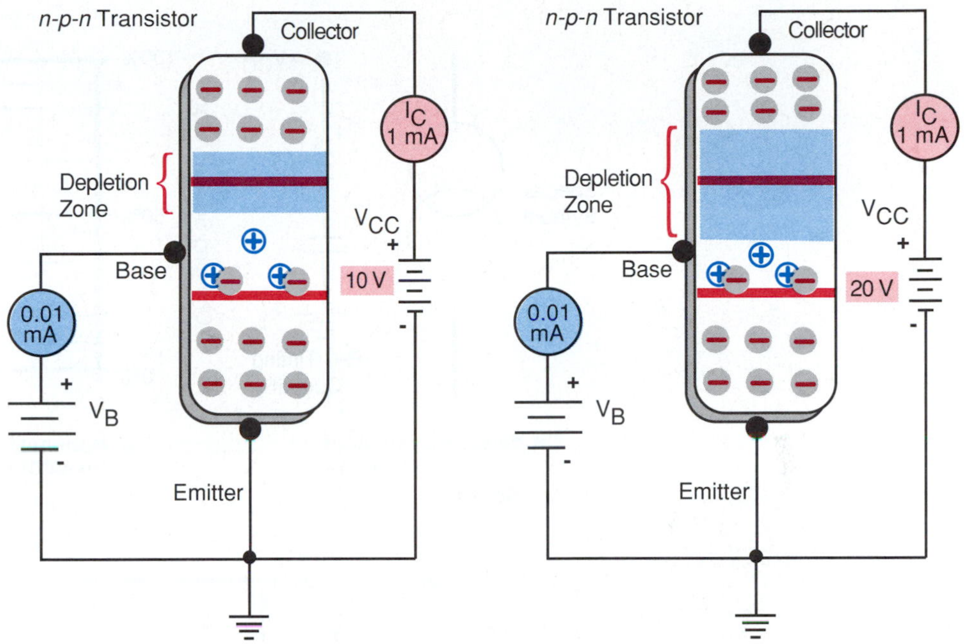

A transistor is frequently used as a current regulator. When a transistor is used for this purpose, it is called a **constant-current source.** It has even been assigned the special symbol shown in Figure 4-18(a) to signify that it is serving in this special capacity. In later chapters, we will examine a number of circuits in which the transistor's dynamic resistance plays a critical role.

The circuit in Figure 4-18(c) illustrates one common application—charging a capacitor through several time constants with a linear charging ramp. If we use an ordinary resistor (as in Figure 4-18(b)) to charge the capacitor, we get the usual nonlinear R-C time-constant curve. But if we replace the timing resistor (R_1) with a transistor, to take advantage of its dynamic resistance, we get the linear charge curve shown in Figure 4-18(c). We can set the transistor's base current at the proper value to get the desired charging current through the transistor, and the capacitor will charge at a constant (linear) rate regardless of the the fact that the capacitor is accumulating a charge voltage.

Normally, the capacitor-charge voltage opposes the supply voltage, reducing the effective collector-emitter voltage. The reduced effective collector-emitter voltage does not affect the charge current, however, because the collector-emitter current is independent of the (effective) collector voltage. With the ordinary timing resistor in Figure 4-18(b), the capacitor charge current is reduced by the opposing capacitor charge voltage, and the charge curve becomes nonlinear.

Output Impedance of a Real Common Emitter Circuit

In transistor amplifier circuits, the dynamic output resistance has a negligible effect on the circuit output resistance. For very practical reasons, external resistance values are so much lower than the transistor's dynamic output

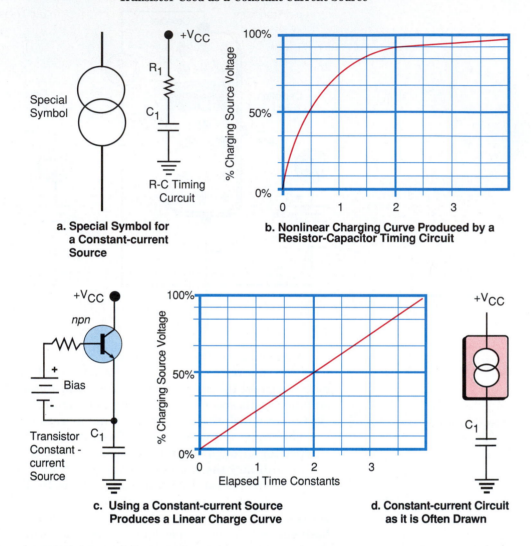

FIGURE 4-18 **Linear R-C Timing Circuit Provided by a Transistor Used as a Constant-Current Source**

+V_CC

R_1

C_1

Special
Symbol

R-C Timing
Curcuit

a. Special Symbol for a Constant-current Source

b. Nonlinear Charging Curve Produced by a Resistor-Capacitor Timing Circuit

+V_CC

npn

Bias

Transistor
Constant-current
Source

C_1

c. Using a Constant-current Source Produces a Linear Charge Curve

+V_CC

C_1

d. Constant-current Circuit as it is Often Drawn

resistance that the external circuit resistances become the dominant resistances in the output circuit.

Figure 4-19(a) shows the transistor output circuit as it appears on the surface. Figure 4-19(b) includes a model of the insides of the transistor, along with the always-needed external load resistance and the internal resistance of the power supply or battery (R_S). Power-supply internal resistance must always be as low as possible. In Figure 4-19(c), the internal battery and the 25-ohm junction resistances have been dropped because they are too small to bother with. The essential model boils down to the 1-megohm transistor dynamic output resistance in parallel with the 1 kilohm external load resistance. Circuit resistances of 1 megohm in parallel with 1 kilohm have a working value very close to 1 kilohm. So it seems that the circuit output resistance is controlled almost entirely by the value of the external load resistance.

Like any other machine, a transistor circuit is of little value if it does no meaningful work. The circuit in Figure 4-20 puts the circuit to work control-

FIGURE 4-19   Output Resistance in a Real Transistor Circuit

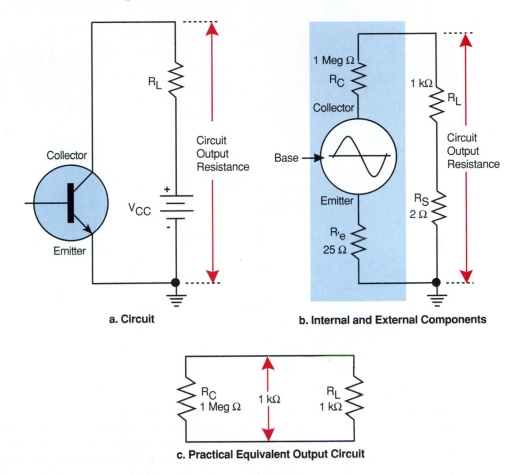

a. Circuit

b. Internal and External Components

c. Practical Equivalent Output Circuit

ling the brightness of a lamp. We will call the lamp the **working load**; and the working load determines the required operating current for the transistor. If the lamp is a 1-amp automotive taillight bulb, the transistor must be capable of maintaining a 1-amp operating current. The base current controls the collector current, so varying the base current causes a beta-times-larger variation in the collector current that flows through the lamp.

If we replaced the 1-amp bulb with a bulb that requires only 100 mA of current, we could probably use the same transistor, but we would have to reduce the base current to a considerably lower value. The 1-amp transistor would then be operating at only about 10% of its maximum current capacity, which is just fine. The working load determines what current the transistor must be capable of handling. The transistor may be capable of handling more current without causing any ill effects, but it must be able to handle at least as much as the working load demands.

Now suppose that we want the normal condition of the 1-amp taillight bulb to be half-brightness, and that we want to be able to increase or decrease the brightness whenever we choose. Suppose, too, that the transistor has a beta value of 100. To set the lamp brightness to one-half, we need a 500-mA

FIGURE 4-20   Using a Transistor to Control Lamp Brightness

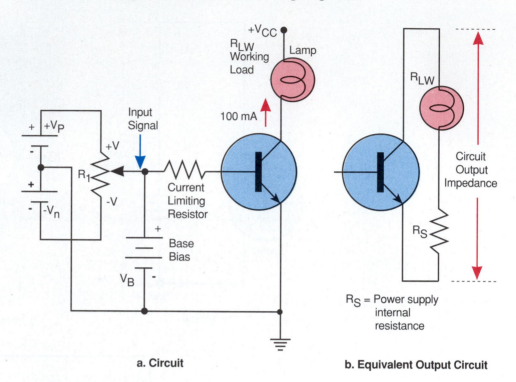

a. Circuit   **b. Equivalent Output Circuit**

idling current. The idling collector current is called the *quiescent collector current* (QI_C). To get 500 mA of quiescent collector current, we need to establish a base current of 5 mA, since 5 mA × 100 [beta] = 500 mA. The base (input) current used to establish the quiescent (idling) collector current is called a **bias current.** The voltage used to force the bias current through the base–emitter circuit is called the **bias voltage.**

If we use some kind of signal to vary the lamp brightness, it must consist of a voltage that either adds to or subtracts from the bias voltage, thus changing the base current and the collector current through the lamp. When the signal voltage is reduced to zero, the bias voltage still drives 5 mA into the base, producing 500 mA of quiescent collector current, and maintaining the lamp at half-brightness. The circuit is now resting at the desired quiescent collector current.

Notice that, in this case, an increase in base bias (or signal-driven base current) causes an increase in lamp brightness. The circuit in this mode is a noninverting current amplifier. It is also a power amplifier, because the voltage current product (Power = $V \times I$) consumed by the lamp is greater than the product of the base voltage and the base current. With the lamp at half-brightness, the current that flows through it is 0.5 amp, and the voltage drop across it is approximately 6 volts. The power delivered to the load (lamp) by the transistor is as follows:

$$\text{Load Power} = \text{Load Voltage} \times \text{Load Current}$$
$$P_{RL} = V_L \times I_L$$

$$= 6 \text{ V} \times 0.5 \text{ A}$$
$$= 3 \text{ watts}$$

The input (base-emitter circuit) power is the product of the base voltage and the base current:

$$\text{Base Circuit (Input) Power} = \text{Base Voltage} \times \text{Base Current}$$
$$P_{in} = V_B \times I_B$$
$$= 0.7 \text{ V} \times 0.005 \text{ A}$$
$$= 0.0035 \text{ W} \quad \text{or} \quad 3.5 \text{ mW}$$

The input power is 0.0035 watts. The output power is 3.0 watts.

The transistor has amplified the input power of 0.0035 watts to a value of 3.0 watts. Thus the input (base circuit) power has been multiplied (amplified) by a factor of 857 to produce an output (load) power 857 times larger. The amount of power amplification (in this case, the number 857) is called the *power gain.*

The symbol for power gain (power amplification) is A_P, where the A stands for amplification, and the subscript P stands for power. Mathematically, power gain is defined as:

$$\text{Power Gain} = \text{Power Delivered to Load/Input Power}$$
$$A_P = P_{RL}/P_{in}$$

or alternately as:

$$\text{Power Gain} = \text{Power Output/Power Input}$$
$$A_P = P_O/P_{in}$$

You can argue, correctly, that voltage amplification also occurs in the circuit in Figure 4-20, because the variations in voltage across the load are larger than the base input voltage variations. However, the circuit in Figure 4.20 is not the best model for a discussion of voltage amplification.

Output Resistance and Voltage-driven Working Loads

The example in Figure 4-20 involved a current-driven or current-controlled working load. Although current-controlled working loads are not unusual, voltage-driven working loads are much more common. Figure 4-21(a) shows a **dummy load** connected in series with the transistor to form a voltage divider. The working load is connected to the midpoint of the voltage divider. This situation is illustrated by the simplified model in Figure 4-21(b).

We can increase the transistor's base–emitter bias current in part (a) to cause enough collector current to flow through the dummy load (R_L) to produce a voltage drop across the dummy of half the supply voltage (V_{CC}). The voltage drops across the dummy load and across the transistor are each equal to half the supply voltage. The voltage across the transistor is then 6 volts, and the 12-volt lamp in parallel with the transistor is driven by half its rated voltage and is therefore at half-brightness.

We can vary the signal voltage to increase or decrease the base current, causing a change in collector current; and this, in turn, varies the voltage

FIGURE 4-21   Voltage-driven Working Load

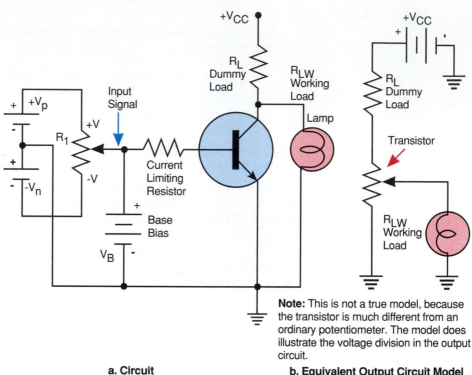

Note: This is not a true model, because the transistor is much different from an ordinary potentiometer. The model does illustrate the voltage division in the output circuit.

a. Circuit   b. Equivalent Output Circuit Model

across the lamp. Figure 4-22 shows the effective output resistance of the complete circuit. The 1-megohm output dynamic resistance (R_C) of the transistor is too high to have much influence on the equivalent parallel resistance, and the 2-ohm internal power supply resistance and the junction resistance (R_e') are too small to have much influence. The dummy load (R_L) and the working load (R_{Lw}) combine to dominate the equivalent circuit output resistance. In most **ac** small-signal amplifiers, the output working load is connected to the collector through a capacitor (or occasionally through a transformer). The capacitor keeps the relatively high **dc** collector voltage from affecting the working load, which is often another transistor circuit. The capacitor blocks the **dc** collector voltage, while allowing the **ac** signal to pass freely to the working load.

The capacitor must be large enough to have a reactance at the lowest expected **ac** signal frequency of near 0 ohm. When a capacitor is used as a coupling device, as illustrated in Figure 4-22, we must talk about output impedance, because the capacitor is a reactive device and is frequency-dependent. An equivalent circuit model that uses a coupling capacitor (or other coupling device) is illustrated in Figure 4-22(b).

Note: A *dummy load* is a load intended to take the place of a working load or to supplement the working load. The dummy load does no useful work (except to provide a voltage drop and to produce some heat). For example, you might connect an 8-ohm resistor to your stereo in place of the 8-ohm speaker. You could then conduct tests with the stereo operating under a normal load, but without people complaining about the noise you are making.

The following definitions apply to Figure 4-22:

FIGURE 4-22   **Using an Output Coupling Capacitor**

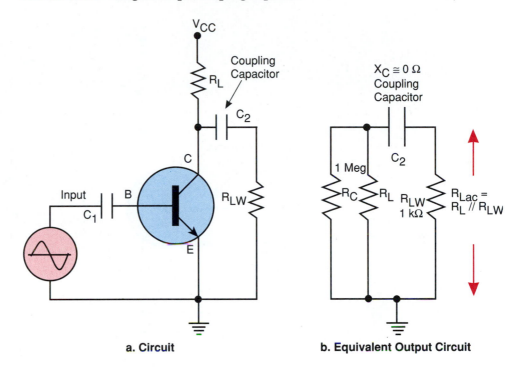

a. Circuit   **b. Equivalent Output Circuit**

R_L is the dummy load.

R_{LW} is the working load.

R_{Lac} is the total **ac** signal output impedance, consisting of R_L and R_{LW} connected in parallel.

R_C is the transistor's dynamic collector–base junction resistance; it is normally two orders of magnitude higher than R_{LW} and R_L, and it has no practical influence on most real circuits.

Voltage Gain in the Common Emitter Circuit

Voltage gain is defined as the ratio of the signal output voltage to the signal input voltage. The symbol for voltage gain (voltage amplification) is A_V, where the A stands for amplification, and the subscript V stands for voltage. The input voltage is applied to the base–emitter circuit. The output voltage is developed across the working load. Thus we have

$$\text{Voltage Gain} = \text{Output Voltage/Input Voltage}$$
$$A_V = V_O/V_{in}$$

The output voltage is the signal voltage developed across the working load.

Suppose that we run through a simple, mathematical analysis of voltage gain, mostly to prepare for the explanation of how the real circuit output impedance affects voltage gain. Voltage gain will be an ongoing subject of discussion. The circuit in Figure 4-23 is missing a number of practical circuit elements that greatly influence the voltage gain numbers, so we won't spend a lot

FIGURE 4-23   Circuit for the Voltage Gain Analysis

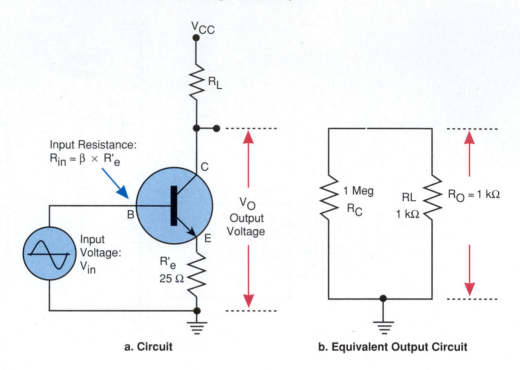

a. Circuit   b. Equivalent Output Circuit

of time discussing the voltage gain of this primitive circuit version. The standard textbook analysis presented here is accurate as far as it goes, but it doesn't go far enough to handle practical circuits.

Refer to Figure 4-23 as we go through the following example.

Example 4.5

Voltage Gain Analysis

Identify the components of the voltage gain for the circuit shown in Figure 4-23.

Solution

Our voltage gain analysis proceeds as follows:

1   $A_V = V_O/V_{in}$ (definition of voltage gain)
2   $V_O = I_C \times R_L$ (Ohm's law)
3   $I_C = \beta \times I_B$ (current gain)
4   $V_O = \beta \times I_B \times R_L$ (substitution)
5   $V_{in} = I_B \times R_{in}$ (Ohm's law)
6   $R_{in} = \beta \times R_e'$ (input resistance looking into the base)

7   $A_V = \dfrac{V_{out}}{V_{in}} = \dfrac{\beta \times I_B \times R_L}{I_B \times \beta \times R_e'}$ (substitution)

Beta and the base current appear in both the numerator and the denominator, so they cancel. Hence, the voltage gain equation can be written as:

$$A_V = R_L/R_e'$$

Keep in mind that this voltage gain equation is only valid because the input resistance is equal to $\beta \times R_e'$. The equation is valid for a **dc** input voltage as well as for an **ac** signal, if no reactive coupling devices are used. The equation is also valid for any operating current, because no current value is involved in the voltage gain equation.

Example 4.6

Calculating Voltage Gain and Output Voltage

For the circuit shown in Figure 4-24, calculate the voltage gain and peak–peak (P–P) output voltage.

Solution

Let $R_e' = 25$ ohms, and let $R_L = 1$ kilohm. With an **ac** signal input of 0.01 volts P–P, find the voltage gain and the output voltage (V_O).

Our solution consists of two parts:

1   $A_V = R_L/R_e' = 1000/25 = 40$
2   $V_O = A_V \times V_{in} = 40 \times 0.01 \text{ V} = 0.4 \text{ V } P–P$

FIGURE 4-24   **Circuit for the Voltage Gain Example (Example 4-4)**

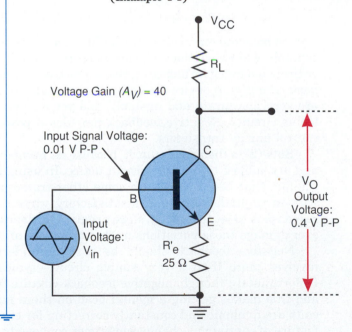

Voltage Gain $(A_V) = 40$

Input Signal Voltage: 0.01 V P-P

Input Voltage: V_{in}

R'_e 25 Ω

V_{CC}

R_L

C

B

E

V_O Output Voltage: 0.4 V P-P

4.4   Negative Feedback in Transistors

The introduction many engineering students get to negative feedback and its power is strictly mathematical, and both the mathematics and the power of feedback are likely to seem almost magical. But what is actually going on here? In spite of its awesome power, negative feedback is really nothing more than applied Ohm's law. In the following sections, we will look at negative feedback from this simple point of view. Later, we will study some general and specific feedback equations. The mathematics is not at all difficult, but it should become even easier once you have developed an intuitive feel for what the equations are telling you about negative feedback circuits.

Although the following sections deal with negative feedback as it occurs in the simplest of transistor circuits, the concepts introduced apply equally well to all other feedback situations. Because negative feedback involves Ohm's law, we can discuss the circuit conditions from three different perspectives: voltage, current, and resistance.

Whichever perspective we take, the final outcome is the same, but the process and the way of looking at things may be quite different when viewed from the different perspectives. It is important to understand as early as possible how things work from the various perspectives. In the future, you will have to read technical explanations of circuits that include negative feedback, and the authors of those explanations may use whichever viewpoint best fits the particular circuits involved. Often, one approach is easier than the others.

The material in the following sections is presented in small sequential steps. It is easy to understand, but the details are crucial and fully comprehending the explanation requires some serious concentration.

Why Transistors Require the Use of Negative Feedback

The semiconductor devices we build electronics around tend to have unacceptably wide variations in manufacturing tolerances, and temperature has entirely too much influence on them. Electronics depends on transistors as the basic amplifying device for both discrete and integrated circuits. Transistors are very inaccurate and unstable, and yet we need accurate and stable electronic circuits. Negative feedback provides a powerful tool to discipline and control unruly transistors.

Beta (ß) is the prime culprit, because its numbers are extremely unreliable and are subject to change without notice. In using feedback, we are accepting the fact that beta (along with some other transistor parameters) is going to vary in unpredictable and unsatisfactory ways that we can't prevent. Our response is to let them vary, since negative feedback will compensate for whatever damage those variations might otherwise inflict.

Negative feedback should be easy to understand, because it seldom involves more than a very simple circuit—typically, a couple of resistors. Unfortunately, though, negative feedback circuits function as moving targets, and this makes getting a mental bead on them more difficult. Feedback circuits are dynamic and constantly correcting for errors in the system, and this feature of continuous change sometimes conveys the impression that two contradictory conditions exist at the same time.

The real negative feedback action has two components:

1   Detect an error condition.
2   Correct the detected error to reestablish the original "no error" condition.

Error detection and correction happen almost instantaneously—so fast that we are usually unaware that anything is happening in a working feedback circuit. Even our instruments seldom show that anything is going on, even though the feedback circuit may be very busy detecting and correcting.

Negative feedback is capable of correcting very large errors, but it doesn't wait for an error to get large before correcting it with spectacular efficiency. Large errors (if any) are corrected on circuit power-up, and the circuit is never allowed to stray far off the mark from then on. Negative feedback performs its task without our being aware of its busy operations in a working circuit.

Because negative feedback requires an error quantity to work, it can't be perfect; but it can be very close. There is always some small resolution error, even after correction. The amount of resolution error depends mostly on the amount of amplification available. In the case of the transistor, higher beta numbers yield better resolution and a smaller remaining error after correction than do lower beta numbers. This is a basic characteristic of negative feedback. The higher the amplifier's gain (voltage gain or current gain), the closer the corrected value of the controlled quantity is to the reference quantity value. The difference between the value of the reference and controlled quantities, after correction, is termed the **resolution.**

FLASHBACK: **Elements of a Negative Feedback Circuit**

1   There must be a controlled quantity.

2   There must be a reference quantity, to establish the desired value of the controlled quantity.

3   There must be an error detector, to compare the controlled quantity value with the reference value and to produce an error signal if these values differ.

4   There must be an amplifier, to amplify the error signal so that the system can respond to small errors.

5   There must be an error corrector, to correct any differences between the reference and controlled quantity values.

Negative Feedback in Transistors

To enable a transistor to work in an analog (or linear) circuit, we must first establish an idling collector–emitter current called the *quiescent collector current (QI$_C$)*. The term *quiescent collector current* is often shortened to *Q-point current, QI$_C$,* or simply *Q-point.* The idling current is established by providing a fixed base–emitter current called **bias current.** This base–emitter bias current appears as a beta-times-larger current in the collector circuit.

Three quantities are especially important in work with transistor currents:

Base–Emitter Current = Bias Current   (I_B)

Collector–Emitter Current = Quiescent Collector Current   (QI_C)

Quiescent Collector Current = Beta × Bias Current ($QI_C = ß \times I_B$)

The idling current (QI_C) is essential to proper transistor operation. Once the designer has established the correct value for QI_C, every remaining parameter of the circuit design is based on that value. If the circuit is to work as designed, the quiescent collector current must stay at the value the circuit was designed around. If the quiescent current varies much during the circuit's normal operation, the circuit will not perform as intended. Too great a deviation from the specified quiescent collector current will cause the circuit to quit working altogether. In short, the quiescent collector current must not be allowed to vary.

The problem is that $QI_C = ß \times I_B$; and although we can hold the base current steady by using a steady bias voltage, we can't depend on ß to stay anywhere near constant. If beta varies, so must the Q-point current. But if we add negative feedback to the transistor circuit and make the quiescent collector current the controlled quantity (to maintain a constant value for QI_C), we no longer have to worry about beta's influence on the collector current. Beta will vary, and there is nothing we can do about it; but the feedback circuit can correct any collector current errors the beta variations may create.

Some Preliminary Ideas

Before we begin to construct a negative-feedback circuit to control the quiescent collector current, let's review the underlying principles involved in real negative-feedback circuits.

Since we are interested in stabilizing the transistor's Q-point current, our controlled quantity is a current value. Figure 4-25 shows a simple series circuit that uses two voltage sources and a resistor. One voltage source (V_S) is the main power supply for the circuit. The second voltage source (V_{adj}) is adjustable and is connected in series in opposition to the main power-supply voltage. The adjustable voltage is used to correct any errors in the 0.5-amp controlled-quantity current.

Refer to Figure 4-25 as we go through the following explanation.

Initial Conditions

The first circuit (Figure 4-25(a)) defines the conditions that the circuit is intended to have. The controlled quantity is the 0.5-amp current, which must be held constant. The main supply voltage (V_S) and the 10-ohm resistor are the only two elements that can cause an error in the 0.5-amp current, if either or both change values.

The power-supply voltage might change as a result of power-line variations and so on. The resistor might change value because you just replaced it with one that has a different value. The variations might also be caused by temperature changes. Ultimately, however, the cause is unimportant; the error must be corrected.

Because the adjustable voltage source is connected to oppose the main power-supply voltage, the effective voltage in the circuit is

$$V_{eff} = V_S - V_{adj}$$
$$= 10\,\text{V} - 5\,\text{V} = 5\,\text{V}$$

FIGURE 4-25 **Open-loop Control Circuit (No Feedback)**

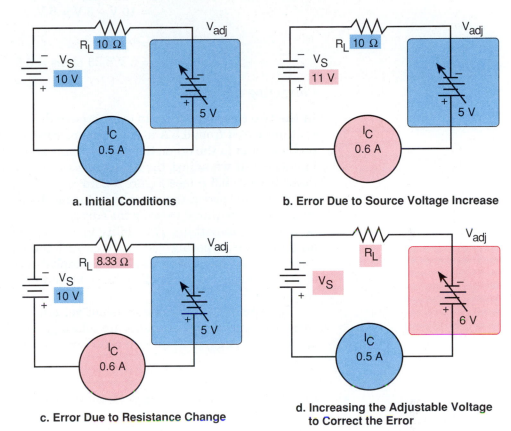

a. Initial Conditions

b. Error Due to Source Voltage Increase

c. Error Due to Resistance Change

d. Increasing the Adjustable Voltage to Correct the Error

With an effective supply voltage of 5 volts and a resistance of 10 ohms, the current is 0.5 A ($I = V_{eff}/R$). These are the conditions for the correct (no error) current.

If the Main Power-supply Voltage Changes

If the main power-supply voltage increases to 11 volts, as in Figure 4-25(b), the current must increase to 0.6 A:

$$V_{eff} = V_S - V_{adj}$$
$$= 11\ V - 5\ V = 6\ V$$

$$I = V_{eff}/R$$
$$= 6\ V/10\ \Omega = 0.6\ A$$

This results in an error, since the required value is 0.5 A.

If the Resistance Value Changes

If the resistance value changes to 8.33 ohms, as in Figure 4-25(c), it produces exactly the same amount of error:

$$V_{eff} = V_S - V_{adj}$$
$$= 10 \text{ V} - 5 \text{ V} = 5 \text{ V}$$

$$I = V_{eff}/R$$
$$= 5 \text{ V}/8.33 \text{ ohms} = 0.6 \text{ A}$$

Correcting the Error

In the two cases in which an error occurred, in our ongoing example, the current increased from 0.5 A to 0.6 A. In one case, the resistor varied to cause the problem; and in the other, the power-supply voltage varied to cause the change in current. If we adjust the adjustable voltage upward to 6 volts, the current drops back to 0.5 A (see Figure 4-25(d)).

Thus, we correct the error by reducing the effective voltage (V_{eff}) to whatever value is required to bring the current back down to 0.5 A. The actual numbers are not important. The ability to correct any error is. This example uses an adjustable opposing voltage to correct the error. Alternatively, a variable resistor could have been used to accomplish the same result, but nearly all electronic negative-feedback circuits depend on an opposing voltage to do the correcting.

The circuit in this example is not yet a negative-feedback circuit, because some of the necessary elements are missing. The circuit has a controlled quantity (the 0.5-A current) and an error corrector (the adjustable voltage), but it

FIGURE 4-26   Negative Feedback Circuit in Which You Are a Critical Part of the Loop

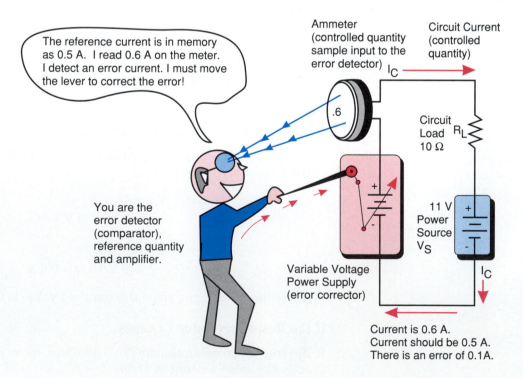

has no reference quantity, no amplifier, and no error detector. When any of these elements is missing, the circuit is said to have an open feedback loop.

Figure 4-26 closes the loop—in a somewhat impractical way—illustrating the elements we need to add to close the negative-feedback loop.

4.5   Applying Negative Feedback to a Transistor Circuit

We must hold the quiescent collector current steady in spite of variations in the elusive beta. And now we will see just how to accomplish that, very simply, using negative feedback.

Outlining the Resistance Perspective

The transistor is a current amplifier. An input (base) current appears in the output (collector) circuit as a current that is beta times larger than the input (base) current. If the transistor has a beta (current multiplier) value of 50 and a base (input) current of 1 mA, it will produce a collector current of 50 mA (1 mA × ß).

Beta is also known as the current gain of the transistor; but remember that the term *gain* is always used to mean multiplication—never addition, as it is often interpreted in ordinary conversation.

Earlier in this chapter, we went through a Kirchhoff loop analysis to prove that the input resistance of the transistor is always beta times the junction resistance:

$$R_{in} = ß \times R_e'$$

It appears that the transistor is not only a current multiplier, but also a resistance multiplier at the same time.

In case the loop equation did not give you a true feeling about what happens when an unexpectedly high input resistance is encountered, let's look at why the input resistance is multiplied by ß in a slightly different way.

The transistor is a current multiplier that multiplies the base current times ß to produce the collector–emitter current. Figure 4-27(a) shows a transistor model we will call a **current-multiplier block.** This particular current-multiplier block has a multiplying factor of 100, corresponding to a beta value of 100 (if this were a transistor). If we input a current of 0.01 mA to the current multiplier, we get 100 times that much current out of the current multiplier block:

$$\text{Current Out} = 0.01 \text{ mA} \times 100 = 1.0 \text{ mA}$$

Simple Illustration Using Ohm's Law

Figure 4-27(b) consists of a 1.0-volt battery connected to a 1-kilohm resistor. Ohm's law ($I = V/R$) indicates that 1.0 mA is flowing through the 1-kilohm resistor, and that the battery is delivering that same 1.0 mA of current to the circuit. In addition, there is (as there must be) a 1.0-volt drop across the 1-kilohm resistor. Ohm's law doesn't get much more basic than this.

FIGURE 4-27   **Current Multiplier Block**

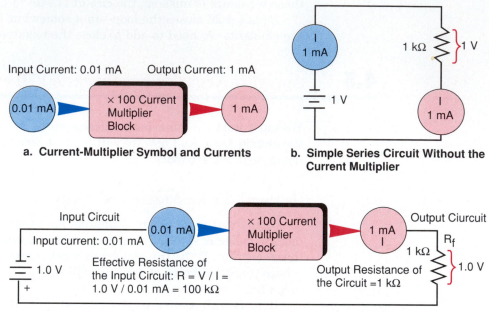

a. **Current-Multiplier Symbol and Currents**

b. **Simple Series Circuit Without the Current Multiplier**

c. **Adding the Times-100 Current Multiplier**

Installing the Current-multiplier Block

In Figure 4-27(c), we have inserted the current-multiplier block into the circuit. A 1.0-mA current still flows through the 1-kilohm resistor. The difference now is that the input battery must deliver only 0.01 mA to the current-multiplier block to cause a 1.0-mA current to flow through the 1-kilohm resistor. If we again apply Ohm's law, this time using the battery voltage and the current it is delivering to the circuit, we get

$$R = V/I = 1.0 \text{ V}/0.01 \text{ mA} = 100 \text{ kilohms}$$

Because of the current-multiplier block, the battery "sees" a resistance of 100 kilohms, even though the real physical value of the sole resistance in the circuit is only 1 kilohm. The current-multiplier block proves to be a resistance-multiplier block as well: the 1-kilohm resistance has also been multiplied by 100. This is exactly what happens in a transistor when the emitter resistor value is multiplied by the transistor's current gain, ß.

If you go back to the Kirchhoff's loop analysis in Example 4.2, you will see that the current-multiplier example here is just another way of approaching the loop analysis. In fact, they say exactly the same thing about this important concept in two slightly different ways.

Replacing the Current Multiplier with a Transistor

In Figure 4-28(a), we have replaced the current-multiplier block with a real current multiplier—a transistor. Notice that the battery voltage has been increased to 1.7 volts. This adjustment was necessary because a 0.6- to 0.7-volt drop always occurs across the forward-bias base–emitter junction. Here, we

FIGURE 4-28 Negative Feedback Corrects for Beta Variations

a. Initial Conditions
Beta (β) = 100
Emitter current is OK.
Calculate
1. Input Resistance: $R_{in} = \beta R_f =$
   $100 \times 1\ k\Omega = 100\ k\Omega$
2. Input (Base) Current:
   $I_B = 1.0\ V / 100\ k\Omega = 0.01\ mA$
3. Emitter Current: $I_E = \beta \times I_B =$
   $100 \times 0.01\ mA = 1.0\ mA$
4. Feedback Resistor Voltage:
   $V_{Rf} = I_E \times R_f = 1\ mA \times 1\ k\Omega = 1.0\ V$

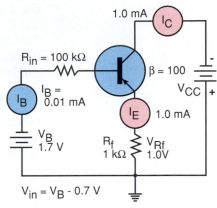

a. Initial Conditions

b. Error
Beta Increases
Beta (β) = 110
Emitter Current is:
1.1 mA before correction. Should be
1.0 mA.
Calculate
1. Input Resistance: $R_{in} = \beta R_f =$
   $100 \times 1\ k\Omega = 100\ k\Omega$
2. Input (Base) Current:
   $I_B = 1.0\ V / 100\ k\Omega = 0.01\ mA$
3. Emitter Current: $I_E = \beta \times I_B =$
   $110 \times 0.01\ mA = 1.1\ mA$
4. Feedback Resistor Voltage:
   $V_{Rf} = I_E \times R_f = 1.1\ mA \times 1\ k\Omega = 1.1\ V$

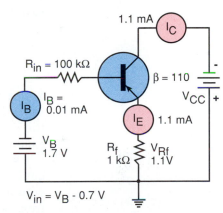

b. Beta Increases-Error

c. Error Corrected
Input resistance *increases* as a result of
the higher emitter current.
Emitter current is OK (corrected).
Beta (β) = 110
Calculate
1. Input Resistance: $R_{in} = \beta R_f =$
   $110 \times 1\ k\Omega = 110\ k\Omega$
2. Input (Base) Current:
   $I_B = 1.0\ V / 110\ k\Omega = 0.00909\ mA$
3. Emitter Current: $I_E = \beta \times I_B =$
   $110 \times 0.00909\ mA = 1.0\ mA$
4. Feedback Resistor Voltage:
   $V_{Rf} = I_E \times R_f = 1\ mA \times 1\ k\Omega = 1.0\ V$

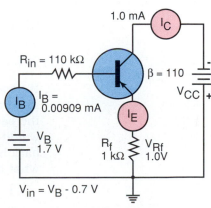

c. After Correction

assumed it to be 0.7 V. Now that we have added the 0.7 volt (and immediately used it up as junction voltage drop), let's forget about it and focus instead on the 1 volt we actually have left to work with.

Figure 4-28(b) ignores the fact that the emitter current includes both the 0.01-mA base current and the much larger 1.0-mA collector current. Accordingly, we will consider only the collector current, and assume that the collector and emitter currents are the same. The resulting error is trivial, and the shortcut will make our computations much easier.

We will assume that the feedback resistor (R_f) is so much larger than the junction resistance (R_e') that it is safe to ignore (R_e'). This is normally a valid assumption.

The battery voltage is 1 volt, and 1.0 mA of current is flowing through the emitter resistor, which we will designate as R_f because it is the feedback resistor. With 1 mA flowing through the 1-kilohm R_f, we get a voltage drop of 1 volt. The transistor (current multiplier) has a beta value of 100, so the input (base) current is 0.01 mA, as it was in the previous example. Again, using the base voltage and the base current to calculate the input resistance, we get an input resistance (looking into the base) of 100 kilohms:

$$R_{in} = V_B/I_B = 1.0 \text{ V}/0.01 \text{ mA} = 100 \text{ kilohms}$$

Negative Feedback at Work—From a Resistance Perspective

So far, we have used a Q-point current value of 1.0 mA; so let 1.0 mA be our controlled quantity, the desired Q-point current. We want the feedback to hold the collector current at a constant 1.0 mA, no matter what other factors may come to bear on the collector current.

We have used a value for beta of 100 in all of the previous examples, but beta is not to be trusted: we expect it to change without notice. And of course, beta is the prime factor we can expect to cause the Q-point current to deviate from our required 1.0-mA quiescent value.

Let's suppose that Beta does change—from 100 to 110, as illustrated in Figure 4-28(b). But if beta rises from 100 to 110, the Q-point current must also rise, because the collector current is the base current times ß. (Assume that the base current remains what it was before, 0.01 mA.) Then we have

$$QI_C = ß \times I_B$$
$$= 110 \times 0.01 \text{ mA} = 1.1 \text{ mA}$$

It appears that our controlled quantity (the quiescent collector current) is now in error because ß has increased to 110—perhaps because of a temperature increase. If we could actually measure the new collector current, however, we would find that it is still the required 1.0 mA; it didn't change. Negative feedback has already made the correction, but we missed seeing it happen. Here is what occurred:

1   The value of beta increased from 100 to 110.

2   This drove the quiescent collector current up from 1.0 mA to 1.1 mA.

3   The input resistance increased when beta went up, because $R_{in} = ß \times R_f$. The new input resistance is

$$R_{in} = 110 \text{ [the new ß]} \times 1 \text{ k}\Omega = 110 \text{ k}\Omega$$

4   We did not change the base-bias battery voltage; it is still 1.0 volt. But if we now recalculate the base (input) current, using the new input resistance of 110 kilohms, we get the new input (base) current:

$$I_B = V_B/R_{in}$$
$$= 1.0 \text{ V}/110 \text{ kilohms} = 0.00909 \text{ mA}$$

5   Because beta really did increase to 110, we must use that figure to calculate the new quiescent collector current, in conjunction with the new base current:

$$\text{Quiescent Collector Current} = \text{Beta} \times \text{Base Current}$$
$$QI_C = \text{ß} \times I_B$$
$$= 110 \times 0.00909 \text{ mA} = 1.0 \text{ mA}$$

The value of beta changed, and this would have altered our controlled quantity, the quiescent collector current, but the increase in input resistance instantly compensated for the change in beta. The circuit returned to its original, required quiescent collector current of 1.0 mA. You can try this example with even larger changes in ß, if you like. The results will be the same.

By adding one negative-feedback resistor (R_f) to the circuit, we have freed ourselves from the dictatorship of the transistor's unreliable beta value. We no longer care whether beta changes with temperature. Just as important, we can replace a defective transistor with one that has a very different beta value, confident that the circuit will work properly with the new transistor!

Negative Feedback at Work—From a Voltage Perspective

We have just examined how the input resistance increase, which results from an increase in beta, corrects the error in the Q-point that results from that same beta increase. Here, we will look at the same action from the perspective of the increased opposing voltage drop across the feedback resistor (R_f).

We have seen mystery-box examples in which the circuit current was reduced by something in the mystery box. The mystery box can contain either a resistor or an opposing voltage; and unless we can take the circuit apart to measure the mystery box by itself (or peek inside) we have no way to determine whether a resistor or an opposing voltage has caused the current reduction. If the mystery box contains a resistor in a series circuit, the result is an increase in total circuit resistance and a consequent decrease in current. If the mystery box contains a series opposing voltage, the result is a decrease in circuit current; and Ohm's law requires that there be an increase in total circuit resistance.

If we know that there is an opposing voltage in the circuit, we have the option of examining the circuit from the perspective of that opposing voltage or from the perspective of the apparent increase in resistance it causes. Either perspective is equally valid, and sometimes one or the other is more useful or convenient. In any case we must select one perspective or the other; we can't view the situation from more than one perspective at a time.

In the case of the transistor circuit in Figure 4-28, we have examined the negative-feedback action from the input resistance perspective, but we know that the voltage drop across the feedback resistor (R_f) is actually an opposing voltage. In fact, the opposing voltage drop across the feedback resistor (R_f)

caused the apparent increase in input resistance. Now let's take a look at the negative feedback action from the opposing voltage perspective.

Holding a Constant Q-point Despite Variations in Beta (A Voltage Perspective Analysis)

Refer to Figure 4-29 as we go through the analysis. This time, let's use some different initial values for ß and for the amount of change in ß. Multiples of 10 make for easy arithmetic, but they sometimes seem to prove that something is true when it really isn't. Multiples of 10 can sometimes be just a little too slick. Consequently, let's adopt the following initial conditions:

Beta = 600

Q-point Collector Current = 2.0 mA

Base Voltage (V_B) = 1.0 V

(Note: The 0.7-volt junction voltage and its source voltage will be ignored.)

1   Calculate the voltage drop across the feedback resistor R_f:

$$V_{Rf} = QI_C \times R_f$$
$$= 2.0 \text{ mA} \times 500 \text{ } \Omega$$
$$= 1.0 \text{ V}$$

2   Calculate the input current:

$$I_B = QI_C/\text{ß}$$
$$= 2.0 \text{ mA}/600$$
$$= 0.0033 \text{ mA}$$

3   Calculate the input resistance (R_{in}):

$$R_{in} = V_B/I_B$$
$$= 1.0 \text{ V}/0.0033 \text{ mA}$$
$$= 303 \text{ k}\Omega$$

4   Assume that beta increases from 600 to 700, to create a Q-point error; and assume that the input current $(I_B = 0.0033 \text{ mA})$ hasn't yet changed. Calculate the new value of QI_C:

$$QI_C = \text{ß} \times I_B$$
$$= 700 \times 0.0033 \text{ mA}$$
$$= 2.31 \text{ mA} \quad \text{(new value)}$$

QI_C has increased to produce an error of about 13%.

5   Now let's see how the Q-point error gets corrected. Calculate the new voltage drop across the feedback resistor, taking into account the increase in QI_C:

$$V_{Rf} = QI_C \times R_f$$
$$= 2.31 \text{ mA} \times 500 \text{ } \Omega$$
$$= 1.155 \text{ V}$$

FIGURE 4-29 Negative Feedback at Work, from a Voltage Perspective: Holding a Constant *Q*-Point in Spite of Variations in Beta

a. Initial Conditions
Beta (β) = 600
Q-Point Collector Current = 2.0 mA
Base Voltage (V_B) = 1.0 V
1. Calculate the voltage drop across R_f:
$V_{Rf} = QI_C \times R_f = 2.0 \text{ mA} \times 500 \text{ }\Omega =$
1.0 V
2. Calculate the Input Current (I_B):
$I_B = QI_C / \beta = 2.0 \text{ mA} / 600 =$
= 0.0033 mA
3. Calculate the input Resistance (R_{in}):
$R_{in} = V_B / I_B = 1.0 \text{ V} / 0.0033 \text{ mA} =$
303 kΩ

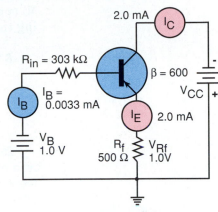

a. Initial Conditions

b. Error-Beta increases from 600 to 700
Beta (β) = 700 (producing a Q-point error
1. Calculate the new value of QI_C:
$QI_C = \beta \times I_B = 700 \times 0.0033 \text{ mA} =$
2.31 mA
2. QI_C has increased for an error of about 13%.

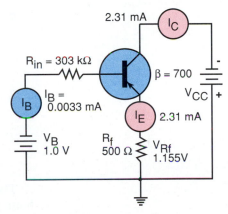

b. Beta Increases-Error

c. How the Q-point Error gets corrected
1. Calculate the new feedback resistor voltage drop:
$V_{Rf} = QI_C \times R_f = 2.31 \text{ mA} \times 500 \text{ }\Omega =$
1.155 V
2. Calculate the effective base bias voltage:
$V_{B (eff)} = V_B - 0.155 \text{ V} =$
1.0 V - 0.155 V = 0.845 V
3. Calculate the new effective base bias current using the new effective base voltage: $I_B = V_{B (eff)} / R_{in} =$
0.845 V / 303 kΩ = 0.0028 mA
4. Calculate the corrected Q-point Current: $QI_C = \beta \times I_B =$
700 × 0.0028 mA = 1.96 mA

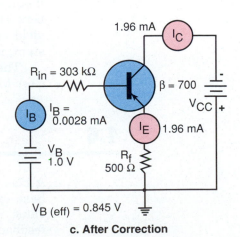

c. After Correction

The voltage drop across the feedback resistor (R_f) has increased as a result of the increased collector current (QI_C), which was caused in turn by the increase in ß from 600 to 700. This additional negative-feedback voltage (0.155 V) opposes the base voltage, canceling part of the base input (bias) and reducing the effective base-bias battery voltage (V_B) by 0.155 volt.

6   Calculate the new effective base (bias) voltage:

$$V_{B(eff)} = V_B - 0.155 \text{ V}$$
$$= 1.0 \text{ V} - 0.155 \text{ V}$$
$$= 0.845 \text{ V}$$

In the previous input resistance–based example, we took the resistance into account and ignored the effect of the increased voltage drop across the feedback resistor. Now we will ignore the input resistance and examine the effect of the negative feedback on the feedback resistor voltage drop. We must select one factor or the other, because they are interdependent variables. We will assume that the input resistance did not change when ß increased.

7   Calculate the base current (I_B), using the new effective base voltage:

$$I_B = V_{B(eff)}/R_{in} = 0.0845 \text{ V}/303 \text{ kilohms}$$
$$= 0.0028 \text{ mA}$$

8   Calculate the corrected Q-point current (QI_C):

$$QI_C = ß \times I_B$$
$$= 700 \times 0.0028 \text{ mA}$$
$$= 1.96 \text{ mA}   \text{(corrected value)}$$

The calculated, corrected value for the Q-point collector current is within about 2% of the original 2.0 mA. In real negative-feedback circuits, we don't expect a perfect correction; but in this example, the difference is largely due to mathematical error associated with rounding off or truncating various numbers. The important conclusion is that both resistance and voltage perspectives yield substantially the same results. The negative feedback holds QI_C constant; it does its job, from whichever perspective we care to view it.

The final corrected value is always slightly imperfect, but it can be very close to the original controlled quantity value (QI_C in this case).

4.6   Using Voltage Divider Bias with Feedback

A bias circuit of some sort is necessary to establish the proper Q-point. We have examined how negative feedback keeps the Q-point at its established value, in spite of changes in beta and other variations. Our initial focus on feedback reflects the fact that the values of the emitter resistor (R_f) and the value of the base-to-ground resistor (which is nearly always a part of any practical bias circuit) are dictated by feedback requirements. Once the values for the base and emitter resistors have been calculated to meet specific feedback requirements, they become part of the bias circuit.

The circuit designer begins by selecting a quiescent collector current that will satisfy the requirements of the working load. The designer's next task is to calculate the required amount of negative feedback and the values for the emitter-feedback resistor and the base-to-ground resistor. Only then does the designer calculate values for the remaining components of the bias circuit.

In the examples we have been studying, we used a 1-volt battery to provide the base-bias current required to establish the desired Q-point collector–emitter current. The battery accomplished its purpose as far as we went, but it is simply not a practical voltage source for supplying bias current. The nearly 0-ohm internal resistance of the battery would short any signal we tried to amplify directly to ground. As a result, none of the signal current, music, or whatever, would get to the transistor input circuit to be amplified. We must have some reasonable resistance value between base and ground, or we find ourselves stuck with a stable but useless circuit.

Implications of Adding a Base-to-Ground Resistor

In previous feedback discussions, where the circuits were powered by a 1-volt bias battery, we realized very nearly perfect corrections to any change in the Q-point when beta varied. The basic nature of negative feedback is to correct any error in the controlled quantity (QI_C), without regard to what caused the error in the first place. The nearly perfect (100%) feedback we encountered earlier allowed almost no change in the quiescent collector current when beta varied.

Unfortunately, that same nearly perfect feedback loop would also prevent a music input signal's base current from changing the output collector current. The feedback system would see our music signal currents as just another error and would correct the error, leaving us with no music signal output from the collector circuit.

Relationship Between Q-point and Signal Currents

The quiescent collector current is a resting or idling current. When there is no input signal current, the collector current sits at the Q-point value. When an input signal current is applied to the base, the output current follows the changes in base current to produce a larger-current replica of the input signal current's variations. This is called *current gain;* and in a real circuit, the circuit current gain is something less than the beta value of the transistor. If we apply a sine-wave signal current to the base, the action is as follows (see Figure 4-30):

1   The sine-wave signal pushes the collector current up above the Q-point value on the positive half-cycle.

2   As the sine wave passes through zero, it passes through the Q-point current value.

3   The sine wave drives the collector current below the Q-point current value on the negative half of the cycle.

FIGURE 4-30　**Establishing a Q-point Current**

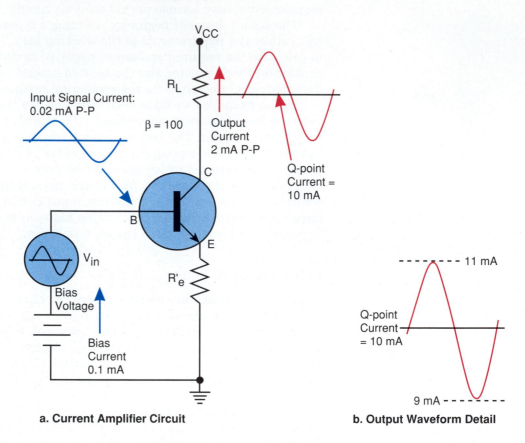

Input Signal Current:
0.02 mA P-P

$\beta = 100$

Output
Current
2 mA P-P

Q-point
Current =
10 mA

R_L

C

B

E

R'_e

V_{in}

Bias
Voltage

Bias
Current
0.1 mA

V_{CC}

11 mA

Q-point
Current
= 10 mA

9 mA

a. Current Amplifier Circuit

b. Output Waveform Detail

4　The *Q*-point current value only exists when no signal input current is present, or when the input signal waveform is passing through zero. When there is no input signal, the collector current always returns to the *Q*-point value.

We must have a stable *Q*-point, but at the same time we must also allow the collector current to vary with a signal applied to the base of the transistor. Since the two conditions tend to be mutually exclusive, we must strike some compromise to allow both a reasonably stable *Q*-point and some latitude for the output (collector current) to vary with a signal.

In reality, we don't need perfect *Q*-point stability. We can tolerate up to about a 10% variation in the *Q*-point without getting into any significant trouble. This means we can add a base-to-ground resistor to permit some signal current to enter the base circuit to get amplified. The base-to-ground resistor shunts some of the signal to ground. R_{B_2}, the base-to-ground resistor, completes the current-mode feedback circuit path, so it is an important part of the feedback loop.

The following figures are approximate but close enough for practical applications. The feedback resolution may now be as much as 10%; that is, we could still have as much as a 10% error in the *Q*-point after correction. On a positive note, we still have about 90% feedback—enough for most situations. On a neg-

FIGURE 4-31 Adding a Base-to-Ground Resistor

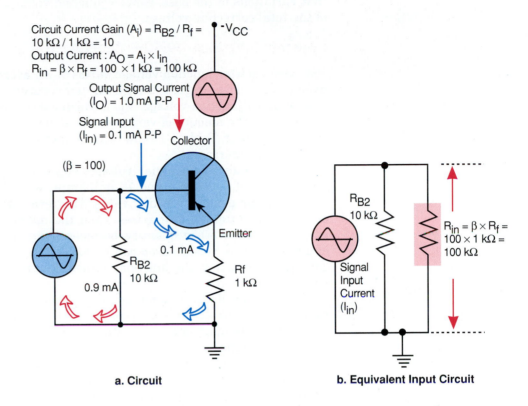

a. Circuit   b. Equivalent Input Circuit

ative note, the 90% feedback leaves only 10% for signal variations. Thus, 90% of our input signal will be shunted to ground through the base-to-ground resistor, and only about 10% of the signal current will enter the base, to appear as a beta-times-larger collector current.

Figure 4-31(a) illustrates the compromise situation. Fortunately, as we will see a little later, the loss of so much of our input signal current through the base-to-ground resistor is not as serious as it might seem now.

We can lower the value of the base-to-ground resistor for the sake of greater Q-point stability (closer to 100% feedback), but at the expense of more signal current loss. Fortunately for us, some practical rules for dealing with this compromise have been established and are almost universally used by circuit designers. Figure 4-32 defines these rules. If we know these rules, we are in a good position to troubleshoot transistor feedback and bias circuits, because we know what to expect.

The rule concerning the minimum voltage drop across the feedback resistor is based on the fact that the base–emitter junction voltage also varies with temperature at an average rate of about 2.5 mV per Celsius degree. Any voltage drop across the feedback resistor of less than 1 volt prevents the feedback loop from adequately compensating for variations in the base–emitter junction voltage. The problem is that the base–emitter junction voltage is in series with the feedback resistor and acts as part of the correction voltage. The last thing we need is to have part of the feedback correction voltage be unstable!

The cure is to make the voltage drop across the feedback resistor so large that variations in the base–emitter junction voltage become a very small part of the total correction voltage.

Swamping the Base–Emitter Junction Voltage Variations

The base–emitter junction voltage varies at an average rate of about 2.5 millivolts per Celsius degree. Increasing temperature causes a decrease in the base–emitter junction voltage (V_{BE}), which causes the quiescent collector current to rise. The amount of variation also depends on the value of the Q-point collector current. Table 4-1 shows the temperature variations in V_{BE} at various operating currents.

Because most small-signal (voltage) amplifiers operate at quiescent collector currents of between 1 and 10 milliamps, we will use 2.5 mV per Celsius degree as our general working value. The term **swamping** is often used to describe the situation that arises when the magnitude of one quantity is so great that some other quantity becomes insignificant. Adding an external emitter-feedback voltage drop can be viewed as swamping the temperature-caused variations in the junction voltage. Figure 4-32 provides basic design rules in this regard.

Because feedback is dynamic and always busy making corrections, let's take a closer look at the swamping concept with feedback. Figure 4-33 illustrates the swamping concept. The drawing in Figure 4-33(a) includes an unstable battery, representing the unstable base–emitter junction voltage. The circuit current is 1.0 mA, with a battery-supply voltage of 0.1 volt.

Now suppose that the unstable battery voltage increases to 0.2 volt. By Ohm's law, we would expect the circuit current to double to a value of 2.0 mA.

If our only interest is in maintaining a constant 1.0 mA of circuit current, in spite of the existence of the unstable battery, we can proceed as follows.

In Figure 4-33(b), a stable voltage source has been added that is much larger than the unstable voltage. Some extra resistance has also been added, to get the desired 1.0-mA circuit current.

Now let the unstable battery again increase from 0.1 volt to 0.2 volt. Ohm's law will tell us how much the current in our new (swamping) circuit will change from the desired 1.0 mA:

$$
\begin{aligned}
I &= V/R \text{ (Ohm's law)} \\
&= 1.2 \text{ V}/1100 \ \Omega \qquad \text{(our swamping circuit)} \\
&= 1.09 \text{ mA} \qquad \text{(a change of only 0.09 mA)}
\end{aligned}
$$

TABLE 4.1
Base–Emitter Junction Voltage Variations

QI_C	ΔV_{be}
0–0.1 mA	−3.0 mV/C°
0.1–1.0 mA	−2.75 mV/C°
1.0–10.1 mA	−2.5 mV/C°
10.0–100 mA	−2.0 mV/C°

FIGURE 4-32 Basic Circuit Design Rules

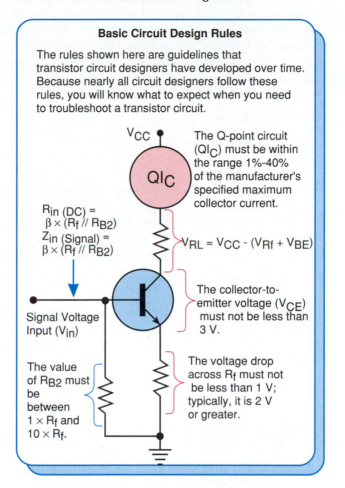

Basic Circuit Design Rules

The rules shown here are guidelines that transistor circuit designers have developed over time. Because nearly all circuit designers follow these rules, you will know what to expect when you need to troubleshoot a transistor circuit.

V_{CC}

QI_C

The Q-point circuit (QI_C) must be within the range 1%-40% of the manufacturer's specified maximum collector current.

R_{in} (DC) = $\beta \times (R_f // R_{B2})$
Z_{in} (Signal) = $\beta \times (R_f // R_{B2})$

$V_{RL} = V_{CC} - (V_{Rf} + V_{BE})$

Signal Voltage Input (V_{in})

The collector-to-emitter voltage (V_{CE}) must not be less than 3 V.

The value of R_{B2} must be between $1 \times R_f$ and $10 \times R_f$.

The voltage drop across R_f must not be less than 1 V; typically, it is 2 V or greater.

FIGURE 4-33
Swamping Example

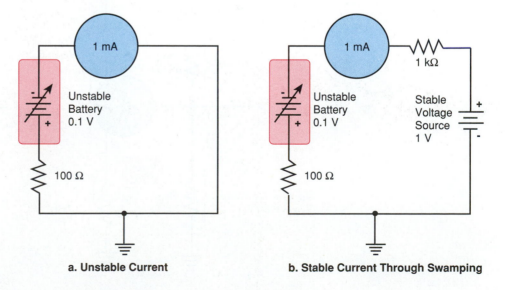

1 mA

Unstable Battery 0.1 V

100 Ω

a. Unstable Current

1 mA

1 kΩ

Unstable Battery 0.1 V

100 Ω

Stable Voltage Source 1 V

b. Stable Current Through Swamping

Without the added swamping voltage source (and resistor), a 100% increase in the unstable battery voltage (actually, V_{BE}), yielded a 100% change in the total circuit current. With the stable voltage source added, a 100% change in the unstable voltage (V_{BE}) resulted in only a 9% change in the total circuit current. The unstable voltage behavior was swamped by the stable voltage source.

Stability Factor S

The **stability factor S** is a figure of merit used to define just how stable the Q-point is expected to be. The stability factor is often employed in parametric design and analysis procedures. Even though we will be using beta-independent procedures, you should know what it means, because it often shows up in the literature.

In the parametric realm, S is a very complex quantity, with an equation to match. However, when all of the unacceptable stability conditions are eliminated by the use of negative feedback, the equation reduces to

$$S = R_{B_2}/R_f$$

This equation is basically the same as the equation for a transistor circuit's current gain, and that is not a coincidence. The stability factor places some practical limits on circuit current gain, depending on the degree of stability required for a given application.

Useful stability factors range from 1 to 10, with 1 being best. Mathematically, it is possible to come up with stability factors that are less than 1, but values less than 1 offer no improvement over a factor of 1. Any S value larger than 10 will not be sufficiently stable. The only exception to this rule involves a special transistor pair, called a **Darlington pair,** which is a

FIGURE 4-34   Adding the Final Bias Resistor R_{B1}: Circuit Voltage Distribution

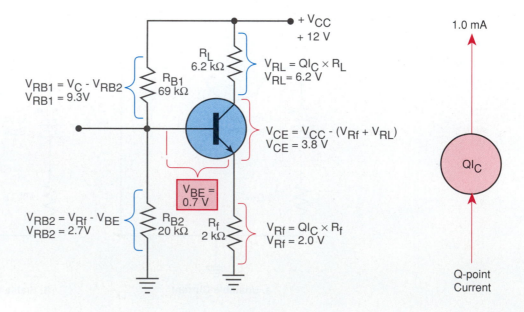

composite device made of two direct-coupled transistors. Each of the transistors in the pair has its own stability factor, and, thus, it may appear that the device's stability factor number is too high. We will study the Darlington super-beta transistor in a later chapter.

Adding the Final Bias Resistor (R_{B1})

All that remains for us to do to complete the voltage-divider bias circuit is to get the desired base voltage to establish the required bias current to set the Q-point. Figure 4-34 shows the added bias resistor R_{B1} and the voltage distributions in a complete circuit. With a Q-point collector current of 1 mA, a 2-volt drop occurs across the feedback resistor. Adding the 0.7-V junction voltage to the feedback resistor voltage drop, we find that the voltage across resistor R_{B2} is 2.7 volts. We have a power supply voltage of 12 volts, so the voltage drop across R_{B1} must be what is left of that 12 V after we subtract the voltage drop of R_{B2}. Because of the designer's rule stating that R_{B2} must fall into the range from $1 \times R_f$ to $10 \times R_f$, and because beta does exist, we can normally treat the voltage divider (R_{B1}, R_{B2}) as an unloaded voltage divider.

4.7   Signal Parameters in the Common Emitter Voltage-divider Biased Circuit

The following section discusses the most important signal parameters of the base-biased, common emitter circuit. Other signal parameters exist, but they are unimportant, or not within our control, or both.

Input Resistance

As Figure 4-34 indicates, R_{B_1} (the bias resistor) is now part of the input circuit. Less obvious is the fact that R_{B_1} and R_{B_2} are effectively in parallel from the point of view of input resistance. This is true because the top of R_{B_1} is connected to ground through the battery (or power supply). The internal resistance of the battery (or power supply) must be very low for it to serve as an acceptable power source. If we connect a 0.1-ohm internal power-supply resistance from the V_{CC} point to ground, the top end of R_{B_1} is effectively grounded, placing R_{B_1} and R_{B_2} in parallel. Because we are now working with a system that is functionally independent of beta and because of the design rules in place, we can usually ignore the input resistance of the transistor itself. The input resistance of the complete circuit is

$$R_{in} \; (Z_{in}) = (R_{B_1} \times R_{B_2})/(R_{B_1} + R_{B_2}) \qquad \text{(resistances in parallel)}$$

Voltage Gain

So far, most of our discussion has revolved around current amplification (gain) because we have been concentrating on collector current (Q-point) stabilization; but any transistor amplifier can yield current gain, voltage gain, or power

FIGURE 4-35   **Voltage Gain Model**

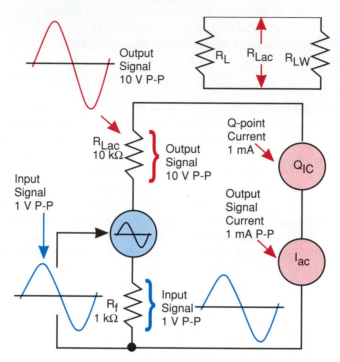

gain. Normally, all three kinds of gain are available to some extent, but we are usually only interested in voltage gain or current gain or the product of the two (power gain). The characteristics of the working load (R_{LW}) determine which kind of gain we need and influence how much gain we can get from a given circuit.

In our earlier discussion of the design rules, we noted that we must never set the Q-point at more than 40% of the manufacturer's specified maximum-rated collector current. For amplifiers that we want to provide current or power amplification, the near 40% figure might be appropriate; but when we want voltage gain, a figure of 1% or 2% of maximum would be more appropriate. We almost always find such low values for the Q-point current when the amplifier is intended to work as a voltage amplifier. The circuit model in Figure 4-35 illustrates how voltage gain is obtained.

Why Voltage Gain Occurs in a Common Emitter Circuit

We can understand the occurrence of voltage gain by breaking the process down into five steps:

1   Assume a Q-point current of 1 mA, and assume that the input (**ac**) signal is causing the collector current to swing 0.5 mA above and 0.5 mA below the 1-mA Q-point. The total signal collector current is 1 mA peak-to-peak (*P–P*).

2   With a 1-mA *P–P* current, the signal voltage across the 1-kΩ feedback resistor (R_f) will be 1 V *P–P* (Ohm's law).

3   The same 1-mA *P–P* signal current flows through the 10-kΩ collector load and produces a signal voltage drop across R_{Lac} of 10 V *P–P*. R_{Lac}

is the working load (R_{LW}) in parallel with the dummy load (R_L). Both resistances will be present in real circuits. We have shown only the combined result, R_{Lac}, in the simplified circuit in Figure 4-35.

4   Ten times more signal voltage is developed across the collector load (R_{Lac}) than is developed across the feedback resistor (R_f), because the resistance ratio is 10 to 1. Remember, we assume that the collector current is equal to the emitter current. If 1 V *P–P* of signal voltage is developed across the feedback resistor (R_f), the input (base–emitter) signal voltage must also be 1 V *P–P*.

5   *Voltage gain* is defined as the output signal voltage divided by the input signal voltage:

$$A_V = V_O/V_{in}$$

From the model, you can see that the *voltage gain* may also be defined in terms of the resistance ratio, R_{Lac}/R_f:

$$A_V = R_{Lac}/R_f$$

Where R_{Lac} is the total **ac** signal load, consisting of the working load (R_{LW}) in parallel with the Dummy load (R_L).

The 0.7-volt base–emitter junction voltage does not get into the act because on half of the signal cycle it adds 0.7 volt to the signal voltage, and on the other half it subtracts 0.7 volt from it. The result is a 0.7-volt shift in the signal's zero reference point, but no change in the *P–P* voltage across R_f.

Now, let's examine the preceding explanation from a mathematical standpoint (see Figure 4-36).

FIGURE 4-36   Circuit for the Voltage Gain Derivation

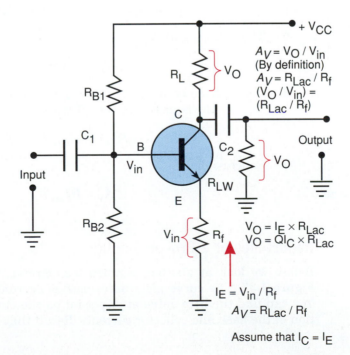

$A_V = V_O / V_{in}$
(By definition)
$A_V = R_{Lac} / R_f$
$(V_O / V_{in}) = (R_{Lac} / R_f)$

$V_O = I_E \times R_{Lac}$
$V_O = QI_C \times R_{Lac}$

$I_E = V_{in} / R_f$
$A_V = R_{Lac} / R_f$

Assume that $I_C = I_E$

Example 4.7

Voltage Gain Equals Resistance Ratios

Prove that voltage gain (A_V) is equal to the collector load resistance (R_{Lac}) divided by the feedback resistance (R_f):

$$A_V = R_{Lac}/R_f$$

Proof

In our proof, we will assume that $I_C = I_E$ (Collector Current = Emitter Current). Now we can proceed as follows:

1 The emitter current is

$$I_E = V_{in}/R_f \qquad \text{(Ohm's law)}$$

2 The output voltage is

$$V_O = QI_C \times R_L \qquad \text{(Ohm's law)}$$

because

$$I_E = I_C \qquad \text{(primary assumption)}$$

3 The output voltage is

$$V_O = I_E \times R_{Lac} \qquad \text{(substitution)}$$

4 Since the emitter current is

$$I_E = V_{in}/R_f \qquad \text{(Step 1)}$$

the output voltage is

$$V_O = (V_{in}/R_f)\, R_{Lac} \qquad \text{(substitution)}$$

5 The voltage gain is

$$A_V = V_O/V_{in} \qquad \text{(by definition)}$$

6 If we divide both sides of the V_O equation in Step 4 by V_{in}, and then refer to the equation in Step 5, we get

$$(V_O/V_{in}) = (R_{Lac}/R_f) = A_V$$

7 Conclusion:

$$A_V = R_{Lac}/R_f$$

Importance of a Proper *Q*-point

Before we look at another common bias circuit, let's examine the curve in Figure 4-37. The curve plots base–emitter current against collector current. You can find the beta value at any point on the curve by finding the intersection of the base and collector currents. Recall that beta varies somewhat with

FIGURE 4-37   *Q*-point, Cut-off, and Saturation

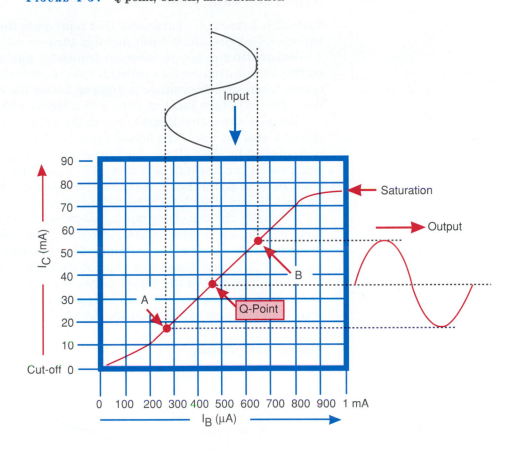

collector current. At point A on the graph, IC = 18 mA, with a base current of about 275 microamps (0.275 mA). The beta value at point A is

$$ß = I_C/I_B$$
$$= 18 \text{ mA}/0.275 \text{ mA}$$
$$= 65.45$$

Thus, beta is = 65.45 at point A.

Similarly, the beta value at point B is

$$ß = I_C/I_B$$
$$= 55 \text{ mA}/0.645 \text{ mA}$$
$$= 85.27$$

Thus, beta is 85.27 at point B.

If it were not for the use of feedback, these differences in beta could cause some distortion of the input waveform, which in the example swings the collector current between these two points. The two endpoints of the curve are called *cut-off* and *saturation*.

Cut-off and Saturation

Cut-off is a transistor parameter that represents the condition where the base voltage falls below the 0.7 volt junction turn-on voltage.

Saturation is not an inherent transistor parameter, because it depends on the voltage drop across a collector load resistance. The saturation condition occurs when so much voltage is dropped across the external load resistor that the collector-to-base junction slips into a forward-bias condition.

Because a real circuit has a load in the collector circuit, saturation is a real collector current limit. The actual value of current at which saturation occurs is controlled by the collector load resistance; but beyond saturation, the transistor no longer responds to increases in base current. This saturation mode is often used to advantage in switching transistor circuits, but it constitutes a problem in any other kind of transistor circuit.

Normally, the transistor must always be biased to obtain a Q-point collector current at some place on the curve where the signal can't drive the current beyond cut-off or saturation. Figure 4-38 illustrates a condition where the transistor has been biased to a Q-point too high on the curve.

The upper part of the signal drives the collector current into the saturation region, causing distortion and clipping. A similar problem can occur if the transistor is biased to a Q-point too close to the cut-off part of the curve, where

FIGURE 4-38   Clipping

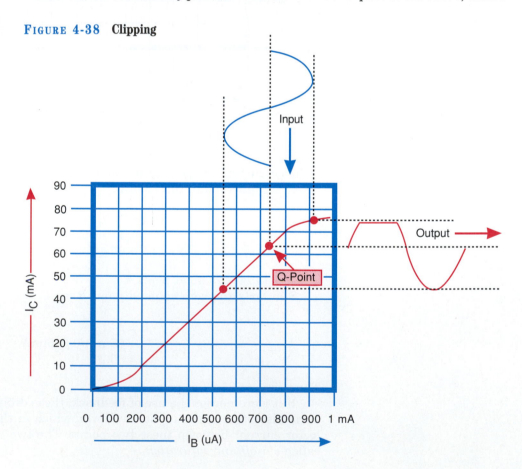

the lower part of the signal is clipped. There is always some nonlinearity near saturation and cut-off, and those areas should be avoided even if actual clipping is not expected.

Finally, it is possible to center the Q-point, as shown in Figure 4-37, and still cause clipping on both halves of the output waveform. This symmetrical clipping will occur if the input signal is large enough to swing the collector current beyond both cut-off and saturation limits. An amplifier is said to be *over-driven* if cut-off or saturation distortion results.

4.8 Emitter Bias Circuit

There are two basic bias circuits: emitter bias and voltage-divider base bias (which we just finished studying). The **emitter bias circuit** requires two power supply voltages, but it allows the higher emitter-feedback resistor voltage drops required by the most critical circuits. The integrated-circuit operational amplifier—the analog workhorse—is one critical circuit that demands the best available Q-point stability, and operational amplifiers normally rely on emitter bias.

The emitter bias arrangement also permits higher input resistances and a more flexible input circuit. **Voltage-divider base bias** can be used satisfactorily for most general-purpose capacitor-coupled transistor amplifiers. The major advantage of voltage-divider base bias is that it requires only a single power-supply voltage. Coupling capacitors must nearly always be used when voltage-divider base bias is used.

The emitter bias circuit is almost universally used when the transistor circuit input must be directly coupled to the driving source without using coupling devices.

A third common bias form is a variant of the voltage-divider base bias circuit, with a different feedback mode called **voltage-mode feedback.** We will investigate this circuit later.

Emitter Biasing

Figure 4-39 illustrates emitter biasing. The quiescent collector current for the circuit shown can be calculated as

$$Q_{IC} = V_{bias}/R_f$$
$$= 10.83 \text{ V}/10,000 \text{ }\Omega = 1.08 \text{ mA}$$

The stability factor is

$$S = R_B/R_f$$
$$= 20 \text{ kilohms}/10 \text{ kilohms} = 2$$

The resulting stability factor of 2 is very close to the best possible S factor of 1.

The circuit current gain (A_i) is equal to the stability factor (S):

FIGURE 4-39   Emitter-biased Common Emitter Circuit

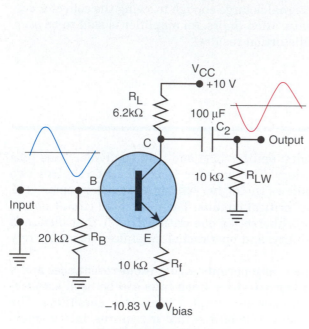

a. Emitter-biased Circuit

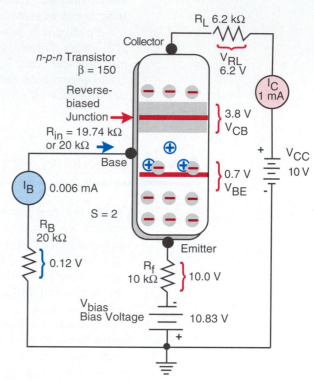

b. Voltage Distribution

$$A_i = S$$

The input resistance of the circuit in Figure 4-39, given a beta of 150, is

$$R_{in} = R_B//(\beta \times R_f)$$
$$= 20 \text{ kilohms}//(150 \times 10 \text{ kilohms})$$
$$= 20 \text{ kilohms}//1.5 \text{ megohms}$$
$$= 19.74 \text{ kilohms}$$

Ignoring ß, we get the following (simpler) calculation for input resistance:

$$R_{in} = R_B = 20 \text{ kilohms} \qquad (R_B = 20 \text{ kilohms in Figure 4-39})$$

Again, ß has a trivial effect on the circuit parameters—19.74 kilohms with ß included, as compared to 20 kilohms with ß excluded.

Notice that we did not include the circuit's R'_e of approximately 25 ohms, because it is so small in comparison to the value of R_f (10 kilohms).

Common Collector Circuit

The **common collector circuit** in Figure 4-40 takes the output signal across the feedback resistor, so the feedback resistor serves as the dummy load resistor as well. There is usually no resistor in the collector circuit, and the collec-

FIGURE 4-40 Common Collector Circuit

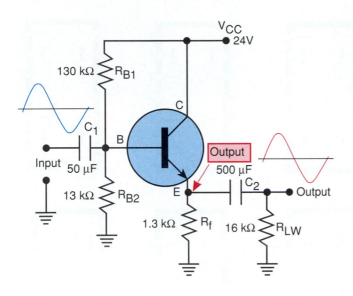

FIGURE 4-41 Common Base Circuit

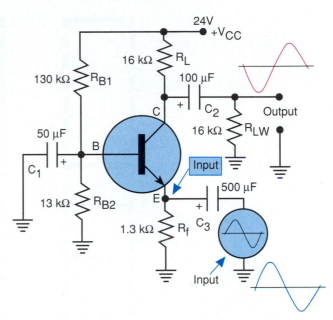

tor is connected directly to V_{CC}. The common collector circuit is used as a buffer circuit in special applications. It has a very high input impedance and a very low output impedance. It has a voltage gain of unity (that is, a 1-volt input yields 1-volt output). In certain special applications these characteristics are ideal, and we will see more of this circuit later.

Common Base Circuit

The **common base circuit** shown in Figure 4-41 has characteristics almost identical to those of the common emitter circuit, with one huge exception. In the common base circuit, the input signal is applied to the emitter instead of to the base, which results in an input impedance that is so low it makes the circuit almost useless. The common base circuit can only be used in a few very special cases. A transformer is almost always necessary to drive such a low impedance, and this makes the circuit useful only at radio frequencies, where a transformer's cost can be justified.

Some Common Multiple-transistor Packages

Various packages contain several transistors. In some of these packages, each transistor is independent; in other cases, some internal connections are made between transistors. Figure 4-42 shows some sample packages from among the many varieties available.

FIGURE 4-42 Multiple-transistor Packages

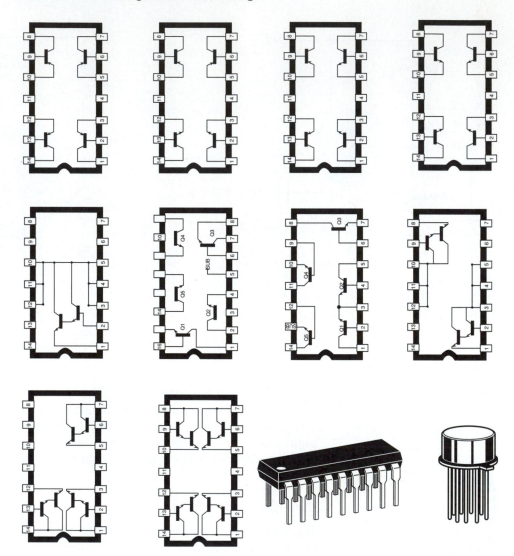

Troubleshooting and Testing Transistors

All of the other troubleshooting procedures described in this text are designed to spot the defective part while it is still physically installed in the circuit. The procedures in the rest of the text are termed *noninvasive*, because they do not involve any risk of damage to the circuit board. Don't guess at which part is defective and remove it for testing! Guess-and-try troubleshooting destroys circuit boards. Guess-and-try methods may make you eligible for unemployment insurance, but otherwise they don't have much to recommend them.

Note: The following tests assume that the transistor is not yet installed in a circuit board.

Using an Ohmmeter

You can use an ohmmeter to test the forward and reverse resistance of the base–emitter and collector–base junctions. Such testing often picks up shorted or open transistors, and it is quick and easy, but it is not very accurate. The test treats the two junctions as back-to-back diodes, as shown in Figure 4-44, and this is where the test falls short: a transistor is more complex than two diodes connected back-to-back.

Some digital multimeters include built-in transistor test circuitry and have a socket to plug the transistor into. If your meter has this feature, it is a far better test to use than the simple ohmmeter test, because it makes the transistor actually work as a transistor during the test.

Constructing a Simple Test Circuit for Testing Transistors with a Voltmeter

Figure 4-45 shows a simple circuit you can construct for testing transistors with any digital voltmeter set on a voltage range. The circuit biases the transistor to a QI_C of about 10 mA, to yield a collector voltage of about 5 volts, depending on actual part values (which are not very critical).

FIGURE 4-43 Pocket Multimeter with Transistor-testing Provisions

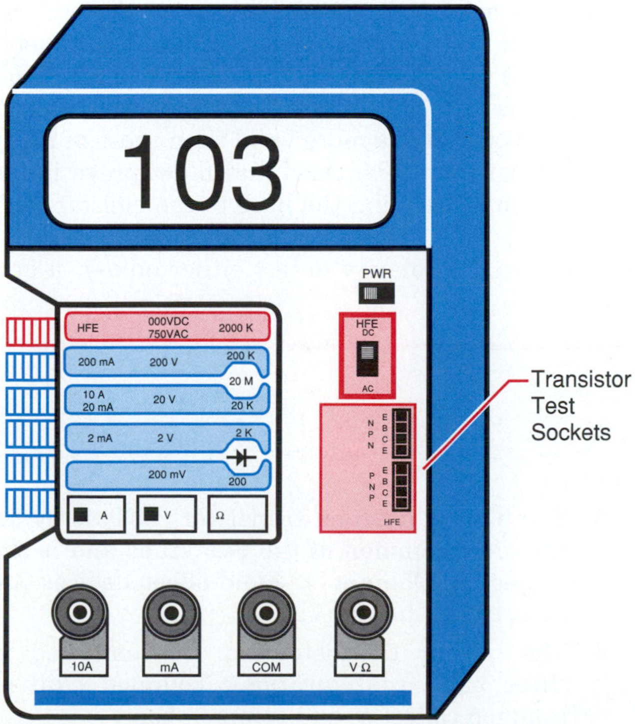

FIGURE 4-44 Testing Transistors with an Ohmmeter

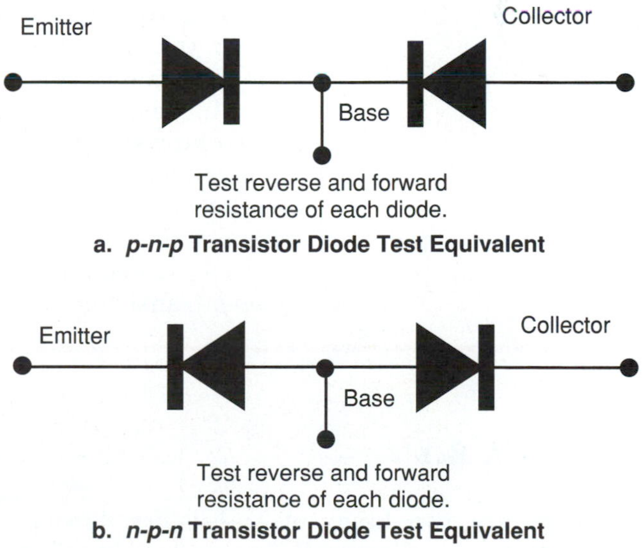

Test reverse and forward resistance of each diode.

a. *p-n-p* Transistor Diode Test Equivalent

Test reverse and forward resistance of each diode.

b. *n-p-n* Transistor Diode Test Equivalent

FIGURE 4-45 **Simple Add-on Circuit That Allows You to Use Your Multimeter to Test Transistors**

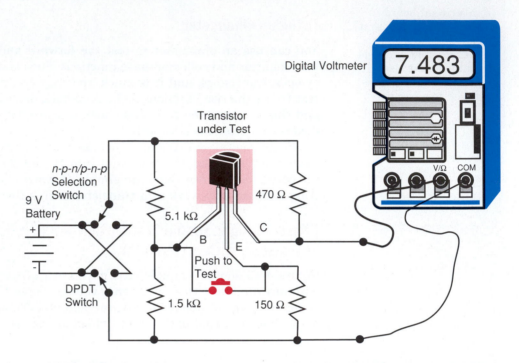

Test a few transistors to get the actual reading. It should be about the same for all transistors. The main test forces the transistor's collector current to swing from QI_C down to near 0 mA by shorting the base to the emitter. When you push the button, the output voltage on the digital multimeter should swing to V_{CC}, indicating that the transistor has been turned off.

This test is basically the same as tests we will use later to test transistors in working systems. The results of this test are more valid than those of many simple beta tests, because the transistor is exercised and must prove it can work over a range of collector currents. Shorting the base to the emitter can't damage the transistor.

The switch allows you to reverse the polarity to test either an *n–p–n* or a *p–n–p* transistor.

SUMMARY

1 There are two kinds of transistors: bipolar junction transistors (BJTs) and field-effect transistors (FETs).

2 The two kinds of transistors perform many of the same functions, but they differ greatly in operation.

3 The bipolar junction transistor (BJT) is by far the more common of the two kinds and is the subject of Chapter 4. Field-effect devices are covered in Chapter 7.

4 The bipolar transistor can be connected in three basic configurations: common emitter, common collector, and common base.

5 The common emitter circuit is the most versatile and frequently used of the three configurations.

6 The BJT has three elements: base, emitter, and collector.

7 The base is nearly always the signal input or control element. The collector is nearly always the output element. Occasionally, the emitter may be used as either the signal input (control) element or the output element.

8 In normal operation, the collector–base junction is reverse-biased, and the base–emitter junction is forward-biased.

9 There is always a 0.6- to 0.7-volt drop across any forward-biased silicon base–emitter junction. This voltage may vary slightly outside these limits with temperature and current extremes.

10 The voltage across the collector–base junction is normally several volts, but it differs in different circuit designs. It is not a constant.

11 Maximum collector voltage, maximum collector current, and maximum power dissipation are manufacturers' specifications that must not be exceeded.

12 Other manufacturers' specifications may represent only rough approximations for any individual transistor.

13 Manufacturing tolerances tend to be very broad, and specification values shift considerably with temperature changes. Operating currents also act to alter many basic transistor specifications.

14 Given these factors, we can never (in practice) know the functional specification values for a given transistor with any reasonable accuracy.

15 Since we can't depend on transistor specifications, we must use negative feedback to compensate for variations in key parameters.

16 Beta (ß) is the most important of the transistor's key parameters. Variations in beta cause the most trouble of any parameter variations in transistor circuits.

17 Beta is the current gain of the transistor and has a more or less typical value of 100. Beta is the factor by which any input (base) current is amplified to appear in the output (collector) as a larger current.

18 The output (collector) current is equal to beta times the input (base) current. If beta is 100, a 1-mA base current will produce a 100-mA collector current.

19 If a transistor circuit is to amplify a signal properly, a specific resting (idling) collector current must be established. This resting current is called the quiescent collector current (QI_C).

20 The quiescent collector current must maintain its specified value even in the face of large temperature changes. If the quiescent current drifts off its design value, the circuit will not work as intended, or it may quit working.

21 The quiescent collector current is set by providing a fixed base current. If beta is 100, a 1-mA fixed base current will produce a quiescent collector current of 100 mA.

22 If 100 mA is the required quiescent collector current, all is well until beta changes—and it will change.

23 Nothing can be done to keep beta from changing, but negative feedback can be used to detect and correct any error in the quiescent collector current (QI_C) resulting from changes in beta.

24 A single resistor installed in series with the emitter can provide automatic negative feedback detection and correction for any unwanted changes in the quiescent collector current.

25 The process of setting the quiescent collector current while using a fixed base current or voltage is called *biasing* the transistor.

26 The base–emitter junction resistance is an important parameter whose value totally depends on the quiescent collector current. The base—emitter junction resistance (R'_e) is equal to 25 divided by the quiescent collector current (in milliamperes); that is, ($R'_e = 25/QI_C$). This is called the Shockley relationship, and it is the same equation we used for forward-bias diode junction resistance.

27 The Shockley relationship is an empirical and statistical relationship. It does not support very accurate junction resistance predictions, but it is the best approximation available.

28 The transistor current gain beta (ß) is also a statistical measure and is not necessarily valid for an individual transistor.

QUESTIONS AND PROBLEMS

4.1 In what three basic configurations can a bipolar transistor be connected?

4.2 Which of the three configurations mentioned in Problem 4.1 is the most often used?

4.3 What are the three connections (elements) of the BJT?

4.4 Which of the three connections (elements) mentioned in Problem 4.3 is generally used for signal input? Which is generally used for signal output?

4.5 In normal operations, is the base–emitter junction forward-biased or reverse-biased? In normal operations, is the collector-base junction forward-biased or reverse-biased?

4.6 What voltage would you expect to measure across the base–emitter junction of a normally operating transistor?

4.7 What voltage would you expect to measure across the base–collector junction of a normally operating transistor?

4.8 What three absolute maximum manufacturers' specifications (ratings) must never be exceeded?

4.9 Can you obtain accurate specifications for a particular transistor in the manufacturer's data manual?

4.10 Why is it difficult to know exact specifications for a particular transistor?

4.11 If it is difficult to know exact specifications, how can we design reliable circuits?

4.12 What single transistor parameter (specification) is the most variable and the most likely to cause problems with temperature changes?

4.13 If we can't depend on the accuracy of transistor specifications, how can we make transistor circuits that have dependable specifications?

4.14 What circuit specification must be maintained at a constant value, in spite of anything that may act to alter its value?

4.15 Why is QI_C (the quiescent collector current) such an important circuit specification?

4.16 What is used to stabilize the value of QI_C, once the circuit designer has established it?

4.17 Define beta (ß). What is a typical beta value?

4.18 If beta is 100 and the base–emitter current is 1.0 mA, what is the collector current?

4.19 Why is the quiescent collector current so important to normal transistor amplifier operation?

4.20 Why must negative feedback be used to maintain a constant value for QI_C?

4.21 How is the quiescent current (QI_C) established in a circuit?

4.22 Once you have established a QI_C of 100 mA, what transistor parameter is most likely to cause the QI_C to drift away from that value?

4.23 What can be done to keep beta from changing?

4.24 What components must be installed in a circuit to produce negative feedback?

4.25 What is the process of establishing a quiescent collector current, using a fixed base current or voltage, called?

4.26 Identify the proper classification for each of the three circuits in Figure 4-46.

4.27 What is the hybrid parameter equivalent of beta for the circuits in Figure 4-46?

4.28 Which of the circuits in Figure 4-46 is the most popular of the three?

FIGURE 4-46   Circuits for
Problems 4.26 Through 4.28

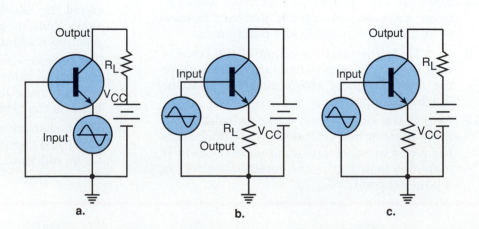

a.   b.   c.

Refer to the circuit in Figure 4-47 for Problems 4.29 through 4.41.

FIGURE 4-47   **Figure for Problems 4.29 Through 4.41**

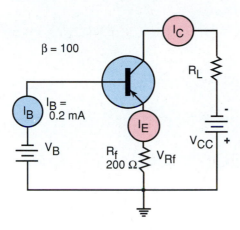

FIGURE 4-48   **Graph for Problems 4.42 Through 4.46**

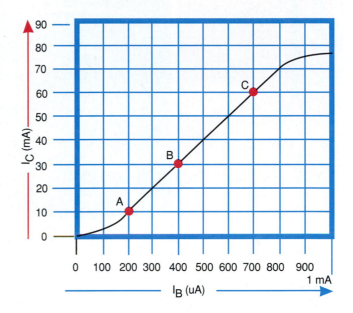

4.29 What is the beta (ß) value of the transistor?

4.30 What is the total emitter current?

4.31 What is the voltage drop across the emitter (feedback) resistor?

4.32 What is the input resistance of the circuit?

4.33 What is the junction resistance (R'_e) of the circuit?

4.34 What is the voltage gain (A_V) in the circuit?

4.35 If the base current in the circuit is increased from 0.2 mA to 0.4 mA, what is the new collector current (I_C)?

4.36 If the base current (I_B) is decreased from 0.2 mA to 0.1 mA, what is the new collector current (I_C)?

4.37 If the base voltage (V_B) is decreased to 0.4 V, what is the new collector current (I_C)?

4.38 Why is your answer to Problem 4.37 correct?

4.39 If the collector voltage is doubled, what is the new collector current (I_C)?

4.40 If you said that doubling the collector voltage doubles the collector current, your answer is wrong. What is the correct answer?

4.41 Which component in the circuit shown is responsible for providing negative feedback?

Refer to the forward current transfer curve in Figure 4-48 for Problems 4.42 through 4.46.

4.42 Find the value of beta at points *A*, *B*, and *C* on the curve.

4.43 Mark the place on the graph that corresponds to a beta value of approximately 85.

4.44 Mark the place on the graph that corresponds to a beta value of approximately 66.

4.45 Mark the place on the graph that corresponds to a quiescent collector current (QI_C) of approximately 40 mA.

4.46 Mark the place on the graph that corresponds to a base current of approximately 0.5 mA.

4.47 For any transistor, what specific transistor current provides the starting value around which the remaining transistor values are calculated?

4.48 Why must we use negative feedback in all analog transistor circuits? List at least three reasons.

4.49 Why can a vendor sell a handful of transistor types that can replace thousands of available type numbers?

4.50 What is meant by a *beta-independent circuit,* and how does a circuit become independent of a very unreliable beta?

4.51 In its simplest form, what is negative feedback?

4.52 Why is negative feedback so important in transistor circuits?

4.53 Why is a dummy load connected in series with a transistor intended for voltage amplification?

4.54 Define the symbol R_{Lac}.

4.55 What is a working load, and how does it differ from a dummy load?

4.56 What is the importance of QI_C?

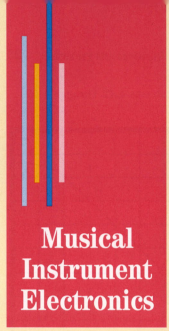

Musical Instrument Electronics

Anyone interested in music and music technology might find musical instrument (MI) electronics servicing a worthwhile career choice. The range of technology in MI electronics covers everything from simple audio amplifiers to computers in sophisticated digital synthesizers and from patch cords to public address systems. Areas of specialization within this field include (in addition to the typical amplifiers and audio mixers) lighting control systems, recording equipment, and even mechanical repairs and alterations to the musical instruments themselves.

Besides performing the usual warranty and nonwarranty work associated with most consumer electronics, MI technicians test and repair rental and lease equipment used by performers. Since this equipment receives rough treatment on the road, failures are frequent. Technicians rarely lack work in stores that rent equipment. Many music stores have their own electronics shop and employ their own technicians. Others contract out their service to independent service companies.

With such a variety of equipment to work on, the electronics technician must be versatile. As in many service situations, the equipment ranges from old to new. In this area, however, some of the old technology is still being built into the latest equipment. For example, this is one of the few areas where tube circuits are still used.

Few technical schools teach the theory and operation of tube circuits anymore, but guitar amplifiers still use vacuum tubes extensively. Many of the larger guitar amps use the old 6L6 power tubes in their output stages, with solid-state input stages. Some use tubes throughout.

The reason for keeping this old technology around relates to distortion. A well-designed solid-state amplifier is less expensive, easier to build, and more reliable than a typical tube amp, but it is just too "clean"—at least in terms of harmonic distortion. Many musical purists maintain that tube amplifiers produce a better, more distinctive sound. The "smooth tube sound" that guitarists like to talk about results from the harmonic distortion produced in the tube circuits.

That is not to say that musical instrument electronics is mired in old technology. During the 1970s, experimentation with electronically synthesized sounds sparked a revolution in the music field. Since then, new developments in microprocessor chips and systems, sampling, analog and digital signal processing, and MIDI (musical instrument digital interface) have brought music performance, production, and recording into the high-tech era.

MIDI is a field unto itself. MIDI is a serial data interface standard that allows synthesizers, rhythm units, and related equipment to be networked. This facilitates various kinds of control of a music system from one or more sources. Personal computers can be interfaced to the system; and with the appropriate software, on-screen editing of musical scores can be done with ease.

Using MIDI, a musician with a couple of synthesizers, a drum synth, and a personal computer can create a digital multitrack recording studio. Each instrument is played in real time, while the computer stores the MIDI information in memory. The computer automatically creates a musical score from the data and corrects for small timing discrepancies. The track can then be played back and transposed, and mistakes can be edited out on-screen. The right software can even create harmony parts to go with the original track. Subsequent tracks can be laid down using other instruments. Nonelectronic instruments such as guitars, brass, and woodwinds can be sampled and their sound-waves converted into MIDI signals for use within the system.

One of the newest and most fascinating areas of MI electronics involves the use of digital signal processing to produce effects. Reverb, echo, flanging, and other effects have been around for years through the use of tape, spring, or analog delay lines. But recent developments in the use of micro-controllers and fast analog-to-digital converters have made a far greater variety of effects possible. Small foot switches or rack-mountable effects boxes provide forms of sound modification that were undreamed of just a few years ago.

In digital signal processing, mathematical algorithms are performed on the digital data obtained from the analog-to-digital converter. The correct algorithm can produce filtering, waveshaping, and other results. Once the digital data are in memory, they can also be manipulated in other ways. For example, they can be delayed and then remixed with the original signal.

Electronics technicians working in the MI field can expect to provide more than just service on this kind of equipment. MIDI gives average musicians the opportunity to produce high-quality music if they know how to use the equipment. As the local expert, the technician may be the one assigned to get the customer "up to speed." Training and technical sales support are important services to the customer. The technician who can provide them becomes an invaluable employee.

A career in musical instrument electronics probably requires the technician to have some background in music, and such a background would certainly make the field more interesting. Musicians often have a unique perspective and commitment to their art but find themselves lost in the tangle of technical details associated with it. As the distinction between musician and technician becomes increasingly blurred, the musical instrument technician can provide the practical expertise that customers need.

CHAPTER
5

Practical Transistor Circuit Analysis

OBJECTIVES

Upon completion of this chapter, you should be able to:

1 Analyze each of the following voltage-divider base-biased circuits:
   a Common emitter circuit with signal feedback.
   b Common emitter circuit without signal feedback.
   c Common collector circuit.
   d Common base circuit.

2 Analyze each of the following emitter-biased circuits:
   a Common emitter circuit without signal feedback.
   b Differential amplifier circuit.

3 Explain the Miller effect and its relationship to circuit input impedance.

4 Analyze the common emitter amplifier with voltage-mode feedback.

Analysis In Brief

Analysis Procedure

The analysis procedure is a simple step-by-step method that uses a special flowchart. The math involved is no more difficult than that of Ohm's law. There are two flowcharts for current-mode circuits and two for voltage-mode circuits. In each case, the first sheet calculates **dc** voltages and currents, and the second sheet deals with signal performance values (voltage gain, etc.). The flowchart procedures are greatly simplified by incorporating a few common-sense assumptions that make very little difference in the final results.

ASSUMPTIONS

1 The collector current is always equal to the emitter current. (The error here is the 1% or less base.)

2 The beta (ß) value of every general-purpose transistor ($I_C < 200$) is 100. All power transistors have a beta of 60. (Because we use negative feedback, any reasonable number can be used for beta).

3 Coupling and bypass capacitors have a reactance near 0 Ω at the lowest frequency of interest. (The degree of accuracy of this is a cost consideration.)

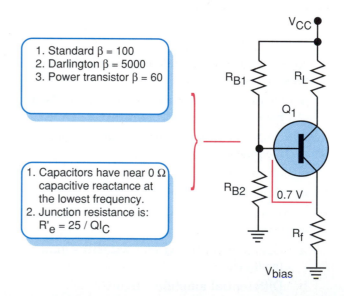

1. Standard β = 100
2. Darlington β = 5000
3. Power transistor β = 60

1. Capacitors have near 0 Ω capacitive reactance at the lowest frequency.
2. Junction resistance is:
   $R'_e = 25 / QI_C$

4 The base–emitter junction voltage is 0.7 V.

5 We will use the Shockley relationship ($R'_e = 25/QI_C$) to calculate the base–emitter junction voltage. (This is the least reliable of the assumptions named, but it is the best we have. Fortunately, it only affects voltage gain and input Z in some circuits.)

Analysis Worksheet, Sheet 1: Output Circuit Calculations

ANALYSIS OF BASE-BIASED CIRCUITS

In all analysis procedures, the first thing we need to know is the quiescent collector current (QI_C). Finding the value of QI_C usually takes several steps.

Step 1

In the base-bias circuit, we find the voltage drop across R_{B2} by using the standard unloaded voltage-divider equation. Generally, we leave out the transistor input resistance ($\beta \times R_f$), which loads the voltage divider. Because of certain stability requirements, the error in ignoring ($\beta \times R_f$) is limited to between 1% and 10%. This is not serious in most troubleshooting situations, but you can include ($\beta \times R_f$) if unusual accuracy is required.

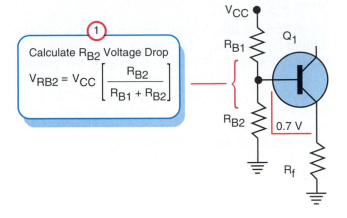

① Calculate R_{B2} Voltage Drop

$$V_{RB2} = V_{CC}\left[\frac{R_{B2}}{R_{B1} + R_{B2}}\right]$$

Step 2

As soon as we know the voltage drop across R_{B2}, we can calculate the feedback resistor (R_f) voltage drop, because we know that the junction voltage is always 0.7 volt. If we subtract 0.7 volt from the R_{B2} voltage, we get the voltage drop across the feedback resistor (V_{Rf}).

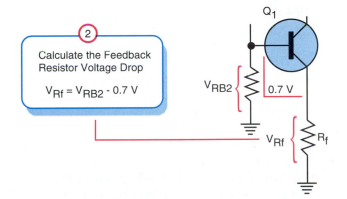

② Calculate the Feedback Resistor Voltage Drop

$$V_{Rf} = V_{RB2} - 0.7\ V$$

Step 3

Here, we take advantage of the assumption that the collector current and the emitter current are equal. We now know the voltage drop across R_f, and the resistance value of R_f. Ohm's law gives us the emitter current and, thus, the quiescent collector current (QI_C).

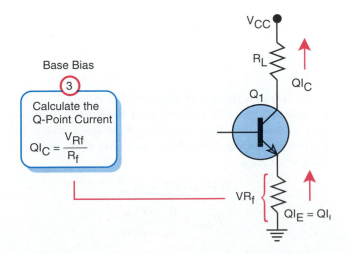

Base Bias

③ Calculate the Q-Point Current

$$QI_C = \frac{V_{Rf}}{R_f}$$

ANALYSIS OF EMITTER-BIASED CIRCUITS

Steps 1–2

When emitter bias is used, we can skip the voltage divider calculations, because a $-V_{bias}$ voltage is applied directly to the emitter. Instead, the preliminary step in the emitter-bias branch states that the entire $-V_{bias}$ voltage is dropped across the feedback resistor ($V_{R_f} = V_{bias}$). This is not quite true, because there is a 0.7-V junction voltage drop plus some small voltage drop across R_{B2}.

Step 3

Since the purpose of emitter bias is to allow for bias voltages of 10 volts or higher, the error in QI_C is normally less than 1%.

UNITED ANALYSIS OF EITHER TYPE OF BIASED CIRCUIT

Step 4

Now we calculate the junction resistance

$$R'_e:$$
$$R'_e = 25/QI_C$$

Step 5

In this step, Ohm's law ($V = IR$) gives us the voltage drop across the dummy load resistor (R_L):

$$V_{RL} = QI_C/R_L$$

Step 6

This step is a bit out of place on the **dc** part of the worksheet, since it only applies to signal parameters shown on sheet #2. We calculate it here to reduce crowding. Calculate the value of R_L in parallel with R_{LW}:

$$\textbf{ac Load } Z = R_{Lac}$$

$$R_{Lac} = \frac{R_L \times R_{LW}}{R_L + R_{LW}}$$

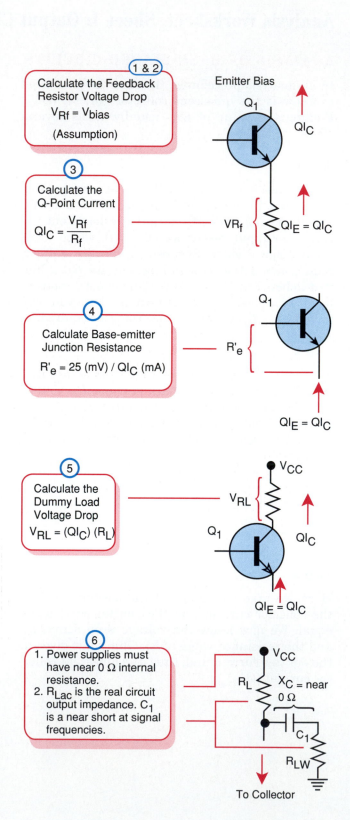

1 & 2
Calculate the Feedback Resistor Voltage Drop
$V_{Rf} = V_{bias}$
(Assumption)

Emitter Bias

Q_1 QI_C

3
Calculate the Q-Point Current
$QI_C = \dfrac{V_{Rf}}{R_f}$

VR_f { $QI_E = QI_C$

4
Calculate Base-emitter Junction Resistance
$R'_e = 25$ (mV) / QI_C (mA)

Q_1
R'_e {
$QI_E = QI_C$

5
Calculate the Dummy Load Voltage Drop
$V_{RL} = (QI_C)$ (R_L)

V_{CC}
V_{RL} {
Q_1 QI_C
$QI_E = QI_C$

6
1. Power supplies must have near 0 Ω internal resistance.
2. R_{Lac} is the real circuit output impedance. C_1 is a near short at signal frequencies.

V_{CC}
R_L X_C = near 0 Ω
C_1
R_{LW}
To Collector

Analysis Worksheet, Sheet 2: Signal Parameters

Note: The following values calculated on sheet #1 will be required by equations on Sheet #2: QI_C, R'_e, V_{CE}, and R_{Lac}.

Step 1

Three input paths to ground make up the input impedance. Resistor R_{B_2} is obvious. Resistor R_{B_1} is less obvious, but any good power supply has a very low internal impedance. For all practical purposes, R_{B_1} is connected to ground.

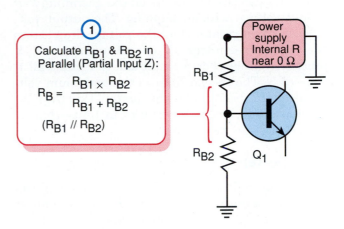

1

Calculate R_{B1} & R_{B2} in Parallel (Partial Input Z):

$$R_B = \frac{R_{B1} \times R_{B2}}{R_{B1} + R_{B2}}$$

$(R_{B1} \mathbin{/\!/} R_{B2})$

Step 2(a)

Our standard beta (ß) is 100. Therefore, if we look into the base, we see any impedance in the emitter as 100 times its actual value. The feedback resistor in most circuits is bypassed with a capacitor whose reactance is near 0 Ω at all frequencies of interest. For signal frequencies, the emitter is directly grounded. The only impedance left in the emitter circuit is the junction resistance (R'_e). The emitter circuit's contribution to the input Z is $100 \times R'_e$.

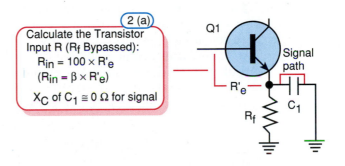

2 (a)

Calculate the Transistor Input R (R_f Bypassed):
$R_{in} = 100 \times R'_e$
$(R_{in} = \beta \times R'_e)$
X_C of $C_1 \cong 0\ \Omega$ for signal

Step 2(b)

Without the capacitor, the emitter resistance is $R'_e + R_f$. Because R'_e is usually around 25 Ω, while R_f is typically 1 kΩ or higher, we can normally ignore R'_e. Looking into the base, we see $100 \times R_f$, base-to-ground resistance.

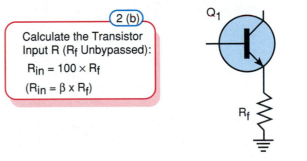

2 (b)

Calculate the Transistor Input R (R_f Unbypassed):
$R_{in} = 100 \times R_f$
$(R_{in} = \beta \times R_f)$

Step 3

The final step in calculating the input impedance is to calculate the equivalent parallel resistance of R_B ($R_{B_1} /\!/ R_{B_2}$) and the bypassed or unbypassed transistor input resistance (ß $\times R'_e$ or ß $\times R_f$).

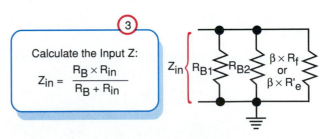

3

Calculate the Input Z:

$$Z_{in} = \frac{R_B \times R_{in}}{R_B + R_{in}}$$

Step 4

There are two possible limits on the maximum P–P output voltage without clipping. The output voltage can be no greater than the voltage drop across the collector–emitter (V_{CE}) minus the 0.7 V junction voltage. For a small safety factor, call it $V_{CE} - 1$ V. The second limit is the maximum voltage drop (IR drop) across the dummy load (R_L). We must use whichever limit kicks in first—that is, the lower of the two values.

Step 5(a)–(b)

The voltage gain is simply the ratio of the real signal output impedance (R_{Lac}) to the actual emitter impedance. The emitter impedance is R'_e (if the bypass capacitor shorts out R_f at signal frequencies) or $R_f + R'_e$ (if there is no bypass capacitor). Since R'_e is so small compared to R_f, we usually ignore it for unbypassed circuits (which are not very common). The emitter (input voltage) and the collector (output voltage) have the same ratio of voltages as of impedances, because the emitter and collector currents are equal. The signal voltage gain is a ratio:

$$A_V = V_{R_{Lac}} \text{ [Output } V]/V_{emitter} \text{ [Input } V]$$

And since $I_E = I_C$,

$$A_V = R_{Lac} \text{ [Output } R]/R_{emitter} \text{ [Input } R]$$

Step 6

This step is not necessary except in power amplifiers. Skip it for small-signal voltage amplifiers. The equation is simply a version of the common power equation: $P = V^2/R$.

Step 7

The maximum P–P output current is something you don't often need to know, but just in case:

$$I = V/R$$

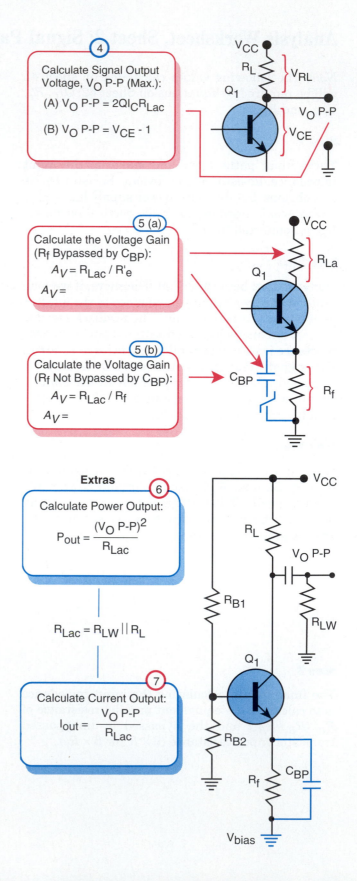

④ Calculate Signal Output Voltage, V_O P-P (Max.):

(A) V_O P-P $= 2QI_C R_{Lac}$

(B) V_O P-P $= V_{CE} - 1$

5 (a) Calculate the Voltage Gain (R_f Bypassed by C_{BP}):

$A_V = R_{Lac} / R'e$

$A_V =$

5 (b) Calculate the Voltage Gain (R_f Not Bypassed by C_{BP}):

$A_V = R_{Lac} / R_f$

$A_V =$

Extras

6 Calculate Power Output:

$$P_{out} = \frac{(V_O \text{ P-P})^2}{R_{Lac}}$$

$$R_{Lac} = R_{LW} \| R_L$$

7 Calculate Current Output:

$$I_{out} = \frac{V_O \text{ P-P}}{R_{Lac}}$$

The Differential Amplifier

DIFFERENTIAL AMPLIFIER COMMON-MODE REJECTION RATIO (CMRR)

The common-mode rejection ratio is the basic figure of merit for differential amplifiers. The CMRR is the ratio of normal differential voltage gain to unwanted common-mode voltage gain:

$$CMRR = 20 \log (A_D/A_{CM})$$

COMMON-MODE VOLTAGE GAIN (A_{CM})

The circuit on page 234 shows the setup for measuring common-mode voltage gain, whose equation is

$$A_{CM} = \Delta V_{out}/\Delta V_{CM}$$

The common-mode voltage gain would be zero if the two sides of the amplifier were perfectly balanced; that is, there would be no output signal if the same signal were applied to both inputs. The smaller the common-mode voltage gain, the better the amplifier.

DIFFERENTIAL VOLTAGE GAIN (A_D)

The method for measuring the differential voltage gain is shown at right. This is the second factor you must know to determine the CMRR. The equation for differential voltage gain is

$$A_D = \Delta V_{out}/\Delta V_D$$

Differential Amplifier Circuit

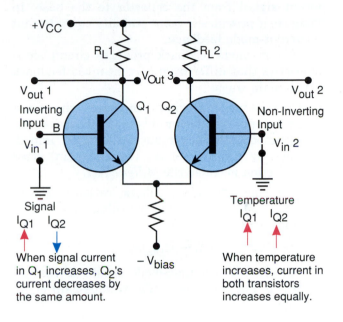

When signal current in Q_1 increases, Q_2's current decreases by the same amount.

When temperature increases, current in both transistors increases equally.

Circuit for Common-mode Voltage Gain

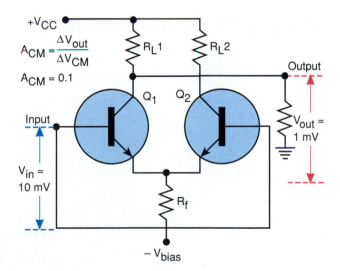

Voltage-mode Feedback

Voltage-mode feedback returns a fraction of the output signal from the collector to the base. In doing so, it provides Q-point stability equal to that of current-mode feedback.

Voltage-mode feedback produces circuit characteristics that differ from current-mode feedback in two main ways:

1   Current-mode feedback increases the input impedance by a factor of beta (ß), whereas voltage-mode feedback reduces the input impedance by a factor of A_V (voltage gain). This reduction is called the *Miller effect*.

2   The current-mode feedback resistor (R_f) (if unbypassed) is a part of the voltage-gain equation:

$$A_V = R_{Lac}/R_f$$

whereas the voltage-mode feedback resistor (R_f) is not involved in the voltage-gain equation:

$$A_V = R_{Lac}/R_e'$$

Circuit for Differential Voltage Gain

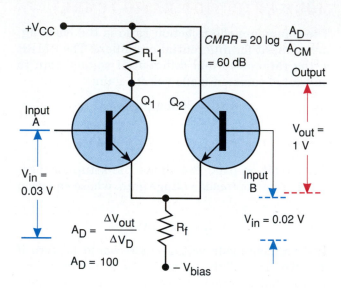

Voltage-mode Feedback and the Miller Effect

EXAMPLE OF THE MILLER EFFECT

When voltage-mode feedback is used, the Miller effect lowers the input impedance by the factor A_V. This is illustrated by the example at right.

Input Resistance for Miller Effect Example:
$$R_{in} = (R_f/A_v)//(\beta \leftrightarrow R_e')$$
$$= (100 \text{ k}\Omega/100)//(100 \leftrightarrow 25)$$
$$= 1 \text{ k}\Omega//2.5 \text{ k}\Omega = 714 \text{ }\Omega$$

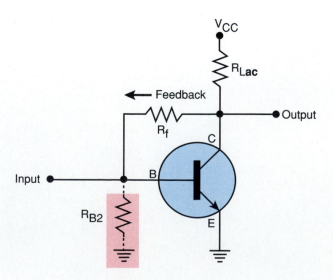

COLLECTOR–BASE CAPACITANCE AND THE MILLER EFFECT

The Miller effect also lowers the effective capacitive reactance of the collector-to-base junction by the factor A_V (voltage gain). This is equivalent to multiplying the junction capacitance value by the factor A_V.

Any external capacitor connected between the collector and the base also becomes A_V microfarads larger as a result of the Miller effect. The capacitor value is multiplied by A_V.

Any capacitance between collector and base has its value multiplied by A_V ▶

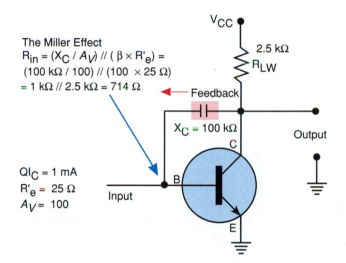

BYPASSING THE FEEDBACK VOLTAGE

The voltage-mode feedback resistor is often split into two parts, and a capacitor is added. The capacitor bypasses the feedback voltage to ground at signal frequencies, but it does not alter the ability of the feedback to stabilize the Q-point. With the signal feedback bypassed, the input impedance is significantly higher, and the circuit is much more useful.

PURPOSE OF R_{B2}

The feedback resistor (R_f) also serves as a source of bias current for the base–emitter junction. In most cases, if the value of R_f is low enough to provide the desired amount of feedback, it delivers far too much bias current. Resistor R_{B2} bleeds off the excess current, leaving the desired bias current. Resistor R_{B2} also reduces the total base-to-ground resistance, yielding a lower stability factor number.

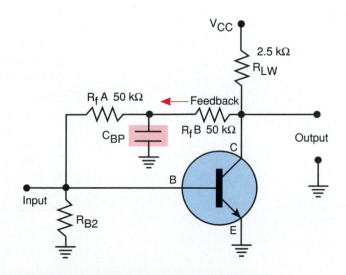

5.1   Introduction to the Universal Analysis Process

Although transistor circuits are called by different names and appear to be quite different, they are really variations on one of two basic circuits that have different feedback modes. Current-mode feedback circuits are by far the more common, with voltage-mode circuits accounting for only a small percentage of the total.

We have two options for studying practical transistor circuits: we can study each circuit variation as though it were unique, with little or nothing in common with other transistor circuits; or we can study the primary circuit and make allowances for the little circuit changes that account for the differences among the various circuits.

The two basic transistor circuits (defined by the way in which negative feedback is obtained) are current-mode feedback circuits and voltage-mode feedback circuits. They can be further subdivided into base-biased and emitter-biased circuits.

Under the two fundamental classifications, we can list the rest of the circuit configurations as minor variations on a theme. Thus we only need two analysis procedures—one for current-mode circuits, and one for voltage-mode circuits. With two simple procedures, we can analyze almost any analog transistor circuit. We don't need to learn a dozen or more different procedures for analyzing a dozen or more slightly different circuits.

5.2   Transistor Circuit Classifications

The classification system for transistor circuits that we will adopt in this chapter fits the analysis procedures we will be working with. It can be outlined as follows:

I. Current-mode Feedback Circuits

    A. Voltage-divider Base-biased Circuits

        1. Common emitter circuits

            a. With signal negative feedback

            b. Without signal negative feedback

        2. Common collector circuits

        3. Common base circuits

    B. Emitter-Biased Circuits

        1. Common emitter circuits

            a. Without signal negative feedback

        2. Common collector circuits

        3. Common base circuits

        4. Differential amplifier circuits

Note: The emitter-biased circuit is not used with signal feedback because voltage gains are too low to be useful.

II. Voltage-mode Feedback Circuits
   1. Common emitter circuit
   2. Special feedback oriented circuits

As you can see from the circuit classifications, current-mode circuits account for most of the circuits and their variations. Voltage-mode circuits account for only 5% or so of the real-world circuits you are likely to encounter. Because current-mode feedback circuits are more important and more common, we will study them first.

First Look at the Universal Transistor Circuit Analysis Worksheets

The analysis procedure for current-mode feedback circuits consists of completing two modified flowcharts called *worksheets*—one for **dc** voltages and the other for signal parameters. The worksheets incorporate a step-by-step approach, with alternate boxed equations to deal with circuit variations. Blank samples of the two current-mode worksheets are shown in Figure 5-1. The worksheets allow you to determine all of the appropriate **dc** voltages and signal parameters, such as voltage gain, for any transistor circuit. To begin, you need to know only the resistor values and the power-supply voltage.

In troubleshooting, you may need to know what voltages and signal parameters you should find in a circuit in order to judge whether they are correct or not. If you have a technical manual for the unit you are working on, you may find that it can supply these details. In many situations, however, you will not have access to a manual, or it will not provide the information you need. The worksheet procedure is based on ideas developed in the previous chapter, and the arithmetic is easy. You will find that your calculations yield results that are more than adequate for troubleshooting purposes.

How to Deal with Beta

Once we have ensured the presence of adequate feedback in the circuit, variations in beta cease to be a problem, but beta does exist as some value. There are two schools of transistor-circuit design and analysis. The parametric philosophy makes design and analysis depend heavily on having accurate values for all parameters, including beta. The beta-independent philosophy tends to ignore transistor parameters, including beta. Both procedures yield adequate results if done properly, but parametric procedures are vastly more difficult to use. We will use a slightly modified version of the beta-independent process, in which we will adopt a standard (fixed) beta value of 100 for certain calculations.

Table 5-1 shows the measured results for the circuit in Figure 5-2, compared to calculated values obtained by using various values of beta and by ignoring beta. As you can see, there is not a lot of difference. In a few special circuits, ignoring beta altogether can yield unacceptable errors. Our adoption of a standard beta value of 100 is intended to avoid those errors.

FIGURE 5-1   Analysis Worksheet Sample

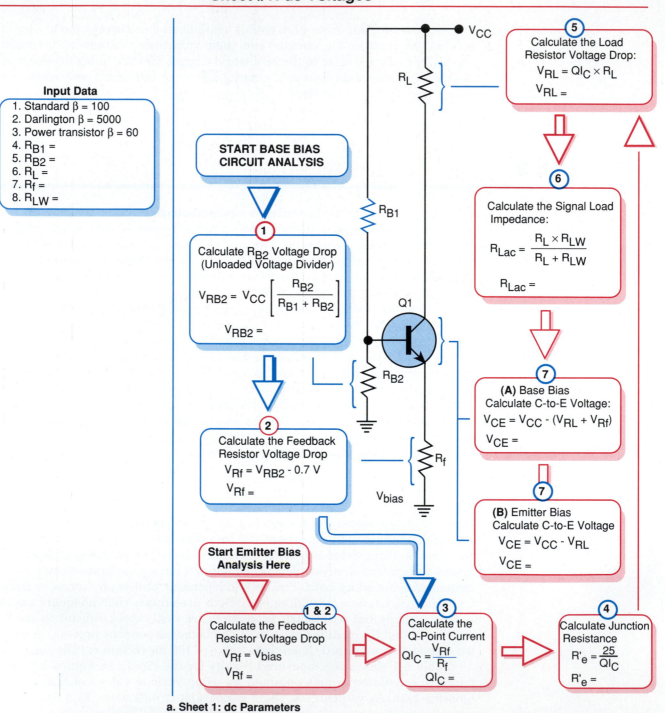

a. Sheet 1: dc Parameters

Sheet #2: Signal Parameters

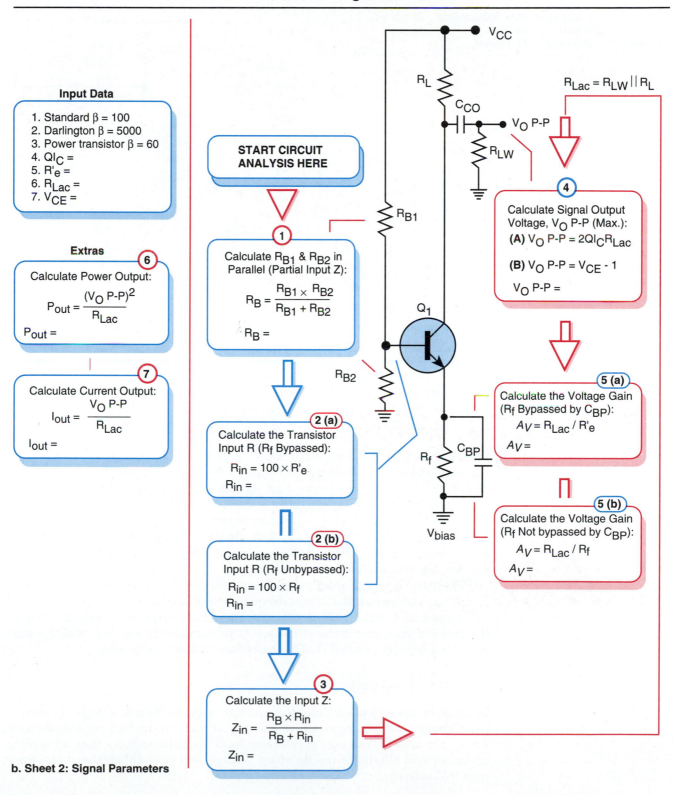

Input Data

1. Standard $\beta = 100$
2. Darlington $\beta = 5000$
3. Power transistor $\beta = 60$
4. $QI_C =$
5. $R'_e =$
6. $R_{Lac} =$
7. $V_{CE} =$

Extras

6 Calculate Power Output:

$$P_{out} = \frac{(V_O \text{ P-P})^2}{R_{Lac}}$$

$P_{out} =$

7 Calculate Current Output:

$$I_{out} = \frac{V_O \text{ P-P}}{R_{Lac}}$$

$I_{out} =$

START CIRCUIT ANALYSIS HERE

1 Calculate R_{B1} & R_{B2} in Parallel (Partial Input Z):

$$R_B = \frac{R_{B1} \times R_{B2}}{R_{B1} + R_{B2}}$$

$R_B =$

2 (a) Calculate the Transistor Input R (R_f Bypassed):

$R_{in} = 100 \times R'_e$

$R_{in} =$

2 (b) Calculate the Transistor Input R (R_f Unbypassed):

$R_{in} = 100 \times R_f$

$R_{in} =$

3 Calculate the Input Z:

$$Z_{in} = \frac{R_B \times R_{in}}{R_B + R_{in}}$$

$Z_{in} =$

$R_{Lac} = R_{LW} \| R_L$

4 Calculate Signal Output Voltage, V_O P-P (Max.):

(A) V_O P-P $= 2QI_C R_{Lac}$

(B) V_O P-P $= V_{CE} - 1$

V_O P-P $=$

5 (a) Calculate the Voltage Gain (R_f Bypassed by C_{BP}):

$A_V = R_{Lac} / R'_e$

$A_V =$

5 (b) Calculate the Voltage Gain (R_f Not bypassed by C_{BP}):

$A_V = R_{Lac} / R_f$

$A_V =$

b. Sheet 2: Signal Parameters

TABLE 5-1

Parameter	Parametric Method-Calculated	Beta Independent Method-Calculated	Measured Values
V_{RB1}	8.3 V	8.0 V	7.92 V
V_{RB2}	1.7 V	2 V	2.09 V
V_{Rf}	1.1 V	1.4 V	1.5 V
QI_C	2.2 mA	2.8 mA	2.6 mA
V_{RL}	2.2 V	2.8 V	2.6 V
V_{CE}	6.7 V	5.8 V	5.9 V
A_V	2	2	1.83
Z_{in}	3.1 k	3.76 k	3.4 k

FIGURE 5-2   Circuit Used to Compile Table 5-1

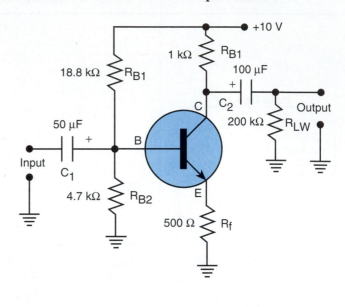

5.4   Capacitors in Transistor Circuits

Capacitors serve several important functions in practical transistor circuits.

Input Coupling Capacitors

In base-biased circuits, an input coupling capacitor is nearly always required to isolate the voltage-divider bias resistors. Nearly any input device, when connected to the amplifier input, has the potential to upset the bias voltage divider, unless a coupling capacitor is used.

The input capacitor is unnecessary in emitter-biased circuits, which is one reason to use emitter bias in the first place. Voltage-divider base bias is the most common bias arrangement because it requires only one power-supply voltage, but a capacitor is required.

Theoretically, the capacitive reactance (X_C) of the input coupling capacitor should be 0 ohms at the lowest frequency we intend to amplify. The actual value used in a real circuit is not critical, and its value is usually decided on the basis of economic rather than scientific considerations. For analysis purposes, we usually pretend that its reactance is essentially zero.

Zero Reference Shift

Because of the way most transistor circuits must be biased, a large **dc** component almost always accompanies the signal voltage. In this regard, **dc** sine waves are typical. You may recall from basic capacitor theory that a coupling capacitor will eliminate the **dc** offset voltage and reestablish the sine wave's true 0-V reference.

Output Coupling Capacitors

An output coupling capacitor is usually used to prevent the working load (R_{LW}) from altering the collector-side **dc** voltages. Again, we assume that the capacitive reactance is insignificant at signal frequencies.

Signal-feedback Bypass Capacitors

The two most significant circuit variations in the common emitter circuit depend on the amount of negative feedback used for the signal. The difference between the two variations depends on whether a (bypass) capacitor is connected across the feedback resistor. If there is no bypass capacitor, the feedback resistor provides the same negative feedback for both Q-point stability and the signal.

Unbypassed Configuration

The large amount of feedback in the unbypassed configuration provides very low distortion, highly predictable voltage gain, and higher input impedance. The unbypassed circuit is superior to the bypassed circuit on all counts, except as regards its very low voltage gain. A maximum voltage gain of about 10 is the best we can expect from the unbypassed configuration in practical circuits.

Bypassed Configuration

Adding a bypass capacitor across the feedback resistor effectively shorts the feedback resistor for signal frequencies. The capacitor provides a reactance of near 0 ohms at the lowest signal frequency. The feedback resistor is effectively shorted (bypassed) at signal frequencies, but the feedback resistor is still fully functional for handling the slow changes involved in temperature changes that cause beta to vary. Q-point stability is not affected by the addition of the bypass capacitor, because the capacitive reactance appears to be nearly infinite (an open circuit) at the very low frequencies represented by the rate at which temperature changes occur.

As illustrated in Figure 5-3, bypassing the feedback resistor converts the feedback resistor into the equivalent of a wire at signal frequencies, leaving only the junction resistance (R'_e) to provide a (very) little negative feedback for the signal. The voltage gain with the feedback resistor bypassed becomes a function of the total collector load and the junction resistance (R'_e); the feedback resistor is no longer in the picture. The voltage gains for the two configurations are illustrated in the following equations (see Figure 5-3).

For unbypassed voltage gain:

$$A_V = R_{Lac}/R_f$$

FIGURE 5-3   Common Emitter Circuits

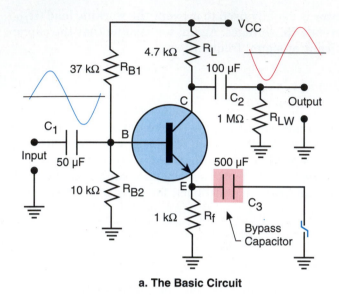

a. The Basic Circuit

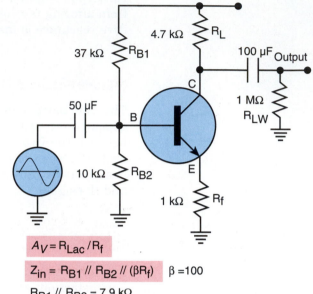

$A_V = R_{Lac}/R_f$

$Z_{in} = R_{B1} \mathbin{/\!/} R_{B2} \mathbin{/\!/} (\beta R_f)$ $\beta = 100$

$R_{B1} \mathbin{/\!/} R_{B2} = 7.9\ \text{k}\Omega$

$7.9\ \text{k}\Omega \mathbin{/\!/} \beta \times 1\ \text{k}\Omega = 7.9\ \text{k}\Omega \mathbin{/\!/} 100\ \text{k}\Omega = 7.3\ \text{k}\Omega$

b. The Bypass Capacitor Removed

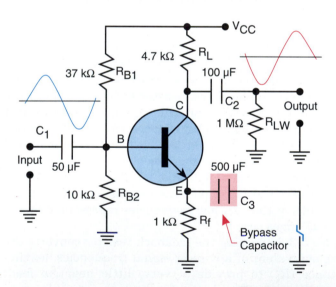

c. The Bypass Capacitor Connected

$Z_{in} = R_{B1} \mathbin{/\!/} R_{B2} \mathbin{/\!/} (\beta R'_e)$ $\beta = 100$

$A_V = R_{Lac}/R'_e$

$R_{B1} \mathbin{/\!/} R_{B2} = 7.9\ \text{k}\Omega$

$7.9\ \text{k}\Omega \mathbin{/\!/} \beta \times 25\ \Omega = 7.9\ \text{k}\Omega \mathbin{/\!/} 2.5\ \text{k}\Omega = 1.9\ \text{k}\Omega$

For bypassed voltage gain:

$$A_V = R_{Lac}/R'_e$$

Example 5.1 | **Voltage Gain in Unbypassed and Bypassed Circuit Configurations**

For this example, assume a junction resistance (R'_e) 25 Ω, a total **ac** (signal) collector load resistance (R_{Lac}) of 4.7 kΩ and a feedback resistor value of 1 kΩ. Find the voltage gain for the unbypassed and bypassed circuit configurations.

Solution   We will proceed separately for the unbypassed and bypassed cases.

Unbypassed

$$A_V = R_{Lac}/R_f$$
$$= 4.7 \text{ k}\Omega/1 \text{ k}\Omega = 4.7$$

The voltage gain is $A_V = 4.7$.

Bypassed

$$A_V = R_{Lac} / R'_e$$
$$= 4.7 \text{ k}\Omega/25 \text{ }\Omega$$

The voltage gain is $A_V = 188$.

Although the voltage gain in the example is some 40 times higher in the bypassed configuration than in the unbypassed version, it is not as predictable. Further, we can expect much greater errors in calculating the voltage gain in the bypassed configuration than we will expect when calculating the voltage gain in the unbypassed circuit.

5.5   Kinds of Worksheets

There are two different kinds of analysis worksheets. The first kind is designed to analyze circuits that use current-mode negative feedback. The second kind is designed to analyze circuits that use voltage-mode negative feedback. Each analysis worksheet is really two sheets—one used for calculating **dc** currents and voltages, and the other used for analyzing signal parameters.

Current-mode Feedback Worksheets

The current-mode worksheet is used for several circuit variations, including both base-bias and emitter-bias circuits. Certain blocks in the worksheet may be skipped or slightly modified for some specific circuits. Specifically, the same basic current-mode worksheet is used to analyze the following circuits:

1   The base-biased common emitter circuit.
2   The base-biased common base circuit.
3   The base-biased common collector circuit.
4   The emitter-biased common emitter circuit.
5   The emitter-biased differential amplifier circuit.

You should be warned that it is possible for circuits 2, 3, and 4 to be emitter-biased instead of base-biased, but you will see how it is done without any examples.

We will start our examples section (section 5.6) with the base-bias common emitter circuit. You will find that the **dc** calculations and the numerical values are identical for several of the circuits, but we won't repeat the same things unnecessarily. In general, a minor variation or two determine what kind of circuit we are looking at.

The fact that all of these circuits are so much alike makes the "one size fits all" kind of worksheets we are using possible. We will try to avoid repetition unless it is necessary to make a point.

Voltage-mode Feedback Worksheets

The voltage-mode circuit is not frequently used in voltage amplifiers, but it does have some advantages in power-amplifier circuits. We will analyze a voltage-mode feedback circuit in section 5.11.

5.6   Current-mode Feedback Analysis Procedures

The following sections consist of worked examples of how to use the current-mode transistor circuit analysis worksheets to analyze the five types of circuits listed in the preceding section. We begin with the base-biased common emitter circuit.

FIGURE 5-4   **Common Emitter Circuit for Worksheet Analysis**

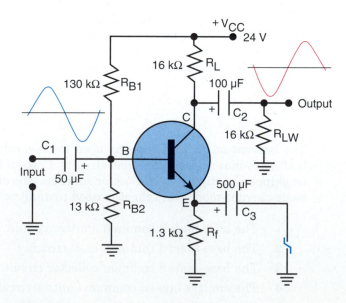

FIGURE 5-5   Worksheet for the Common Emitter Circuit, Sheet #1

Sheet #1: dc Voltages

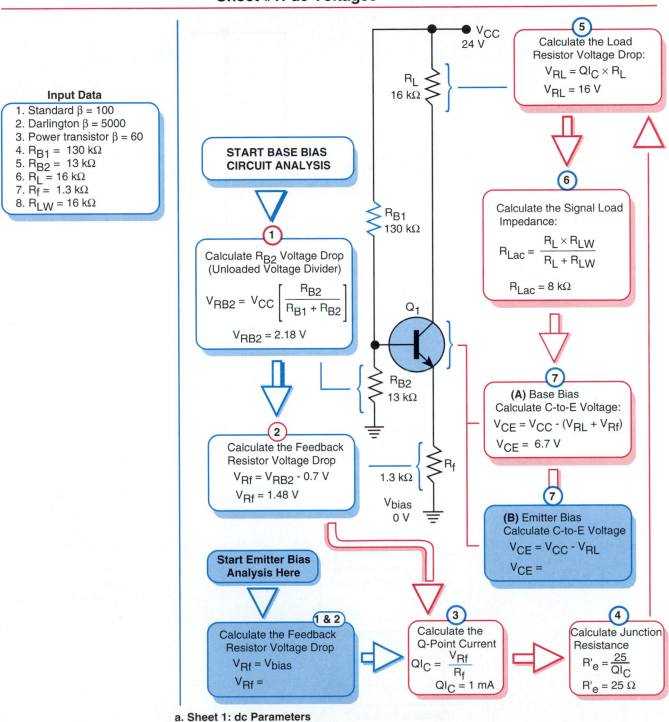

Input Data

1. Standard $\beta = 100$
2. Darlington $\beta = 5000$
3. Power transistor $\beta = 60$
4. $R_{B1} = 130\ k\Omega$
5. $R_{B2} = 13\ k\Omega$
6. $R_L = 16\ k\Omega$
7. $R_f = 1.3\ k\Omega$
8. $R_{LW} = 16\ k\Omega$

START BASE BIAS CIRCUIT ANALYSIS

1 Calculate R_{B2} Voltage Drop (Unloaded Voltage Divider)

$$V_{RB2} = V_{CC}\left[\frac{R_{B2}}{R_{B1} + R_{B2}}\right]$$

$$V_{RB2} = 2.18\ V$$

2 Calculate the Feedback Resistor Voltage Drop

$$V_{Rf} = V_{RB2} - 0.7\ V$$

$$V_{Rf} = 1.48\ V$$

Start Emitter Bias Analysis Here

1 & 2 Calculate the Feedback Resistor Voltage Drop

$$V_{Rf} = V_{bias}$$

$$V_{Rf} =$$

3 Calculate the Q-Point Current

$$QI_C = \frac{V_{Rf}}{R_f}$$

$$QI_C = 1\ mA$$

4 Calculate Junction Resistance

$$R'_e = \frac{25}{QI_C}$$

$$R'_e = 25\ \Omega$$

V_{CC} 24 V

R_L 16 kΩ

R_{B1} 130 kΩ

Q_1

R_{B2} 13 kΩ

R_f 1.3 kΩ

V_{bias} 0 V

5 Calculate the Load Resistor Voltage Drop:

$$V_{RL} = QI_C \times R_L$$

$$V_{RL} = 16\ V$$

6 Calculate the Signal Load Impedance:

$$R_{Lac} = \frac{R_L \times R_{LW}}{R_L + R_{LW}}$$

$$R_{Lac} = 8\ k\Omega$$

7 **(A)** Base Bias Calculate C-to-E Voltage:

$$V_{CE} = V_{CC} - (V_{RL} + V_{Rf})$$

$$V_{CE} = 6.7\ V$$

7 **(B)** Emitter Bias Calculate C-to-E Voltage

$$V_{CE} = V_{CC} - V_{RL}$$

$$V_{CE} =$$

a. Sheet 1: dc Parameters

FIGURE 5-6   Worksheet for the Common Emitter Circuit, Sheet #2

Sheet #2: Signal Parameters

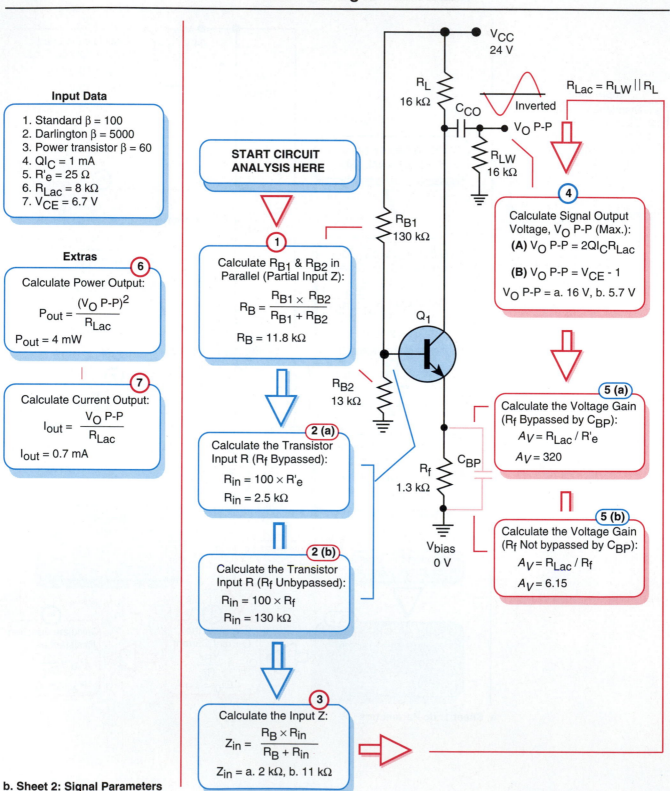

Input Data

1. Standard $\beta = 100$
2. Darlington $\beta = 5000$
3. Power transistor $\beta = 60$
4. $QI_C = 1$ mA
5. $R'_e = 25\ \Omega$
6. $R_{Lac} = 8$ kΩ
7. $V_{CE} = 6.7$ V

Extras

6 Calculate Power Output:

$$P_{out} = \frac{(V_O\ P\text{-}P)^2}{R_{Lac}}$$

$P_{out} = 4$ mW

7 Calculate Current Output:

$$I_{out} = \frac{V_O\ P\text{-}P}{R_{Lac}}$$

$I_{out} = 0.7$ mA

START CIRCUIT ANALYSIS HERE

1 Calculate R_{B1} & R_{B2} in Parallel (Partial Input Z):

$$R_B = \frac{R_{B1} \times R_{B2}}{R_{B1} + R_{B2}}$$

$R_B = 11.8$ kΩ

2 (a) Calculate the Transistor Input R (R_f Bypassed):

$R_{in} = 100 \times R'_e$

$R_{in} = 2.5$ kΩ

2 (b) Calculate the Transistor Input R (R_f Unbypassed):

$R_{in} = 100 \times R_f$

$R_{in} = 130$ kΩ

3 Calculate the Input Z:

$$Z_{in} = \frac{R_B \times R_{in}}{R_B + R_{in}}$$

$Z_{in} = $ a. 2 kΩ, b. 11 kΩ

V_{CC}
24 V

R_L
16 kΩ

C_{CO}

Inverted

V_O P-P

R_{LW}
16 kΩ

$R_{Lac} = R_{LW} \| R_L$

R_{B1}
130 kΩ

Q_1

R_{B2}
13 kΩ

R_f
1.3 kΩ

C_{BP}

V_{bias}
0 V

4 Calculate Signal Output Voltage, V_O P-P (Max.):

(A) V_O P-P $= 2QI_C R_{Lac}$

(B) V_O P-P $= V_{CE} - 1$

V_O P-P $= $ a. 16 V, b. 5.7 V

5 (a) Calculate the Voltage Gain (R_f Bypassed by C_{BP}):

$A_V = R_{Lac} / R'_e$

$A_V = 320$

5 (b) Calculate the Voltage Gain (R_f Not bypassed by C_{BP}):

$A_V = R_{Lac} / R_f$

$A_V = 6.15$

b. Sheet 2: Signal Parameters

Analyzing the Base-biased Common Emitter Circuit

As you read the description of the analytical procedure here, refer to Figure 5-4 for a depiction of the circuit involved and to Figures 5-5 and 5-6 for filled-in versions of the two worksheet pages.

Introduction to Sheet #1

Resistors R_{B1} and R_{B2} form a voltage divider to provide the bias voltage to set the Q-point collector current. We usually treat the voltage divider as an unloaded voltage divider. Once in a while, the transistor input circuit represents a significant load for the voltage divider, or we may have reason to want more accurate values. In such cases we then perform Step 1A. Most of the time, we skip it.

Because ß does exist, the factor $ß \times R_f$ can be significant in some circuits. Feedback frees us from having known ß accurately, but we do need *some* value for ß. Hence, we adopt a value of 100 as a standard ß. Any reasonable value would do, but 100 is a good value and makes the arithmetic easier.

Step 1: **Finding the Voltage Drop Across R_{B2}**

$$V_{RB2} = V_{CC}\left(\frac{R_{B2}}{R_{B1} + R_{B2}}\right)$$

$$= 24\text{ V}\left(\frac{13\text{ k}\Omega}{130\text{ k}\Omega + 13\text{ k}\Omega}\right)$$

$$= 2.18\text{ V}$$

Step 2: **Calculating the Voltage Drop Across the Feedback Resistor (R_f)**

The voltage drop across the feedback resistor is always 0.6 to 0.7 volts less than the voltage drop across the base-to-ground resistor (R_{B_2}). In this step, we subtract the 0.7-volt junction voltage from the voltage across R_{B_2}.

$$V_{R_f} = V_{RB2} - 0.7\text{ V}$$

$$= 2.18\text{ V} - 0.7\text{ V}$$

$$= 1.48\text{ V}$$

Step 3: **Calculating the Quiescent Collector Current (QI_C)**

Here we assume that the collector current and the emitter current are equal. We can find the emitter current by using Ohm's law, because we know the voltage drop across the feedback resistor and its resistance. The emitter current flows through the feedback resistor R_f. Thus, we have

$$Q_{IC} = V_{R_f} / R_f$$
$$= 1.48 \text{ V}/1.3 \text{ k}\Omega = 1.0 \text{ mA}$$

Note: The calculator says 1.1384. . . . Ordinarily, I would call it 1.0 mA.

Step 4: Calculating the Base–Emitter Junction Resistance (R_e')

We need the value of the junction resistance in order to calculate the voltage gain for one of the circuit variations on sheet #2 (signal parameters) of the worksheet. To obtain this, we use the Shockley relationship:

$$R_e' = 25/QI_C$$
$$= 25 \text{ mV}/1.0 \text{ mA} = 25 \text{ } \Omega$$

Step 5: Calculating the Dummy-load Voltage Drop

The dummy load R_L has a voltage drop (V_{RL}) that is a function of its resistance and the current through it. By Ohm's law, we get

$$V_{RL} = QI_C \times R_L$$
$$= 1.0 \text{ mA} \times 16 \text{ k}\Omega = 16.0 \text{ V}$$

Step 6: Calculating the ac Signal-load Impedance R_{Lac}

This step simply calculates the equivalent value of the dummy-load (R_L) and the working-load (R_{LW}) impedances, connected in parallel. For **dc** calculations, the usual output coupling capacitor blocks the **dc** component from reaching the working-load impedance (R_{LW}). Because the two impedances (R_L and R_{LW}) are only really in parallel for signal frequencies, the calculation should have been saved for sheet #2 of the worksheet, which deals specifically with **ac** signal parameters. The only reason we do it here is to keep sheet #2 from getting more crowded and confusing. Our calculation is

$$R_{Lac} = \frac{R_L \times R_{LW}}{R_L + R_{LW}}$$

$$= \frac{16 \text{ k}\Omega \times 16 \text{ k}\Omega}{16 \text{ k}\Omega + 16 \text{ k}\Omega}$$

$$= 8 \text{ k}\Omega$$

Step 7: Calculate the Collector-to-Emitter Voltage (V_{CE})

In this case, we have two options: base bias or emitter bias. Since our circuit is base-biased by voltage divider R_{B1}/R_{B2}, we use the base-bias option. We know by Kirchhoff's law that

$$V_{RL} + V_{CE} + R_f = V_{CC} \text{ [24 V]}$$

Therefore,

$$V_{CE} = V_{CC} - (V_{RL} + V_{Rf})$$
$$= 24 \text{ V} - (16 \text{ V} + 1.3 \text{ V})$$
$$= 6.7 \text{ V}$$

Introduction to Sheet #2

The bypass capacitor (C_{BP}) across the feedback resistor (R_f) significantly lowers the circuit input impedance, but it also causes a drastic increase in the circuit voltage gain. In general, the lowered input impedance is considered a good trade-off for the greatly improved voltage gain. As a result, most circuits are bypassed.

In a few applications, accurate voltage gain predictability and low signal distortion are more important than lots of (not very predictable) voltage gain. In those cases, the bypass capacitor is not used. The capacitor effectively eliminates the negative feedback at signal frequencies, but it has no effect on the negative feedback for **dc** or stability parameters.

For this example, we will calculate the input impedance for both conditions—with and without the bypass capacitor (C_{BP})—so that we can compare the two easily. This should give you some idea of the relative difference, although the absolute amount of difference depends on the specific circuit design. This circuit is pretty typical.

Step 0: Copying Data from Sheet #1

The four pieces of relevant data are

$$QI_C = 1.0 \text{ mA}$$
$$R'_e = 25 \text{ } \Omega$$
$$V_{R_{Lac}} = 8 \text{ k}\Omega$$
$$V_{CE} = 6.7 \text{ V}$$

Step 1: Calculating R_B, the Parallel Combination of R_{B1} and R_{B2}

Resistors R_{B1} and R_{B2} are a part of the input impedance, whether or not the bypass capacitor is installed, so let's get that calculation out of the way first. Resistor R_{B2} is obviously part of the input impedance because it is directly grounded. Resistor R_{B1}'s path to ground is a little less obvious, but R_{B1} is effectively grounded through the power supply via V_{CC}. An acceptable power supply must have an internal impedance of near 0 ohms. We calculate R_b as follows:

$$Z_{in} = \frac{R_{B2} \times R_{B1}}{R_{B2} + R_{B1}}$$
$$= \frac{13 \text{ k}\Omega \times 130 \text{ k}\Omega}{13 \text{ k}\Omega + 130 \text{ k}\Omega}$$
$$= 11.8 \text{ k}\Omega$$

For troubleshooting purposes, we could call Z_{in} equal to 11.8 kΩ and be close enough most of the time. But in some settings—particularly those involving power amplifiers (IC = 250 mA or more)—you should go on with the rest of the steps.

Step 2(a): Calculating the Transistor Input Impedance (R_{in}) If R_f Is Bypassed

$$R_{in} = 100 \times R_e' \qquad [100 = ß]$$
$$= 100 \times 25 \ \Omega = 2.5 \ k\Omega$$

Step 2(b): Calculating the Transistor Input Impedance (R_{in}) If R_f Is Not Bypassed

$$R_{in} = 100 \times R_f \qquad [100 = ß]$$
$$= 100 \times 1.3 \ k\Omega = 130 \ k\Omega$$

Step 3: Calculating the Final Total Input Impedance (Z_{in})

Let's do step 3 twice: once for a bypassed R_f, and once for an unbypassed R_f. Then we can easily compare the two values to see how much difference there is with and without the bypass capacitor. We will do the same thing with the voltage gain calculations when we get to those steps in the procedure.

Bypassed:

$$Z_{in} = \frac{R_B \times R_{in}}{R_B + R_{in}}$$
$$= \frac{11.8 \ k\Omega \times 2.5 \ k\Omega}{11.8 \ k\Omega + 2.5 \ k\Omega}$$
$$= 2 \ k\Omega \qquad \text{(bypassed)}$$

(Calculated: 2.06 kΩ)

Unbypassed:

$$Z_{in} = \frac{R_B \times R_{in}}{R_B + R_{in}}$$
$$= \frac{11.8 \ k\Omega \times 130 \ k\Omega}{11.8 \ k\Omega + 130 \ k\Omega}$$
$$= 11 \ k\Omega \qquad \text{(unbypassed)}$$

(Calculated: 10.8 kΩ)

Step 4: Calculating the Maximum P–P Output Voltage (V_O) Without Clipping

The traditional equations used to predict the maximum undistorted P–P signal output voltage are very difficult to use and require accurate values for several variables. Not coincidentally, the traditional equations usually yield disappointing results in real circuits.

The equations presented here constitute a useful alternative, based on testing many hundreds of circuits. Statistically based empirical equations are not known for great precision, but these work well and are sufficiently conservative to ensure that you will always get at least as much signal output voltage as you calculate. You may get a larger maximum output voltage than your math predicts, but this is never a problem. Not having enough output signal is the condition to avoid.

Two **dc** voltages primarily limit the signal output voltage: $V_{R_{Lac}}$ ($2QI_C \times R_{Lac}$) and V_{CE}. Normally, one of these reaches its limit before the other. And the first one to run out of voltage starts to clip the waveform, introduces distortion, and represents the real limit. Clipping is only sometimes symmetrical.

Logic might lead us to wonder why one of the equations is $V_O = V_{CE} - 1$ instead of $2V_{CE} - 1$, if we expect such symmetrical action. But hundreds of measurements tell us that the expected symmetry isn't there. $V_O = V_{CE} - 1$ is correct. Because one of the two factors starts clipping before the other, we must calculate both, and then select the smaller of the two values.

In our example, we must select the lesser of

$$V_O P\text{–}P_{(max)} = 2QI_C \times R_{Lac}$$
$$= 2 \, (1 \text{ mA}) \times 8 \text{ k}\Omega$$
$$= 16 \text{ V } P\text{–}P$$

and

$$V_O P\text{–}P_{(max)} = V_{CE} - 1$$
$$= 6.7 \text{ V} - 1 \text{ V}$$
$$= 5.7 \text{ V } P\text{–}P$$

Obviously, our choice in this case must be

$$V_O P\text{–}P_{(max)} = 5.7 \text{ V } P\text{–}P$$

Step 5: Calculating Voltage Gain (A_V) With and Without the Bypass Capacitor Installed

If the bypass capacitor is installed, it provides an effective short across the feedback resistor (R_f) for signal frequencies. The only signal frequency emitter and feedback impedance is the junction resistance R'_e. In this example, that resistance is only 25 ohms. In the bypassed circuit, the voltage gain is a function of the emitter resistance (R'_e) and the collector impedance (R_{Lac}).

With the Bypass Capacitor:

$$A_V = R_{Lac} / R'_e$$
$$= 8 \text{ k}\Omega/25 \text{ }\Omega$$
$$= 8000/25 = 320$$

Without the Bypass Capacitor:

$$A_V = R_{Lac} / R_f$$
$$= 8 \text{ k}\Omega/1.3 \text{ k}\Omega$$
$$= 8000/1300 = 6.15$$

The voltage gain of 320 with the bypass capacitor is much greater than the voltage gain of 6.15 without it. Adding the bypass capacitor also lowered the circuit input impedance, but to a much lesser extent. The input impedances were 2 kΩ bypassed and only 11 kΩ unbypassed. The much larger voltage gain makes the bypassed configuration by far the more popular of the two versions.

Step 6: Calculating the Maximum Circuit Power Output (P_{out})

This is a standard power formula: $P = V^2/R$. The V part of the equation is the maximum P–P output voltage. Thus, we have

$$P_{out} = V_O \text{ P--P}/R_{Lac} = (5.7 \text{ V})^2/8 \text{ k}\Omega = 32.5/8 \text{ k}\Omega = 0.004 \text{ W (4 milliwatts)}$$

Step 7: Calculating the Maximum Output Current (I_O P–P)

This is directly obtainable from Ohm's law ($I = V/R$):

$$(I_O \text{ P--P}) = (V_O \text{ P--P})/R_{Lac} = 5.7 \text{ V}/8 \text{ k}\Omega = 0.0007 \text{ A} \qquad (0.7 \text{ milliamps})$$

Analyzing the Base-biased Common Base Circuit

Figure 5-7 shows the schematic diagram of a common base circuit. It is infrequently used, and the analysis of the circuit differs very little from that of the common emitter circuit, which we just went through in some detail.

You can use the same analysis sheets. Indeed, R_{B_1}, R_{B_2}, R_L, R_f, R_{LW}, and the power supply voltage are the same here as in the previous common emitter example. There is nothing unusual about the similarity in values, because most designers use the same approach for both circuits.

The only real difference between the common base and common emitter circuits involves where the input signal is connected. The common emitter circuit uses the base as the input element. The common base circuit applies the signal to the emitter, which results in a normally impractically low signal input impedance.

Capacitor C_1 serves as a signal bypass, so no signal voltage develops across R_{B_2}(and R_{B_1}). It must be there, but it doesn't change the analysis procedure. We will not repeat the sheet #1 calculations for this example, because they are identical to those in the common emitter example. We will start with worksheet sheet #2, the **ac** signal parameters (see Figure 5-8).

FIGURE 5-7

Common Base Circuit for Worksheet Analysis

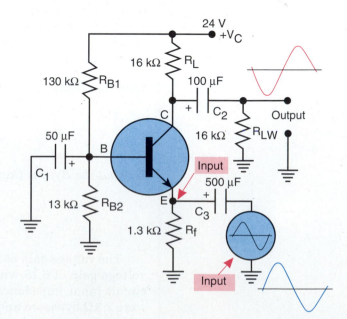

FIGURE 5-8 Worksheet for the Common Base Circuit, Sheet #2

Sheet #2: Signal Parameters

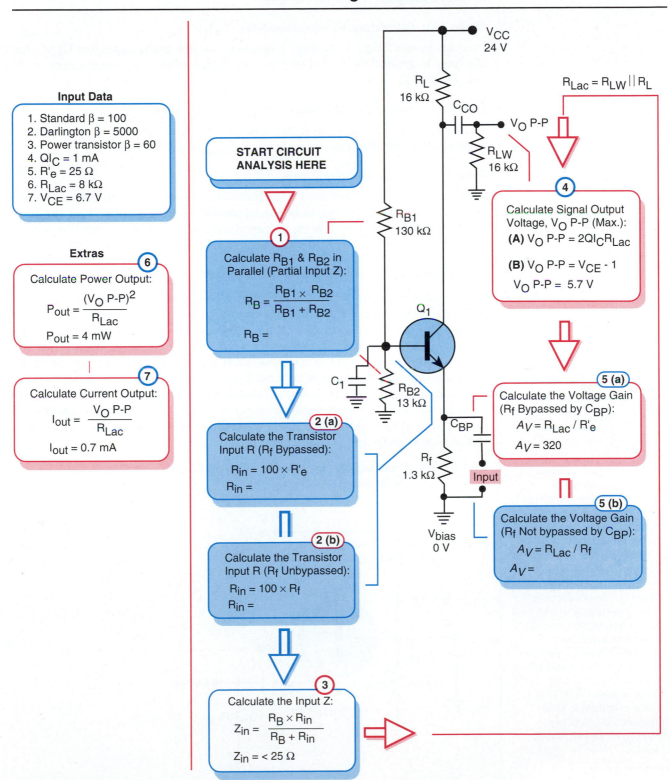

Input Data

1. Standard $\beta = 100$
2. Darlington $\beta = 5000$
3. Power transistor $\beta = 60$
4. $QI_C = 1$ mA
5. $R'_e = 25$ Ω
6. $R_{Lac} = 8$ kΩ
7. $V_{CE} = 6.7$ V

Extras

6 Calculate Power Output:

$$P_{out} = \frac{(V_O \text{ P-P})^2}{R_{Lac}}$$

$P_{out} = 4$ mW

7 Calculate Current Output:

$$I_{out} = \frac{V_O \text{ P-P}}{R_{Lac}}$$

$I_{out} = 0.7$ mA

START CIRCUIT ANALYSIS HERE

1 Calculate R_{B1} & R_{B2} in Parallel (Partial Input Z):

$$R_B = \frac{R_{B1} \times R_{B2}}{R_{B1} + R_{B2}}$$

$R_B =$

2 (a) Calculate the Transistor Input R (R_f Bypassed):

$R_{in} = 100 \times R'_e$

$R_{in} =$

2 (b) Calculate the Transistor Input R (R_f Unbypassed):

$R_{in} = 100 \times R_f$

$R_{in} =$

3 Calculate the Input Z:

$$Z_{in} = \frac{R_B \times R_{in}}{R_B + R_{in}}$$

$Z_{in} = < 25$ Ω

V_{CC} 24 V

R_L 16 kΩ

C_{CO}

V_O P-P

R_{LW} 16 kΩ

$R_{Lac} = R_{LW} \| R_L$

R_{B1} 130 kΩ

Q_1

C_1

R_{B2} 13 kΩ

C_{BP}

R_f 1.3 kΩ

Input

V_{bias} 0 V

4 Calculate Signal Output Voltage, V_O P-P (Max.):

(A) V_O P-P $= 2QI_C R_{Lac}$

(B) V_O P-P $= V_{CE} - 1$

V_O P-P $= 5.7$ V

5 (a) Calculate the Voltage Gain (R_f Bypassed by C_{BP}):

$A_V = R_{Lac} / R'_e$

$A_V = 320$

5 (b) Calculate the Voltage Gain (R_f Not bypassed by C_{BP}):

$A_V = R_{Lac} / R_f$

$A_V =$

Step 0: Copying Data from Sheet #1

This step and the values in it are identical to those in the previous common emitter example.

Steps 1–3: Calculating the Input Impedance (Z_{in})

You can ignore Steps 1 through 3 because you are using capacitor C_1 to short the base to ground for signal frequencies. If you wish to make an entry in this box, enter 0 Ω.

Step 4: Calculating the Maximum *P–P* Signal Output Voltage

This step and the values in it are identical to those in the previous common emitter example. The end value is V_O $P–P_{(max)}$ = 5.7 V *P–P*.

Step 5: Calculating the Voltage Gain

This step and the values in it are identical to those in the previous common emitter example. However, only the bypassed version applies, and the input source must have a very low input impedance. This usually raises the need for a transformer of some sort. The voltage gain is 320 (see Figure 5-8).

Analyzing the Base-biased Common Collector Circuit

Figure 5-9 shows the schematic diagram of a common collector circuit, sometimes called an *emitter-follower*. This circuit is used fairly often, even though it always has a voltage gain of 1 (unity). The common collector circuit is often used as a buffer amplifier, because it features a high input impedance that won't load critical circuits or components. It also has a lower output impedance than the common emitter circuit, and it can tolerate heavier working load current demands than can the common emitter circuit.

The value of the working-load (R_{LW}) must not have a resistance (impedance) value of less than $0.1 \times R_f$. Lower values of R_{LW} cause the voltage gain to fall below unity, and introduce distortion.

FIGURE 5-9   Common Collector (Emitter-follower) Circuit for Worksheet Analysis

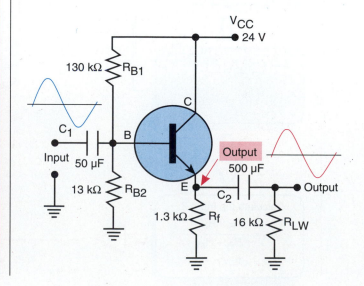

You can use the same analysis sheets we used in the previous analysis examples (see Figures 5-10 and 5-11). Again, R_{B1}, R_{B2}, R_f, R_{LW}, and the power-supply voltage in Figure 5-7 are the same as those in the previous common emitter circuit.

In this circuit, however, R_L (the dummy load) is missing. Because we are not taking any output signal from the collector, R_L serves no purpose. The current and power that R_L would normally waste as heat are now available as output current and power.

Once in a while, you will see a common-collector circuit with a dummy load. The circuit designer probably had a little higher V_{CC} than was needed, and R_L was added to get rid of some surplus voltage. This is not a very common design solution.

The only real difference between the common collector circuit and the common emitter circuit is that the common emitter circuit uses the collector as the output terminal, whereas the common collector circuit takes the output signal from the emitter. Because the signal voltage on the base and the signal voltage on the emitter are almost exactly the same, the voltage gain is very close to 1 (unity). The details of the analysis are spelled out on worksheet sheet #2 (Figure 5-11), particularly Steps 4 through 7.

5.7   Emitter-biased Circuits

Bias current for the base-to-emitter circuit can also be provided by a battery or other voltage source connected between the emitter and ground, as illustrated in Figure 5-12(a) and (b). A battery or other source could be installed between the base and ground, but a voltage divider is cheaper for base bias. Additional voltage supplies connected in the base circuit are generally a poor way to go, because the extra voltage source interferes with the input signal source.

The emitter-biased circuit in Figure 5-12 offers a number of advantages over voltage-divider base bias. First, it leaves the base free to use either a base-to-ground resistor, as in part (a), or to use a direct-coupled signal source, as in part (b).

A signal source cannot be connected directly to the base in a voltage-divider base-biased circuit, because the internal resistance of the signal source would then be in parallel with R_{B2}. Additional parallel resistance would alter the base bias and change the Q-point. Voltage-divider base-biased circuits require an input coupling capacitor, and this places a lower limit on the frequency response of the amplifier.

The emitter-biased circuit in Figure 5-12(b) allows the signal source to be connected directly to the base, permitting the frequency response to go all the way down to 0 Hz **(dc).** The signal source can have any internal resistance that does not exceed ten times the value of the emitter-bias/feedback resistor (R_f). This is a Q-point stability requirement and is almost never a problem to meet.

The second advantage of emitter bias is that we can make the emitter feedback resistor value as large as we wish. Highly stable circuits are possible without further circuit compromises, because the bias circuit is almost totally

FIGURE 5-10   Worksheet for the Common Collector Circuit, Sheet #1

Sheet #1: dc Voltages

Input Data

1. Standard β = 100
2. Darlington β = 5000
3. Power transistor β = 60
4. R_{B1} = 130 kΩ
5. R_{B2} = 13 kΩ
6. R_L = 0 Ω
7. R_f = 1.3 kΩ
8. R_{LW} = 16 kΩ

Special Considerations

1. The working load is connected to the emitter.
2. No dummy load collector resistor is required. The collector current is independent of collector voltage.
3. Voltage gain always = 1.

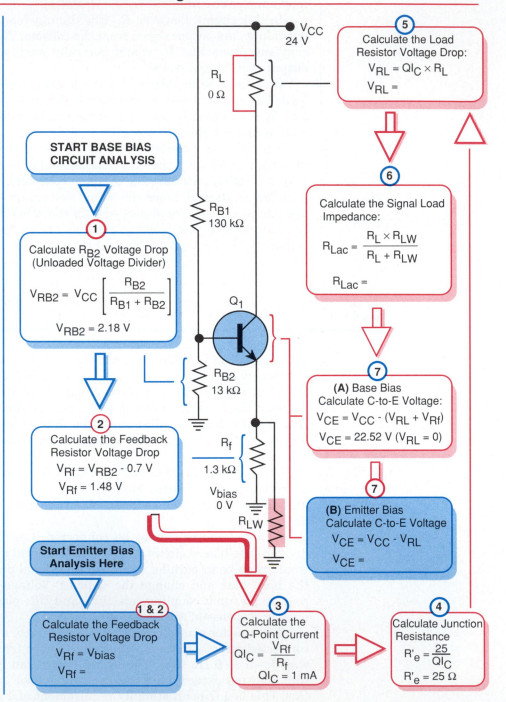

START BASE BIAS CIRCUIT ANALYSIS

1 Calculate R_{B2} Voltage Drop (Unloaded Voltage Divider)

$$V_{RB2} = V_{CC} \left[\frac{R_{B2}}{R_{B1} + R_{B2}} \right]$$

$$V_{RB2} = 2.18 \text{ V}$$

2 Calculate the Feedback Resistor Voltage Drop

$$V_{Rf} = V_{RB2} - 0.7 \text{ V}$$

$$V_{Rf} = 1.48 \text{ V}$$

Start Emitter Bias Analysis Here

1 & 2 Calculate the Feedback Resistor Voltage Drop

$$V_{Rf} = V_{bias}$$

$$V_{Rf} =$$

V_{CC} 24 V

R_L 0 Ω

R_{B1} 130 kΩ

Q_1

R_{B2} 13 kΩ

R_f 1.3 kΩ

V_{bias} 0 V

R_{LW}

5 Calculate the Load Resistor Voltage Drop:

$$V_{RL} = QI_C \times R_L$$

$$V_{RL} =$$

6 Calculate the Signal Load Impedance:

$$R_{Lac} = \frac{R_L \times R_{LW}}{R_L + R_{LW}}$$

$$R_{Lac} =$$

7 **(A)** Base Bias Calculate C-to-E Voltage:

$$V_{CE} = V_{CC} - (V_{RL} + V_{Rf})$$

$$V_{CE} = 22.52 \text{ V } (V_{RL} = 0)$$

7 **(B)** Emitter Bias Calculate C-to-E Voltage

$$V_{CE} = V_{CC} - V_{RL}$$

$$V_{CE} =$$

3 Calculate the Q-Point Current

$$QI_C = \frac{V_{Rf}}{R_f}$$

$$QI_C = 1 \text{ mA}$$

4 Calculate Junction Resistance

$$R'_e = \frac{25}{QI_C}$$

$$R'_e = 25 \text{ }\Omega$$

FIGURE 5-11 Worksheet for the Common Collector Circuit, Sheet #2

Sheet #2: Signal Parameters

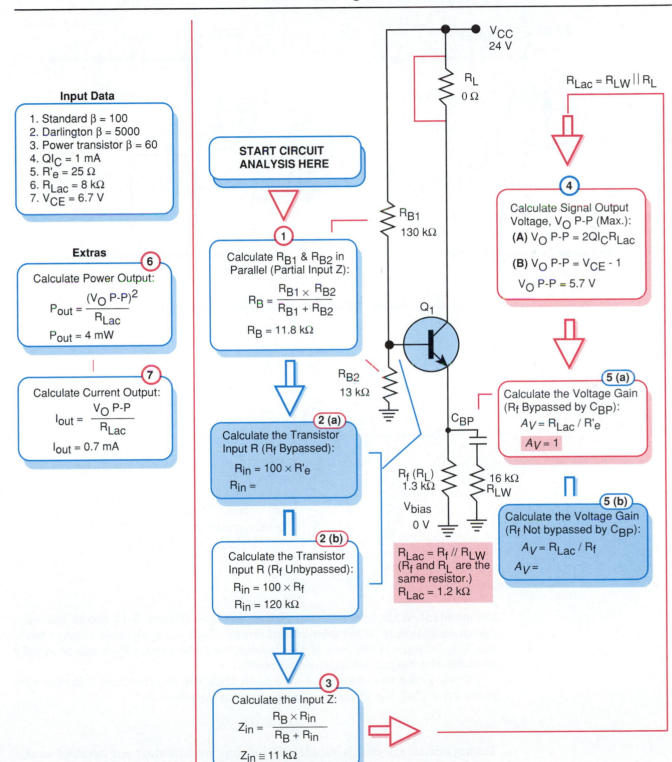

Input Data

1. Standard β = 100
2. Darlington β = 5000
3. Power transistor β = 60
4. QI_C = 1 mA
5. R'_e = 25 Ω
6. R_{Lac} = 8 kΩ
7. V_{CE} = 6.7 V

Extras

6 Calculate Power Output:

$$P_{out} = \frac{(V_O \text{ P-P})^2}{R_{Lac}}$$

P_{out} = 4 mW

7 Calculate Current Output:

$$I_{out} = \frac{V_O \text{ P-P}}{R_{Lac}}$$

I_{out} = 0.7 mA

START CIRCUIT ANALYSIS HERE

1 Calculate R_{B1} & R_{B2} in Parallel (Partial Input Z):

$$R_B = \frac{R_{B1} \times R_{B2}}{R_{B1} + R_{B2}}$$

R_B = 11.8 kΩ

2 (a) Calculate the Transistor Input R (R_f Bypassed):

$R_{in} = 100 \times R'_e$

R_{in} =

2 (b) Calculate the Transistor Input R (R_f Unbypassed):

$R_{in} = 100 \times R_f$

R_{in} = 120 kΩ

3 Calculate the Input Z:

$$Z_{in} = \frac{R_B \times R_{in}}{R_B + R_{in}}$$

$Z_{in} \cong$ 11 kΩ

V_{CC} 24 V

R_L 0 Ω

R_{B1} 130 kΩ

R_{B2} 13 kΩ

Q_1

C_{BP}

$R_f (R_L)$ 1.3 kΩ

16 kΩ R_{LW}

V_{bias} 0 V

$R_{Lac} = R_f // R_{LW}$
(R_f and R_L are the same resistor.)
R_{Lac} = 1.2 kΩ

$R_{Lac} = R_{LW} || R_L$

4 Calculate Signal Output Voltage, V_O P-P (Max.):
(A) V_O P-P = $2QI_C R_{Lac}$
(B) V_O P-P = $V_{CE} - 1$
V_O P-P = 5.7 V

5 (a) Calculate the Voltage Gain (R_f Bypassed by C_{BP}):
$A_V = R_{Lac} / R'_e$
A_V = 1

5 (b) Calculate the Voltage Gain (R_f Not bypassed by C_{BP}):
$A_V = R_{Lac} / R_f$
A_V =

FIGURE 5-12   Emitter-biased Common Emitter Circuit

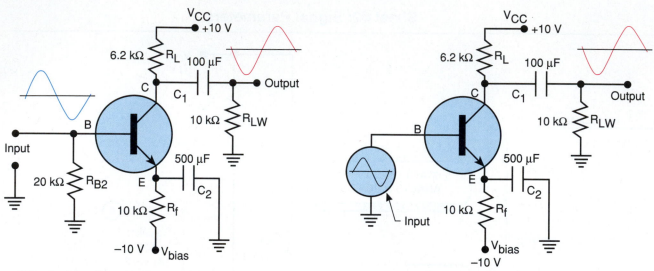

a. Circuit with a Base-To-Ground Resistor

b. Using the Internal Resistance of the Signal Source as R_{B2}

FIGURE 5-13   Voltage Distribution in an Emitter-biased Circuit

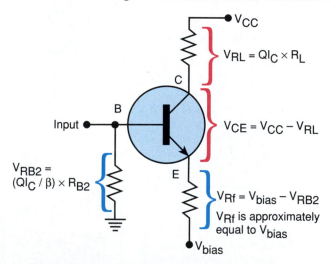

independent of the other output circuit voltages. Figure 5-13 shows the voltage distribution in an emitter-biased circuit. There is a division of labor here: the V_{bias} supply takes care of the biasing, and the main (V_{CC}) power supply provides the output voltage and current.

If we make the common assumption that the emitter and collector currents are equal, the Q-point collector current equation is:

$$QI_C = V_{bias}/R_f$$

Taking advantage of this simple relationship, we will start our detailed analysis of sample emitter-biased circuits (in section 5.9) by calculating QI_C.

FIGURE 5-14   Circuit for the Emitter-biased Loop Analysis

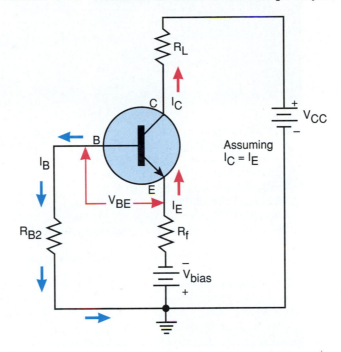

FOR THE MATHEMATICALLY INCLINED

The following series of calculations represents the loop analysis for the emitter-biased circuit (see Figure 5-14).

Earlier, we discussed the fact that the value of the feedback resistor in the emitter circuit is magnified by the factor beta ($R_{in} = ß \times R_f$). There is another side to this coin. If we measure the value of the base-to-ground resistor (R_{B_2}) from the emitter circuit, the value of R_{B_2} appears to be R_{B_2} divided by beta. This should not be too surprising, since we have a current in the emitter circuit that is greater by a factor of beta than the current in the base circuit.

Calculating the Quiescent Collector Current, Using Emitter Bias

Suppose that we write a Kirchhoff's loop equation (see Figure 5-14). Kirchhoff's base–emitter loop equation for emitter biased circuits is

$$V_{bias} - I_B R_{B_2} - I_E R_f - V_{BE} = 0$$

Assuming that $I_C = I_E$, we can write

$$V_{bias} - I_B R_{B_2} - I_C R_f - V_{BE} = 0$$

Because $I_B = I_C / ß$, we can substitute

$$V_{bias} - (I_C R_{B_2} / ß) - I_C R_f - V_{BE} = 0$$

Rearranging the equation, we obtain

$$I_C R_f + (I_C R_{B_2} / ß) = V_{bias} - V_{BE}$$

(continued)

Factoring out I_C, we get

$$I_C \, [R_f + (R_{B2}/\beta)] = V_{bias} - V_{BE}$$

Solving for I_C, we get

$$I_C = (V_{bias} - V_{BE})/R_f + (R_{B2}/\beta)$$

Notice the term R_{B2}/β. This implies that R_{B2} appears to be divided by the factor beta with respect to resistances on the emitter side of the loop. The reason is that beta times as much current is flowing on the emitter side. What this means for the bias loop is that a very small voltage drop occurs across the base-to-ground resistor, relative to voltage drops on the emitter side of the bias loop.

We can simplify the equation for calculating the quiescent collector by making a couple of normally valid assumptions. If we assume that V_{bias} is much larger than the 0.7 V junction voltage and that (R_B/β) is much less than the feedback resistor R_f, the collector current (in this case, QI_C) reduces to

$$QI_C = V_{bias}/R_f$$

The assumptions generally are valid, since a large voltage drop across the feedback resistor must be used; otherwise, there is no point in using the emitter-biased approach. The second assumption must be valid because of mandatory stability requirements.

Analyzing the Emitter-biased, Common Emitter Circuit

We now undertake our first worksheet analysis of an emitter-biased circuit.

Introduction to Sheet #1

The circuit we will use for this example appears in Figure 5-15. Because this circuit is emitter-biased, we can ignore Steps 1 and 2 on the base-bias branch of worksheet sheet #1 (see Figure 5-16). Step 1 for the emitter-bias branch simply tells you where to start the procedure. Step 2 reminds you that we are assuming that the emitter and collector currents are equal.

Step 3: Calculating the Quiescent Collector Current (QI_C)

The actual process starts here:

$$QI_C = \frac{V_{R_f}}{R_f}$$

$$= \frac{10 \text{ V}}{10 \text{ k}\Omega}$$

$$= 1.0 \text{ mA}$$

FIGURE 5-15 Emitter-biased Common Emitter Circuit for Worksheet Analysis

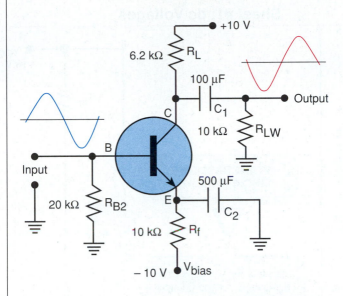

Step 4: Calculating the Shockley Relationship Junction Resistance (R_e')

$$R_e' = \frac{25}{QI_C}$$

$$= \frac{25\ (\text{mV})}{1.0\ \text{mA}}$$

$$= 25\ \Omega$$

Step 5: Calculating the Voltage Drop Across the Dummy-load Resistor (R_L)

$$V_{R_L} = QI_C \times R_L$$
$$= 1.0\ \text{mA} \times 6.2\ \text{k}\Omega$$
$$= 6.2\ \text{V}$$

Step 6: Calculating the Total Signal–load Impedance ($R_{L\text{ac}}$)

$$R_{L\text{ac}} = \frac{R_L \times R_{LW}}{R_L + R_{LW}}$$

$$= \frac{6.2\ \text{k}\Omega \times 10\ \text{k}\Omega}{6.2\ \text{k}\Omega + 10\ \text{k}\Omega}$$

$$= 3.8\ \text{k}\Omega$$

Step 7 (Option B): Calculating the Collector-to-Emitter Voltage (V_{CE})

$$V_{CE} = V_{CC} - V_{R_L}$$
$$= 10\ \text{V} - 6.2\ \text{V}$$
$$= 3.8\ \text{V}$$

FIGURE 5-16   Worksheet for the Emitter-biased Common Emitter Circuit, Sheet #1

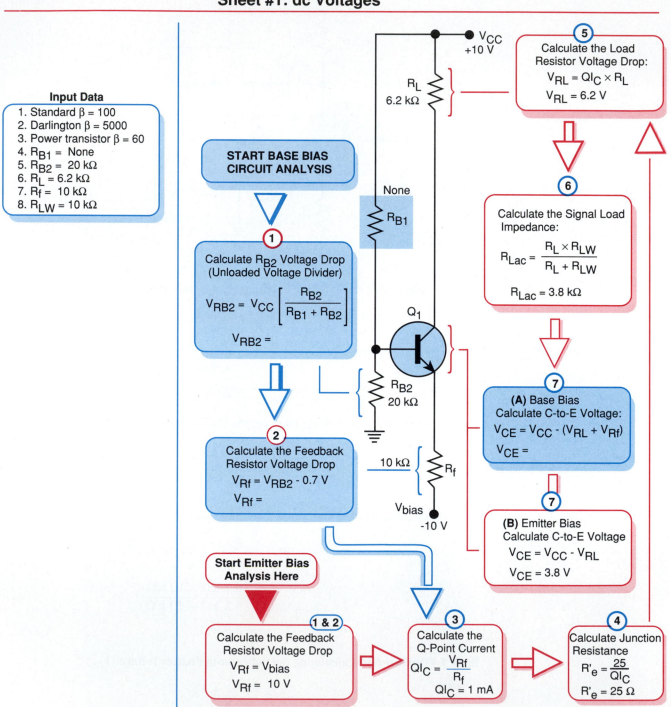

Sheet #1: dc Voltages

Input Data
1. Standard β = 100
2. Darlington β = 5000
3. Power transistor β = 60
4. R_{B1} = None
5. R_{B2} = 20 kΩ
6. R_L = 6.2 kΩ
7. R_f = 10 kΩ
8. R_{LW} = 10 kΩ

START BASE BIAS CIRCUIT ANALYSIS

(1) Calculate R_{B2} Voltage Drop (Unloaded Voltage Divider)
$$V_{RB2} = V_{CC}\left[\frac{R_{B2}}{R_{B1} + R_{B2}}\right]$$
$V_{RB2} =$

(2) Calculate the Feedback Resistor Voltage Drop
$V_{Rf} = V_{RB2} - 0.7\text{ V}$
$V_{Rf} =$

Start Emitter Bias Analysis Here

(1 & 2) Calculate the Feedback Resistor Voltage Drop
$V_{Rf} = V_{bias}$
$V_{Rf} = 10\text{ V}$

(3) Calculate the Q-Point Current
$$QI_C = \frac{V_{Rf}}{R_f}$$
$QI_C = 1\text{ mA}$

(4) Calculate Junction Resistance
$$R'_e = \frac{25}{QI_C}$$
$R'_e = 25\ \Omega$

V_{CC} +10 V

R_L 6.2 kΩ

None R_{B1}

Q_1

R_{B2} 20 kΩ

10 kΩ R_f

V_{bias} -10 V

(5) Calculate the Load Resistor Voltage Drop:
$V_{RL} = QI_C \times R_L$
$V_{RL} = 6.2\text{ V}$

(6) Calculate the Signal Load Impedance:
$$R_{Lac} = \frac{R_L \times R_{LW}}{R_L + R_{LW}}$$
$R_{Lac} = 3.8\text{ k}\Omega$

(7) **(A)** Base Bias Calculate C-to-E Voltage:
$V_{CE} = V_{CC} - (V_{RL} + V_{Rf})$
$V_{CE} =$

(7) **(B)** Emitter Bias Calculate C-to-E Voltage
$V_{CE} = V_{CC} - V_{RL}$
$V_{CE} = 3.8\text{ V}$

FIGURE 5-17   Worksheet for the Emitter-biased Common Emitter Circuit, Sheet #2

Sheet #2: Signal Parameters

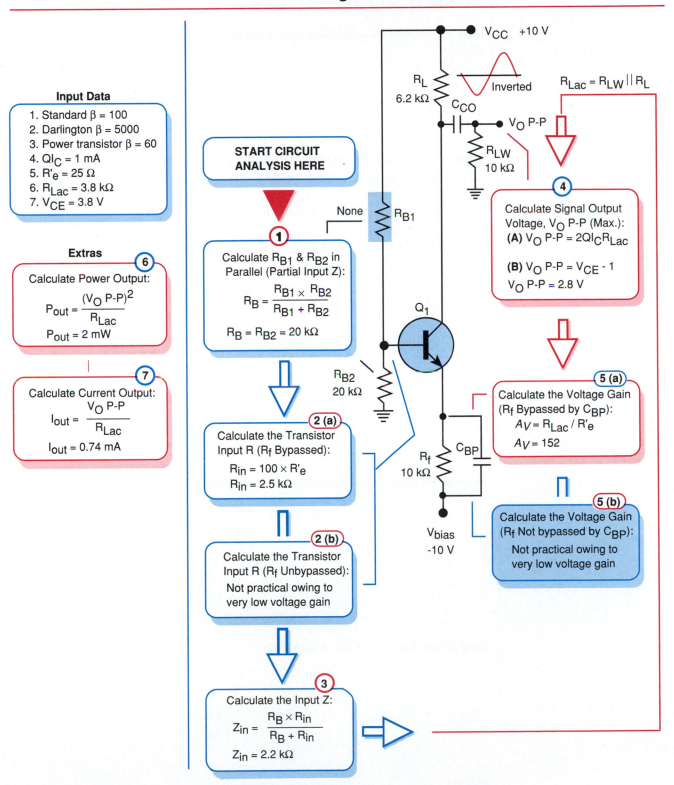

Input Data

1. Standard $\beta = 100$
2. Darlington $\beta = 5000$
3. Power transistor $\beta = 60$
4. $QI_C = 1$ mA
5. $R'_e = 25\ \Omega$
6. $R_{Lac} = 3.8\ k\Omega$
7. $V_{CE} = 3.8$ V

Extras

6
Calculate Power Output:
$$P_{out} = \frac{(V_O\ P\text{-}P)^2}{R_{Lac}}$$
$P_{out} = 2$ mW

7
Calculate Current Output:
$$I_{out} = \frac{V_O\ P\text{-}P}{R_{Lac}}$$
$I_{out} = 0.74$ mA

START CIRCUIT ANALYSIS HERE

1
Calculate R_{B1} & R_{B2} in Parallel (Partial Input Z):
$$R_B = \frac{R_{B1} \times R_{B2}}{R_{B1} + R_{B2}}$$
$R_B = R_{B2} = 20\ k\Omega$

2 (a)
Calculate the Transistor Input R (R_f Bypassed):
$R_{in} = 100 \times R'_e$
$R_{in} = 2.5\ k\Omega$

2 (b)
Calculate the Transistor Input R (R_f Unbypassed):
Not practical owing to very low voltage gain

3
Calculate the Input Z:
$$Z_{in} = \frac{R_B \times R_{in}}{R_B + R_{in}}$$
$Z_{in} = 2.2\ k\Omega$

V_{CC} +10 V

R_L 6.2 kΩ

Inverted

$R_{Lac} = R_{LW} \| R_L$

C_{CO}

V_O P-P

R_{LW} 10 kΩ

None R_{B1}

Q_1

R_{B2} 20 kΩ

R_f 10 kΩ

C_{BP}

V_{bias} -10 V

4
Calculate Signal Output Voltage, V_O P-P (Max.):
(A) V_O P-P $= 2QI_C R_{Lac}$
(B) V_O P-P $= V_{CE} - 1$
V_O P-P $= 2.8$ V

5 (a)
Calculate the Voltage Gain (R_f Bypassed by C_{BP}):
$A_V = R_{Lac} / R'_e$
$A_V = 152$

5 (b)
Calculate the Voltage Gain (R_f Not bypassed by C_{BP}):
Not practical owing to very low voltage gain

Introduction to Sheet #2

Our procedures on the parameter analysis worksheet (sheet #2) remain much the same as they were for the base-bias circuits we analyzed earlier.

Step 0: Copying Essential Data from Sheet #1

$QI_C = 1.0$ mA
$R_e' = 25$ Ω
$R_{Lac} = 3.8$ kΩ
$V_{CE} = 3.8$ V

Step 1: Calculating the Equivalent Parallel Value of R_L and R_{LW} (R_{Lac})

Since there is no R_{B_1} resistor when we use emitter bias, we go on to Step 2(a).

Step 2(a): Calculating the Transistor Input Resistance (R_{in})

We will always use the bypassed mode in this circuit. Without the bypass capacitor, there is no voltage gain and no point to the circuit. Since we have agreed to adopt a value of 100 for beta, our input resistance calculation becomes

$$R_{in} = 100 \times R_e' = 2.5 \text{ k}\Omega$$

Step 3: Calculating the Input Impedance (Z_{in})

$$Z_{in} = \frac{R_B \times R_{in}}{R_B + R_{in}}$$

$$= \frac{20 \text{ k}\Omega \times 2.5 \text{ k}\Omega}{20 \text{ k}\Omega + 2.5 \text{ k}\Omega}$$

$$= 2.2 \text{ k}\Omega$$

Step 4: Calculating the Maximum P–P Output Signal Voltage (V_O)

$$\textbf{a } V_O\, P\text{–}P = 2QI_C \times R_{Lac}$$
$$= 2 \text{ mA} \times 3.8 \text{ k}\Omega$$
$$= 7.6 \text{ V } P\text{–}P$$

$$\textbf{b } V_O\, P\text{–}P = V_{CE} - 1$$
$$= 3.8 \text{ V} - 1 \text{ V}$$
$$= 2.8 \text{ V } P\text{–}P$$

Step 5(a): Calculating the Voltage Gain

$$A_V = \frac{R_{Lac}}{R_e'}$$

$$= \frac{3.8 \text{ k}\Omega}{25 \text{ Ω}}$$

$$= 152$$

Step 6: Calculating the Output Power (P_{out}) [Extra]

$$P_{out} = \frac{(V_O\ P\text{--}P)^2}{R_{Lac}}$$

$$= \frac{(2.8\ V\ P\text{--}P)^2}{3800\ \Omega}$$

$$= 2\ mW$$

Step 7: Calculating the Maximum Output Current ($I_O\ P\text{--}P$) [Extra]

$$I_O\ P\text{--}P = \frac{V_O\ P\text{--}P}{R_{Lac}}$$

$$= \frac{2.8\ V\ P\text{--}P}{3.8\ k\Omega}$$

$$= 0.74\ mA$$

5.8   Darlington-pair Super-beta Transistor

Figure 5-18 shows a pair of transistors connected in an arrangement known as the **Darlington-pair** configuration. The Darlington pair is generally treated as a single super-beta transistor. Darlington-pair power transistors are available in a single case. The most common version consists of a small signal transistor for Q_1 and a power transistor for Q_2, as shown in Figure 5-18. Because

FIGURE 5-18   **Super-beta Darlington Pair**

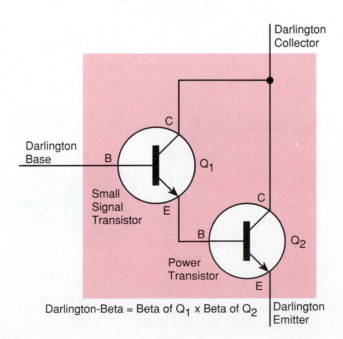

Darlington-Beta = Beta of Q_1 x Beta of Q_2

the collector current of Q_1 is the base current for Q_2, a small signal transistor can comfortably drive a fairly large power transistor.

The Darlington pair is called a *super-beta transistor* because the total beta of the pair is the product of the individual betas ($\beta_{Q1} \times \beta_{Q2}$). If we assume a beta of 100 for both Q_1 and Q_2, the composite beta would be 100×100, for a Darlington beta of 10,000.

In practice, the input transistor (Q_1) is often operated at an abnormally low collector current (QI_C), because Q_1's collector current is Q_2's base current. We can expect the beta of Q_1 to drop somewhat at the lower collector currents, so we will derate our standard beta of 100 to 85. This value seems to reflect reality pretty well.

The power transistor will probably operate at normal collector currents, and a beta of 60 is a good standard for power transistors. Figure 5-19 shows a Darlington pair with a Darlington compound ß of about 5000.

The super-beta of 5000 exerts its primary influence on the input resistance of the circuit, and that is the main reason to use a Darlington pair instead of a single transistor. In Figure 5-19, the input impedance of the Darlington is about 50 kilohms, with an emitter resistor of 10 ohms in the emitter of Q_2. If we used Q_2 alone, we would have an input resistance of only 600 ohms.

$$R_{in} = \text{ß} \times R_f$$
$$= 60 \times 10 \ \Omega = 600 \ \Omega$$

This is a very inexpensive way to increase the input resistance from 600 ohms to 50 kilohms.

The Darlington pair has two junctions. Therefore, it introduces two junction voltage drops of 0.7 volt each, for a total of 1.4 volts. There are also two junction resistances in series, the value of each depending on QI_C values.

FIGURE 5-19 **Characteristics of the Super-beta**
Darlington Pair

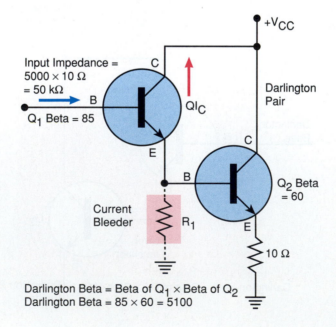

Input Impedance =
$5000 \times 10 \ \Omega$
$= 50 \ k\Omega$

Q_1 Beta = 85

QI_C

Darlington Pair

Q_2 Beta = 60

Current Bleeder R_1

$10 \ \Omega$

Darlington Beta = Beta of $Q_1 \times$ Beta of Q_2
Darlington Beta = $85 \times 60 = 5100$

Using a Darlington pair in a bypassed common emitter circuit produces a lower voltage gain than would a single transistor. The voltage gain equation becomes

$$A_V = R_{Lac} / (R_e' 1 + R_e' 2)$$

In some cases, an additional current bleeder resistor (R_1) is added to allow Q_1 to operate at a higher QI_C in situations where beta may be higher.

5.9 Differential Amplifier

In the electronics world of compromise, no single circuit is best for all occasions, but the **differential amplifier** comes very close to that ideal. The differential amplifier is also a natural for integrated-circuit technology, because its remarkable performance depends on close matching of transistors and other components, which is easy to do with integrated circuits. Accurate absolute values are not critical to the differential amplifier, so it doesn't matter that accurate absolute values are hard to obtain on an integrated circuit chip.

The differential amplifier's frequency response ranges from 0 Hz (**dc**) to as high a frequency as required. It is inherently very stable, provides high voltage gains, and requires no coupling or bypass capacitors. It also features a number of other performance characteristics not available in any other common amplifier form. The differential amplifier is not very popular in discrete circuits, where individual transistors are used, because of the matched-component requirement, but it is the king of the integrated-circuit world.

The circuit in Figure 5-20 represents the basic differential amplifier. The differential amplifier consists of a balanced pair of transistors that share a common emitter (feedback) resistor. This feature makes it quite different from any other transistor circuit. The differential amplifier is usually emitter-biased for reasons we will soon see. The circuit has two inputs and two outputs, but in most cases only one of the possible outputs is actually used. The circuit in Figure 5-20 has a collector load only in the output that has been selected for use. Working in harmony with that single output, the two inputs provide one inverting and one noninverting input. The circuit in Figure 5-20 is the most common arrangement of inputs and outputs; but other combinations are sometimes used, as shown in Figure 5-21.

Quiescent Collector Current Stability

The differential amplifier has an inherently stable Q-point as a result of current-mode feedback in a shared emitter resistor. Because the circuit's stability depends on identical temperature behavior in the two transistors, a matched pair is critical. There must also be almost perfect temperature tracking between the two transistors. It is virtually impossible to meet these two requirements with separate transistors, although matched-pair transistor arrays can be used. The differential amplifier, however, is best used in integrated circuit technology. The transistors share the same chip and, if they are

FIGURE 5-20   Differential Amplifier Circuit

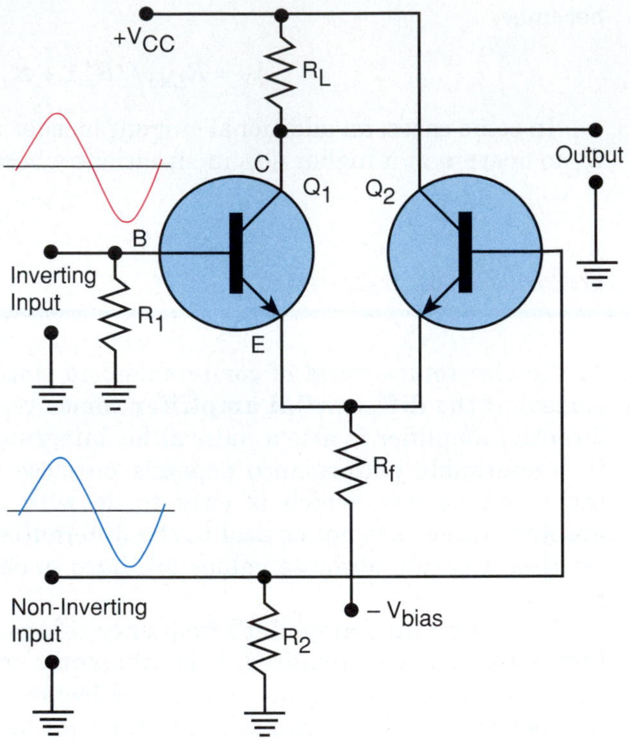

accurately matched, should both run at the same temperature. Let's examine the circuit behavior when the temperature increases.

Differential Amplifier and Temperature Stability

Here's what happens when the temperature rises:

1   As the temperature increases, beta also increases, resulting in an identical increase in collector current in both transistors. Although two collector currents are involved, the current-mode feedback works in the same way as in a single transistor. The increased voltage drop across the common feedback resistor opposes the bias voltage, reducing the collector current to its original value.

2   If the temperature increases, the junction voltage decreases, causing an identical increase in collector current in both transistors. Again, the current-mode feedback voltage developed across the common emitter resistor adjusts the Q-point back to normal.

3   In previous circuits, we had to use either a bypass capacitor or a very small value for the feedback resistor to get large voltage gains. In the differential amplifier, we can use as large a feedback resistor as we want, without its having any effect on the voltage gain. We can use as

FIGURE 5-21   How the Differential Amplifier Circuit Works

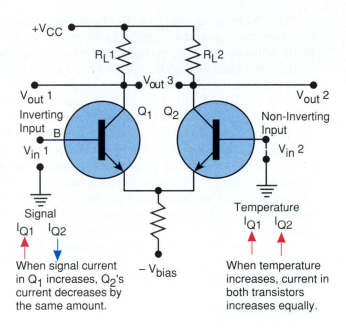

Signal
I_{Q1} I_{Q2}

When signal current
in Q_1 increases, Q_2's
current decreases by
the same amount.

Temperature
I_{Q1} I_{Q2}

When temperature
increases, current in
both transistors
increases equally.

much negative feedback as we want without paying a voltage gain penalty or using bypass capacitors.

Differential Amplifier Operation with a Signal

The differential amplifier circuit shown in Figure 5-21 consists of a symmetrical combination, in which a common emitter circuit (Q_1) also acts as a common collector circuit to drive a common base circuit (Q_2). When the signal is applied to the input of Q_2, it too acts as both a common emitter circuit to amplify the signal and a common collector circuit to drive Q_1 as a common base circuit. This tends to be a bit confusing, so let's examine what actually happens step-by-step when we apply an input signal to the base of Q_1 (see Figure 5-21). Assume the base of Q_2 is connected through a resistor to ground, as indicated by the dotted lines in Figure 5-21; and assume that the input signal drives the base of Q_2 more positive. Here's what happens.

1   The base current of Q_1 increases, causing the collector current of Q_1 to increase.

2   The emitter of Q_1 also goes more positive (emitter-follower action).

3   The emitter of Q_2 (which is connected to Q_1's emitter) also goes more positive with respect to the base of Q_2.

4   This reduces the forward-bias base-emitter voltage of Q_2, which reduces the forward-bias base-emitter current of Q_2.

5   When the base-emitter current of Q_2 is reduced, its collector current goes down proportionately.

The increase in positive (Q_1) base voltage has caused an increase in Q_1's collector current and an identical decrease in Q_2's collector current. In a perfectly balanced circuit, no net change occurs in the total current through the shared (emitter) feedback resistor, because the increase in Q_1's emitter current is exactly counterbalanced by the decrease in Q_2's emitter current.

But if the current through the (emitter) feedback resistor does not vary with signal voltage variations, there is no negative feedback action for signal variations. Consequently, the differential amplifier produces a lot of negative feedback for temperature-induced collector-current changes, but no negative feedback at all for signal-induced collector-current changes. Temperature increases cause both transistors to increase their collector currents by the same amount. The increased voltage drop across the (emitter) feedback resistor corrects for the Q-point increase. As far as the signal is concerned, the feedback resistor does not exist and appears as an open circuit.

The ultimate result is all the negative feedback we want for Q-point stability, with no negative feedback for the signal. The voltage gain is no longer influenced by the value of the feedback resistor.

Output Signals

The output signal can be taken from either collector, with respect to ground. Assume that the signal is taken from the collector of Q_1 (Figure 5-21). If the signal is input to the base of Q_1, the resulting output signal is inverted. If the signal is input to the base of Q_2, the resulting output signal is noninverted in phase. Because the circuit is symmetrical, the output signal could just as well be taken from the collector of Q_2. In that case, the base of Q_2 becomes the inverting input, and the base of Q_1 becomes the noninverting input. It is also possible—although rarely done—to take the output between the two collectors. Most loads require one side to be grounded.

Differential Input Mode

In addition to its inverting and noninverting input modes, the differential amplifier has a third input mode: the differential mode. If input signals are applied to both inputs at the same time, the circuit amplifies the algebraic difference between the two input signals. The differential mode is unique to the differential amplifier circuit and has a multitude of applications.

Common-mode Feedback Signal (CMS)

Because a perfect balance on both sides of the circuit is not possible, some slight change in feedback resistor current always occurs as a result of signal variations. This small difference in balance causes a small signal voltage to develop across the shared emitter feedback resistor (see Figure 5-22). This unwanted signal voltage is called the *common mode feedback signal (CMS)*.

The feedback signal voltage can be measured directly in a differential amplifier, but it is not accessible when the differential amplifier is part of an operational amplifier integrated circuit. As a result, we must use a different approach to measure the degree of imbalance in differential amplifiers integrated into an operational amplifier. An operational amplifier figure of merit called the *common-mode rejection ratio (CMRR)* is generally used for this purpose.

Common-mode Rejection Ratio (CMRR)

If we apply identical ground-referenced signals to both inputs, a perfectly balanced differential amplifier produces no signal at the output. This arrangement is called the **differential connection,** where the signal applied to the inverting input is subtracted from the signal applied to the noninverting input, and the difference voltage is amplified. If the amplifier is perfectly balanced and the two input voltages are identical, the subtraction yields 0 volts, and 0 volts times any amount of voltage gain is still 0 (see Figure 5-22).

In the real world, things are never perfect and some output voltage is produced. Indeed, for the common-mode rejection ratio to be meaningful, some reference value that includes the voltage gain of the amplifier is required. The **differential voltage gain (A_{CM}),** is measured by applying two known but unequal voltages to the two inputs and measuring the output voltage. The differential voltage gain is then calculated as

Differential Voltage Gain = Output Voltage/[(Voltage in A) − (Voltage in B)]

The differential voltage gain applies to a normal amplifying mode and is illustrated in Figure 5-23(b).

The two input voltages are then made equal, and the **common-mode voltage gain** is determined by measuring the signal output voltage in the

FIGURE 5-22 Common Mode Signal (CMS)

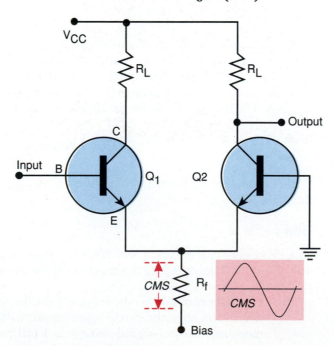

FIGURE 5-23   Common-mode Rejection Ratio (CMRR)

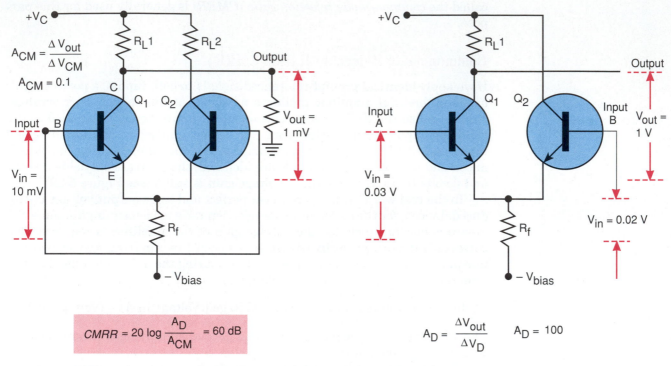

a. Circuit for the Common-Mode Voltage Gain

b. Circuit for the Differential Voltage Gain

same way that the differential voltage gain is determined. The conditions for the common-mode voltage gain are illustrated in Figure 5-23(a). The **common-mode rejection ratio (CMRR)** is the ratio of the differential voltage gain to the (unwanted) common-mode voltage gain, expressed in decibels. CMRR is defined by the following equation:

$$\text{CMRR} = 20 \log \frac{A_D}{A_{CM}}$$

Where A_D is the differential voltage gain and A_{CM} is the common-mode voltage gain.

How to Do It

The process of calculating the CMRR for a particular circuit (in this case, the one shown in Figure 5-23(a)) consists of three steps:

1   Find the common-mode voltage gain for the circuit in Figure 5-23(a), where the common-mode input signal voltage is 10 millivolts and the measured output signal voltage is 1 millivolt:

$$A_{CM} = \frac{\Delta V_{out}}{\Delta V_{CM}}$$

$$= \frac{1 \text{ mV}}{10 \text{ mV}} = 0.1$$

2   Find the differential voltage gain, where the input voltage on input A is 0.03 volts and the signal in input B is 0.02 volts, and the two signals are in phase. Because the amplifier amplifies the difference between the two input signal voltages, the effective input signal voltage is

$$A_D = \frac{\Delta V_{out}}{\Delta V_D}$$

$$= \frac{1 \text{ V}}{0.01 \text{ V}} = 100$$

3   Find the common-mode rejection ratio (CMRR):

$$\text{CMRR} = 20 \log \frac{A_D}{A_{CM}}$$

$$= 20 \log \frac{100}{0.1}$$

$$= 60 \text{ dB}$$

Input Impedance of the Differential Amplifier

Refer to Figure 5-24 as you read the following explanation. Because Q_1's signal currents are canceled out in the feedback resistor by the opposite changes in Q_2's currents, the shared feedback resistor appears to be an open circuit to input signals from either input. If the input voltage is applied to the base of Q_1, the signal path passes through the junction resistance of Q_1, and through

FIGURE 5-24   Input Impedance and Voltage Gain

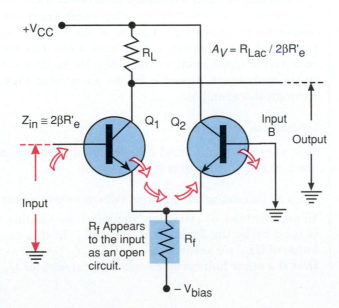

the junction resistance of Q_2, before reaching ground. The input resistance (impedance) is beta times the junction resistance, as usual. But now there are two junctions in series. The input resistance equation is

$$Z_{in} = 2\text{\ss}R_e'$$

Voltage Gain in the Differential Amplifier

The input signal path (Figure 5-24) is from the base of Q_1, through the base-emitter junction of Q_2, to ground via the base of Q_2. The two junction resistances in series constitute the only feedback resistance in the circuit. For signal purposes, the emitter feedback resistor might as well be disconnected and discarded. The emitter feedback resistor is still essential to maintain the proper Q-point and to provide bias.

Because the amplifier includes two junctions in series, the theoretical voltage gain equation is

$$A_V = R_{Lac}/2R_e'$$

But since the junction resistances are not ordinary resistors, the real situation is somewhat different. A practical equation for the voltage gain is

$$A_V = R_{Lac}/2.5R_e'$$

Analyzing the Differential Amplifier Circuit

The differential amplifier is the most important analog amplifier in today's analog world (see Figure 5-25). The differential amplifier is the heart of nearly all modern operational and other integrated-circuit amplifiers. In some cases, a special differential-pair of transistors is used to make a discrete circuit you can analyze.

The current-mode feedback analysis worksheets are suitable for analyzing this circuit, once a few minor allowances are made in the process to take into account the symmetrical nature of the circuit. For this example, some additional notes have been added, and some unused blocks (steps) have been removed to make room for the notes. No fundamental changes have been made in the worksheet, however. Follow along with Figures 5-26 and 5-27 as we go through the example.

Introduction to Sheet #1

It is very unusual to find a base-bias differential amplifier circuit, so we will start at the emitter-bias track (step 3).

Step 3: **Calculating the Quiescent Collector Bias Current (QI_C)**

In emitter bias, as elsewhere (usually), we assume that the entire bias voltage is used up by the feedback resistor (R_f). In this case, though, once we find the value of QI_C, we assume that half of QI_C becomes collector current for Q_1, and that the other half becomes collector current for Q_2. Thus (by assumption), the

FIGURE 5-25   **Differential Amplifier Circuit for Worksheet Analysis**

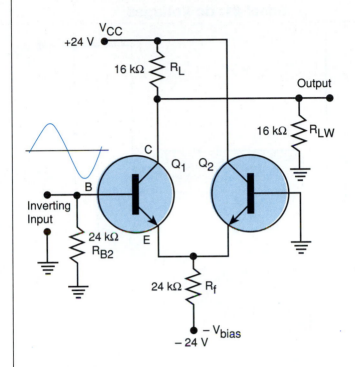

current splits equally between Q_1 and Q_2. Consequently, we must use one-half the current (calculated in Step 3) for each calculation that involves the collector current value. According to Sheet #1 (Figure 5-26), we calculate a QI_C of 1.0 mA. The collector–emitter circuit of Q_1 gets 0.5 mA, and that of Q_2 gets the other 0.5 mA.

Step 4:   Calculating the Junction Resistance (R_e')

We are concerned about the junction resistance of each transistor, so we must take into account the amount of QI_C at which the transistor actually runs. In each case, this is one-half the total 1 mA, or 0.5 mA. Hence, our twin equations are

$$Q_1 \text{ Junction Resistance} = 25/0.5 \text{ mA} = 50 \ \Omega$$
$$Q_2 \text{ Junction Resistance} = 25/0.5 \text{ mA} = 50 \ \Omega$$

Step 5:   Calculating the Dummy-load (R_L) Resistor Voltage Drop

$$V_{RL} = QI_C \times R_L$$
$$= 0.5 \text{ mA} \times 16 \text{ k}\Omega$$
$$= 8.0 \text{ V}$$

Step 6:   Calculating R_{Lac}, the Equivalent Value of R_L and R_{LW} in Parallel

If $R_L = 16 \text{ k}\Omega$ and $R_{LW} = 16 \text{ k}\Omega$, then R_{Lac} must be 8 kΩ.

FIGURE 5-26 Worksheet for the Differential Amplifier Circuit, Sheet #1

Sheet #1: dc Voltages

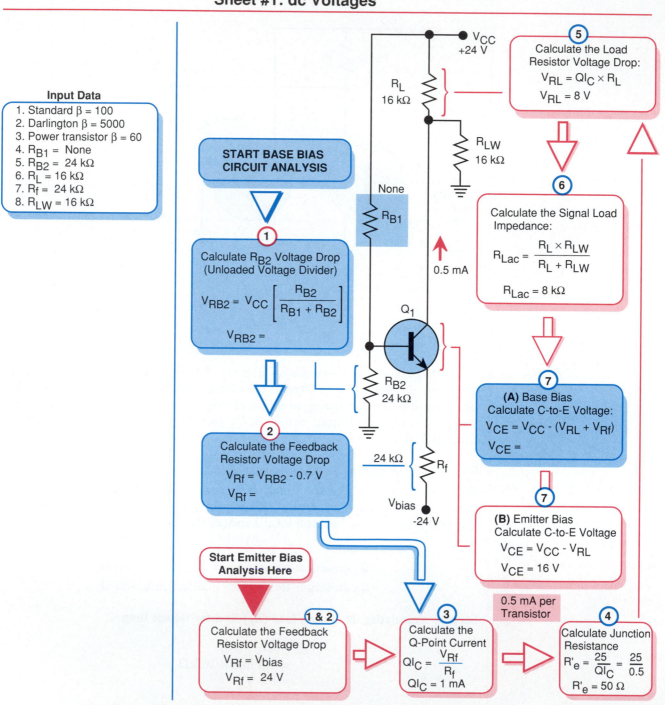

Input Data
1. Standard $\beta = 100$
2. Darlington $\beta = 5000$
3. Power transistor $\beta = 60$
4. $R_{B1} = $ None
5. $R_{B2} = 24\ k\Omega$
6. $R_L = 16\ k\Omega$
7. $R_f = 24\ k\Omega$
8. $R_{LW} = 16\ k\Omega$

START BASE BIAS CIRCUIT ANALYSIS

1 Calculate R_{B2} Voltage Drop (Unloaded Voltage Divider)
$$V_{RB2} = V_{CC}\left[\frac{R_{B2}}{R_{B1} + R_{B2}}\right]$$
$V_{RB2} = $

2 Calculate the Feedback Resistor Voltage Drop
$V_{Rf} = V_{RB2} - 0.7\ V$
$V_{Rf} = $

Start Emitter Bias Analysis Here

1 & 2 Calculate the Feedback Resistor Voltage Drop
$V_{Rf} = V_{bias}$
$V_{Rf} = 24\ V$

3 Calculate the Q-Point Current
$$QI_C = \frac{V_{Rf}}{R_f}$$
$QI_C = 1\ mA$

4 Calculate Junction Resistance
$$R'_e = \frac{25}{QI_C} = \frac{25}{0.5}$$
$R'_e = 50\ \Omega$

V_{CC} +24 V
R_L 16 kΩ
R_{LW} 16 kΩ
None
R_{B1}
0.5 mA
Q_1
R_{B2} 24 kΩ
24 kΩ R_f
V_{bias} -24 V

5 Calculate the Load Resistor Voltage Drop:
$V_{RL} = QI_C \times R_L$
$V_{RL} = 8\ V$

6 Calculate the Signal Load Impedance:
$$R_{Lac} = \frac{R_L \times R_{LW}}{R_L + R_{LW}}$$
$R_{Lac} = 8\ k\Omega$

7 **(A)** Base Bias Calculate C-to-E Voltage:
$V_{CE} = V_{CC} - (V_{RL} + V_{Rf})$
$V_{CE} = $

7 **(B)** Emitter Bias Calculate C-to-E Voltage
$V_{CE} = V_{CC} - V_{RL}$
$V_{CE} = 16\ V$

0.5 mA per Transistor

Step 7(b): Calculating the Collector-to-Emitter Voltage (V_{CE})

$$V_{CE} = V_{CC} - V_{RL}$$
$$= 24 \text{ V} - 8 \text{ V} = 16 \text{ V}$$

Introduction to Sheet #2

Please follow along with sheet #2 of the analysis worksheet (Figure 5-27) for the remaining steps of this example.

Step 0: Copying Essential Data from Sheet #1

$QI_C = 0.5$ mA for each transistor (1.0 mA total)
$R'_e = 50 \ \Omega$
$R_{Lac} = 8$ kΩ
$V_{CE} = 16$ V

Step 1: Calculating the Base-to-Ground Resistance (R_B)

Because the circuit is emitter-biased, there is no R_{B_1}, so $R_B = R_{B_2}$

$$R_B = 24 \text{ k}\Omega$$

Step 2(b): Calculating the Transistor Input Resistance (R_{in})

The differential amplifier gets a little peculiar here. It includes an unbypassed emitter feedback resistor; but in this circuit, balanced emitter currents make it appear as an open circuit. For all practical (signal) purposes, resistor R_f does not exist.

When a signal is applied to the input (base) of transistor Q_2, part of the signal current flows through R_{B_2} to ground. The signal path for the rest of the signal current is as follows:

1   The signal current enters the base of Q_1, and passes through Q_1's base–emitter junction resistance (R'_e).

2   The signal current then flows to the emitter of Q_2, through Q_2's junction resistance R'_e.

3   The signal current then flows from Q_2's base to ground (through any base-to-ground resistance in the circuit).

The input resistance is approximately

$$R_{in} = \text{ß} \times 2R'_e$$
$$= 100 \times 2 \times 50 \ \Omega = 10 \text{ k}\Omega$$

Step 3: Calculating the Input Impedance (Z_{in})

Here, we find the parallel value of R_{in} and R_B.

$$Z_{in} = (R_B \times R_{in})/(R_B + R_{in})$$
$$= (24 \text{ k}\Omega \times 10 \text{ k}\Omega)/(24 \text{ k}\Omega + 10 \text{ k}\Omega)$$
$$= 7 \text{ k}\Omega$$

FIGURE 5-27   **Worksheet for the Differential Amplifier Circuit, Sheet #2**

Sheet #2: Signal Parameters

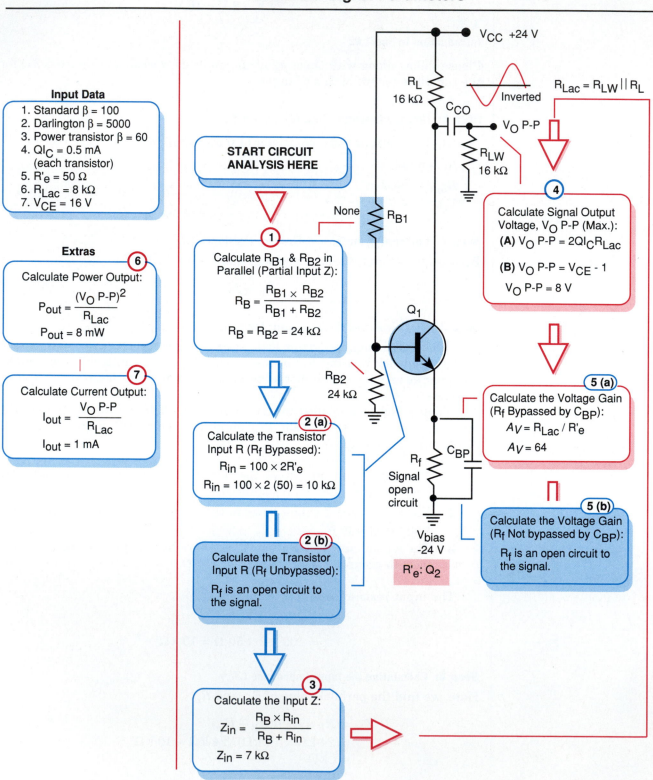

Input Data
1. Standard $\beta = 100$
2. Darlington $\beta = 5000$
3. Power transistor $\beta = 60$
4. $QI_C = 0.5$ mA (each transistor)
5. $R'_e = 50\ \Omega$
6. $R_{Lac} = 8\ k\Omega$
7. $V_{CE} = 16$ V

Extras

6 Calculate Power Output:
$$P_{out} = \frac{(V_O\text{ P-P})^2}{R_{Lac}}$$
$$P_{out} = 8\text{ mW}$$

7 Calculate Current Output:
$$I_{out} = \frac{V_O\text{ P-P}}{R_{Lac}}$$
$$I_{out} = 1\text{ mA}$$

START CIRCUIT ANALYSIS HERE

1 Calculate R_{B1} & R_{B2} in Parallel (Partial Input Z):
$$R_B = \frac{R_{B1} \times R_{B2}}{R_{B1} + R_{B2}}$$
$$R_B = R_{B2} = 24\ k\Omega$$

2 (a) Calculate the Transistor Input R (R_f Bypassed):
$$R_{in} = 100 \times 2R'_e$$
$$R_{in} = 100 \times 2\,(50) = 10\ k\Omega$$

2 (b) Calculate the Transistor Input R (R_f Unbypassed):
R_f is an open circuit to the signal.

3 Calculate the Input Z:
$$Z_{in} = \frac{R_B \times R_{in}}{R_B + R_{in}}$$
$$Z_{in} = 7\ k\Omega$$

V_{CC} +24 V

R_L 16 kΩ

C_{CO}

Inverted

$R_{Lac} = R_{LW} \| R_L$

V_O P-P

R_{LW} 16 kΩ

None R_{B1}

Q_1

R_{B2} 24 kΩ

R_f Signal open circuit

C_{BP}

V_{bias} -24 V

R'_e: Q_2

4 Calculate Signal Output Voltage, V_O P-P (Max.):
(A) V_O P-P = $2QI_C R_{Lac}$
(B) V_O P-P = $V_{CE} - 1$
V_O P-P = 8 V

5 (a) Calculate the Voltage Gain (R_f Bypassed by C_{BP}):
$$A_V = R_{Lac} / R'_e$$
$$A_V = 64$$

5 (b) Calculate the Voltage Gain (R_f Not bypassed by C_{BP}):
R_f is an open circuit to the signal.

This comparatively low input impedance is not a problem. Negative feedback, Darlington-pair input circuits, and field-effect transistors can all be used to increase the input impedance when necessary.

Step 4: Calculating the Maximum *P–P* Output Signal Voltage

a $\quad V_{Out}\ P\text{–}P = 2 \times (0.5\ \text{mA}) \times 8\ \text{k}\Omega = 8\ \text{V}\ P\text{–}P$

b $\quad V_{Out}\ P\text{–}P = V_{CE} - 1 = 16\ \text{V} - 1 = 15\ \text{V}\ P\text{–}P$

The smaller of the two values (and therefore the allowable maximum) is 8 V *P–P*.

Step 5: Calculating the Voltage Gain

The data for Step 5(a) in Figure 5-27 has been removed to make room for the modified equation. When you use the analysis sheets, you should ignore any data that don't fit the particular problem.

The differential amplifier is unique in several ways, one of which is the fact that changes in Q_1 and Q_2 collector–emitter currents are always equal and opposite. The result is the cancellation of all signal feedback voltage drops across the feedback resistor. This is equivalent to removing R_f altogether for signal frequencies. R_f still works fine for preventing Q-point drift, but it has no effect on signal frequencies.

In the bypassed common emitter amplifier circuit, we used a large-value capacitor to short out the feedback resistor at signal frequencies, in order to get higher voltage gains. That approach had three problems:

1 An expensive capacitor was required.

2 The voltage gain was frequency-dependent, particularly at very low frequencies.

3 The amplification of **dc** signals was not possible, because of the finite reactance (X_C) of the capacitor.

Because the differential amplifier does not depend on capacitors to eliminate signal feedback, it can continue to obtain the same high voltage gains from **dc** up to some very high frequency. The voltage gain in the differential amplifier depends only on the values of the signal load impedance (R_{Lac}) and the base–emitter junction resistance (R'_e). This is almost the same situation as we had with the bypassed common emitter circuit, except that we have two transistors in the differential circuit. And because we have two transistors, we also have two junction resistances $(2 \times R'_e)$. Theoretically, the voltage gain equation should be

$$A_V = R_{Lac}/2R'_e$$

In the real world, however, the equation

$$A_V = R_{Lac}/2.5\ R'_e$$

seems to yield better predictability. Remember, junctions are not just simple resistors; they often behave a bit strangely.

For this example,

$$A_V = 8 \text{ k}\Omega/125 \ \Omega$$
$$= 64$$

Step 6: Calculating the Maximum Power Output [Extra]

The power output is 8 milliwatts *P–P*. Remember that the circuits we have been working with in this chapter are intended to be voltage amplifiers, so we have no reason to expect much output power. The power output (as well as the current output) is just extra information.

Step 7: Calculating the Maximum Output Current [Extra]

The maximum output current is 1.0 mA *P–P*.

5.10 Voltage-mode Feedback and the Miller Effect

The voltage-mode feedback connection can provide stability and most circuit parameters equal to those provided by current-mode feedback. One factor, however, makes current-mode feedback generally more popular than voltage-mode feedback in single-transistor amplifiers. The problem involves a phenomenon known as the Miller effect, which drastically lowers the circuit's input impedance.

The circuit in Figure 5-28 shows the voltage-mode feedback connection. This simple version typically does not work as a practical circuit, because the feedback resistor also supplies the base-bias current. A feedback resistor must be selected to provide the proper amount of negative feedback, and the value selected nearly always delivers too much base-bias current. The way to cure the problem is to add a current-bleeding resistor from base to ground, and we will look at that remedy shortly.

The negative feedback applied in this mode reduces the input resistance to

$$R_f/A_V$$

The current-mode feedback circuit increases the value of R_f by the factor beta. This circuit has one characteristic that can be either an advantage or a disadvantage, depending on the circuit's intended use. Good stability is not possible unless the collector-to-emitter voltage (V_{CE}) is small compared to the voltage across the dummy-load resistor (V_{RL}); at the same time, the circuit can operate with V_{CE} values of 1 volt or less, where a minimum of 2 to 3 volts (V_{CE}) is required for current-mode circuits.

The ability to operate with low V_{CE} values is very useful in power-amplifier circuits, where the largest possible output voltage swing is desired. Voltage-mode feedback circuits can serve as quite respectable voltage amplifiers for small signals (low-current signals), but they are more commonly found in power-amplifier circuits, because negative feedback around several stages is possible with voltage-mode feedback. Feedback around several stages can take advantage of the stability of low-power stages to help in stabilizing the power-

FIGURE 5-28   Common Emitter Circuit with
Voltage-mode Feedback

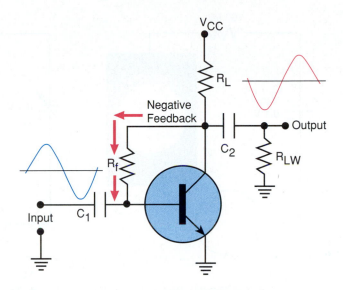

amplifier stages. Power stages are more difficult to stabilize because of the higher temperatures at which they operate.

Voltage-mode feedback is often used to enclose an entire multistage amplifier in a common negative-feedback loop. Generally, each of the several stages has its own negative-feedback circuit (often current-mode feedback), and the overall voltage-mode feedback is added on top of the existing individual stage feedback. We will see some common examples of overall feedback as we go along.

The Miller Effect

The voltage-mode feedback connection causes the effective value of the feedback resistor (R_f) to decrease to R_f/A_V. This reduction of input resistance by negative feedback is also called the **Miller effect.** The term *Miller effect* originally applied to interelectrode capacitive feedback in vacuum-tube amplifiers, but it is now used more broadly. Again, we will use a black box analogy to see how it works.

If you examine the mystery box in Figure 5-29(a) and do a little arithmetic, you must conclude that something is wrong with the current indicated by the current meter in the circuit. By Ohm's law, it is not possible! The current should be 1 microamp, not the 0.1 mA indicated.

Let's assume that the mystery box revealed (Figure 5-29(b)) represents an amplifier with a voltage gain of 100. The dotted-line linkage between V_{in} and V_O simulates the voltage-gain condition, where any change in the input voltage causes a 100-times-larger change in the output voltage. The result is a 100-times-larger change in the circuit current than would be expected, given the input voltage and the obvious circuit resistance. If we calculate the effective circuit resistance with the output voltage included, we get:

$$R = 0.1 \text{ V}/0.1 \text{ mA} = 1 \text{ k}\Omega$$

FIGURE 5-29   Miller-effect Mystery Box

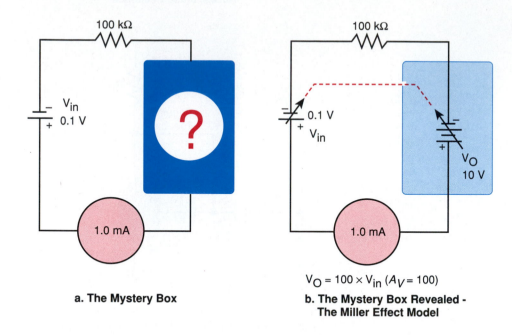

a. The Mystery Box

b. The Mystery Box Revealed -
The Miller Effect Model

$$V_O = 100 \times V_{in} \ (A_V = 100)$$

The effective resistance (because of V_O) is only 1 kilohm. We ignored the opposing 0.1-volt input voltage in the calculation. Ignoring it leaves us with a small (1%) error, but including it adds more to the arithmetic than it is worth.

The output voltage (V_O) gets involved because this is a case of negative feedback, where a portion of the output voltage is fed back to the input circuit. In this analogy, we returned 100% of the output voltage back to the input. The common collector transistor circuit and the voltage follower operational-amplifier (op-amp) circuits do use 100% feedback, but in most cases only a fraction of the output voltage is returned to the input. The factor by which the physical input resistance is reduced depends on the feedback fraction and the voltage gain of the amplifier.

The Miller effect tends to reduce the amplifier input resistances by factors of 400 or less in transistor circuits. Real operational amplifiers have open-loop voltage gains (without the feedback connected) of from 100,000 to 2 million, which accounts for the near zero ohms effective input resistance of nearly 0 kilohms called a virtual ground (almost ground) at the inverting input of the op-amp in Figure 5-30(a). For many circuits, this virtual ground is a necessary feature, and we use the inverting input (−) when we need ultra-low input resistance. In the case of the operational amplifier in Figure 5-30(a), the feedback signal provides a backdoor bootstrapping that greatly increases the input impedance of the noninverting (+) input. We will see how that happens later in this chapter. Op-amps are mentioned here (prematurely) only to make the point that voltage-mode feedback behaves in the same way regardless of the kind of amplifier involved.

In the case of transistor circuits, the low input resistance (impedance) caused by the Miller effect is usually unacceptable from a signal standpoint.

FIGURE 5-30 Drastic Lowering of Input Impedance, Due to the Miller Effect

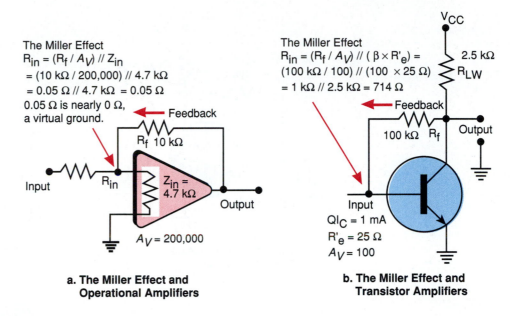

The Miller Effect
$R_{in} = (R_f / A_V) // Z_{in}$
$= (10 \text{ k}\Omega / 200{,}000) // 4.7 \text{ k}\Omega$
$= 0.05 \Omega // 4.7 \text{ k}\Omega = 0.05 \Omega$
0.05Ω is nearly 0Ω,
a virtual ground.

Rf 10 kΩ ← Feedback

Input R_{in} $Z_{in} = 4.7 \text{ k}\Omega$ Output

$A_V = 200{,}000$

**a. The Miller Effect and
Operational Amplifiers**

The Miller Effect
$R_{in} = (R_f / A_V) // (\beta \times R'_e) =$
$(100 \text{ k}\Omega / 100) // (100 \times 25 \Omega)$
$= 1 \text{ k}\Omega // 2.5 \text{ k}\Omega = 714 \Omega$

V$_{CC}$

2.5 kΩ
R_{LW}

← Feedback

100 kΩ R$_f$ Output

Input
Q$I_C = 1$ mA
$R'_e = 25 \Omega$
$A_V = 100$

**b. The Miller Effect and
Transistor Amplifiers**

The negative feedback is essential to stabilize the circuit, and there are ways to get around the too-low input impedance problem.

Formal Miller Effect Derivation

This explanation is based on the circuit and the two models in Figure 5-31.

Assumptions

1 The voltage gain of the amplifier and the models is 100:

$$A_V = 100$$

2 We will ignore small-order values of 2% or less.

Voltage Gain Equivalent Forms for the Example There are four voltage gain equivalents to consider in this example:

$$A_V = 100 \qquad A_V = V_O/V_{in} \qquad V_O = 100\ V_{in}$$

and most important,

$$A_V = \Delta V_O / \Delta V_{in}$$

where ΔV_O means a change in V_O, and ΔV_{in} means a change in V_{in}.

Because the feedback involved in the Miller effect is negative, and because we can't really deal with feedback unless we deal with changes in the variables, we will use the formula

FIGURE 5-31 **Circuit Models of the Miller Effect**

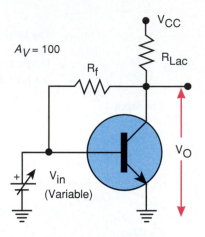

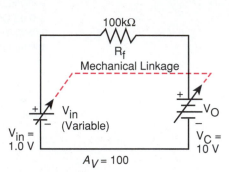

a. **Voltage-Mode Feedback**
   Transistor Circuit

b. **Electromechanical Model**

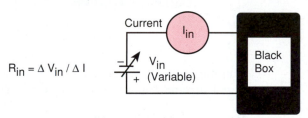

c. **Measuring Input Resistance**

$$A_V = \Delta V_O / \Delta V_{in}$$

to define the voltage gain dynamically.

Procedure Now, let's suppose that we have the input source (V_{in}) and a black box whose dynamic input resistance (or impedance) we want to measure. How do we go about it? We begin by varying the input voltage (V_{in}) and calling that variation ΔV_{in}. The variation ΔV_{in} causes a variation in the circuit current, which we record as ΔI_{in}. Then, using Ohm's law, we calculate the effective dynamic resistance:

$$R_{in} = \Delta V_{in} / \Delta I_{in}$$

In this case, R_{in} is actually resistor R_f in Figure 5-30(a), for which we are trying to find the effective value. Let's try a step-by-step calculation on the model (Figure 5-31(b)):

1 If V_{in} changes by 0.01 volt, and the voltage gain is 100, then

$$\begin{aligned} \Delta V_O &= A_V \times \Delta V_{in} \\ &= 100 \times 0.01 \text{ V} \\ &= 1.0 \text{ V} \end{aligned}$$

The total change in the circuit voltage is actually 1.01 volt, but we will ignore the 0.01 volt because it is a small-order effect (less than 2% of the total). Thus, our initial conclusion is that

$$\Delta V_O = 1 \text{ V}$$

2 The current through R_f is:

$$I_{R_f} = \Delta V_O/R_f \quad \text{Or} \quad I_{RF}$$
$$= 1 \text{ V}/100 \text{ k}\Omega$$
$$= 1 \times 10^{-5} \text{ (0.01 mA)}$$

So far, we have proved that when we change V_{in} by 0.01 volt, the current changes by 1×10^{-5} A or 0.01 mA. This is what actually happened. It happened in our paper model and it would happen in a real circuit. We know the value of ΔV_{in} because we chose it. We know what the change in circuit current is, including the effect of the voltage gain on that current. Because we now know ΔV_{in} and ΔI_{in}, we can solve for the effective input resistance of R_f with the effects of the voltage gain included. We will assume that the internal resistances of the two voltage sources (V_{in} and V_O) are negligible.

3 Solving for the effective resistance of R_f, we get

$$R_{f(effective)} = \Delta V_{in}/\Delta I_{in} \quad \text{(Ohm's law)}$$
$$= (1 \times 10^{-2} \text{ V})/(1 \times 10^{-5} \text{ A}) = 1 \times 10^{3} \text{ }\Omega$$
$$= 0.01 \text{ V}/0.01 \text{ mA} = 1 \text{ k}\Omega$$

What has happened here is that the voltage gain, in concert with the way in which the feedback resistor R_f is connected in the circuit, has lowered the effective value of R_f from its color code value of 100 kilohms to a real or effective value of 1 kilohm. The value of R_f has effectively been lowered by a factor of 100. It is no coincidence that the factor by which the value of R_f has been lowered and the voltage gain are the same. This is what the Miller effect is all about.

4 Because of the Miller effect, R_f in the circuit in Figure 5-31(a) appears as

$$R_{f(Miller)} = R_f/A_V$$

The effective value of R_f is always reduced by the factor A_V.

Miller Effect and Collector–Base Junction Capacitance

The circuit in Figure 5-32 shows a capacitor connected from collector to base. This is a voltage-mode feedback connection, and the Miller effect applies. Suppose that the capacitor has a reactance of 100 kilohms at some frequency, and that the amplifier has a voltage gain of 100. The Miller effect works for reactance in the same way as it does for resistance: it reduces the effective reactance to 1 kilohm, just as it did the effective resistance in the previous example.

The picture gets a little more complex when a capacitor is involved, because of phase shift and so on, but the result is essentially the same: the

FIGURE 5-32   **Miller Effect and Collector-base Junction Capacitance**

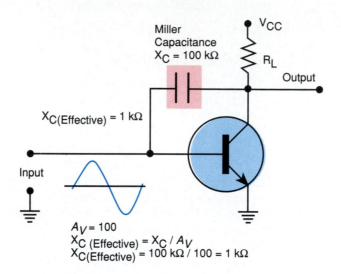

$A_V = 100$
$X_{C \text{ (Effective)}} = X_C / A_V$
$X_{C(\text{Effective})} = 100 \text{ k}\Omega / 100 = 1 \text{ k}\Omega$

reactance is reduced by the factor A_V. If the voltage gain is 100, the reactance is divided by 100, yielding $X_C/100$. If X_C is reduced by a factor of 100, the effective capacitance value is increased by the factor of 100; the capacitor then appears to be 100 times larger than its marked value.

The voltage-mode feedback acts as a capacitance multiplier, and for this reason the circuit is sometimes called a **capacitance multiplier** or *capacitance amplifier*. Some circuits take advantage of the Miller effect's ability to simulate a large capacitor by using a relatively small real capacitor value.

When Miller discovered the effect, back in the early days of electronics, it was nothing but a problem. Interelectrode capacitance in vacuum tubes, coupled with the voltage-mode feedback's Miller effect, yielded such a large effective capacitance that it put a lid on the frequency of the tuned circuit Miller was trying to use. The problem was so serious that special vacuum tubes with extra electrodes had to be built to reduce the Miller effect capacitance and to allow the construction of higher-frequency radio transmitters.

Collector–Base Junction Capacitance

We don't escape the Miller effect capacitance problem in transistors either. Recall that the collector–base junction is normally reverse-biased. The depletion zone is an insulator (dielectric) because it has no carriers for current flow. The areas on each side of the depletion zone are rich in carriers, and are therefore conductors. A capacitor consists of two conductors separated by an insulator, and a reverse-bias junction fits the description. Although the real junction capacitance is quite small, the Miller effect can make it quite significant.

Output Capacitance

Because the collector–base junction capacitance is part of the output circuit in common emitter circuits, it plays a part in the circuit output impedance. It does not, however, get amplified by the Miller effect in the output. For most practical purposes, only the small actual junction capacitance is part of the output. If the circuit voltage gain is much less than 10, the Miller effect can be a problem in certain unusual circumstances.

Cascode Amplifier Circuit

In some very-high-frequency amplifiers, the Miller effect and the junction capacitance can present a problem. A two-transistor circuit has been designed to get around that problem; it effectively places two Miller-effect capacitances in series. The **Cascode circuit** shown in Figure 5-33 consists of a common emitter circuit driving a common base circuit. The two transistors are in series, and this places the two Miller capacitors in series as well.

The input to the compound circuit is connected to the common emitter stage, so the Cascode circuit has the characteristics of a common emitter

FIGURE 5-33 **Cascode Circuit**

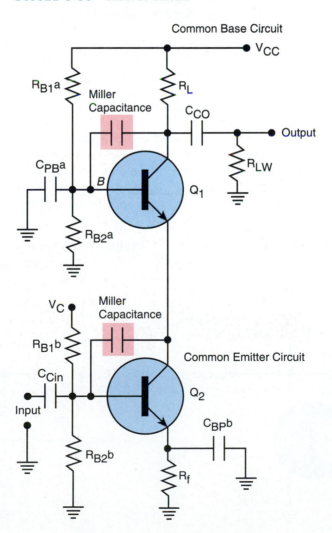

amplifier. The common base circuit serves only to provide an added series Miller capacitance. Equal-value capacitors in series yield an equivalent of half the capacitance of either capacitor.

Stability Factor S in Voltage-mode Amplifiers

The stability factor S and the approximate circuit current gain for the voltage-mode feedback amplifier in Figure 5-34 are equal to

$$S = A_i = R_f/R_L$$

Most circuits include a base-to-ground resistor in addition to R_f as part of the bias network. The value of R_f must be dictated by feedback considerations, and the base-to-ground resistor serves as a current bleeder to shunt the extra base current to ground. The value of R_f cannot exceed 10 times the value of R_L without sacrificing Q-point stability, since stability factors must range from 1 to 10. More often than not, a value for R_f that is small enough for good Q-point stability provides far too much base current for proper biasing. A base-to-ground resistor (R_{B2}) bleeds off the excess bias current.

Bypassing the Feedback at Signal Frequencies

So far, the circuits we have discussed have used negative feedback, which affected the signal characteristics as well as the circuit stability. In current-mode circuits, the negative feedback required for stability also held voltage-gain values quite low.

In voltage-mode feedback circuits, negative feedback produces the Miller effect, which lowers the input resistance to (often) unacceptable values. Q-point stability is the goal that necessitates the use of negative feedback; and Q-point variations caused by temperature changes occur very slowly compared to changes caused by variation in the input signal. Temperature changes

FIGURE 5-34   **Circuit for the Stability Factor**

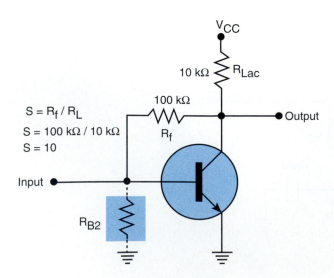

FIGURE 5-35   **Bypassing the Negative Feedback at Signal Frequencies**

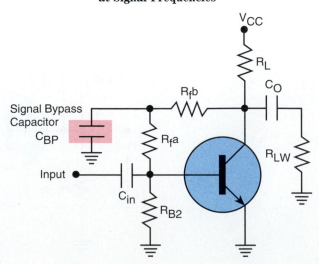

may take many minutes or hours, while signal changes typically occur at rates in excess of 30 or 40 times a second (30 or 40 Hz).

We can use a capacitor with a low reactance at all signal frequencies to bypass negative feedback voltages to ground before they can reach the input. A capacitor that has a low reactance to signal frequencies still has nearly infinite reactance at the very low Q-point drift–equivalent frequency (rate of change).

If we connect capacitor C_{BP} as shown in Figure 5-35, it acts as a near open circuit at Q-point drift change rates; as a result, the feedback signal reaches the input just as it would if the capacitor were left out. The added capacitor has no effect on the slowly changing **dc** parameters.

However, the capacitor is a virtual short at all signal frequencies, and the feedback signal is shunted to ground. The signal-frequency feedback-signal voltages never get to the input, so no feedback occurs at signal frequencies. Normal feedback still exists at Q-point change rates.

Comparing Common Emitter Circuits Using Current-mode Feedback to Common Emitter Circuits Using Voltage-mode Feedback

Our comparison will proceed in two main parts, based on the feedback mode involved. Assume R_{B2} is the same for both circuits, so leave it out of the calculations.

Current-mode Circuit

The voltage gains and input impedances for the current-mode circuit in Figure 5-36(a) can be analyzed as follows.

Voltage Gain

1 With the capacitor not connected:

$$A_V = R_{Lac}/R_f$$
$$= 6.2 \text{ k}\Omega/2 \text{ k}\Omega$$
$$= 3.1$$

2 With the capacitor connected:

$$A_V = R_{Lac}/R'_e$$
$$= 6.2 \text{ k}\Omega/25 \text{ }\Omega \qquad \text{(assuming } QI_C = 1 \text{ mA)}$$
$$= 248$$

Input Impedances

1 With the capacitor not connected:

$$Z_{in} = \beta \times R_f$$
$$= 100 \times 2 \text{ k}\Omega$$
$$= 200 \text{ k}\Omega \qquad \text{(not including } R_{B_1} \text{ and } R_{B_2} \text{ in parallel)}$$

FIGURE 5-36   **Comparing Circuits with Voltage-mode and Current-mode Feedback**

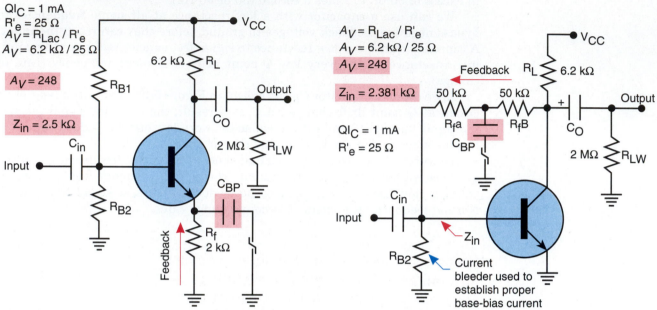

a. **Common-Emitter Circuit with Current-Mode Feedback**   b. **Common-Emitter Circuit with Voltage-Mode Feedback**

2   With the capacitor connected:

$$Z_{in} = \beta \times R'_e$$
$$= 100 \times 25 \ \Omega$$
$$= 2.5 \ \text{k}\Omega$$

Voltage-mode Circuit

The voltage gains and input impedances for the voltage-mode circuit in Figure 5-36(b) can be calculated as follows.

Input Impedance

1   With capacitor C_{BP} not connected:

$$Z_{in} = (R_f/A_V)//(\beta \times R'_e)$$
$$= (100 \ \text{k}\Omega/100)//(100 \times 25 \ \Omega)$$
$$= 714 \ \Omega$$

2   With capacitor C_{BP} connected:

$$Z_{in} = R_{fa}//(\beta \times R'_e)$$
$$= 50 \ \text{k}\Omega//2.5 \ \text{k}\Omega$$
$$= 2.381 \ \text{k}\Omega$$

Voltage Gain

1   With capacitor (C_{BP}) not connected:

$$A_V = R_{Lac}/R_e'$$
$$= 6.2 \text{ k}\Omega/25$$
$$= 248$$

2   With capacitor (C_{BP}) connected:

$$A_V = R_{Lac} / R_e'$$
$$= 6.2 \text{ k}\Omega/25 \ \Omega$$
$$= 248$$

In voltage-mode circuits, the capacitor does not affect the voltage gain.

5.11   Voltage-mode Feedback Analysis Procedures

Circuits that use current-mode feedback are all variations of the same basic circuit; and the same analysis procedure, with variations, can be used for all of them. Voltage-mode feedback circuits are very different from current-mode circuits and require a different analysis process. There is only one voltage-mode circuit, but sometimes some secondary current-mode feedback is added. Voltage-mode feedback is nearly always dominant when both are used, and the voltage-mode worksheet applies.

Analyzing the Common Emitter Circuit with Voltage-mode Feedback

For the circuit in Figure 5-37, we will use the voltage-mode version of the universal transistor circuit analysis worksheet to calculate the appropriate **dc** voltages, the collector current, and the signal parameters. Please follow along with the two example worksheets in Figures 5-38 and 5-39.

Introduction to Sheet #1

Sheet #1 of the two-page universal worksheet deals with **dc** voltages and collector current that must be calculated.

Step 1: **Calculating the Current Through R_{B2} (I_{RB2})**

$$I_{RB_2} = 0.7 \text{ V}/R_{B2} \qquad \text{(given that the base–emitter junction voltage is 0.7 V)}$$
$$= 0.7 \text{ V}/6 \text{ k}\Omega$$
$$= 0.1 \text{ mA} \qquad \text{(actually, 0.1167 mA)}$$

Step 2: **Calculating the Quiescent Collector Current (QI_C)**

$$QI_C = \frac{(V_{CC} - 0.7 \text{ V}) - (I_{RB2}R_f)}{R_L + (0.01 \times R_f) + R_{FE}} = 1 \text{ mA}$$

Note: $0.01 \times R_f = R_f /\beta$, using our standard beta of 100.

FIGURE 5-37   **Common Emitter, Voltage-mode Circuit for Worksheet Analysis**

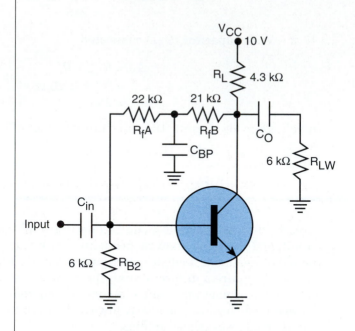

Step 3: **Calculating the Secondary Current-mode Resistor Voltage Drop (V_{RFE})**

Skip this step, since this circuit has no secondary current-mode feedback resistor.

Step 4: **Calculating the Base–Emitter Junction Resistance (R_e')**

$$R_e' = 25 \text{ mV}/QI_C$$
$$= 25 \text{ mV}/1 \text{ mA}$$
$$= 25 \ \Omega$$

Step 5: **Calculating the Voltage Drop Across the Dummy-load Resistor (V_{R_L})**

$$V_{R_L} = QI_C \times R_L$$
$$= 1 \text{ mA} \times 4.3 \text{ k}\Omega$$
$$= 4.3 \text{ V}$$

Step 6: **Calculating the ac Signal–load Impedance (R_{Lac})**

$$R_{Lac} = (R_L \times R_{LW})/(R_L + R_{LW})$$
$$= (4.3 \text{ k}\Omega \times 6 \text{ k}\Omega)/(4.3 \text{ k}\Omega + 6 \text{ k}\Omega)$$
$$= 25.8 \text{ k}\Omega/10.3 \text{ k}\Omega$$
$$= 2.5 \text{ k}\Omega$$

Step 7: **Calculating the Collector-to-Emitter Voltage Drop (V_{CE})**

$$V_{CE} = V_{CC} - (V_{R_L} + V_{R_{FE}})$$
$$= 10 \text{ V} - (4.3 \text{ V} + 0 \text{ V})$$
$$= 5.7 \text{ V}$$

FIGURE 5-38   Worksheet for the Common Emitter Circuit with Voltage-mode Feedback, Sheet #1

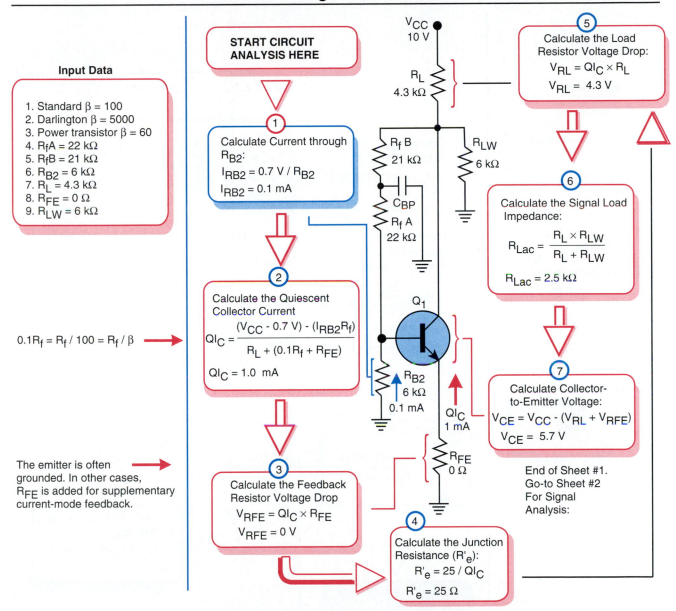

Sheet 1: dc Voltages and Currents

Input Data

1. Standard $\beta = 100$
2. Darlington $\beta = 5000$
3. Power transistor $\beta = 60$
4. $R_fA = 22\ k\Omega$
5. $R_fB = 21\ k\Omega$
6. $R_{B2} = 6\ k\Omega$
7. $R_L = 4.3\ k\Omega$
8. $R_{FE} = 0\ \Omega$
9. $R_{LW} = 6\ k\Omega$

$0.1R_f = R_f\ /\ 100 = R_f\ /\ \beta$ →

The emitter is often grounded. In other cases, R_{FE} is added for supplementary current-mode feedback. →

START CIRCUIT ANALYSIS HERE

1 Calculate Current through R_{B2}:
$$I_{RB2} = 0.7\ V\ /\ R_{B2}$$
$$I_{RB2} = 0.1\ mA$$

2 Calculate the Quiescent Collector Current
$$QI_C = \frac{(V_{CC} - 0.7\ V) - (I_{RB2}R_f)}{R_L + (0.1R_f + R_{FE})}$$
$$QI_C = 1.0\ mA$$

3 Calculate the Feedback Resistor Voltage Drop
$$V_{RFE} = QI_C \times R_{FE}$$
$$V_{RFE} = 0\ V$$

4 Calculate the Junction Resistance (R'_e):
$$R'_e = 25\ /\ QI_C$$
$$R'_e = 25\ \Omega$$

V_{CC} 10 V

R_L 4.3 kΩ

$R_f\ B$ 21 kΩ

C_{BP}

$R_f\ A$ 22 kΩ

R_{LW} 6 kΩ

Q_1

R_{B2} 6 kΩ 0.1 mA

QI_C 1 mA

R_{FE} 0 Ω

5 Calculate the Load Resistor Voltage Drop:
$$V_{RL} = QI_C \times R_L$$
$$V_{RL} = 4.3\ V$$

6 Calculate the Signal Load Impedance:
$$R_{Lac} = \frac{R_L \times R_{LW}}{R_L + R_{LW}}$$
$$R_{Lac} = 2.5\ k\Omega$$

7 Calculate Collector-to-Emitter Voltage:
$$V_{CE} = V_{CC} - (V_{RL} + V_{RFE})$$
$$V_{CE} = 5.7\ V$$

End of Sheet #1. Go-to Sheet #2 For Signal Analysis:

Introduction to Sheet #2

This sheet of the universal transistor-circuit analysis worksheet covers the signal parameters for the voltage-mode circuit.

Step 0: Copying Relevant Data from Sheet #1 (See Sheet #2)

FIGURE 5-39 Worksheet for the Common Emitter Circuit with Voltage-mode Feedback, Sheet #2

Sheet #2: Signal Parameters

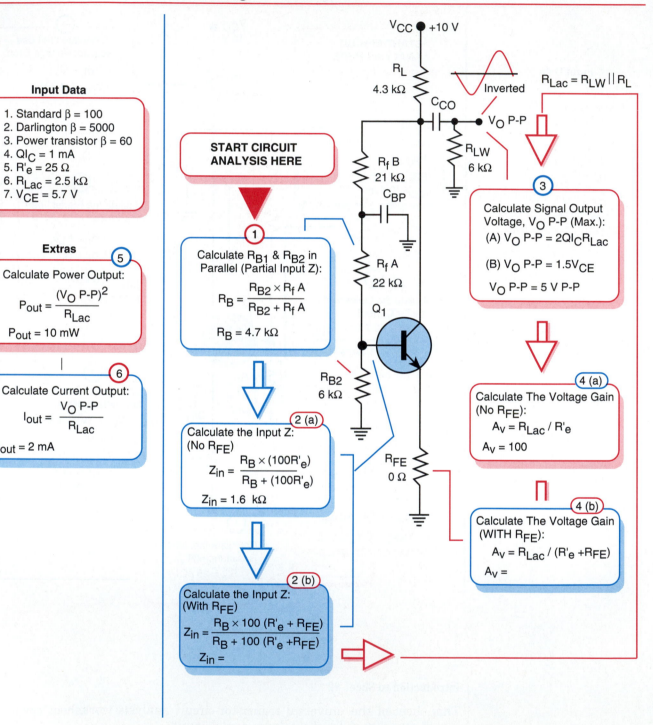

Step 1: Calculating the Equivalent Parallel Impedance (Resistance) of the Current-bleeder Resistor (R_{B2}) and the R_fA Half of the Feedback Resistor

The bypass capacitor has been selected to have nearly 0 ohms of capacitive reactance at the lowest expected signal frequency. For signal frequencies, C_{BP} might as well be a wire. The procedure for this step is just a simple parallel-resistance calculation:

$$R_B = (R_{B2} \times R_fA)/(R_{B2} + R_fA)$$
$$= (6 \text{ k}\Omega \times 22 \text{ k}\Omega)/(6 \text{ k}\Omega + 22 \text{ k}\Omega)$$
$$= 132 \text{ k}\Omega/28 \text{ k}\Omega$$
$$= 4.7 \text{ k}\Omega$$

Step 2(a): Calculating Z_{in}

Step 2 on the worksheet has two parts. Part (a) consists of finishing the input impedance calculation, while part (b) is for use if R_{FE} exists.

To perform part (a), we must find the value of R_B (from the previous step) in parallel with the transistor's input resistance. The transistor's usual input resistance is simply beta times the junction resistance (R_e'). Thus we have

$$R_B = 4.7 \text{ k}\Omega$$
$$R_e' = 100 \times 25 \text{ }\Omega = 2.5 \text{ k}\Omega$$

$$Z_{in} = [R_B \times (100 \text{ } R_e')]/[R_B + (100 \text{ } R_e')]$$
$$= (4.7 \text{ k}\Omega \times 2.5 \text{ k}\Omega)/(4.7 \text{ k}\Omega + 2.5 \text{ k}\Omega)$$
$$= 1.6 \text{ k}\Omega$$

Some circuits combine both current-mode and voltage-mode feedback, in which case we must include the extra current-mode feedback resistor in our input impedance calculations. Here, however, no current-mode feedback resistor is included, so we can skip Step 2(b).

Step 3: Calculating the Maximum Undistorted *P–P* Output Signal Voltage

The maximum output signal voltage is the lower of the values calculated in parts (a) and (b).

a
$$V_O P\text{–}P = 2QI_C \times R_{Lac}$$
$$= 2 \times 1 \text{ mA} \times 2.5 \text{ k}\Omega = 5 \text{ V } P\text{–}P$$

b
$$V_O P\text{–}P = 1.5 \times V_{CE}$$
$$= 1.5 \times 5.7 \text{ V} = 8.55 \text{ V } P\text{–}P$$

We therefore select the part (a) product [$V_O P\text{–}P = 5 \text{ V } P\text{–}P$] as our maximum output signal voltage.

Note: The equations used in Step 3 are derived empirically, based on measurement experience with hundreds of real circuits. The mathematically correct derivations that are commonly used seem to be poor predictors in the real world.

Step 4(a): Calculating the Voltage Gain (A_V) for a Circuit Without a Current-mode Feedback Resistor (R_{FE})

$$A_V = R_{Lac}/R'_e$$
$$= 2.5\ \Omega/25\ \Omega$$
$$= 100 \qquad \text{(for this example)}$$

Step 4(b): Calculating the Voltage Gain (A_V) for a Circuit with a Current-mode Feedback Resistor (R_{FE})

$$A_V = R_{Lac}/(R'_e + R_{FE})$$

Step 5: Calculating the Maximum Undistorted Output Power (P_{out}) [Extra]

Don't expect much power in this example; the circuit is primarily a voltage amplifier.

$$P_{out} = (V_O\,P\text{–}P)^2\,/\,R_{Lac}$$
$$= 25\ \text{V}\,/\,2.5\ \text{k}\Omega$$
$$= 0.01\ \text{W} = 10\ \text{mW}$$

Step 6: Calculating the Maximum Undistorted Output Current (I_O) [Extra]

$$I_O\,P\text{–}P = V_O\,P\text{–}P\,/\,R_{Lac}$$
$$= 5\ \text{V}\,/\,2.5\ \text{k}\Omega$$
$$= 2\ \text{mA}$$

Note: The power output and current output calculations are generally of greater interest when you are using voltage-mode feedback than when you are using current-mode feedback. Voltage-mode feedback is more likely to be used in power amplifiers than in voltage amplifiers. Naturally, if we are working with power amplifiers, power-output and current-output capabilities are both probably of interest to us.

SUMMARY

1 Transistor circuits can be classified according to the two ways in which negative feedback can be obtained.

2 The two classifications are current-mode feedback and voltage-mode feedback.

3 This chapter is primarily concerned with current-mode circuits.

4 The current-mode feedback voltage is developed across an emitter resistor as the result of the emitter current flowing through it.

5 In the analysis procedures, we assume that the collector and emitter currents are equal.

6 The base to ground resistor is part of the feedback circuit and must have a value of between 1 and 10 times the value of the feedback resistor in the emitter circuit.

7 The upper bias resistor, from the power supply to the base, is calculated by the designer only after the two feedback resistor values have been determined.

8 The circuits in this chapter are intended to produce voltage gain. Current gain and power gain are by-products, and we are only slightly concerned with them.

9 The analysis process involves two phases: **dc** voltages and quiescent collector current; and signal parameters.

10 The purpose of a transistor circuit is to process the signal to produce the desired voltage gain and other signal parameters.

11 However, the required signal parameters cannot be obtained unless the **dc** voltages and Q-point currents are correct.

12 The analysis process used in the chapter is beta-independent, but only in the sense that we need not have an accurate value for it. Negative feedback accounts for variations in beta, but beta does exist as some value.

13 For purposes of our analysis procedures, we have adopted a value for beta of 100, and we call that value our standard beta value.

Negative feedback takes care of the fact that an individual transistor may have a different value of beta.

14 Adding a bypass capacitor across the feedback resistor acts as a short circuit for signal frequencies, greatly increasing the voltage gain.

15 Negative feedback reduces the voltage gain. Bypassing the emitter feedback resistor eliminates the negative feedback for signal frequencies, resulting in a much larger voltage gain.

16 Because temperature changes are slow compared with signal frequencies, the bypass capacitor has no effect on the circuit's Q-point stability, and this is the purpose of adding negative feedback. The bypass capacitor does not interfere with Q-point stability.

17 The total voltage gain for a string of cascaded transistor amplifier stages is the product of the gains of the individual stages.

Troubleshooting Transistor Circuits

CAUTION
Always verify that the power supply voltages are correct, before starting your search for other problems.

In this section, we look at a series of comprehensive noninvasive tests to enable you to find the trouble in transistor circuits. The base-biased common emitter circuit is used in the illustrations, but the techniques work equally well with other circuit configurations. The tests normally don't require that you analyze the circuit first.

If these quick tests don't pinpoint the trouble, you may need to know fairly accurate values for the circuit parameters to go further. At that point, you must put your circuit analysis skills to work. In the vast majority of cases, these quick tests will find the problem, but a small percentage of tough problems (called *dogs*) will not yield to the easy way.

Base–Emitter Junction Voltage

This is one of the simplest and most useful tests you can perform. The base-to-emitter junction voltage must be in the range from 0.5 to 0.8 volts, and most often should fall between 0.58 and 0.78 volts. This is true for any operating silicon transistor, no matter what circuit it is in or what kind of transistor it is.

You can measure the junction voltage directly if you are using a battery-operated voltmeter. If you are using a line-operated instrument, you must measure the emitter-to-ground voltage and the base-to-ground voltage and subtract the emitter voltage from the base voltage:

$$\text{Junction Voltage} = \text{Base Voltage} - \text{Emitter Voltage}$$

If the junction voltage is incorrect, you have found a problem. Typically, if there is a problem, the junction voltage will not even be close to the correct (0.6–0.7 V) value. After checking for mechanical problems, solder joints, broken wires, and so on, if there is enough voltage on the base to bias the junction on, the transistor is probably faulty. Figure 5-40 illustrates the technique.

Collector-to-Emitter Voltage

Measuring the collector-to-emitter voltage is another valuable quick check. The voltage (V_{CE}) can have virtually any value, but the ideal is to have a collector-to-emitter voltage equal to roughly one-half the supply voltage. You will not often encounter the ideal, but it is a ballpark figure.

A voltage of less than 1 volt for V_{CE} almost always indicates that the transistor is overbiased or defective. A V_{CE} voltage of 0.2 to 0.8 volt usually indicates saturation and an excessive base bias condition. A reading of 0 V most likely indicates a shorted transistor. Any V_{CE} voltage of less than about 3 volts merits further investigation.

Exercising the Transistor

If the base-to-emitter junction voltage is okay, and V_{CE} seems to be okay, but the circuit is still questionable, there may still be a problem in the collector circuit. The collector-to-base junction is reverse-biased and can have a range of acceptable voltage values.

The trick in this test is to force the transistor's collector voltage to change, to see if the base actually has control over the collector current. Connecting a temporary jumper from base to emitter, as shown in Figure 5-41, turns the base current off. If the collector current turns off, as it should, the collector voltage will rise to approximately the power-supply voltage. If this doesn't happen, the transistor is probably defective.

In most cases, this test will tell the story, and you won't need to use the slightly more risky procedure that follows.

The second transistor exercising procedure, shown in Figure 5-42, pulls the base up to the full supply voltage, driving the transistor toward—or all the way into—saturation. The current-limiting resistor should be selected to be about one-tenth the value of the upper bias resistor (R_B). In the case of an emitter-biased circuit, the resistor should be about equal to the base-to-ground resistor value. Some technicians use a plain jumper wire for this test, but in doing so they risk damaging the transistor.

This test is usually not necessary; the turn-off test discussed earlier is safer and is probably sufficient. In Chapter 6, we will examine some more sophisticated tests, using a scope-and-function generator.

FIGURE 5-40 **Testing the Base-to-Emitter Junction Voltage**

FIGURE 5-41 **Quick-test Pull-down Test for Transistor Operation**

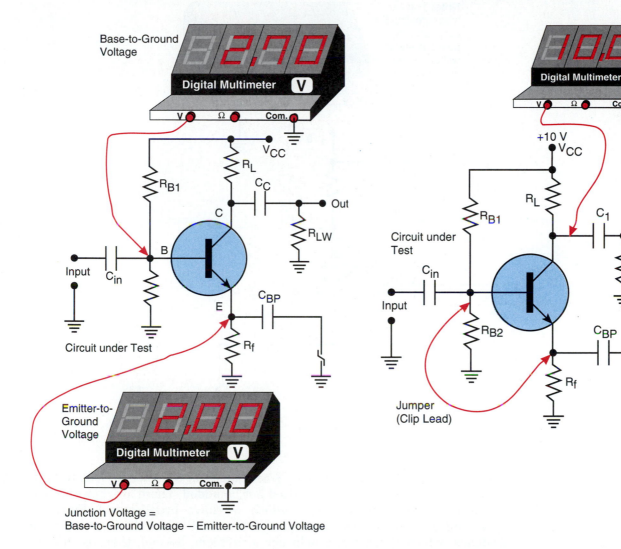

Junction Voltage =
Base-to-Ground Voltage − Emitter-to-Ground Voltage

The Digital Connection: Switching Transistor Circuits

The switching mode in transistors differs considerably from the linear operating mode we have been discussing. In the switching mode, we are not concerned with stability, since the transistor is either turned fully on or turned all the way off. It passes through the linear-operating part of the curve in a few

FIGURE 5-42 **Quick-test Pull-up Test for Transistor Operation**

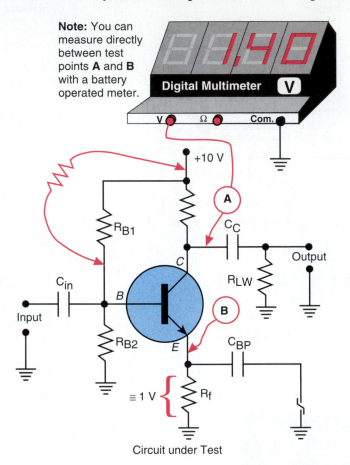

Note: You can measure directly between test points **A** and **B** with a battery operated meter.

Circuit under Test

nanoseconds to a few microseconds. Since quiescent collector-current stability is not a factor, negative feedback need not be added. There are, however, two modes of internal feedback over which we have little control. First, the base–emitter junction resistance provides a small amount of current-mode feedback, which is neither a help nor a problem. Second, there is internal voltage-mode feedback through the base–collector junction capacitance, which can be a significant problem. We normally think of capacitance coupling as involving some frequency, with the capacitive reactance being low only for high frequencies. A high frequency, however, represents a rapid rate of change in some signal voltage. In switching transistor circuits, a very rapid rate of change occurs when the transistor switches from off to on or from on to off. The rate of change in switching circuits is normally so fast that even the small collector–base junction capacitance can feed a significant inverted signal voltage from the collector back to the base.

The transistor does not exhibit voltage gain in either the full-off condition or the full-on condition, but to make the change from one of those conditions to the other, the transistor must pass through the linear part of the operating curve. The transistor exhibits a high voltage gain during the time it is chang-

FIGURE 5-43 **Switching Time Delay Factors**

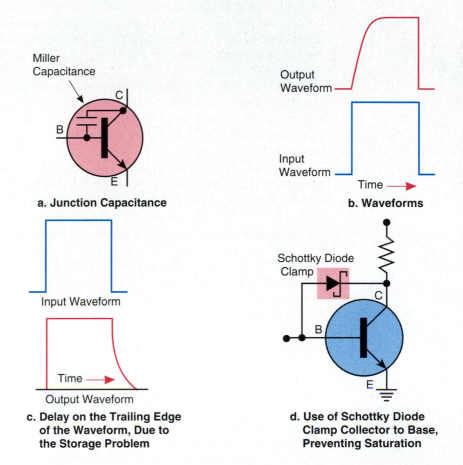

a. Junction Capacitance

b. Waveforms

c. Delay on the Trailing Edge of the Waveform, Due to the Storage Problem

d. Use of Schottky Diode Clamp Collector to Base, Preventing Saturation

ing from one state to the other. When voltage-mode feedback and voltage gain occur, a phenomenon called the *Miller effect* arises. The Miller effect causes the capacitance of the collector–base junction to appear to be much larger than it really is.

When the input signal switches to turn the transistor on, the transistor can't start to change states until the Miller capacitor has been charged. Unfortunately, it takes time to charge the capacitor, which slows down the switching action and distorts the output waveform. Figure 5-43(a) shows the junction capacitance "connection," and Figure 5-43(b) illustrates the distorting effect of the input capacitance, which produces rounding of the square wave input on the leading edge of the waveform.

We can minimize the input capacitance problem by using a small-value collector-load resistor. During a switching transition, the voltage gain is

$$A_V = R_L/R_e'$$

Lowering the voltage gain reduces the Miller effect, reducing the effective value of the input capacitance. A low-impedance drive source can deliver more current, to charge the capacitor quicker.

If the transistor is in the full-on state, it is in saturation ($V_{CE(sat)}$) and the collector–base depletion zone is full of carriers (holes/electrons). The depletion zone is the dielectric of the capacitor in the reverse-bias junction condition; but a forward-biased junction has no dielectric, because the depletion zone has become a conductor. As a result, no input capacitance exists when the transistor is in saturation.

The good news is that there is no junction capacitance to slow down the on-to-off switching action. The bad news is that it takes time to clear the carriers out of the depletion zone, and this slows down the switching action. The depletion zone must be re-formed before normal transistor action can turn the transistor off.

For slower-speed switching circuits, the depletion-zone carrier-storage phenomenon may not be a problem; but in high-speed digital circuits, it can be deadly. The only cure is to prevent the transistor from ever going into saturation. This is normally accomplished by using a diode to clamp the collector to the base just prior to allowing the collector–base junction to switch to a forward-bias condition. A Schottky diode is usually used as a clamp, because it has no carrier storage or capacitance problems of its own. Figure 5-43(c) shows the effect of depletion-zone carrier storage on the input waveform, and Figure 5-43(d) shows a transistor equipped with a Schottky diode clamp to eliminate the problem.

Some Practical Switching Circuit Considerations

The circuit in Figure 5-44(a) is a basic transistor switching circuit. Notice that the input voltage is +5 V (equal to the collector supply voltage). The resistor in series with the base limits the base current to a safe value.

In order to saturate the transistor in the on condition, we must drop the entire 5 volts (V_{CC}) across R_L. Ohm's law tells us that we need a collector current of 5 mA to obtain a 5-V drop across a 1-kΩ collector-load resistor. Consequently, we need a base current of 5 mA/ß. If beta is 100, we need 0.005 mA of base current. Since the input voltage in the high condition is 5 volts, we can again use Ohm's law to calculate the value of the input resistor value:

$$R = 5 \text{ V}/0.005 \text{ mA} = 100 \text{ k}\Omega$$

This sort of calculation makes a nice textbook exercise, but it is not very useful in practice. The input resistor combines with the Miller input capacitance to yield a very long R-C time constant, making the circuit unacceptably slow.

Ideally, the input resistor should have as low a value as possible. In practice, an overdrive base current of 50 to 100 times the value calculated in the example is used. The base–emitter junction can tolerate 5 to 10 mA comfortably, and there is no reason not to overdrive the base by that amount in the interest of obtaining faster switching action. A value of somewhere between 4.7 kΩ and 1 kΩ have become typically accepted values for the input resistor in this circuit.

Darlington-pair Switching Circuit

The Darlington-pair circuit in Figure 5-44(b) offers some significant advantages in switching service. The input capacitance is greatly reduced because

FIGURE 5-44   **Transistor Switching Circuits**

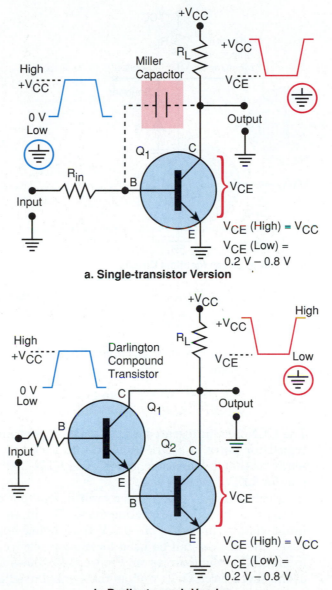

a. Single-transistor Version

b. Darlington-pair Version

the input transistor is basically an emitter-follower (common collector) stage that has a voltage gain of 1. Because the input capacitance is effectively multiplied by the voltage gain, the Darlington circuit eliminates the input transistor's Miller effect problem. A voltage gain of 100 makes the input capacitance appear to be 100 times as large as it really is, slowing the switching speed proportionately.

Even if the complete pair is used in a common emitter circuit (which does have a high voltage gain), there is no input transistor Miller effect. There actually is Miller-effect feedback in the output transistor of the pair, but the input of the second transistor is directly driven by the emitter of the input transistor,

FIGURE 5-45   **ULN-2003 Darlington Transistor Array**

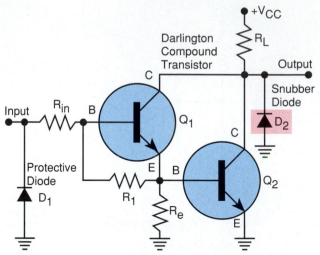

a. Schematic Diagram of One of Seven Identical Circuits

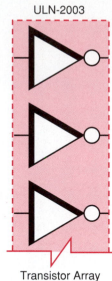

Transistor Array

b. Symbol (for Three of the Seven Circuits)

and this direct drive does not involve an added input resistor to increase the R-C time constant.

ULN-2003, Seven-circuit Darlington Transistor Array

The ULN-2003 integrated circuit package contains seven separate Darlington transistor pairs in a DIP (dual in-line package). The device is typical of the many available transistor arrays. The ULN-2003 was designed for switching and for LED driver service.

Each Darlington pair has a built-in snubber diode for inductive switching service and an input protective diode. The input diode prevents any negative input voltage in excess of –0.7 V from being applied to the input transistor base. The devices can be used for analog service, if care is taken to keep the input diode from turning on. Figure 5-45 shows the internal schematic of one of the seven circuits, as well as the symbol used in systems diagrams.

ANALYSIS PROBLEMS

5.1   Analyze the base-biased common emitter circuit (see Figure 5-46) by calculating the values for each of the following (using the universal transistor circuit analysis worksheet provided):

FIGURE 5-46 **Circuit for Problem 5.1**   **FIGURE 5-47** **Circuit for Problem 5.2**

FIGURE 5-46 **Circuit for Problem 5.1**   **FIGURE 5-47** **Circuit for Problem 5.2**

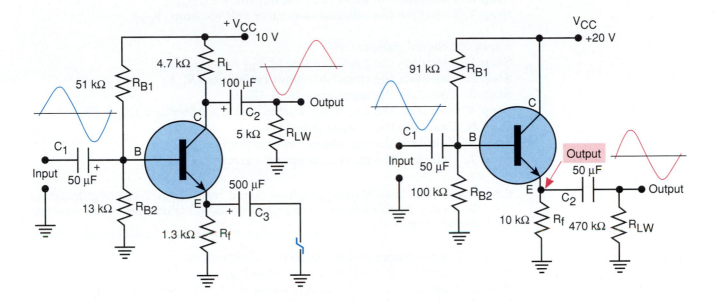

SHEET #1: **dc** Voltages and the Q-point Collector Current
Step 1. Calculate the voltage drop across R_{B2}.
Step 2. Calculate the voltage drop across the feedback resistor (V_{R_f}).
Step 3. Calculate the quiescent collector current (QI_C).
Step 4. Calculate the base–emitter junction resistance (R'_e).
Step 5. Calculate the dummy-load voltage drop (V_{R_L}).
Step 6. Calculate the **ac** signal-load impedance (R_{Lac}).
Step 7. Calculate the collector-to-emitter voltage drop (V_{CE}).

SHEET #2: Signal Parameters
Step 1. Calculate the parallel value of R_{B1} and R_{B2} (R_B).
Step 2. Calculate the transistor input resistance (R_{in}).
Step 3. Calculate the input impedance (Z_{in}).
Step 4. Calculate the maximum P–P output signal voltage (V_O P–P).
Step 5. Calculate the voltage gain (A_V).
Step 6. Calculate the maximum power output (P_{out}).
Step 7. Calculate the maximum output current (I_O P–P).

5.2 Analyze the base-biased common collector circuit (see Figure 5-47) by calculating the values for the following (using the universal transistor circuit
analysis worksheet provided):

SHEET #1: **dc** Voltages and the Q-point Collector Current
Step 1. Calculate the voltage drop across R_{B2}.
Step 2. Calculate the voltage drop across the feedback resistor (V_{R_f}).
Step 3. Calculate the quiescent collector current (QI_C).
Step 4. Calculate the base–emitter junction resistance (R'_e).

Step 5. Calculate the dummy-load voltage drop (V_{R_L}).
Step 6. Calculate the **ac** signal-load impedance ($R_{L_{ac}}$).
Step 7. Calculate the collector-to-emitter voltage drop (V_{CE}).

SHEET #2: Signal Parameters
Step 1. Calculate the parallel value of R_{B1} and R_{B2} (R_B).
Step 2. Calculate the transistor input resistance (R_{in}).
Step 3. Calculate the input impedance (Z_{in}).
Step 4. Calculate the maximum P–P output signal voltage (V_O P–P).
Step 5. Calculate the voltage gain (A_V).
Step 6. Calculate the maximum power output (P_{out}).
Step 7. Calculate the maximum output current (I_O P–P).

5.3   Analyze the base-biased paraphase circuit (see Figure 5-48) by calculating the values for the following (using the universal transistor circuit analysis worksheet provided):

SHEET #1: **dc** Voltages and the Q-point Collector Current
Step 1. Calculate the voltage drop across R_{B2}.
Step 2. Calculate the voltage drop across the feedback resistor (V_{R_f}).
Step 3. Calculate the quiescent collector current (QI_C).
Step 4. Calculate the base-emitter junction resistance (R_e').
Step 5. Calculate the dummy-load voltage drop (V_{RL}).
Step 6. Calculate the **ac** signal-load impedance (R_{Lac}).
Step 7. Calculate the collector-to-emitter voltage drop (V_{CE}).

SHEET #2: Signal Parameters
Step 1. Calculate the parallel value of R_{B1} and R_{B2} (R_B).
Step 2. Calculate the transistor input resistance (R_{in}).

FIGURE 5-48   **Circuit for Problem 5.3**

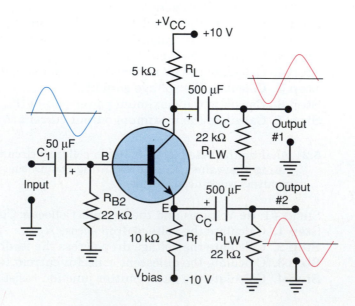

Step 3. Calculate the input impedance (Z_{in}).
Step 4. Calculate the maximum P–P output signal voltage (V_O P–P).
Step 5. Calculate the voltage gain (A_V).
Step 6. Calculate the maximum power output (P_{out}).
Step 7. Calculate the maximum output current (I_O P–P).

5.4 Analyze the emitter-biased common-emitter circuit (see Figure 5-49) by calculating the values for the following (using the universal transistor circuit
analysis worksheet provided):

SHEET #1: **dc** Voltages and the Q-point Collector Current
Step 1. Start emitter-biased circuit here.
Step 2. Enter feedback voltage value.
Step 3. Calculate the quiescent collector current (QI_C).
Step 4. Calculate the base-emitter junction resistance (R_e').
Step 5. Calculate the dummy-load voltage drop (V_{R_L}).
Step 6. Calculate the **ac** signal-load impedance (R_{Lac}).
Step 7. Calculate the collector-to-emitter voltage drop (V_{CE}).

SHEET #2: Signal Parameters
Step 1. Calculate the parallel value of R_{B1} and R_{B2} (R_B).
Step 2. Calculate the transistor input resistance (R_{in}).
Step 3. Calculate the input impedance (Z_{in}).
Step 4. Calculate the maximum P–P output signal voltage (V_O P–P).
Step 5. Calculate the voltage gain (A_V).
Step 6. Calculate the maximum power output (P_{out}).
Step 7. Calculate the maximum output current (I_O P–P).

FIGURE 5-49 **Circuit for Problem 5.4**

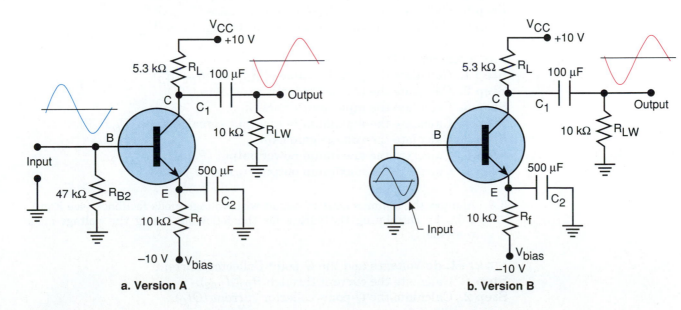

a. Version A   **b. Version B**

FIGURE 5-50   **Circuit for Problem 5.5**

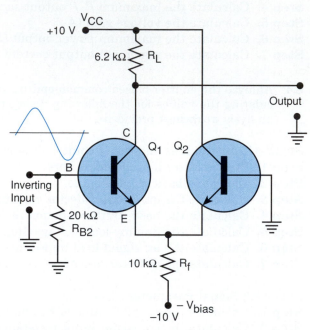

5.5   Analyze the differential amplifier circuit (see Figure 5-50) by calculating the
values for the following (using the universal transistor circuit analysis worksheets provided):

SHEET #1: **dc** Voltages and the Q-point Collector Current
Step 1.   Start emitter-biased circuit here.
Step 2.   Enter feedback voltage value.
Step 3.   Calculate the quiescent collector current (QI_C).
Step 4.   Calculate the base–emitter junction resistance (R'_e).
Step 5.   Calculate the dummy-load voltage drop (V_{R_L}).
Step 6.   Calculate the **ac** signal-load impedance (R_{Lac}).
Step 7.   Calculate the collector-to-emitter voltage drop (V_{CE}).

SHEET #2: Signal Parameters
Step 1.   Calculate the parallel value of R_{B1} and R_{B2} (R_B).
Step 2.   Calculate the transistor input resistance (R_{in}).
Step 3.   Calculate the input impedance (Z_{in}).
Step 4.   Calculate the maximum P–P output signal voltage (V_O P-P).
Step 5.   Calculate the voltage gain (A_V).
Step 6.   Calculate the maximum power output (P_{out}).
Step 7.   Calculate the maximum output current (I_O P–P).

5.6   Analyze the common emitter circuit with voltage-mode feedback (see Figure
5-51) by calculating the values for the following (using the voltage-mode
worksheet):

SHEET #1: **dc** Voltages and the Q-point Collector Current
Step 1.   Calculate the current through R_{B2} (I_{RB2}).
Step 2.   Calculate the Q-point collector current (QI_C).

FIGURE 5-51 Circuit for Problem 5.6

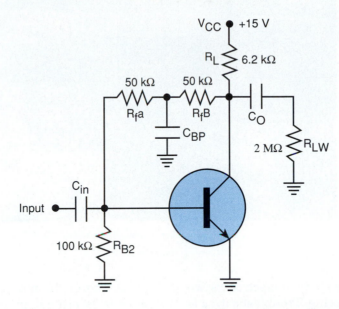

Step 3. There is no supplementary current-mode feedback.
Step 4. Calculate the base–emitter junction resistance (R_e').
Step 5. Calculate the dummy-load voltage drop (V_{R_L}).
Step 6. Calculate the collector-to-emitter voltage drop (V_{CE}).

SHEET #2: Signal Parameters
Step 0. Copy essential data from sheet #1.
Step 1. Calculate the value of R_{B2} and R_fA in parallel (R_B).
Step 2. Calculate the input impedance (Z_{in}).
Step 3. Calculate the maximum P–P output voltage (V_O P–P).
Step 4. Calculate the voltage gain (A_V).
Step 5. Calculate the maximum power output (P_{out}).
Step 6. Calculate the maximum output current (I_O P–P).

Trouble-shooting

As we proceed through our study of electronics, we talk a lot about troubleshooting. Troubleshooting is an important responsibility of electronics technicians. But what is troubleshooting, really? Is it merely the process of making measurements in faulty electronic equipment, or of replacing modules or components until the equipment starts working again? Or does troubleshooting hinge on the technician's ability to analyze a schematic and determine exactly what should be happening in a fully functional circuit?

Talk to a dozen experienced technicians, and you will probably get a dozen different definitions of *troubleshooting*. And all of them may well contain important elements of truth. Technicians develop their own style and approach to solving problems—some possibly more effective than others—but a few common threads can be identified in everybody's personal approach to troubleshooting.

First, you will find that there is no substitute for a solid understanding of electronic circuit theory. Ohm's law, Kirchhoff's voltage and current laws, and the voltage divider rule are like the oxygen in air—indispensable. You have to know the basics if you want to understand circuits. A good grasp of series and parallel circuits and of the characteristics of resistive, capacitive, and inductive circuits is likewise essential. You need them to understand sensors, transmission lines, output devices and every other aspect of electrical circuits.

A thorough understanding of semiconductor devices is critical in troubleshooting electronics, even if you are only replacing modules. The input and output characteristics of electronic modules determine how they can be applied, how and where they may fail under certain conditions, and how subsequent failures can be avoided. A technician trying to troubleshoot the circuitry of a module without understanding transistor biasing or typical amplifier configurations is like a navigator trying to steer by dead reckoning without knowing which stars to use.

Next, the ability to use various types of test equipment is important. And it's not enough to just know how to make measurements. You have to know whether the measurement is accurate and reliable and what the measurement means. Every electronics technician should be able to measure voltage, current, and resistance on a digital multimeter. Taking oscilloscope measurements is another mandatory skill. The ability to use power supplies, capacitance meters, frequency counters, and many other specialized types of test equipment is also important.

But the most important piece of test equipment available to you is your mind. Effective troubleshooting depends on how you approach problem solving and how you apply your knowledge of electronics and your skills with test equipment.

Many technicians, if asked what kind of thinking skills are important in troubleshooting, would

Two-stage Amplifier Schematic

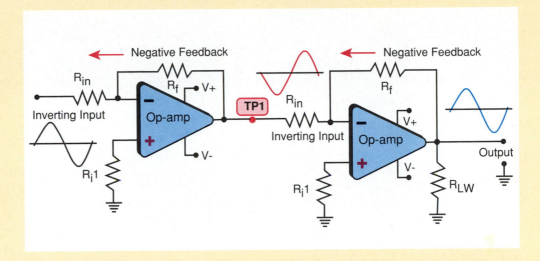

answer: logical thinking skills. This is true, but only partly so. Certainly, troubleshooting does require logic. Consider a simple example: In a two-stage amplifier circuit, we have a signal entering stage one but nothing registering at the output. Logically, we "divide and conquer." We use an oscilloscope to check for signal at the output of the first stage. If there is signal there, we can logically assume that the problem must be in the second stage, so we look more closely at that area.

In real-life troubleshooting, however, the situation is usually more complex. Rather than encountering a problem that we can isolate by means of a logical, linear process, we often find that problems and obstacles prevent us from moving step by step toward the solution.

For example, in another type of amplifier circuit, negative feedback is used to control the gain and to minimize distortion levels. Again there is no output, even though an input has been applied. In this case, the negative feedback complicates mat-

ters, and the simple "divide and conquer" method does not work as well. Other kinds of tests must be devised. Perhaps you turn off the power and check all the semiconductors for shorts and opens. Possibly a bias voltage check will reveal the problem.

A logical process is still involved, but it is not as linear as before. It also incorporates an intuitive element, since the technician must devise new ways of testing and new approaches to the problem. This can be thought of as moving laterally, taking another approach, and then proceeding as far as possible again in a linear direction.

The most valuable aspect of developing excellence in troubleshooting is that the skill is transferable to almost any other pursuit. Whether you apply it to electronic systems, to some other kind of system, to software debugging, to customer service, or to personnel development, the ability to solve problems creatively will always be a marketable skill.

CHAPTER 6

Multistage and Power Amplifiers

OBJECTIVES

Upon completion of this chapter, you should be able to:

1 Analyze multistage R-C-coupled amplifiers.

2 Explain the operation of the inductor–capacitor-coupled amplifier.

3 Explain why transformer coupling can yield very high voltage gains.

4 Explain the operation of tuned radio-frequency amplifiers.

5 Explain how each of the following turbocharging techniques is used:

 a The Darlington pair.

 b The constant-current device.

 c Bootstrapping.

6 Explain how the differential amplifier can be turbocharged.

7 Explain the operational differences between class-A, class-B, and class-AB amplifiers.

8 Identify and compare the percentage efficiencies of class-A and class-B amplifiers.

9 Explain the cause and cure of crossover distortion.

10 Draw the schematic diagram of a two-stage class-A power amplifier and driver, with voltage-mode feedback around both stages.

11 Identify complementary-symmetry power-amplifier circuits of various types and levels of complexity.

12 Draw the schematic diagram of a basic two-transistor differential amplifier.

13 Draw the schematic diagram of a transistor differential amplifier, with a current mirror in place of the shared emitter resistor.

14 List the outstanding characteristics of the differential amplifier.

15 Explain what happens when voltage-mode feedback is added to a differential amplifier.

16 Draw the schematic diagram of a current-mirror constant-current source.

17 Define class-A amplifier operation.

18 Define class-B amplifier operation.

19 Define class-C amplifier operation.

Analysis In Brief

Multistage & Power Amplifiers in Brief

Coupling Techniques

R-C-COUPLED MULTISTAGE AMPLIFIERS

R-C-coupled amplifiers use a capacitor to provide isolation between stages for **dc** voltages, while passing **ac** signal voltages from one stage to the next. Usually, all stages are identical and each stage (except the last one) has a maximum voltage gain of about 66. The input impedance of the driven stage is the signal load (R_{LW}) for the driving stage. The voltage gain of several R-C coupled stages is equal to $A_{V1} \times A_{V2} \times A_{V3} \cdots$

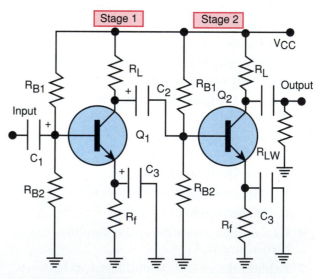

R-C coupled Amplifier

INDUCTOR–CAPACITOR COUPLING

Inductor–capacitor coupling is used in much the same way as R-C coupling, except that the inductor serves as the collector load. The collector-load impedance is higher at higher frequencies, so the voltage gain of the amplifier peaks at the high-frequency end. The inductor is sometimes used to balance out stray capacitance, which tends to lower the gain at higher frequencies.

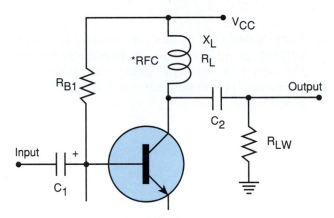

Inductor–Capacitor Coupling

TRANSFORMER COUPLING

Transformer coupling can provide voltage gains of up to 800 per stage, compared to a maximum of 66 for R-C-coupled stages. The input impedance of the driven stage still serves as the signal working load for the driving stage, but that impedance is stepped up by the turns ratio of the transformer. The voltage gain of the transformer coupled stage is $A_V = n \times R_{Lac}/R'_e$ where n is the transformer's turns ratio.

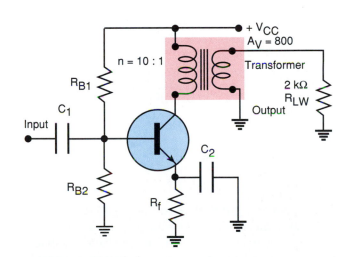

Transformer Coupling

Turbocharging Techniques For Transistor Amplifiers

Three common techniques can drastically improve the performance of transistor amplifiers, simply and inexpensively. Many integrated-circuit amplifiers use all three methods.

THE DARLINGTON SUPER-BETA TRANSISTOR PAIR

The Darlington pair uses two direct-coupled transistors to form a composite transistor with a ß

value of from 2000 to 5000. This very high ß affects circuit performance in two primary ways:

1   Any resistance in the emitter is elevated by the super-ß of 2000 to 5000, as compared to the factor of 100 for a single transistor. The resulting very high input impedance yields much higher voltage gains in situations where several stages are coupled, as well as a light load on any signal source.

2 Power gain is the product of voltage gain times current gain. The super-ß directly increases the circuit's current gain and indirectly increases its voltage gain—a double boost in power amps.

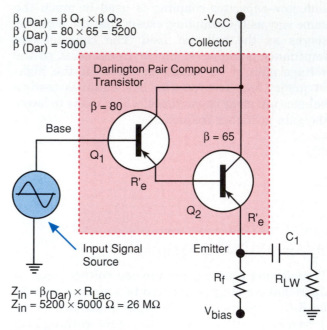

$$\beta \text{ (Dar)} = \beta\, Q_1 \times \beta\, Q_2$$
$$\beta \text{ (Dar)} = 80 \times 65 = 5200$$
$$\beta \text{ (Dar)} = 5000$$

$$Z_{in} = \beta_{(Dar)} \times R_{Lac}$$
$$Z_{in} = 5200 \times 5000\ \Omega = 26\ M\Omega$$

Darlington Super-β Pair

POSITIVE FEEDBACK

When properly limited, positive feedback can be used to cancel real impedances, making them appear to be open circuits. In the positive-feedback (bootstrapping) model, opposing voltages cancel. No current flows through the resistor. As viewed from either generator, the resistor appears to be an open circuit. In real situations, voltages partly cancel. The resistor appears to have a much higher value than its color code indicates.

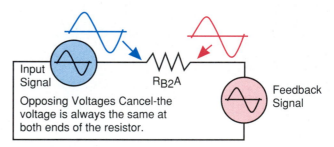

Bootstrapping Model, Using Positive Feedback

THE CONSTANT-CURRENT DEVICE

When a transistor is biased on to some specific collector current, that current value is exclusively a function of ß and of the base-bias current. We can change the collector voltage, but the collector current doesn't change. When we increase the collector voltage, the collector–base depletion zone widens. The wider depletion zone has a higher dynamic resistance, however, which cancels the expected increase in collector current. This effect only involves collector voltage changes, not steady state **dc** conditions. There are two resistances: one **dc** resistance and one resistance to any change.

When we use a biased-on transistor in a circuit, it behaves as a 1-megohm (or so) resistor for signal- or temperature-induced collector voltage variations, but it exhibits **dc** voltage drops that correspond to much lower values of resistance.

Because the transistor's collector current value remains nearly the same, no matter what voltage is applied, the collector current is called a *constant-current circuit*.

A transistor can be used to replace an ordinary 1-megohm resistor for signal purposes. A real 1-megohm resistor has a voltage drop of 1000 V, with a current of 1 mA. A transistor can simulate a 1-megohm resistor for signal voltages, but with a **dc** voltage drop of only a few volts.

In the following circuit, what happens to the collector current if we change the collector voltage from 10 to 20 volts? Answer: Nothing. The equation for collector current is $QI_C = ß \times I_B$ where I_B is the base–emitter current. The collector voltage has almost no effect on collector current.

Question: What happens to the collector current if we change V_{CC} from 10 V to 20 V?

Answer: Nothing! Collector current is independent of collector voltage.

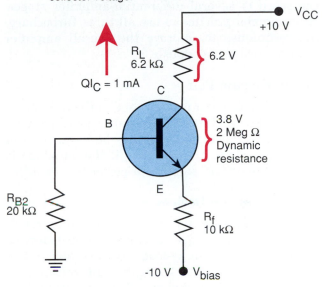

Circuit with Changed Collector Voltage

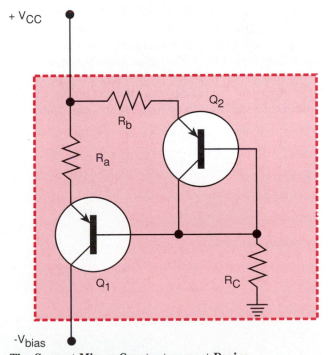

The Current Mirror Constant-current Device

TURBOCHARGING THE DIFFERENTIAL AMPLIFIER

Most modern integrated-circuit amplifiers are composed of several differential-amplifier stages. It is common practice to use all three turbocharging techniques to improve the overall amplifier performance tremendously.

The Darlington Pair

The Darlington pair is often used in the first stage, to provide a high-input impedance for the signal source. The Darlington pair may be used again in the output stages, particularly if the amplifier is intended to deliver significant power to a load.

Constant-current Devices

It is not unusual for a multistage integrated-circuit amplifier to use constant-current (sources) for all collector dummy-load resistors, and to use constant-current (sinks) to replace all emitter (current-mode) feedback resistors. These constant-current devices not only provide improved performance, but actually cost less to manufacture on the IC chip.

Bootstrapping and the Differential Amplifier

When voltage-mode negative feedback is added, (usually around several stages), the inverting input becomes a very low input impedance because of the Miller effect. Positive feedback sneaks in the back door to bootstrap the noninverting input, causing it to have a very high input impedance.

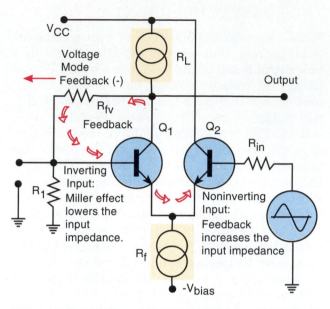

Differential Amplifier with Added Voltage-mode Feedback and Constant-current Devices Replacing the Collector and Emitter Resistors

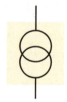

Constant-current Device Symbol

Voltage-mode Feedback

VOLTAGE-MODE FEEDBACK AND POWER TRANSISTORS

One special advantage of voltage-mode feedback is that it can be used to enclose several stages in the feedback loop. This allows designers to include a hard-to-stabilize power transistor in a feedback loop with an easily stabilized low-power transistor. The combination results in a circuit almost as easy to stabilize as a single low-power transistor.

Voltage-mode feedback is infrequently used to stabilize a single transistor stage.

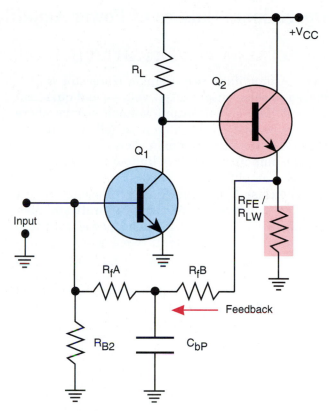

Voltage-mode Feedback Used to Stabilize Power Transistor Q_2 and Small-signal Transistor Q_1 in the Same Feedback Loop

VOLTAGE-MODE FEEDBACK AND MULTISTAGE IC AMPLIFIERS

Many integrated circuit amplifiers, including the operational amplifier, have provisions for external voltage-mode feedback.

In most cases, negative feedback is used on-board the IC chip to provide Q-point stability; but external feedback can be added to improve stability or to control gains and impedances.

MULTISTAGE CURRENT-MODE FEEDBACK

Multistage current-mode feedback is not a common feature of multistage and power amplifiers.

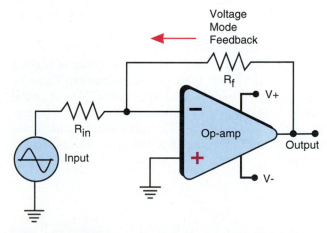

Voltage-mode Feedback Used as External Feedback for an IC Amplifier

Operational Classes of Power Amplifier

THE CLASS-A POWER AMPLIFIER

Class-A amplifiers use a single transistor that is biased to some output offset voltage and quiescent current. The signal output voltage swings above and below that offset value as it follows a signal. Class A is the most common class of operation for voltage amplifiers, but it is limited to 5 watts or so in power amplifiers.

The class-A amplifier has an efficiency of only about 25%, which means that a 25-watt amplifier would deliver 25 watts of useful power to a load and waste 75 watts as heat. When there is no signal to amplify, the quiescent collector current still causes power to be produced as heat. Using a 100-watt power supply for a 25-watt amplifier is not economically smart.

Input Waveform

Transistor turns off at 0.7 V

Crossover Distortion

Output Waveform

Part of signal is lost

Transistor turns on at - 0.7 V

Crossover Distortion

THE CLASS-B AMPLIFIER

The class-A amplifier uses two power transistors—one for each half of a sine-wave signal. One transistor amplifies the positive half of the waveform, while the other transistor rests. Then the second transistor amplifies the negative half of the cycle, while the first transistor rests. No power is wasted in heat when there is no signal to amplify. The efficiency of a class-B (sometimes called a *push–pull*) power output circuit is about 78%. Thus, a 78-watt amplifier would deliver 78 watts to the load, while wasting only 22 watts of a 100-watt power supply as heat.

The Class-B (Push-Pull) Complementary-symmetry IC Power Amplifier

6.1   Introduction

In this chapter, we will examine methods of coupling amplifier stages, using capacitors, inductors, transformers, and direct coupling. We will look at the different circuits that are used, and assess their advantages and disadvantages. We will study three techniques that drastically improve the performance of basic transistor amplifier circuits, and we will investigate how these turbocharging techniques can be applied to the differential amplifier.

We will then examine power amplifier circuits, and see how a simple class-A amplifier with dual negative-feedback paths has evolved into the modern circuit forms used in power-amplifier integrated circuits.

Finally, we will examine several common commercial integrated-circuit power amplifiers.

6.2   Capacitor-coupled Multistage Amplifiers

Capacitor coupling is one of the most popular methods of coupling transistor amplifier stages. One of its primary advantages is that the driven stage loads the driving stage only for the **ac** signal, without affecting the quiescent collector current or the stability of the driving stage. The signal input impedance of the driven stage becomes the working load (R_{LW}) for the driving stage. Other than that, the two stages are completely independent of each other. This makes analyzing the circuit easy, because you can use the universal transistor circuit analysis worksheet on one stage by itself, as you would for a lone transistor circuit.

Capacitor coupling is generally preferred when an **ac** signal is involved. Direct coupling must be used for **dc** or very-low-frequency signals, but direct coupling demands some compromises and generally yields less gain per stage than does capacitor coupling.

When several capacitor-coupled stages are cascaded, all stages are usually identical. There is no necessity behind this, but it proves to be the best approach in terms of total voltage gain. The term *cascading stages* means that the output of one stage drives the input of the next. The voltage gain of three stages in cascade is

$$A_{V\,(total)} = A_{V\,(stage\ 1)} \times A_{V\,(stage\ 2)} \times A_{V\,(stage\ 3)} \cdots$$

The voltage gain is always the product of the several individual stage voltage gains. The maximum possible voltage gain per stage in a chain of identical stages is about 66. The **ac** signal input impedance of each driven stage is the working load (R_{LW}) for the driving stage, and R_{LW} largely determines the maximum possible gain of the driving stage. It turns out that, no matter how the circuit is designed, 66 is about the limit. In general, circuits are designed for the maximum available voltage gain, so a voltage gain of 60 to 66 is about what you can expect, most of the time.

Figure 6-1 shows a two-stage capacitor-coupled amplifier circuit. This circuit will be the focus of Example 6-1.

Example 6.1

Calculate the Voltage Gain of Stage 1 of the R-C-coupled Amplifier in Figure 6-1.

Solution

We begin by finding the input impedance (Z_{in}) of stage 2, which forms the working load for stage 1. We must know that value if we are to calculate the voltage gain of the first stage in Figure 6-1(a).

The equivalent circuit for stage 1 is shown in Figure 6-1(b). Let us proceed to calculate the voltage gain of stage 1, using its equivalent in Figure 6-1, and Step 1 in the universal transistor circuit analysis worksheet, sheet #2 (Figure 6-2(b)).

Step 1: Calculating R_B

$$R_B = (R_{B1} \times R_{B2})/(R_{B1} + R_{B2})$$
$$= (50 \text{ k}\Omega \times 10 \text{ k}\Omega)/(50 \text{ k}\Omega + 10 \text{ k}\Omega)$$
$$= 500 \text{ k}\Omega/60 \text{ k}\Omega = 8.33 \text{ k}\Omega$$

Step 2(a): Calculating R_{in} (Bypassed)

$$R_{in} = 100 \times R'_e$$
$$= 100 \times 25 = 2.5 \text{ k}\Omega$$

Step 3: Calculating the Input Impedance (Z_{in})

$$Z_{in} = (R_B \times R_{in})/(R_B + R_{in})$$
$$= (8.33 \text{ k}\Omega \times 2.5 \text{ k}\Omega)/(8.33 \text{ k}\Omega + 2.5 \text{ k}\Omega) = 20.825/10.83$$
$$= 2.0 \text{ k}\Omega$$

Because Z_{in} of stage 2 is the working load (R_{LW}) for stage 1, and $R_{LW} = 2.0$ kΩ, we can calculate the voltage gain from stage 1. Go back to the analysis worksheet sheet #1 (Figure 6-2(a)), and calculate the value of the **ac** signal load (R_{Lac}).

Sheet #1, Step 6: Calculating the ac signal-load impedance (R_{Lac})

If $R_{LW} = Z_{in}$ (stage 2), then

$$R_{Lac} = (R_L \times Z_{in})/(R_L + Z_{in})$$
$$= (10 \text{ k}\Omega \times 2 \text{ k}\Omega)/(10 \text{ k}\Omega + 2 \text{ k}\Omega) = 20 \text{ k}\Omega/12 \text{ k}\Omega = 1.66 \text{ k}\Omega$$

Now we are ready to return to sheet #2.

Step 5: Calculating the Voltage Gain (Bypassed)

$$A_V = R_{Lac}/R'_e$$
$$= 1.66 \text{ k}\Omega/25\Omega = 66$$

Remember that the emitter feedback resistor in this circuit is bypassed for signal frequencies, so the voltage gain completely depends on R'_e and on beta. Because neither of these factors is very predictable, any voltage gain calculated for this circuit can only represent an approximation. If we measured the voltage gain of this circuit in a real situation, and found a two-stage gain of

FIGURE 6-1   **Two-stage Current-mode, Common-emitter, R–C-coupled Amplifier for Worksheet Analysis**

Total Voltage Gain = A_V (stage 1) $\times$ A_V(stage 2) = 66 $\times$ 66 = 4356

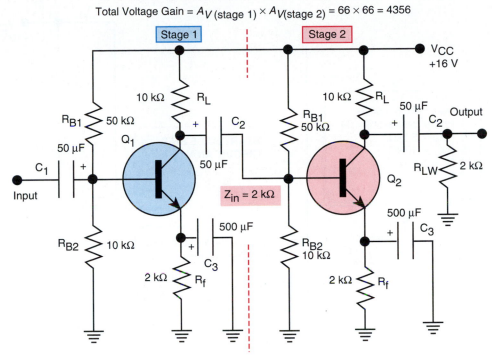

a. Complete circuit for Example 6-1

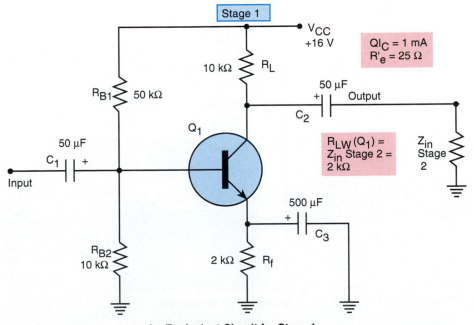

b. Equivalent Circuit for Stage 1

FIGURE 6-2 Worksheet for Two-stage Current-mode, Common Emitter, R–C-coupled Amplifier

Sheet #1: dc Voltages

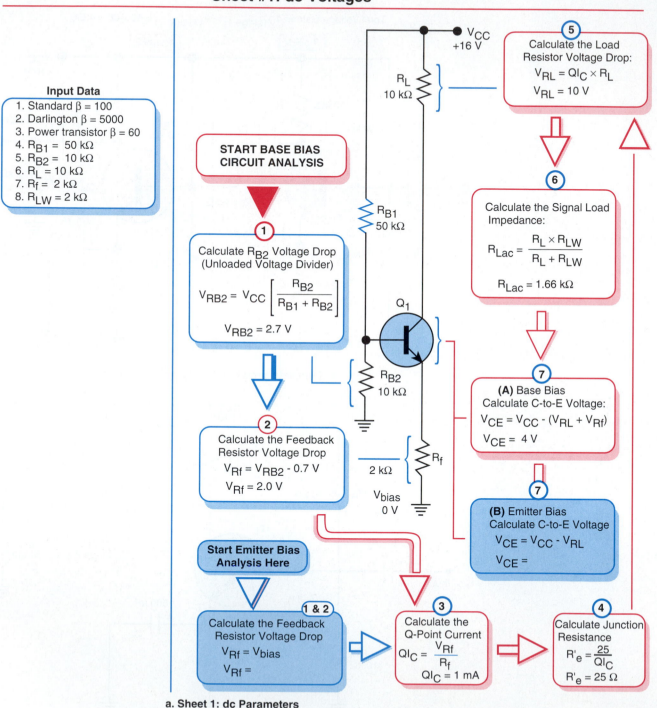

a. Sheet 1: dc Parameters

FIGURE 6-2   Continued

Sheet #2: Signal Parameters

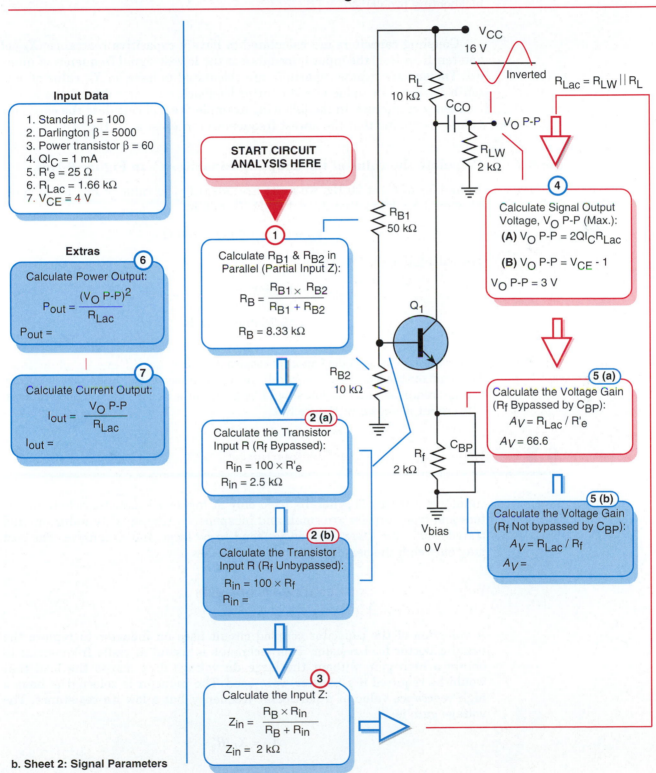

Input Data

1. Standard $\beta = 100$
2. Darlington $\beta = 5000$
3. Power transistor $\beta = 60$
4. $QI_C = 1$ mA
5. $R'_e = 25 \ \Omega$
6. $R_{Lac} = 1.66$ kΩ
7. $V_{CE} = 4$ V

Extras

6 Calculate Power Output:
$$P_{out} = \frac{(V_O \text{ P-P})^2}{R_{Lac}}$$
$$P_{out} =$$

7 Calculate Current Output:
$$I_{out} = \frac{V_O \text{ P-P}}{R_{Lac}}$$
$$I_{out} =$$

START CIRCUIT ANALYSIS HERE

V_{CC} 16 V

R_L 10 kΩ

C_{CO}

Inverted $R_{Lac} = R_{LW} \parallel R_L$

V_O P-P

R_{LW} 2 kΩ

4 Calculate Signal Output Voltage, V_O P-P (Max.):
(A) $V_O \text{ P-P} = 2QI_C R_{Lac}$
(B) $V_O \text{ P-P} = V_{CE} - 1$
$V_O \text{ P-P} = 3$ V

R_{B1} 50 kΩ

1 Calculate R_{B1} & R_{B2} in Parallel (Partial Input Z):
$$R_B = \frac{R_{B1} \times R_{B2}}{R_{B1} + R_{B2}}$$
$$R_B = 8.33 \text{ k}\Omega$$

Q_1

R_{B2} 10 kΩ

2 (a) Calculate the Transistor Input R (R_f Bypassed):
$R_{in} = 100 \times R'_e$
$R_{in} = 2.5$ kΩ

R_f 2 kΩ C_{BP}

V_{bias} 0 V

5 (a) Calculate the Voltage Gain (R_f Bypassed by C_{BP}):
$A_V = R_{Lac} / R'_e$
$A_V = 66.6$

5 (b) Calculate the Voltage Gain (R_f Not bypassed by C_{BP}):
$A_V = R_{Lac} / R_f$
$A_V =$

2 (b) Calculate the Transistor Input R (R_f Unbypassed):
$R_{in} = 100 \times R_f$
$R_{in} =$

3 Calculate the Input Z:
$$Z_{in} = \frac{R_B \times R_{in}}{R_B + R_{in}}$$
$$Z_{in} = 2 \text{ k}\Omega$$

b. Sheet 2: Signal Parameters

3000 or 6000, we could reasonably conclude that the circuit was working just fine. If the two-stage gain measured only 80 or 800, however, we would know that we had a serious problem. A ballpark figure is usually sufficient for troubleshooting problems.

Coupling capacitors are calculated to have a capacitive reactance (X_C) of one-tenth (or less) the input impedance at the lowest signal frequency of interest. The emitter bypass capacitors are calculated to have an X_C value of one-tenth (or less) the value of the emitter feedback resistor value at the lowest frequency of interest. In the following example, we will calculate the capacitor values, assuming that the lowest frequency of interest is 20 Hz.

Example 6.2

Calculate the value of the coupling capacitor C_2 in Figure 6-1(a).

Solution

To find X_C of C_2 at 20 Hz, we must first identify the value of Z_{in}. From sheet #1, step 6, we know that Z_{in} = 1.6 kΩ. Therefore,

$$X_C = 0.1 \times 1.6 \text{ k}\Omega = 160 \text{ }\Omega$$

The calculation for C_2 itself is

$$\begin{aligned} C_2 &= 1/(2\pi f X_C) \\ &= 1/(6.28 \times 20 \times 160) \\ &= 49.76 \text{ microfarads} \end{aligned}$$

Emitter-biased Version of the Capacitor-coupled Amplifier
The emitter-biased circuit is rarely used, because of the dual power-supply voltage requirement, but you should be aware of it in case you do encounter it. The circuit is shown in Figure 6-3.

6.3 Inductive Coupling Techniques

Inductive coupling is generally used only at higher frequencies, where inductors and transformers are small and inexpensive. High-quality inductors and transformers for lower frequencies tend to be large and expensive. The cost may outweigh the benefits at audio frequencies.

Inductor–Capacitor Coupling (A Common Video Amplifier Method)

A variation of the capacitor coupled circuit uses an inductor to replace the usual collector load resistor. This technique is useful at radio frequencies to obtain a high gain without the large **dc** voltage drop across the load that would be required if a resistor were used. The inductor is selected to have a high reactance value at a particular frequency, but a low **dc** resistance. The voltage gain is:

$$A_V = X_L/R'_e$$

FIGURE 6-3   Emitter-biased Two-stage R–C-coupled Amplifier

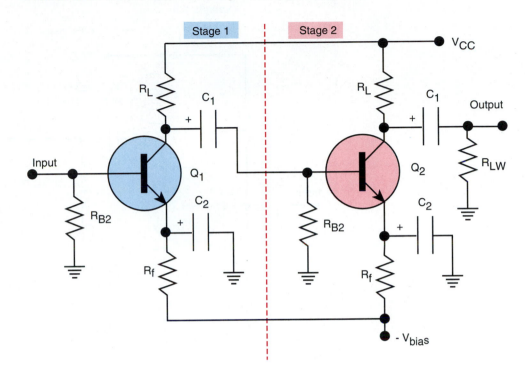

Inductive coupling is used in radio frequency amplifiers where a single frequency is involved. It may also be used in video amplifiers where it is desirable to have higher voltage gains at higher frequencies. In the video amplifier case, the inductor is often called a **peaking coil.** The increasing peaking coil reactance tends to peak the gain at the higher frequencies, where capacitor effects tend to cause a loss in voltage gain. The peaking coil helps to compensate for those capacitor problems.

Example 6.3

Calculating the Inductance Value and Voltage Gain

Figure 6-4 shows a circuit with inductance–capacitance coupling. Let's suppose we want a 20-kilohm load at a frequency of 1 MHz. Find (a) the inductance value required; (b) the voltage gain of the amplifier at 1 MHz.

Solution

a We begin by stating the inductive reactance equation:

$$X_L = 2\pi f L$$

Solving for L, we get

$$L = X_L/2\pi f$$
$$L = 20 \text{ k}\Omega/(6.28 \times 1 \times 10^6) = 3.2 \text{ mH}$$

b To find the voltage gain of the amplifier at 1 MHz, we must first find the value of the **ac** load at 1 MHz. R_{Lac} at 1 MHz is equal to X_L (which in this case is 20 kΩ) at 1 MHz in parallel with R_{Lw}. Thus, we calculate

FIGURE 6-4   Inductance–Capacitance Coupling, for
Example 6.3

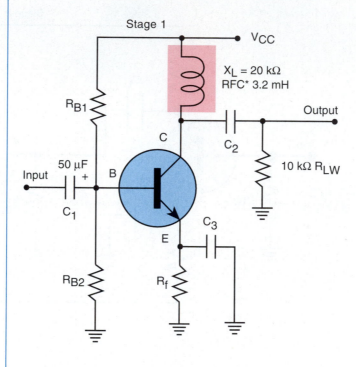

*RFC = Radio Frequency Choke (Inductor)

$$R_{Lac} = (R_L \times R_{Lw})/(R_L \times + R_{Lw})$$
$$= (20 \text{ k}\Omega \times 10 \text{ k}\Omega)/(10 \text{ k}\Omega + 20 \text{ k}\Omega)$$
$$= 6.7 \text{ k}\Omega$$

Now we are ready to find the voltage gain at 1 MHz, given that QI_C at 1 mA is equal to 25 Ω (the Shockley relationship).

$$A_V = R_{Lac}/R_e'$$
$$= 6.7 \text{ k}\Omega/25 \text{ }\Omega$$
$$= 268 \text{ (bypassed)}$$

Ideally, we would consider the R_{Lac} impedance as a vector combination with an X_L component and an R component. On the other hand, the fact that the voltage gain depends on the value of R_e' means that the voltage gain calculation can only be approximate anyway. Using a vector solution for R_{Lac} might or might not produce greater accuracy; and because we can't know whether it is worthwhile or not, we might as well take the easy way.

Untuned Transformer Coupling

Transformer coupling has some significant advantages in RF circuits, because RF transformers are cheap and make much higher gains per stage possible. In audio-frequency circuits, we can still get a greater gain per stage by using transformer coupling than by using R-C coupling, but audio-frequency transformers present some problems.

Audio-frequency transformers must use iron cores, which are subject to saturation problems, which in turn cause distortion. Good audio transformers tend to be expensive, bulky, and not very efficient. The voltage-gain advantages must always be weighed against the cost, space demands, and other undesirable features.

Maximum Power Transfer

Some of the available literature implies that the sole function of transformer coupling is to obtain maximum power transfer by matching input and output impedances. This is only occasionally true in electronics. In most cases, fairly large impedance mismatches are quite acceptable. Even when we are dealing with a power amplifier, where delivering power to a load is our goal, reasonable impedance mismatches are tolerable.

When we use transformers to couple voltage amplifiers, we are interested in reflected impedances—not in impedance matching or power transfer. Table 6-1 illustrates why fairly large impedance mismatches seldom pose a problem.

TABLE 6-1
Power Loss vs. Impedance Mismatch

Ratio of Load to Source Z	Percentage Mismatch	Percentage of Power Loss to Load
10:1	1,000%	83%
4:1	400%	68%
2:1	100%	56%
1.5:1	50%	14%
1:1	0%	0%

Example 6.4

Comparing Transformer and R-C Coupled Circuits

Calculate and compare the voltage gain of an R-C-coupled common emitter amplifier with the voltage gain of a similar circuit that uses transformer coupling. Figure 6-5 shows the two circuits.

Solution

We will begin with the R-C-coupled circuit. Assume that both circuits are operating at a quiescent collector current of 1 mA. This means that both circuits have a junction resistance (R_e') of 25 ohms. Let's find the voltage gain of the capacitor-coupled circuit in Figure 6-5(a).

Step 1: Calculating the Value of the Total ac Signal Load (R_{Lac}) for the R-C-coupled Circuit

$$R_{Lac} = (R_L \times R_{LW})/(R_L + R_{LW})$$
$$= (10 \text{ k}\Omega \times 2 \text{ k}\Omega)/(10 + 2 \text{ k}\Omega)$$
$$= 1.66 \text{ kilohms}$$

FIGURE 6-5 Comparing R–C and Transformer Coupling

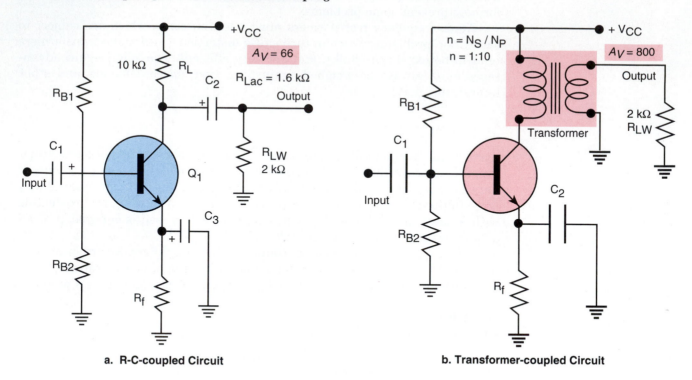

a. R-C-coupled Circuit

b. Transformer-coupled Circuit

Step 2: Calculating the Voltage Gain (A_V) for the R-C-coupled Circuit

$$A_V = R_{Lac}/R'_e$$
$$= 1.66 \text{ k}\Omega/25\ \Omega$$
$$= 66$$

The voltage gain of the R-C-coupled amplifier, with a 2-kΩ working load (R_{LW}) is 66. Now let's look at the transformer-coupled version, with the same final working-load value of 2 kΩ for R_{LW}.

Step 3: Calculating the Total Reflected ac Load Impedance ($R_{Lac\ (reflected)}$) for the **Transformer-coupled Circuit**

If we use a transformer with a 10:1 ratio, as shown in Figure 6-5(b), we can make the collector see a reflected impedance that is much larger than the actual 2-kilohm value of the working load (R_{LW}). A higher value for R_{LW}—real or reflected—increases the **ac** load impedance (R_{Lac}) and consequently the voltage gain. Let's look at some numbers.

Elementary transformer theory tells us that the impedance reflected from the secondary back into the primary is

$$R_{Lac(reflected)} = n^2 \times R_{LW}$$

where $R_{Lac(reflected)}$ is the effective load reflected into the transformer primary at the collector, and n is the turns ratio of the transformer (in this case, 10:1).

If we use the numbers from the circuit in Figure 6-5(b), we get

$$R_{\mathbf{L}ac(reflected)} = 10^2 \times 2 \text{ k}\Omega$$
$$= 200 \text{ k}\Omega$$

Step 4: **Calculating the Voltage Gain (A_V) for the Transformer-coupled Circuit**

Because of the 200 kilohm collector load, we get an apparent voltage gain of

$$A_V = R_{\mathbf{L}ac(reflected)}/R'_e = 200 \text{ k}\Omega/25 \text{ }\Omega = 8000$$

A voltage gain of 8000 is a whopping increase, but don't get too excited yet. Remember, we are using a 10:1 step-down transformer, which means that the signal voltage developed in the transformer primary is reduced by a factor of 10 in the secondary. Our apparent voltage gain of 8000, divided by 10 (the step-down ratio), gives us a real voltage gain of 800. Still, a voltage gain of 800 is a lot larger than the voltage gain of 66 provided by the capacitor-coupled circuit.

The following general voltage gain equation for transformer-coupled circuits takes all of the factors into account:

$$A_V = (n^2 \times R_{\mathbf{L}ac})/(n \times R'_e) = n \times R_{\mathbf{L}ac}$$

Dividing by n (the turns ratio), we can simplify the general equation to

$$A_V = (n \times R_{\mathbf{L}ac})/R'_e$$

Multistage Transformer-coupled Circuits

Figure 6-6 shows the schematic diagrams for emitter-biased and base-biased two-stage transformer-coupled amplifiers. The two circuits shown are typical of common emitter/transformer-coupled voltage amplifiers. The total voltage is, again, the product of the individual voltage gains. These amplifiers are best-suited for high frequencies, where input signal voltages may be in the micro-volt region and for which transformers are cheap. Transformer coupling can significantly reduce the circuit's cost in such cases. Broadband video amplifiers are one example of such an application.

Tuned Radio-frequency Amplifiers

Radio-frequency circuits offer fertile ground for transformer coupling because good-quality radio-frequency transformers are inexpensive and commonly available. In contrast, good-quality audio-frequency transformers are expensive and not commonly available. When poor fidelity and high distortion are acceptable, cheap audio-frequency transformers are a viable option.

Most radio-frequency circuits require the selectivity provided by tuned resonant circuits, so tuned radio-frequency inductors and coupling transformers are used. The circuit in Figure 6-7 uses a resonant tuned circuit as a collector load. Most radio-frequency amplifiers are intended to amplify a single frequency or a narrow band of frequencies. The **tank** (tuned circuit) in the collector is made resonant at the desired frequency.

FIGURE 6-6 Two-stage Transformer-coupled Amplifiers

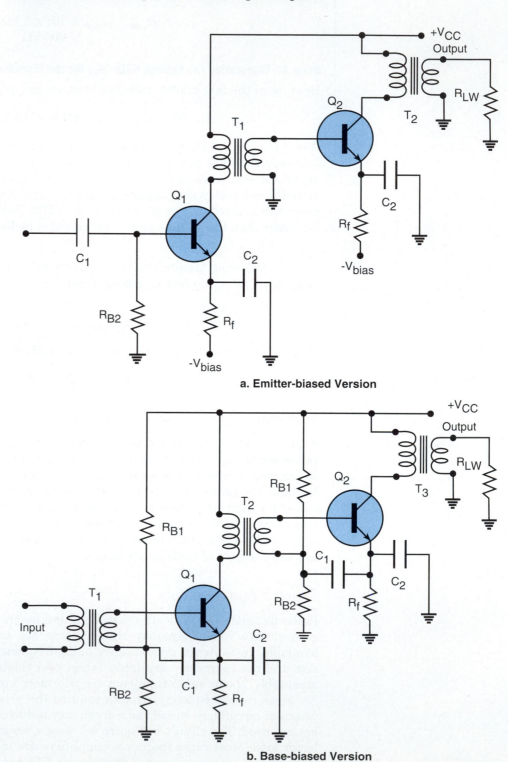

a. Emitter-biased Version

b. Base-biased Version

FIGURE 6-7 Emitter-biased Tuned Radio-frequency Amplifier

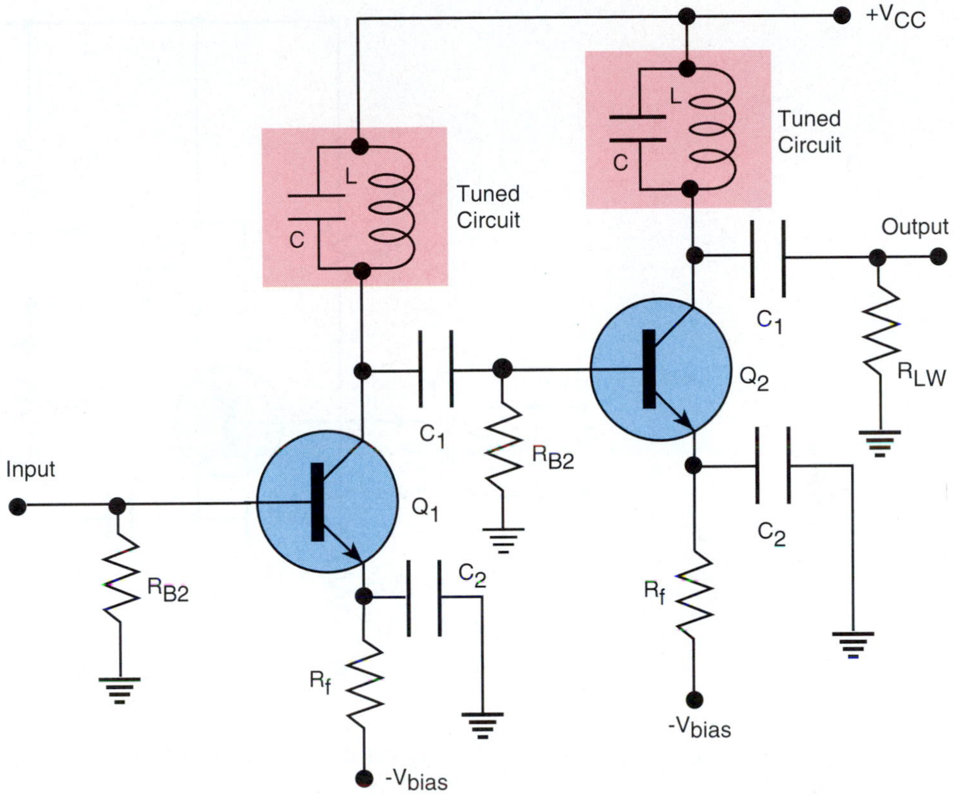

The tank is parallel-resonant, so it exhibits a very high impedance at the resonant frequency. The high impedance at resonance provides a very high effective value for the collector-load impedance, yielding a high voltage gain at the resonant frequency. At frequencies off resonance, the impedance of the tank is comparatively low, yielding a low voltage gain for all frequencies except the resonant frequency. The amplifier is thus frequency-selective. The circuit in Figure 6-7 couples the two tuned stages with a capacitor.

The circuit in Figure 6-8 uses a tuned transformer instead of a capacitor to couple the stages. The circuit is superior to the circuit in Figure 6-7 because the input impedance of the second stage is increased as a result of the trans-former-reflected impedance. Figure 6-9 shows two commonly used radio-frequency transformer configurations—tapped primary and autotransformer.

6.4   Turbocharging Transistor Circuits

In this section, we will examine some techniques that greatly improve the performance of transistor amplifier circuits. You will find most of these tricks used in integrated circuits to boost their performance. Specifically, we will examine the following three techniques:

FIGURE 6-8   Base-biased Tuned Radio-frequency Transformer-coupled Amplifier

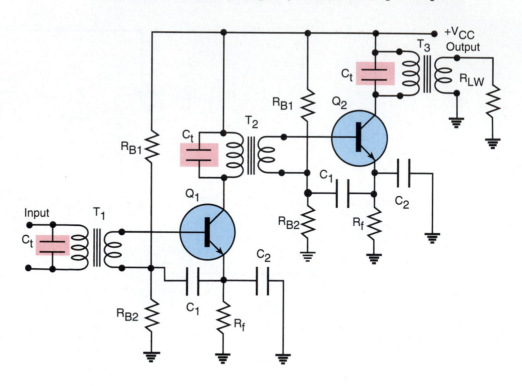

FIGURE 6-9   Typical Radio-frequency Transformer Configurations

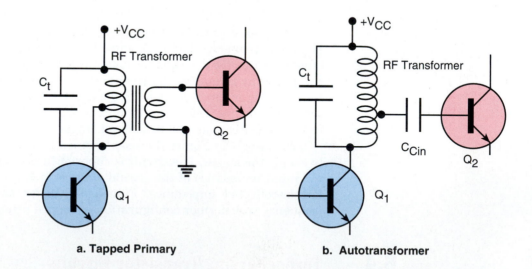

a. Tapped Primary   b. Autotransformer

1   Boosting the input impedance and the current gain with a Darlington pair.

2   Using a transistor as a constant-current source or sink to replace current-mode emitter feedback and collector-load resistors.

3   **Bootstrapping**—using a limited amount of positive feedback to make impedances seem much larger than their real value.

Nearly all operational amplifiers—the most frequently used of all analog integrated circuits—use constant-current transistors to enhance their performance. Most operational amplifiers are designed around the differential amplifier circuit, and a version of bootstrapping provides the impedance boost for the noninverting input. Many op-amps also use the Darlington pair to produce higher input impedance for the amplifier.

The Darlington pair is a favorite turbocharging method in power amplifiers. Power amplifiers are also likely to take advantage of the positive-feedback bootstrapping technique. The **super-beta** provided by the Darlington pair makes much larger power gains possible, with almost no extra circuitry.

We will look at the nature of each of the turbocharging techniques here; and later in the chapter, we will put them to work in real circuits.

Darlington-pair Compound Transistor

The **Darlington pair** is a simple, two-transistor direct-coupled circuit that behaves like a single super-beta transistor. The circuit is limited to two transistors, because any collector-current drift in the input transistor is multiplied by the beta value of the output transistor. Since only two transistors are involved, this is not a problem. The beta of the Darlington pair is the product of the two betas ($\beta_{Q_1} \times \beta_{Q_2}$). For example, if Q_1 has a beta value of 80 and Q_2 has a beta value of 65, the beta value of the Darlington compound transistor is 5000. Figure 6-10 shows the emitter-biased version of the Darlington pair, connected in the common collector configuration.

FIGURE 6-10   **Emitter-biased Darlington Pair**

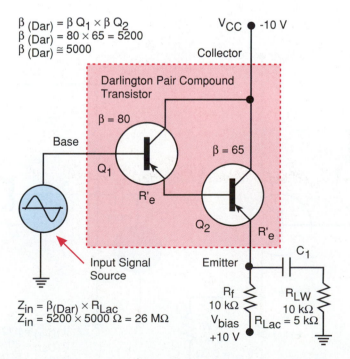

$\beta_{(Dar)} = \beta Q_1 \times \beta Q_2$
$\beta_{(Dar)} = 80 \times 65 = 5200$
$\beta_{(Dar)} \cong 5000$

$Z_{in} = \beta_{(Dar)} \times R_{Lac}$
$Z_{in} = 5200 \times 5000\ \Omega = 26\ M\Omega$

The main advantage of the Darlington circuit is its ability to deliver very high input impedances (resistances), without any sacrifice in circuit stability. The circuit in Figure 6-10 has a compound beta of 5000. The working load and the emitter feedback resistor (R_f) are both 10 kilohms, yielding a total load (R_{Lac}) value of 5 kilohms. The input impedance is

$$\beta_{Dar} \times R_{Lac} = 5000 \times 5000 = 25 \text{ megohms}$$

If the Darlington pair were replaced by a single transistor with a beta of 100, the input resistance would only be about 0.5 megohm.

Darlington Pair Beta Stability Factor

The 25-megohm input impedance in the previous example is higher than in most circuits. Typically, circuits require a base-to-ground resistor (R_{B2}) with a low enough value to satisfy the stability factor requirements. For a single transistor, the base-to-ground resistor in Figure 6-11 could have a maximum value of 10 times the value of R_f, or 10×5 kΩ = 50 kΩ.

Fortunately, the Darlington circuit has two separate stability factors, which allows the overall Q-point stability to be excellent (a stability factor 10 or less) even though the Q_1 base-to-ground resistance is several hundred times greater than the value of R_f.

Voltage-Divider Base Bias

The higher values allowable for the base-to-ground resistance mean that voltage-divider base bias, as shown in Figure 6-12, can be used with an R_{B2} value

FIGURE 6-11   Circuit for the Stability Analysis   **FIGURE 6-12   Base-biased Darlington Pair**

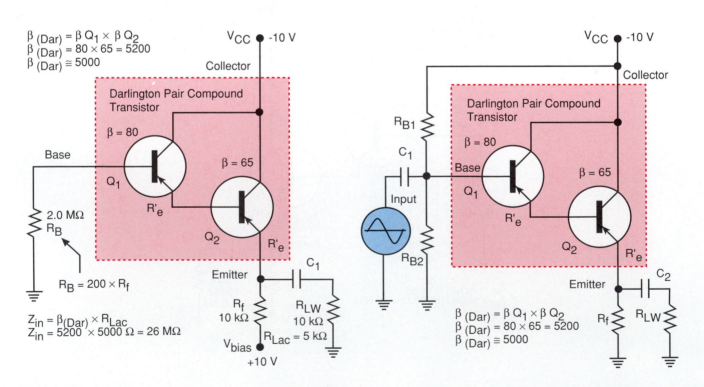

of up to 200 times the value of the emitter-feedback resistor (R_f). Bias resistor R_{B1} can also be scaled up. So, even with base bias, the Darlington pair can elevate the normal input resistance to somewhere between 50 and 200 times more than the input resistance we could get from a single transistor.

Darlington Pair and Voltage Gain

The Darlington pair can also be used in the common emitter configuration, but you must remember that there are two junction resistances in series. The result is that the voltage gain with a bypassed circuit is cut in half. On balance, you can multiply the input resistance by a factor of 100, at the cost of dividing the voltage gain by 2.

Using a Constant-current Device as a Collector Load

FLASHBACK

We first encountered the concept of using a constant-current device as a collector load in Chapter 4. Now we are going to to put the phenomenon to work as a circuit supercharger. As you may recall, the formula for collector current is beta times the base current ($QI_C = ß \times I_B$). Notice that the collector-current equation does not mention the collector voltage. If we change the collector voltage in Figure 6-13 from 10 volts to 20 volts (a 100% increase), the collector current remains unchanged at 1.0 mA.

Whatever happened to trusty old Ohm's law, which you worked so hard to learn? Ohm's law says that, if we change only the circuit source voltage (in this

FIGURE 6-13 **Relationship of Collector Current to Collector Voltage**

Question: What happens to the collector current if we change
V_{CC} from 10 V to 20 V?

Answer: Nothing! Collector current is independent of
collector voltage.

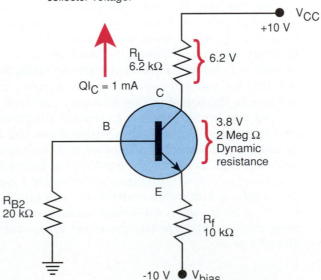

case, doubling it), the current must also change (double). But it doesn't happen here; so is this an exception to the law? No, there are no exceptions. Something else must be going on—something that automatically increases the collector–base junction's dynamic resistance in proportion to any changes in the collector voltage.

The collector–base junction is normally reverse-biased, and the width of the depletion zone determines the dynamic resistance of the junction. A wide depletion zone has a higher effective resistance than a narrow one. If we double the reverse-bias voltage (in this case, V_{CC}), the width of the depletion zone also doubles.

In fact, the effective resistance doubled at the very same time that the collector voltage doubled, leaving the collector current unchanged. There is no violation of Ohm's law here, because the increase in dynamic resistance is equal to the increase in reverse-bias voltage. The increased collector voltage is almost exactly balanced out by the wider depletion zone's increased resistance.

The balance is not perfect, however, or the collector–base junction would act as if it were an open circuit, with an infinite dynamic resistance. In practice, a very small increase in collector current occurs if we double the collector voltage (V_{CC}). This small increase amounts to the Ohm's law dynamic equivalent of about 1 megohm. Although this dynamic resistance value is quite variable, it is very high, which suffices to make it useful.

Notice that Figure 6-13 has a collector-to-emitter voltage drop of 3.8 volts, with a QI_C of 1 mA. By Ohm's law, this works out to an equivalent resistance of 3.8 kilohms for static **dc** conditions. This is the static resistance and applies as long as V_{CC} stays constant.

A second resistance exists only when the collector voltage changes and alters the depletion zone. The changing collector voltage is usually some kind of signal voltage or some temperature-induced change in V_{CC}. We have an effective resistance for signals of 1-megohm, but only a 3.8 kilohm resistance for constant **dc.**

We call a transistor that is used in this way, with no provision for signal inputs or outputs, a **constant-current device.** The current through the transistor doesn't vary, even though V_{CC} changes.

We can use the constant-current device as a substitute for either a 1-megohm collector-load resistor or a 1-megohm emitter current-mode feedback resistor. When used as a collector-load resistor it is called a **constant-current source.** When it is used as an emitter current-mode feedback resistor, it is called a **constant-current sink.** Because some people work with conventional current instead of with electron flow, you may find some inconsistency in the use of the terms *source* and *sink.*

But why would we want to replace a common resistor with a more expensive transistor? A 1-megohm resistor has the same static resistance as its dynamic resistance. Its resistance is 1-megohm for either **dc** or signal voltages. The transistor in Figure 6-13 has a 1-megohm signal resistance and a **dc** resistance of only 3.8 kilohms, which yields a **dc** voltage drop of 3.8 volts at 1 mA of QI_C. A 1-megohm resistor at 1 mA QI_C would have a voltage drop ($V = I \times R$) of 1000 volts—not a very useful power-supply voltage. We can get the same results with a constant-current transistor and a much more reasonable 10 volts or so.

As a practical matter, we generally assume a (very conservative) dynamic resistance of 500 kilohms, just to acknowledge that we never really know the precise value.

Using a Constant-current Source as a Collector-load Resistor

Theoretically, we should get very high voltage gains if we use a very high value for the collector-load resistance (R_L). For example, if we have an R_e' value of 25 ohms and an R_L of 500 kilohms we should get a (very theoretical) voltage gain of 500,000/25, or 20,000. In practice, however, we can't get a value that high, because 500 kilohms has about the same order of magnitude as the reverse-biased collector–base junction's dynamic resistance.

Refer to Figure 6-14 as we go through the following discussion. If we assume a collector–base dynamic resistance of 500 kilohms, the effective collector load value becomes 250 kilohms, bringing the gain down to 10,000. Once we add a working load, we pull the voltage gain down still lower.

Even with these considerations, voltage gains of several thousand are theoretically possible with a 0.5-megohm collector load value. But there is one little practical problem. In the preceding example, we used a value for R_e' of 25 ohms, indicating a quiescent collector current of 1 mA. A 1-mA current through a 500-kilohm resistor produces a voltage drop of 500 volts. But most transistors have maximum collector voltages of 30 to 50 volts; and given a collector supply voltage of over 500 volts, a very small reduction in collector current would produce a voltage drop that exceeded the transistor's maximum collector voltage. In addition to that, a 500-volt power supply is anything but cheap and convenient.

This constant-current source trick, which is often used in integrated-circuit amplifiers, allows us to have our electronic cake and eat it too. The trick is to use the dynamic collector–base junction resistance of another transistor as a collector load resistor (R_L). This produces an effective collector-load resistance approaching 1 megohm, with a voltage drop across the transistor of only a few volts.

The circuit in the dotted box in Figure 6-14 is called a **current mirror.** Transistor Q_1 serves as the collector load (R_L). Transistor Q_2 biases transistor Q_1 in relation to the desired Q-point, and it uses the base–emitter junction of Q_1 to track and compensate for temperature changes in the base–emitter junction of Q_2. Figure 6-14(b) gives the details of what is happening. Without Q_2, temperature-induced changes in the base–emitter junction of the synthetic resistor would cause QI_C to drift. The current-mirror circuit adds transistor Q_2 to stabilize the base–emitter junction voltage.

Figure 6-15 shows two versions of the current mirror, as well as its symbol. Packaged constant-current devices, many of them adjustable, are commercial products.

Bootstrapping

The bootstrapping technique uses positive feedback to produce the same voltages at both ends of a resistor or other impedance. If the voltages at both ends of a resistor are equal in polarity and magnitude, then effectively no voltage

FIGURE 6-14   Using a Constant-current Device as a Collector Load

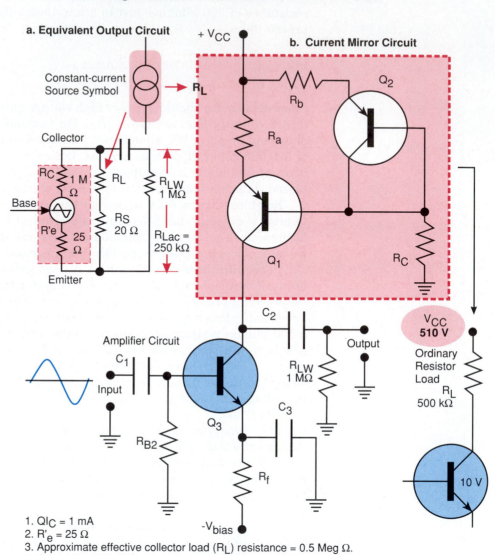

1. $QI_C = 1$ mA
2. $R'_e = 25\ \Omega$
3. Approximate effective collector load (R_L) resistance = 0.5 Meg Ω.
4. $R_{Lac} = R_C \mathbin{//} R_L \mathbin{//} R_{LW} = 1$ Meg $\Omega \mathbin{//} 0.5$ Meg $\Omega \mathbin{//} 1$ Meg $\Omega = 250$ kΩ
5. $A_V = R_{Lac} / R'_e = 250$ k$\Omega / 25\ \Omega = 10{,}000$

exists and no current flows through the resistor. If no current flows, the resistor appears to be an open circuit to both voltage sources. The idea is to make a resistor (or other impedance) effectively vanish. It remains in place, but since it demands no current, it seems not to be there as far as the voltage sources are concerned.

Figure 6-16 shows a simplified model of the concept. The input signal voltage and the feedback signal voltage are in phase opposing; as a result, each point in the sine curve of one signal has a nullifying counterpart in the sine curve of the other signal, and the two points cancel each other. Figure 6-17 shows a practical application of the bootstrapping principle. The signal voltage

FIGURE 6-15   **Current Mirror Constant-current Device, Circuits, and Symbol**

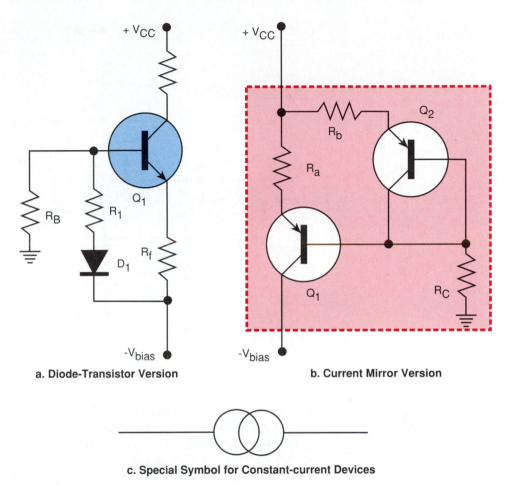

a. Diode-Transistor Version

b. Current Mirror Version

c. Special Symbol for Constant-current Devices

FIGURE 6-16   **Simplified Bootstrapping Model**

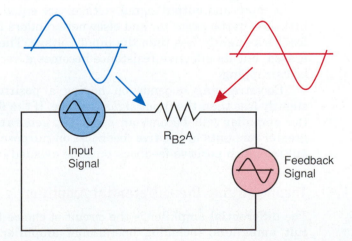

FIGURE 6-17   Bootstrapping the Common Collector Circuit

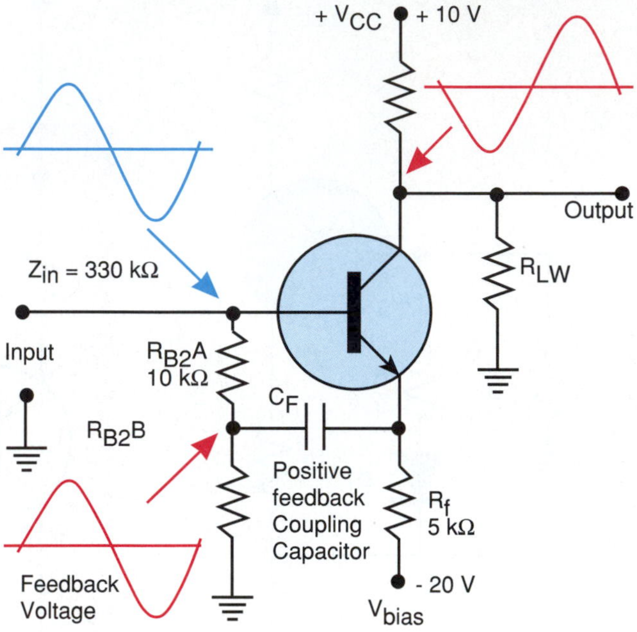

is applied to one end of the upper half of the base-to-ground resistor $R_{B2}A$, and an exact duplicate of the input signal voltage is fed back to the other end. The generator providing the signal delivers no current through R_{B2} to ground, so it functions as if R_{B2} doesn't exist. R_{B2} appears to be an open circuit to the input signal generator. Removing R_{B2} by bootstrapping the circuit raises the input impedance of the common collector circuit in Figure 6-17 from less than 10 kilohms to over 100 kilohms.

Remember, the output voltage of the common collector is in phase with the input voltage. And because the common collector circuit has a voltage gain of 1, the input and output signal voltages are equal. Again, nothing is perfect; a little loss in the capacitor and elsewhere renders the amplitude of the positive feedback slightly less than the input voltage. The resistor $R_{B2}A$ is not totally erased, but its effective resistance becomes a very high value because of the bootstrapping.

Bootstrapping depends on having a positive-feedback voltage that is slightly less than the input signal voltage. If the feedback voltage is too great, the amplifier will become an oscillator, generating its own signal. We use greater amounts of positive feedback, on purpose, to make oscillator circuits; but too much positive feedback must be avoided in amplifier circuits.

Turbocharging the Differential Amplifier

The differential amplifier is the circuit of choice for nearly all integrated circuit amplifiers, including operational amplifier and power amplifier ICs. Figure 6-18 shows a differential amplifier that uses a constant-current source

FIGURE 6-18 **Differential Amplifier, Using a Constant-current Source to Replace the Collector Load Resistor, and a Constant-current Sink to Replace the Emitter Current-mode Feedback Resistor**

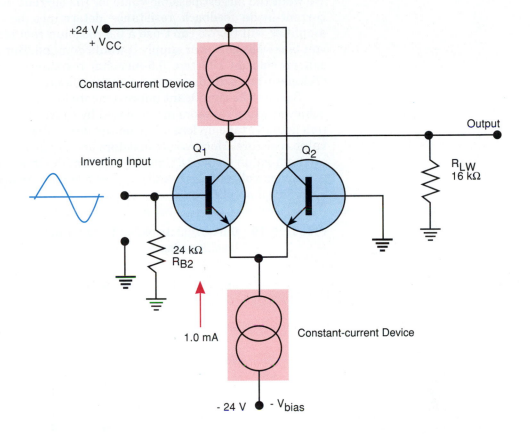

to replace the collector dummy-load resistor and a constant-current sink to replace the emitter resistor.

Constant-current Collector Load

Using a constant-current source to replace the collector-load resistor (R_L) yields an effective signal impedance of 0.5 megohm or so. This particular circuit is a converted form of a circuit that used a 16-kilohm load resistor. In that circuit, we had a voltage gain of 128 at 1 mA. The value of R_{Lac} in Figure 6-18 consists of the 0.5-megohm constant-current source in parallel with the 16-kilohm working load (R_{LW}). The two junction resistances of 25 ohms each at 1 mA provide a total resistance of 50 ohms. The voltage gain equation is

$$A_V = R_{Lac}/2.5\ R'_e$$
$$= 16\ \text{k}\Omega/62.5\ \Omega$$
$$= 256$$

The effective collector-load resistance of the constant-current source is so high that the voltage gain is determined entirely by the value of the working load (R_{LW}), in most applications.

Using the Constant-current Sink to Replace the Emitter Feedback Resistor

We want the largest possible value for the current-mode feedback resistor. The current-mode feedback resistance determines how temperature-stable the amplifier will be. We can't use a 0.5-megohm real resistor because of the 500-volt bias-voltage power supply it would demand. But the constant-current sink can provide an effective 0.5-megohm resistance for any changing voltages. (Temperature drift demands a voltage change.)

Besides the significant improvements in voltage gain and stability that are achieved when resistors are replaced by constant-current devices, the transistors actually take up less space on the integrated-circuit chip than would high-value resistors. Moreover, transistors are cheaper to fabricate on the chip than are high-value resistors. Constant-current devices are almost universally used to replace collector-load and emitter-feedback resistors in integrated-circuit differential amplifiers.

FIGURE 6-19   Differential Amplifier, Using Darlington Pairs to Increase Input Impedance

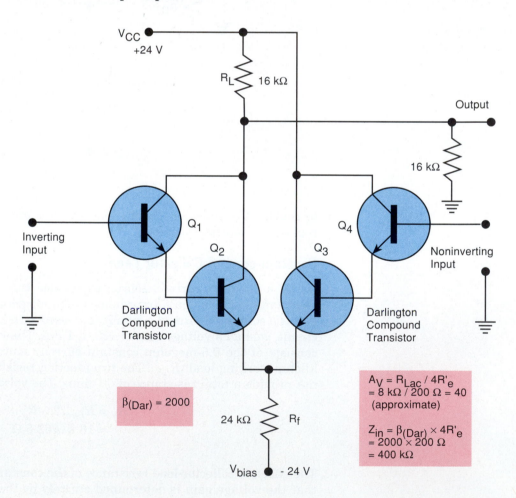

$A_V = R_{Lac} / 4R'_e$
$= 8\ k\Omega / 200\ \Omega = 40$
(approximate)

$Z_{in} = \beta_{(Dar)} \times 4R'_e$
$= 2000 \times 200\ \Omega$
$= 400\ k\Omega$

$\beta_{(Dar)} = 2000$

Using Darlington Pairs to Increase the Input Resistance

Darlington pairs are often included at the input stage of multistage differential amplifier circuits, such as op-amps. In the case of integrated circuits, the added cost is almost nil. Because the input transistor of the pair must operate at a rather low collector current, its beta value is likely to be a little lower than that of Darlington pairs using a small signal transistor to drive a power transistor. Figure 6-19 shows how Darlington pairs are used in a differential amplifier.

Adding Voltage-mode Feedback to a Differential Amplifier

The following subsections explain why operational amplifiers have one low-impedance input and one high-impedance input. Operational amplifiers are extremely important integrated-circuit devices. Nearly all op-amps are made from several stages of differential amplifiers. Operational amplifiers (almost) always add voltage-mode feedback to the overall circuit (there is only one exception).

Refer to Figure 6-20 as we go through the discussion. There are no new ideas in this discussion, but several old ones are used in a new combination.

FIGURE 6-20 Adding Voltage-mode Feedback to a Differential Amplifier, Drastically Altering the Input Impedance of Both Inputs

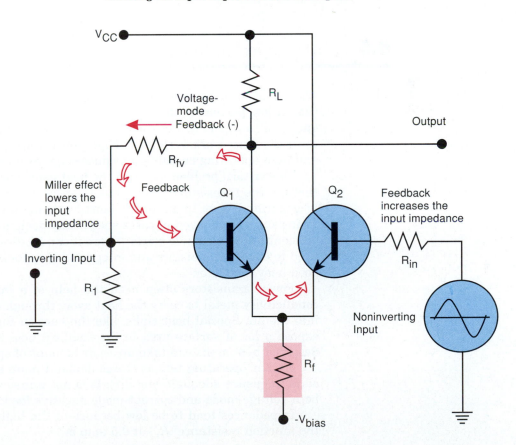

Low-Impedance Inverting Input   First, the inverting input receives a voltage-mode negative-feedback voltage, which always involves the Miller effect. The Miller effect lowers the effective input resistance of the inverting input in proportion to the voltage gain of the amplifier. This reduction in input resistance is drastic when a multistage differential amplifier has voltage gains of 100,000 or more. In fact, the Miller-effect input resistance in op-amps is often reduced to a fraction of an ohm—so close to zero that it is called a **virtual ground.** A real ground would be 0 ohms. We can still use that input by placing a resistor or capacitor in series with it, to make the input resistance tolerable to the input signal source.

High-impedance Noninverting Input   The feedback signal path is traced by the arrows in Figure 6-20. The voltage-mode feedback voltage is positive feedback applied to the emitter of Q_2. This a case of backdoor bootstrapping. The junction resistance of Q_2 appears as a near-infinite value—almost an open circuit—to the signal input source. The signal source sees a very high input impedance.

Remember, R_f is not part of the signal input impedance because the two opposite current changes make R_f appear to be an open circuit as well. R_f is not a bootstrap concept; it is simply a case of equal and opposite currents canceling, and it behaves the same way whether or not voltage-mode feedback is used.

6.5 Introduction to Power Amplifiers

The circuits we have studied so far have worked the transistors at no more than a few percent of their capacity. It is as though you had a job where you only had to report to work one day each month. Power-transistor circuits, in contrast, must work very hard. High power implies large operating currents, and high operating currents introduce some new problems.

One critical problem involves the heat generated by large currents flowing through the transistor's internal resistances. We must get rid of the internally generated heat to keep the junction temperatures at acceptable levels. Transistors are housed in cases specifically designed to conduct internal heat efficiently to the outside world, where it can be dissipated. One material often used to encase transistors and integrated circuits is a special plastic that is a good heat conductor.

Power transistors often need the help of a large area of high-thermal-conductivity metal to carry the heat away through conduction and convection into the air. Special heat sinks, like the ones in Figure 6-21, are designed to expose a lot of surface area in the smallest possible volume. Even properly designed heat sinks can take up a fair amount of space.

Higher operating temperatures demand more sophisticated feedback circuits to ensure adequate stability. It is not unusual in power circuits to find both voltage-mode and current-mode negative feedback in the same circuit.

Impedances tend to be low because of the higher currents. For example, the junction resistance (R_e') at 0.5 amp is

FIGURE 6-21   Common Heat Sinks and Mounting Arrangements

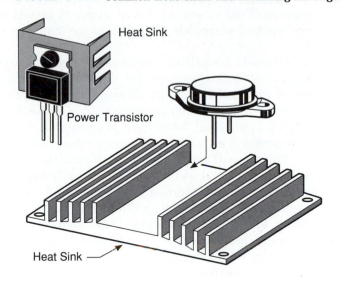

$$R'_e = 25 \text{ mV}/QI_C \text{ mA}$$
$$= 25 \text{ mV}/500 \text{ mA} = 0.05 \ \Omega$$

The input resistance in this case, assuming a beta of 65 (which is a typical value for a power transistor), is:

$$\beta \times R'_e = 100 \times 0.05 = 5 \ \Omega$$

Previous circuits have had something like 2 kilohms of input resistance. Such low input resistances bring up some new problems to deal with. Does the Darlington pair come to mind as a way to boost the low input impedance?

We will also lean more heavily on voltage-mode feedback for stability. In lower-power circuits, we are not concerned with power loss in the emitter-feedback resistors; but in power amplifiers, those power losses can be intolerable. Voltage-mode feedback is far less wasteful of power and can be used around several stages. Power amplifiers often use a combination of current-mode feedback and voltage-mode feedback, with the voltage-mode feedback dominant.

Classes of Operation

The different modes of amplifier operation are designated class A, class B, class C, class D, and so on. The most common of these are class A and class B. Classes A and B deliver power to the load during the entire sine wave and are classed as linear amplifiers. Audio amplifier measurements are normally made using a 1-kHz sine wave. The sine wave is the only waveform that has no extra harmonic frequencies (because it is the only pure waveform). The sine wave may not be a very good substitute for music; but until recently, music was far too complex to measure. The following classifications are all based on a sine wave.

In **class-A amplifiers,** a single transistor delivers power for the full 360 degrees of the sine-wave power output. **Class-B amplifiers** use two transistors, each of which delivers power to the load for 180 degrees of the output sine wave.

Class-C amplifier operation delivers power to the load for only 120 degrees or less. Consequently, class-C operation can only be used to ring a tuned resonant circuit, which completes the cycle by using the energy stored in the resonant circuit's capacitor and inductor. Class-C amplifiers are not considered to be linear amplifiers, even though the end result they produce may be a perfect sine wave at one specific frequency.

The **class-D amplifier** is a pulse-width-modulated power amplifier, which is as much digital as it is analog. Class-D amplifiers are not practical except when they use modern digital integrated circuits. Class-D amplifiers also require a tuned resonant circuit or some other filter to fill in the gaps between pulses. We will examine pulse-width modulation in Chapter 12. Other classes also exist, but these are too specialized for this book.

In class-A operation, a single transistor amplifies the entire waveform. The circuits we have examined so far have all been class-A amplifiers. A class-B amplifier uses two transistors, one of which amplifies the positive half-cycle of the signal, and the other of which amplifies the negative half of the cycle. Each transistor works only half the time, allowing it some cooling time while it is at rest. Class-B amplifiers are sometimes called **push-pull amplifiers.**

Crossover Distortion and Modified Class-B Operation

Transistors must operate in a modified class-B mode, because of the 0.6 to 0.7 volt required to turn the input base–emitter junction on. This junction voltage requirement causes a glitch in the waveform when it crosses the zero reference line. One transistor turns off at + 0.7 volt, and the other does not turn on until the signal reaches -0.7 volt—a 1.4 volt gap. The resulting distortion is called **crossover distortion** and is illustrated in Figure 6-22.

The cure for crossover distortion is to bias both transistors slightly on— just enough to keep the base–emitter junction slightly forward–biased when the signal approaches zero volts. When a little bias is added, the class of operation is called *modified class B*. The amount of anti-crossover distortion bias is so small that it has little effect on power output or efficiency, but the distortion without the extra bias is intolerable.

Power and Efficiency

Power wasted in small-signal circuits is not a significant consideration. **Efficiency** is the ratio of power used to power dissipated as heat. An efficiency of 25% is no problem when only a few milliwatts are involved. In power amplifiers, however, where power levels range from several watts to several hundred watts, efficiency is a critical issue.

Class-A Power Amplifiers

Class-A power amplifiers are often biased to a *Q*-point of about 40% of their maximum rated current, and the signal swings the collector current from near

FIGURE 6-22 **Crossover Distortion**

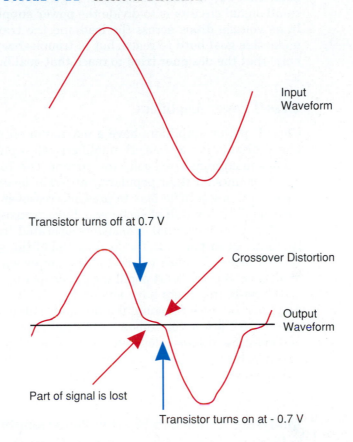

zero to near 90% of maximum. Because the transistor works very hard and idles at a fairly high current, it generates a lot of heat. The class-A amplifier is only 25% efficient (maximum), as measured by the equation

$$E_{ff} = P_{O\ (max)}/P_S$$

where $P_{O(max)}$ is the maximum signal power output to the load, and P_S is the power demanded from the power supply. The difference between the two power numbers represents the amount of power wasted as heat. The maximum transistor power dissipation (internal heat) occurs at the Q-point current, so the transistor generates the most heat when no signal is being amplified and the transistor is at rest. The equation for class-A power dissipation is

$$P_{D\ (max)} = QI_C \times V_{CE}$$

In power amplifiers, V_{CE} is usually equal to about one-half the power-supply voltage, and QI_C is equal to about 40% of the transistor's maximum current rating. Therefore, the transistor's quiescent operating current must be 40% or less of its maximum rating, and the transistor must be capable of a power dissipation of at least twice the rated maximum (from the manual). This contrasts sharply with the small-signal class-A amplifiers we have studied so far, in which the Q-point collector currents were only 1 to 10% of the transistor's maximum rated collector current. A general design goal in both power and small-signal circuits is to divide the power supply voltage more or less equally, as voltage drops across the load and the transistor (V_{CE}). Various factors make this goal hard to reach, but in troubleshooting you can assume (reasonably) that the designer tried to reach that goal but probably didn't quite make it.

Class-B Power Amplifiers

Class-B power amplifiers have a maximum efficiency of 78%. The maximum power dissipation in class-B amplifiers takes place at about 64% of the transistor's maximum rated collector current. The 78% efficiency of class-B amplifiers accounts for their popularity at power levels above about 5 watts. Other than retaining a little bias to prevent crossover distortion, each transistor is turned off during half of each cycle. Each transistor works only half the time.

A pair of particular transistors operated in class-B mode can deliver 5 times as much power to the load as one of the same transistors operating in class-A mode. Because the cost of the power supply can be significantly lowered, class B is the most popular class of operation in higher-power amplifiers.

Class-B amplifiers have an additional advantage when acting as a music amplifier, because the amplifier dissipates almost no power when no signal is present. All music contains a lot of dead-time and soft elements. The amount of dead-time in music depends on the kind of music involved; and it is highly variable, but it is always far more than we would suspect on the basis of experience alone.

Analyzing A Common Emitter Power-amplifier Circuit with Voltage-mode and Supplementary Current-mode Feedback

Given the circuit shown in Figure 6-23, use the voltage-mode version of the universal transistor circuit analysis worksheet (parts (a) and (b) of Figure 6-24) to calculate the appropriate **dc** voltage, the quiescent collector current, and the signal parameters.

The circuit under investigation is a class-A power amplifier circuit that is commonly used as a basic module for more sophisticated power amplifiers. This particular circuit is popular because of its remarkable performance. It has been a mainstay in the author's laboratory for years. Students are impressed with its performance and with the feedback circuit's ability to adapt to a wide range of junk-box part values. Being a class-A amplifier, it is not very efficient and constitutes a poor candidate for power applications beyond 5 watts or so.

Solution

We will use the same detailed analytical procedure for this example that we used in the analysis section of Chapter 5, based on the universal worksheet.

Introduction to Sheet #1

As in Chapter 5, sheet #1 of the universal transistor circuit analysis worksheet focuses on calculations of **dc** voltages and collector current. We will proceed through this sheet step by step.

FIGURE 6-23 **Common Emitter Class-A Power Amplifier, with a Darlington Pair, Voltage-mode Feedback, and Secondary Current-mode Feedback, for Worksheet Analysis**

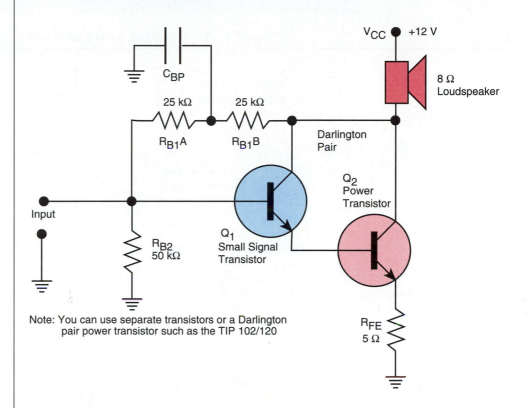

Note: You can use separate transistors or a Darlington pair power transistor such as the TIP 102/120

Step 1: **Calculating the Current Through R_{B2} ($I_{R_{B2}}$)**

The Darlington pair has two base–emitter junction voltages of 0.7 V each, for a total of 1.4 V. Therefore, our equation is

$$I_{R_{B2}} = 1.4 \text{ V}/R_{B2}$$
$$= 1.4 \text{ V}/50 \text{ k}\Omega$$
$$= 0.028 \text{ mA}$$

Step 2: **Calculating the Quiescent Collector Current (QI_C)**

In the previous example, we used a simplified version of the following equation:

$$QI_C = \frac{V_{CC} - V_{BE} - (I_{R_{B2}} R_f)}{R_L + (R_f/\beta) + R_{FE}}$$

where $R_f/\beta = 0.01 \times R_f$, using our standard β of 100, and where V_{BE} is always 0.7 V. Because we are using a Darlington pair, a β value of only 100 is not realistic here. A β of 80 for Q_1 and a β of 65 for the Q_2 power transistor, for a total

FIGURE 6-24   Universal Circuit Analysis Worksheet, Voltage Mode, for the Power Amplifier Example

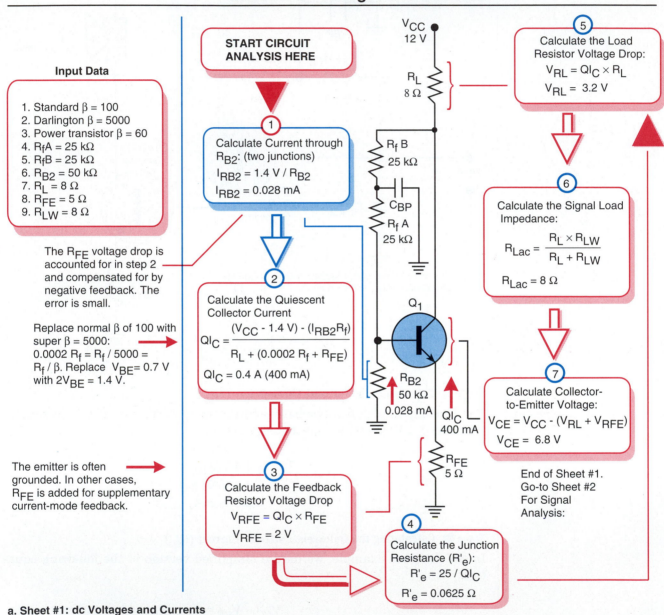

a. Sheet #1: dc Voltages and Currents

FIGURE 6-24 Continued

Sheet #2: Signal Parameters

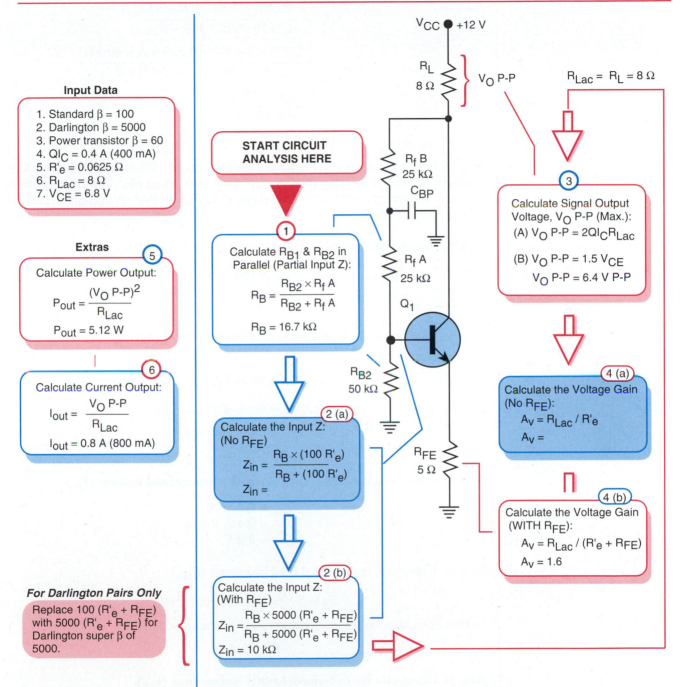

Input Data

1. Standard β = 100
2. Darlington β = 5000
3. Power transistor β = 60
4. QI_C = 0.4 A (400 mA)
5. R'_e = 0.0625 Ω
6. R_{Lac} = 8 Ω
7. V_{CE} = 6.8 V

Extras ⑤

Calculate Power Output:

$$P_{out} = \frac{(V_O \, P\text{-}P)^2}{R_{Lac}}$$

$$P_{out} = 5.12 \text{ W}$$

⑥ Calculate Current Output:

$$I_{out} = \frac{V_O \, P\text{-}P}{R_{Lac}}$$

$$I_{out} = 0.8 \text{ A (800 mA)}$$

For Darlington Pairs Only

Replace 100 (R'_e + R_{FE}) with 5000 (R'_e + R_{FE}) for Darlington super β of 5000.

START CIRCUIT ANALYSIS HERE

① Calculate R_{B1} & R_{B2} in Parallel (Partial Input Z):

$$R_B = \frac{R_{B2} \times R_f A}{R_{B2} + R_f A}$$

$$R_B = 16.7 \text{ k}\Omega$$

② (a) Calculate the Input Z: (No R_{FE})

$$Z_{in} = \frac{R_B \times (100 \, R'_e)}{R_B + (100 \, R'_e)}$$

$$Z_{in} =$$

② (b) Calculate the Input Z: (With R_{FE})

$$Z_{in} = \frac{R_B \times 5000 \, (R'_e + R_{FE})}{R_B + 5000 \, (R'_e + R_{FE})}$$

$$Z_{in} = 10 \text{ k}\Omega$$

V_{CC} ● +12 V

R_L 8 Ω V_O P-P

R_f B 25 kΩ

C_{BP}

R_f A 25 kΩ

Q_1

R_{B2} 50 kΩ

R_{FE} 5 Ω

R_{Lac} = R_L = 8 Ω

③ Calculate Signal Output Voltage, V_O P-P (Max.):

(A) V_O P-P = $2QI_C R_{Lac}$

(B) V_O P-P = 1.5 V_{CE}

V_O P-P = 6.4 V P-P

④ (a) Calculate the Voltage Gain (No R_{FE}):

$$A_V = R_{Lac} / R'_e$$

$$A_V =$$

④ (b) Calculate the Voltage Gain (WITH R_{FE}):

$$A_V = R_{Lac} / (R'_e + R_{FE})$$

$$A_V = 1.6$$

b. Sheet #2 Signal Parameters

($\beta_1 \times \beta_2$) of 5000 is a good value. Recognizing that 1/5000 = 0.0002, we can modify the simplified equation for a β value of 5000 as follows:

$$QI_C = \frac{(V_{CC} - 1.4 \text{ V}) - (I_{R_{B2}}R_f)}{R_L + (0.0002 \times R_f) + R_{FE}}$$

$$= \frac{(12 \text{ V} - 1.4 \text{ V}) - (0.028 \text{ mA} \times 50 \text{ k}\Omega)}{8 + (0.0002 \times 50{,}000) + 5}$$

$$= \frac{10.6 - 1.4 \text{ V}}{8 + 10 + 5}$$

$$= 0.4 \text{ A}$$

Notice that the quiescent collector current is almost 1/2 ampere (400 mA). This is much larger than the QI_C in any previous example. Power amplifiers demand larger currents and (sometimes) higher collector voltages than do voltage amplifiers.

Step 3: Calculating Secondary Current-mode Resistor Voltage Drop ($V_{R_{FE}}$)

$$V_{R_{FE}} = QI_C \times R_{FE}$$
$$= 0.4 \text{ A} \times 5 \text{ }\Omega$$
$$= 2.0 \text{ V}$$

Step 4: Calculating the Base–Emitter Junction Resistance (R'_e)

Our general equation is

$$R'_e = 25 \text{ mV}/QI_C$$

and we know that $QI_C = 0.4$A, or 400mA. Thus, we have

$$R'_e = 25 \text{ mV}/400 \text{ mA}$$
$$= 0.0625 \text{ }\Omega$$

Step 5: Calculating the Voltage Drop Across the Dummy Load Resistor (V_{RL})

$$V_{RL} = QI_C \times R_L$$
$$= 0.4 \text{ A} \times 8 \text{ }\Omega$$
$$= 3.2 \text{ V}$$

Step 6: Calculating the ac Signal Load Impedance (R_{Lac})

$$R_{Lac} = (R_L \times R_{LW})/(R_L = R_{LW})$$

In this case, the 8-ohm loudspeaker is the total load:

$$R_{Lac} = 8 \text{ }\Omega$$

Step 7: Calculating the Collector-to-Emitter Voltage Drop (V_{CE})

$$V_{CE} = V_{CC} - (V_{R_L} + V_{R_{FE}})$$
$$= 12 \text{ V} - (3.2 \text{ V} + 2.0 \text{ V})$$
$$= 6.8 \text{ V}$$

Introduction to Sheet #2

Again, as in Chapter 5, the second sheet of the universal transistor circuit analysis worksheet is devoted to the signal parameters.

Step 0: Copying the Relevant Data from Sheet #1

$QI_C = 0.4$ A   [400 mA]

$R'_e = 0.0625$ Ω

$R_{Lac} = 8$ Ω

$V_{CE} = 6.8$ V

Step 1: Calculating the Equivalent Parallel Impedance of the Current-bleeder Resistor (R_{B2}) and the R_{fA} Half of the Feedback Resistor

The bypass capacitor has been selected to have nearly 0 Ω of capacitive reactance at the lowest expected signal frequency. For signal frequencies, C_{BP} might as well be a wire. The procedure is just a simple parallel resistance calculation:

$$
\begin{aligned}
R_B &= (R_{B2} \times R_{fA})/(R_{B2} + R_{fA}) \\
&= (50 \text{ k}\Omega \times 25 \text{ k}\Omega)/(50 \text{ k}\Omega + 25 \text{ k}\Omega) \\
&= 1250 \text{ k}\Omega/75 \text{ k}\Omega \\
&= 16.7 \text{ k}\Omega
\end{aligned}
$$

Step 2(b): Calculating Z_{in}

The circuit in this example combines both current-mode feedback and voltage-mode feedback, so we must include the extra current-mode feedback resistor (which has a resistance of 5 Ω) in our input impedance calculations. Our general equation is

$$Z_{in} = [R_B \times 100(R'_e + R_{FE})]/[R_B + 100(R'_e + R_{FE})]$$

with $R_B = 16.7$ kΩ. This circuit offers a rare instance where we can't assume the standard ß value of 100. The Darlington pair requires us instead to use the Darlington's super-beta value of 5000. Let's rewrite the Z_{in} equation with the 5000 super-ß:

$$
\begin{aligned}
Z_{in} &= [R_B \times 5000(R'_e + R_{FE})]/[R_B + 5000(R'_e + R_{FE})] \\
&= [16.7 \text{ k}\Omega \times 5000(0.0625 + 5)]/[16.7 \text{ k}\Omega + 5000(0.09 + 5)] \\
&= (16.7 \text{ k}\Omega \times 25 \text{ k}\Omega)/(16.7 \text{ k}\Omega + 25 \text{ k}\Omega) \\
&= 417.5 \text{ k}\Omega/41.7 \text{ k}\Omega \\
&= 10 \text{ k}\Omega
\end{aligned}
$$

Step 3: Calculating the Maximum Undistorted *P–P* Output Signal Voltage

At this step, we will select the lower of the two values calculated in parts (a) and (b).

a
$$
\begin{aligned}
V_O \, P\text{–}P &= 2QI_C \times R_{Lac} \\
&= (2 \times 0.4 \text{ A}) \times 8 \, \Omega = 6.4 \text{ V } P\text{–}P \\
&= 6.4 \text{ V } P\text{–}P
\end{aligned}
$$

b
$$V_O\,P\text{--}P = 1.5 \times V_{CE}$$
$$= 1.5 \times 6.8 \text{ V}$$
$$= 10.2 \text{ V } P\text{--}P$$

Since the value in part (a) is lower than the value in part (b), we select: $V_O\,P\text{--}P$ = 6.4 V *P–P.*

Step 4(b): **Calculating the Voltage Gain (A_V) when a Current-mode Feedback Resistor (R_{FE}) Is Present**

$$A_V = R_{L\mathbf{ac}}/(R'_e + R_{FE})$$
$$= 8 \,\Omega/5.0625 \,\Omega$$
$$= 1.6$$

A voltage gain of 1.6 is nothing to write home about, but we are dealing with a power amplifier here. From a power amplifier, we want power gain, which is the product of the voltage gain and the current gain ($A_P = A_V \times A_i$). This amplifier has a current gain of 2000 or so. You will be asked to calculate it and the power gain in one of the end-of-chapter student problems.

Step 5: **Calculating the Maximum Undistorted Output Power (P_{out}) [Extra]**

Since this is a power amplifier, we are particularly interested in power output. Our general equation is

$$P_{out} = (V_{O\,(P\text{--}P)})^2/R_{L\mathbf{ac}}$$
$$= 40.96 \text{ V}/8 \,\Omega$$
$$= 5.12 \text{ W}$$

Step 6: **Calculating the Maximum Undistorted Output Current (I_O) [Extra]**

$$I_O\,P\text{--}P = V_O\,P\text{--}P/R_{L\mathbf{ac}}$$
$$= 6.4 \text{ V}/8 \,\Omega$$
$$= 800 \text{ mA} \quad [0.8 \text{ A}]$$

Direct-coupled Class-A Emitter-follower Power Amplifier

The emitter-follower (common collector) circuit is nearly always preferred in power-amplifier output stages. In the common emitter circuit, the input resistance tends to be too low to drive comfortably, and the circuit is difficult to stabilize against temperature problems.

Power gain requires a product of voltage gain and current gain. The output emitter-follower stage can't produce voltage gain, so all of the required voltage gain must occur in earlier driver stages. Figure 6-25 presents the simplest version of the circuit.

The next two power-amplifier circuits are operated as class-A amplifiers and are limited to a few watts of output power. The push–pull modified class-B amplifier uses the same feedback techniques, however, and is really just a modification of the two class-A circuits that follow.

FIGURE 6-25   Class-A Power Amplifier with Two-stage Negative Feedback

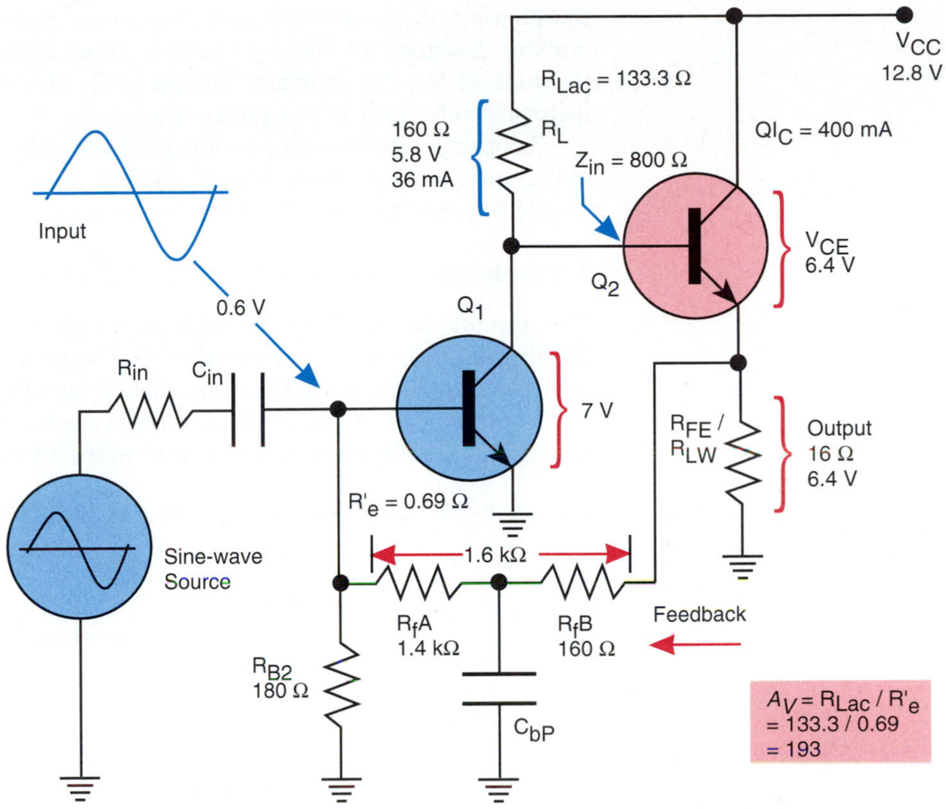

Notice that the power supply voltage in Figure 6-25 is split about equally between the transistor's collector-to-emitter voltage (V_{CE}) and the 16-ohm load at the quiescent collector current. This yields the greatest possible output power.

The base-to-ground voltage of the power output transistor must be 0.6 to 0.7 volt higher than the emitter-to-ground voltage to turn the output transistor on. In this case, the voltage from base to ground is 7 volts (6.4 V + 0.6 V). The 0.6-volt value is used to simulate worst-case temperature conditions more closely, but 0.7 volt could be used instead. Because the collector of the driver transistor (Q_1) is directly coupled to the base of the power output transistor, the voltage between its collector and its emitter must also be 7 volts.

Voltage-mode Feedback Loop

The driver transistor (Q_1) could be a standard voltage-mode feedback circuit with the feedback resistors R_{fA} and R_{fB} connected to its own collector. In this case, it is connected to the emitter of the power output transistor. Because the power output transistor is an emitter follower, with a gain of 1, it makes no difference to Q_1 whether the feedback is derived from its own collector or is connected as shown in Figure 6-25.

Connecting the Q_1 feedback network to the emitter of the power transistor, however, makes the power transistor part of the feedback loop. Any collector current drift in the power transistor causes a change in voltage at its emitter. Because any change in **dc** voltage across the load causes a change in the bias of Q_1, the collector voltage of Q_1 also changes, correcting for the collector-current drift in the power transistor.

Adopting the two-stage voltage-mode scheme is far more effective than using independent feedback for each transistor, and it is standard practice. The circuit in Figure 6-25 is capable of a power output of about 780 mW.

Turbocharged Version of the Class-A Power-amplifier Circuit

The circuit in Figure 6-26 adds a single low-power transistor to form a Darlington-pair output transistor. Darlington power pairs are commercially available for little more than the cost of an ordinary power transistor.

Replacing an ordinary power transistor with a Darlington increases the voltage gain of Q_1 by a factor of 10 or more, by increasing Q_2's input resistance

FIGURE 6-26   Improved Class-A Power Amplifier Circuit

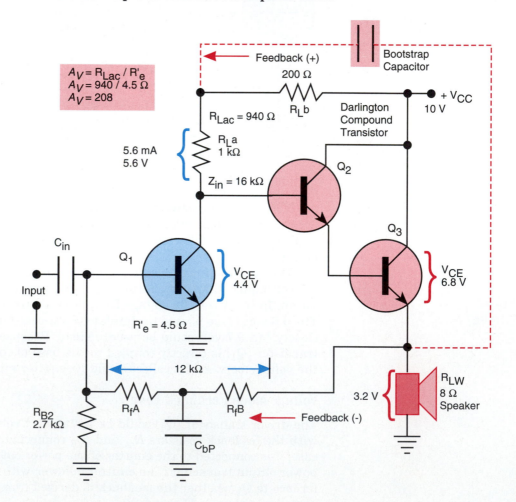

and the effective value of the working load. The total current gains and power gains increase significantly as well. The total output power increases, and the overall stability greatly improves. The Q_1 input impedance is also raised, because higher values for the feedback-loop resistors can be used. The Darlington pair provides at least a tenfold improvement in all of the desired amplifier characteristics, at little or no added cost.

Bootstrapping Q_1's Collector Load

A second optional improvement involves adding $R_L b$ to the Q_1 collector load and bootstrapping it to the output. The bootstrapping greatly increases the effective resistance of R_L and drastically raises the voltage gain of Q_1. Again, the output stage produces no voltage gain, so if we want high power gain we must obtain a high voltage gain from Q_1. The bootstrap is optional because it provides positive feedback and adds to the existing distortion; the additional distortion may or may not be acceptable.

Complementary-symmetry Push–Pull Class-B Power Amplifier

Some variation of the complementary-symmetry class-B push–pull circuit has become the industry standard for audio amplifiers and most other modern power amplifiers. It is also the preferred circuit for most integrated-circuit power amplifiers. Back in the days of vacuum tubes, every audio engineer dreamed of having a high-fidelity amplifier that did not use a transformer to couple the power tubes to the loudspeaker. The output transformer was always the weak link in the system. The transistor made the dream come true, in the form of the complementary-symmetry power amplifier.

Figure 6-27 shows the circuit for a basic modified class-B complementary-symmetry power amplifier. The driver stage (Q_1) and the feedback loop in the complementary-symmetry circuit are very similar to those used in the class-A circuit we discussed previously. Both output transistors—one n–p–n and one p–n–p—are connected as common collector circuits.

The collector load for Q_2 is a bit more involved, because it includes diode D_1 and resistor R_D, in addition to the usual split collector-load resistor. Diode D_1 and resistor R_D produce a voltage drop to bias both output transistors slightly on, producing a few mA of quiescent collector current in both. The bias is just enough to ensure that neither transistor ever turns off completely. This small amount of bias eliminates the crossover distortion (Figure 6-22); once it is added, the circuit is called a *modified class-B* or a **class-AB amplifier.**

The diode and resistor together produce 1.2 to 1.4 volts to forward-bias the junctions of Q_2 and Q_3. Some circuits use two diodes instead of the resistor–diode combination. Diodes are used to track the temperature changes in the transistors' base–emitter junctions. If the temperature rises, the junction voltage of the transistors decreases, which normally would cause the quiescent collector current to increase. Meanwhile, however, the voltage drop across the bias diodes also decreases, lowering the bias current and preventing the collector current from rising.

Two complementary transistors—one p–n–p and one n–p–n—form the power output stage. The n–p–n transistor is turned on for one half of the cycle,

FIGURE 6-27 **Complementary-symmetry Push–Pull Power Amplifier**

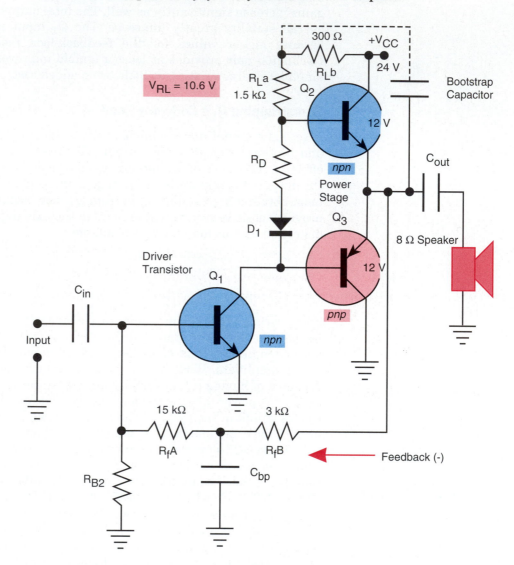

and the *p–n–p* transistor works during the opposite half cycle (see Figure 6-28). During the part of the cycle where the driver transistor (Q_1) is conducting very lightly, the base voltage of Q_2 rises, causing the base current to increase, which in turn causes a heavy current to flow through the load (speaker) and charge the output coupling capacitor.

When the collector current of Q_1 is high, it pulls the voltage at the base lower. Q_2 turns nearly off and the output coupling capacitor starts discharging through Q_3, driving the speaker in the opposite direction. The output coupling capacitor must be fairly large because it serves as the power supply for the *p–n–p* transistor (Q_3). Capacitor coupling is a common practice for audio applications. Other circuits may be **dc**-coupled, but two power-supply voltages are

FIGURE 6-28 **Bias Network Model**

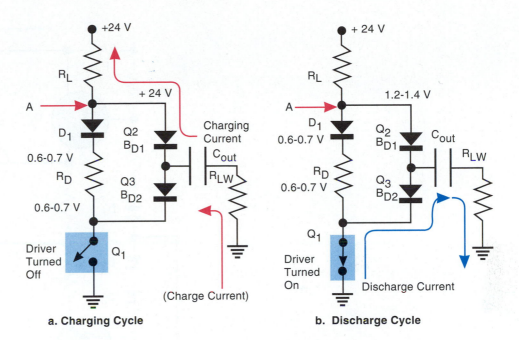

a. Charging Cycle   b. Discharge Cycle

required. Figure 6-29 shows a version of the circuit that contains two power-supply voltages: a direct-coupled load and Darlington output transistors.

6.6 Introduction to the Operational Amplifier's Internal Circuitry

The following analysis applies to any multistage differential amplifier circuit, but the particular circuit under examination is typical of those used in integrated-circuit operational amplifiers. You will probably never have to analyze this kind of circuit, but it is sometimes an advantage to know something about what goes on in the integrated circuit, when tough troubleshooting problems arise. The circuit is shown in Figure 6-30, along with the simplified version we will use to study the details.

Analyzing the Input Stage (Q_1 and Q_2)

1 The current through transistors Q_1 and Q_2 is 0.5 mA per transistor, because the current mirror is biased to 1 mA.

2 We know that the current through the 14.8-kΩ collector-load resistor is 0.5 mA, so we can use Ohm's law to calculate a voltage drop of 7.4 V.

3 The collector–emitter voltage drop for Q_1 (and for Q_2) can be found by subtracting the 7.4-V drop across the load from the 10.6-V positive supply voltage. We get 3.2 V.

FIGURE 6-29 Turbocharged Complementary-symmetry Push–Pull Power Amplifier

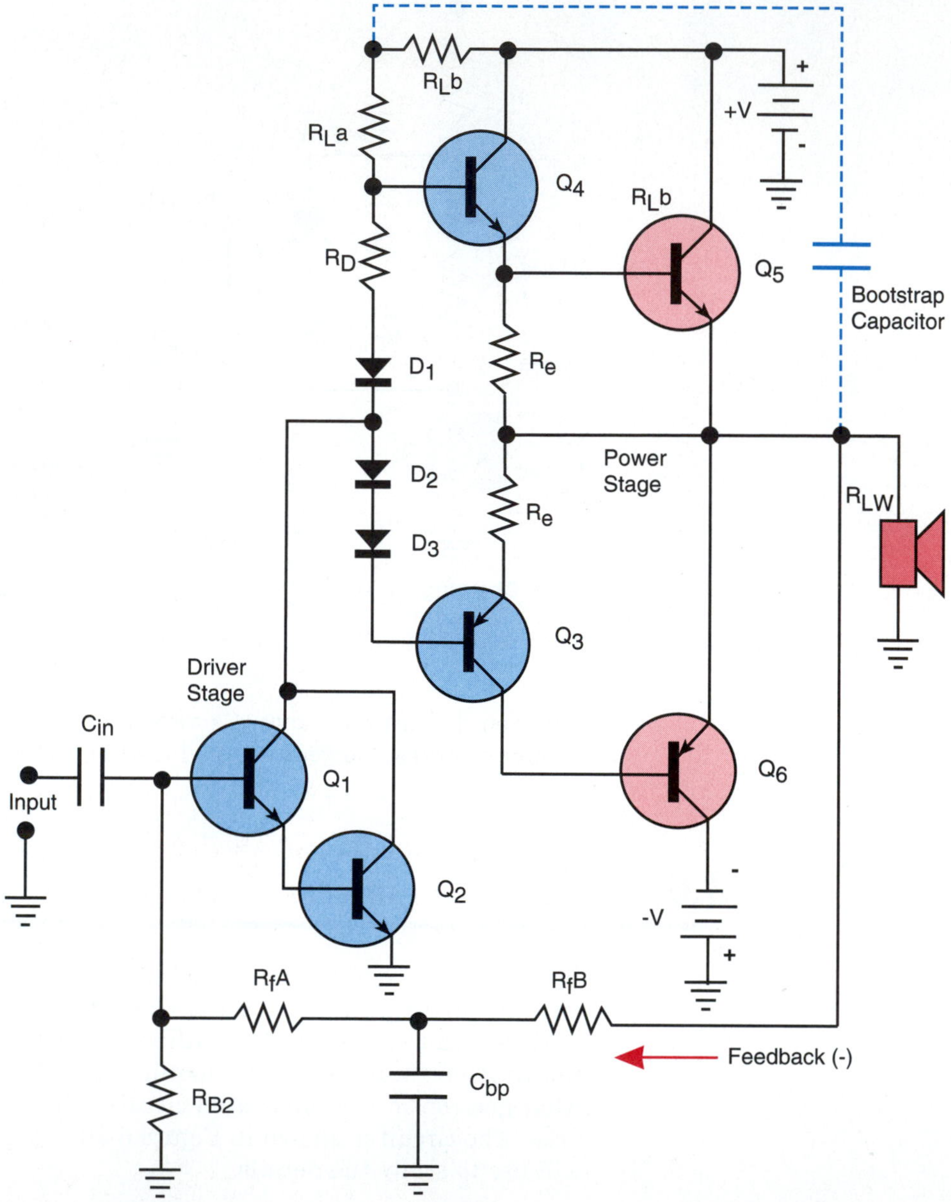

Analyzing the Second Stage (Q_3 and Q_4)

1   The base-to-ground voltage for Q_3 and Q_4 is 3.2 V.
2   If we subtract the 0.6-V junction voltage from 3.2 V, we get the feed-back resistor voltage drop of 2.6 V.

FIGURE 6-30
**Multistage Differential
Amplifier Analysis**

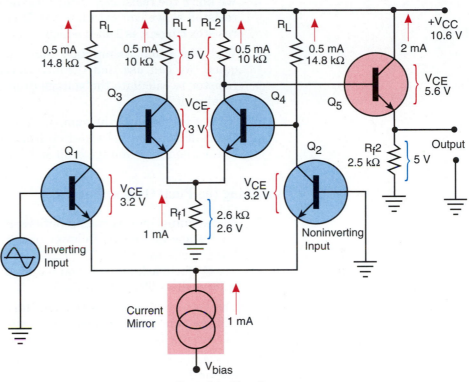

a. Complete Circuit

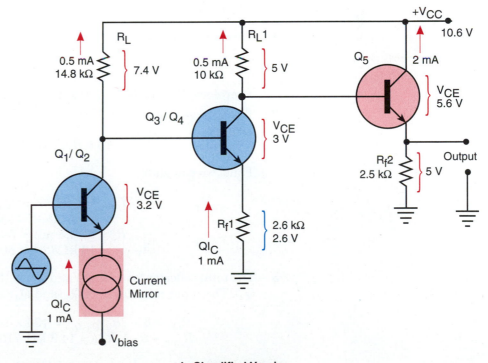

b. Simplified Version

3 We know the resistance value and the voltage drop across the feedback resistor, so we can calculate the Q-point for Q_3 and Q_4. The feedback resistor current is 1 mA, which splits into 0.5 mA for each transistor.

4 With the usual assumption that collector and emitter currents are equal, we can use that 0.5-mA current and the 10-kΩ collector-load resistor to calculate the voltage drop across the collector load. We get a value of 5 V.

5 Now, if we subtract the sum of the feedback resistor voltage drop and the collector-load voltage drop from the supply voltage, we get a collector-to-emitter voltage drop of 3.0 V for Q_3 and Q_4.

Analyzing the Final Output Stage

1 In designing operational amplifiers, it is common practice to make the final stage an emitter-follower (common collector) stage, to provide a low-impedance output. The base-to-ground voltage for Q_5 is the sum of the Q_3 feedback resistor voltage drop plus the Q_3 collector-to-emitter voltage drop. The total is 5.6 V.

2 The voltage drop across Q_5's emitter resistor is the Q_5 base-to-ground voltage minus the 0.6-V junction voltage. The voltage across Q_5's emitter resistor is 5 V.

3 The emitter current (and collector current) for Q_5 is equal to 5 V/2.5 kΩ, or 2 mA.

Analyzing the Signal Parameters

1 The input resistance for Q_5 is the working load for Q_4:

$$Z_{in} = \beta \times R_f = 100 \times 2.5 \text{ k}\Omega = 250 \text{ k}\Omega$$

Therefore, R_{LW} for Q_4 is 250 kΩ.

2 The voltage gain of Q_3–Q_4 is

$$A_V = R_{Lac}/2.5\,R'_e$$

Thus, we calculate

$$R_{Lac} = R_{L(Q4)}//Z_{in}\,(Q_5)$$
$$= 10 \text{ k}\Omega//250 \text{ k}\Omega = 10 \text{ k}\Omega$$
$$R'_e = 25/0.5 \text{ mA} = 50\ \Omega$$
$$A_{V\,(Q4)} = 10 \text{ k}\Omega/(2.5 \times 50) = 80$$

3 The total collector load for Q_1 (and Q_2) is the input impedance of Q_3 (and Q_4). The input impedance of Q_3 (and Q_4) is equal to $\beta \times 2R'_{e\,(Q_3)(\text{and } Q_4)}$.

$$R_{LW} = \beta \times 2R'_{e\,(Q_3-Q_4)} = 100 \times 100 = 10 \text{ k}\Omega$$
$$R_{Lac} = R_L\ //\ R_{LW} = 14.8 \text{ k}\Omega\ //10 \text{ k}\Omega = 6 \text{ k}\Omega$$

4 The differential voltage gain of $Q1$–$Q2$ deserves a little preliminary explanation. The voltage gain of an amplifier driven in the differential mode uses only one of the junction resistances in the voltage-gain calculation, because both bases are driven by the input signal:

$$A_{V_D} = R_{L\mathbf{ac}}/R'_e$$

The voltage gain for Q_1–Q_2 is thus

$$R_{L\mathbf{ac}}/R'_e = 6 \text{ k}\Omega/100 \text{ }\Omega$$

and

$$A_V = 60$$

The total amplifier voltage gain is expressed as

$$A_{V\,(total)} = A_{V\,(stage\ 1)} \times A_{V\,(stage\ 2)}$$
$$= 80 \times 60 = 4800$$

6.7   Integrated-circuit Power Amplifiers

Many types of integrated-circuit audio power amplifiers are available, and new ones are introduced frequently. In this section, we will look at some common (and representative) types. For power output levels below 20 watts or so, complete audio-amplifier ICs are available that require little more than a volume control, a speaker, and a power source.

The power-level limit for any integrated circuit depends on the circuit's ability to get rid of internally generated heat. There is no theoretical limit to a semiconductor device's power level, if the heat can be carried away from the junctions fast enough to keep the temperature from rising to a destructive level. But there are many practical limitations on how well heat can be conducted away from the junctions, and manufacturers are constantly making little improvements in thermal conductivity to allow for higher power levels for a given package. At higher power levels, several design possibilities exist:

1   An outboard power-booster transistor on a proper heat sink can be added to a low-power integrated-circuit amplifier.

2   Hybrid packages may include the low-power-driver IC amplifier and separate power-booster transistors in the same package.

3   There are special high-power-driver ICs that produce no significant power of their own but contain all of the bias and stability circuits, feedback circuits, and so on for a high-power amplifier circuit. The only external parts required are the high-power transistors themselves and some components to allow the user to set gain and feedback parameters.

Circuit Configuration

In spite of the many different type numbers, power level specifications, and case styles, nearly all integrated-circuit power amplifiers use the same basic circuit configuration.

TABLE 6-2

Typical Integrated-circuit Power Amplifiers with Low- to Medium-power Integrated Circuits

Device Type Number	Power Output	Power Supply	Package Style	Monaural/ Stereo
LM-380/ ULN2280B	2.5 W	8.0 to 26 V	14-pin DIP	Mono
ULN-2283B (Low voltage)	1.2 W	3.0 to 18 V	8-pin DIP	Mono
LM-383/ TDA2002/ CA2002/ ULN-3701Z	10 W	8.0 to 18 V	TO-220	Mono
LM-384/ ULN-3784 (Can also replace LM-380 and ULN-2280B)	4.0 W	9.0 to 28 V	14-pin DIP	Mono
TDA-2003 ULN-3703Z	10 W	8.0 to 18 V	TO-220	Mono
TDA-2008/ 3702Z	12 W	8.0 to 26 V	TO-220	Mono
ULN-3705M (Low voltage)	600 mW	1.8 to 9.0 V	8-pin DIP	Mono
ULN-3783 (Low voltage)	520 mW	2.4 to 9.0 V	14-pin DIP	Stereo
LM-1877 (Replaces older LM-377)	2 W	6.0 to 24 V	14-pin DIP	Stereo
LM-1875	20 W	xxxx	5-pin TO-220	Mono
TDA-1515/ 24WBTL	Dual 12 W	xxxx	11-pin SIP	Stereo
TDA-1020U Car Radio	12 W	12 V	9-pin SIP	Mono
TDA-1520U Hi-Fi	20 W	xxxx	9-pin SIP	Mono

Power Stages

The power output stage is nearly always a complementary-symmetry power amplifier circuit using Darlington (or inverted Darlington) output transistors. With minor variations, the standard power-amplifier IC output stage matches the complementary-symmetry circuit we studied earlier in this chapter.

TABLE 6-3

Audio Amplifier Power Transistor Drivers

Device Type Number	Power Output (with external power transistors)	Supply Voltage	Package Style
LM391	100 W	100 V or +50, −50	16-pin DIP
MC3321P	10 W	12 to 30 V	8-pin
LM-12CLK	150 W	xxxx	4-pin TO-3

Driver Stage

The IC driver stage is normally a differential amplifier with current mirrors, for both the collector-load and emitter-feedback resistors. The differential amplifier would replace the Q_1 Darlington driver in Figure 6-29. There are also some minor variations, in the form of improved stability circuits, modified feedback circuits, and the like, but the circuits we have studied in discrete form represent the core circuits used in nearly all IC power amplifiers.

Some Samples

Tables 6-2 and 6-3 list some common examples of the various types of power amplifiers available. This is anything but a comprehensive list of type numbers. More than any other IC classification, the audio power-amplifier IC has seen an unusual proliferation of new versions and the extinction of older versions. You almost need a scorecard to keep up.

The LM-384, ULN-3784, 4-watt IC Audio Amplifier

The LM-384 (shown in Figure 6-31) is a 4-watt integrated-circuit audio amplifier in a special 14-pin DIP package. The LM-384 replaces the LM-380 and other similar devices. Three pins on each side of the LM-384's 14-pin DIP package are combined into a wide tab to serve as a heat sink (see Figure 6-31(a)).

The actual power output available from any of these ICs depends on how well the heat is carried away from the chip. The two heat-sink tabs can be used as the only heat sink in free air at lower power levels. Higher power levels can be obtained if the tabs are soldered or bolted to a large pad of copper on the printed circuit board. If some kind of special, more efficient heat sink is bolted to the tabs, an even higher power output level can be obtained.

The 4-watt rating is a nominal rating. The actual power capability of an IC is extremely heat-sink-dependent. For example, if the two tabs are soldered to a circuit board's copper pad (with an area of 6 square inches), the amplifier will be capable of a power output of 3.6 watts. Just the tabs, in free air, permit a power output of only 1.5 watts. The manufacturers' data manuals normally provide a significant amount of information about detailed heat-sink requirements, for various power levels, for each of their products.

FIGURE 6-31   LM-384/ULN-3784 Integrated-circuit Audio Amplifier

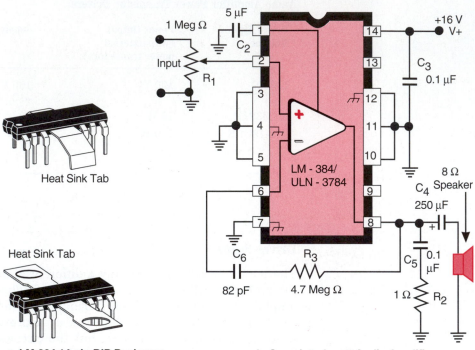

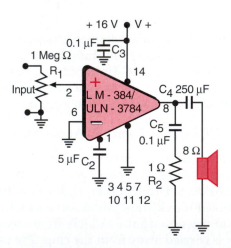

a. LM-384 14-pin DIP Packages

Heat Sink Tab

Heat Sink Tab

b. Complete 4-watt Audio Amplifier, Noninverting Version

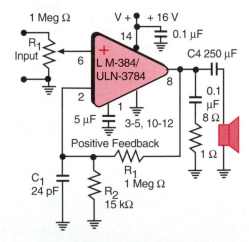

c. 4-watt Amplifier Circuit from Part b, Drawn in an Alternate Form

d. Inverting Amplifier Version with Positive Feedback to Increase Gain

Note: This is another case of backdoor bootstrapping of a differential amplifier input circuit.

The circuit in Figure 6-31(b) shows one way of drawing the circuit, with external parts included. Parts (c) and (d) of Figure 6-31 show inverting and noninverting versions of the complete amplifier. Both inverting and noninverting circuits are normally possible as a result of using a differential amplifier driver on the chip. The circuits in Figure 6-31(c) and (d) are drawn by using an alternative method to the approach used in Figure 6-31(a). The two drawing methods are more or less equally common.

What the External Components Do

The two circuits in Figure 6-31(c) and (d) use only a handful of external components. Let's examine the function of each component. You will find that the same external components and networks are used in nearly all monolithic audio amplifier ICs. The same external parts are required in most of these ICs because most of them use the same basic internal circuit.

C_4: the Loudspeaker Coupling Capacitor

The 250- to 500-microfarad capacitor (C_4) allows the amplifier to be operated from a single power supply. This capacitor charges to the power-supply voltage during the positive half of the output signal, and it discharges to provide the entire power for operating the amplifier during the negative half of the output cycle.

C_5 and R_2: the Stability Network

The stability network is a high-frequency filter used to counteract the normal tendency of high-gain negative-feedback amplifiers to become unstable and break into oscillation at some very high frequency. The network may not always be necessary, but it is generally installed to ensure that oscillation can't happen. In Chapter 8, we will look at the concept of high-frequency roll-off in considerable detail. This kind of unwanted high-frequency oscillation is a common problem in most high-gain negative-feedback amplifiers.

C_2: Bypass Capacitor

Because a power amplifier requires large excursions in current from the power supply, resulting changes in power-supply voltage can cause problems. A sudden, heavy peak-current demand by the amplifier can strain the power supply's ability to keep its output voltage up to normal. These peak current demands don't last very long, however, so we can use a capacitor to store enough energy reserve to handle the peaks, without demanding the extra current directly from the power supply.

In addition, large current demands that are felt on the power-supply bus can affect other circuits that use the same power supply. The bypass capacitor also helps to decouple (isolate) the power-supply bus from radical changes in the power amplifier's current demands. The capacitor makes the power-supply requirements less critical and, consequently, less expensive.

C_3: High-frequency Bypass Capacitor

Capacitor C_3 is also a bypass capacitor, but it is designed to bypass changes that occur at higher frequencies. Large electrolytic capacitors tend to have relatively large inductances, which make them less effective in bypassing higher frequencies. Again, the purpose of C_2 is to decouple the amplifier from the power-supply bus.

R_1, R_2, and C_1: Positive Feedback Network (Figure 6-31(d))

The amplifier has a built-in negative-feedback loop that sets the voltage gain to 34. Most IC power amplifiers have provisions for external negative feedback, so the voltage gain can be set to any desired value. In the case of the LM-384/ULN-3784, however, the voltage gain is fixed at 34. If a higher voltage gain is required, positive feedback can be used to bootstrap the input to increase the gain. When positive feedback is used to increase the gain, it also increases the distortion; so positive feedback is used where economy is more important than quality.

Bridge Configuration

Figure 6-32 shows a simple way to increase the output power of monolithic integrated-circuit audio amplifiers. The two amplifiers are connected in such a way that the inputs of the two amplifiers are cross-coupled; thereafter, one amplifier output goes positive with a positive input signal, while the output of the other one moves in the opposite direction, toward ground. The effective result is delivery of twice as much peak-to-peak voltage to the speaker as could be delivered from a single amplifier.

$$P = V^2/R_L$$

Assuming the same loudspeaker in both cases, doubling the output voltage produces four times the power output available from a single amplifier.

FIGURE 6-32   **Using Two Integrated-circuit Amplifiers in a Bridge Connection to Increase the Output Power by a Factor of Four**

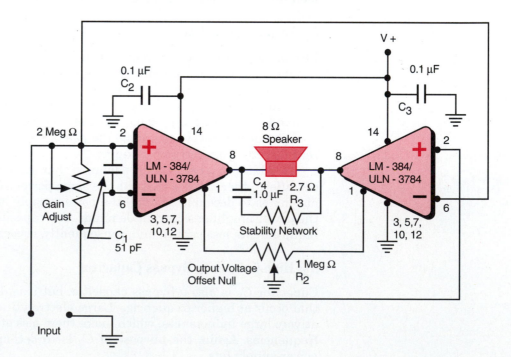

Low-voltage, Battery-operated Audio Amplifiers

Several monolithic low-voltage, low-power, audio amplifiers have been designed specifically for battery-operated portable equipment. The ULN-3705M, with its 1.8- to 9-volt supply, and the ULN-3783, with its 2.4- to 9-volt supply, are typical examples.

Figure 6-33 shows the basing diagram and the complete circuit for the ULN-3705M low-voltage audio amplifier. The amplifier is housed in a modified 8-pin DIP package, with pins 2 and 3 and pins 6 and 7 formed into heat-sink tabs. The amplifier is capable of generating up to 600 milliwatts, with an adequate heat sink. The 500–μF bypass capacitor serves the same power-supply

FIGURE 6-33 3705 Low-voltage 500- to 600-mW Audio Amplifier Circuit

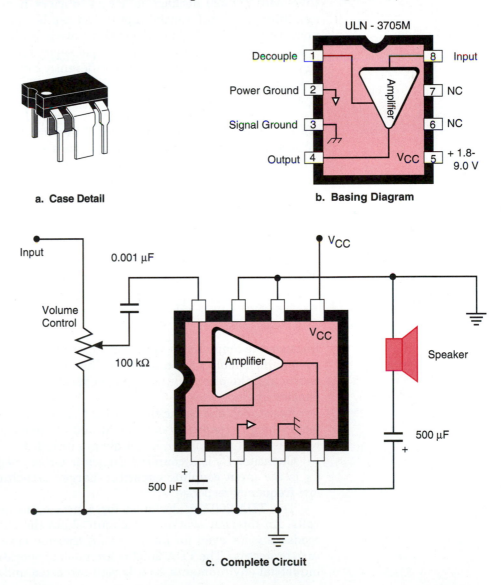

a. Case Detail

b. Basing Diagram

c. Complete Circuit

decoupling function as it did in the LM-384 we just examined. Similarly, the 500-μF output coupling capacitor in this circuit provides the amplifier power for the negative half of the audio output cycle. The 0.001-μF input coupling capacitor prevents the volume control and signal sources from altering the input bias.

TDA-2002, TDA-2003, and TDA-2008 10- and 12-watt Monolithic Power Amplifiers

The TDA-2002, -2003, and -2008 monolithic power amplifiers are capable of from 10 to 12 watts of output power with a modest heat sink. The modified (five-lead) TO-220 package is a very efficient heat-removing package. These amplifiers have differential inputs and can be operated in either inverting or noninverting input mode. External negative feedback is normally required, as it is in most higher-power amplifier packages. Power-supply decoupling is also required, but it is handled a little differently here than in the LM-384 class of devices. A high-frequency roll-off filter (stability network) is also required, as it was in the LM-384 circuits.

Figure 6-34 shows the complete circuit for a TDA-2002 (LM-383/CA-2002/ULN-3701Z) 10-watt audio amplifier circuit.

Capacitors C_3 and C_4 are the bypass capacitors, used to decouple the amplifier from the power-supply bus. The larger capacitor handles the large current demands, while the smaller capacitor takes care of the higher-frequency decoupling.

Resistor R_4 and capacitor C_6 provide protection against high-frequency oscillation or instability. The values shown in Figure 6-34 are useful for output-load (speaker) impedance values of between 2 and 16 ohms.

Resistors R_2 and R_3 form a voltage-mode feedback voltage divider, which controls the voltage gain of the amplifier. The voltage-gain equation is

$$A_V = (R_2 + R_3)/R_3$$

Because the amplifier is not internally compensated for gains of less than 20, such gains should not be used. In Chapter 8, we will study internal compensation and the effects of negative feedback on bandwidth and instability tendencies in amplifiers.

Resistor R_f and capacitor C_f form a secondary negative-feedback loop. This loop, which is frequency-dependent and designed to limit the high-frequency response of the amplifier, is not always included.

Capacitor C_2 is a coupling capacitor for the negative feedback. Capacitor C_1 is the input coupling capacitor. Larger capacitance values provide better low-frequency response.

The TDA-2002 is designed for a maximum power-supply voltage of 18 volts. An internal overvoltage circuit shuts the chip down if the voltage exceeds 28 volts, even for an instant. A thermal cut-out provides overtemperature protection. The TDA-2002 is intended to operate in harsh automotive and industrial environments, so it is well-protected and rugged.

FIGURE 6-34   LM-383, TDA-2002, CA-2002, ULN-3701Z 10-watt Audio IC Amplifier

LM-383, TDA-2002,
CA 2002, ULN-3701Z
10-watt Audio
IC Amplifier

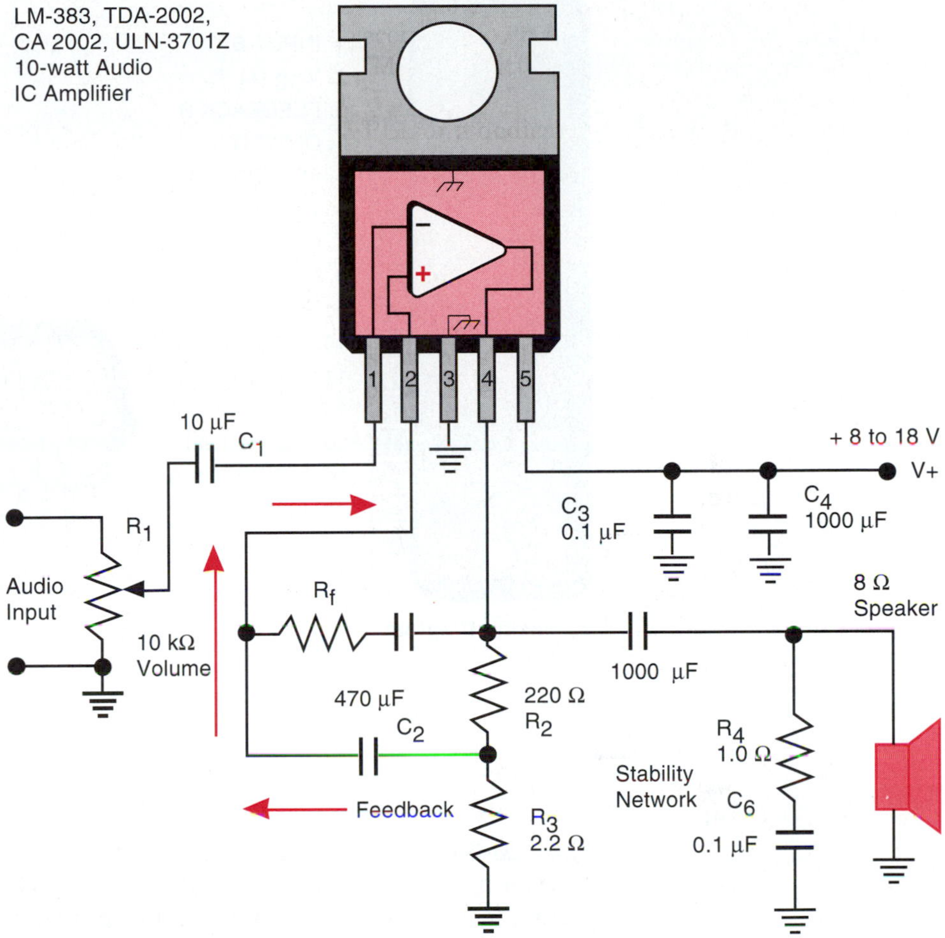

The TDA-2008 is identical to the TDA-2002 except that the overvoltage protection circuit is disabled. This allows the device to be operated at 28 volts (maximum), for a power output of 12 watts. The TDA-2008 should not be used with power supplies that are likely to produce large voltage spikes.

The TDA-2002 and -2008 have both inverting and noninverting inputs and can be connected in the bridge configuration for higher power levels.

LM-1875 Dual 20-watt Hybrid Audio Amplifier

Figure 6-35 shows the basing diagram and case style for the LM-1875. This device consists of monolithic driver circuits coupled to external power transistors, but with everything included in a single case that is designed to bolt directly to a heat sink. This seems to be the fastest-growing group of new analog devices, so expect to find a variety of type numbers for similar devices.

FIGURE 6-35   LM-1875 Dual 20-watt Power Amplifier

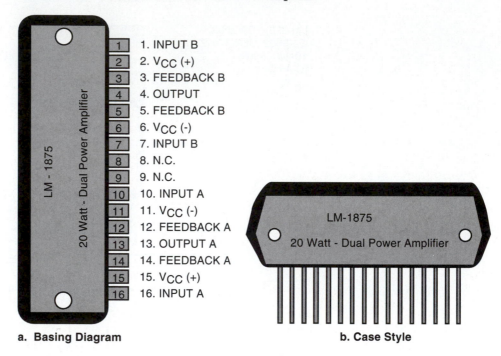

a. Basing Diagram   b. Case Style

LM-391 Power Driver for 100-watt Power Amplifiers

The LM-391 power-amplifier driver IC (shown in Figure 6-36) represents another approach to high power levels. In this approach, all of the circuits necessary to bias, stabilize, and provide proper signal-drive voltages for high-power transistors are included on the IC chip. All that must be added are the power transistors, a heat sink, and a few resistors and capacitors. The complete circuit is shown in Figure 6-36(b).

Troubleshooting Multistage and Power Amplifiers

Many function generators have a provision for a **dc** offset voltage, and that can cause test and measurement problems. If the **dc** offset voltage is not zero, it can seriously alter the transistor bias, causing the stage to operate poorly or not at all. In addition, many function generators have low output resistances, which are equally likely to upset the bias circuit. A capacitor, of about the same value as the amplifier's coupling capacitor, when used in series with the generator, can eliminate both problems.

FIGURE 6-36   LM-391 Power Driver for 100-watt Power Amplifiers

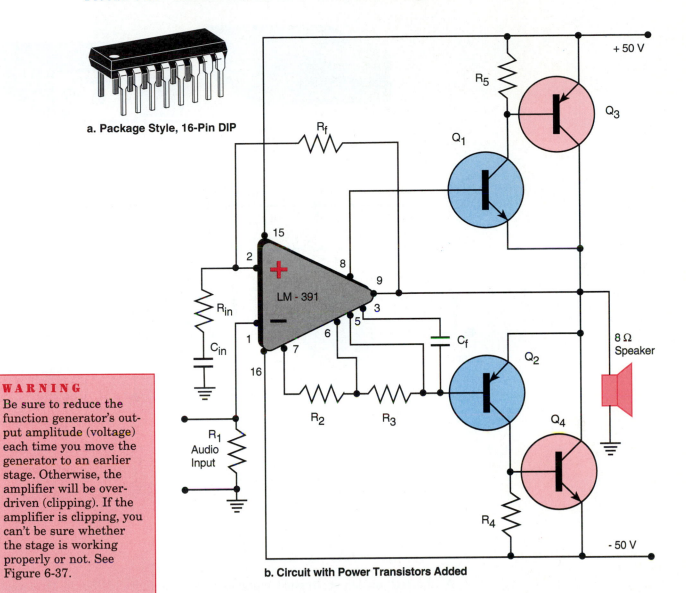

a. Package Style, 16-Pin DIP

b. Circuit with Power Transistors Added

Capacitor Problems

If the collector-to-emitter voltage in the second stage (V_{CE}) is too low—and typically, less than 2 volts is too low—the transistor (Q_2) is overbiased. The most common cause of this problem is a leaky coupling capacitor, which may add extra unwanted bias current to Q_2. Electrolytic capacitors are prone to leakage, and this problem is a common one.

An open feedback bypass capacitor results in the much lower voltage gain of an unbypassed circuit. This is easy to detect and doesn't happen often.

FIGURE 6-37   Multistage Troubleshooting Techniques

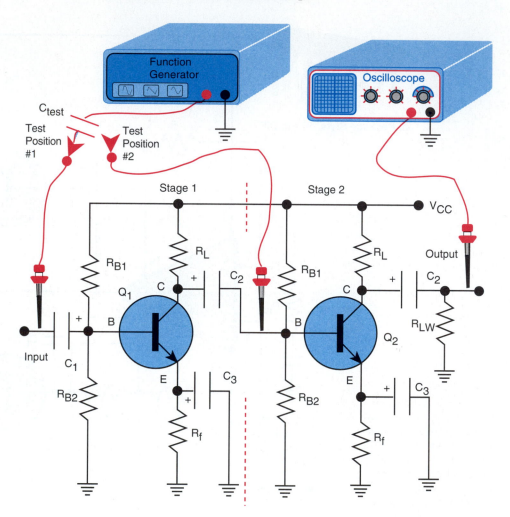

A shorted or leaky bypass capacitor can be more difficult to detect. Leakage reduces the effectiveness of the feedback, making the circuit less stable and less likely to work well in hot weather. A large amount of leakage or a short often causes overbiasing and a low collector-to-emitter voltage. Enough current is available in the emitter circuit to cause significant heating in a bad capacitor, so your sense of touch may be useful. If the feedback bypass capacitor is shorted or leaky, the **dc** voltage drop across the feedback resistor will be incorrect; but you may have to do some calculating to find out what the voltage should be.

QUESTIONS AND PROBLEMS

6.1 What is the typical voltage gain per stage in a multistage R-C-coupled amplifier?

6.2 What is the purpose of the inductor in an inductor–capacitor coupled amplifier?

6.3 Why does proper transformer coupling yield very high voltage gains?

6.4 Explain the operation of tuned radio-frequency amplifiers.

6.5 Why can voltage gains be higher in tuned RF amplifiers than for similar untuned transformer-coupled amplifiers?

6.6 How can the super-β of a Darlington pair be used to improve the performance of a transistor amplifier?

6.7 How can the dynamic resistance of a reverse-biased transistor's collector–base junction be used to improve the performance of a transistor amplifier?

6.8 What is the name of the device described in Problem 6.7?

6.9 How can bootstrapping be used to improve the performance of a transistor amplifier?

6.10 How can the differential amplifier be turbo-charged?

6.11 How do the class-A, class-B, and class-AB classes of amplifier operation differ?

6.12 Identify and compare the percentage efficiencies of class-A and class-B amplifiers.

6.13 What causes crossover distortion? What cures it?

6.14 Draw the schematic diagram of a two-stage class-A power amplifier and driver with voltage-mode feedback around both stages.

6.15 Identify the circuit in Figure 6-38.

6.16 Draw the schematic diagram of a basic two-transistor differential amplifier.

6.17 Draw the schematic diagram of a transistor differential amplifier, with an added current mirror in place of the shared emitter resistor.

6.18 List the outstanding characteristics of the differential amplifier.

6.19 What happens when voltage-mode feedback is added to a differential amplifier?

FIGURE 6-38   **Circuit for Problem 6.15**

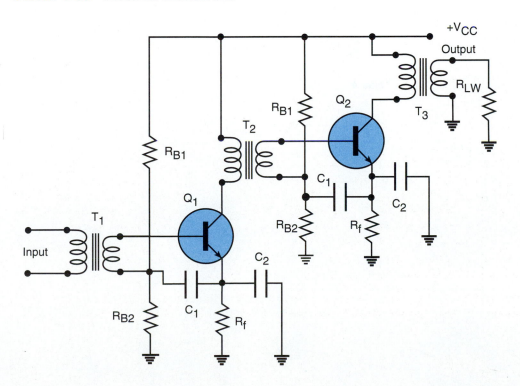

6.20 Draw the schematic diagram of a current-mirror constant-current source.

6.21 Define *class-A amplifier operation.*

6.22 Define *class-B amplifier operation.*

6.23 Define *class-C amplifier operation.*

6.24 What are the advantages of using a constant-current device to replace an emitter resistor or collector-load resistor?

6.25 Why must most higher-power IC audio amplifiers usually use external power transistors?

6.26 Describe the three kinds of high-power IC amplifiers.

6.27 Identify the circuit in Figure 6-39.

6.28 Identify the circuit in Figure 6-40.

6.29 Identify the circuit in Figure 6-41.

FIGURE 6-39   Circuit for Problem 6.27

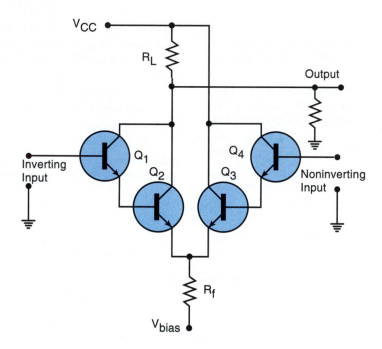

FIGURE 6-40
Circuit for Problem 6.28

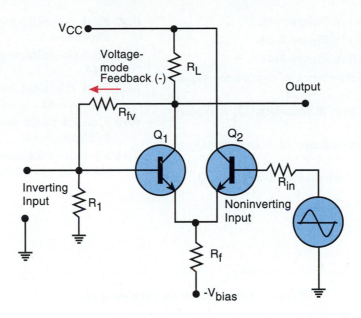

FIGURE 6-41
Circuit for Problem 6.29

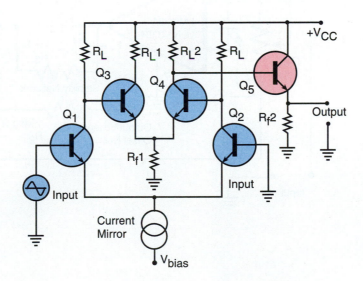

6.30 Identify the circuit in Figure 6-42.

6.31 Identify the circuit in Figure 6-43.

6.32 Identify the circuit in Figure 6-44.

6.33 Find the input impedance of transistor Q_2 in Figure 6-44.

6.34 Find the voltage gain of transistor Q_1 in Figure 6-44.

6.35 Find the input impedance of transistor Q_1 in Figure 6-44.

6.36 Find the collector current of transistor Q_1 in Figure 6-44.

6.37 Find the collector current of transistor Q_2 in Figure 6-44.

6.38 Find the voltage gain of transistor Q_2 in Figure 6-44.

6.39 Find the total voltage gain of both stages in Figure 6-44.

6.40 Find the junction resistance (R_e') of transistor Q_1 in Figure 6-44.

6.41 Find the maximum P–P output voltage of transistor Q_2 in Figure 6-44.

FIGURE 6-42 Circuit for Problem 6.30

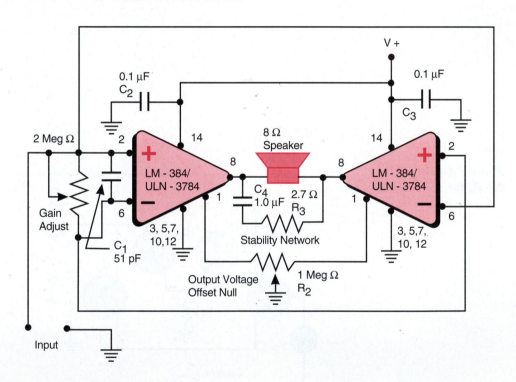

FIGURE 6-43
Circuit for Problem 6.31

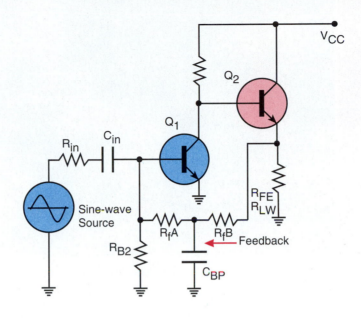

FIGURE 6-44
Circuit for Problems 6.32
Through 6.41

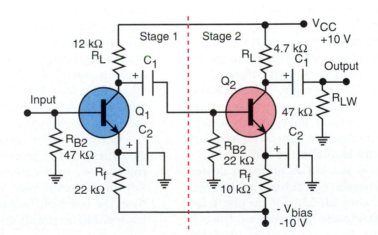

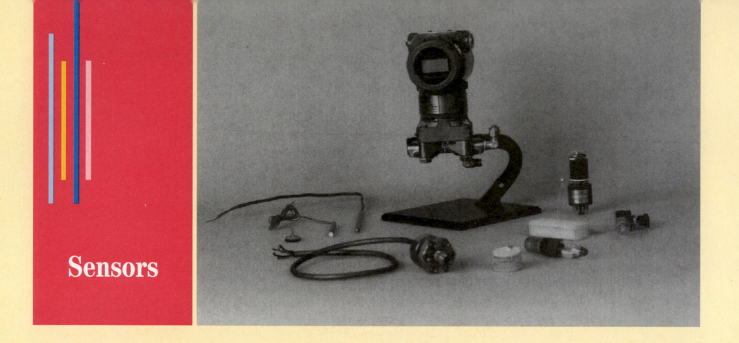

Sensors

To be useful, electronic equipment of any kind must have some kind of input and output. The attachments involved may be as simple as a switch to activate the circuit and a lamp to indicate that it is working. Audio amplifiers receive **ac** signals in the audio range and output a replica of the signal to a load such as a speaker. A computer receives input from its keyboard, mouse, disk drive, or modem and sends output data to a screen, or back to the disk drive or modem for storage or transmission.

A transducer is a device that converts energy from one form into another. Sometimes the energy undergoes more than one conversion before reaching the desired form. For example, a loudspeaker uses a movable coil around a permanent, fixed magnet to convert electrical energy into kinetic energy. This moves the speaker cone, which in turn creates sound pressure levels in the air.

When a transducer is used as an input device, we call it a *sensor*. There are hundreds of different kinds of sensors, but they all operate on just a few basic principles. In electronic applications, the output of a sensor must be some kind of electrical parameter that can be processed and used in an electronic circuit. Again, some sensors convert energy through several forms before creating the electrical signal.

Ultimately, only three electrical parameters can be manipulated to produce a sensor output: voltage, current, and resistance. And even so, both current and resistance are usually converted into voltage at some point in the input circuitry of an electronic device. Of course, voltage and current can be created as **dc** or **ac,** and sensors may produce signals that vary from microvolts (or even less, in some cases) to several volts. In addition, **ac** signals can vary in terms of frequency, waveshape, and phase relationship (to some reference).

Some electronics technicians work as technologists or engineering assistants in designing, modifying, and prototyping new equipment. A good understanding of the principles associated with sensor signals and transducer operation is invaluable in successfully performing these functions. This is no less true for the bench or field service technician, who must know what kind of signal is being handled by the equipment under repair.

Sensors can be categorized in a variety of ways: According to the parameter they measure; according to their application; or according to their measurement principle. The following paragraphs describe some of the sensor types that an electronics technician might encounter.

Self-generating Sensors

Many sensors produce their own voltage or current signal. One example of a self-generating sensor is the thermocouple: connecting two dissimilar wires together creates a millivoltage signal at the junction of the wires that is proportional to the temperature of the junction. These sensors are used to

Sensors

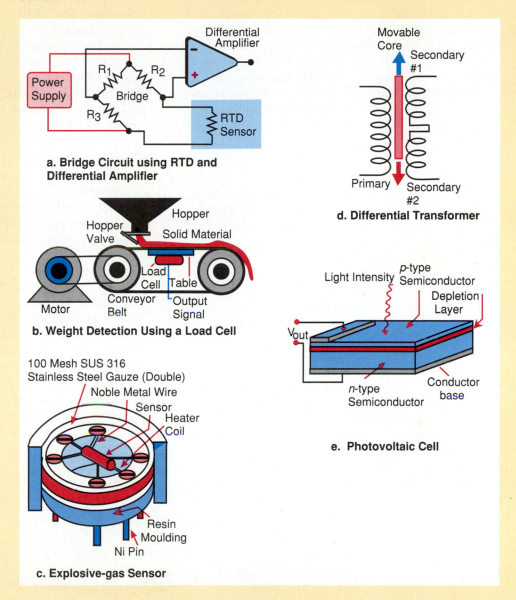

a. Bridge Circuit using RTD and
Differential Amplifier

b. Weight Detection Using a Load Cell

c. Explosive-gas Sensor

d. Differential Transformer

e. Photovoltaic Cell

measure temperature over a wide range in industrial and commercial applications.

Another example of a self-generating transducer is the piezo-electric crystal. Piezo-electrics are used in many applications, including vibration monitoring, weighing systems, level measurements, and clock signal generation. Piezo-electric crystals create a voltage across the faces of the crystal in response to the mechanical stress applied to them.

Faraday's law of magnetic induction is the principle underlying velocity sensors and other inductive motion sensors. As magnetic lines of force cut a conductor (usually a coil), a voltage is generated that is proportional to the velocity of movement of the magnetic field. These kinds of sensors are used in vibration monitoring, tachometers, and machine speed monitors.

Other self-generating transducers include photocells (which convert light energy into electrical current) and electrochemical cells (which convert energy released during chemical reactions into electrical energy). Electrochemical cells are used to measure parameters such as the oxygen content of air and water, the pH (acidity, or alkalinity) of water, and the presence of other ions dissolved in water. Photocells are used in a wide variety of applications to detect mechanical position and to sense the intensity of light generated in chemical reactions.

Passive Sensors

Passive sensors change an electrical parameter in response to some stimuli, but they require an excitation voltage or current in order to operate. One example of a passive sensor is the ubiquitous strain gauge, which is used in measuring weight, pressure, force, vibration, and many other variables. Strain gauges are resistive devices and are usually incorporated into a Wheatstone bridge configuration. The Wheatstone bridge is excited by either **ac** or **dc** voltage or current, and the differential voltage across the bridge is then amplified.

Resistive sensors abound. Thermistors adjust resistance as temperature changes; potentiometers indicate position; dissolved ion content is correlated with the conductivity of the solution.

Capacitance is often used as a variable in passive sensors. The capacitance of a probe is incorporated into an **ac** Wheatstone bridge (excited with **ac** voltage). When plate area, plate distance, or the dielectric constant is varied, capacitance (and therefore, the differential voltage across the bridge) changes.

Variable inductors are sometimes used in the same way. One variation involves the linear variable differential transformer (LVDT), which is used to sense mechanical position. The **ac** excitation voltage is applied to the primary of the transformer. Dual secondaries connected in opposition produce an algebraic total of the output voltage, depending on the position (and therefore, the magnetic coupling) of the core.

Semiconductors are also used as sensors. Because semiconductors are current-controlled devices, the voltage applied in the bias circuitry causes a current to flow. Variation in the amount of light striking the semiconductor material (as in phototransistors and photodiodes) varies the current. The same is true for semiconductor temperature sensors. As the temperature changes, the current through the device changes.

Many other sensors exist as well, but most are variations on the same principles. To work with any sensor, you must understand the principle on which it operates and have access to information on its input-to-output specifications. But electronic design, prototyping, and servicing requires more than just an understanding of devices and circuits. The technician must also know the overall purpose of the circuits under repair in order to solve application and service problems effectively.

CHAPTER 7

Field-effect Transistors

Objectives

Upon completion of this chapter, you should be able to:

1. Explain how the structure of field-effect transistors differs from the structure of BJTs.

2. Explain how the J-FET channel control operates, and describe the channel characteristics.

3. Interpret the J-FET family of drain curves.

4. Use the transconductance curves for the J-FET.

5. Describe the method of J-FET bias, and identify the feedback connections.

6. Determine J-FET signal parameters for the various configurations.

7. Explain the theory of operation for the depletion/enhancement MOS-FET.

8. Interpret the depletion/enhancement MOS-FET family of drain curves.

9. Use the depletion/enhancement MOS-FET transconductance curve.

10. Explain how the enhancement (only) MOS-FET works.

11. Explain the operation of C-MOS field-effect devices.

12. Discuss the principle of operation of the analog/transmission gate.

13. Define the characteristics and properties of analog/transmission gates.

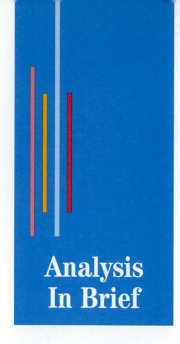

Analysis In Brief

Field-effect Transistors in Brief

FIELD-EFFECT TRANSISTORS

The bipolar junction transistor (BJT) uses an input current (base) to control the output (collector) current. The BJT is a current amplifier. The field-effect transistor (FET) uses an input (gate) voltage to control an output (drain) current. Because a "voltage" controls a "current," FETs are called *transconductance amplifiers*. Field-effect transistors are most common in a digital logic family called *complementary metal-oxide semiconductors* (C-MOS). C-MOS also finds application in some analog circuits because of its very high input impedence. Otherwise, FETs are not used much in linear analog circuits.

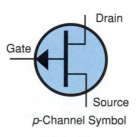

p-Channel Symbol

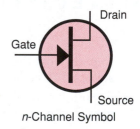

n-Channel Symbol

J-FET Symbols

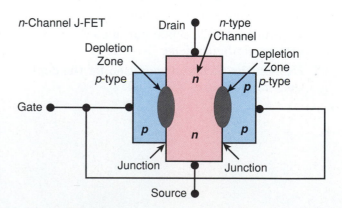

The J-Fet Structure

HOW THE J-FET WORKS

The J-FET (junction field-effect transistor) was the first of the field-effect transistors. The device is actually two diode junctions that share a common element called a *channel*. The channel may be either *p*-type or *n*-type. Each of the two junctions are reverse-biased, creating a depletion zone on each side of the channel. Increasing the reverse-bias voltage widens both depletion zones abutting the common source-to-drain channel. A narrow channel allows only a small source/drain-channel current. With a high enough reverse bias, the two depletion zones expand to meet in the center of the channel, pinching off the channel current. A wider channel allows a larger source-to-drain current.

FET TRANSCONDUCTANCE CURVES

There are two primary reasons why field-effect transistors are infrequently used in analog circuits. First, they are not capable of very high gains. Typical voltage gains with bypass capacitors may be around 5. With such low gains, too many transistors are needed to get what we want. And even worse, there is no spare gain to trade off for necessary negative feedback. The transconductance curves are nonlinear and cause distortion. Negative feedback could fix this problem, but there is no spare gain for feedback. The extremely high input impedance is the FET's one redeeming feature, and the FET is used as an input stage in some circuits for that reason. The variable-resistance action is also quite useful in analog gates, where no amplification is involved. Field-effect transistors can be used in common source, common drain, or common gate configurations.

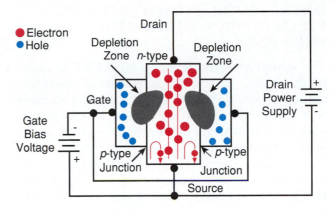

How the J-FET Works

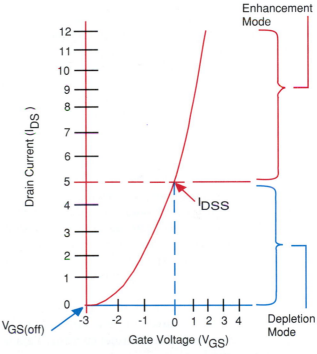

$$I_D = I_{DSS}[1 - (V_{GS}/V_{GS(off)})]$$

Transconductance Curve for a Depletion/Enhancement MOS-FET

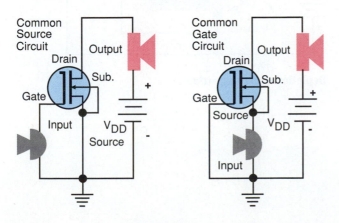

FET Common Gate Circuit

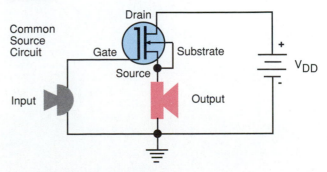

FET Common Drain Circuit

METAL-OXIDE-SEMICONDUCTOR FIELD-EFFECT TRANSISTORS (MOS-FETS)

The structure of the common MOS-FET is shown to the right. It has three modes of operation: enhancement, depletion, and resistive. The resistive function is sometimes used as a voltage-variable resistor with no amplication involved. In the depletion mode, channel current is normally flowing and is reduced by a control signal. In the enhancement mode, a control signal must be used to start channel-current flow. A bias voltage is used to set the mode.

HOW THE MOS-FET WORKS

All field-effect transistors use some method to control the amount of current flowing through an *n*-type or *p*-type silicon channel. The two ends of the channel are called *source* and *drain,* similar to *emitter* and *collector* in the BJT. The difference between the various types of FETs is the method used to control the channel current and the design of the channel.

In the MOS-FET family, a metal (or silicon) plate is insulated from the channel. The insulator is usually silicon dioxide, commonly known as glass. A positive gate voltage attracts electrons upward from the *n*-substrate against the underside of the glass. This yields an increased number of electrons available for conduction through the channel that flows along the underside of the glass. This is the enhancement mode. Using a negative bias voltage forces electrons out of the small channel and deep into the substrate. The reduced number of electrons means less channel current. This is the depletion mode.

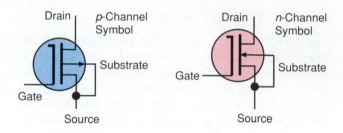

MOS-FET Symbols

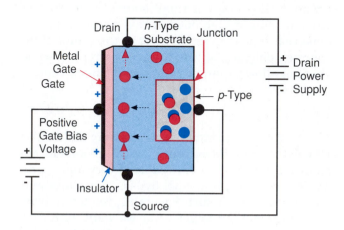

Enhancement-mode Operation

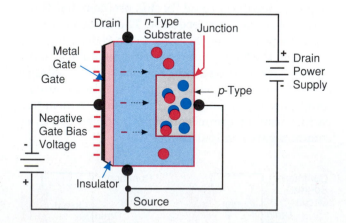

Depletion-mode Operation

Switching MOS-FETs

In the enhancement mode, MOS-FET transistors make nearly ideal switches and form a popular digital circuit family called *complementary* metal oxide semiconductors, or C-MOS. The C-MOS family uses one *n*-channel device and one *p*-channel device to form a single C-MOS switch.

The Analog Gate

Perhaps the single most important FET application involves a family of devices called *analog transmission gates*. The C-MOS transmission gate is commonly used in audio electronic volume controls, photoreceptor selectors in videocamera imaging chips, analog memory selectors in recording devices, and a multitude of other applications.

The digital C-MOS version of the transmission gate is the heart of sophisticated multiplexers/demultiplexers, and even the ever-present "D" flip-flop. The transmission gate is a very useful device in both technologies. The transmission gate also finds employment in analog-to-digital and digital-to-analog converters.

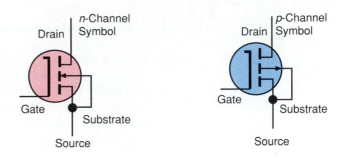

Enhancement-mode (only) Symbols

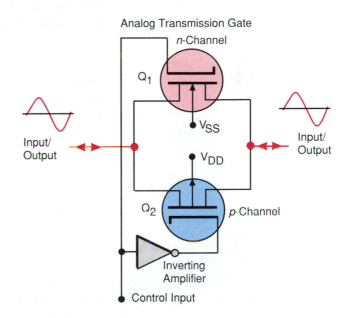

Analog Gate Circuit

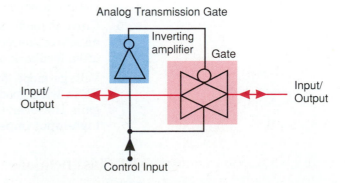

Analog Gate Symbol

7.1   Introduction

In this chapter, we explore field-effect transistors. This reflects the fact that bipolar and field-effect devices are very different animals. Bipolar transistors are by far the more popular of the two device classifications in analog circuits; but in some special applications, field-effect transistors are superior to bipolar devices. Specifically, field-effect devices are often preferred under the following conditions:

1   Where very high input impedances are required.

2   Where oscillators and radio frequency–tuned circuits must maintain a high *Q*. (The high FET input impedance can then be used to avoid loading the resonant circuit.)

3   Where low noise is required.

4   In mixer circuits, where a nonlinear amplifier is required.

5   Where the FET's special operating region (not available in the bipolar devices) can enable it to operate as a voltage-controlled variable resistor. (This operating mode can be used for both analog and digital circuits.)

6   In switching circuits, where some FETs can provide a lower output voltage in the off condition than can bipolar devices.

FETS also have some disadvantages:

1   Only comparatively low voltage gains are available from FETS.

2   The transfer curve in a FET is parabolic. This nonlinearity limits the input signal swing and the peak-to-peak output voltage to fairly small values for nearly linear operation. Only a short segment of the curve can be used if linear operation is required.

3   The FET is less temperature-sensitive than are bipolar transistors, but its manufacturing tolerances are even broader.

4   Replacing defective FETs is more difficult than replacing bipolar transistors, because FET circuits are much more device-dependent.

5   Current-mode feedback can be used with depletion-mode devices, but an active source resistor is sometimes needed to produce enough negative feedback and still meet bias requirements.

6   Voltage-mode feedback is commonly used with enhancement-mode FETs with good results, because there is more than enough current gain. However, the Miller effect in this mode can sometimes defeat the high-input impedance advantage.

FET Classifications

There are two major classifications of field-effect transistors: junction field-effect transistors (J-FETs), and insulated-gate field-effect transistors

(IGFETs). Both types of devices are available in n- and p-channel versions, which correspond roughly to n–p–n and p–n–p bipolar transistors.

The junction field-effect device operates in depletion mode. Depletion-mode devices are normally fully on and must be biased toward the off condition. The bipolar transistor is an enhancement-mode device because it is normally off and must be biased toward the on condition. Insulated-gate field-effect transistors are available for both enhancement- and depletion-mode operations. IGFETs are also available for combination depletion–enhancement-mode operation. IGFETs were originally composed of a metal gate and a silicon dioxide insulator. Some of them are still made that way, but others use silicon gates with silicon oxide insulators. IGFETs are often referred to as *MOS (metal-oxide semiconductors)*.

7.2   Junction Field-effect Transistors (J-FETs)

Figure 7-1 offers a simplified representation of p-channel and n-channel **junction field-effect transistor** (J-FET) structures, together with their schematic symbols. The transistor consists of a p or n doped silicon channel and two oppositely doped areas that form two junctions. The two junctions are connected and act in unison to control channel current. The two **gate** blocks serve as the control element (roughly equivalent to the base in bipolar devices). The **drain** connected to one end of the channel is roughly equivalent to the collector in bipolar transistors, and the **source** connected to the other end of the

FIGURE 7-1   **Structure of the Junction Field-effect Transistor**

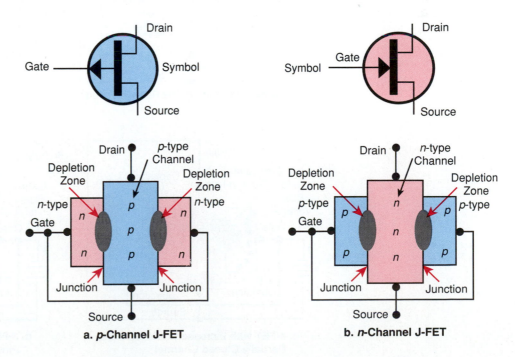

a. *p*-Channel J-FET   b. *n*-Channel J-FET

channel serves the same function as the emitter in bipolar devices. The controlled current flows through the channel from source to drain. When the silicon crystal is formed, a transition (depletion) zone is formed at the channel–gate, p–n boundary. A junction is then formed just as it is in bipolar transistors.

How the FET Amplifies

Refer to Figure 7-2 for the following discussion. When the drain-to-source power supply is connected, current begins to flow through the channel. The two gate junctions are intrinsically reverse-biased by the normal silicon junction voltage of 0.6 V. Part of the depletion zones lies within the channel and very slightly restricts the current through the channel. If 0 volts is applied between gate and source, the current flowing through the channel is called I_{DSS} (where I stands for current, and the subscript DSS stands for drain-to-source). The final subscript S indicates that the measurement is made with the gate shorted to the source.

FIGURE 7-2   **How the Channel Current Is Controlled by the Gate Voltage**

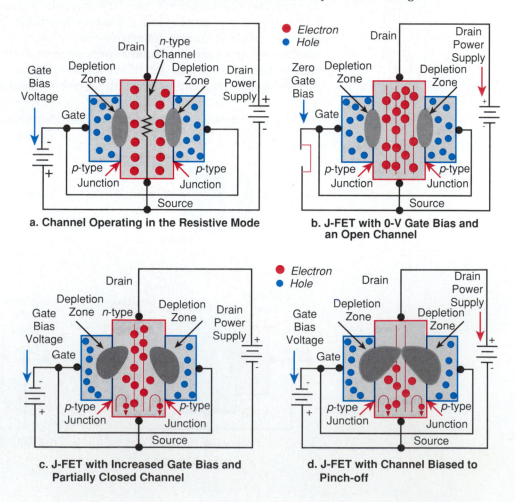

a. Channel Operating in the Resistive Mode

b. J-FET with 0-V Gate Bias and an Open Channel

c. J-FET with Increased Gate Bias and Partially Closed Channel

d. J-FET with Channel Biased to Pinch-off

The gate–channel junctions are always operated in the reverse-bias condition; and as a result, they have a very high impedance. Because of the high impedance of the reverse-biased junctions, static charges can collect on the gate. The gate is shorted to the source prior to I_{DSS} measurements, to eliminate this problem and to ensure that the voltage between gate and source is actually 0 volts.

Figure 7-2(a) shows a resistor along the channel. When the source-to-drain voltage falls below a certain threshold, the channel behaves as if it were an ordinary resistor, except that its resistance can be varied by varying the gate voltage. This is not the normal amplifying mode, but it can be useful when a voltage-controlled variable resistor is required.

When the drain-source voltage (V_{DS}) is increased beyond the threshold voltage, the channel is said to be **saturated.** When the channel is in saturation, the FET is in its normal amplifying mode, and the channel no longer acts as an ordinary resistance. In saturation, the channel has a very high dynamic impedance and behaves very much like the reverse-biased collector-to-base junction in a bipolar transistor.

Once the drain-to-source voltage is high enough to get the channel into saturation, the source-to-drain channel current (I_{DS}) becomes independent of the channel voltage. Once in saturation, the channel current is completely under the control of the gate-to-source voltage (V_{GS}).

Figure 7-2(b) shows the conditions that prevail with a gate-to-source bias voltage of 0 volts, with the gate shorted to the source to ensure that the gate bias is zero.

The gate–channel junction is reverse-biased only by the intrinsic junction voltage of 0.6 volt. The channel is as wide open as it can get, allowing the maximum channel current to flow from source to drain.

I_{DSS} Defined

The channel current with zero gate bias is an important FET parameter. The channel current (I_{DS}) does not depend on the channel voltage; it only depends on the reverse-bias voltage (0.6 volt, in this case) applied between the gate and the source.

Varying the Gate Bias to Control I_{DS}

If we increase the reverse-bias voltage of the two gate–channel junctions, the two depletion zones widen, closing off more of the channel and reducing the channel current. Increasing the bias voltage results in a proportional reduction in channel current (see Figure 7-2(c)).

Channel Pinch-off

Figure 7-2(d) shows the conditions that prevail when the gate-to-source bias voltage closes the channel almost completely. This is called the *gate cut-off voltage* ($V_{GS\,(off)}$) or the **pinch-off voltage (V_P),** and it is the point at which virtually no channel current exists.

If the gate-to-channel reverse-bias voltage is increased further (beyond $V_{GS\,(off)}$), the junction Zener voltage will be reached and the junction will go into Zener conduction and risk triggering a destructive avalanche. Some devices come with built-in Zener diodes to prevent the gate–channel junction from becoming forward-biased.

The Zener breakdown voltage has the specification BV_{DSX}, where X is the reverse-bias Zener voltage. The Zener voltage depends to some extent on the drain–source voltage. The total specification is written as (for example) BV_{DS3} = 20, meaning that the Zener voltage is 3 volts when the drain–source voltage is 20 volts.

Specifications Summary

The following symbols and terms are key elements in any work with FET parameters:

V_{DS} = Drain–source voltage

I_{DS} = Drain–source current

I_{DSS} = Drain–source current (with gate shorted to source; 0 volts gate bias)

V_{GS} = Gate-to-source bias voltage

BV_{DSX} = Reverse-bias breakdown (Zener) voltage for the gate–channel junction

Saturation = Channel condition for normal amplifier service, where channel current is independent of channel voltage

Unsaturated Channel = Channel condition where the channel behaves like an ordinary resistor; a change in channel voltage results in a corresponding change in channel current, obeying Ohm's law.

Maximum Drain Current

Field-effect transistor manufacturers do not generally provide a specification for maximum drain current. However, they usually provide a maximum power dissipation specification (P_D). If you know P_D and the drain–source voltage you can calculate the maximum drain current at any particular drain–source voltage.

$$I_{DS\,(max)} = P_{D\,(max)}/V_{DS}$$

$P_{D\,(max)}$ is rated at 25°. For temperatures higher than this, the transistor must be derated by about 2 milliwatts per additional Celsius degree.

Quiescent Drain–source Current (QI_{DS})

The quiescent drain–source current (channel current) is determined by the gate–source bias voltage (V_{GS}). As is the case with bipolar transistor collector current, the quiescent drain current must be adequate to supply the working load. The most common bias point in J-FET is about 40% of the maximum drain current for the particular drain–source voltage.

Bias and Signal

The J-FET is normally operated with some value of gate voltage (V_{GS}) that produces a quiescent drain current. The quiescent drain current is then increased or decreased by signal variations in gate voltage (V_{GS}). Figure 7-3

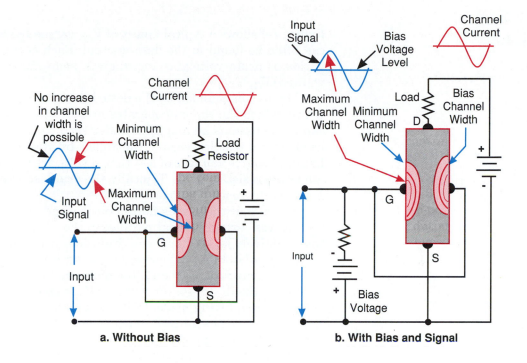

FIGURE 7-3   J-FET with Bias and an Input Signal

a. Without Bias

b. With Bias and Signal

illustrates how bias and signal voltages are related. Figure 7-3 implies that operation with zero bias is impossible, but that is not quite true. The J-FET has an intrinsic reverse bias of 0.6 to 0.7 volt, allowing the channel to widen sightly. Zero-bias operation is possible, but the junction must never be allowed to go into forward bias. This necessitates using a relatively small input signal—a few millivolts *P–P,* in practice. The J-FET is not frequently operated with zero bias. Zero-bias operation is also a poor choice in terms of temperature sensitivity.

Leakage Current

Leakage current in FETs poses only a minor problem, since the current is not amplified as it is in bipolar transistors. The very high input impedance of the J-FET is influenced by temperature, because the input to the transistor is a reverse-biased junction, and leakage current is the only thing that prevents the input impedance from being infinite. As it is, the leakage current amounts to only a few pico-amperes and is not much of a problem.

Current Gain

Current gain is equal to $\Delta I_{DS}/\Delta I_{GS}$. The current gain is almost infinite because ΔI_{GS} is so tiny for any value of ΔI_{DS}. The current gain is not very important in itself; but the high current gain means a high input impedance, and high input impedance is one of the most common reasons for using a field-effect transistor.

J-FET Family of Drain Voltage (V_{DS}) versus Drain Current (I_{DS}) Curves

Figure 7-4 shows a typical family of V_{DS} versus I_{DS} curves. Curves of this kind can often be found in the data manual. Each curve shows how drain current varies as drain voltage changes, given some fixed value of gate–source bias voltage (V_{GS}).

The region bounded by the dotted line on the left of Figure 7-4 is the area within which the channel operates in a strictly resistive mode. The transistor cannot be used as an amplifier under such conditions, but the channel can be used as a variable resistor controlled by the gate-to-source voltage (V_{GS}). Drain-to-source voltages of less than 4 volts are required for the J-FET to operate in the resistive mode. Typically, the drain–source voltage remains at about 1 volt in the resistive mode.

Notice that the curves flatten out beyond about $V_{DS} = 4$ volts. Thereafter, the "curves" extend almost parallel to the x-axis. Closer examination reveals that relatively large changes in drain-to-source voltage (V_{DS}) cause almost no change at all in the drain current (I_{DS}), with V_{GS} held constant. We must operate in the saturation region if the device is to act as an amplifier.

When the drain–source voltage (V_{DS}) exceeds 4 or 5 volts, the transistor enters its amplifying mode. In its normal amplifying mode, drain-source current (I_{DS}) becomes nearly independent of drain–source voltage (V_{DS}) and becomes strictly a function of gate–source voltage (V_{GS}).

Two particularly important specifications are shown in Figure 7-4:

1   I_{DSS} is the drain–source current with 0 volts of gate bias ($V_{GS} = 0$). The measurement is taken with the gate shorted to the source to ensure against static-charge accumulation on the gate.

2   $V_{GS\,(off)}$, also called the *pinch-off voltage,* is the gate-to-source voltage (V_{GS}) required to reduce the channel current to approximately 0 amps.

Notice that the value of I_{DSS}—the drain current with a gate bias of 0 volts—is just slightly less than 3.75 mA. Although the set of curves in Figure 7-4 is quite impressive and seems very scientific, the value of I_{DSS} plotted there is only a statistical average. According to most of the literature, real transistors can deviate from the value shown (3.75 mA) over a range from 1.5 mA to 7.5 mA. In practice, the variation seems to be a bit greater. I_{DSS} is one of those variable parameters that resist being pinned down with any degree of certainty.

Notice that the $V_{GS} = -4$ V curve is only visible when it reaches the Zener knee; this is because most of it is congruent to the $I_{DSS} = 0$ mA axis. $V_{GS\,(off)}$ can also (according to the literature) vary by a factor of as much as ±3 from the value specified on the family of I_{DS} curves, marking it as another somewhat uncertain parameter. I_{DSS} defines the maximum channel current, which occurs at $V_{GS} = 0$ V. $V_{GS\,(off)}$ defines the gate bias voltage (V_{GS}) required for minimum channel current (I_{DS}).

The family of I_{DS} curves provides a limited amount of information, because conditions are shown for only five values of V_{GS}. Suppose that we are interested in I_{DS} at a V_{GS} value of -1.5 volts. If we can identify I_{DSS} and $V_{GS\,(off)}$ from the family of drain–source curves or from a listing in the data manual,

FIGURE 7-4 Family of J-FET Drain Curves

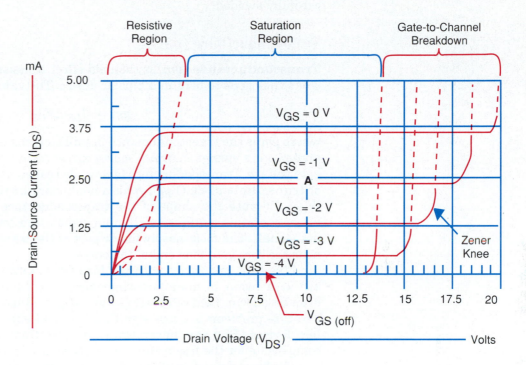

we can calculate the drain current (I_{DS}) for any value of gate voltage (V_{GS}) by using the following equation:

$$I_{DS} = I_{DSS} \times [1 - (V_{GS}/V_{GS\ (off)})]^2$$

For this particular FET, we can get I_{DSS} and $V_{GS\ (off)}$ from Figure 7-4. Thus, we have

$$I_{DSS} = 3.75\text{ mA}$$

$$V_{GS\ (off)} = -4\text{ V}$$

To find I_{DS} for a V_{GS} of 1.5 V, we calculate

$$\begin{aligned}
I_{DS} &= 3.75 \times [1 - (1.5/4)]^2 \\
&= 3.75 \times (0.625)^2 \\
&= 3.75(0.391) \\
&= 1.466\text{ mA}
\end{aligned}$$

An examination of Figure 7-4 confirms that our calculations tend to agree with the curves, although the equation gives us far more exact values than we could get by interpolation of the data shown in Figure 7-4. The nearly vertical dotted lines at the far right of Figure 7-4 indicate where the gate–channel junction goes into Zener or avalanche conduction. Notice that the gate–channel breakdown voltage (Zener voltage) depends on both the drain–source voltage (V_{DS}) and the gate–source voltage (V_{GS}). This breakdown area must be carefully avoided.

Transconductance (*gm*)

Transconductance (*gm*) is the field-effect transistor's figure of merit, somewhat analogous to beta in a bipolar device. The value of *gm* is defined as:

$$gm = \Delta I_{DS}/\Delta V_{GS}$$

where *gm* is the transconductance in millisiemens (mS) or microsiemens (μS), and ΔI_{DS} is a change in drain–source current resulting from ΔV_{GS}, a change in gate voltage. Transconductance (*gm*) is also called *forward transfer admittance* or **transadmittance (Y_{FS})**; *gm* is a nonlinear function of the gate bias voltage (V_{GS}). Because the shape of the transconductance curve is parabolic, near-linear operation requires small input signals and consequently small *P–P* output signals. The most nearly linear part of the transconductance curve occurs near the zero gate bias end of the curve.

Like beta, *gm* varies with the bias level, with temperature, and (most of all) as a result of manufacturing tolerances. Manufacturing tolerances are even broader in field-effect transistors than in bipolar transistors, while temperature problems are less significant in field-effect transistors than in bipolar devices. But the uncertainties in *gm* from other factors more than compensate for the less acute temperature problems.

The basic definition of *gm* tells us nothing about the transistor parameters that determine the numerical value of *gm*. They are $V_{GS\,(off)}$ or pinch-off (V_P) and I_{DSS}, the channel current with 0 volts of gate–source bias. The numerical value of *gm* also depends on the actual bias value of V_{GS}. Once all of these variables are taken into account, the *gm* relationship is

$$gm = [2I_{DSS}/V_{GS(off)}] \times [1 - (V_{GS}/V_{GS(off)})]$$

The circuit designer selects the bias voltage (V_{GS}) but must live with the existing values for I_{DSS} (the zero-bias channel current) and V_{GS} (the channel pinch-off voltage).

In the equation

$$gm = [2I_{DS}/V_{GS\,(off)}] \times [1 - (V_{GS}/V_{GS(off)})]$$

the first term defines *gm* at zero bias ($V_{GS} = 0$):

$$gm_0 = 2I_{DS}/V_{GS(off)}$$

Thus, the equation for *gm* at other values of V_{GS} can be written as

$$gm = gm_0 \times [1 - (V_{GS}/V_{GS(off)})]$$

An approximate transconductance curve can be derived from the family of drain curves in Figure 7-4 (see Figure 7-5). Since there are only a few data points to work with, the transconductance curve is not a very accurate representation of a parabolic transconductance curve, but it can still tell us a number of things. The poor accuracy of the transconductance curve is not really a

FIGURE 7-5 **Transconductance (*gm*) Curves, Derived from the Family of Drain Curves**

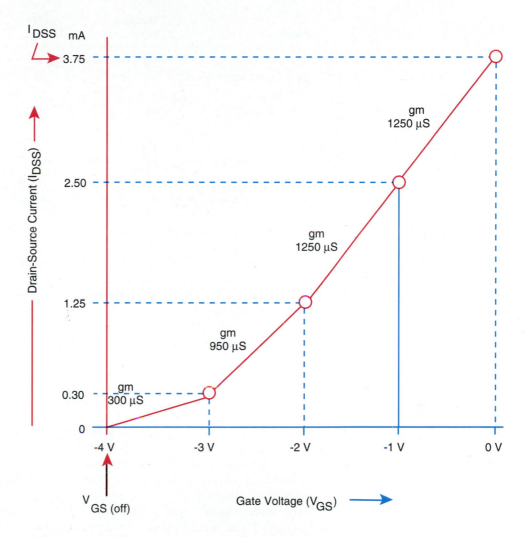

problem when we consider that the family of curves from which it was derived is itself a statistical set of curves, and that the individual FET we are using may differ considerably from any of these.

We derived the transconductance curve from *A* in Figure 7-4, which represents a drain–source (V_{DS}) voltage of 10 volts. Any voltage (V_{DS}) within the saturation area of the curve would do as well, because *gm* does not depend significantly on drain–source voltage. The *gm* curve in Figure 7-5 indicates that *gm* has its highest value and is at its most nearly linear in the upper half of the curve, where gate bias voltage (V_{GS}) values range from 0 to –2 volts. Therefore, for the highest and the most linear *gm,* operation in the upper half of the curve is preferable.

Certain circuits, such as mixers in receiver circuits, operate well into the lower half of the curve, in order to obtain the nonlinearity and attendant distortion there. These circuits are the exception, however, not the rule. Generally,

TABLE 7-1
Sample from the Manufacturer's Data Manual

Parameter	Minimum Value	Typical Value	Maximum Value	Units
Zero-gate-voltage Drain Current (I_{DSS})	2.5	3.0	3.75	mA
Gate Cut-off Voltage ($V_{GS\,(off)}$)	-2	-3	-4	V_{dc}
Transconductance (gm)	1875	2000	2500	μS

more linear operation is desirable. The gm in the curve shown in Figure 7-5 is the slope (tangent) of the curve at any point on the curve.

Let's examine a realistic situation where the manufacturer gives us maximum and minimum values for $V_{GS\,(off)}$ and I_{DSS}. Table 7-1 offers a representative sample of values from a manufacturer's data manual. Suppose that we plot two transconductance curves—one for the minimum values of $V_{GS\,(off)}$ and I_{DSS}, and one for the maximum values. We can find I_{DS} for each plot point by using the equation

$$I_{DS} = I_{DSS}\,[1 - (V_{GS}/V_{GS\,(off)})]^2$$

The results are shown in Figure 7-6. Keep in mind that these two curves are for the same transistor type number, not for two different ones. Notice that curve B features higher gm values than does curve A, but that it allows a narrower range of possible bias (V_{GS}) voltages. A V_{GS} of -2 volts would be an acceptable value for a transistor with curve A, but it would amplify only half of the input signal (sine wave) for a transistor with curve B. A bias voltage (V_{GS}) of -1 volt would place us near the midpoint ($^1/_2\,I_{DSS}$) of curve A but the same -1 volt would put us low on curve B, where gm begins to drop off. The curves in Figure 7-6 indicate that a bias voltage of about -0.6 to -0.7 volt would be optimum. A bias voltage of -0.7 volt is shown in Figure 7-6. We will see how that value was obtained in Example 7.1.

In reality, neither minimum nor maximum values are encountered very often. The values exhibited by the vast majority of transistors fall somewhere between. The distribution of parameters follows a curve that resembles the standard bell-curve often used for classroom grading purposes and other statistical applications. In mass-producing FET circuits, manufacturers are wise not to design to the extremes. Generally, the bias voltage is based on a geometric average, and the most convenient geometric average is the gm average:

$$gm_{0\,(average)} = \sqrt{gm_{0\,(max)} \times gm_{0\,(min)}}$$

We will see how to apply the gm average in the next section.

FIGURE 7-6   Maximum (*A*) and Minimum (*B*) Transconductance Curves

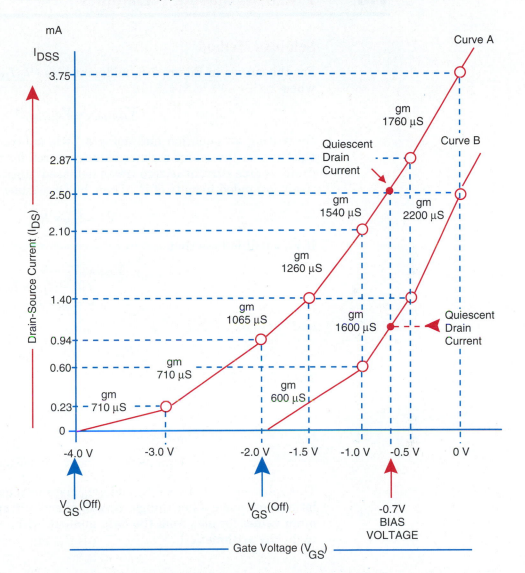

There are two common bias methods: self-bias and voltage-divider bias. **Self-bias** uses the source resistor voltage drop to bias the FET, and it provides negative feedback. The source resistor is primarily for bias purposes. Any negative feedback is incidental. Self-bias depends on the high input impedance of the gate. The gate impedance is almost infinite, so any generator or other gate-to-ground resistance (R_g) is negligible and can be ignored. Voltage-divider bias is easier to control, but it does reduce the circuit input impedance to a few megohms. Self-bias is used when the generator must look into the near infinite input impedance. Voltage-divider bias is preferred for most applications.

7.3   Practical Biasing Techniques

Self-bias Method

For the curves in Figure 7-6, midpoint bias ($\frac{1}{2}\,I_{DSS}$), occurs (approximately) when

$$V_{GS(mid)} = V_{GS\,(off)}/4$$

By writing an equation and doing a little subtracting, we can get a simple equation for the source resistor value required for midpoint bias. Assume that drain–source current is zero (reverse-biased junction), making the gate effectively grounded in spite of R_g. Then, by Ohm's law, we have

$$V_S = I_{DS} \times R_s$$

If $V_G = 0$, it follows that

$$
\begin{aligned}
V_{GS} &= (V_G - V_S) \\
&= [0 - (I_{DS} \times R_s)] \\
&= -I_{DS} \times R_s
\end{aligned}
$$

and

$$R_S = V_{GS\,(off)}/2I_{DSS}$$

Then, for $\frac{1}{2}\,I_{DSS}$,

$$gm_0 = 2I_{DSS}/V_{GS\,(off)}$$

which is equal to $1/R_S$. Therefore,

$$1/gm_0 = R_S \qquad \text{and} \qquad 1/R_S = gm_0$$

This relationship between R_s and gm_0 gives us a quick solution for bias calculations and some other things. We can almost always get maximum and minimum values for gm_0 from the data manual, without having to plot any graph or do any arithmetic.

| Example 7.1 | **Self-bias Example** |

Calculate the value of the self-bias resistor R_s and the voltage drop across R_s for the circuit in Figure 7-7 and the data in Table 7-1.

Solution

1   From Table 7-1 we get

$$
\begin{aligned}
gm_{0\,(min)} &= 1875 \text{ microsiemens} \\
gm_{0\,(max)} &= 2500 \text{ microsiemens}
\end{aligned}
$$

2   The average gm_0 is

$$
\begin{aligned}
gm_{0\,(average)} &= \sqrt{gm_{0\,(min)} \times gm_{0\,(max)}} \\
&= 1875 \times 2500 \\
&= 2165 \text{ microsiemens}
\end{aligned}
$$

FIGURE 7-7   **J-FET Circuit with Self-bias**

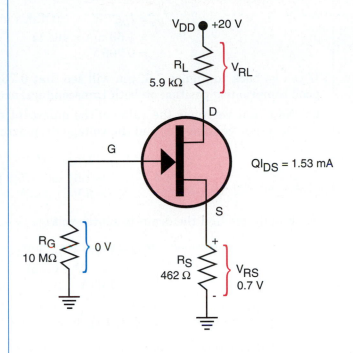

3   The value of the self-bias resistor is

$$R_s = 1/gm_{0\ (average)}$$
$$= 1/2165 \times 10^{-6}$$
$$= 462\ \Omega$$

4   Because the gate is 0 volts, the gate–source voltage (V_{GS}) is equal to the voltage drop across R_s. We now have a value for R_s, but we don't know the current through it or the voltage drop across it. To find the midpoint current, we first find the geometric average I_{DSS} corresponding to the minimum and maximum gm values:

$$I_{DSS\ (average)} = \sqrt{I_{DSS(max)} \times I_{DSS\ (min)}}$$

From Table 7-1 we get

$$I_{DSS\ (max)} = 3.75\ \text{mA}$$
$$I_{DS\ (min)} = 2.5\ \text{mA}$$
$$I_{DSS\ (average)} = \sqrt{3.75\ \text{mA} \times 2.5\ \text{mA}}$$
$$= 3.06\ \text{mA}$$

The midpoint is therefore

$$3.06\ \text{mA}/2 = 1.53\ \text{mA}$$

5 Now we can calculate the voltage drop across R_S:

$$V_{RS} = I_{DS} \cdot R_s \qquad \text{(Ohm's law)}$$
$$= 1.53 \text{ mA} \times 462 \text{ } \Omega$$
$$= 0.706 \text{ V}$$

If you go back to Figure 7-6, you will see that 0.76 volts bias (V_{GS}) falls in a good compromise position on both transconductance curves.

6 Now that we know the value of the quiescent drain current, we can find the other voltages. To find the voltage drop across R_L, we use

$$V_{R_L} = QI_{DS} \times R_L$$
$$= 1.53 \text{ mA} \times 5.9 \text{ k}\Omega$$
$$= 9 \text{ V}$$

7 Finally, we find the drain-to-source voltage (V_{DS}):

$$V_{DS} = V_{DD} - (V_{R_L} + V_{RS})$$
$$= 20 - (9 + 0.71)$$
$$= 10.3 \text{ V}$$

Voltage-divider Bias and Negative Feedback

The self-bias circuit not only provides the appropriate bias; it also tends to compensate for variations in transistor parameters and temperature drift. If I_{DS} increases, it causes a larger voltage drop across the source (feedback) resistor (R_S/R_f). The increased voltage drop across R_f represents an increase in the bias voltage V_{GS}, which tends to decrease the drain–source current. This form of negative feedback is analogous to current-mode feedback in a bipolar transistor and is a very effective way of maintaining the desired quiescent drain current in spite of variations in transistor parameters. The source resistor (R_S/R_f) serves two functions: setting the proper bias voltage for the desired quiescent drain current, and stabilizing the quiescent drain current with negative feedback.

The problem with self–bias is that good stability generally demands a larger voltage drop across R_S/R_f than can be produced by the desired bias voltage. The larger the voltage drop across R_S/R_f, the greater the amount of feedback, and the more tolerant the circuit will be of wide variations in transistor parameters. Fortunately, the problem has an easy solution. We can select a voltage drop across R_S/R_f that is large enough to provide the desired stability (negative feedback), and then we can provide a positive offset voltage to the gate to get the bias voltage V_{GS} back down to the desired voltage. The circuit is called voltage-divider bias and is shown in Figure 7-8.

Example 7.2 | **Voltage Divider Bias Example**

Suppose that we want to maintain the voltage of 0.7 volt and a QI_{DS} of 1.53 mA from the previous example, but we want a voltage drop of 5 volts or more across R_f for stability purposes. How can we accomplish this?

FIGURE 7-8   J-FET Voltage-divider Bias Circuit

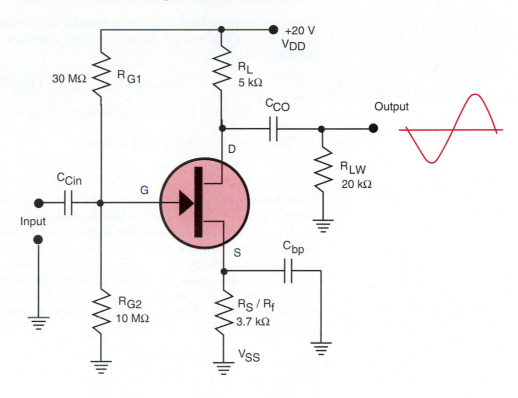

Solution

In Figure 7-8, the voltage divider provides an offset voltage of +5 volts:

$$V_{RG2} = V_{DD} \times [R_{G2}/(R_{G1} + R_{G2})] \quad \text{(the classical voltage divider equation)}$$
$$= (V_{(gate)} = 5 \text{ V}$$

where $V_{(gate)}$ is the gate-to-drain voltage.

If R_f is equal to 3.7 kilohms and QI_{DS} is equal to 1.53 mA, then

$$V_{Rf} = 3.7 \text{ k}\Omega \times 1.53 \text{ mA}$$
$$= 5.66 \text{ V}$$

which may be approximated as 5.7 V. Since we now have values for ($V_{(gate)}$ and (V_{R_f}), we can caluclate

$$V_{GS} = V_{Rf} - V_{gate}$$
$$= 5.7 \text{ V} - 5.0 \text{ V}$$
$$= 0.7 \text{ V} \quad \text{(effective gate bias voltage)}$$

The gate is 0.7 volt less positive than the source, so we have an effective bias voltage of −0.7 volt. We have the same quiescent operating current and the same bias voltage as we did in the previous example; but we have increased the stability voltage drop across R_f by a factor of about 8.

Output Resistance in the Resistive Region

The source–drain conductance (channel conductance) when source–drain voltages V_{DS} are low enough to permit the transistor to operate in the resistive region is

$$G_{DS \, (on)} = gm_0 \times [1 - (V_{GS}/V_P)]$$

(*Note:* In this case, G is the conductance symbol, and not the symbol for the transistor gate.)

Therefore, the channel resistance is

$$R_{DS \, (on)} = 1/G_{DS \, (on)}$$

since resistance is the reciprocal of the conductance. The value for $G_{DS \, (on)}$ when $V_{GS} = 0$ is often listed in the data manual.

$G_{DS \, (on)}$ actually has nothing to do with gm, because no transconductance occurs in the resistive mode. But gm_0 is a convenient point on the V_{DS}–I_{DS} curve. The rest of the equation looks familiar because the curve of $R_{DS \, (on)}$ versus V_{GS} has the same parabolic shape as the transconductance curve. If you look back at Figure 7-4, you will see a dotted curve with a parabolic shape that marks the border of the resistive operating region.

Output Resistance in the Amplifying (Transconductance) Mode

Output admittance Y_{OS} is the reciprocal of the output resistance:

$$R_O \quad [\text{or } R_D] = 1/Y_{OS}$$

The output admittance is the parameter most generally available from data manuals. Because we are normally interested in the output resistance R_O (or R_D), we must take the reciprocal of Y_{OS}. Y_{OS} is given in millisiemens or microsiemens, and is generally measured at 1 kHz.

Example 7.3

Calculating FET Circuit Parameters

If $Y_{OS} = 10$ microsiemens, find the output resistance R_O, the total drain load R_{Lac}, the voltage gain A_V, and the input impedance $Z_{in.}$

Solution

$$R_O = 1/(10 \times 10^{-6}) = 100 \text{ k}\Omega$$

We need to find the total load resistance, which consists of three resistances in parallel:

$$R_L // R_{Lw} // (1/Y_O)$$

The transistor's output resistance (R_O) may or may not be significant, depending on the values of R_L and R_{Lw}. In bipolar transistors, the output resistance of the transistor is from 10 to 50 times greater than the typical transistor's output-load resistances, and we generally ignore R_O. We cannot always ignore R_O in field-effect transistors, however. Let's look at the situation in Figure 7-9.

FIGURE 7-9   **Circuit for the Signal Parameter Discussion**

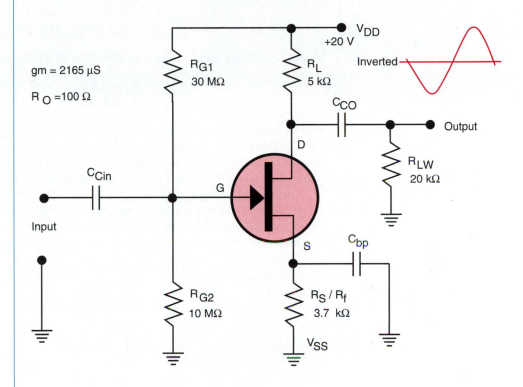

For the circuit in Figure 7-9, we have the following drain-load values:

$$R_L = 5\,\text{k}\Omega$$
$$R_{LW} = 20\,\text{k}\Omega$$
$$R_O = 100\,\text{k}\Omega$$

The total drain load is

$$R_{Lac} = R_L // R_{LW} // R_O$$
$$= 5\,\text{k}\Omega // 20\,\text{k}\Omega // 100\,\text{k}\Omega$$
$$= 3.85\,\text{k}\Omega$$

If we leave R_O out, we get

$$R_{Lac} = 5\,\text{k}\Omega // 20\,\text{k}\Omega = 4\,\text{k}\Omega$$

For practical purposes, the difference here is not significant. As a rule of thumb, we can ignore R_O whenever

$$R_O = 10 \times (R_L // R_{LW})$$

Voltage Gain for the Circuit in Figure 7-9

Without the bypass capacitor (C_{bp}), resistor R_F in the circuit in Figure 7-9 adds negative feedback that reduces the voltage gain but makes it stable and independent of the uncertainties of *gm* values. The voltage gain without the bypass capacitor is

$$A_V = R_{Lac}/R_f$$

Using R_{Lac} from the previous calculation of the total drain load, we get

$$A_V = 4 \text{ k}\Omega/3.7 \text{ k}\Omega$$
$$= 1.08$$

The voltage gain is hardly impressive, and if voltage gain were the circuit's only purpose, the circuit would be useless. However, we could be using this stage just to obtain the very high input impedance available from a field-effect transistor. Then we could obtain the necessary voltage gain from successive FET or bipolar stages.

If we add the bypass capacitor, we can short out the negative feedback at signal frequencies and achieve a significant increase in voltage gain. With the bypass capacitor added (and assuming that *gm* = 2165 µS), the voltage is

$$A_V = gm \times R_{Lac}$$
$$= 2165 \times 10^{-6} \times 4 \times 10^3$$
$$= 8660 \times 10^{-3}$$
$$= 8.66$$

The voltage gain of a field-effect transistor circuit is generally much lower than that of a similar bipolar circuit. The transistor used in Figure 7-9 is not a particularly high-*gm* device, but even high-transconductance devices rarely allow voltage gains of more than 20.

Input Impedance of the Circuit in Figure 7-9

The input impedance of the reverse-biased gate–channel junction is typically much greater than 100 megohms. In nearly all practical FET circuits, the input impedance is controlled by the values of the external gate resistors.

In the case of the circuit in Figure 7-9, two resistors are included in the voltage divider. The upper resistor (R_{G1}) is grounded through the negligible impedance of the drain power supply, so the two resistors (R_{G1}) and (R_{G2}) are effectively in parallel.

The input impedance of the circuit in Figure 7-9 is

$$Z_{in} = R_{G1}//R_{G2}$$
$$= (30 \text{ M}\Omega \times 10 \text{ M}\Omega)/(30 \text{ M}\Omega + 10 \text{ M}\Omega)$$
$$= 7.5 \text{ M}\Omega$$

Common Drain (Source-follower) Circuit

Figure 7-10 shows the **common drain circuit,** which is analogous to the common collector circuit in bipolar technology. The input impedance of the circuit in Figure 7-10 is identical to that of the common source circuit we have previously discussed:

$$Z_{in} = R_{G1}//R_{G2}$$

FIGURE 7-10 Common Drain (Source-follower) Circuit

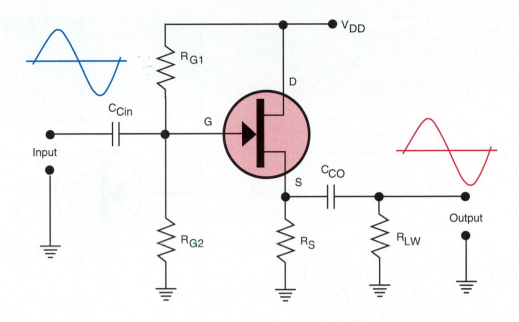

The voltage gain of the common drain circuit is always approximately 1. The output impedance is

$$Z_{out} = R_L /\!/ (1/gm)$$

All of the previous discussions of gm, biasing considerations, and so on apply to the common drain circuit.

Common Gate Circuit

Figure 7-11 shows the **common gate circuit,** which is analogous to the common base circuit in bipolar technology. The voltage gain and all gm and bias considerations previously discussed for the common source circuit apply with equal validity to the common gate circuit.

The input impedance for the common gate circuit is quite low (as it is in the equivalent bipolar common base circuit):

$$Z_{in} = R_f /\!/ (1/gm)$$

7.4   Depletion/Enhancement-mode MOS-FET

The MOS-FET operates on the same basic principal as the J-FET, except that the control element (gate) that controls the channel is a metal plate insulated from the channel by a very thin layer of silicon dioxide (glass). Figure 7-12

FIGURE 7-11 **Common Gate Circuit**

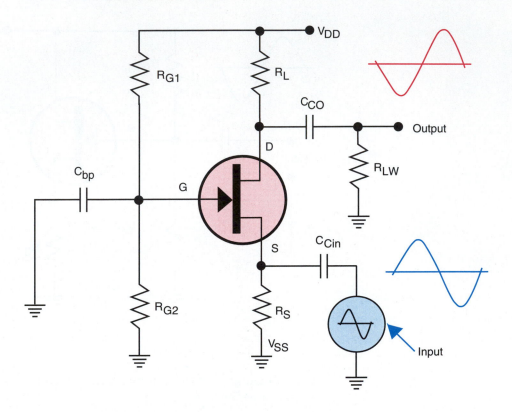

FIGURE 7-12 **Structure of the Depletion/Enhancement MOS-FET**

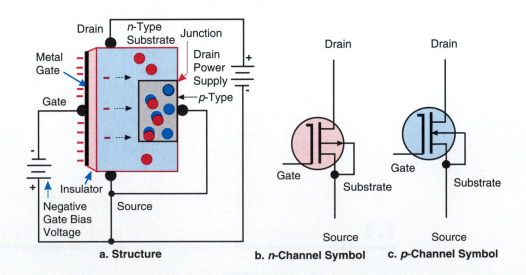

a. Structure   b. *n*-Channel Symbol   c. *p*-Channel Symbol

shows the structure of the depletion/enhancement MOS-FET, and the appropriate schematic symbols for this design.

Depletion-mode Operation of a MOS-FET

In Figure 7-13(a), the gate is in its zero–bias condition, and electrons flow freely through the *n*-type silicon channel.

In Figure 7-13(b), a potential difference of several volts of negative bias is applied to the gate, driving the free electrons into the *p*-type substrate. The electrons are captured by holes in the substrate, making them unavailable for channel conduction. The source-to-drain (channel) current is cut off.

Between these two extremes—full conduction and cut-off—is a range in which varying numbers of electrons remain in the channel while

FIGURE 7-13 Depletion-mode Operation in a MOS-FET

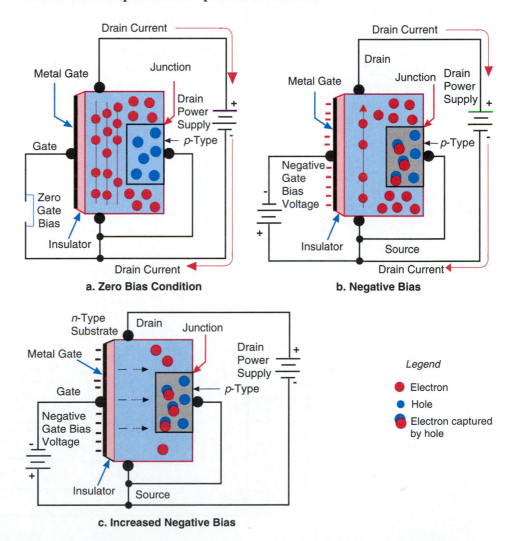

a. Zero Bias Condition

b. Negative Bias

c. Increased Negative Bias

Legend

● Electron

● Hole

◑ Electron captured by hole

others are driven into the substrate by gate bias. When many electrons are in the channel, a relatively high source-to-drain current flows. A high negative gate voltage drives more electrons into the substrate, reducing the channel current proportionally. The partially conducting channel condition is shown in Figure 7-13(b).

Enhancement Mode

If the polarity of V_{GS} is reversed to make the gate positive with respect to the source, electrons are drawn out of the p-region, increasing the normally available channel current. This mode of operation is called **enhancement.** All MOS-FET devices are capable of some enhancement-mode operation, but many are tailored for a fairly symmetrical operation in both modes.

Figure 7-14 shows a family of drain curves for a MOS-FET that is intended for combination depletion/enhancement-mode operation. This family of MOS-FET curves is very similar to the family of J-FET drain curves shown earlier (Figure 7-4). The main difference is that the MOS-FET family includes I_{DS} values for both positive and negative gate voltages. In addition, the resistive region is a bit more pronounced in the MOS-FET family.

FIGURE 7-14 **Family of Drain Curves for a Depletion/Enhancement MOS-FET (*n*-Channel)**

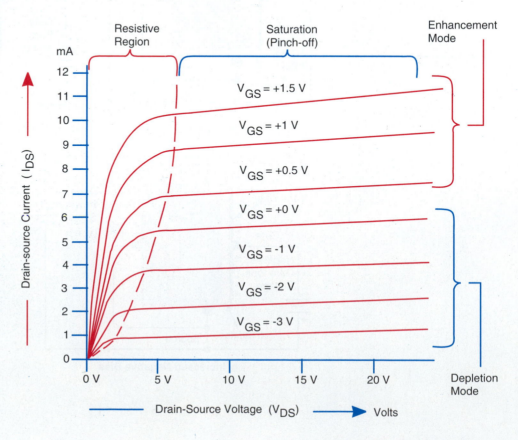

FIGURE 7-15 Transconductance Curve for a Depletion/Enhancement MOS-FET

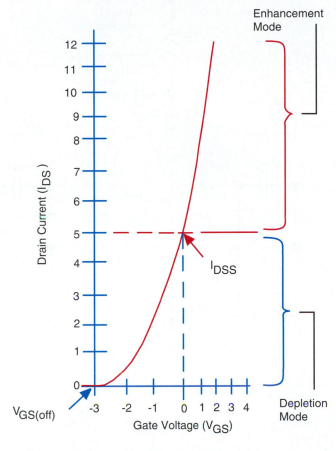

$$I_D = I_{DSS} [1 - (V_{GS} / V_{GS(off)})]$$

Figure 7-15 shows the transconductance curve for a typical depletion/enhancement MOS-FET. Again, other than the fact that both polarities of gate voltage are involved, this curve is similar to the transconductance curve for a J-FET device. Like the J-FET transconductance curve, the MOS-FET transconductance curve is parabolic. All of the mathematical relationships involving I_{DS}, I_{DSS}, gm, and so on that we used with the J-FET apply to the MOS-FET. However, MOS-FETs designed for depletion/enhancement-mode operation are often operated at $V_{GS} = 0$, whereas J-FETs are rarely operated at zero bias.

Figure 7-16 shows the MOS-FET version of the three basic configurations: common source, common drain, and common gate.

Enhancement-mode-only MOS-FET

Another kind of MOS-FET is a mainstay of digital integrated circuits. It is an enhancement-mode-only device and is normally off. Like the bipolar transistor, the enhancement MOS-FET is normally off and requires a finite gate-bias voltage to initiate conduction. And like the bipolar transistor, the same polar-

FIGURE 7-16 **MOS-FET Versions of the Three Configurations**

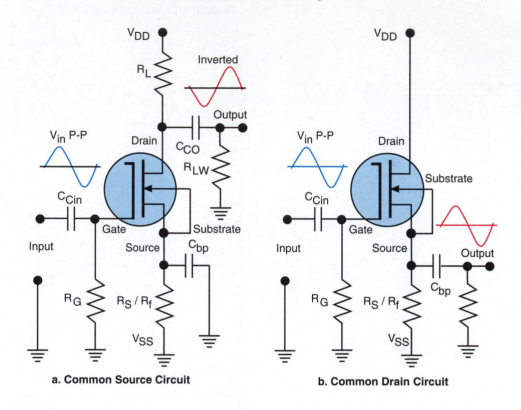

a. **Common Source Circuit**

b. **Common Drain Circuit**

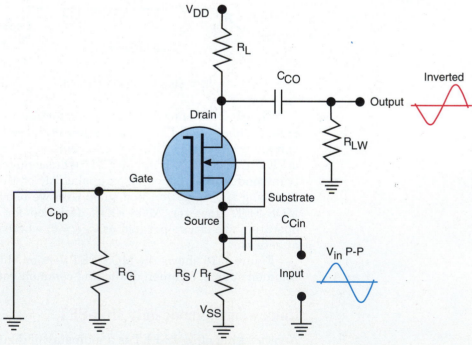

c. **Common Gate Circuit**

FIGURE 7-17   Enhancement-only MOS-FET

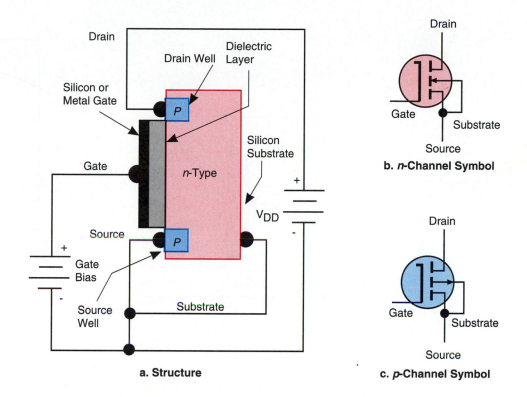

a. Structure

b. *n*-Channel Symbol

c. *p*-Channel Symbol

ity as the drain (collector) is required to turn the device on. The enhancement-mode MOS-FET has no built-in channel, and its two junctions must be forward-biased before source–drain current can flow.

The voltage required to turn the device on is called the **threshold voltage (V_T)**. Once the threshold voltage is exceeded, the device behaves like a depletion/enhancement-mode device that is operating in the enhancement mode.

Figure 7-17(a) shows the structure of an *n*-channel enhancement MOS-FET and identifies its operating polarities. Parts (b) and (c), respectively, of Figure 7-17 show *n*-channel and *p*-channel enhancement-mode MOS-FET schematic symbols. The *n*-channel device shown in Figure 7-17 actually has no built-in channel; but the channel, once formed, is the equivalent of an *n*-channel. Figure 7-18 illustrates how the device works.

At first, as the gate voltage is increased (see Figure 7-18(a)), no source–drain current flows: there is no channel through which current can flow. But the gate voltage continues to increase until the threshold voltage is reached, at which point the drain-to-substrate and source-to-substrate diode junctions become forward-biased. The two junctions and the electric field structure form a complex system that can be tailored to threshold voltages ranging from a little less than 1 volt to several volts.

As soon as the two junction voltages have been overcome, the positive charge on the metal gate draws electrons up against the silicon dioxide insulator, where they form a layer of *n*-type carriers (see Figure 7-18(b)). This layer

FIGURE 7-18   How the Enhancement MOS-FET Switching Amplifier Works

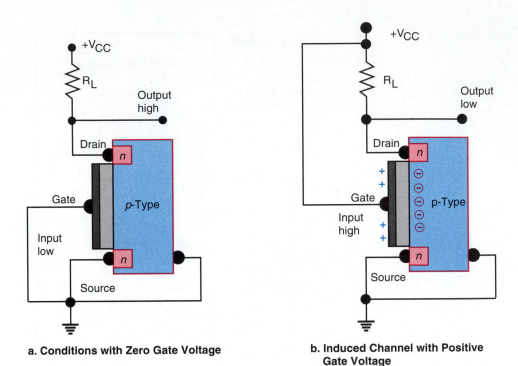

a. Conditions with Zero Gate Voltage

b. Induced Channel with Positive Gate Voltage

of electrons becomes a source–drain channel. As the positive voltage on the gate increases, more electrons are drawn into the induced n-type channel. The amount of current that can flow between source and drain is proportional to the number of electrons available in the induced channel. The number of available channel electrons, in turn, is proportional to the positive voltage on the gate.

The transconductance curve for enhancement-only devices has the typical field-effect parabolic shape. However, zero-bias drain current (I_{DSS}) and $V_{GS\,(off)}$ have no meaning for the enhancement-only MOS-FET.

We need to develop some new equations for the enhancement-only MOS-FET. Two important ones are the equations for drain current (I_D) and for *gm*:

$$I_D = k \times [(V_{GS} - V_T)^2\,]$$
$$gm = 2k(V_{GS} - V_T)$$

where I_D is the source–drain current at a specific gate-to-source voltage (V_{GS}), V_T is the threshold voltage from the data manual, *gm* is the transconductance at a particular gate voltage (V_{GS}), and k is a constant (expressed in mA/V^2) from the data manual.

The enhancement-only MOS-FET can be used for either analog or digital–switching applications. For analog service, enough gate bias voltage must be applied to the FET to bias it above the threshold voltage. The enhancement-only MOS-FET is ideal for digital and other switching applications because the threshold voltage provides a minimum turn-on gate voltage, which provides some natural immunity to unwanted noise signals.

7.5   Complementary MOS (C-MOS) Devices

The use of complementary pairs of MOS-FET—one n-channel and one p-channel device—has produced a class of devices known as **analog transmission gates.** There is no bipolar equivalent of the transmission gate, which is commonly used in both analog and digital technologies. A complete family of digital complementary MOS (C-MOS) logic exists, including large-scale integrated circuits such as microprocessors.

In the realm of analog circuits, the transmission gate is a rare kind of semiconductor device, because it is truly bidirectional: signal current can flow through the transmission gate in both directions. The transmission gate offers a wide range of control for either an **ac** or a **dc** signal. In digital circuits, transmission gates are used to form latches and flip–flops, and to serve as data-selector switches and bidirectional computer-bus drivers.

Figure 7-19 shows the C-MOS analog (or digital) transmission-gate circuit and its symbol. The circuit consists of a pair of complementary MOS transistors, with an inverting (MOS-FET) amplifier that permits both n-channel and p-channel transistors to be controlled by a single control voltage.

Transmission gates normally come several to an integrated-circuit package. Figure 7-20 shows two of the several available IC package configurations. The quad analog gate in part (a) is the electronic version of a mechanical rotary selector switch. The selector can be used to select from among various audio signals in a recording studio, for example. In addition, each of the four transmission gates can operate as a variable resistance controlled by its con-

FIGURE 7-19   Complementary MOS Analog Transmission Gate

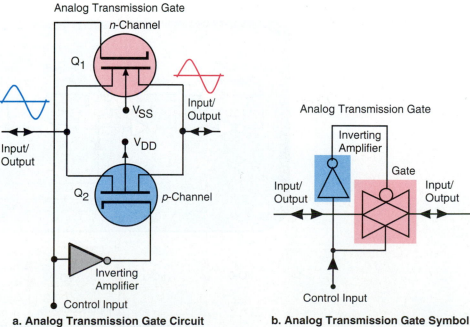

a. Analog Transmission Gate Circuit   b. Analog Transmission Gate Symbol

FIGURE 7-20
Integrated-circuit Quad
Analog Transmission Gate

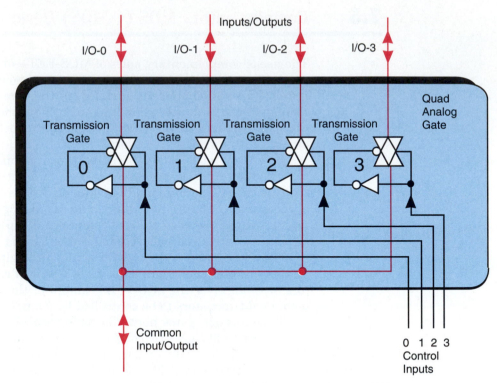

a. Quad Analog Gate with Common Input/Output (I/O)

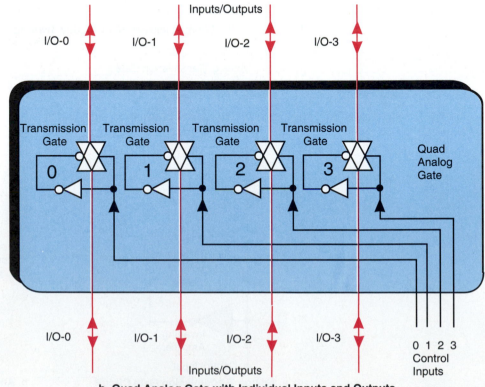

b. Quad Analog Gate with Individual Inputs and Outputs

trol input. The second configuration (shown in part (b)) has four completely separate analog gates. Analog gates make excellent audio-control devices. The noise-free operation is a boon to audio systems, whether the analog gate is used as an on/off switch or as a voltage-controlled variable resistor.

The power supply voltage applied to an analog gate may be a single voltage, with V_{SS} grounded; or V_{SS} may be connected to a negative supply voltage. With V_{SS} grounded, the peak-to-peak signal voltage to be controlled by the transmission gate must not exceed V_{DD}. If a negative supply voltage is used for V_{SS}, the peak-to-peak signal voltage must not exceed the sum of $V_{DD} + V_{SS}$. In a practical circuit, V_{DD} should be $1/2$ to 1 volt higher than the signal peak. A similar allowance should be made for V_{SS}. We will encounter a number of transmission-gate applications in later chapters.

Troubleshooting Field-effect Transistor Circuits

You will not have many opportunities to test field-effect transistor circuits. They are not often used in analog applications. Frequently, when FETs are used, they are part of an integrated circuit, so you can't get at the FET to test it. The test set-up in Figure 7-21 is a quick and easy way to make the test if you should ever have the need.

As usual, you should first test power-supply voltages before looking for other problems.

Once again, the transistor test involves making the output voltage change by altering the gate bias with a clip lead. Try both clip-lead positions indicated on the drawing. If either test causes a significant output voltage change, the FET is probably okay.

If neither clip-lead connection yields any change in the output voltage, the FET is either bad or some voltage is missing or incorrect.

The Digital Connection: Field-effect Transistors

In this section, we will look at complementary MOS (C-MOS), which consists of an n-channel and a p-channel pair of MOS-FETs in a kind of class-B mode. There are no transistor-transistor-logic (TTL) equivalents for some C-MOS devices. The analog gate is an important device for which there is no TTL equivalent. Analog gates can be used for either analog or digital signal control.

C-MOS Logic Family

Modern C-MOS digital devices are now providing viable competition to the venerable TTL logic families. For many years, C-MOS digital logic was slow,

FIGURE 7-21   Set-up for Testing Field-effect Transistor Circuits

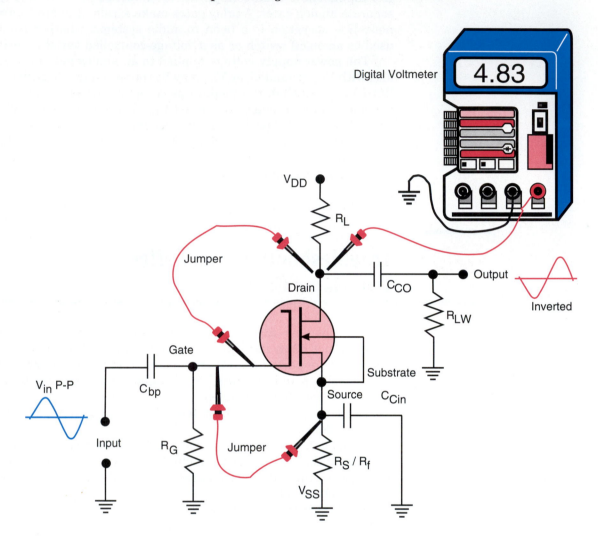

delicate, and subject to a vexing latch-up problem. Modern digital C-MOS is as fast as TTL, uses far less power, is no longer so fragile (electrically), and has eliminated the latch-up problem. The C-MOS logic family is composed of enhancement-only MOS-FETs. The turn-on threshold is set at about $\frac{1}{2} V_{DD}$, which provides excellent noise immunity. Figure 7-22 shows the circuits and the switching-circuit equivalents for the digital logic inverter and NAND gates.

Digitally Controlled Analog Gates

A number of digitally controlled analog-gate integrated-circuit packages are available. Some of these are designed to control digital signals only. Other devices are intended for either digital or analog signals. The 4052 dual-quad analog gate is an example of a digitally controlled selector switch that can switch either analog or digital data. Analog/digital gates can serve as analog-signal selectors or distributors. In the digital world, the same functions are called *data selector/multiplexer* and *data distributor/demultiplexer.*

FIGURE 7-22 **Complementary MOS-FET (C-MOS) Digital Logic Circuits: Inverter and NAND Gate**

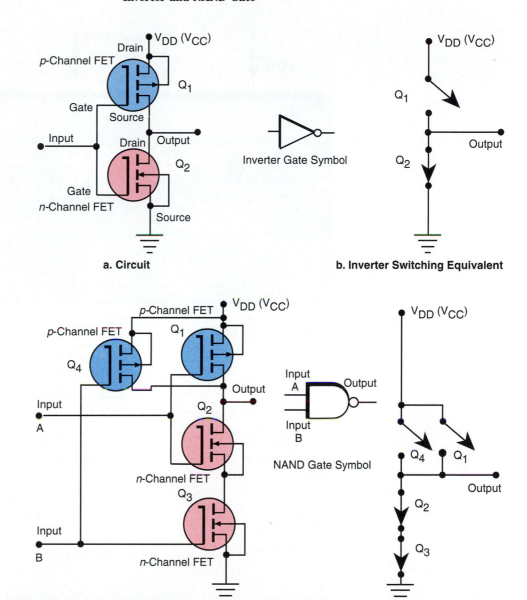

a. Circuit

Inverter Gate Symbol

b. Inverter Switching Equivalent

NAND Gate Symbol

c. NAND Gate Circuit

d. NAND Gate Switching Equivalent

Digitally selected analog/digital switches, like the one illustrated in Figure 7-23, use the binary values 0 and 1—represented by ground and +5 volts respectively—to determine the input/output selection. The switches themselves can be connected to +5 volts (V_{DD}) and ground (V_{SS}) to switch digital signals.

Analog signals, including **ac** voltages, may be connected to the switches, provided that the voltages for V_{DD} and V_{SS} are appropriate. For example, an

FIGURE 7-23   **Digitally Selected Analog Gates for Both Analog and Digital Applications**

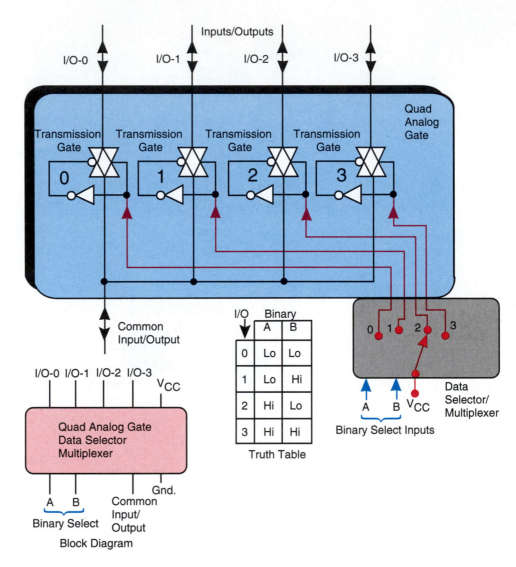

I/O	Binary	
	A	B
0	Lo	Lo
1	Lo	Hi
2	Hi	Lo
3	Hi	Hi

Truth Table

ac voltage of 10 volts peak-to-peak can be switched, provided that V_{DD} is at least +10 volts and V_{SS} is at least −10 volts. The circuit in Figure 7-23 illustrates a simplified version of a typical digitally controlled analog gate.

The value +5 volts is represented on the truth table as "High," corresponding to a binary 1. A "Low" on the truth table represents ground, 0 volts, or a binary 0. A binary 0 (00) selects I/O-0; a binary 1 (01) selects I/O-1; a binary 2 (10) selects I/O-2; and a binary 3 (11) selects I/O-3.

Summary

1 The field-effect transistor has input impedances of 1000 to 100,000 times greater than the input impedance of a bipolar transistor in a similar circuit.

2 The output voltage of a field-effect transistor often swings closer to both V_{CC} and ground than does that of a bipolar device. This allows a larger signal output swing in linear circuits, and it permits more nearly perfect logic levels in switching circuits than can usually be had from bipolar transistors.

3 Field-effect transistors are capable of a mode of operation in which the output (source–drain) circuit can act as an ordinary resistor. The resistance, in this mode, can be varied by altering the input (gate–source) voltage. In this mode, the field-effect transistor acts as a voltage-variable resistor.

4 Bipolar transistors are not capable of operating in this voltage–variable resistor mode.

5 Field-effect transistors can be operated in a controlled resistive mode to create a device known as an analog gate. There is no bipolar counterpart of the analog gate; it is available only in field-effect technology.

6 The preceding five statements identify the major strengths and features of field-effect transistors. If none of these five characteristics is important in a particular application, a bipolar transistor will probably be used instead of a field-effect transistor.

7 There are two basic kinds of field-effect transistors; the junction (or J-FET) and the metal-oxide semiconductor (or MOS-FET).

8 The MOS-FET is sometimes called an *insulated gate field-effect transistor* (or IGFET).

9 The field-effect transistor is called a *transconductance amplifier*, because an input voltage is used to control an output current. The symbol for transconductance is *gm*.

10 Transconductance in a FET is defined as $gm = I_{DS}/V_{GS}$, where *gm* is the transconductance in microsiemens, V_{GS} is the gate-to-source voltage, and I_{DS} is the drain-to-source current.

11 The transconductance (*gm*) in a FET is analogous to beta (β) in a bipolar transistor.

12 The source in a FET is analogous to the emitter in a bipolar transistor.

13 The drain in a FET is analogous to the collector in a bipolar transistor.

14 The gate in a FET is analogous to the base in a bipolar transistor.

15 A FET can be connected in one of three configurations: common source, common drain, and common gate. The common source circuit is the most common of these.

16 At lower drain-to-source voltages, FETs have an operating mode in which they act not as an amplifier, but simply as a resistor. This operating mode allows the FET to be used as a voltage-variable resistor, or electronic switch. A family of FET-based devices called *analog gates* (or *transmission gates*) operate in this mode.

17 When the FET is in its resistive mode, the channel resistance between source and drain is controlled by varying the voltage on the gate.

18 There is no operating mode similar to resistive mode in bipolar transistors, so bipolar analog gates do not exist.

19 The second FET operating mode is amplifying mode. Transconductance, the amplifying figure of merit, applies only to amplifying mode.

20 In general, FETs are not very good amplifiers, since they do not produce much gain. They make excellent switches, however, and are more common in digital systems for that reason.

21 FETs are generally used as the input stage of an amplifier, with the rest of the amplifier constructed from bipolar transistors. In some instances, power FETs are used as output stages, where they do have some advantages over bipolar power transistors, particularly at radio frequencies.

22 Negative feedback is not as important in FET circuits as it is in bipolar circuits, because FETs need less negative feedback to operate

properly. On the other hand, FETs normally don't deliver enough gain to have much extra to trade for the benefits of feedback. There just isn't much gain to spare.

23   FET circuits are much more standardized than bipolar circuits, with fewer variations. This makes FET-circuit troubleshooting comparatively easy.

QUESTIONS AND PROBLEMS

7.1 Define the terms *enhancement mode* and *depletion mode* as they apply to field-effect transistors.

7.2 Are J-FETs enhancement-mode or depletion-mode devices?

7.3 In what ways does an enhancement-only device differ from a depletion/enhancement-mode device?

7.4 Can a J-FET be operated with zero bias? If so, what are the limitations on zero-bias operation?

7.5 If a J-FET were operated at zero bias, would you expect a relatively low or a relatively high transconductance value?

7.6 If a depletion/enhancement-mode device were operated at zero bias, would you expect a relatively low or a relatively high transconductance value?

7.7 If an enhancement-only device were operated at zero bias, would you expect a relatively low or a relatively high transconductance value?

7.8 What is pinch-off in a J-FET?

7.9 How does pinch-off relate to gate voltage and drain current?

7.10 Define I_{DS}.

7.11 Define I_{DSS}.

7.12 Define V_{GS}.

7.13 Define gm.

7.14 Define $V_{GS\,(off)}$.

7.15 Define *transconductance*.

7.16 Write the equation for J-FET transconductance when $V_{GS} = 0$.

7.17 Write the equation for transconductance when V_{GS} has some bias value other than zero.

7.18 If you know I_{DSS}, how can you find the drain–source current at some value of V_{GS}? Write the appropriate equation.

7.19 If you know I_{DSS} and $V_{GS\,(off)}$, how can you find the value of gm at $V_{GS} = -2$ volts?

7.20 If you know the transconductance of a J-FET at $V_{GS} = 0$ and the value of $V_{GS\,(off)}$, how can you find the value of gm at $V_{GS} = -2$ volts?

7.21 Write the two equations for the voltage gain of a J-FET common source amplifier—one with the feedback resistor bypassed, and one without. Include the working-load and the transistor output impedance in each equation.

7.22 Write the two equations for the voltage gain of a J-FET common gate amplifier—one with the feedback resistor bypassed, and one without. Include the working-load and the transistor output impedance in each equation.

7.23 Write the input impedance equation for a common gate FET amplifier.

7.24 What are the input impedances of the common source and the common drain FET amplifiers?

7.25 Write the equation for the transconductance of an enhancement-only MOS-FET.

7.26 Write the equation for I_{DS} for an enhancement-only MOS-FET.

7.27 What two purposes does a source resistor used in a depletion/enhancement-mode MOS-FET common source amplifier serve?

7.28 Write the equation for the channel resistance of a FET that is operating in resistive mode.

7.29 What are the characteristics of an FET channel when the FET is operating in resistive mode?

7.30 What is the typical range of FET dynamic channel resistance when the channel is in saturated mode?

Use the family of drain–source curves in Figure 7-24 to answer Problems 7.31 through 7.37.

FIGURE 7-24 **Family of Drain Curves for Problems 7.31 Through 7.37**

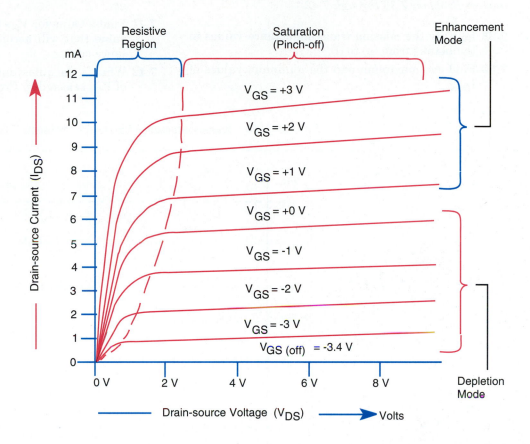

7.31 Find the value of I_{DSS} at V_{DS} = 4 volts.

7.32 Find the value of I_{DS} when V_{GS} = –2 volts (V_{DS} = 4 V).

7.33 Find the value of I_{DS} when V_{GS} = –2 volts (V_{DS} = 8 V).

7.34 Find the value of I_{DS} when V_{GS} = +2 volts (V_{DS} = 4 V).

7.35 Find the transconductance when V_{GS} varies from 0 to –2 volts.

7.36 Find the transconductance when V_{GS} varies from 0 to +2 volts.

7.37 Find the value of $V_{GS(off)}$.

Use the transconductance curve in Figure 7-25 to answer Problem 7.38 through 7.42.

7.38 Identify the missing transconductance values in the blanks from (a) to (i).

7.39 Find the maximum and the minimum values of I_{DSS}.

7.40 Find the maximum and the minimum values of gm.

7.41 Find a value for V_{GS} for a suitable midpoint bias value that will accommodate both transconductance curves.

7.42 What is the quiescent drain current at the value of V_{GS} selected in Problem 7.41?

FIGURE 7-25   **Transconductance Curves for Problems 7.38 Through 7.42**

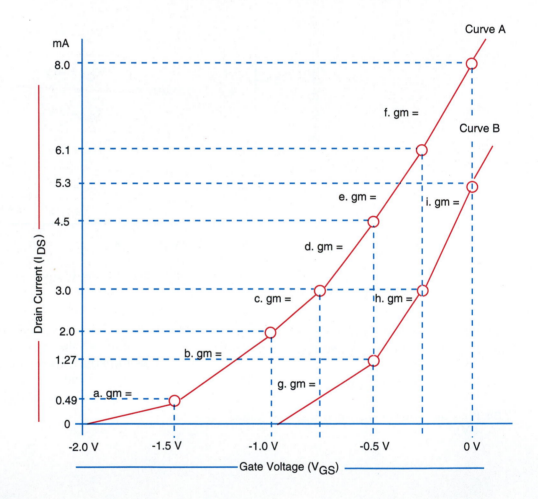

7.43 Using the transistor defined in Figure 7-25, calculate the value of R_S in Figure 7-26 that is required to achieve the quiescent drain current in Problem 7.42.

7.44 With the value of R_S obtained in Problem 7.43, calculate the voltage gain of the circuit in Figure 7-26, with and without the bypass capacitor.

FIGURE 7-26   Circuit for Problems 7.43 and 7.44

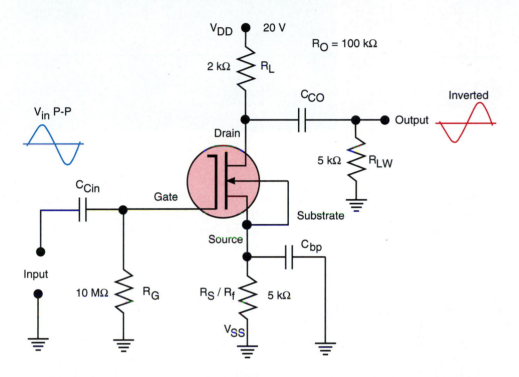

Soldering

Anybody can solder, right? A little heat, a little solder, and you're done. What's the big deal?

Soldering is a skill that is sometimes overlooked in the plethora of theoretical courses offered by technical schools. And many people assume that it is as simple as picking up a soldering iron and melting solder onto a connection. The truth is that good soldering technique demands knowledge and practice: knowledge of the correct materials, tools, and methods to use; and lots of practice to perfect the technique.

Soldering involves more than just joining two conductors with molten metal. It involves properly preparing component leads and surfaces, deoxidizing contacts, applying the proper amount of heat in the correct place, using the correct alloy, and stabilizing the solder joint while the molten solder cools and reverts to a solid state. Component lead dress and board cleaning are important, too.

A related area (and one that is often more difficult to master) involves the process of desoldering. Electronics technicians must be able to remove faulty components quickly and without damaging the printed circuit board. This is no small feat, since many circuit boards now use very light gauges of copper tracing. Too much heat and the wrong technique can result in removal of the trace along with the component lead.

These concerns are multiplied when the technician is working on surface-mount boards. Small components and minute lead spacings make successful soldering and desoldering very difficult. A whole new set of tools and techniques has been developed for this area of practice. Many electronics shops bring in surface-mount soldering equipment and have the manufacturer supply extensive training on how to use it.

But why are all these details so important? Perhaps you have soldered hundreds of joints already with no problems. Quality control is the answer. Manufacturers and end users alike want their electronic equipment to be reliable. Even a small percentage of warranty returns cuts into the manufacturer's profits, not to mention its reputation. Every consumer wants to have a reliable product; but some customers, such as the military and NASA, require very high reliability. The possibility of cold solder joints on circuit boards in the space shuttle represents an unacceptable risk.

The details of proper soldering and desoldering techniques cannot be fully described here, but several basic considerations can be mentioned.

First, you should use the right tools for the job. Electronic soldering requires a temperature-controlled soldering iron with a grounded tip. The temperature control ensures that sensitive components are not heated beyond a damaging temperature. The grounded tip is just one aspect of good electrostatic discharge precautions. Using a tip of the correct size and shape is also important for efficient soldering. Most soldering irons have replaceable tips so that the correct one can be selected for

Resin Core Solder

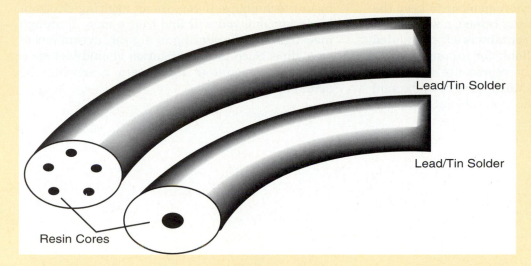

Lead/Tin Solder

Lead/Tin Solder

Resin Cores

each job. If you keep the tip clean and tinned with solder, you will ensure good heat transfer and better solder joints.

For desoldering, there are several options. In situations where a lot of printed-circuit rework is required, the best approach is to use a special soldering station that includes a desoldering iron with a hollow tip and vacuum. This equipment is expensive but very effective. For less frequent desoldering situations, a separate "solder sucker" can be used with a conventional soldering iron. The solder joint is heated until molten, and then the "solder sucker" is quickly positioned over the joint and triggered. A spring-actuated plunger creates a vacuum which sucks the molten solder out of the joint. Alternatively, solder wick, a product that soaks molten solder into a copper braid, can be used.

Other tools, such as small side-cutters, needlenose pliers, and wire strippers, are useful in soldering. It is important to get good-quality tools of the right size and to use them correctly. With all of these tools, professional-quality skills require time to develop.

The correct choice of solder is important. Always use resin-core solder. In plumbing applications, solid-core solder and acid-flux solder paste are used to deoxidize the metal to be soldered. This works fine for plumbing, but the acid flux is deadly to electronic components and printed circuit boards. Electronic solder has a resin flux embedded in the solder itself. Resin flux only becomes active when it is heated above a certain temperature. While it is active, it removes oxidation from the parts to be soldered and allows the solder to flow and bond with the metal parts. When the resin cools, it becomes inactive and will not harm the components.

Separate liquid resin flux is often useful for soldering. Although electronic solder contains flux, brushing a layer of flux onto the leads, circuit pads, or wire prior to soldering ensures better deoxidation. For tinning stranded wire or for creating large solder joints, this is especially helpful. Surface-mount soldering uses solder flux and pretinned printed circuit pads. The component leads and the circuit board are heated (usually with hot air) and the component is positioned on the pads. The flux helps ensure that the solder flows properly.

Armed with this information, you will find that your best course of action is to practice soldering and desoldering. A few hints may speed your progress, though.

Heat rises. Whenever possible, you should heat the solder joint from below. When heating a joint that is to be soldered (especially if liquid flux has not been preapplied), you should heat both parts of the joint at the same time and aid the heat conduction further by melting a drop of solder onto the tip of the iron itself. Always remember to keep the soldering iron tip cleaned and tinned; a wet sponge is useful for this. Small containers of granulated solder with resin are available for tinning tips.

This information only represents a cursory introduction to soldering. Much more could be said about lead bending, specific kinds of joints and connections, varieties of solder, and different surface-mount techniques. If you are an aspiring electronics technician, you will find that a more thorough investigation of soldering is a good investment of your time. Such an investigation should include a significant amount of time spent practicing the actual hand skills.

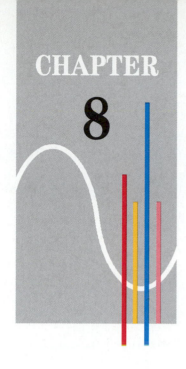

CHAPTER 8

Introduction to Operational Amplifiers

OBJECTIVES

Upon completion of this chapter, you should be able to:

1. Identify each of the following op-amp configurations on sight:
   a. The inverting amplifier.
   b. The noninverting amplifier.
   c. The voltage follower.
   d. The summing amplifier.
   e. The differential amplifier.

2. Write the equations for each of the following op-amp parameters:
   a. The voltage gain and input impedance of an inverting amplifier.
   b. The voltage gain of a noninverting amplifier.
   c. The voltage gain of a voltage follower.
   d. The output voltage of a summing amplifier.
   e. The output voltage of a differential amplifier.

3. List the characteristics of the ideal operational amplifier.

4. Compare the characteristics of the ideal operational amplifier with those of real operational amplifiers.

5. Explain the op-amp **dc** output offset voltage, and what can be done about it.

6. Explain the Miller effect as it applies to operational amplifiers.

7. Explain how the Miller effect alters the input impedance on the inverting input.

8. Explain how the Miller effect alters the input impedance on the noninverting input.

9. Define the term *virtual ground*.

10. Explain and write the equation for the input impedance of the inverting input.

11. Explain and write the equation for the input impedance of the noninverting input.

12. Draw a schematic diagram showing how an op-amp can be biased to operate with a single power supply.

13. Describe the purpose and limitations of using positive feedback bootstrapping with an op-amp.

Op-amps in Brief

THE STANDARD OP-AMP SYMBOL

The inverting input is indicated by the minus sign and the noninverting input is indicated by the positive symbol. Most op-amps require two power-supply voltages. The two power-supply inputs are not always shown as part of the symbol.

Standard Op-Amp Symbol
$A_{OL} > 100,000$ (Typical) ►

Open Loop Voltage Gain: $A_{OL} > 100,000$ (Typically)

Inverting Input
Noninverting Input

V+
Op-amp
V-
Out

Operational Amplifier Symbol

THE COMPARATOR

The comparator is the only op-amp circuit that does not use negative feedback. It functions as a switching amplifier and is a key circuit in analog-to-digital converters. It also serves in many timing and waveform-generating circuits.

The Comparator
$A_V = A_{OL} > 100,000$ (Typical) ►
$Z_{in} = 1$ Meg Ω (Typical)

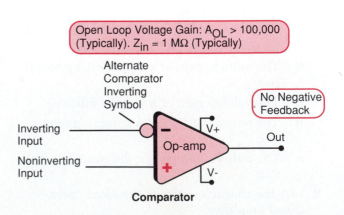

Open Loop Voltage Gain: $A_{OL} > 100,000$ (Typically). $Z_{in} = 1$ MΩ (Typically)

Alternate Comparator Inverting Symbol

No Negative Feedback

Inverting Input
Noninverting Input

V+
Op-amp
V-
Out

Comparator

THE VOLTAGE FOLLOWER

The voltage follower is the simplest amplifier circuit. It produces a voltage gain of 1 (unity), and features a very high input impedance and a very low output impedance. It is used as a buffer between a load and a circuit that can't drive that particular load.

The Voltage Follower
$A_V = 1$ (Unity) ➤
Negative Feedback = 100%

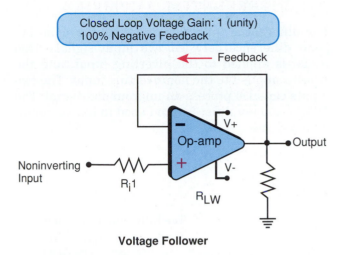

Voltage Follower

THE INVERTING AMPLIFIER

The inverting amplifier is a general-purpose voltage amplifier that provides an output signal 180° out of phase with the input signal. The circuit's voltage gain is determined by the ratio of the value of the feedback resistor to the value of the input resistor. The amplifier's input impedance is equal to the value of the input resistor.

The Inverting Amplifier
$A_V = -R_f/R_{in}$ ➤
$Z_{in} = R_{in}$

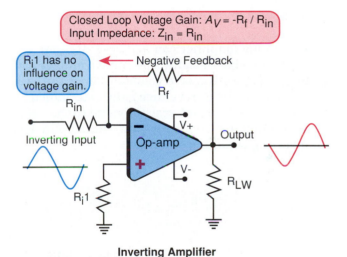

Inverting Amplifier

THE NONINVERTING AMPLIFIER

The noninverting amplifier is a general-purpose voltage amplifier that provides an output signal in phase with the input signal.

The circuit voltage gain is still determined by the ratio of the value of the feedback resistor to the value of the input resistor, which is connected to the inverting input.

This amplifier features a very high input impedance.

The Noninverting Amplifier
$A_V = (R_f/R_{in}) + 1$ ➤
$Z_{in} = > 500$ kΩ (Higher with Increased Feedback)

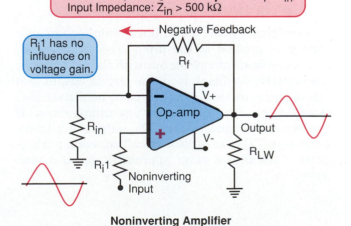

Noninverting Amplifier

THE DIFFERENTIAL AMPLIFIER

The differential amplifier amplifies the point-by-point difference between two input signals. One signal is applied to the inverting input, and the other is applied to the noninverting input. The two inputs can also process an ungrounded signal. The differential amplifier is often used in instrumentation systems.

The Differential Amplifier
$$V_{out} = A_V (V_{in2} - V_{in1})$$ ►
Zin = > 500 kΩ (Higher with Increased Feedback)

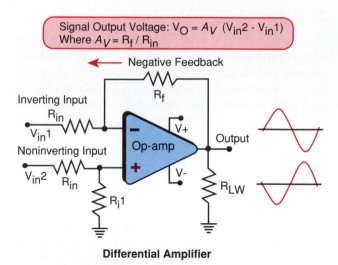

Differential Amplifier

THE SUMMING AMPLIFIER

The summing amplifier can add several input signal voltages algebraically. The input voltages can be negative, positive, or a combination of the two. Each input has its own voltage gain, determined by the ratio of the feedback resistor value to the value of each individual input resistor. The amplifier is an inverting summing amplifier.

The Summing Amplifier
$$V_{out} = -R_f [(V_1/R_1) + (V_2/R_2) + (V_3/R_3)]$$ ►

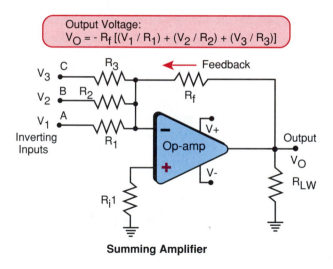

Summing Amplifier

THE INTEGRATOR CIRCUIT

The integrator generates a linear ramp when a **dc** voltage is applied to the input. The circuit is an R-C time-constant circuit; but with the capacitor in the negative-feedback loop, the circuit output is a linear ramp instead of the typical nonlinear R-C charge curve. The output voltage ramps down. If the input is a square wave, the output is a triangular wave. The integrator circuit converts a triangular wave into a close approximation of a sine wave.

The Integrator Circuit
$$V_{out} = -V_{in} \leftrightarrow (1/R_{in}C_f) \leftrightarrow \text{Time}$$ ►

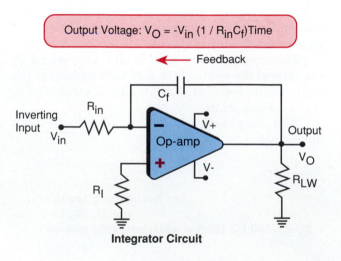

Integrator Circuit

8.1   Introduction

The **operational amplifier** is a universal amplifier: it can be programmed to perform a multitude of different tasks by adding a few external components. The op-amp can be programmed to serve any ordinary amplifier function if one or more input resistors and a negative-feedback resistor are added. The addition of positive feedback allows the amplifier to be task-programmed as an oscillator or function generator. Configured as a comparator, the op-amp can perform a variety of timing and control functions.

The op-amp can also be programmed as a current amplifier, voltage regulator, linear integrator, or differentiator. This is not a complete list, but it gives you some idea of the scope of the operational amplifier's capabilities. We will study these and other applications here and in subsequent chapters.

Nearly every op-amp circuit uses negative feedback to control the operating characteristics of the circuit. Negative feedback is the key player in nearly all of the op-amp circuits we will encounter in this text. External feedback components have almost total control over the characteristics of each op-amp circuit, and they are the primary agents that allow us to program the op-amp to perform its multitude of possible tasks.

Figure 8-1 provides an overview of the op-amp circuits we will study.

8.2   Negative Feedback in Operational Amplifiers

We can look at negative feedback in an op-amp in a number of ways (as usual). We will look at it first as a generic negative-feedback self-regulating system. Then we will look at some more formal feedback equations. Figure 8-2 shows the simplest op-amp circuit—the voltage follower—and indicates the standard feedback circuit control elements.

How the Feedback Loop in Figure 8-2 Operates

1   A reference voltage, which is actually the input signal, is applied to the noninverting input. Assume a 1-volt positive **dc** input (reference) voltage.

2   The output voltage is the controlled quantity. The voltage gain of the amplifier, which may be very high, produces an output voltage that is higher than the input voltage.

3   Because of the high open-loop voltage gain, the output voltage tries to rise rapidly. However, the output voltage is applied (fed back) to the inverting input of the differential amplifier, and that feedback voltage opposes the 1-volt input (reference) voltage.

4   The feedback voltage opposes any attempt on the part of the output voltage (the controlled quantity) to rise beyond the 1-volt reference (input) voltage. Any difference between the input (reference) voltage

FIGURE 8-1 **Sample of Common Op-amp Circuits**

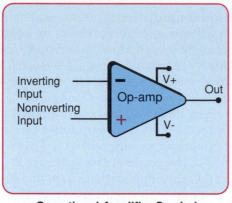

a. Operational Amplifier Symbol

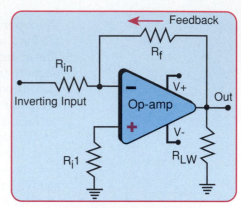

b. Inverting Amplifier

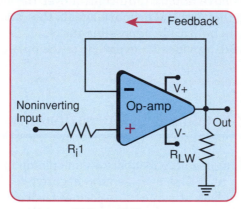

c. Voltage Follower

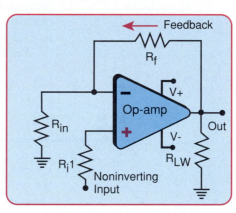

d. Noninverting Amplifier

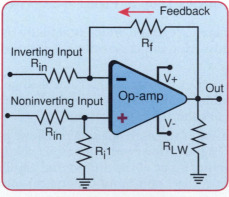

e. Differential Amplifier

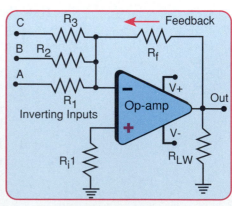

f. Summing Amplifier

FIGURE 8-1 Continued

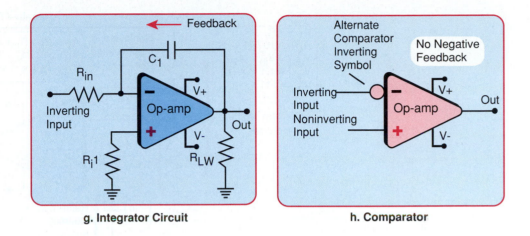

g. Integrator Circuit

h. Comparator

FIGURE 8-2 Negative Feedback: The Voltage-follower Circuit

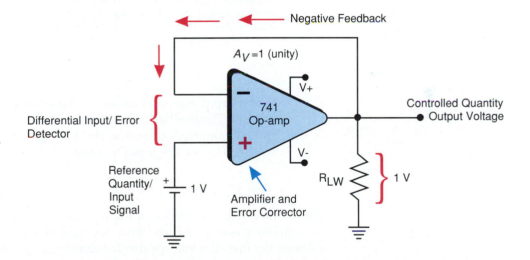

and the output (controlled quantity) voltage constitutes an error voltage.

5   When the voltages on the two inputs are equal, there is no error voltage, and no further correction of the output voltage takes place. The feedback loop is balanced and stable at 1 volt input, with a 1-volt output.

6   The voltage gain of the amplifier, with the feedback connected, is 1 (unity). This is true even though the amplifier has a potential open-loop gain of 200,000 or more. The actual circuit voltage gain is totally controlled by the negative feedback.

FIGURE 8-3 **How to Get Desired Voltage Gains with Feedback: The Noninverting Amplifier**

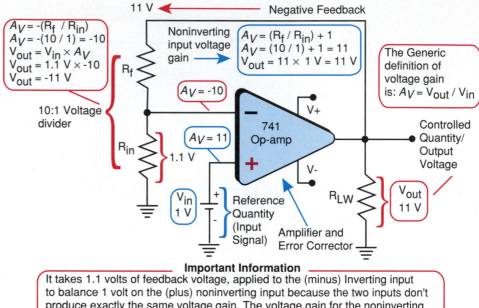

7 If the input signal changes, it represents a change in the reference voltage. The feedback circuit instantly corrects the output voltage (controlled quantity) to match the new input voltage value. The output voltage follows the input voltage.

How to Get Higher Voltage Gains with Negative Feedback

The noninverting amplifier in Figure 8-3 can be programmed to provide any reasonable value of voltage gain. Instead of feeding back the entire output voltage, we install a voltage divider that consists of a feedback resistor and a resistor to ground. We trick the inverting input into behaving as though the output voltage were only a fraction of its actual value.

For example, assume an input (reference) voltage of 1 volt. If the voltage divider divides the output voltage by a factor of 10, the output voltage can rise to 10 volts before the two input voltages are equal at 1 volt. When the actual amplifier output voltage is 10 volts, 1 volt will be applied to the inverting input, matching the 1-volt reference voltage on the noninverting input. There is no error voltage, but the output voltage is 10 times the input voltage. The amplifier circuit with feedback now has a voltage gain of 10. You can program any voltage gain you want by selecting the proper ratio of feedback and inverting input-to-ground resistor. If the input voltage changes, it represents a change in the reference voltage, so the feedback adjusts the output voltage; but it is still 10 times the new input voltage.

FIGURE 8-4   **Noninverting Op-amp Circuit**

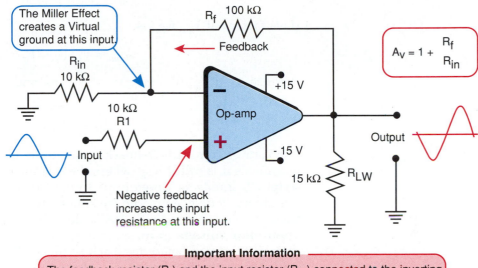

___ **Important Information** ___

The feedback resistor (R_f) and the input resistor (R_{in}) connected to the inverting input always control the voltage gain. Resistors connected to the noninverting input serve to balance input circuit currents, but they exert no control whatsoever over the circuit's voltage gain: $A_V = 1 + R_f / R_{in}$.

8.3   Formal Analysis of the Noninverting Amplifier

The noninverting amplifier uses the noninverting input of the op-amp for the signal input. The noninverting input is a high-impedance input because the input stage is a differential amplifier and is bootstrapped by the feedback voltage.

Voltage Gain

Now that we know how the feedback circuit controls the voltage gain, let's look at the formal equations we use to predict the voltage gain in a working op-amp circuit. The Miller effect influences our analysis of the noninverting input in two ways: it places the inverting input at virtual ground, and it effectively lowers the resistance between the two inputs to near 0 ohms. At the same time, the input resistance as seen by signal source has been greatly increased by the backdoor bootstrap. Refer to Figure 8-4 as you read the following analysis.

If we assume that the Miller effect has reduced the resistance between the two inputs to effectively 0 ohms, we can write

$$V_{R_f} = I_{Rf} R_f = V_{in} (R_f/R_{in})$$

where V_{R_f} is the voltage drop across the feedback resistor (R_f), I_{R_f} is the current through the feedback resistor, and V_{in} is the voltage drop across R_{in}. Now, we can write

$$V_{out} = V_{in} + [V_{in}\,(R_f/R_{in})]$$

If we factor out V_{in}, we get

$$V_{out} = V_{in}\,[1 + (R_f/R_{in})]$$

Because $A_V = V_{out}/V_{in}$, we can substitute and rearrange the preceding equation as follows:

$$A_V = 1 + (R_f/R_{in})$$

The voltage gain with feedback is called the **closed-loop voltage gain,** and in symbols it is called A_{CL} (where the A stands for amplification, and the subscript CL stands for closed loop). In most cases, we are only interested in the closed-loop voltage gain, and we simply use A_V. In cases where we must contrast the closed-loop (with feedback) voltage gain with the amplifier's raw **open-loop voltage gain,** we use A_{CL} and A_{OL}. A_{OL} is the amplifier's wide-open voltage gain before feedback is added. The most popular integrated-circuit op-amp, the 741, has an A_{OL} of about 200,000. A third voltage gain value, called simply the **loop gain** (A_L), is defined as:

$$A_L = A_{OL}/A_{CL}$$

Noninverting Amplifier Input Impedance

The input impedance of the noninverting amplifier with feedback is

$$Z_{in} = Z_{iin} \times A_L$$

where Z_{in} is the input impedance as seen by the signal source with feedback applied (this is the actual working circuit input impedance), and Z_{iin} is the intrinsic input impedance of the particular op-amp in question. The value for intrinsic input impedance can often be obtained from the manufacturer's data manual.

Noninverting Amplifier Output Impedance

The output impedance (Z_{out}) of the noninverting amplifier with negative feedback is

$$Z_{out} = Z_{iout}/A_L$$

where Z_{out} is the output impedance with feedback, and Z_{iout} is the intrinsic output impedance of the particular op-amp, as found in the manufacturer's data manual.

Example 8.1 | **Calculating Voltage Gain**

a Find the voltage gain (A_V) with feedback for the noninverting op-amp circuit shown in Figure 8-4.

b Find the input and output impedances for the noninverting op-amp circuit shown in Figure 8-4.

Solution

a
$$A_V = A_{CL}$$
$$= 1 + (R_f/R_{in})$$
$$= 1 + (100 \text{ k}\Omega/10 \text{ k}\Omega)$$
$$= 11$$

b To find the input and output impedances, we must first find the amplifier's open-loop (intrinsic) voltage gain. This value is a statistical typical value we can get from the manufacturer's data manual. The 741 op-amp is made by a number of manufacturers and has a typical open-loop voltage gain of 200,000. If we use this value for the open-loop gain (A_{OL}), and if we use the closed-loop voltage gain (A_{CL}) of 11 that we just calculated in part (a), we can find the loop gain (A_L). Calculating A_L is a necessary first step:

$$A_L = A_{OL}/A_{CL}$$
$$= 200,000/11$$
$$= 18,181$$

The manufacturer's data manual gives us a figure of 2 megohms for the intrinsic input resistance (Z_{iin}). Calculating Z_{in}, we get

$$Z_{in} = Z_{iin} \times A_L$$
$$= 2 \text{ Meg } \Omega \times 18,181$$
$$= 36.36 \text{ Meg } \Omega$$

The manufacturer's data manual gives us a value of 75 ohms for the typical intrinsic output impedance of the 741 op-amp. We can now calculate the output impedance with feedback:

$$Z_{out} = Z_{iout}/A_L$$
$$= 75 \text{ } (\Omega)/18,181$$
$$= 0.004 \text{ } \Omega$$

In practice, these input and output impedance values are far from exact, but the input impedance is very high and the output impedance is very low with feedback added. It is usually enough to know that the input impedance is high enough for nearly all purposes and that the output impedance is low enough for most practical purposes. Only rarely do we need to be concerned with actual values of input and output impedances.

8.4   Inverting Amplifier

The inverting amplifier uses the inverting input of the op-amp for the signal input. The inverting input is a low-impedance input because of the Miller effect produced by the feedback voltage. In practical situations, the amplifier's input terminal impedance is nearly at ground potential. The circuit's input resistance is equal to the value of the input resistor (R_{in}). The voltage gain differs only slightly from that of the noninverting configuration.

FIGURE 8-5   Inverting Op-amp Input Impedance and the Miller Effect

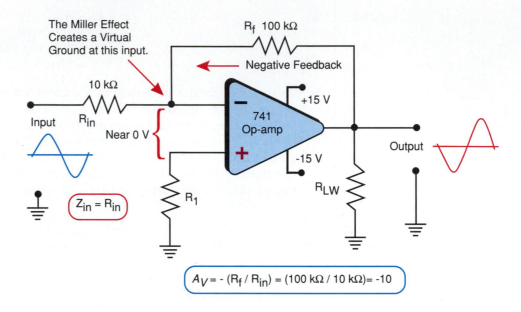

$$A_V = -(R_f / R_{in}) = (100\ k\Omega / 10\ k\Omega) = -10$$

Input and Output Impedances

The inverting configuration in Figure 8-5 can be analyzed in the same way as the noninverting version, but it is actually much simpler because of the Miller effect. The output impedance is the same for both inverting and noninverting versions. The input impedance is the same as the value of the input resistor (R_{in}), because the end of the resistor that is connected to the amplifier input terminal is virtually grounded. Thus we have

$$Z_{in} = R_{in}$$

Voltage Gain

Because of the virtual ground, the currents through R_f and R_{in} are equal, so the voltage gain is simply the ratio of the feedback resistor (R_f) value to the input resistor (R_{in}) value. The voltage gain equation for the inverting amplifier is

$$A_V = -R_f/R_{in}$$

The resistor R_1, connected between the noninverting input and ground, is used to balance slight input offset currents. It should have a value of

$$R_1 = R_{in}//R_f$$

The noninverting input may be connected directly to ground when input offset currents need not be balanced.

FIGURE 8-6   Differential Amplifier

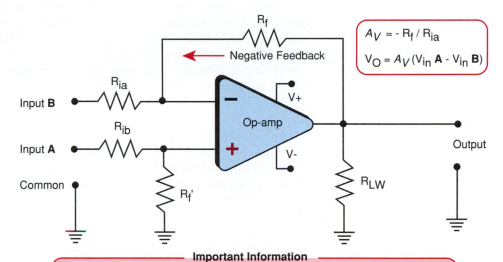

$$A_V = -R_f / R_{ia}$$
$$V_O = A_V (V_{in} \text{A} - V_{in} \text{B})$$

Important Information

In many op-amp circuits, adding a resistor from the noninverting input to ground is optional or included for a specific voltage divider function. R_f' in the differential amplifier configuration is essential to an acceptable *CMRR* (common-mode-rejection-ratio), and acceptable input circuit current balance. R_f' is actually part of the negative feedback circuit.

8.5   Differential Amplifier

The **differential amplifier** in Figure 8-6 uses both inputs as signal inputs, often with different signals on each input. The two inputs can also be connected to floating devices or circuits. A **floating device** or **floating circuit** is one in which neither connection is connected to ground. The output of a Wheatstone bridge circuit is an example of a floating circuit. Certain transducers (used to measure temperature, pressure and so on) may be connected to the two inputs of an operational amplifier instead of to one input and to ground. The differential amplifier amplifies the algebraic difference between the voltages applied to the two inputs.

The output voltage equation is

$$V_{out} = (V_{in}A - V_{in}B) \times A_V$$
$$= (V_{in}A - V_{in}B) \times (R_f/R_{in})$$

Resistor $R_{in}B$ is optional and is used to balance input offset currents. If $R_{in}B$ is included, it should have an approximate value of

$$R_{in}B = R_f//R_{in}A$$

In any op-amp circuit, this resistor may or may not be required, but it is more likely to be present when the absolute value of R_f or of R_{in} exceeds 0.5

megohm. The higher the resistance value(s), the more offset voltage is produced.

Noise pickup from the power line and from other sources can be a serious problem in many industrial and biomedical instrumentation systems. The signal levels from transducers and biomedical instrumentation can be very small and easily overwhelmed by stray electrical noise pickup. Even shielded cables often fail to provide sufficient protection against unwanted trash pickup. When signals must be transmitted over wires for a considerable distance, the problem grows even worse. One of the most useful applications of the differential amplifier involves canceling such noise components, while leaving the signal unaltered.

Figure 8-7 shows the differential amplifier connected to remove noise from a remote signal source. The wire from the signal source carries the signal, but it also picks up electrical noise. The signal source is connected to the noninverting input, so both signal and noise are applied to the noninverting input. The inverting input is connected to a wire that follows the same physical path as the signal wire, so it picks up the same noise signal as the signal wire. The wire from the inverting input is connected to a dummy signal source, which simulates the characteristics of the signal source but does not produce a signal.

The noninverting input receives the desired signal plus the noise signal. The inverting input receives only the noise signal, which it subtracts from the identical noise signal on the noninverting input. The noise signal is thus canceled at the differential input, but the desired signal is preserved. We will look at other differential amplifier applications in other sections of the text.

FIGURE 8-7   Using the Differential Amplifier to Remove Unwanted Noise

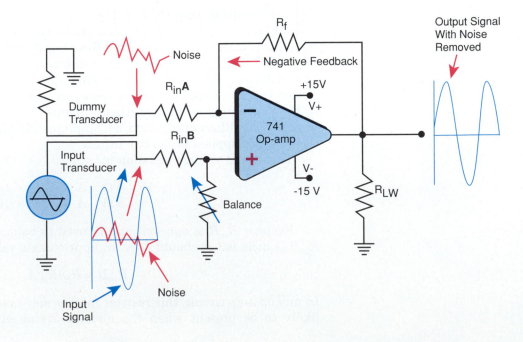

FIGURE 8-8   Inverting Summing Amplifier

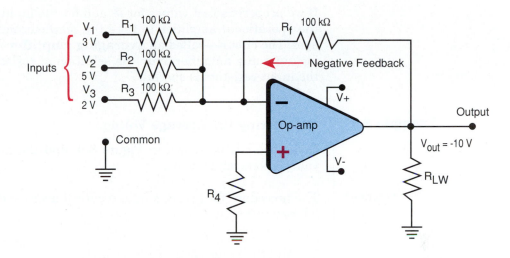

8.6   The Inverting Summing Amplifier (Adder)

The **inverting summing amplifier** in Figure 8-8 is capable of adding and amplifying several voltages. The input to each input resistor (R_1, R_2, R_3, and so on) has its own voltage gain ($A_V = R_f/R_{in}$). This allows the input voltages to be scaled—that is, individually multiplied or divided by some desired factor. The equations that follow illustrate the scaling possibilities of the summing amplifier. The minus sign in the equations indicates that we are dealing with an inverting amplifier.

The inverting input makes an almost ideal summing node, because it is a virtual ground; and the individual input resistor currents are entirely a function of the resistor values. The individual input resistor currents add in the amplifier input. The virtual ground ensures that component values in the amplifier input circuit have little or no influence over the summing operation.

The full equation for output voltage in this setting is

$$V_{out} = -[V_1 \times (R_f/R_1)] - [(V_2 \times (R_f/R_2)] - [V_3 \times (R_f/R_3)]$$

If we factor out R_f, we get

$$V_{out} = -R_f \times [(V_1/R_1) + (V_2/R_2) + (V_3/R_3)]$$

For the special case where all resistor values are equal ($R_1 = R_2 = R_3 = R_f$), the equation becomes

$$V_{out} = -(V_1 + V_2 + V_3)$$

Averaging Amplifier

If we properly select values for R_f and for the input resistors, we can program the operational amplifier to take the mathematical average of several voltages. The circuit—called an **averaging amplifier**—is the same as the inverting summing amplifier; only the selection of resistance values for R_f and for the input resistors is special.

Example 8.2

Calculating the Average Voltage

Given the circuit shown in Figure 8-9, find the average of three voltages: 5 volts, 3 volts, and 9 volts.

Solution

The procedure for taking a mathematical average of a group of numbers is as follows:

1   Add the values of all the numbers.
2   Divide the resulting sum by the number of numerical entries used to get the sum.

In our example, there are three digits to average. First, we get the sum of the three digits: 5 + 3 + 9 = 17. Second, we divide by the number of values to be averaged (three, in this case): 17/3 = 5.66 . . .

FIGURE 8-9   **Averaging Amplifier**

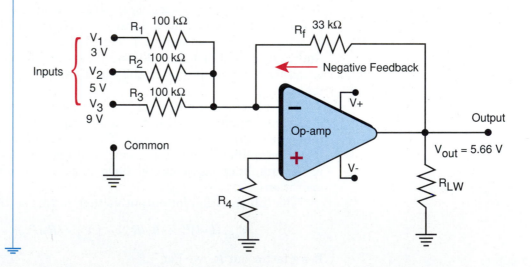

To implement the averaging amplifier circuit with an op-amp, we must have an op-amp configuration that is capable of division. Voltage gain is a multiplication operation. Dividing an input voltage by some factor is called **attenuation**. An operational amplifier will attenuate an input voltage if the value of the feedback resistor (R_f) is lower than the value of the input resistor.

Example 8.3

Dividing by Three

Suppose that we have a 17-volt input and a 100-kilohm feedback resistor, and we wish to divide by three. What value of input resistor (R_{in}) is required?

Solution

Let's start by reviewing the general voltage-gain equation involved and then substitute it into the equation for voltage output:

$$A_V = -R_f/R_{in}$$

$$\begin{aligned}
V_O &= -V_{in} \times (R_f/R_{in}) \\
&= -V_{in} \times [R_f/[R_f/(3 \times R_f)]] \\
&= 17\,\text{V} \times (100\,\text{k}\Omega/300\,\text{k}\Omega)
\end{aligned}$$

Thus, we need an input resistor value of 300 kilohms.

The averaging circuit in Figure 8-9 requires a value for R_f that is equal to one-third the value of each input resistor: $R_f = R_{in}/3$.

The general equations for the output voltage of an averaging amplifier is

$$V_{out} = -(V_{inA} + V_{inB} + V_{inC})/n$$

where n is the number of inputs, and where

$$R_{inA} = R_{inB} = R_{inC}.$$

8.7   Noninverting Summing Amplifier

The noninverting input can also be used as a summing input with the restriction that the voltage gain must be made equal to the number of inputs. Because the noninverting input is a high-impedance input, it is less than ideal as a summing node. The voltage gain restriction makes the circuit somewhat inflexible, so it is not often used. The circuit for the **noninverting summing amplifier** is shown in Figure 8-10.

8.8   Standard Op-amp Power Connections

The standard op-amp power connection uses two power-supply voltages, as shown in Figure 8-11(a). The output voltage of the op-amp, when the input voltage is zero, is 0 volts. The positive voltage provides power for the positive-going signal, and the negative power-supply voltage supplies power for negative-going parts of the waveform. The two power-supply voltages are generally derived from a pair of IC linear regulators—one positive and one negative. One pair of regulators can supply power for several op-amps (probably as many as there are in the system).

Figure 8-11(b) illustrates the use of two protective diodes to prevent damage to the op-amp in the event that the power-supply connections get reversed.

FIGURE 8-10   **Noninverting Summing Amplifier**

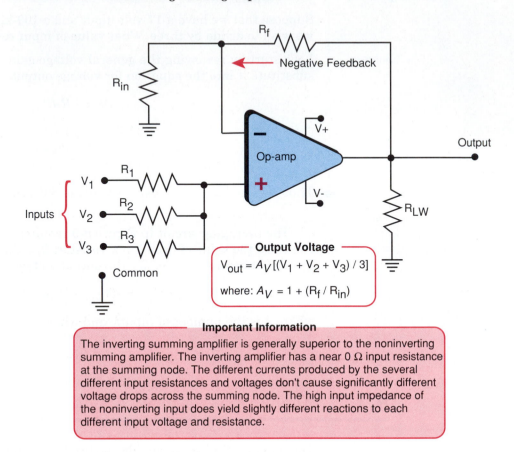

Output Voltage

$$V_{out} = A_V [(V_1 + V_2 + V_3) / 3]$$

where: $A_V = 1 + (R_f / R_{in})$

Important Information

The inverting summing amplifier is generally superior to the noninverting summing amplifier. The inverting amplifier has a near 0 Ω input resistance at the summing node. The different currents produced by the several different input resistances and voltages don't cause significantly different voltage drops across the summing node. The high input impedance of the noninverting input does yield slightly different reactions to each different input voltage and resistance.

FIGURE 8-11   **Standard Power-supply Connections**

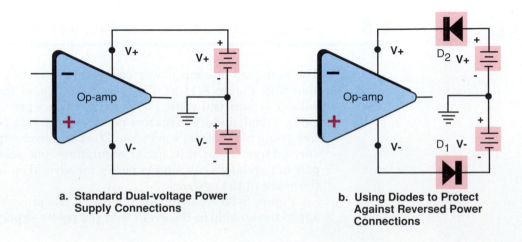

a. Standard Dual-voltage Power Supply Connections

b. Using Diodes to Protect Against Reversed Power Connections

Reverse polarity is often fatal for the op-amp(s). The diodes are particularly useful in battery-operated circuits, where polarity reversal is a real possibility. Including a couple of diodes in the circuit can be a cheap insurance policy.

8.9 Biasing Op-amps for Use with a Single-power Supply

A small number of op-amps are specifically designed for single-power-supply operation, but most op-amps are intended for dual-power-supply operation. Here are some examples of op-amps designed for single-supply operation:

Quad operational amplifiers: LM-124/224/324, LM-2902.
Dual operational amplifiers: LM-158/258/358, LM-2904.
Quad Norton operational amplifiers: LM-3900, LM-3301/3401.

All of the amplifiers listed are low-power devices and are particularly useful in battery-operated equipment.

Operating Conventional Op-amps with a Single-supply Voltage

Any op-amp can be biased for use with a single power supply if we are willing to accept the following conditions:

1   An input coupling capacitor must be used. The circuit will be an **ac**-only amplifier.
2   The output voltage with no signal input will be equal to one-half the supply voltage. It may be necessary to use an output coupling capacitor.
3   Frequency response limitations will be imposed by the introduction of coupling capacitors.
4   The bias resistors will lower the input impedance of the noninverting input.

When we ground the negative power-supply terminal on an op-amp, we no longer have bias for the normally emitter-biased differential amplifier transistors. We can no longer amplify negative-going signals. Even positive-going signals must overcome two junction voltages before we get an output signal. This means that positive input voltages of less than 1.4 volts are lost, including any part of a waveform that is less than 1.4 volts in amplitude.

To make the amplifier work properly, we must add a voltage-divider base-bias circuit. Figure 8-12(a) shows the transistor differential amplifier circuit with added base bias. Remember, the input circuit of the op-amp is a transistor differential amplifier, implemented as part of an integrated circuit. Figure 8-12(b) shows a noninverting op-amp biased for single-supply operation. Figure 8-12(c) shows an inverting op-amp biased for single-supply operation.

FIGURE 8-12 Using a Single Power-supply Voltage with Op-amps

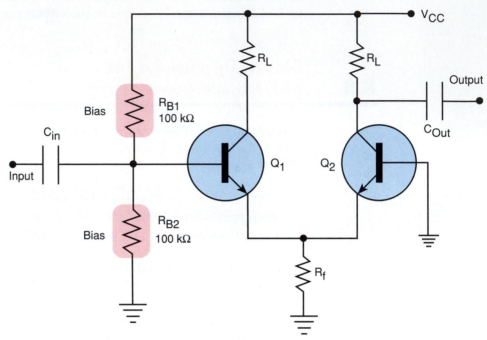

a. Biasing the Differential Amplifier Using Voltage-divider Base Bias

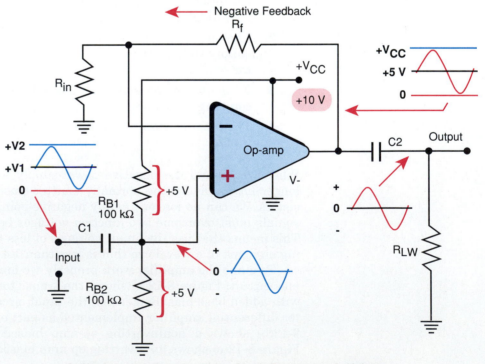

b. Noninverting Op-amp with a Single Power-supply Voltage

FIGURE 8-12   Continued

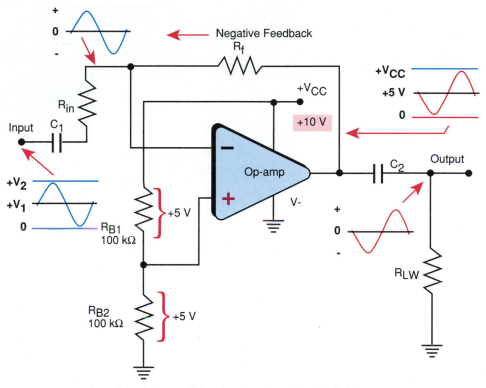

c. Inverting Op-amp with a Single Power-supply Voltage

Since we are using voltage-divider bias, we can use the familiar voltage-divider equation to calculate the output offset voltage. The output offset voltage is the quiescent output voltage, and an input signal will cause the output voltage to swing above and below the offset voltage value. The equation is

$$V_{out\,(\mathbf{dc})} = V+ \times [(R_{B2})] / (R_{B1} + R_{B2})]$$

It is customary to make R_{B1} and R_{B2} equal values, even though this practice results in a slightly off-center quiescent output voltage. The maximum peak-to-peak output voltage is approximately 2 volts less than the power-supply voltage (+V).

Voltage Follower Using a Single Power Supply

The **voltage follower** can be used with a single power supply, as shown in Figure 8-13(a). The problem here is that the high input impedance of the non-inverting input is one of the main reasons for using the voltage follower, and the bias resistors significantly lower the input impedance. Because the circuit is already an **ac**-only amplifier, and since coupling capacitors are allowed, we can use our old bootstrapping trick to get the input impedance back up. Figure 8-13(b) shows how.

Figure 8-13 ac Voltage Followers

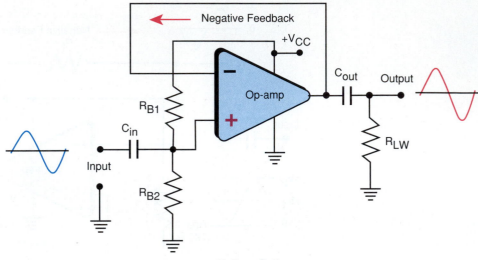

a. ac Voltage Follower

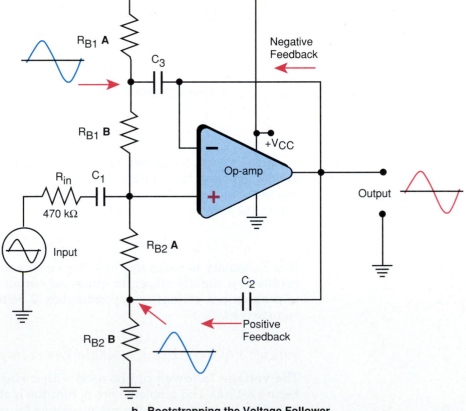

b. Bootstrapping the Voltage Follower

8.10   Adding Gain Control to Op-amps

There are several ways to provide user control over the effective voltage gain of an op-amp. A simple voltage divider can be used to divide the input signal voltage before it reaches the amplifier. The voltage divider attenuator is not a true gain control, but it serves the same purpose. True gain control requires control of the feedback loop. A special nonlinear potentiometer is generally used in audio systems to match the dB characteristics of the human ear more closely. Linear gain control is used in most other applications.

Voltage-divider Signal Attenuator

The traditional method for controlling input signal amplitude is to use a common voltage-divider potentiometer as an attenuator for the signal. The attenuated signal is applied to the input of the op-amp, allowing volume control in audio amplifiers and amplitude control in other circuits. Potentiometers are available with linear tapers and log tapers used for audio volume controls. The **log-tapered potentiometer** generates a curve that follows the logarithmic decibel (dB) scale, which is based on the sound-level response of the human ear. Log-tapered potentiometers are called *audio-tapered potentiometers*. The voltage-divider attenuator is shown in Figure 8-14(a). This method does not vary the gain of the amplifier; it simply controls the amplitude of the signal delivered to the amplifier input.

Voltage-gain Control Using the Feedback Voltage

The circuit shown in Figure 8-14(b) controls the voltage gain of the amplifier instead of attenuating the input signal. A potentiometer is used to control the percentage (fraction) of the output voltage used to supply the negative-feedback voltage loop. The potentiometer may be a linear- or an audio-tapered control. This circuit is a true variable voltage-gain control.

Audio-taper Control Using a Linear Potentiometer

The circuit in Figure 8-14(c) shows an audio gain/attenuator circuit in which an audio attenuation curve is produced with a linear-tapered potentiometer. The curve generated is a good approximation of the dB curve. The attenuation curve for an audio-tapered device is described by the following equation:

$$\text{Attenuation [dB]} = 20 \log (V_{in}/V_{out})$$

Linear potentiometers are less expensive than audio-tapered potentiometers; but just as important, high-quality linear potentiometers are easier to find than high-quality audio-tapered potentiometers. Placing the control in the feedback loop helps minimize the noise generated by the sliding element in the potentiometer.

FIGURE 8-14 **Gain-control Techniques**

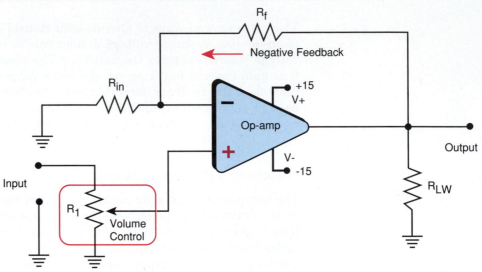

a. Simple Input Voltage Divider to Control Input Voltage

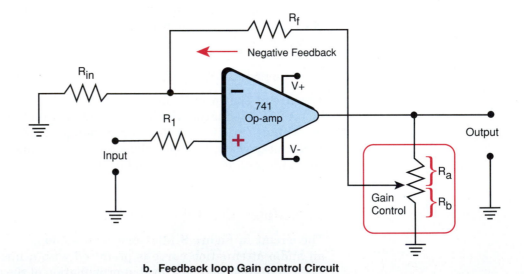

b. Feedback loop Gain control Circuit

Using Analog Gates for Remote Or Noise-free Gain Control

It is common practice on professional audio mixing boards to use an analog gate as a variable resistor to control amplifier gain and attenuation. The advantages of the analog gate are virtually noise-free operation, no mechanical wear, and the ability to control the amplifier remotely. A potentiometer is still used to control the analog gate, but the noise-free analog gate controls the

FIGURE 8-14 Continued

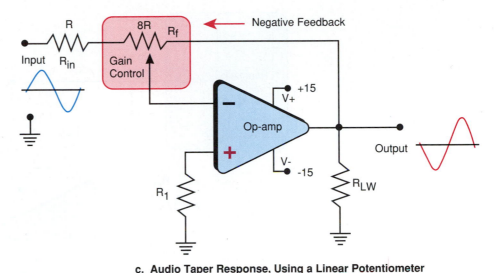

c. Audio Taper Response, Using a Linear Potentiometer

amplifier gain. Potentiometer noise does not get into the system. Because the control voltage is **dc,** it is relatively easy to keep power line noise out of the system, even for long remote-control lines.

The 4016/66 quad analog gate can control analog signals with peak-to-peak voltages of up to $V_{DD} + V_{SS}$. The circuits in Figure 8-15 show two ways of connecting the analog gain/attenuator control. In Figure 8-15(a), the analog gate is connected in the feedback loop, and it functions as a variable feedback resistance. In this case, the gain is controlled by varying the feedback resistance.

In the circuit in Figure 8-15(b), the analog gate substitutes for the input resistor (R_{in}) and serves as a variable input resistance.

8.11   Output Offset Voltage: Its Causes and Cures

Imperfections in the balance of the input circuits of a differential amplifier, together with finite input impedances, give rise to small unwanted input currents and voltages when there is no signal present. The amplifier treats these unwanted input voltages and currents as valid **dc** input signals. The op-amp amplifies the unwanted false input signals, and they appear in the output as a nonzero output voltage. Ideally, when no signal is applied to the amplifier input, the output voltage should be exactly 0 volts. The small nonzero output voltage, with no signal, is called **offset voltage.** In many cases, the small offset voltage is tolerable and presents no problem. In other op-amp applications, the offset voltage is a problem and must be eliminated.

FIGURE 8-15   Noise-free Gain Control Using an Analog Gate

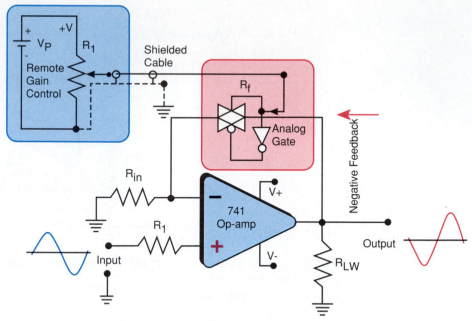

a. Using the Analog Gate as a Feedback Resistor

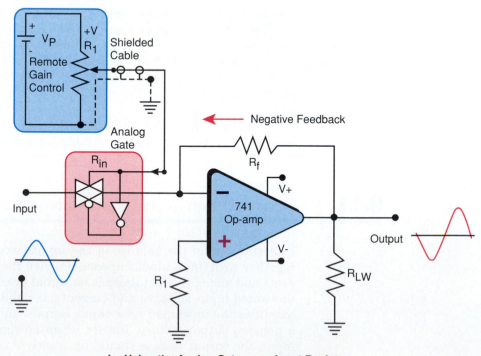

b. Using the Analog Gate as an Input Resistor

Three Causes of Output Offset Voltage

There are three main causes of output offset voltage: input bias current, input offset current, and input offset voltage. We will look at each of these factors separately in the subsections that follow.

Input Bias Current (I_{ib})

The intrinsic input impedance of commercial integrated-circuit op-amps is very high, but it is not infinite. And because the input resistance is not infinite, some small current flows into both inverting and noninverting inputs. The average of the inverting and noninverting input currents is called the **bias current** (see Figure 8-16).

The equation for input bias current is

$$I_{ib} = (I_{bias}I + I_{bias}N)/2$$

where $I_{bias}I$ is the inverting input current and $I_{bias}N$ is the noninverting input current.

When there is no input signal, the bias current flows through the input and feedback resistors. The bias current acts as a small but unwanted input signal, which is then amplified and appears in the output as an offset voltage. The output offset voltage resulting from the bias current is defined by the following equation:

$$V_{OS} = I_{ib} \times R_f$$

Input Offset Current (I_{iO})

Input offset current is another contributor to the unwanted output offset voltage. The input offset current (I_{iO}) is the difference between the inverting input

FIGURE 8-16 **Circuit for the Input Offset Voltage Discussion**

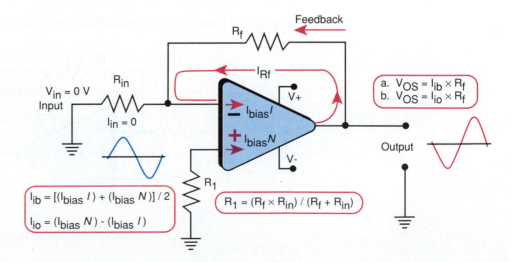

and the noninverting input bias currents. The input offset current is defined by the equation

$$I_{iO} = (I_{bias}N) - (I_{bias}I)$$

where $(I_{bias}N)$ is the noninverting input bias current, and $(I_{bias}I)$ is the inverting input bias current.

The output offset voltage resulting from the input offset current is defined by the following equation:

$$V_{OS} = I_{iO} \times R_f$$

Because the input offset currents can push the output voltage in either the negative or the positive direction and can either add to or subtract from other offset factors, the offset effect is not predictable in practice. We can, however, reduce the bias-current-produced output offset voltage to near zero by adding a resistor to the noninverting input. The resistor value should be approximately equal to the parallel combination of R_f and R_{in}:

$$R_1 = R_f // R_{in}$$

(see Figure 8-15).

Input Offset Voltage (V_{iO})

The input offset voltage is caused by a voltage imbalance in the input circuit and can be simulated by the battery shown in Figure 8-17. Balancing the external input resistors is no help in reducing the input offset voltage and its resulting output offset voltage. We must either live with the output offset voltage or provide some kind of external balancing circuit. We will look at some balancing techniques next.

The output offset voltage is caused by a combination of the factors we have already discussed. The magnitude of the offset voltage is the sum of the results

FIGURE 8-17   Circuit for the Input Offset Voltage Discussion

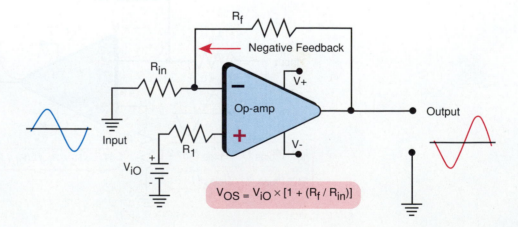

FIGURE 8-18   **Offset Null Circuit for the 741 Op-amp**

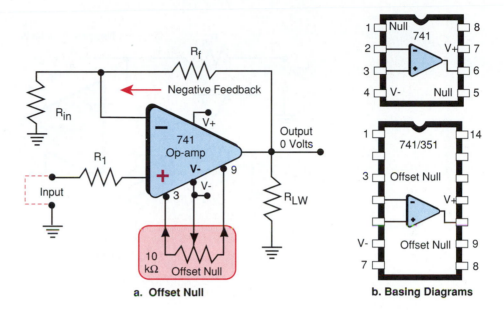

a. **Offset Null**

b. **Basing Diagrams**

of the three factors. The amount of output offset voltage resulting from the input offset voltage is defined by the equation

$$V_{OS} = V_{iO} \times [1 + (R_f/R_{in})]$$

Because the system is linear, the total offset voltage is the sum of the three individual output offset voltages resulting from I_{ib}, I_{iO}, and V_{iO}.

Offset Null and Balance Compensation Techniques

Operational amplifiers such as the 741 make the addition of a balance control easy. The only additional element required is an inexpensive trimming potentiometer. Any unwanted output offset voltage, regardless of its source, can be nullified by adjusting the potentiometer. In the case of the 741 op-amp circuit in Figure 8-18, the potentiometer allows an adjustment range of ±15 millivolts—more than enough to nullify any normal output offset voltage.

Integrated-circuit op-amps that don't have special provisions for an offset null potentiometer can be balanced by using add-on circuits like the ones shown in Figure 8-18. The add-on circuits require extra resistors in addition to the adjustment potentiometer. The extra resistors are needed to limit the adjustment range of the potentiometer to 10 to 15 millivolts. Otherwise, the adjustment would be too sensitive to adjust properly.

It is good practice to use the add-on circuit specified by the manufacturer for the particular op-amp being used. The examples shown in Figure 8-19 represent typical add-on null circuits.

FIGURE 8-19   Add-on Offset Null Circuits

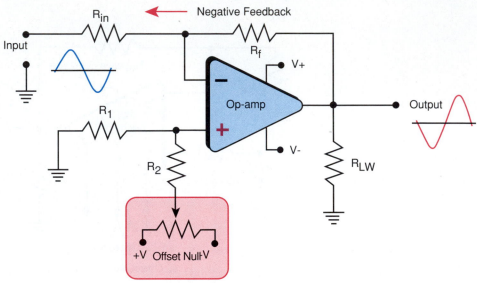

a. Inverting Amplifier Version

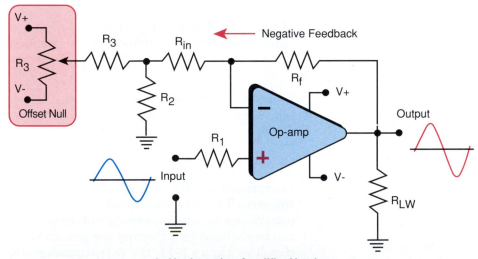

b. Noninverting Amplifier Version

8.12   Norton Current-differencing Operational Amplifier

The **Norton operational amplifier** is a very different kind of op-amp from the traditional differential-amplifier based op-amp. The Norton op-amp does not use the conventional transistor differential-amplifier circuit, and it has characteristics very different from those of the conventional op-amp. The

FIGURE 8-19 **Continued**

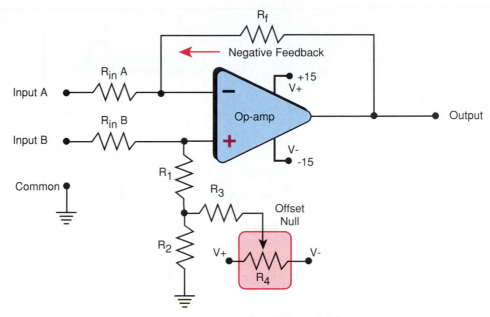

c. Offset Null Add-on for the Differential Amplifier

Norton amplifier is primarily designed for single-power-supply, low-power operation, but it can be used with dual power supplies.

Although most ordinary op-amp circuits can be implemented with the Norton amplifier, the Norton is not considered a general replacement for the conventional op-amp. Many of the Norton task-programmed circuits appear to be the same as the circuits for conventional op-amps, but resistance values in Norton circuits tend to be much higher, and sometimes they serve a different purpose than do their counterparts in conventional op-amps circuits.

The Norton's ability to operate on a single low-voltage, low-current power supply makes it ideal for many battery-operated systems. Unlike conventional op-amps biased for single-supply operation, the Norton can operate from a single power-supply voltage that ranges from 4 volts to 36 volts, with a peak-to-peak output voltage that is about 1 volt less than the supply voltage.

Let's examine the simplified Norton circuit in Figure 8-20(a). The inverting input consists of a single common emitter amplifier with a grounded emitter (Q_2). The common emitter inverting input amplifier uses a constant-current source as a collector load (R_L). The feedback resistance for Q_2 is only R'_e—a few tens of ohms—and the constant-current source provides a high collector-load resistance (R_L). The common-collector output stage (Q_3) also uses a constant-current source (R_f) as an emitter-load/feedback resistance, which makes the input impedance of the output stage very high as well.

The single gain stage (Q_2) in the Norton amplifier yields a voltage gain of about 3000 (70 dB), as compared to the conventional 741 op-amp's voltage gain of 200,000 (110 dB). The lower gain of the Norton is nonetheless adequate for the kinds of tasks the Norton amplifier is normally assigned.

FIGURE 8-20   Norton Differencing Op-amp

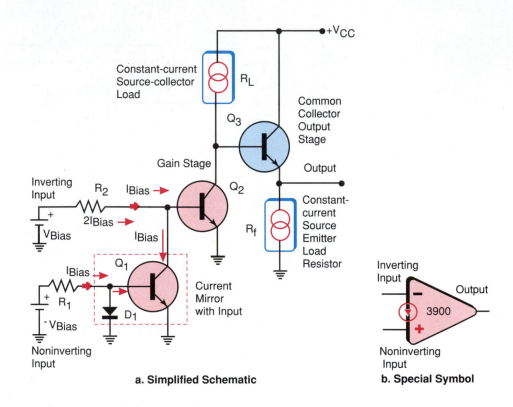

a. Simplified Schematic   b. Special Symbol

So far, we have identified nothing very unusual about the circuit except the fact that constant-current sources are used in place of the dummy collector load and current-mode feedback resistors. Most conventional op-amps also use some constant-current sources for collector loads and emitter-feedback resistances.

How the Noninverting Input Is Formed

Every op-amp must have both inverting and noninverting inputs, and the way in which the Norton amplifier derives its noninverting input is what makes the Norton circuit unique. To begin with, the gain stage (Q_2) does not have a current-mode feedback resistor in the emitter circuit. The Q-point of Q_2 is stabilized by Q_1, which acts as a current mirror (constant-current circuit). Diode D_1 tracks temperature variations to stabilize Q_1. Both transistors (Q_1 and Q_2) must be biased to an appropriate Q-point on the strength of bias current provided for the base–emitter junctions of Q_1 and Q_2. The only source for the bias current is the inverting input for Q_2 and the noninverting input for Q_1. Bias resistors must be provided outside the IC. The two inputs serve as both bias and signal inputs.

The Norton amplifier circuit is called a **current-differencing amplifier** to distinguish it from the more common differential amplifier. Unlike in the differential amplifier, the Norton's two inputs have very low input resistances, which makes them current inputs rather than voltage inputs.

Input Circuit Conditions

The external bias circuit, consisting of a bias voltage and bias resistors R_1 and R_2, must first overcome the 0.6 to 0.7-volt junction voltages in transistor Q_1 and diode D_1. Once the positive diode D_1 is turned on, it turns on transistor Q_1 and clamps the noninverting input to 0.6 to 0.7 volt. Likewise, when Q_2's base–emitter junction is turned on, it clamps the inverting input to 0.6 to 0.7 volt. Because both inputs are forward-biased junction diodes, the maximum input signal voltage swing is limited to a few tens of millivolts, peak-to-peak.

Because both inputs in Figure 8-20(a) are diode junctions, a high-value current-limiting resistor must be placed in series with the bias supply voltage and each input. A similar high-value resistor is nearly always placed between the amplifier input and the signal source. In both cases, the added input resistors convert the Norton's current inputs into voltage inputs with relatively high input resistances. Figure 8-20(b) shows the Norton amplifier symbol. The arrow always points to the noninverting input and indicates that the inputs are current-operated.

How the Current Differencing Works

The gain stage (Q_2) in Figure 8-20 always amplifies the difference between the two input currents. For any current flowing into the inverting input, a part of that inverting input current is bypassed to ground through Q_1 as long as Q_1 is turned on.

The amount of current bypassed through Q_1's collector–emitter circuit depends only on Q_1's base–emitter current; it is independent of any action on Q_2's part.

Any increase in Q_1's base–emitter current causes a reduction in Q_2 base current and a consequent reduction in Q_2's collector–emitter current.

The additional current demanded by Q_1 must be drawn (and subtracted) from the inverting input current. This subtraction from the inverting input current, in response to variations in the noninverting input current, gives the circuit the *current differencing* part of its name.

Transistor Q_2 is biased to a desired Q-point by selecting a value for R_1, as in Figure 8-20. The collector current of the gain stage (transistor Q_2) can then be altered by having some of its base current drawn off. Once the bias values for both Q_1 and Q_2 have been established, an **ac** signal can be capacitor-coupled into either or both inputs.

Figure 8-21 shows a circuit model of a noninverting Norton amplifier. Notice that the value of bias resistor R_2 is half the value of bias resistor R_1. We have established some bias current ($2I_{bias}$), then biased the noninverting input to bleed off half of the original inverting input bias current. The two input bias currents are set at I_{bias}.

Analysis Equations for the Norton Amplifier

Let's examine the basic equations for the Norton amplifier. Then we will use the values in Figure 8-22 as examples of how to use the equations. We will specify the 3900 quad Norton integrated-circuit op-amp, but the equations are generic and apply to any Norton amplifier. The 3900 is the industry standard,

FIGURE 8-21   Norton Amplifier Input Circuit Bias and Signal

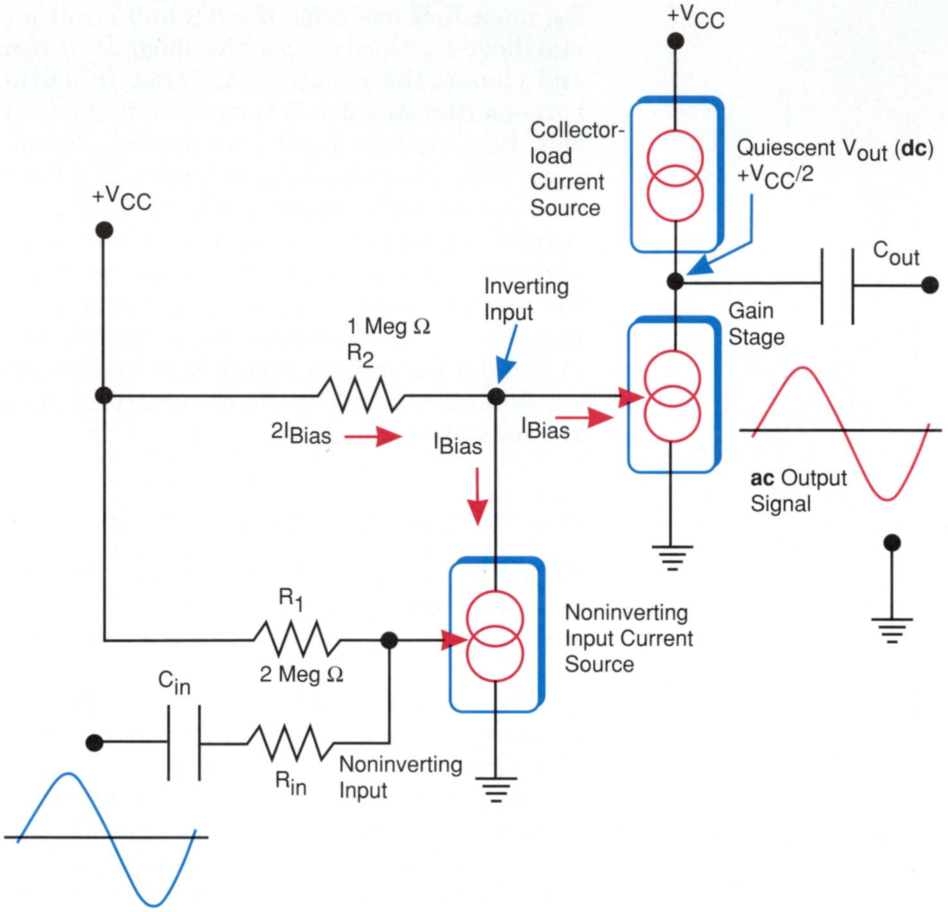

and there are not many other choices. The relatively low power dissipation and the simplicity of the Norton circuit allow the manufacturer to put four independent Norton op-amps in a single 14-pin DIP package. The four amplifiers share the two power pins (V+ and ground) but are otherwise independent.

Finding the Quiescent dc Output Voltage ($V_{out(\text{dc})}$)

$$V_{out(\text{dc})} = +\text{V} \ (R_f/R_{bias})$$
$$= +12 \text{ V} \ (1 \text{ Meg}\Omega/2 \text{ Meg}\Omega)$$
$$= +12 \text{ V} \times 0.5$$
$$= +6 \text{ V}$$

Biasing the amplifier for **ac** signal amplification nearly always uses midpoint biasing, where the quiescent **dc** output voltage is set to one-half the power-supply voltage. This equation applies to both noninverting (Figure 8-22 (a)) and inverting (Figure 8-22(b)) versions of the circuit.

FIGURE 8-22 Practical Norton Op-amp Circuits

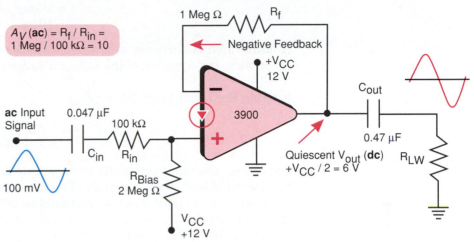

A_V (ac) = R_f / R_{in} =
1 Meg / 100 kΩ = 10

a. **Noninverting Norton Op-amp Circuit**

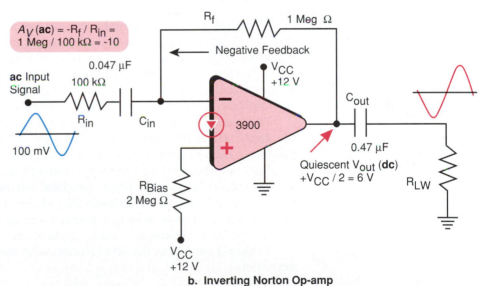

A_V (ac) = -R_f / R_{in} =
1 Meg / 100 kΩ = -10

b. **Inverting Norton Op-amp**

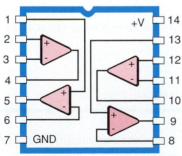

c. **3900 Norton Op-amp Basing Diagram**

Important Information

Coupling capacitor values are typically much lower than those used in bipolar op-amps. For input capacitors, the capacitive reactance should be about one-tenth the value of the circuit input impedance at the lowest frequency you intend to amplify.

Output coupling capacitors are often less certain, because we don't always know the impedance of the working load. A good rule-of-thumb is to make the output capacitor value ten times the value of the input capacitor.

Finding the ac Voltage Gain (A_V)

$$A_V = R_f/R_{in}$$
$$= 1 \text{ Meg}\Omega/100 \text{ k}\Omega$$
$$= 10$$

A minus sign is normally used to indicate the gain in an inverting amplifier. Otherwise, the equation is the same for inverting and noninverting versions.

Finding the Input Impedance (Z_{in})

$$Z_{in} = R_{in}$$
$$= 100 \text{ k}\Omega$$

Remember, the input terminals to the Norton amplifier are connected directly to forward-biased diode junctions. The input resistance of the junctions is so low that we can call it 0 ohms, compared to 100,000 ohms. In all real voltage-amplifier circuits, R_{in} will be high enough for this assumption to be valid.

The Miller effect is active in this circuit, with the negative feedback driving the input impedance of the inverting input terminal to virtual ground. The virtual ground doesn't alter the situation much, however, since the input resistance was already so small.

When we discussed summing amplifiers a few pages back, we noted that the noninverting input was not very useful as a summing junction. The noninverting input in conventional op-amps is a high-impedance input. In the case of the Norton, both inputs are low-impedance inputs, so either input can serve as a summing junction. An audio mixing-board manufacturer devised a very clever mixer/fader circuit by using both inputs of a Norton as summing inputs.

The Norton amplifier sounds pretty good, and it is—as long it is used in the applications for which it is best suited. A major drawback is the Norton's relatively high output impedance of about 8 kilohms, as compared to the low 75 ohms of output impedance of the conventional 741 op-amp. The Norton's 8 kilohm output impedance limits practical output drive currents to something less than 5 mA, and working loads (R_{Lw}) to values of 10 kilohms or higher. The Norton is a low-power device, which is a strength for battery operation and so on but can be a weakness in other applications.

Table 8-1 compares the 3900 Norton with the 741 conventional operational amplifier.

8.13   Operational-amplifier Frequency Response

There are actually two sets of op-amp frequency response characteristics: one for small-amplitude signals, and another for large-amplitude signals.

When **ac** output signals are less than about 1 volt *P–P*, a common frequency-response graph of voltage gain versus frequency defines the amplifier's upper frequency limits. In addition, noise levels may become important at these low-level output signal voltages. The upper frequency limit for an op-amp is primarily determined by internal capacitances—often a 30-picofarad

TABLE 8-1

3900 and 741 Op-amps Compared

	Open-loop Voltage Gain	Input Impedance	Output Impedance	Normal Power-supply Voltage
3900	3000	near 0 Ω to 1 Meg Ω with input resistor	8 kΩ	4 to 36 V or ±2 to ±18 V
741	200,000	2 Meg Ω	75	±9 to ±15 V

(typical value) capacitor deliberately put on the chip to prevent unwanted oscillations from occurring under certain conditions. The addition of this capacitor is called **frequency compensation,** and some form of frequency compensation is required in nearly all op-amp circuits.

When the op-amp's output voltage must make large voltage excursions—roughly in excess of 2 volts—a second factor called **slew rate** becomes the dominant upper frequency limit. The term *slew* means a very rapid movement from one output voltage to another. For example, you might slew around if startled or frightened by a sound behind you. When the op-amp output voltage changes (as it follows a sine waveform, for example), the output voltage must be able to change fast enough to follow each change in the waveform's direction or slope. In other words, the output voltage must be able to slew at a fast enough rate to keep up.

Slew-rate Concept

Let's look at the example in Figure 8-23. The waveform shown in part (a) has an amplitude of 1 volt and a period for the complete waveform of 1 microsecond. In order for the op-amp to produce an output waveform, its output voltage must change, faithfully following the waveshape. The arrows show the changes in output voltage. Arrow 1 indicates an output voltage rise from 0 volts to 1 volt. After a stable period R, arrow 2 describes a second 1-volt transition, from +1 volt to 0 volts. After another stable period R, the waveform makes a third 1-volt transition, as indicated by arrow 3.

The output voltage has changed three times, each time by 1 volt. Since it has changed by a total of 3 volts during a 1-microsecond time period, its rate of change is identified as 3-volts per microsecond.

The waveform shown in part (b) of Figure 8-23 has a frequency twice that of the waveform in part (a), so obviously the output voltage must change more times during the same 1-microsecond time period. In fact, five 1 volt changes occur during the 1-microsecond time period, so the output voltage has changed by a total of 5 volts during the microsecond. This represents a rate of change of 5 volts per microsecond. The output voltage must be able to change even faster if the frequency is increased.

The waveform shown in part (c) of Figure 8-23 has the same frequency as the waveform in part (b), but with an amplitude of 2 volts instead of 1 volt. The

Fɪɢᴜʀᴇ 8-23   **Slew Rate Concept**

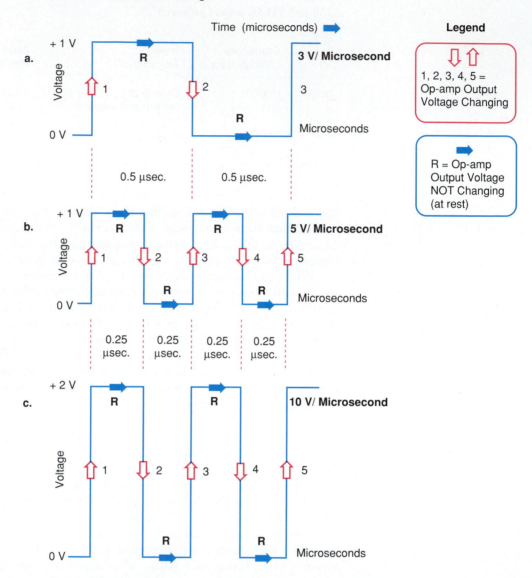

op-amp output voltage must be able to slew faster for the 2v waveform than for the 1v waveform. The output voltage still makes five changes, but now each change is 2 volts. The total voltage change that the output voltage must undergo in 1 microsecond is now 10 volts. This represents a rate of change of 10 volts per microsecond.

Figure 8-23 illustrates the concept of slew rate and demonstrates how frequency and peak-to-peak output voltage relate to one another. In practice, the amplifier's slew rate must be fast enough to follow the rise and fall of each cycle. If the amplifier's slew rate is too slow to trace accurately the rise and fall of a waveform, the waveform becomes distorted, as illustrated in Figure 8-24. The manufacturer of an op-amp normally uses a square or rectangular wave

FIGURE 8-24   **What Happens to a Sine Wave When the Slew Rate Is Too Slow**

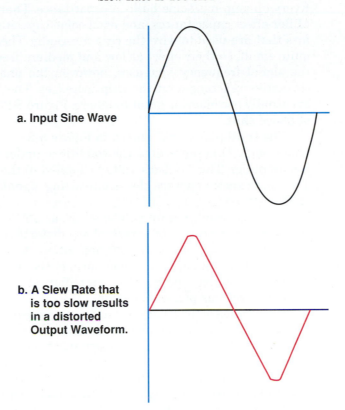

a. Input Sine Wave

b. A Slew Rate that is too slow results in a distorted Output Waveform.

to measure the device's slew rate. It is very hard to come up with standard measurement parameters for slew rate when a sine wave is used. The rise and fall of a square or rectangular waveform are often referred to as **step functions.**

The 741 op-amp has a slew-rate specification of 0.5 volt per microsecond. The maximum frequency at which we can expect a sine-wave signal to remain undistorted by slew-rate limitations can be calculated by using the formula

$$F_{max} = \text{Slew Rate}/(6.28 \times V_{out})$$

where F_{max} is the maximum undistorted frequency in (Hz), V_{out} is the peak output voltage, and slew rate is expressed in volts per microsecond.

Example 8.4 | **Maximum Undistorted Frequency**

What is the maximum undistorted frequency for a peak output voltage swing of 10 volts? The slew rate is specified as 0.5 volts per microsecond.

Solution

$$F_{max} = \text{Slew Rate}/(6.28 \times V_{out})$$
$$= [0.5 \text{ V}/10^{-6} \text{ sec}]/(6.28 \times 10)$$
$$= 8 \text{ kHz}$$

Phase Shift in an Op-amp

Every op-amp has some built-in capacitance. There are junction capacitances, Miller-effect capacitances, and even some capacitances between microconductors that are used to wire the chip internally. These capacitances can be kept quite small, so their effect at low and medium frequencies is insignificant. As the signal frequency increases, however, the phase shift resulting from the capacitive reactances on the chip increases. The total amount of phase shift eventually becomes too great to ignore. Figure 8-25 illustrates the significance of phase shift.

The third phase shift shown in Figure 8-25 is the near 0° op-amp internal phase shift. This represents the conditions under which we normally operate the amplifier. The feedback voltage applied to the inverting input is inverted 180° and exactly opposes the noninverting signal, as is required for negative feedback.

The next condition up is labeled "Near 90°" on Figure 8-25. Actually, this represents a condition of severe phase distortion, because the negative feedback signal is no longer in 180° opposition. As a result, the feedback signal opposes the input signal in some parts of the waveform but not in others. We ordinarily do not operate the amplifier at a frequency high enough to encounter serious phase distortion.

If we increase the frequency still more, we eventually reach the point where the total internal op-amp phase shift amounts to a full 180° shown in the top of Figure 8-25. If we were to input a signal at this frequency into the noninverting input, the input signal would be inverted 180° by the the time it got through the amplifier. If we then fed that signal back to the inverting input, the signal would get inverted 180° again. The total phase shift would be 360°, so the feedback signal would add to the input signal instead of opposing it.

If the feedback signal adds to the input signal, the output signal voltage tends to increase with each cycle as the feedback voltage is added to the input signal voltage. The output signal, fed back through the feedback resistor, becomes its own input signal and the operation becomes self-sustaining, even if the original input signal that started the process is removed. This phenomenon is called **oscillation.** The output signal is fed back to the input in phase and is continuously recirculated. We might set up such a positive-feedback oscillator circuit on purpose if we needed a signal generator.

Oscillation is fine when we need it for an application; but unwanted oscillation, over which we have little or no control, is nearly always disastrous to what we are trying to accomplish.

In the case of the 180° internal op-amp phase-shift situation, we might try to be very careful not to use any input signal with a high enough frequency to reach the 180° phase shift required to start the oscillation, but our efforts ultimately won't help. If there is any energy in the system—from any source—at the frequency needed to start oscillation, the circuit will oscillate. There is always some electrical noise in the system, due to random electron motion, and that electrical noise contains at least a little energy at all frequencies. Noise energy is enough to start the oscillation; and once started, it continues.

If unwanted oscillation does occur, you are most likely to see it when you try to evaluate the desired output signal. Figure 8-26(a) shows approximately what output signal you can expect to see when unwanted oscillation and your

FIGURE 8-25   Capacitance and Phase Shift in an Op-Amp

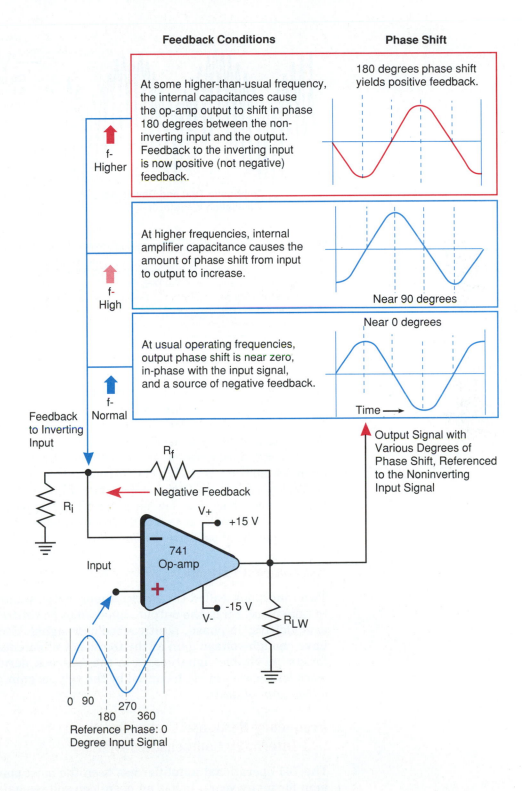

Feedback Conditions

f-Higher

At some higher-than-usual frequency, the internal capacitances cause the op-amp output to shift in phase 180 degrees between the non-inverting input and the output. Feedback to the inverting input is now positive (not negative) feedback.

f-High

At higher frequencies, internal amplifier capacitance causes the amount of phase shift from input to output to increase.

f-Normal

At usual operating frequencies, output phase shift is near zero, in-phase with the input signal, and a source of negative feedback.

Phase Shift

180 degrees phase shift yields positive feedback.

Near 90 degrees

Near 0 degrees

Time →

Feedback to Inverting Input

Output Signal with Various Degrees of Phase Shift, Referenced to the Noninverting Input Signal

R_f

R_i

Negative Feedback

$V+$ +15 V

741 Op-amp

Input

$V-$ −15 V

R_{LW}

0   90
180   270   360
Reference Phase: 0 Degree Input Signal

FIGURE 8-26   Unwanted Oscillations in an Op-amp

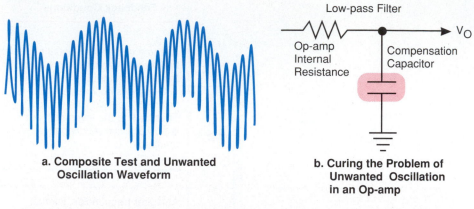

a. Composite Test and Unwanted Oscillation Waveform

b. Curing the Problem of Unwanted Oscillation in an Op-amp

Important Information

The standard audio test frequency is 1 kHz, while the unwanted oscillations are likely to happen at frequencies in the MHz range. Because of the difference in the two frequencies, you will probably not be able to see both waveforms with the same horizontal time base setting on the oscilloscope.

desired signal are both present. The desired signal has a much lower frequency than the unwanted oscillations, so the oscillations may at first appear on the scope as a blur. If you change the time base setting on the scope, you will be able to measure the oscillation frequency.

The standard cure for unwanted oscillations is simply to filter out all frequencies above a certain value. An added capacitor, called a *frequency compensation capacitor,* forms a low-pass filter network (see Figure 8-26(b)). Frequencies above the low-pass cut-off are bypassed to ground. Those frequencies don't make it to the feedback resistor, so there is no energy to feed back at potential oscillating frequencies.

Barkhausen Criteria

Two conditions, called the **Barkhausen criteria,** are required for oscillation to take place. First, the output signal must be returned to the input as a voltage-additive, in-phase, positive-feedback signal. Second, the amplifier must have enough voltage gain at the frequency of potential oscillation to make up for the inevitable signal losses in the feedback network. If the feedback network losses are exactly balanced by the voltage gain, the circuit is said to have a *loop gain of unity.*

Frequency Response Curves for the 741 Internally Compensated Op-amp

The 741 operational amplifier has been the most popular general-purpose op-amp for many years. It has an open-loop voltage gain for **dc** (and up to a few

FIGURE 8-27 Open-loop Frequency Response for the 741 Op-amp

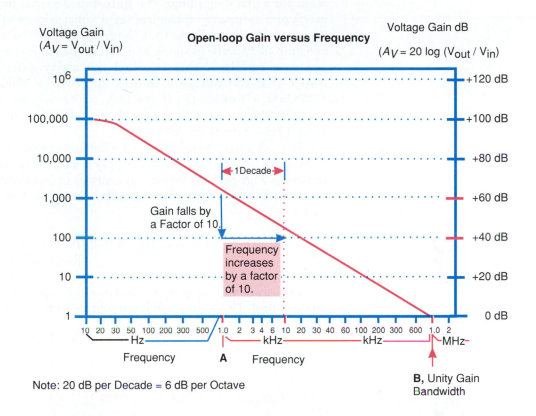

Note: 20 dB per Decade = 6 dB per Octave

hertz) of about 200,000. Figure 8-27 is a graph of the open-loop voltage gain versus frequency for the 741 op-amp. The 200,000 voltage-gain figure is only a statistical value, and it is specified as a **dc** value.

Assume that a particular 741 has a **dc** open-loop gain of 200,000. If we begin to apply an **ac** signal to the input and slowly increase the frequency, the voltage gain will begin to go down as the frequency increases. The point at which the voltage gain drops to $0.707 \times 200,000$, is termed as the **break point** between **dc** and **ac** operation of the amplifier. The break point is defined as the frequency at which the voltage gain is 3 dB below the open-loop **dc** voltage gain. The break-point frequency is typically around 5 Hz.

Notice that the voltage gain, as shown in Figure 8-27, decreases at a rate of 20 dB per decade. The slope of the graph in Figure 8-27 is 20 dB per decade, and the slope is the result of the 741's internal frequency-compensation capacitor. The compensation capacitor is part of a voltage divider (see Figure 8-26 (b)), and we should expect the output voltage to go down as the capacitive reactance goes down.

If the frequency increases by a factor of ten, the capacitive reactance decreases by a factor of 10 and the amplifier output voltage drops by a factor of ten. If we do not change the input voltage of the amplifier (when we increase the frequency), the drop in output voltage is really a drop in voltage gain. The fact that an increase in frequency results in proportional drop in voltage gain leads to the concept of the *gain-bandwidth product*.

The product of the voltage gain times the frequency always yields a constant for a given amplifier. The **gain-bandwidth product** is a basic figure of merit and is usually measured at a voltage gain of unity (1). If you look at Figure 8-27, you will see that the frequency response graph crosses the voltage gain of 1 (0 dB) baseline at 1 MHz. The frequency is 1 MHz when the voltage gain equals 1. This is called the **unity-gain bandwidth** (1,000,000). Notice that the voltage gain at 1 kilohertz is 1000, and 1000 Hz × 1000 = 1,000,000. Alternatively, if you look at the voltage gain at a frequency of 10 kHz (10,000 Hz), on Figure 8-27, you will find a voltage gain of 100, and 10,000 Hz × 100 = 1,000,000.

Not all op-amp specification sheets provide a value for the unity-gain bandwidth, nor do they always supply a frequency response graph like the one shown in Figure 8-27. When the unity-gain bandwidth is not listed, it can be calculated from a specification called the *transient response rise time (unity gain):*

Unity-gain Bandwidth (in MHz) = 0.35/Transient

Response Rise Time (in microseconds)

The transient response rise time for the 741 op-amp is listed as 0.35 microsecond. The unity-gain bandwidth can be calculated as

Unity-gain bandwidth = 0.35/0.35 = 1 megahertz

Closed-loop Frequency Response Curve

The closed loop (with negative feedback) frequency response curve is shown in Figure 8-28. Notice that the voltage gain is constant from 0 Hz to 1 kilohertz. Once feedback is added, the voltage gain over the closed-loop small-signal bandwidth is a function of the ratio of the feedback resistance (R_f) to the input resistance (R_{in}), and not a function of the amplifier's characteristics:

$$A_V = R_f / R_{in}$$

The closed-loop small-signal bandwidth still depends on the basic amplifier figure of merit, the gain-bandwidth product. The gain-bandwidth product for the 741 is 1 MHz (1,000,000). In Figure 8-28, the voltage gain (with feedback) is 1000 and the small signal bandwidth is 1 kilohertz. A voltage gain of 1000 × 1000 Hz = 1,000,000. If we change the feedback values for a voltage gain of 100, the voltage gain remains constant at 100 until we reach a frequency of 10 kHz (100 × 10,000 = 1,000,000), at which point it begins to roll off at a rate of 20 dB per decade.

The 741 op-amp is internally compensated to prevent it from oscillating under any condition. The internal frequency compensation capacitor ensures that the frequency response rolls off to a voltage gain of unity at 1 megahertz. Because the frequency at which 180° of phase shift occurs is several megahertz, and because the voltage gain is less than 1 at any frequency above 1 megahertz, no oscillation can occur. The Barkhausen gain criterion requires that there be some positive voltage gain to compensate for losses in the feed-

FIGURE 8-28 **Closed-loop Frequency Response
for the 741 Op-amp**

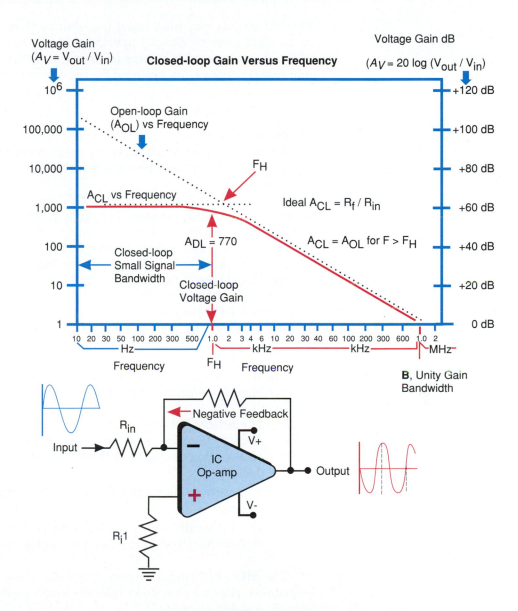

back circuit; but there is no such voltage gain available above 1 megahertz. Therefore oscillations cannot be sustained.

Some op-amps are not internally frequency-compensated, and must be compensated externally. Internally compensated op-amps are far more convenient, but uncompensated amplifiers allow the circuit designer to squeeze a little extra bandwidth out of the amplifier by adding only the amount of capacitance needed for the specific operating conditions.

8.14   Bi-FET and J-FET Op-amps

Sometimes you may need input impedances that are higher than conventional op-amps provide. Special hybrid op-amps are available that combine field-effect transistors with bipolar transistors. A FET is used for the input stage, and bipolar transistors are used for the remaining circuits.

Bi-MOS Op-amps

Field-effect transistors do not make good general-purpose transistors, so there has been no attempt to manufacture op-amps composed entirely of field-effect devices. Several bipolar op-amps, however, use a field-effect transistor as the input stage.

The RCA 3140 family uses MOS-FET input stages. The 3140 is pin-for-pin compatible with the common 741 op-amp. The 3140's circuitry is similar to that of the compensated bipolar circuit used in the 741, except in relation to the MOS-FET input stage, which yields input impedances 1 million times higher than those of the 741 and input bias and offset currents 1000 times smaller than those of the 741. The 3140 serves as an excellent and inexpensive replacement for the 741 when these input circuit characteristics are required.

When a frequency higher than the 741 can manage is involved, the 3130 Bi-MOS op-amp can be used. The 3130 is basically an uncompensated bipolar 301 op-amp with a MOS-FET input stage. The 3130 is useful even at frequencies slightly beyond 15 MHz.

Bi-MOS op-amps should not be considered a universal replacement for similar bipolar op-amps, for the following reasons:

1   The MOS-FET input stage is electrically delicate. Static charges, transient voltages, and so on can punch through the gate insulator. Diode protection is included in the IC, but it is still not as rugged as bipolar devices.

2   The output offset voltage tends to be significantly higher in Bi-MOS op-amps.

3   The internally generated noise in Bi-MOS amplifiers is significantly higher than that in the 741 and similar bipolar amplifiers.

The MOS-FET input circuit shines for low-current amplifiers, for high-impedance sources that can't tolerate much loading, and for integrator and sample-and-hold circuits.

Bi-FET Op-amps

The Bi-FET op-amp is a bipolar circuit with a J-FET input stage. The Bi-FET has input characteristics very similar to those of the Bi-MOS. Offset and bias currents are a little higher, and the input impedance is a little lower, than in Bi-MOS amplifiers. Although the J-FET's input stage is electrically nearly as rugged as a bipolar stage, it does suffer from noise generation problems. The Bi-FET op-amps 351, 353, and 357 are pin-for-pin replacements for the 741, the 747 dual 741, and the 748 quad 741 op-amps, respectively.

Table 8-2 compares technical data for several popular commercial op-amps.

TABLE 8-2
Comparison of Common Operational Amplifiers

Technical Parameter	Bipolar			Bi-MOS		Bi-FET	
	741C	301A	LM-11C	3140	3130	LF-351	LF-356
Open-loop Voltage Gain	100,000	100,000	100,000	100,000	100,000	50,000	100,000
Input Impedance	1 Meg	4 Meg	10^5M	10^6M	10^6M	10^6M	10^6M
Unity-gain Bandwidth	1 MHz	4 MHz	5 MHz	4.5 MHz	15 MHz	4 MHz	4 MHz
Common-mode Rejection Ratio	90 dB	90 dB	110 dB	90 dB	90 dB	100 dB	100 dB
Output Current	25 mA	20 mA	15 mA	20 mA	20 mA	20 mA	25 mA
Power Dissipation	500 mW	500 mW	500 mW	600 mW	600 mW	500 mW	570 mW
Type of Compensation	Int.	Ext.	Ext.	Int.	Ext.	Int.	Int.

Troubleshooting Operational Amplifiers

The following procedures are simple and noninvasive. The tests can be made without removing any components from the PCB (printed circuit board).

Finding the Faulty Stage

Locate the faulty amplifier stage, as illustrated in Figure 8-29. Start at the final output stage and work your way back toward the input stage. You must decrease the input signal amplitude as you move the function generator toward the input stage.

FIGURE 8-29   Locating the Faulty Stage

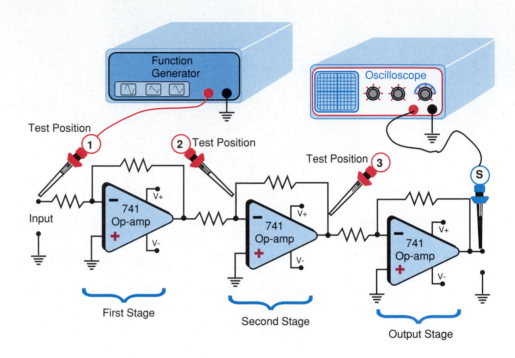

FIGURE 8-30   Testing at the *IC* Pins

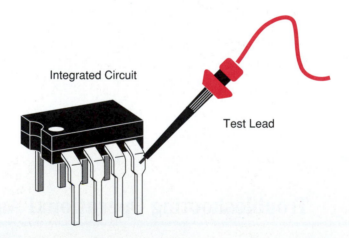

Testing the Suspect Stage

1   As usual, the first thing to do is to make sure that all power-supply
     voltages are normal. If you suspect a particular IC, it is a good idea to
     measure voltages (and signals) directly on the IC pins, as shown in
     Figure 8-30. This ensures that the voltage or signal is not lost in a
     socket or IC solder joint.

2 Verify the op-amp's **dc** output offset voltage. If the op-amp uses a dual supply, the offset should not be more than a few millivolts. If the amplifier is powered by a single supply, the **dc** offset should be about one-half the supply voltage.

Causes of Excessive Offset Voltage

Excessive offset voltage may be due to any of three underlying causes:

1 A **dc** input voltage in a working circuit. Measure the **dc** input voltage. If there is an input voltage to account for the **dc** offset, the **dc** offset does not signify an op-amp problem. Look at input for a **dc** input offset voltage.

2 An open feedback loop. Connect a temporary external feedback resistor, as shown in Figure 8-31. The existing feedback network may be more complex, but we don't care. If the excessive output offset voltage vanishes when you connect the temporary feedback resistor, the circuit has an open feedback loop.

3 A defective op-amp IC. Before you change the IC, consider the single-stage signal tests that are described next.

FIGURE 8-31 **dc Offset and Feedback Loop Test**

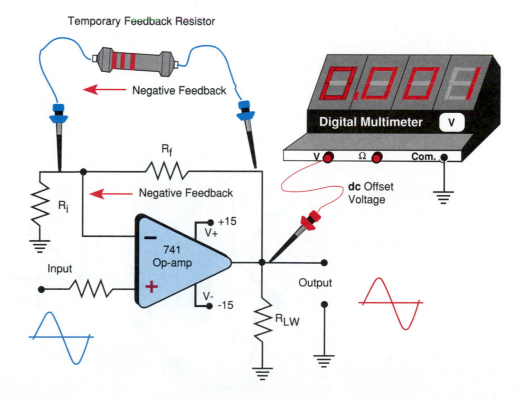

Single-stage Tests

In-circuit testing of op-amps is quick and easy in all but a few cases. The quick forcing tests described in the following section will detect nearly all cases of op-amp failure. On occasion, the problem is poor performance rather than a dead stage. A different and more complex procedure is required for troubleshooting these comparatively rare problems. We will also examine procedures for diagnosing problems involving poor performance.

A Quick Dead-or-Alive Response Test

A simple test can be used to verify that an op-amp stage is at least alive. If the circuit passes the test, it tells you only that the stage is not totally dead; it doesn't tell you anything about how well the stage is working. On the other hand, assuming that the IC has power, if it fails the test, you can be fairly certain that the IC is bad. Figure 8-32 shows how the test is done. Simply short the input temporarily to V+, and then to V−, and observe the **dc** output voltage. The output voltage should swing from near V− to near V+. Do not leave the short connected for long. Do not use an input test voltage that is higher than the supply voltage applied to the amplifier. This is a brute-force test, since you are forcing the amplifier all the way into saturation in both directions. The test is not subtle, but it won't damage the IC.

A More Sophisticated Test

Figure 8-33 shows how to make a more accurate evaluation of a stage. In this test, you take a real voltage-gain measurement. The procedure involves adding a temporary extra input resistor to create a summing amplifier circuit. The

FIGURE 8-32   Dead-or-Alive Quick Test

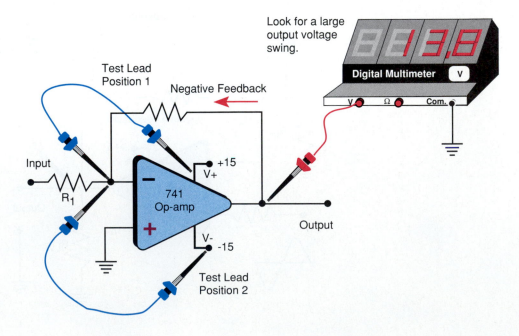

FIGURE 8-33   **Voltage-gain Test**

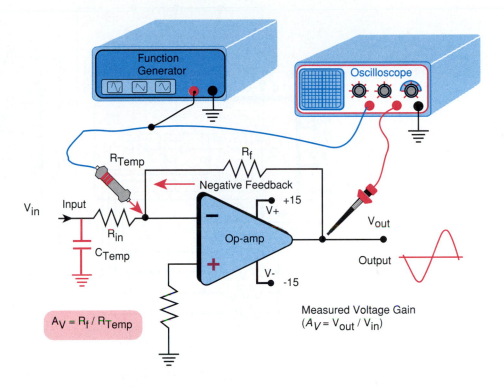

voltage gain of your newly created circuit has a predicted voltage gain of

$$A_V = R_f/R_{temp}$$

Your added summing input is independent of any other signals, but you may have to disable any existing signal to measure the gain of your temporary circuit. A temporary capacitor can be used to shunt other signals to ground. If the amplifier is a noninverting amplifier, the temporary capacitor should be connected from the noninverting input to ground. If the measured voltage gain ($A_V = V_{out}/V_{in}$) approximates the predicted voltage gain, the amplifier itself must be okay.

The Digital Connection: Op-amps and Analog Gates

It is sometimes necessary to control the voltage gain of an op-amp from the output of a computer or other digital circuit. An analog gate with digital input/output selection can be used to allow digital control of an analog circuit. The second part of this section discusses some ways in which C-MOS digital IC's can be used as analog amplifiers.

FIGURE 8-34 **Control of Op-amp Gain by Binary Digital Values**

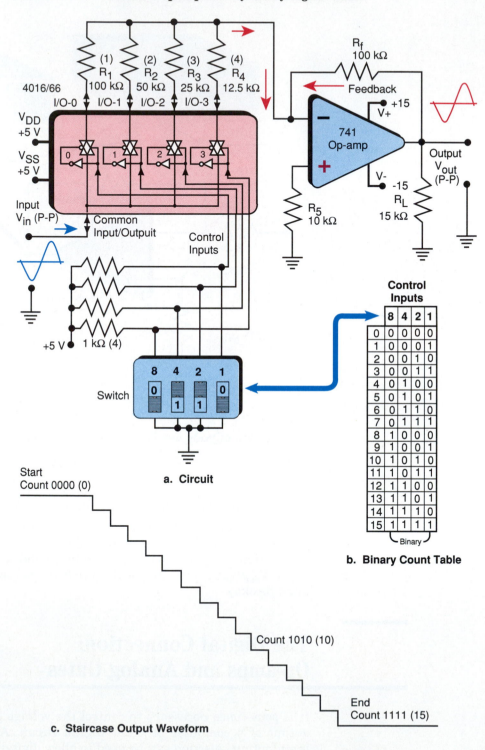

	Control Inputs			
	8	**4**	**2**	**1**
0	0	0	0	0
1	0	0	0	1
2	0	0	1	0
3	0	0	1	1
4	0	1	0	0
5	0	1	0	1
6	0	1	1	0
7	0	1	1	1
8	1	0	0	0
9	1	0	0	1
10	1	0	1	0
11	1	0	1	1
12	1	1	0	0
13	1	1	0	1
14	1	1	1	0
15	1	1	1	1

Binary

a. Circuit

b. Binary Count Table

c. Staircase Output Waveform

Binary Control of Op-amp Gain

Figure 8-34 illustrates one of many techniques for controlling an op-amp by using digital signals. In the example in Figure 8-34, the op-amp is configured as a summing amplifier, designed to produce input voltage gains in binary values of 1, 2, 4, and 8. Because the circuit is a summing amplifier, all of the binary combinations needed to count in binary from 0 to 15 are available.

In the example, a manual switch is used to step through the binary counts, according to the binary-count table shown in the figure. The switches can be set to produce op-amp voltage gains from 0 to 15. Each of the switches is connected to the control input of an analog gate, which turns its respective input resistor on/off. A similar scheme can be used to control feedback-resistor values.

The manual switch can be replaced by binary output signals from a digital IC or even from a microcomputer output port. In data-acquisiton systems, it is often necessary to use digital circuits to control analog devices.

The circuit in Figure 8-34(a) is also used to convert binary digital values into analog voltages. Assume that a fixed voltage representing a binary 1 (usually +5 V) is applied to the signal input of the analog gate. If we step the manual switch through the first ten combinations on the binary-count table, we get the output voltages shown in Table 8-3 from the summing amplifier.

Theoretically, we could complete the table all the way to a count of 15, but reality is less obliging. Because we have a 15-volt power supply, we theoretically can get 15 volts of output; but in reality we can only expect about 13 volts maximum, before the relationship between the input numbers and the output voltages begins to show some error. We can't run the op-amp too near saturation in either direction without developing some distortion. In this case, any distortion shows up as an error between the binary input numbers and the analog (op-amp) output voltage.

As a practical matter, we can increase the power-supply voltage, limit the binary count to 10 or so, or scale the output to produce an output voltage range of 0.1 V to 1.5 V, instead of from 1 V to 15 V.

If a binary counter is used to step through the binary values automatically and in order, the circuit becomes a staircase generator with a set of op-amp output voltages, as shown in Figure 8-34(b).

TABLE 8-3

Output Voltages Produced by the Summing Amplifier for Different Combinations of Binary Values

Binary Values		Op-amp Output Voltage	Binary Values		Op-amp Output Voltage
0 + 0 + 0 + 0	=	0 volts	0 + 4 + 2 + 0	=	6 volts
0 + 0 + 0 + 1	=	1 volt	0 + 4 + 2 + 1	=	7 volts
0 + 0 + 2 + 0	=	2 volts	8 + 0 + 0 + 0	=	8 volts
0 + 0 + 2 + 1	=	3 volts	8 + 0 + 0 + 1	=	9 volts
0 + 4 + 0 + 0	=	4 volts	8 + 0 + 0 + 2	=	10 volts
0 + 4 + 0 + 1	=	5 volts			

Using C-MOS Digital Gates as Analog Amplifiers

Before we go any farther, let's make sure we understand that the following circuits are not possible with TTL (transistor-transistor-logic) gates. The basic characteristics of the complementary MOS circuits used in C-MOS gates make it possible to bias an otherwise switching-mode circuit into its linear operating region. The 74Cxx and 74HCxx families are good candidates for this service. The 74HCxx family is preferred.

The circuit in Figure 8-35 illustrates how a C-MOS digital IC can be used as a negative-feedback amplifier. The basic inverting negative-feedback amplifier shown in Figure 8-35 cascades three of the six inverters in the 74HC04 digital inverter package. Adding the feedback resistor biases the inverter into the center of the linear part of its operating curve.

There must be an odd number of stages. Each stage has an open-loop gain of about 100 with a 12-volt power supply for a total open-loop gain for the three

FIGURE 8-35 Using Bias and Feedback to Permit Operating C-MOS Digital ICs in Linear (Analog) Circuits

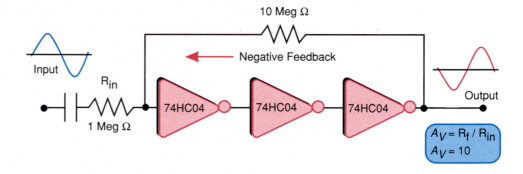

FIGURE 8-36 Summing Amplifier with Linearly Biased C-MOS Digital IC (May Be Used to Sum Analog or Digital Signals)

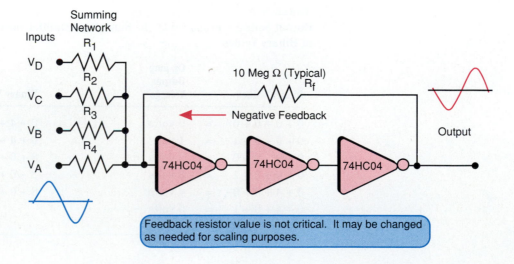

stages of about 1 million. The closed-loop gain is given by the usual inverting op-amp equation:

$$A_V = R_f/R_{in}$$

The amplifier can have a high input impedance, and it uses very little power when no signal is applied. The output can swing to within a few millivolts of the power-supply voltage (either voltage, if a dual supply is used). The 74HC04 is an excellent analog amplifier, particularly in battery-operated devices.

The circuit can be used in place of an op-amp if only an inverting input is needed. The primary advantages of the circuit in Figure 8-35 are its very high input impedance and its ability to operate with a power-supply voltage as low as 3 volts.

Figure 8-36 extends the idea represented in Figure 8-35, showing how a C-MOS digital IC can be used as an analog summing amplifier.

Figure 8-37 illustrates how two C-MOS gates can be used to provide a low-voltage power booster for the LM-308 low-voltage op-amp.

FIGURE 8-37   **Dual Power-supply Power Amplifier with C-MOS Digital Gates in a Linear Mode**

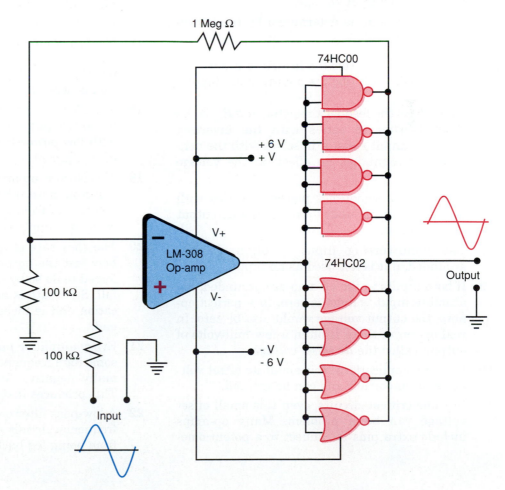

SUMMARY

1 The operational amplifier is a nearly universal amplifier. A single op-amp type can be used for hundreds of different applications.

2 The op-amp's open-loop voltage gain ranges from 100,000 to 2 million.

3 The op-amp's closed-loop gain is totally defined by the value of two feedback resistors, which form a feedback voltage divider.

4 The integrated-circuit (IC) op-amp is composed of differential amplifiers and a low-impedance output stage.

5 Because of its differential amplifiers, an op-amp has one input that produces an inverted output voltage and another that produces a noninverted output voltage.

6 The voltage gain is determined by the following equations:

$$A_V = R_f/R_{in} \qquad \text{(for the inverting input)}$$
$$A_V = (R_f/R_{in}) + 1 \quad \text{(for the noninverting input)}$$

where R_f is the feedback resistor, and R_{in} is the input resistor in series with the inverting input. An input resistor in series with the noninverting input has no effect on the voltage gain.

7 A resistor is sometimes placed in series with the noninverting input to reduce the **dc** output voltage offset.

8 The noninverting input is often directly grounded, if it is not used as an input.

9 If both inputs to the op-amp are grounded, the absolute input voltage is zero. In a perfect op-amp, the output voltage would also be zero. In real op-amps, there is often a few millivolts of output, called the **dc** *offset voltage*.

10 In many applications, this small **dc** offset voltage is unimportant and can be ignored.

11 In some critical circuits, even this small offset voltage can be a problem. Many op-amps include extra pins to connect to a potentiometer that can balance out, or nullify, the **dc** offset voltage to zero.

12 The Miller effect applies to operational amplifiers.

13 The Miller effect decreases the input impedance of the inverting input.

14 The Miller effect increases the input impedance of the noninverting input.

15 The Miller effect causes the inverting input terminal of the amplifier to have a voltage near zero. Because the input voltage has almost the same potential as ground, the input is said to be a *virtual ground*.

16 Because of the virtual ground, the input resistance of the inverting input is equal to the value of the input resistor (R_{in}).

17 Most op-amps are designed to use two power-supply voltages. Any op-amp can be externally biased for single-supply use, if **ac**-only operation is acceptable.

18 The Norton operational amplifier is designed for single-power-supply use and for systems with low power levels, such as battery-operated equipment.

19 The Norton op-amp uses a very different circuit from most other op-amps. It can operate at voltages of from 2 to 36 volts and can be used with two supply voltages if necessary.

20 The slew rate of an amplifier is a measure of how fast the output voltage can change when forced to do so by a square-wave input's rise or fall. Slew rate is an important op-amp specification and is measured in volts per microsecond.

21 Every amplifier produces some phase shift. At some high frequency, phase shift can cause normally negative feedback to become positive. This produces instability and oscillation.

22 A low-pass filter can be used to eliminate high-frequency signals and to prevent those signals from being fed back to the input.

23 Adding the filter causes the amplifier's upper frequency to be limited to a frequency called the *roll-off frequency.*

24 If the filter is built into the op-amp, it is called an *internally compensated amplifier.* If the fil-ter must be added externally, the amplifier is termed *externally compensated.* Both kinds of IC op-amps are available.

Questions and Problems

Given the schematic diagram shown in Figure 8-38, answer the questions in Problems 8.1 through 8.4.

8.1 This circuit is called which of the following?
- **a.** An inverting amplifier.
- **b.** A noninverting amplifier.
- **c.** A voltage follower.
- **d.** A differential amplifier.
- **e.** A bifilter amplifier.

8.2 Write the equation that defines the closed-loop voltage gain of the circuit.

8.3 Based on the part values shown, what is the calculated voltage gain of the circuit?

8.4 Based on the part values shown, what is the calculated input impedance of the circuit?

Given the schematic diagram shown in Figure 8-39, answer the questions in Problems 8.5 through 8.8.

8.5 This circuit is called which of the following?
- **a.** An inverting amplifier.
- **b.** A noninverting amplifier.
- **c.** A voltage follower.
- **d.** A differential amplifier.
- **e.** A summing amplifier.

8.6 Write the equation that defines the closed-loop voltage gain of the circuit.

8.7 Based on the part values shown, what is the calculated voltage gain of the circuit?

8.8 Based on the part values shown, what is the calculated input impedance of the circuit?

FIGURE 8-38   **Circuit for Problems 8.1 Through 8.4**

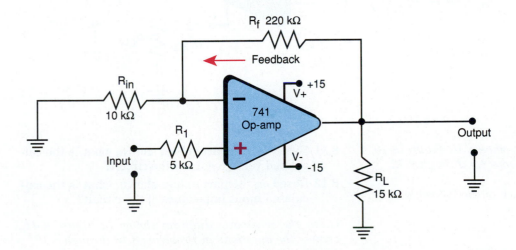

FIGURE 8-39   Circuit for Problems 8.5 Through 8.8

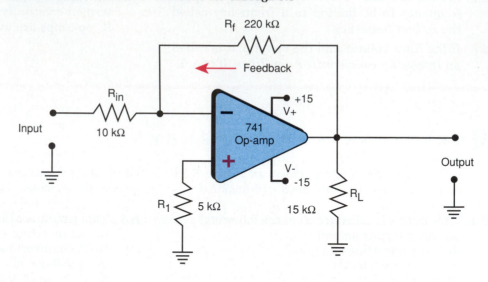

FIGURE 8-40   Circuit for Problems 8.9 Through 8.12

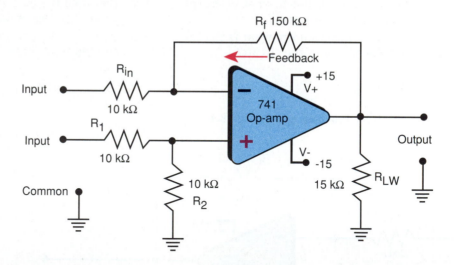

Given the schematic diagram shown in Figure 8-40, answer the questions in Problems 8.9 through 8.12.

8.9 This circuit is called which of the following?
   a. An inverting amplifier.
   b. A noninverting amplifier.
   c. A voltage follower.
   d. A differential amplifier.
   e. A Bi-MOS amplifier.

8.10 Write the equation that defines the closed-loop voltage gain of the circuit.

8.11 Based on the part values shown, what is the calculated voltage gain of the circuit?

8.12 Based on the part values shown, what is the calculated input impedance of the circuit?

Given the schematic diagram shown in Figure 8-41, answer the questions in Problems 8.13 through 8.16.

8.13 This circuit is called which of the following?
   a. An inverting amplifier.
   b. A noninverting amplifier.

FIGURE 8-41 Circuit for Problems 8.13 Through 8.16

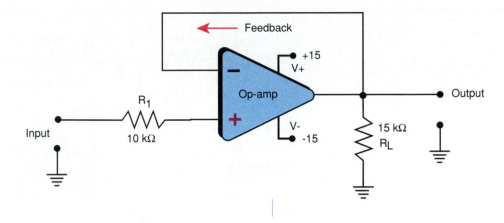

FIGURE 8-42 Circuit for Problem 8.22

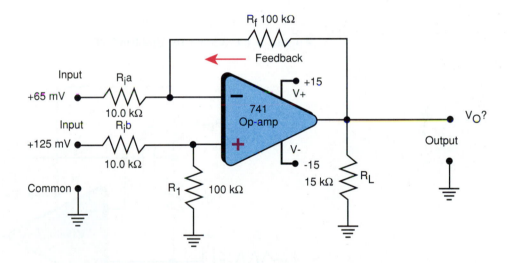

c. A voltage follower.

d. A differential amplifier.

e. A Norton amplifier.

8.14 Write the equation that defines the closed-loop voltage gain of the circuit.

8.15 Based on the part values shown, what is the calculated voltage gain of the circuit?

8.16 Based on the part values shown, what is the calculated input impedance of the circuit?

8.17 Draw a circuit showing how you would connect a potentiometer to a 741 op-amp to nullify the **dc** offset voltage.

8.18 Explain the Miller effect as it applies to operational amplifiers.

8.19 How does the Miller effect alter the input impedance of the inverting input?

8.20 How does the Miller effect alter the input impedance on the noninverting input?

8.21 Define the term *virtual ground*.

8.22 Given the differential amplifier circuit in Figure 8-42, what is the output voltage?

8.23 Given the summing amplifier in Figure 8-43, what is the output voltage?

FIGURE 8-43   **Circuit for Problem 8.23**

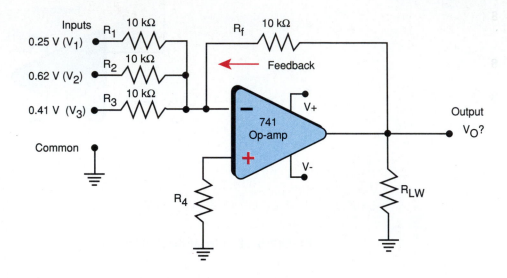

FIGURE 8-44   **Circuit for Problem 8.26**

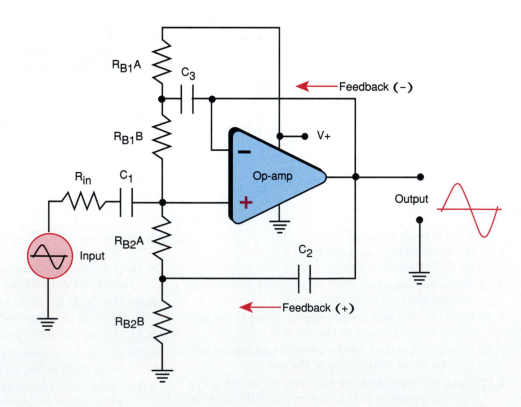

8.24 Draw a schematic diagram showing how an op-amp can be biased to operate with a single power supply.

8.25 If a 741 op-amp is biased to allow single-power-supply operation, what is the **dc** output offset voltage when there is no input signal?

8.26 Given the circuit in Figure 8-44, what is the purpose of capacitors C_2 and C_3?

8.27 Contrast the Norton current-differencing amplifier with the common 741 variety with regard to the following parameters:

a. Kind of input circuit.
b. Input impedance.
c. Output impedance.

8.28 Define *slew rate*.

8.29 Explain the need for frequency compensation in an op-amp.

8.30 Identify the Barkhausen criteria.

8.31 Define *unity-gain bandwidth*.

8.32 Define *gain-bandwidth product*.

Teamwork

Any history of the development of electrical and electronics technology will reveal that this is a field where teamwork counts. Yes, many famous people have made outstanding discoveries and contributions to the field, but each innovator "stood on the shoulders" of their predecessors.

Many developments in electronics in this century were accomplished by teams working for universities or private research laboratories. Consider, for instance, the team that developed the first transistor.

William Shockley, John Bardeen, and Walter Brattain are credited with developing the first bipolar transistor in 1948 at Bell Telephone Labs. In 1929, Brattain joined Bell Labs to do vacuum-tube research, but his background in solid-state physics led him into work on semiconductors. Brattain, again "standing on the shoulders" of other researchers, investigated the motion of electrons in semiconductors. Related work had been done by such scientists as the British physicists Alan H. Wilson, academicians Frenkel and Davydov from the Soviet Union, Schottky in Germany, and Mott in England. William Schockley, another physicist, joined Bell Labs in 1936 and eventually began working with Brattain, experimenting with copper oxide.

Other Bell Labs personnel contributed to the process during this same period. Russel S. Ohl, a chemist, experimented with the properties of silicon. J. H. Schaff and H. C. Theurer, metallurgists, found that silicon, when melted in a vacuum, produced materials that would conduct in one direction but not in the other. This led to the process of doping silicon to produce p- and n-type materials. Ohl's subsequent experiments demonstrated the photoelectric properties of doped silicon materials.

During World War II, Shockley and Brattain's work was interrupted. Meanwhile, however, other researchers, such as Karl LarHorovitz at Purdue University, continued to experiment with another semiconductor: germanium. After the war, Bell Labs resumed its research in this area, building on the data discovered at Purdue. In 1945, a solid-state physics group was created under Schockley and Stanley O. Morgan. A subgroup focusing on semiconductors included Brattain and another physicist, Gerald Pearson. They were later joined by John Bardeen.

Over the next three years, Brattain and Bardeen conducted extensive experiments attempting to find a way to control the current through a semiconductor material by using a third lead. One of their difficulties involved making very fine connections to the semiconductor material. Late in 1947, they found a way to accomplish their goal and were able to build a simple amplifier. Interestingly, the name for the device was coined by yet another Bell Labs employee, John R. Pierce. He suggested the name *transistor* because, unlike the vacuum tube (which exhibits transconductance), the new device exhibited transresistance.

Many improvements were necessary before the new semiconductor device could be produced commercially in quantity and with predictable specifications. Bell Labs, in an effort to promote the development of the transistor, ran seminars and instituted a widespread licensing program. Bell also waived all license fees and royalties on transistors built for use in hearing aids. The result was that many of the needed improvements were accomplished by other companies and research facilities.

The development of the transistor is a good example of how teamwork by many gifted scientists resulted in an important technological development. Teamwork is an effective approach to accomplishing goals whether it is applied in research, engineering design, or service and maintenance. The old adage about the result being greater than the sum of its parts is correct. Assembling a team of technically oriented people to focus on one or more specific goals allows each individual to contribute out of his or her own expertise. At the same time interaction between individuals tends to spark ideas for innovative solutions to problems.

Many employers involved in electronics use the team approach. For example, an engineering team involved in producing new electronics products may consist of the following members: a team leader, who coordinates and sets the direction for the team; a design engineer, who researches and formulates the technical approach to the problem; an associate engineer or technologist, who develops the actual circuit schematic and layout; and a technician, who builds, tests, and troubleshoots the prototype.

In a production facility, several interactive teams may exist, working as part of a larger team. A quality-control group may network with the production team, a test-and-repair group, and the engineering group, to fine-tune the operation for maximum efficiency. In a less complex work environment, the team may be as simple as a salesman, an accounts clerk, and a service technician working together to make sure that the company produces a profit.

The nature of business in our global economy has driven companies, and even whole industries, toward higher levels of efficiency and productivity. Many companies have turned to innovative techniques like total quality management (TQM) in an effort to achieve these goals. TQM attempts to build on the strengths of the individuals in the organization, encouraging their ideas, input, and suggestions. The outstandingly successful technological and manufacturing growth of Japan over the last 40 years has been attributed largely to the application of TQM principles.

More recently, the International Standards Organization, with participants from 91 countries, has developed a set of quality management standards called *ISO 9000*. The 9000 series of standards specifies how businesses, manufacturers, service companies, and others should operate in order to maintain high-quality products and workmanship. Already, many employers have adopted ISO 9000 and delegated many of the responsibilities for implementing it to their engineering and management teams. In the future, many electronics technicians may find themselves required to know and apply these standards in their workplace.

Whether the employer chooses to incorporate TQM or ISO 9000 or to develop its own team concept, electronics technicians—as part of the engineering team—will often be affected. There are very few jobs where the employee's only concern is the technical nature of the work. The current trend toward productivity, quality, and excellence demands that each member of an organization work cooperatively as a team member for the benefit of the whole organization.

CHAPTER

9

Comparators

OBJECTIVES

Upon completion of this chapter, you should be able to:

1 Explain how the comparator's switching action takes place.

2 Define the following comparator terms:

  a *Slew rate.*

  b *Response time.*

  c *Rail voltage.*

  d *Hysteresis.*

  e *Upper trip point.*

  f *Lower trip point.*

  g *Reference voltage.*

  h *Schmitt trigger.*

  i *Jitter.*

3 Describe three ways in which a comparator can be used in conjunction with a reference voltage.

4 Explain why the comparator produces a switchlike output instead of a linear (analog) output signal.

5 Draw the schematic diagram for and explain the operation of each of the following comparator circuits:

  a The zero-crossing detector.

  b The basic comparator circuit with a fixed reference voltage.

  c The comparator-based R-C timing circuit.

  d The basic comparator-based square-wave generator.

  e The window detector.

  f The ladder comparator.

  g Common 311 analog/digital comparator circuits.

6 Explain the operation of a comparator-based pulse-width modulator.

Analysis In Brief

Comparators in Brief

WHAT IS A COMPARATOR?

Comparators are operational amplifiers, used without negative feedback. They are open-loop and have very high voltage gain.

Comparator inputs accept any analog waveform voltage.

Comparator output voltages are digital and binary. The output can be one of only two possible voltages. The possible output voltages correspond to the two power supply voltages, called the *rail voltages.*

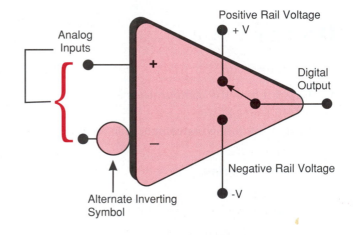

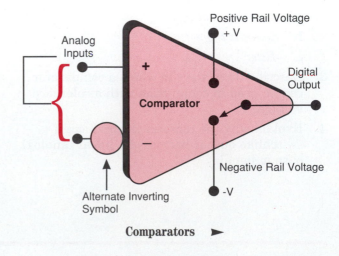

Comparators ▶

THE OUTPUT: WHICH RAIL?

1   Change the sign of the voltage applied to the inverting input. Call this voltage *A*.

2   Call the voltage applied to the noninverting input voltage *B*.

3   Compare voltage *A* to voltage *B*. Use the sign of the larger value.

4   If the sign of the larger value is positive (+), the comparator output will switch to the positive (+) rail.

5   If the sign of the larger value is negative (−), the comparator output will switch to the negative (−) rail.

Note: Rail voltages may consist of a positive and a negative voltage, or a positive voltage and ground (0 volts).

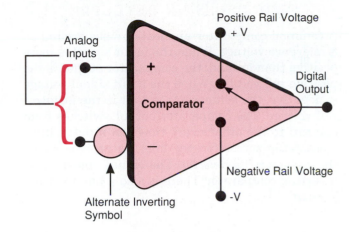

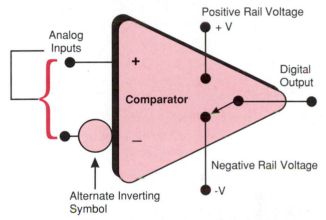

Which Output Rail?   ➤

THE ZERO-CROSSING DETECTOR

A common comparator application uses ground as a reference voltage. When the input signal is more positive than ground, the comparator output is the positive rail voltage. When the input signal voltage is negative, the comparator output is the negative rail voltage. The comparator output switches from one rail to the other every time the input voltage crosses the zero reference line; hence, it is called the *zero-crossing detector*. The circuit shown is an inverting comparator. There is also a noninverting version.

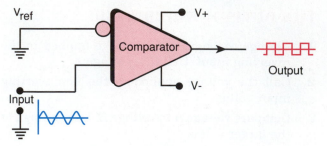

Zero-crossing Detector Circuit

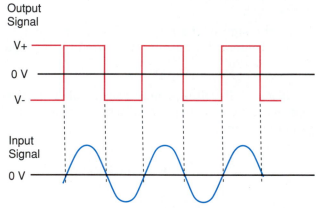

Zero-crossing Detector Waveforms

USING A NONZERO REFERENCE

The comparator can be used to detect the crossing of voltages other than zero. A nonzero reference voltage is usually obtained from a resistive voltage divider. The output voltage switches from one rail to the other each time the input signal voltage crosses the reference voltage. The reference voltage is also called the *trip-point voltage,* where the comparator switches.

The equation for the reference voltage is the standard voltage-divider equation:

$$V_{ref} = V_{CC} \, [R_1/(R_1 + R_2)]$$

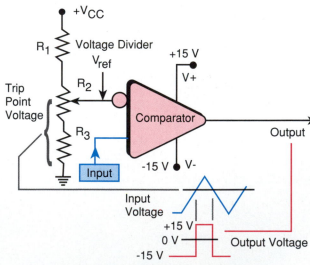

Using a Non-zero Reference Voltage

THE SCHMITT TRIGGER

A comparator becomes a Schmitt trigger when its output is connected to the noninverting input (positive feedback). The comparator output becomes the reference voltage. Because there are two possible output voltages, there are now two reference voltages, and two trip points.

HYSTERESIS

The voltage range between the two trip points is called the *hysteresis voltage*. The hysteresis voltage's range is a dead zone; input voltages within this range cause no change in the comparator's output voltage status.

THE UPPER TRIP POINT

An input voltage starting within the hysteresis band must rise until it crosses the +7.5-volt trip point (in the example). When the input voltage crosses the +7.5-volt reference voltage, it passes the upper trip point. The output thereupon switches to the −15-volt rail. The new trip point voltage is −7.5 volts.

THE LOWER TRIP POINT

An input voltage that starts above the upper trip-point voltage must fall all the way through the hysteresis band and cross the −7.5-volt trip point (in the example) before it will change the output voltage. When the input voltage crosses the −7.5-volt reference voltage, it has crossed the lower trip point. The output switches to the +15-volt rail. The new reference (and trip-point) voltage is +7.5 volts. The input voltage must cross the upper trip point again to switch the comparator once more.

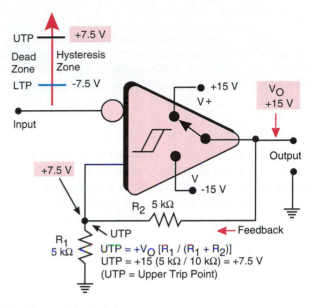

$\text{UTP} = +V_O \,[R_1 \,/\, (R_1 + R_2)]$
$\text{UTP} = +15 \,(5\ \text{k}\Omega \,/\, 10\ \text{k}\Omega) = +7.5\ \text{V}$
(UTP = Upper Trip Point)

The Upper Trip Point

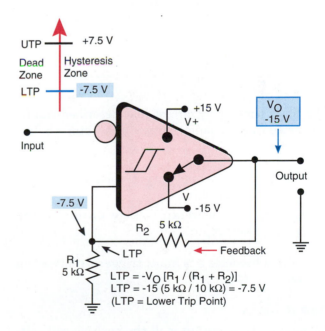

$\text{LTP} = -V_O \,[R_1 \,/\, (R_1 + R_2)]$
$\text{LTP} = -15 \,(5\ \text{k}\Omega \,/\, 10\ \text{k}\Omega) = -7.5\ \text{V}$
(LTP = Lower Trip Point)

The Lower Trip Point

COMPARATOR-BASED TIME-DELAY CIRCUIT

The time-delay circuit establishes a fixed reference voltage—63.2% or less of the capacitor's charging voltage. Here, 63.2% represents a one-time constant limit, to use the most linear part of the capacitor-charging curve.

How It Works

When the pushbutton is opened, capacitor C_1 begins charging through resistor R_1. When the charge voltage on the capacitor crosses the reference voltage, the comparator switches to the upper rail. The LED (b) turns on, indicating that the timing period is complete.

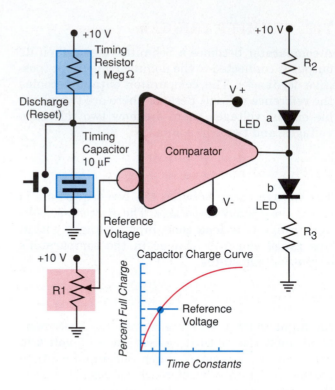

Comparator-based Time-delay Circuit

SCHMITT-TRIGGER SQUARE-WAVE GENERATOR

The basic time-delay circuit just described can be converted into a square-wave generator in two steps:

1   Use the comparator output voltage to charge the timing capacitor.
2   Connect the reference voltage divider to the comparator output.

The circuit becomes a Schmitt trigger with two trip points.

How It Works

Assume that the output is at the positive rail voltage. When the capacitor-charge voltage crosses the UTP, the comparator output switches to the negative rail. The capacitor now begins to charge toward the negative-rail voltage. When the charge voltage crosses the LTP, the comparator output switches to the positive rail. The cycle repeats endlessly.

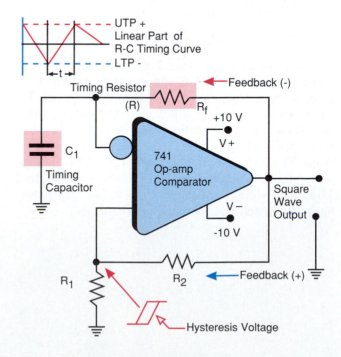

Schmitt Trigger Square-wave Generator Circuit

9.1   Introduction

The **comparator** is an operational amplifier that is used in a differential mode, without any negative feedback. It is the only common op-amp circuit in which the operational amplifier is used with an open feedback loop. The entire open-loop voltage gain of 200,000 or more is used to provide a switching action instead of a linear response.

Any operational amplifier IC can be used as a comparator, if speed is no object. For high-speed comparator applications, particularly when digital circuits are involved, several op-amp ICs have been specially designed for comparator service. These devices are listed in the manufacturers' data manuals as *comparators*. Not only does the comparator circuit serve as an essential element in nearly all analog-to-digital converters, but it is also a principal component in a number of analog integrated circuits. We will examine these analog ICs in subsequent chapters.

The comparator compares two voltages, so it is actually a voltage comparator, although its full title is rarely used.

Figure 9-1 shows some comparator symbols and terminology. The conventional op-amp symbol is often used, and you must look to confirm that a negative-feedback loop that is included in the conventional design is missing, in order to determine that the circuit is a comparator circuit instead of a regular op-amp circuit.

The symbol in Figure 9-1(b) shows a more modern version of the comparator symbol, using a bubble to identify the inverting input. The bubble symbol was taken from the digital world, where it is used to identify a digital inverter.

Because many special comparator ICs were designed to meet digital circuit requirements, the bubble symbol (Figure 9-1(b)) is most often associated with ICs called comparators by the device manufacturer. These special comparator ICs may also be used in analog circuits. The symbols are used pretty much at the discretion of the designer, so don't expect any hard-and-fast rules about which symbol should be used where.

The term **rail voltage** is unique to comparator circuits; it is used as another name for the two power-supply voltages. Since the comparator output voltage is always one or the other power-supply voltage, the term *rail voltage* is a convenient way to refer to the comparator output, without being specific about supply voltages.

The comparator compares two voltages and initiates a specific action based on that comparison. For example, a comparator may be used to monitor a power supply. If the actual power-supply output voltage exceeds some predetermined (reference) voltage, the comparator output voltage switches to the alternate rail voltage. This change in comparator output voltage can be used to sound an alarm, to shut down the power supply, or to initiate some other action.

Whenever we must initiate some action based on the relationship between two voltages, a comparator is the circuit we should use. The comparator finds a place in timing circuits, in voltage-controlled oscillators, in waveshaping circuits, and in various other discrete and integrated circuits.

FIGURE 9-1   **Comparator Symbols**

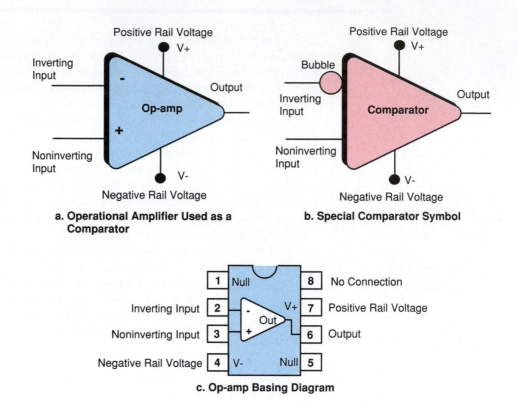

a. **Operational Amplifier Used as a Comparator**

b. **Special Comparator Symbol**

c. **Op-amp Basing Diagram**

9.2   Comparator Model

Let's examine a simplified model of how the comparator works before we look at why it works that way. Refer to Figure 9-2 as we go through the following explanation.

First, notice that the output of the comparator is shown as a switch with two possible positions. In one position, the switch connects the output to the positive rail (the positive 12-volt power-supply terminal). The other switch position connects the output to the negative rail (the negative 12-volt power-supply terminal). This is a good model for the comparator's output circuit behavior. Of course, the switch is electronic and not quite perfect. We will see what actually causes the switching action and how it works electronically, after we finish with the model.

The important idea here is that the comparator's output voltage can be either the positive-rail voltage or the negative-rail voltage. No other output voltage is possible. Now, let's look at the inputs to the comparator in Figure 9-2.

A voltage applied to the noninverting input enters the comparator without a change of sign: a positive voltage enters the comparator as a positive voltage, and a negative voltage enters the comparator as a negative voltage.

FIGURE 9-2 How the Comparator Works When Positive Voltage Dominates

FIGURE 9-3 How the Comparator Works When Negative Voltage Dominates

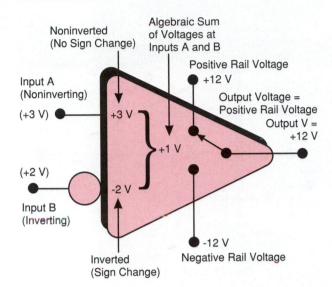

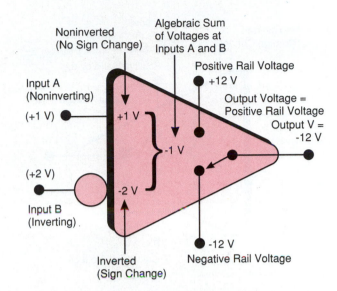

A voltage applied to the inverting input enters the comparator with its sign changed (inverted): a positive voltage enters the comparator as a negative voltage, and a negative voltage enters the comparator as a positive voltage. Once inside the comparator, the larger of the two voltages dominates, and its sign controls which rail the comparator output switches to. If the positive voltage is dominant, the comparator switches to the positive rail. If the negative voltage is dominant, the comparator switches to the negative rail.

Figure 9-2 illustrates the case where the positive voltage dominates. In this case, input voltages of +3 V and –2 V (inverted) yield a difference of +1 V. The positive difference voltage causes the comparator to switch to the +12-V rail. Figure 9-3 illustrates the case where the negative voltage is substantially greater. In this case, input voltages of +1 V and –2 V (inverted) yield a difference voltage of –1 V. The negative difference voltage causes the comparator to switch to the –12-V rail. Figure 9-4 illustrates the case where the difference voltage is very small but positive. In this case, input voltages of +1.02 V and –1.01 V (inverted) yield a difference voltage of +0.01 V. Although quite small, this difference voltage still drives the comparator output to the +12-V positive rail. Figure 9-5 illustrates how a comparator can use a ground (0 volts) for the negative rail. The comparator action—switching from rail to rail—is the same, regardless of the actual rail voltages.

9.3 What Really Happens

The action of the input circuit is really the same as occurs when the op-amp is used as a differential amplifier, because that is exactly what the comparator input circuit is. The model is simply another way of looking at a differential input circuit. The model provides a quick and easy way to determine which of

FIGURE 9-4 How the Comparator Works When
Positive Voltage Is Slightly Greater

FIGURE 9-5 How the Comparator Can Use a Ground
(0 V) as the Negative Rail

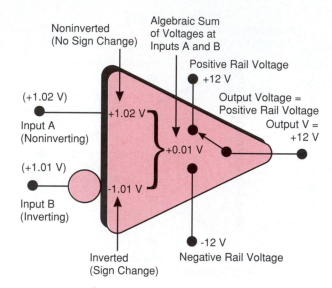

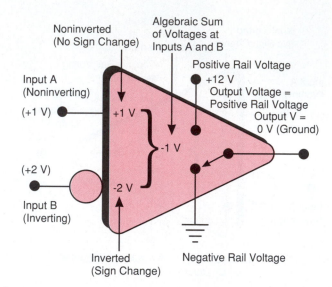

the two possible output voltages should be present for any given set of input conditions. In troubleshooting comparator circuits, it is often necessary to know what output voltage to expect. Now let's look at how the output circuit simulates the mechanical switch shown in the model.

Figure 9-6 shows a comparator with a fixed 1-volt reference voltage on the inverting input. Column A in the associated table contains only 1-volt entries, to reflect the fact that the voltage remains the same throughout the example. Column B lists the different voltages to be applied to the noninverting input for the example. Column C lists the difference voltages that would result from a combination of the constant 1-volt on the inverting input and the voltages applied to the noninverting input, as listed in column B. (Keep in mind that we are talking about an operational amplifier—a special one perhaps, but still an op-amp. And we are using the op-amp in an open-loop configuration, with no negative feedback. Open-loop voltage gains are very high, so the assumption in this example of an open-loop voltage gain of 250,000 is not out of line.) Column D calculates the theoretical output voltage based on the following equation:

$$\text{Voltage Out} = \text{Voltage In} \times \text{Voltage Gain}$$
$$V_O = (V_{in})(A_V)$$

In row R, column C, we find a differential input voltage of +1.0 volt. If we multiply this 1-volt input by the voltage gain of 250,000, we get a calculated theoretical output voltage of 250,000 volts (row R, column D).

In column E, row R, the real output voltage is +12 volts, the positive rail voltage. Because we can't get something for nothing, and 12 volts is all we have, that is the absolute maximum output voltage that the comparator (op-amp) can deliver.

FIGURE 9-6 How the Comparator Output Simulates a Switch

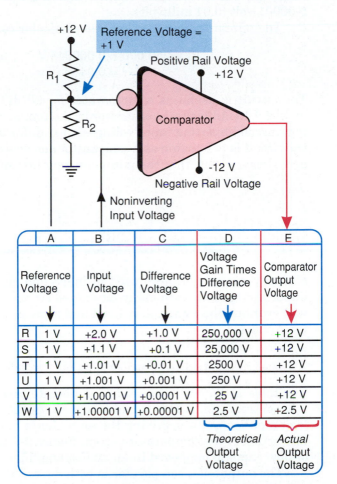

	A	B	C	D	E
	Reference Voltage	Input Voltage	Difference Voltage	Voltage Gain Times Difference Voltage	Comparator Output Voltage
R	1 V	+2.0 V	+1.0 V	250,000 V	+12 V
S	1 V	+1.1 V	+0.1 V	25,000 V	+12 V
T	1 V	+1.01 V	+0.01 V	2500 V	+12 V
U	1 V	+1.001 V	+0.001 V	250 V	+12 V
V	1 V	+1.0001 V	+0.0001 V	25 V	+12 V
W	1 V	+1.00001 V	+0.00001 V	2.5 V	+2.5 V
				Theoretical Output Voltage	*Actual* Output Voltage

In a real comparator or op-amp, the output voltage is always slightly less than the rail (power supply) voltage, because the output transistor in the amplifier will have a small saturation voltage drop ($+V_{sat}$) of a few tenths of a volt. The same is true when the comparator switches to the negative rail: we will have an output voltage of slightly less than −12 volts.

Now let's go back to the table in Figure 9-6. In row *S,* the differential voltage is 0.1 volt, and the theoretical output voltage is 2500 volts. Again, 12 volts is the limit, so we get an output voltage of 12 volts—the rail voltage. The same situation exists for rows *T, U,* and *V:* the differential input voltage times the open-loop voltage gain yields a theoretical output voltage that is greater than the rail voltage. In each case, however, the real output voltage is limited to the rail (power-supply) voltage.

In row *W,* we get both a theoretical and an actual output voltage of +2.5 volts. The circuit is not useful as a comparator in this case, because it no longer simulates the mechanical switch in the model. The comparator does not switch all the way to rail voltage, and it now acts as an ordinary linear operational amplifier. The comparator cannot resolve a differential voltage as small as 0.00001 volt (0.01 millivolt).

The minimum resolution of this particular comparator circuit is

$$12 \text{ V}/250{,}000 = 0.000048 \text{ V} \qquad \text{(differential)}$$
$$= 0.048 \text{ mV}$$

The circuit is useful as a comparator for all differential input voltages that exceed 0.48 millivolts. In other words, the comparator requires that the inverting and noninverting input voltages must differ by at least 0.048 millivolts, if the circuit is to function as a comparator and switch all the way to the rail voltage. Hence we say that the minimum **resolution** is 0.048 millivolts.

9.4   Basic Comparator Circuits

Not only are comparators used in individual circuits, but they also perform important operations in many integrated circuits. In the following sections, we will examine the comparator itself and some of the most important comparator circuits.

Analog Input/Digital Output Characteristics

The table in Figure 9-6 records a deliberate case of overdriving an amplifier into severe clipping. This would produce a dreadful signal distortion in an analog amplifier, but overdriving the amplifier is how we get the desired switch-like action of the comparator output. Figure 9-7 is a flashback, reminding you of how clipping happened in an analog amplifier.

The output of a comparator is both digital and binary, in the sense that it only has two possible states. However, the two specific comparator output voltages that are possible depend on the two power-supply (rail) voltages used in the particular circuit. The two rail voltages may or may not be the standard digital voltages of +5 volts and ground.

Zero-crossing Detector

The input voltages for a comparator are essentially analog, and they may have any polarity or magnitude within the limitations of the integrated circuit. The simplest comparator circuit is the **zero-crossing detector,** which can switch output rail voltages whenever the comparator input crosses the zero reference line (if a sine-wave input is used), as shown in Figure 9-8.

The circuit in Figure 9-8(a) is a noninverting comparator because the signal is applied to the noninverting input, with the reference voltage (0 volts in this case) connected to the inverting input. If the input signal had been connected to the inverting input, the comparator would be called an *inverting comparator.*

The comparator output voltage in Figure 9-8 switches to the positive-rail voltage as soon as the sine wave rises to a few millivolts on the positive side of 0 volts. The comparator output switches to the negative-rail voltage when the

FIGURE 9-7 **Amplifier Clipping**

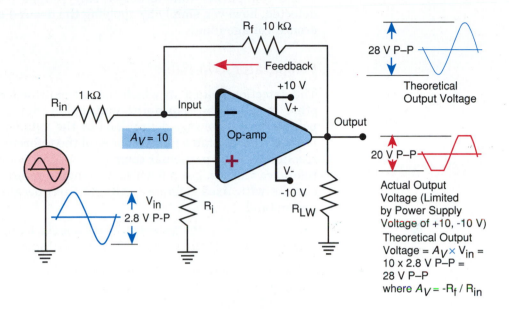

FIGURE 9-8 **Sine-wave Zero-crossing Detector**

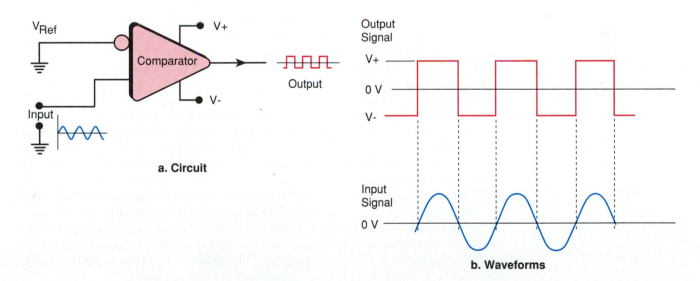

sine wave voltage falls to a few millivolts negative. The reference voltage is set to 0 volts, and the comparator changes output voltage states each time the sine wave crosses the zero reference line. The reference voltage is set to 0 volts (ground) to correspond to the zero baseline, where the sine wave's polarity reverses. This is why it is called a *zero-crossing detector*.

The circuit is often used to convert a sine-wave input into a square-wave output. The zero-crossing detector action is not limited to sine waves; any waveform that crosses zero (any **ac** waveform) can be used (see Figure 9-8(b)).

In the zero-crossing detector, the reference voltage is 0 volts, because that is the voltage we want to detect. Any voltage point on a waveform can be detected, however, simply by applying the desired detection voltage to the reference voltage input.

Comparator Trip Point

The detection voltage at which the comparator switches from one rail to the other is called the **trip-point voltage,** or simply the **trip point.** The trip-point voltage is always determined by the value of the reference voltage, and it always lies within a millivolt or so of the reference voltage (depending on the resolution of the particular comparator). Most comparator circuits use a fixed reference voltage, but a few use a changing reference voltage. We will look at circuits with fixed reference voltages first, because they are much easier to understand.

Comparator Circuits with Nonzero Reference Voltages

Figure 9-9 illustrates how the reference voltage can be set to detect when a power-supply voltage exceeds or falls below specific limits. Some electronic circuits and systems must have accurate power-supply voltages to operate properly. In those cases, it is important to be alerted if the power supply voltage has gone out of tolerance for some reason. An undervoltage or overvoltage comparator-based detector can activate an alarm or a circuit to shut the supply down as soon as it detects that the supply voltage is out of tolerance.

In Figure 9-9(a), a reference voltage of 11 volts is applied to the comparator's noninverting input, representing the power supply's upper voltage tolerance limit. The normal 10-volt power-supply voltage is connected to the inverting input. As long as the power-supply voltage is less than 11 volts, the reference voltage dominates, and the output of the comparator is switched to the +V rail. With +V on both ends of the LED, the LED is dark, indicating that the power-supply voltage is less than the tolerance limit (11 volts).

As long as the power-supply output voltage stays below 11 volts, the comparator output remains at the positive-rail voltage. But if the power-supply voltage rises to a millivolt or two above 11 volts, the inverting input becomes dominant, and the comparator output switches to the negative rail (ground). The LED now has +V on one end and ground on the other, so current flows and the LED lights up. The turned-on LED indicates a power-supply overvoltage.

In the undervoltage circuit in Figure 9-9(b), the reference voltage is connected to the noninverting input, and the power-supply voltage to be monitored is connected to the inverting input. The reference voltage is set to 9 volts, representing the lower allowable power-supply voltage limit.

As long as the power-supply output remains above 9 volts, the comparator output is switched to the positive rail, the LED is dark, and all is well. If the power-supply output voltage drops below 9 volts, however, the inverted reference voltage becomes dominant, the comparator output switches to the negative rail (ground), and this turns the LED on.

Figure 9-9(c) illustrates how the comparator converts **ac** power-line voltage into line-synchronized trigger pulses. The pulse width is controlled by varying the comparator reference voltage, as depicted in Figure 9-9(d).

FIGURE 9-9

Comparators with Nonzero Reference Voltages

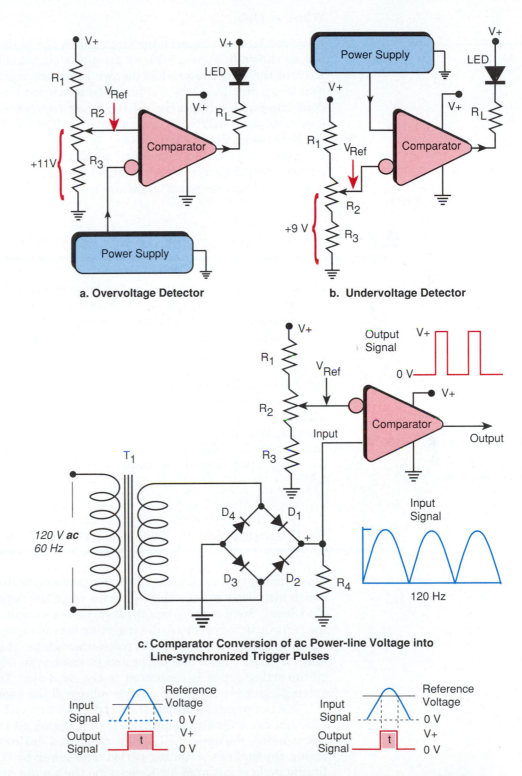

a. Overvoltage Detector

b. Undervoltage Detector

c. Comparator Conversion of ac Power-line Voltage into Line-synchronized Trigger Pulses

Lower reference voltage yields wider pulse

Higher reference voltage yields narrower pulse

d. Controlling Pulse-width by Varying the Reference Voltage

Window Detector

If we combine the overvoltage circuit with the undervoltage circuit in Figure 9-9, as shown in Figure 9-10, we get a new circuit called a **window detector.** Each of the comparators has its own reference voltage and, consequently, its own trip point. Thus, the window-detector circuit has two completely independent trip-point voltages, called the *upper trip point (UTP)* (positive) and the *lower trip point (LTP)* (negative or ground). The window represents the voltage range between the two trip points.

Figure 9-11 shows a modified window-detector circuit, in which the comparator outputs have been combined so that a single output provides both upper and lower trip-point indications.

9.5 Typical Comparator Output Connections

The circuits in Figure 9-12 show two ways in which LED indicators can be connected to the output of a comparator. LEDs are used in the example; other devices can be connected in the same way. A LED or other device can be turned on when the comparator is switched to the upper rail, as shown in Figure 9-12(a).

If you are familiar with TTL logic, you will recall that this connection is not used with TTL logic gates. TTL is capable of producing only very small output currents when the output is +V (high). The connection is just fine, however, for comparators that are capable of the same output current at either rail voltage. The condition shown in Figure 9-12(b) turns the LED on when the comparator is switched to the lower rail. If both LEDs are installed as shown, each of the two comparator output conditions will be indicated by a turned-on LED.

9.6 Comparator-based Timing Circuit

Most analog timing circuits use a **resistor-capacitor (R-C) timing circuit,** which includes a comparator that trips when the capacitor voltage is at a specified level. Such a comparator-based timing circuit is shown in Figure 9-13 (page 514). In Figure 9-13, the inverting input is connected to a potentiometer, which is used to set the desired reference voltage. The reference potentiometer may be fitted with a dial calibrated in seconds to adjust the time period. The noninverting input is connected to the capacitor. The voltage on the noninverting input is always the charge voltage of the capacitor at that instant.

The comparator's output state is defined by two LED indicators. The upper LED indicates that a timing cycle is in progress. At the end of the preset time-delay period, the upper LED will turn off, and the lower LED will turn on, indicating the end of the timing period. The lower LED will stay on until a new timing cycle is initiated by discharging the timing capacitor and allowing it to start a new charging cycle.

FIGURE 9-10   Window Detector Overvoltage Detector

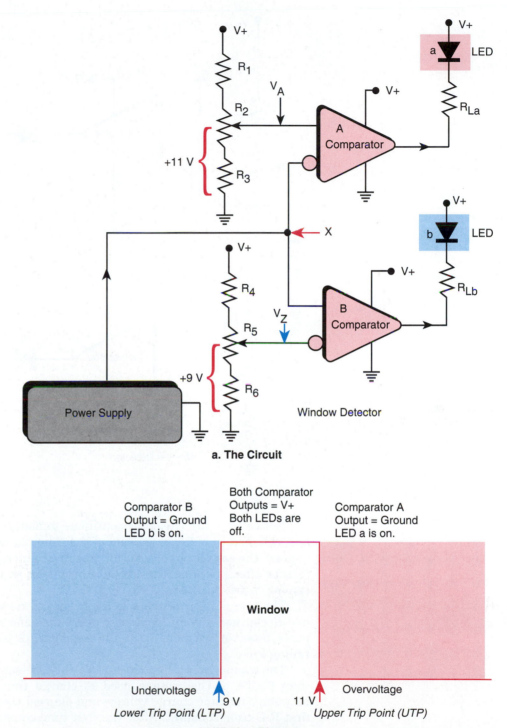

a. The Circuit

b. Window Diagram

FIGURE 9-11   Window Detector with a Common Output

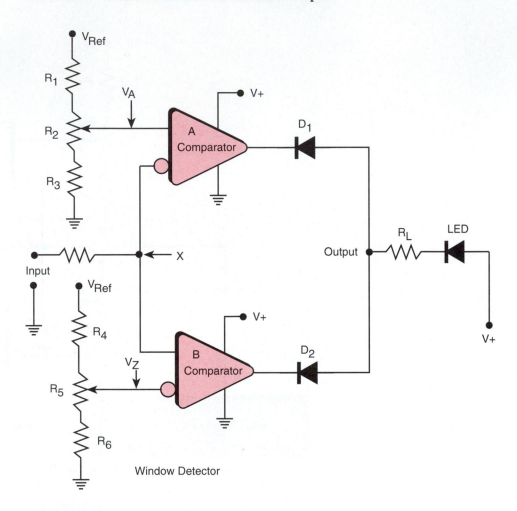

Window Detector

Notice that the capacitor continues to charge even after the comparator has switched to indicate the end of the time period. Once the comparator has tripped, the capacitor voltage can continue to increase; but it will not have any further effect, because the comparator output is already at its upper rail and cannot go any further.

The comparator switches to the lower rail as soon as the capacitor voltage has discharged to a millivolt or so below the reference voltage. For timing purposes, however, we must discharge the capacitor totally before initiating a new timing cycle.

The comparator trip point must always be set at a voltage equal to or less than 63.2% of the voltage used to charge the capacitor. The relationship between capacitor charge voltage and elapsed time is fairly linear during the first R-C time constant. After the first time constant, the nonlinearity of this relationship is considered too pronounced to allow any subsequent measurement of the voltage-to-time relationship. The 63.2% figure represents the charge on the capacitor at the end of the first time constant.

FIGURE 9-12 LED Output Circuit Conditions for the Timing Circuit in Figure 9-13

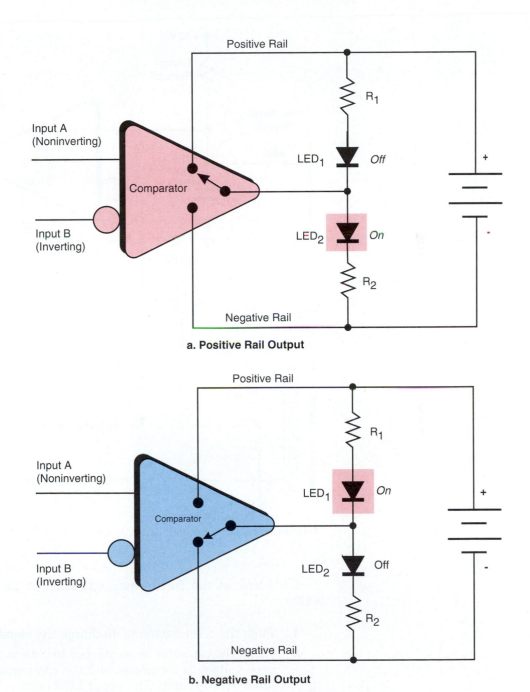

a. Positive Rail Output

b. Negative Rail Output

FIGURE 9-13   A Comparator-based Timing Circuit

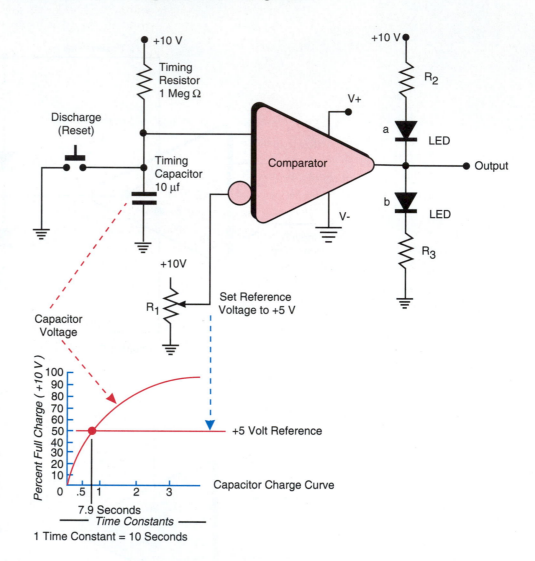

Let's look at the step-by-step operation of the timing circuit in Figure 9-13(a):

1   Push the reset button to discharge the capacitor.

2   When the capacitor is discharged to 0 volts, the inverted +5-volt reference voltage is dominant, and the comparator output switches to the lower rail (ground). The upper LED is on.

3   Release the reset button. At once, the capacitor starts to charge. The comparator output is still at the lower rail. The upper LED is still on.

4   The capacitor continues to charge, but nothing else happens until the charge on the capacitor reaches a millivolt or so above the inverted +5-volt reference voltage.

5   As soon as the capacitor charge voltage exceeds 5 volts by a millivolt or so, the comparator suddenly switches to the upper rail. The lower LED turns on, indicating that the 7.9-second timing period is finished. The upper LED turns off.

9.7   Comparator Ladders

When you watch the bar of LEDs on your stereo expand and contract with the music, you are watching a **comparator ladder** in action. Many of the new portable digital multimeters feature a voltage bar, much like the one used as a stereo output monitor, in addition to the usual digital-number readout. This expanding and contracting bar is usually referred to as an *analog meter*. This kind of expanding-bar (analog) meter has the advantage over the usual digital meter of nearly instantaneous response.

The common digital meter updates its reading every few tenths of a second and requires a little time to make its internal calculations for a new reading. In contrast, if you are trying to adjust a circuit for maximum output voltage in a working circuit, for example, the bar-type analog meter can instantly follow your adjustment changes. The digital meter, because of the relatively long time it takes to calculate a new reading, can be a very unsatisfactory instrument to use when you are trying to make adjustments in a working circuit. Having the analog bar on your digital multimeter is a real advantage in such circumstances.

Although its name suggests otherwise, the analog bar-type meter is really partly analog and partly digital. The circuit consists of a stack of comparators, each with a slightly higher reference voltage than the one immediately below it in the stack. The basic comparator ladder, applied as a fuel gauge, is shown in Figure 9-14.

In the circuit in Figure 9-14, the noninverting inputs are used as reference voltage inputs. Each comparator in the ladder has a reference voltage that is 1-volt higher than the one below it. The reference voltage can be obtained from any source, but the simple and inexpensive voltage-divider resistor is the most common source. The trip point of the bottom comparator is 1 volt. The trip point of the comparator just above it is 2 volts; the trip point of the one above that is 3 volts; and so on. The trip-point voltages increase in increments of 1 volt from the bottom step of the ladder to the top.

The inverting inputs of all comparators are connected to form a common input for the voltage that is to be measured. If we start with an input voltage derived from the linear potentiometer (in Figure 9-14), of less than 1 volt, all four LEDs will be dark. Each of the reference voltages on the noninverting inputs is higher than the input voltages on the inverting inputs, so the comparators all switch to the positive rail. The LEDs have a positive voltage on both ends, so no current flows and the LEDs stay dark.

If the float in the fuel tank rises high enough to cause the linear potentiometer to provide an input voltage of a millivolt or so above 1 volt, the bottom comparator switches to the lower rail. The LED then turns on, indicating a fuel level of one-quarter tank.

FIGURE 9-14   A Ladder Comparator Fuel Gauge

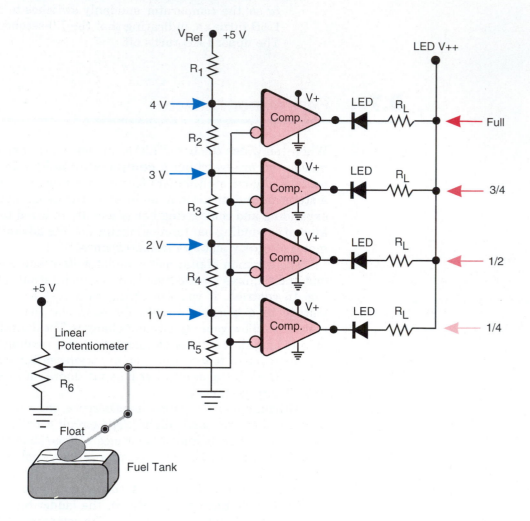

If more fuel is added to the tank and raises the float high enough, the linear potentiometer will deliver sufficient voltage to cross the 2-volt trip point. In response, the second comparator output switches to the negative rail (ground), and the second LED turns on, indicating that the tank is one-half full. The first LED stays on because 2 volts exceeds the 1-volt trip point for the first comparator (the one-quarter tank indicator). As the tank fills, the potentiometer produces an increasing voltage at the inverting inputs. As each comparator's trip point is crossed, the comparator switches, turning its associated LED on. The previous LEDs stay on, and the bar of lit LEDs gets longer as the tank fills.

There is no theoretical limit to how many comparators we can install in the ladder. We can also set the increment between the comparators to any reasonable value. In the example, we supplied 4 volts at the top of the divider string. Four resistors (and four comparators) are involved in the reference voltage divider, so 4 volts divided by four resistors or comparators gives us our increment: 1 volt per comparator. Now suppose that we change the voltage at the top of the divider string from 4 volts to 1 volt. The new increment per comparator becomes

$$1 \text{ (volt)}/4 \text{ (comparators)} = 0.25 \text{ volts per division (increment)}$$

The increment defined here may also be called volts-per-division or volts per LED. If we assume an increment of 1 volt per division, the volts per division is also the **resolution** of the meter. The output LEDs light in 1-volt steps. In-between voltages such as 2.5 volts cannot be displayed and, hence, are not available. An analog input voltage of 2.0 volts will light the 1-volt and 2-volt LEDs, indicating a measured input voltage of 2 volts. However, a 2.5-volt analog input can only be displayed as 2 volts. The system cannot resolve values any closer than 1 volt apart.

Analog versus Digital

The comparator responds to analog input voltages and has an infinite resolution—no limitations. This is the nature of analog signals. The comparator outputs are essentially digital and can respond only in specific increments. Digital data always have a finite resolution limit.

In the preceding example, the comparator ladder could resolve values only in increments of 1 volt. We could, for example, make a string of 1000 comparators with a 1-volt reference voltage at the top of the string. We would then have a meter with an analog input that would provide a 1-volt full-scale reading, with an interior-range resolution based on increments of 0.001 volt. The meter would read a maximum voltage of 1 volt, but its resolution would be 1 millivolt (0.001 volt).

Digital devices always have a definite resolution. But we can make the defining increment (resolution) as small as we wish, if we are willing to use enough comparators or other hardware. Fortunately for digital electronics, integrated technology allows us to put hundreds of thousands to millions of circuits on a chip.

The comparator ladder is essentially a very fine and fast analog-to-digital converter. In its common form (shown in Figure 9-14), the ladder has an analog input and a digital output, but the digital output is not in the binary form required by computers. Integrated circuits are available that can translate the comparator ladder's decimal output into standard binary digital form.

Many high-speed analog-to-digital converters that are used for converting video digital data use the comparator ladder plus binary encoding circuits on an IC chip. The IC is called a *flash analog-to-digital converter,* in recognition of its ability to make almost instantaneous conversions. Many other kinds of analog-to-digital conversion circuits exist; most of them are much slower (but much less expensive) than the flash converter.

Commercial Comparator Ladder Integrated Circuits

The LM-3914 and LM-3915 bar-dot LED display drivers are typical of the available, everything-on-a-chip, comparator ladder devices available. Figure 9-15(a) shows a functional block diagram of the LM-3914/3915. The LM-3914 consists of ten comparators in a ladder structure, with a linear voltage divider for the reference voltage string.

FIGURE 9-15   The LM-3914 and LM-3915 LED Bar-dot Drivers

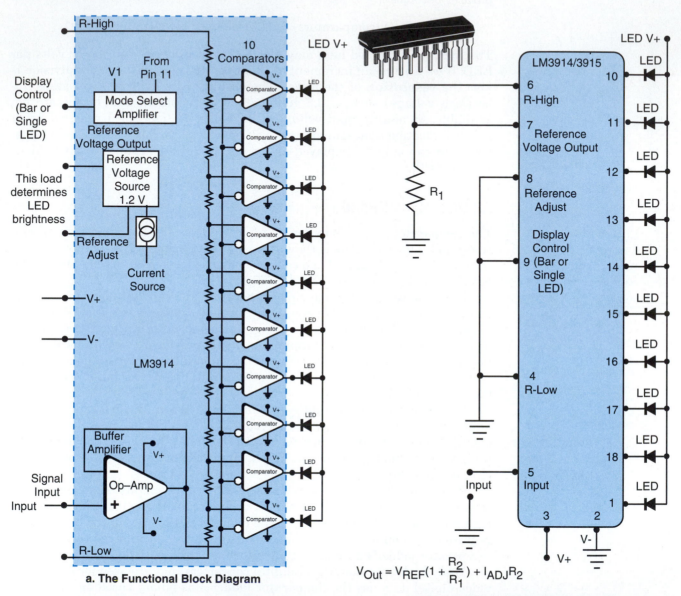

a. The Functional Block Diagram

$$V_{Out} = V_{REF}(1 + \frac{R_2}{R_1}) + I_{ADJ}R_2$$

b. Typical Basic Analog Meter Circuit

The LM-3915 uses resistors with unequal values in the reference voltage string. Using different resistor values creates a ladder with varying increments. If resistor values are selected properly, the LED output representation can be made to approximate the logarithmic decibel (dB) scale. The dB scale, used in the LM-3915, is much better than any constant-increment scale as an audio power output indicator, because it more nearly approximates the logarithmic response of the human ear. Other kinds of nonlinear scales can be created, but they are not normally commercially available in an IC package.

The outputs of all of the comparators in the 3914/15 are connected to the output pins through constant-current sources. This eliminates the need for a current-limiting resistor for each LED. The constant-current sources are actually controlled sources—a feature that allows the current to all of the LEDs to be adjusted simultaneously with a single resistor. LED brightness can easily be set to the desired level, and even automatically adjusted for daylight and dark viewing conditions. Interesting effects, such as a flashing overrange, can be accomplished by using the LED current-control pin.

In the LM3914/15, both ends of the reference voltage-divider string are brought out to pins for maximum flexibility. In most cases, the divider-string voltage is connected to the internal 1.2-volt reference source, to provide a natural full-scale range of 1.2 volts and a resolution of 0.12 volt (125 mV) per division. A resistive voltage divider can be used to alter the full-scale range and the resolution. We can measure larger input voltages by using a voltage divider on the input. Two or more units can be cascaded for twenty or more steps.

The LM3914/15 also contains a mode amplifier, which allows a normal expanding-bar presentation or a single-dot-at-a-time display. The mode input can also be used to aid in cascading units for larger counts. A voltage follower on the input buffers the IC from the external circuits and provides a high input impedance.

Figure 9-15(b) shows a typical circuit application for the LM-3914. Similar devices are available, such as the TSM3934/36 module, which includes a 3914/15-style IC driver and a 10-LED light bar in a single functional package. LED bar-dot drivers for various numbers of steps can also be found.

9.8   Circuits That Use Variable Reference Voltages

Comparators often use variable reference voltages to control the output pulse width of a comparator-based square-wave generator. The variable reference voltage may be a repetitive waveform (to form pulse-width modulator circuits) or it may be a music signal (to form frequency modulation circuits).

Using a Reference Voltage to Control the Pulse Width of a Pulse Shaper

Most common comparator circuits use a fixed reference voltage, but one special (and important) class of comparator circuits uses variable reference voltages. A waveform can be used as a reference voltage when it is necessary to modulate one signal with another.

Figure 9-16(a) shows a schematic diagram of a circuit in which an adjustable reference voltage is used to produce a rectangular output pulse that has a continuously variable pulse width. The circuit depends on two factors: the slope of a triangular waveform into the comparator input, and the value of the reference voltage. In Figure 9-16(b), the reference voltage is set to sightly above 0 volts. As the triangular input voltage rises just slightly above the 0-volt line, it crosses the comparator trip point.

FIGURE 9-16   Controlling Linear Variable Pulse Width by Varying
the Reference Voltage

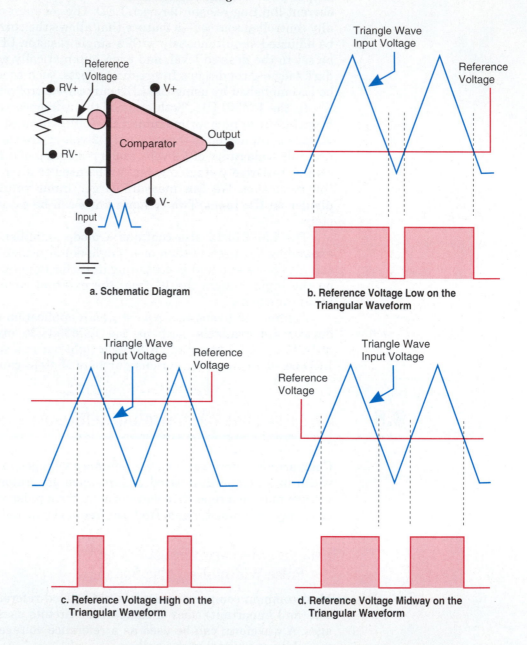

a. Schematic Diagram

b. Reference Voltage Low on the
Triangular Waveform

c. Reference Voltage High on the
Triangular Waveform

d. Reference Voltage Midway on the
Triangular Waveform

The trip point is set by the reference voltage and is just slightly above 0
volts. The comparator trips just as the triangular wave rises above the zero
line. The comparator output stays at the positive rail voltage until the triangular wave falls to near zero, where it crosses the trip point again. The comparator output stays at the positive rail during almost the entire triangular
wave period.

In Figure 9-16(c), the reference voltage sets the trip point very high on the
triangular wave, so the comparator trips and the triangular wave starts down

again almost immediately, very quickly crossing the trip point. The comparator output stays at the positive rail voltage for a very short time, yielding a narrow pulse.

Figure 9-16(d) shows the reference voltage (and the trip point) set to midway up the sawtooth slope, yielding a medium-width pulse. Study the drawing carefully, since this is very difficult to explain in words.

Pulse-width Modulator

The previous circuit was a rather trivial application of the pulse-width control, consisting of varying the reference voltage; but it is necessary to understand the circuit in Figure 9-16(a) if we are to understand what happens in the more complex situation presented in Figure 9-17.

FIGURE 9-17 **Linear Pulse-width Modulator's Use of an Input Signal as a Reference Voltage to Vary the Pulse Width**

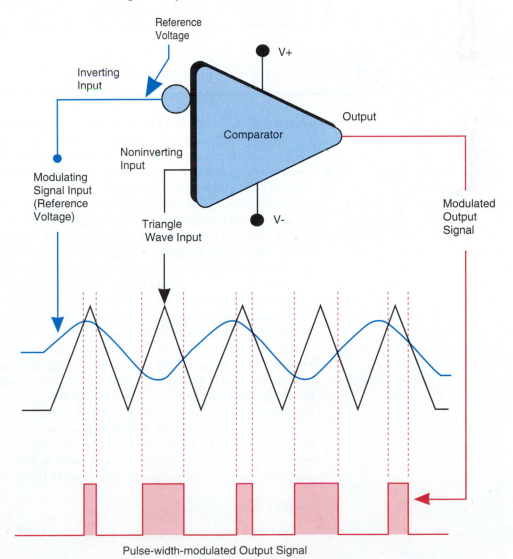

Pulse-width-modulated Output Signal

Pulse-width modulation is a form of frequency modulation that is used in several kinds of communications systems and in switching power-supply circuits. The basic principle is also used in certain kinds of phase detectors. The frequency of the modulating signal is always much lower than the frequency being modulated.

In the circuit in Figure 9-17, a triangular wave is applied to the comparator's noninverting input. The modulation voltage or signal is applied to the inverting input as a varying reference voltage. When the signal voltage rises, it moves the reference voltage and the comparator trip point upward on the triangle, resulting in a wider pulse.

The pulses become narrower as the modulating signal voltage decreases and moves the trip point lower on the triangle. The width of the comparator's output pulse is proportional to the modulating input voltage and its width varies as the modulating signal varies. The modulating signal variations are encoded in the form of pulses of varying width; and with a proper decoding circuit, the original modulating waveform can be recovered. We will see some practical communications applications of pulse-width modulation in Chapter 12, and some power-supply applications elsewhere.

9.9   Schmitt Trigger and Hysteresis

In some comparator applications, the trip point is simply too sensitive to be practical. The output can be erratic—a condition called **jitter,** when viewed on a scope. Fortunately, there is a simple way to provide the comparator with two separate trip-point voltages, counteracting jitter problems and opening the way for some new circuit configurations.

The procedure involves adding positive feedback. As soon as positive feedback (as shown in Figure 9-18(a), page 524) is added, the circuit has two trip points and gets a new title. A comparator circuit with positive feedback is called a **Schmitt trigger.** Besides creating two trip points, the introduction of positive feedback also introduces a new concept, called **hysteresis.**

The hysteresis voltage is the voltage difference between the two trip-point voltages; within this voltage range, no change occurs in the comparator output. The hysteresis voltage area in Figure 9-18(c) thus represents a dead zone—where nothing happens—between the two trip-point voltages.

In Figure 9-18(a), the comparator reference input (noninverting) is connected to the midpoint of the R_1–R_2 voltage divider. Because a fraction of the output voltage is returned to the noninverting input, this is a case of positive feedback. The fact that we are dealing with positive feedback is not yet obvious, but it will be as soon as an output voltage change takes place.

In Figure 9-18(a), the output voltage is +10 volts, and the voltage divider divides it down to +3.19 volts for the reference (noninverting) input. The reference voltage is no longer derived from a fixed voltage, but from the comparator's output. Let's see how the process takes place, step by step.

Step by Step

  1   The comparator output voltage is +10 V.

2   The R_1–R_2 voltage divider reduces the +10 V output voltage to +3.19 V. The reference voltage is +3.19 V.

3   If the input voltage on the inverting input is less than +3.19 V, the comparator output remains at the +10-V rail. If the voltage on the inverting input is increased until it is slightly higher than +3.19 V, the trip point is crossed and the comparator output switches to the –10-V rail.

4   The comparator output voltage switches to –10 V, and we have a new reference voltage of –3.19 V instead of +3.19 V (see Figure 9-18(b)). The reference voltage to the noninverting input is now controlled by the output voltage, so this is positive feedback.

5   Now, to switch the comparator output back to the +10-V rail, we must make the inverting input voltage change all the way to –3.19 V, and we must make the input voltage cross the new –3.19-V trip point. As soon as the –3.19-V trip point is crossed, the comparator output switches back to the +10-V rail, and the reference voltage changes to +3.19 V again.

The comparator circuit now has positive feedback, two trip points (+3.19 V and –3.19 V), a dead zone called the *hysteresis voltage,* and a new name: *Schmitt trigger.* Simply adding two positive-feedback resistors has completely changed the character of the comparator circuit. The hysteresis diagram for the circuit is shown in Figure 9-18(c).

9.10   Schmitt Trigger Comparator-based Square-wave Oscillator

The Schmitt trigger comparator is essential to the operation of the comparator-based square-wave oscillator. We obtain this oscillator (one of the most important of all the comparator-based circuits) by modifying the timing circuit in Figure 9-13 as shown in Figure 9-19(a), page 526. The first modification involves adding positive feedback. This change yields a Schmitt trigger with two trip points. The second modification involves connecting the capacitor charging resistor to the output of the comparator instead of to a fixed charging voltage.

With these changes, the comparator timing circuit switches itself automatically at the end of each charging cycle and starts a new charging cycle in the opposite direction. We'll walk through this process next.

Step by Step

1   The output of the comparator is at the +10-V rail (see Figure 9-19(b)).

    a   The comparator reference voltage is +3.19 V.

    b   The capacitor charging voltage is +10 V.

2   The capacitor starts charging toward the +3.19-V upper trip point (*UTP*).

FIGURE 9-18   Schmitt Trigger, Positive Feedback, and Hysteresis

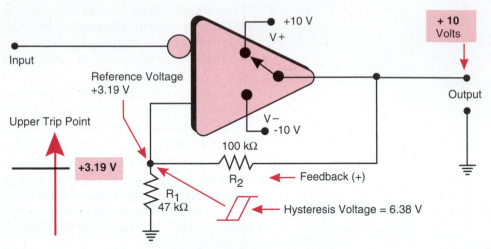

a. Conditions for Steps 1, 2 and 3

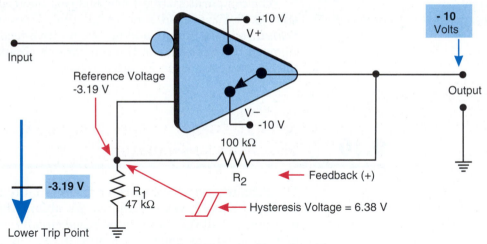

b. Conditions for Steps 4 and 5

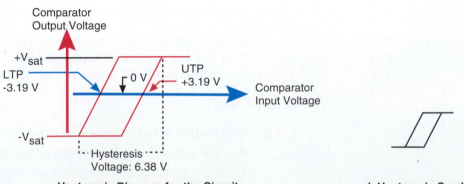

c. Hysteresis Diagram for the Circuit

d. Hysteresis Symbol

FIGURE 9-18   Continued

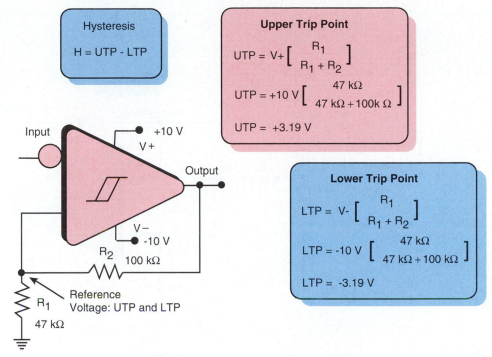

e. Calculating Hysteresis and Trip Points

3 When the capacitor charge voltage crosses the +3.19-V upper trip point (*UTP*), the comparator output switches to the −10-V rail.

4 The output of the comparator is at the −10-V rail (see Figure 9-19(c)).

    a The comparator reference voltage is −3.19 V.

    b The capacitor charging voltage is −10 V.

5 The capacitor starts charging toward the −3.19 V lower trip point (*LTP*).

6 When the capacitor charge voltage crosses the −3.19-V lower trip point (*LTP*), the comparator output switches to the +10-V rail.

7 The output of the comparator is at the +10-V rail (see Figure 9-19(b)).

    a The comparator reference voltage is +3.19 V.

    b The capacitor charging voltage is +10 V.

Then the cycle starts over again; it continues indefinitely, producing a continuous square-wave output. The circuit is called a *square-wave oscillator* or *square-wave generator*.

We will find this Schmitt trigger comparator-based square-wave generator at the heart of an entire family of integrated-circuit voltage-controlled oscillators, phase-locked loops, and similar devices.

FIGURE 9-19 Schmitt Trigger Comparator-based Square-wave Oscillator

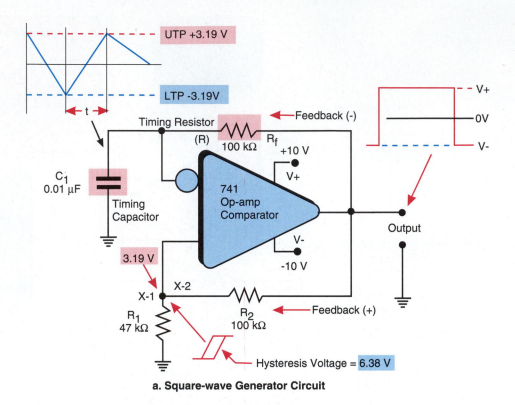

a. Square-wave Generator Circuit

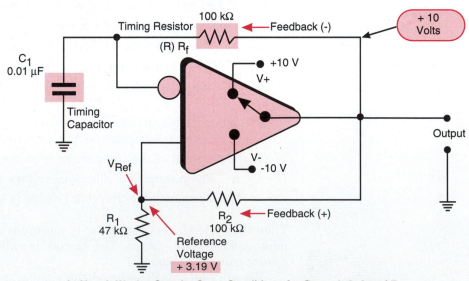

b. How it Works, Step by Step: Conditions for Steps 1, 2, 3 and 7

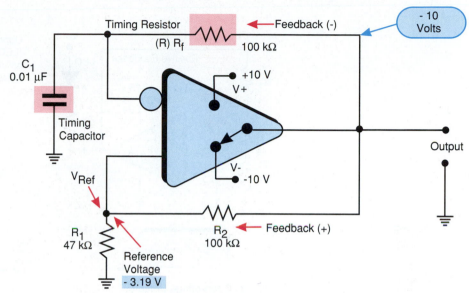

c. How it Works, Step by Step: Conditions for Steps 4, 5 and 6

Schmitt Trigger as a Memory Element

In some applications, it is essential to know whether something happened when nobody was watching. For example, we may need a circuit to light a LED if some overvoltage threshold is exceeded, and to keep the LED lit even though the voltage returns to normal during the night. The power-supply overvoltage monitor circuit shown in Figure 9-20(a) is an example of how a comparator with hysteresis (Schmitt trigger) can be used to detect a transient event and remember that it happened.

How It Works

The circuit in Figure 9-20(a) is designed to monitor a 10-volt power supply, to turn a LED on if the power-supply voltage exceeds 11 volts, and to keep the LED on even if the overvoltage condition goes away. The trick is hysteresis:

1   Adding hysteresis creates two trip points. In this case, the upper trip point is set to 11 volts. The lower trip point is approximately 0 volts, because the negative power-supply rail is grounded instead of being connected to some negative voltage. (The circuit would work just as well with trip points of +11 volts and −11 volts, but a negative voltage is not required. Positive- and negative-rail voltages that have different absolute values can be used to get nonsymmetrical upper and lower trip points, but it is not a very common procedure.)

2   Initially, the comparator output voltage must be at the +15-volt rail.

FIGURE 9-20   **How an Overvoltage Detector with Hysteresis "Remembers" an Overvoltage Condition**

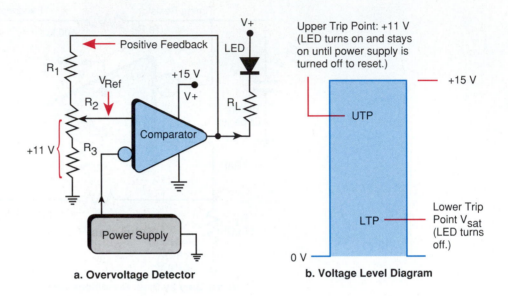

a. Overvoltage Detector   b. Voltage Level Diagram

3   The upper trip point is set by the voltage divider (R_1, R_2, and R_3) to +11 volts. The noninverting input is at a reference voltage of +11 volts.

4   The voltage applied to the inverting input from the power supply that is being monitored is normally +10 volts. The comparator output voltage remains at the +15-volt rail as long as the power-supply output voltage stays below +11 volts.

5   Suppose a sudden voltage surge occurs and the power-supply voltage rises to 12 volts. Since 12 volts is above the upper trip-point voltage of +11 volts, the comparator output will switch to the negative rail, turning the LED on.

6   If the power supply later returns to its normal 10-volt output, the comparator output stays at the negative (ground) rail. In order to reset the comparator and turn the LED off, the power-supply voltage must cross the lower trip point at approximately 0 volts. Each time the power supply is turned off, the overvoltage alarm is reset. Subsequently, the overvoltage alarm is rearmed when the power supply is turned back on.

Bistable Schmitt Trigger Latch

The bistable Schmitt trigger latch in Figure 9-21 is an extension of the idea presented in Figure 9-20. In the bistable version, provision is made for a trigger pulse to force the comparator to the positive rail. Once tripped, the trigger pulse can vanish, but the comparator stays at the upper rail voltage. To reset the circuit, the operator forces the comparator to the lower rail by transmitting a trigger pulse. Hysteresis creates a bistable circuit where the comparator latches in one of two possible states—at the upper rail or at the lower rail.

FIGURE 9-21   Latching Schmitt Trigger Circuit

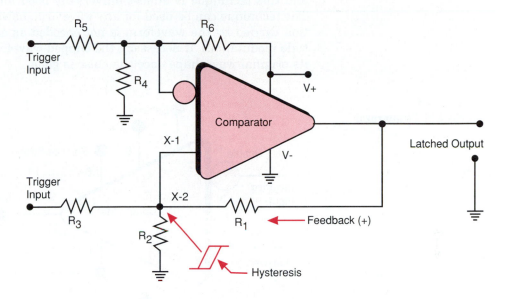

Using a Schmitt Trigger to Remove Noise from a Waveform

Whenever a waveform is transmitted from one location to another, there is always the possibility that stray electrical noise will be added to the original waveform. Noise voltages always add to (or subtract from) the voltage of the original waveform, and thus constitute a form of amplitude modulation. Figure 9-22 illustrates the situation.

An ordinary comparator is of no help here; but if we add hysteresis and a second trip point, as shown in Figure 9-22, the noise can be eliminated. Figure

FIGURE 9-22
Hysteresis and the Noise
Problem

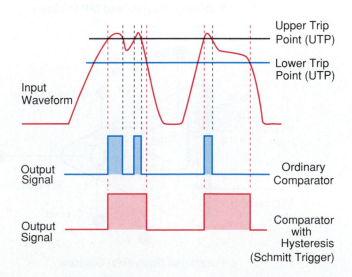

9-22 assumed that the original waveform was a rectangular or square wave, and this technique is almost universally used for cleaning up digital pulses. The technique can be used for any waveform, however, as long as the information carried by the waveform is not encoded as amplitude variations (amplitude modulation). If necessary, the clipped waveform can easily be restored to its original waveshape once the noise is gone.

FIGURE 9-23

Versatile 311 Comparator

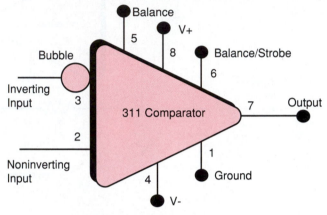

a. Schematic Symbol

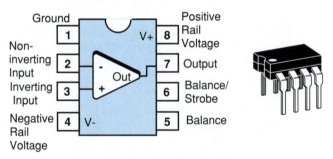

b. Basing Diagram and DIP Package

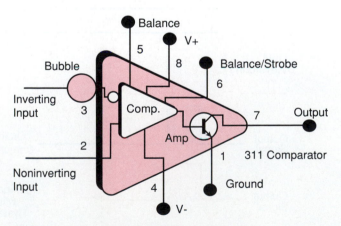

c. Functional Equivalent Diagram

9.11   The Very Versatile 311 Comparator

The 311 comparator is a popular IC comparator because of its versatility. The 311 features high open-loop gain, a very fast slew rate, external balance, and digital strobe terminals. In addition, the 311 can be used to interface an analog circuit to a digital circuit or to another analog circuit when the two circuits require different power-supply voltages. The comparator section and the output stage can use separate single or dual power supplies.

Figure 9-23(a) shows the common 311 symbol. Figure 9-23(b) shows the basing diagram of the most common 8-pin DIP, and the DIP package. Figure 9-23(c) shows a simplified internal functional diagram of the 311 comparator.

Balancing out Offset Voltage

Figure 9-24 shows the 311 comparator using a dual power supply for the comparator section, and using a single power-supply voltage for the output stage. The proper connection is also shown for situations where precise offset voltage balance (null) is essential. If the balance/strobe terminal is not used, it is usually connected to the comparator's positive supply to avoid possible stray noise pickup.

FIGURE 9-24   Balance Circuit

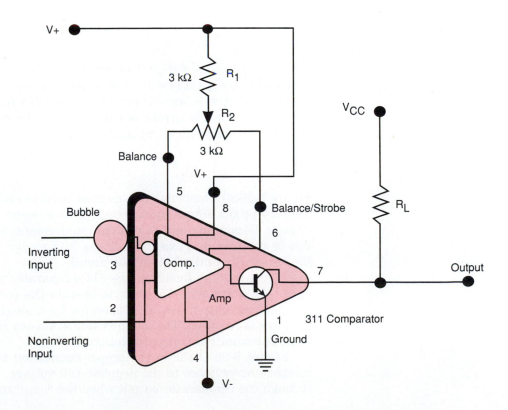

FɪɢᴜʀE 9-25 **Strobe Circuit**

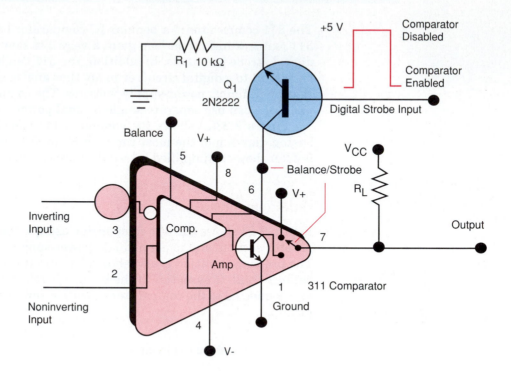

Enable or Disable Control Pin

The strobe allows a digital circuit to enable or disable the comparator, as required by a digital system. The strobe terminal is a current-operated circuit. Anything that will draw a current of 3 to 5 mA from the terminal will disable the comparator. The strobe pin should never be connected directly to ground. Figure 9-25 shows a typical strobe circuit.

311 Output Circuit

Figure 9-26(a) illustrates one method used in ordinary comparators to allow them to produce an output voltage that is some voltage other than the rail voltage. Depending on the voltage requirement, these output add-on circuits can be complex and expensive. Parts (b) through (g) of Figure 9-26 illustrate the versatility of the 311 output circuit.

Figure 9-26(b) illustrates the 311 comparator using two power-supply voltages (+15 volts and −15 volts) to operate the comparator section, while the independent output circuit is operated by a single +10-volt supply. No extra circuitry is required. The load resistor is always required, because the output is an uncommitted transistor collector.

Figure 9-26(c) shows the proper connection to turn on a LED when the comparator switches to the negative-rail voltage. The LED itself is grounded through the comparator output when the comparator is at the negative rail.

Figure 9-26(d) shows a relay connected to close when the comparator section switches to the −12-volt rail. The relay is powered by a separate 24-volt supply. The snubber diode is necessary to prevent the relay's collapsing magnetic field induction from damaging the output transistor.

Figure 9-26(e) shows the comparator and the output circuit connected to two different dual power-supply voltages. In Figure 9-26(f), the same dual power-supply voltages are used to power both the comparator section and the output stage.

Figure 9-26(g) shows the output circuit connected as an emitter follower. These examples don't exhaust the possibilities, but they provide some insight into just how versatile the 311 can be.

FIGURE 9-26 **311 Comparator Output Circuit**

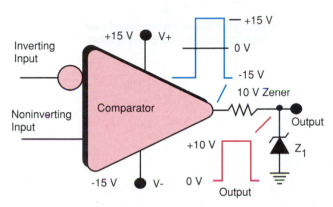

a. Altering the Output Voltage of
a Conventional Comparator

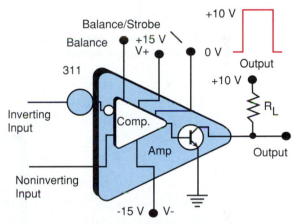

b. Versatile Output Circuit of the 311

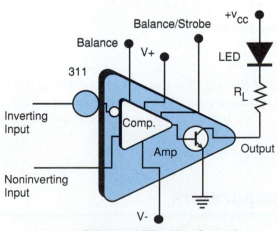

c. Driving a LED with a Ground
Reference

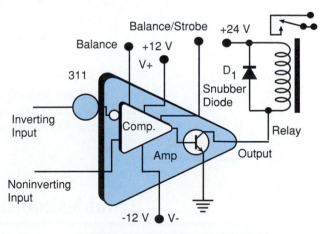

d. Using the 40-V, 50-mA Output to
Drive a Relay

(Fig. 9-26 continues)

FIGURE 9-26   Continued

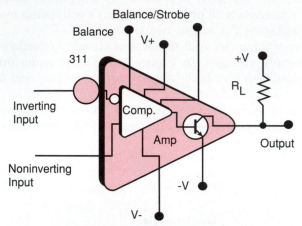

e. Using Different Supply Voltages for Comparator and Output Amplifier

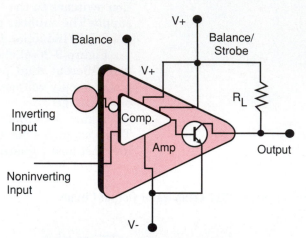

f. Using the Same Power Supply for Comparator and Output Amplifier

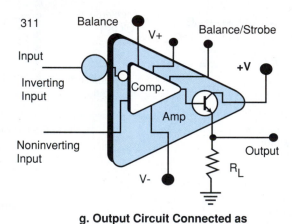

g. Output Circuit Connected as an Emitter Follower

There are several versions of the 311. For example, the A311 is the standard; the AF311 is a very low input current (high input impedance FET) version of the 311. The trade-off for low input current is slower response time.

Troubleshooting Comparators

Both common comparators and Schmitt trigger comparator circuits can be tested quickly without being removed from the printed circuit board. Simple forcing tests are quick and certain.

Quick Check (Noninvasive) for a Common Comparator

Symptom: Comparator circuit doesn't seem to work.

This test (depicted in Figure 9-27) allows you to force a comparator from its present state to the opposite rail. If both comparator output states are possible, the comparator IC and the power-supply voltages are okay. If the comparator passes this test, the problem most likely involves the reference voltage or the input signal. This test can be used with any circuit that uses a voltage divider as a reference-voltage source. It works with nearly any input signal, any combination of rail voltages, and any reference voltage.

Procedure

1   Temporarily connect jumper J_1 or J_2, to force the comparator from its current state to the opposite rail voltage. This will not cause any damage unless the voltage applied to the reference input exceeds the rail voltage (power-supply voltage). This is usually not a problem.

2   If you can't force the comparator into the alternate state, check the power-supply voltages and the voltages in the reference-voltage circuit. Another possibility is that the IC is bad.

3   Verify that the reference voltage is correct. Readjust the reference voltage (assuming that adjustment is possible).

FIGURE 9-27

Quick Test (Noninvasive) for Common Comparator

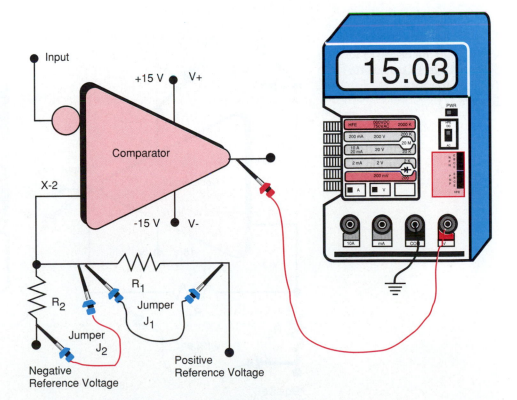

4 If the reference voltage is okay, and if it is possible to force the comparator into the alternate state, then the comparator is okay, and the likely problem is with the input signal.

Quick Check (Noninvasive) for a Comparator with Hysteresis (Schmitt Trigger)

Symptom: Schmitt trigger circuit doesn't seem to work.

This test (depicted in Figure 9-28) allows you to force a Schmitt trigger from its present state to the opposite rail. If both comparator output states are possible, the comparator IC and the power-supply voltages are okay. If the comparator passes this test, the problem most likely involves the positive-feedback circuit-reference voltage or the input signal. This test can be used with any circuit that uses a voltage divider as a reference-voltage source. It works with nearly any input signal, with any combination of rail voltages, and with any reference voltage.

Procedure

1 Temporarily connect jumper J_1 or J_2, to force the comparator from its current state to the opposite rail voltage. This will not cause any damage.

FIGURE 9-28

Quick Test (Noninvasive) for Schmitt Trigger Comparator (Comparator with Hysteresis)

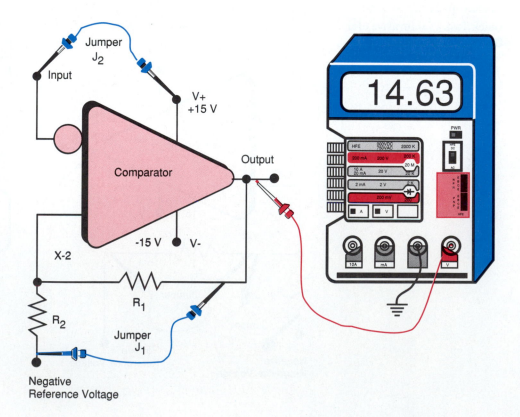

2   If you can't force the comparator into the alternate state, check the power-supply voltages and the components in the positive-feedback circuit. Alternatively, the IC may be bad.

3   If the positive feedback circuit is okay, and it is possible to force the comparator into the alternate state, the comparator is okay and the likely problem is with the input signal.

If the problem seems to be an input signal problem, use an oscilloscope to verify that the waveform has the correct shape and amplitude. You will need the manufacturer's technical data to determine the correct signal waveform and amplitude. If the input signal is nonexistent or incorrect, the problem lies in one of the circuits prior to the comparator.

The Digital Connection: Introduction to Flip-flops

The **flip-flop** is a latching circuit; once set it stays set until it is deliberately reset. Then it remains in the reset condition until it is given another set signal. A flip-flop uses positive feedback and can be constructed from two of any kind of inverting amplifier. Because the flip-flop is so common in digital circuits, digital versions are often used in analog circuits as well. The inverting amplifiers used in flip-flops are operated in a comparator-like switching mode, with (normally) a large input signal. This means that a small voltage gain is adequate for the purpose.

The inverting amplifiers shown in Figure 9-29 operate in a comparator-like switching mode, switching the output from +5 volts to ground. Assume a voltage gain of 100 or so, to ensure solid rail-to-rail switching. When the input is +5 volts, the inverting amplifier produces a 0-volt (ground) output voltage. When the input voltage is 0 volts, the inverted output voltage is +5 volts. In Figure 9-29, op-amp #1 has an input voltage of +5 volts through R_1, causing an inverted output voltage of zero (ground). The output of op-amp #1 is zero/ground, which pulls the input of op-amp #2 to ground.

The input voltage to op-amp #2 is zero, which is inverted for an output voltage of +5 volts. The +5 volts at the input of op-amp #1 is inverted to 0 volts at the output of op-amp #1, and 0 volts is applied to op-amp #2's input. The 0-volt input voltage is inverted and produces a voltage of +5 volts at the output of op-amp #2.

The inversion of +5 volts to zero represents a 180° phase shift, as does the inversion of zero to +5 volts. The total phase shift through the two amplifiers is 360°. The output of op-amp #2 is in phase with the input voltage of op-amp #1.

Flip-flop Using an Inverter

Figure 9-29(b) shows the standard symbol for a digital inverting/switching amplifier called, appropriately, an inverter. We will use this digital inverter in

FIGURE 9-29 **Inverting Switching Amplifiers with Large dc Input Voltages**

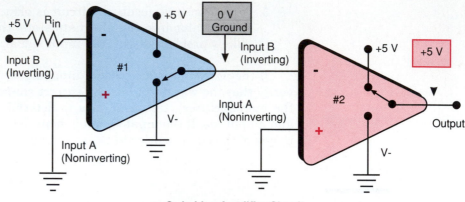

a. Switching Amplifier Circuit

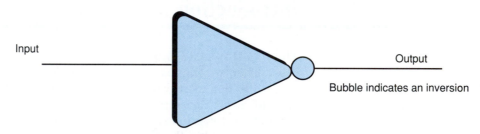

b. Digital Switching Amplifier (Inverter)

the following explanation of how a flip-flop works. Figure 9-30 shows the most common way of drawing the basic flip-flop, as a pair of cross-coupled inverters. Many students find this form of the drawing hard to follow, so we will draw it a little differently for our explanation.

How It Works

First, let's review a theoretical example, as laid out in Figure 9-31.

Phase 1: Initial Conditions   Figure 9-31(a) shows the initial conditions of the circuit. Which of the two inverters comes up with a +5-volt output when the power is turned on is a coin-flip. Let's assume that the initial conditions are as shown on Figure 9-31(a). Because output Q is connected to the input of inverter #2; and the $\overline{Q}$ (Not-Q) output is connected to the output of inverter #2, the two outputs are always in opposite states—one at +5 volts and one at 0 volts.

Phase 2: Setting the Flip-flop   Now suppose that we momentarily push the set button (see Figure 9-31(b)). This momentarily grounds the output of inverter #2, and because of the feedback connection, the input of inverter #1 is also at ground. Because the input of inverter #1 is at ground, its inverted output goes to +5 volts. The input of inverter #2 is at +5 volts, so its inverted output is now at ground.

FIGURE 9-30   **Flip-flop Drawn as Cross-coupled Inverters**

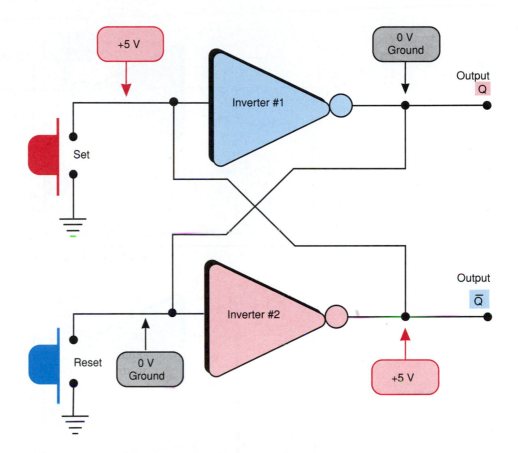

Both the set push-button and the inverter #2 output are at ground. We can release the push-button, and nothing will happen. The Q output is now at +5 volts—the Set condition—and will remain in that state indefinitely, unless we disconnect the power or press the reset button. The flip-flop is now latched at Q = +5 volts, and $\overline{Q}$ (Not-Q) is at 0 volts.

Phase 3: Resetting the Flip-flop   Now suppose that we momentarily push the reset button (see Figure 9-31(c)). This momentarily grounds the output of inverter #1. The input of inverter #2 also goes to ground. Because the input of inverter #2 is at ground, its inverted output goes to +5 volts. The input of inverter #1, by way of the feedback path, is at +5 volts, so its inverted output is now at ground.

Both the reset push-button and the inverter #1 output are at ground. We can release the push-button, and nothing will happen. The Q output is now at 0 volts—the Reset condition—and will remain in that state indefinitely, unless we disconnect the power or press the set button again. The flip-flop is now latched at Q = 0 volts and $\overline{Q}$ (Not-Q) is at +5 V. The flip-flop remembers one of two possible states or conditions.

FIGURE 9-31   Flip-flop Drawn for the Example

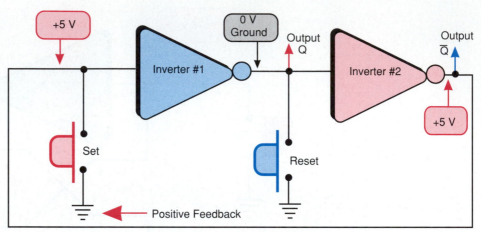

a. Phase 1: Initial Conditions

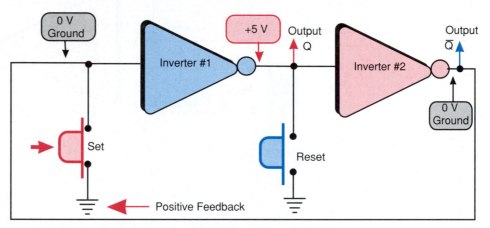

b. Phase 2: Setting the Flip-flop

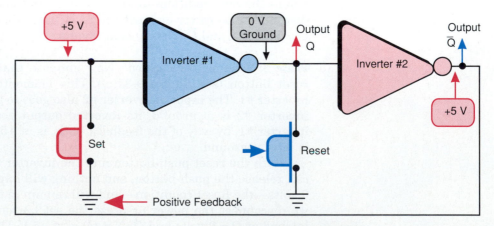

c. Phase 3: Resetting the Flip-flop

So much for theory. To get practical, we must be able to set or reset the flip-flop with an electronic pulse or other electronic signal. Figure 9-32 shows one way to accomplish this. A transistor replaces the set and reset push-buttons. A brief turn-on voltage applied to the transistor base acts in the same way as a momentarily closed push-button, setting or resetting the flip-flop.

Figure 9-33 shows the standard symbol for the set–reset flip-flop.

FIGURE 9-32   **Flip-flop Modified to Use Electronic Set and Reset Signals**

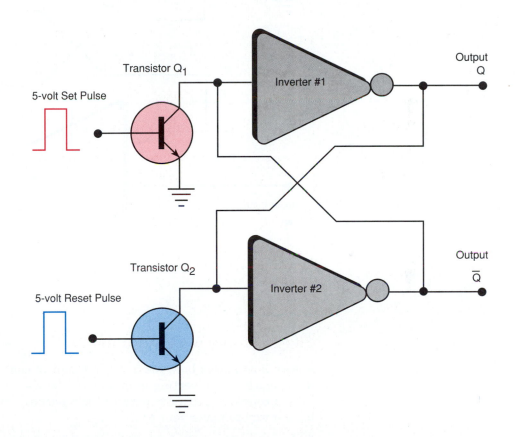

FIGURE 9-33   **Standard Symbol for a Set–Reset Flip-flop**

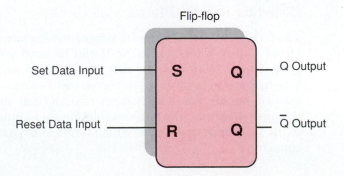

FIGURE 9-34 Flip-flop Using Two Comparators

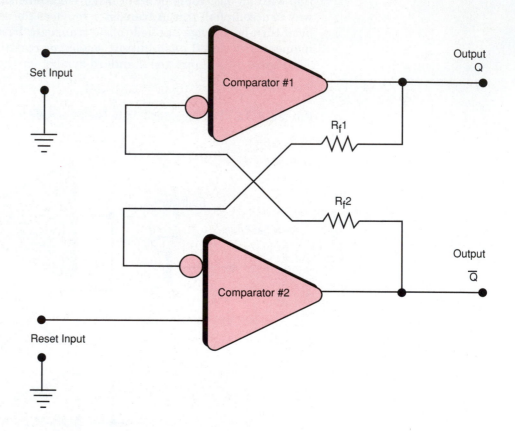

Flip-flop Using Two Comparators

Figure 9-34 shows how you can use a pair of comparators to construct a flip-flop. Nearly any inverting amplifier can be used to make a flip-flop; so if you have a couple of extra comparators in a package and need a flip-flop, you now have a way to make your own.

Comparator-based Square-wave Generator
Based on a Digital Integrated Circuit

The 74HC14 C-MOS Schmitt trigger inverter makes an excellent comparator-based square-wave generator. It can be used for a digital clock or for analog purposes. C-MOS devices can be operated at voltages other than 5 volts. The hysteresis is built-in, so the circuit becomes very simple. The circuit produces a square wave with short rise and fall times, and a nice flat top. TTL Schmitt inverters are not suitable for this application, because of their relatively low input resistance, which tends to bypass the current intended for the capacitor.

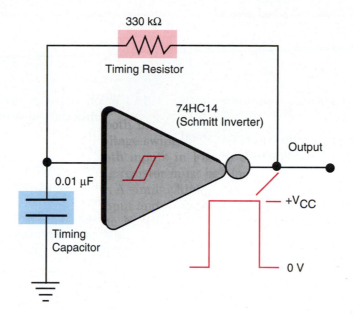

FIGURE 9-35

Schmitt Trigger Devices, Members of the Digital Integrated Circuit Family

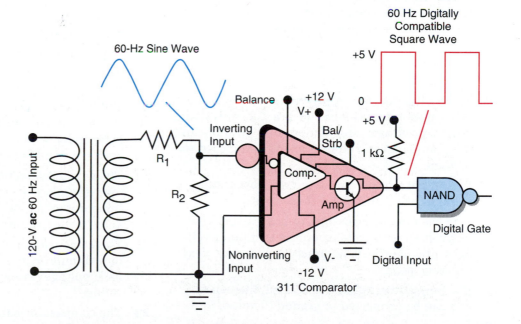

FIGURE 9-36

311 Zero Crossing Detector for Converting an ac Line Voltage Sine Wave into +5/Ground Square Wave That Is Compatible with Digital Circuits

Using a 311 Comparator to Convert Any Analog Waveform into a Digitally Compatible Square Wave

One of the main strengths of the 311 comparator is its ability to have different input and output power-supply voltages. The circuit in Figure 9-36 takes advantage of this property to convert any analog signal into a standard +5-volt digital signal. The circuit shown converts a power-line **ac** sine wave into a digital pulse, but any other analog waveform would work as well.

SUMMARY

1 A comparator is an operational amplifier with no negative feedback.

2 With an open feedback loop, an operational amplifier has a typical voltage gain of 200,000 or more.

3 With such high voltage gains, an operational amplifier will saturate with a very small input signal, resulting in a switchlike action.

4 The comparator compares two voltages, of any magnitude or sign, within the comparator's voltage limits.

5 The comparator is operated in a differential mode to compare the two voltages.

6 If the differential input voltage has a net positive value, the comparator output switches to the positive power-supply voltage.

7 If the differential input voltage has a net negative value, the comparator output switches to the negative power-supply voltage.

8 The positive power-supply voltage and the negative power-supply voltage are the only two possible voltages the comparator output can have. In-between voltages are not supposed to happen.

9 In comparator jargon, the two power-supply voltages are called the *rail voltages*.

10 It is theoretically possible to have a third output voltage of 0 volts when the two input voltages are identical. In practice, this is a very unlikely state.

11 If a zero output voltage is desired, the negative rail can be connected to ground (0 volts).

12 In the real world, the comparator output voltages are slightly less than the rail voltages because of the saturation voltage of the op-amp (comparator) output transistors.

13 Because the input differential voltage required to switch the comparator to the upper rail is almost identical to that required to switch it to the lower rail, the comparator is said to have a single trip point (voltage).

14 When noise is present, the single trip point permits the noise to cause erratic switching by moving the input voltage back and forth across the trip-point voltage.

15 The noise problem can be solved by adding hysteresis, using positive feedback. Adding hysteresis yields two trip-point voltages spaced far enough apart (by a dead band) that noise cannot normally drive the input voltage from one trip point to the other.

16 A comparator with hysteresis added is called a *Schmitt trigger*.

17 If the real signal input voltage is at one trip point, a noise signal would have to be large enough to drive the input all the way to the other trip point to cause erratic action. This is very unlikely.

18 Using positive-feedback hysteresis is a popular way to create two trip points, but the two trip point values are not totally independent of one another.

19 If the circuit's requirements demand complete independence and total control of the two trip points, a window detector circuit fills the bill.

20 The window detector requires two comparators (and therefore is more expensive), with each comparator set to its own desired trip point and sharing a common input signal.

21 The comparator often uses a fixed, predetermined limit or reference voltage. A second (signal) voltage is compared to the reference voltage.

22 The comparator can initiate some action if the signal voltage exceeds (or drops below) the reference voltage. For example, a comparator can sound an alarm or shut the circuit down.

23 The comparator makes an excellent timing circuit. It can detect and switch rails when the charge on a capacitor in an R-C timing circuit reaches a desired value.

24 A comparator with hysteresis can be used to sense the charge on a capacitor in an R-C timing circuit. It can then discharge the capacitor automatically, sense the discharge value, and start it charging again. The result is a square-wave generator.

25 The comparator-based square-wave generator is a popular circuit in its discrete form, and it also serves as a basic element in voltage-controlled oscillators and other integrated circuits.

26 Several comparators can be stacked (to light a series of LEDs) with their trip points set at 1-volt increments (or any other desired increment). An increasing input voltage lights the first LED at 1 volt, the second LED at 2 volts, the third LED at 3 volts, and so on. The circuit is called a *comparator ladder*.

27 Commercial comparator ladder integrated circuits with 10 or 12 comparators are available for use as stereo output-level indicators, as voltmeters, and in other applications.

28 If a triangular waveform is applied to one input of a comparator, and an adjustable **dc** reference voltage is applied to the other input, the comparator will produce a rectangular-wave output. The pulse width can be varied by adjusting the **dc** reference voltage.

29 A variation on the variable pulse-width circuit replaces the adjustable **dc** reference voltage with a signal. The output pulse width then follows the signal, changing width as the signal voltage changes. Because the pulse width is a function of the signal voltage, the circuit is called a *pulse-width modulator*.

30 Although any op-amp can be used as a comparator, op-amps may not be fast enough for critical applications. For that reason, you will find performance specifications in the manufacturers' data manuals under the heading *Comparators*.

31 Comparators are op-amps optimized for comparator service. They may include features useful only for comparator service.

32 The 311 comparator is an example. The 311 has an output stage that can be configured in a variety of ways, and the 311 output circuit can use a power-supply voltage different from that used in the comparator part of the IC.

33 There are a number of special comparator ICs. Most of these are designed to meet the stringent requirements of digital circuits.

QUESTIONS AND PROBLEMS

9.1 In terms of negative feedback, what distinguishes a comparator circuit from other op-amp circuits?

9.2 What are the possible output voltages from a comparator circuit?

9.3 If a 0-volt output is required, how is it best obtained?

9.4 In a comparator without hysteresis, what problem can noise on the signal input cause?

9.5 A comparator without hysteresis has how many trip points?

9.6 A comparator with hysteresis has how many trip points?

9.7 A particular comparator with hysteresis has one trip point at +5 volts and another trip point at −5 volts. What is the hysteresis (dead-band) voltage?

9.8 What is a Schmitt trigger?

9.9 What is one particular advantage of using a Schmitt trigger?

9.10 What is meant by the term *overdrive*?

9.11 Why is overdrive important?

9.12 Define *slew rate*.

9.13 Define *response time*.

9.14 What is the most common method for obtaining a desired reference voltage for a comparator?

9.15 Name three comparator circuits that have more than one trip point.

9.16 Define *rail voltages*.

9.17 Which comparator has an output circuit that can use power-supply voltages different from the comparator rail voltages?

9.18 Suggest three possible applications for a comparator ladder.

9.19 What is the purpose of the strobe input on the 311 comparator?

9.20 If the balance/strobe terminal on the 311 comparator is not used, what should it be connected to?

9.21–9.26 Given the circuit shown in Figure 9-37, complete the appropriate rows of Table 9-1.

9.27–9.32 Given the circuit shown in Figure 9-38, complete the appropriate rows of Table 9-2.

9.33 The circuit in Figure 9-38 has two inputs. One of the inputs is indicated by a bubble symbol. Which input uses the bubble symbol?

9.34 What kind of circuit is used in Figure 9-38 to provide the reference voltage?

9.35 What are the rail voltages in Figure 9-37?

Fɪɢᴜʀᴇ **9-37** **Circuit for Problems 9.21 Through 9.26 and 9.35**

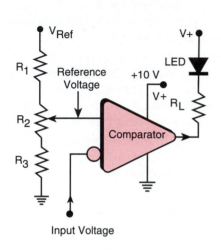

Input Voltage

Tᴀʙʟᴇ **9-1**

Work Table for Problems 9.21 Through 9.26

Problem Number	Reference Voltage	Input Voltage	Output Voltage	LED On/Off
9.21	+1.5 V	−3.1 V		
9.22	+2.01 V	+2.09 V		
9.23	−3.1 V	−2.2 V		
9.24	2 V	4.5 V		
9.25	+6.4 V	−4.4 V		
9.26	+5 V	+5 V		

Fɪɢᴜʀᴇ **9-38** **Circuit for Problems 9.27 Through 9.34 and 9.36**

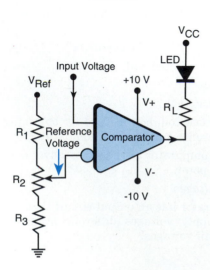

Tᴀʙʟᴇ **9-2**

Work Table for Problems 9.27 Through 9.32

Problem Number	Reference Voltage	Input Voltage	Output Voltage	LED On/Off
9.27	+1.5 V	−3.1 V		
9.28	+2.01 V	+2.09 V		
9.29	−3.1 V	−2.2 V		
9.30	2 V	4.5 V		
9.31	+6.4 V	−4.4 V		
9.32	+5 V	+5 V		

9.36 What are the rail voltages in Figure 9-38?

Given the circuit in Figure 9-39, answer Problems 9.37 through 9.40.

9.37 Identify the circuit.

9.38 What is the circuit's full-scale voltage?

9.39 What is the volts-per-division value for the circuit?

9.40 What voltage would you connect to the top of the voltage divider (V_{Ref}) to get a volts-per-division value of 0.25 V?

9.41 What is the purpose of R_L?

Given the circuit in Figure 9-40, answer Problems 9.42 through 9.45.

9.42 What is the name of the circuit?

9.43 What is the upper trip-point voltage?

9.44 What is the lower trip-point voltage?

9.45 What is the hysteresis voltage?

FIGURE 9-39
Circuit for Problems 9.37 Through 9.40

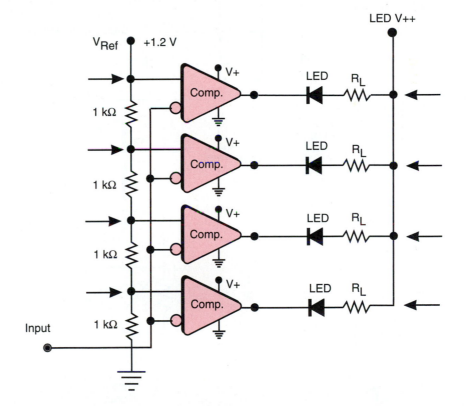

FIGURE 9-40
Circuit for Problems 9.41 Through 9.45

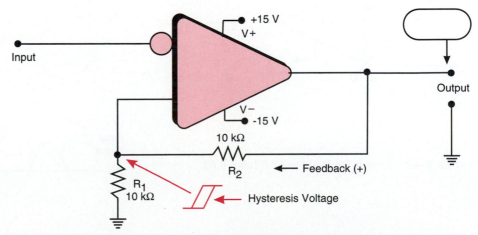

Given the circuit in Figure 9-41, answer Problems 9.46 through 9.51.

9.46 Identify the circuit.

9.47 What is the upper trip-point voltage?

9.48 What is the lower trip-point voltage?

9.49 What is the hysteresis voltage?

9.50 Make a sketch of the output waveform.

9.51 Make a sketch of the waveform across the capacitor.

Given the circuit in Figure 9-42, answer Problems 9.52 and 9.53.

9.52 Identify the circuit.

9.53 Draw the output waveform.

9.54 Identify the circuit in Figure 9-43.

9.55 Identify the circuit in Figure 9-44. What are the two trip-point voltages?

FIGURE 9-41
Circuit for Problems 9.46 Through 9.51

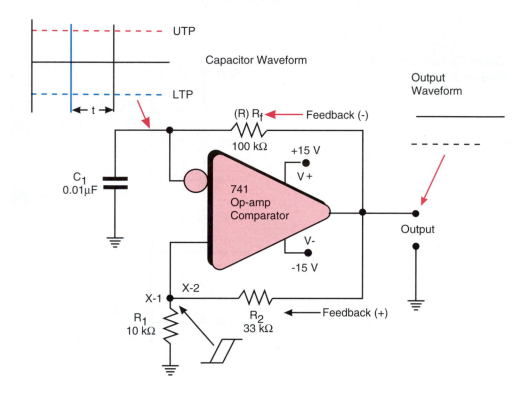

FIGURE 9-42 **Circuit for Problems 9.52 and 9.53**

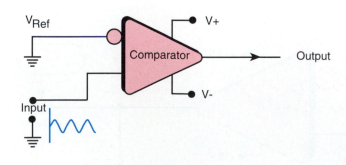

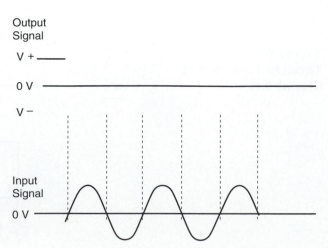

FIGURE 9-43
Circuit for Problem 9.54

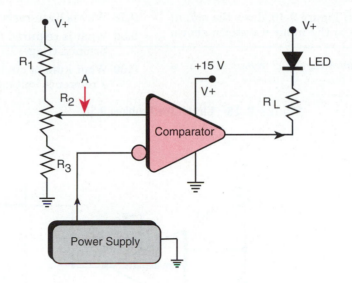

FIGURE 9-44
Circuit for Problem 9.55

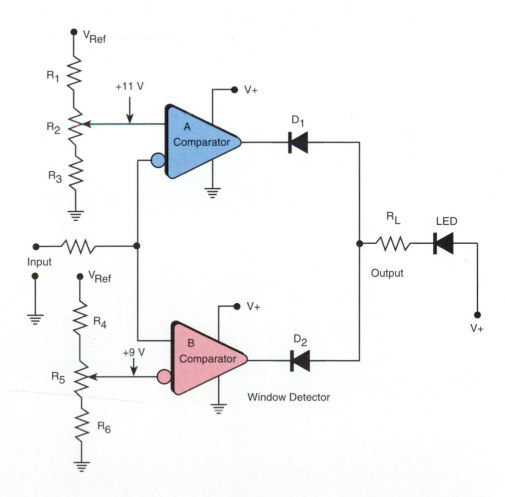

9.56 Given the circuit in Figure 9-45, draw the output waveform relative to the input waveform shown in the drawing.

9.57 What are the advantages of hysteresis in a comparator?

9.58 Why is hysteresis in comparators important?

9.59 What is required for a comparator to qualify as a Schmitt trigger?

9.60 What kind of circuit is most often used to derive a reference voltage for comparator circuits?

FIGURE 9-45   Circuit for Problem 9.56

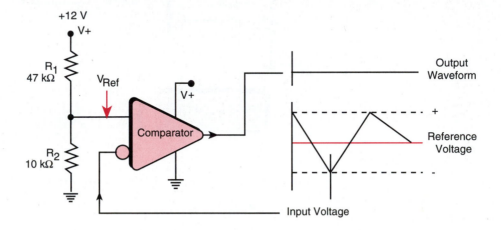

Television Waveforms

Television is so much a part of our lives today that it is rather shocking to realize that the technology behind it was only developed about 50 years ago. Commercial broadcast television became available in the late 1950s; cable and satellite signals, in just the last few years. Today, many people anxiously await the introduction of high-definition television, which promises a wider aspect ratio, more features, and much better picture quality.

In the meantime, we are still using a signal standard that was introduced in 1941. Despite the overall constraints caused by the standard, the television industry has managed to incorporate a surprisingly large number of new features into the existing format.

The original television video signal standard specified several parameters. The signal had to take up no more than 4.24 MHz of frequency bandwidth, each screen was to be composed of 525 horizontal scan lines, and the signal was to update the screen 30 times per second. In addition to the video information itself, some method was needed to synchronize the television circuitry with the video signal coming in. (The video signal itself modulates the transmitter frequency and is again demodulated in the receiver section of the television.)

To reproduce the image correctly on the screen, the television circuitry must create a horizontal ramp waveform that pulls the electron beam across the screen at the appropriate time (see Figure A). After completing each scan line, the sig-

nal reverts to its starting point. This is called *retrace*. The horizontal ramp waveform is created by the horizontal oscillator circuit. As the horizontal waveform repeatedly sweeps the trace across the screen and back, the vertical oscillator must produce another, slower ramp waveform that pulls the trace from the top of the screen to the bottom.

For each vertical ramp, 262.5 horizontal scan lines are traced across the screen (see Figure B). This constitutes one field. The vertical ramp then retraces and starts over, and another 262.5 scan lines are interlaced between the lines of the first field. Two fields make up a complete frame, which occurs 30 times per second.

To synchronize the video signal with the vertical and horizontal oscillators, sync pulses are embedded in the composite video signal. The video signal is a voltage waveform that controls the intensity of the electron bean striking the phosphorus on the inside of the cathode-ray tube. The amplitude of the signal controls the brightness. A 12.5% (of maximum) level should produce the brightest trace, whereas the trace should be black at a 67.5% level (blank).

As Figure A shows, at the end of each frame of video information, the signal goes briefly to a value of 75% of maximum (called the *front porch*). This is the beginning of the horizontal blanking pulse. During this time, the horizontal ramp can retrace without creating a visible trace on the screen. After 1.3 microseconds, the signal goes to 100% for 5

Television Waveforms

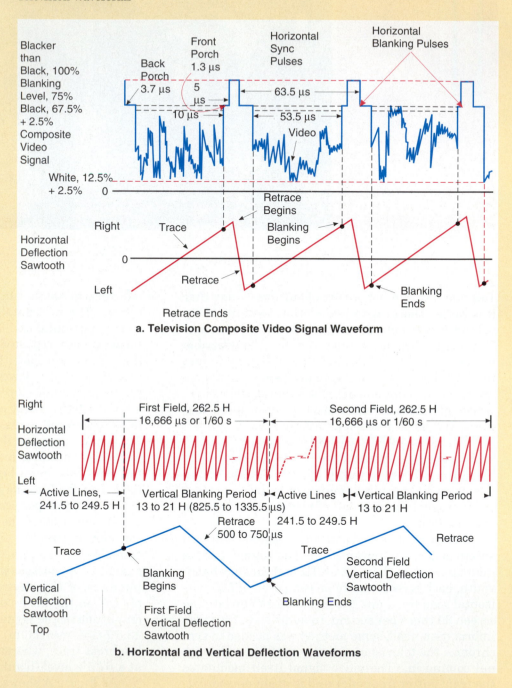

a. Television Composite Video Signal Waveform

b. Horizontal and Vertical Deflection Waveforms

microseconds. This is called the *horizon sync pulse.* Circuitry inside the television detects this pulse and triggers the horizontal oscillator to start another ramp waveform. (The remaining 3.7 seconds of the blanking pulse is called the *back porch;* it has a special use that will be explained later.)

A similar technique is used to trigger the vertical oscillator. After each field (262.5 horizontal lines), the video signal contains a series of pulses. Some of these pulses are used to synchronize the vertical oscillator; others carry special information (some of it optional) for use in the television.

Television Waveforms

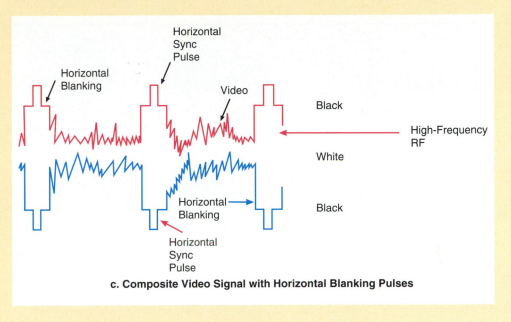

c. Composite Video Signal with Horizontal Blanking Pulses

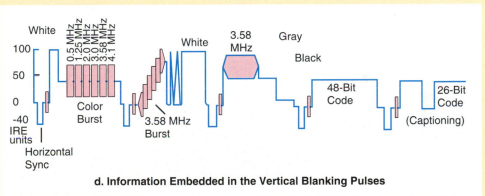

d. Information Embedded in the Vertical Blanking Pulses

Another important signal, which is embedded in the back porch of the horizontal blanking pulse, is a short burst of sine wave at 3.58 MHz. This signal is used to synchronize the receiver color circuits in a color television. (Color signals are decoded by comparing the phase of the 3.58-MHz oscillator with the phase of the chrominance signal.)

Other signals, transmitted during the vertical blanking pulse include VITS (vertical interval test signal), VIR (vertical interval reference), and data for closed captioning.

Without pursuing more of the specifics of television circuitry and operation, it should be obvious that the waveforms found inside a television are complex. Even though the signal standards for television are over 50 years old, they continue to be used today, with several features added to improve the quality of the television picture.

Understanding the complexity of television technology may be important to electronics technicians in deciding the area of technology they may wish to pursue. Just because televisions have invaded practically every home in the nation does not mean that they are simple devices. Television repair and the related areas of repair for VCRs and camcorders can be challenging and rewarding work for the technician who decides to pursue this work as a livelihood.

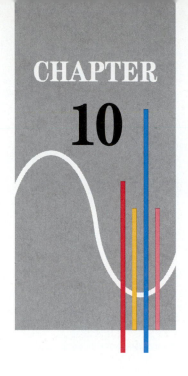

CHAPTER 10

Timing and Waveform Generation

OBJECTIVES

Upon completion of this chapter, you should be able to:

1 Explain the basic problem involved in using resistor-capacitor (R-C) analog timing circuits.

2 Draw a schematic diagram of an op-amp integrator.

3 Explain why the op-amp integrator can be used as a linear timing circuit.

4 Draw the schematic diagram of a hysteresis comparator-based square-wave generator, with a constant-current charging and discharging source and sink.

5 Explain why the constant-current version of the comparator-based square-wave generator is superior to the R-C version.

6 Draw a schematic diagram of a ramp generator in which a comparator is used to control the start and stop of the ramp.

7 Draw the schematic diagram of a practical staircase waveform generator.

8 Explain the operation of the circuit for a staircase waveform generator.

9 Explain how frequency is controlled in a hysteresis comparator-based oscillator that uses

a constant-current source/sink charging circuit.

10 Draw the internal functional block diagram of a 555 timer.

11 Explain the operation of the 555 timer in monostable mode.

12 Draw the external block diagram of the 555 in monostable mode.

13 Explain the operation of the 555 timer in astable mode.

14 Draw the external block diagram of the 555 in astable mode.

15 Explain how very long time delays can be obtained with a 555 timer.

16 Explain how the 555 can be used as a voltage-controlled oscillator (VCO).

17 Draw an internal block diagram of the 566 integrated-circuit voltage-controlled oscillator (VCO).

18 Explain the operation of the 566 IC VCO.

Analysis In Brief

Waveform Generation in Brief

ANALOG TIMING CIRCUITS, THE PROBLEM

The circuit and waveform to the right are for a typical R-C timing circuit. This is the basic circuit used for nearly all analog timing circuits and non-sinusoidal oscillator circuits. The problem is that the capacitor's charge voltage opposes the applied charging voltage, reducing the charging current. This makes the voltage versus time curve nonlinear and makes the timing inaccurate. Noncritical timing circuits can use the part of the curve within the first time constant, because it is approximately linear.

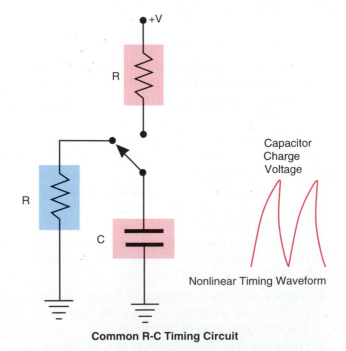

Common R-C Timing Circuit

The Problem: Inaccurate Timing

THE CURE FOR THE NONLINEARITY PROBLEM

The nonlinearity problem is caused by the fact that the charging current decreases as time passes. If we can charge the capacitor at a constant current—no matter what charge voltage it has accumulated—we can obtain a linear charging circuit. In a linear charging situation, the charge voltage is always exactly proportional to the elapsed time.

THE CONSTANT-CURRENT SOURCE

The constant-current device in the figure to the right is the most common solution to the problem of producing a linear timing circuit.

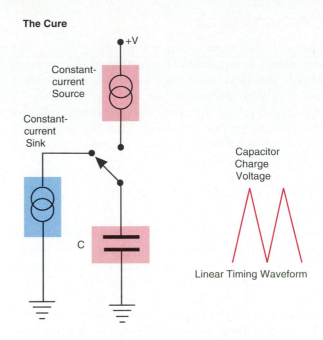

The Cure

The Cure: Charging the Capacitor Through a Constant-current Source

THE OP-AMP INTEGRATOR

The op-amp integrator is another solution to the nonlinear capacitor charging problem.

The op-amp integrator is a circuit in which a capacitor is used as the negative-feedback element. The negative feedback maintains a constant charging current. The constant charging current results in an op-amp output voltage that increases or decreases linearly with time. The slope of the linear output ramp is controlled by the values of R and C. The ramp produced is a linear graph of op-amp output voltage versus time. Resistor R_S is often added to bleed off capacitor charging due to leakage currents in the op-amp input circuit.

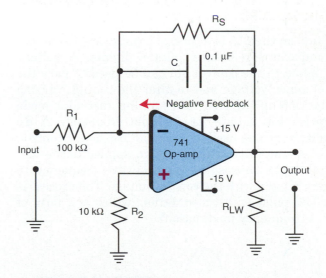

The Op-amp Integrator Linear R-C Timing Circuit

THE 555 TIMER

The 555 timer uses a simple R-C timing circuit, but the circuit is automatically restricted to operation within the first time constant. Two comparators use a voltage divider to set the trip points to $1/3$ V_{CC} and $2/3$ V_{CC}. Because the reference voltages are derived from a voltage divider, the trip points are within the first time constant—no matter what supply voltage is used.

THE 555 TIMER IN ITS OSCILLATOR (ASTABLE) MODE

The 555 can also be connected to an internal transistor which automatically discharges the timing capacitor at the end of the timing cycle. As soon as the timing capacitor has been discharged, the discharge transistor turns off, and the new cycle begins. The output result is a square-wave output. The output frequency can be set to one cycle in many hours or to frequencies up to about 100 kHz. The output frequency is controlled by resistors R_a and R_b and timing capacitor C_1.

THE 555 AS A VOLTAGE-CONTROLLED OSCILLATOR

Pin 5 on the 555 is a control input connected to the comparator reference voltage divider. A voltage applied to the control pin can be used to vary the reference voltage and to alter the timing. Varying the timing, in this free-running (astable) mode varies the square-wave output frequency. A **dc** voltage can be used as a remote frequency control.

A sine wave or other signal applied to the control input causes the square-wave frequency to vary in step with the control voltage. The process is called pulse-width modulation, which is a form of FM (frequency modulation).

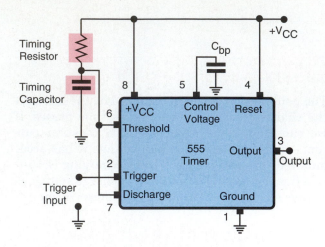

The 555 Timer IC, Connected as a Timer

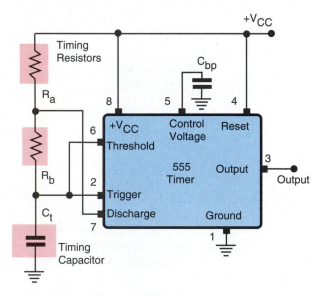

The 555 Rectangular-wave Generator

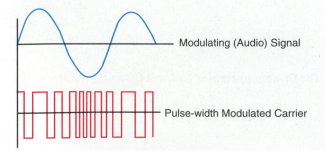

The Pulse-width Modulated Waveform

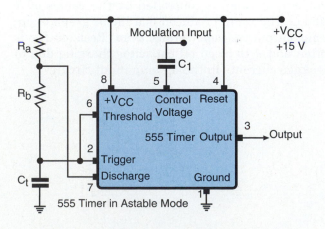

Using the Control Pin for Pulse-width Modulation

558

HOW THE 555 TIMER COMPARATOR TRIP POINTS WORK

The two comparators are arranged as shown in the partial schematic drawing to the right. The reference voltage is connected to the inverting input of one comparator and to the noninverting input of the other.

SUMMARY OF POSSIBLE TRIP-POINT CONDITIONS

The table to the right summarizes how the comparator trip points operate when a changing input signal voltage is applied to the two inputs (threshold and trigger). If the flip-flop is set, it will stay in that condition until a comparator switches to reset it. If the flip-flop is reset, it will stay in that condition until a comparator switches to set it.

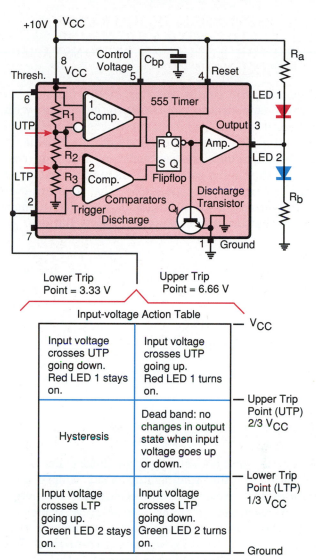

Input-voltage Action Table		
Input voltage crosses UTP going down. Red LED 1 stays on.	Input voltage crosses UTP going up. Red LED 1 turns on.	
Hysteresis	Dead band: no changes in output state when input voltage goes up or down.	
Input voltage crosses LTP going up. Green LED 2 stays on.	Input voltage crosses LTP going down. Green LED 2 turns on.	

Lower Trip Point = 3.33 V

Upper Trip Point = 6.66 V

V_{CC}

Upper Trip Point (UTP) 2/3 V_{CC}

Lower Trip Point (LTP) 1/3 V_{CC}

Ground

The 555 Timer and Its Trip-points

Using the Control Pin to Vary the Frequency of the Square-wave-Generator (Pulse-width Modulation)

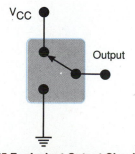

555 Equivalent Output Circuit

THE HYSTERESIS COMPARATOR OSCILLATOR WITH THE TIMING RESISTOR REPLACED BY CONSTANT-CURRENT SOURCES

Replacing the timing resistor in the classic hysteresis comparator square-wave generator with constant-current sources produces a precision oscillator. The constant-current sources provide an easily controlled, precise, constant charging current to the timing capacitor. This makes the timing very accurate and repeatable. The constant-current sources also minimize timing (frequency) errors that arise from power-supply variations.

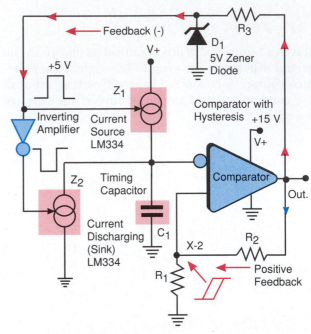

Hysteresis Comparator Square-wave Generator with Constant-current Sources

COMMERCIAL IC VOLTAGE-CONTROLLED OSCILLATORS

Most IC voltage-controlled oscillators use some variation of the comparator-based square-wave generator, using constant-current sources. Constant-current sources are actually transistor collector-to-base junctions, so the transistor base can be used to vary the constant-current value. The base can vary the collector current, which varies the capacitor's charging current, the timing, and the frequency.

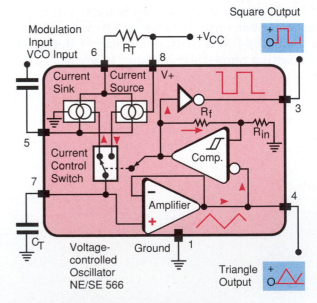

Commercial Voltage Controlled Oscillator

10.1   Introduction

This chapter will be devoted to examining circuits that perform timing and signal-generation operations, special timing circuits that generate nonsinusoidal waveforms, and circuits that control or modify nonsinusoidal waveforms. We will look at a number of methods for providing improved timing accuracy and frequency control. We will examine some very useful integrated circuits and will look at several specific applications, including a couple of fairly complex subsystems.

Most common analog timing circuits use a resistor-and-capacitor combination to form the basic timing element, in the form of an R-C time-constant circuit. The R-C timing circuit is used with an operational amplifier or comparator to make active timing and wave-generating circuits. We will look at some discrete timing and waveform-generating circuits, and then we will see how many of those elements and operating principles are combined in the circuits and performance characteristics of some of the most popular timing and waveform-generating integrated-circuit types.

10.2   R-C Time-constant Nonlinearity Problem, and Some Solutions

Most useful timing and waveform-generating circuits demand linear timing action. The R-C time-constant circuit is an excellent, simple, and inexpensive timing circuit; but it is not linear enough for most critical timing and waveform-generating applications.

FLASHBACK

Figure 10-1(a) shows a typical R-C timing circuit, and Figure 10-1(b) presents a standard graph of how a capacitor charges through a resistor as time elapses. Notice that the charge does not accumulate at the same rate for the full 5 or so equal time intervals leading to full charge. Instead, the capacitor charges relatively rapidly at first and then slows down. Notice that the capacitor's charging current also decreases as time progresses, and the capacitor accumulates a charge voltage that opposes the supply voltage trying to charge it.

The idea behind the R-C timing circuit is that there should be a direct and linear relationship between elapsed time and accumulated charge voltage on the capacitor. For example, if the capacitor is charging at the linear rate of 1 volt per second, at the end of 5 seconds we should have a capacitor charge voltage of 5 volts. If we can measure the voltage, we also measure the time. Unfortunately, the relationship between time and charge voltage is not linear. One fairly linear segment on the curve—during the first time constant—offers reasonable timing accuracy for noncritical operations; but the rest of the curve is not even close to linear.

FIGURE 10-1   Capacitor Charging, Nonlinearity Problem

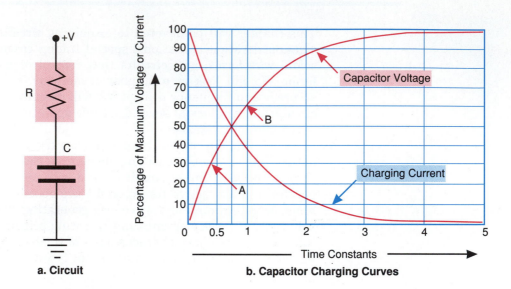

a. Circuit

b. Capacitor Charging Curves

There are two common solutions to the nonlinearity problem. The first is illustrated in Figure 10-1 by points *A* and *B*, where only a short and fairly linear segment of the charging curve is used. This solution requires that we begin the timing process at point *A* and end it at point *B*. Using a comparator circuit with a lower trip point at point *A* and an upper trip point at point *B* is the preferred method of restricting the timing to the part of the curve between points *A* and *B*.

The segment of the curve between about $1/3$ and $2/3$ of full charge is fairly standard. The upper part of the curve is much too nonlinear, and voltage levels too near 0 volts can sometimes cause trouble with transistor junction turn-on voltage near the 0.6 to 0.7-volt junction voltage. If some transistor in the system doesn't turn on until some time after the official timing period has started, it could throw off the timing accuracy of the system. In addition, capacitors can exhibit problems of retaining small residual voltages after discharge that may affect the timing near zero. We will soon see how comparator circuits actually accomplish the task of restricting operation to part of the curve.

The second solution to the linearity problem is to throw out the timing resistor and replace it with a constant-current device. The linearity problem is caused by the accumulated charge voltage, which opposes the applied charging voltage, reducing the total effective charging voltage and consequently reducing the charging current and the charging rate. The constant-current capacitor-charging circuit shown in Figure 10-2 provides a constant charge or discharge current no matter what charge voltage the capacitor has accumulated. If the charging current is constant, the rate of charge must also be constant.

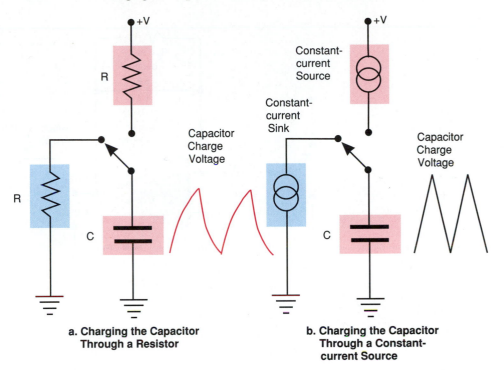

FIGURE 10-2 · Charging the Capacitor Through a Constant-current Source

a. Charging the Capacitor
Through a Resistor

b. Charging the Capacitor
Through a Constant-
current Source

10.3   Operational Amplifier Integrator

The integrating op-amp circuit shown in Figure 10-3 provides another method of supplying a constant charging current to a capacitor. The simple R-C timing circuit is often called an *integrator circuit,* and it does accumulate a gradually increasing voltage over time. But it is an imperfect integrator because its voltage does not increase (or decrease) at a constant rate over the time period.

If we place the timing capacitor in the feedback loop of an op-amp, we get a nearly constant charging current and, consequently, a linear op-amp voltage output change. The inverting input is a virtual ground because of the Miller effect, so it can serve as a current-summing junction. The higher initial capacitor-charging current opposes the current provided by the fixed input voltage. As the charging current decreases, the input source current becomes increasingly effective, essentially maintaining a constant charging current until the amplifier saturates. The constant charging (or discharging) current produces a linear ramp output voltage, as illustrated by the waveform in Figure 10-3.

When the input is switched to the V+ position, the output voltage ramps down at a linear rate until the amplifier reaches the negative saturation voltage. The output stays at that level until the input voltage is changed. When the input is switched to V−, the output voltage ramps up linearly until it

FIGURE 10-3   **Basic Op-amp Integrator Circuit**

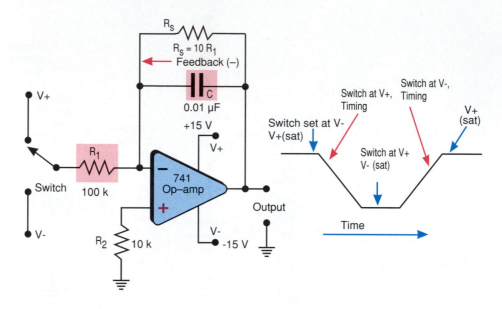

reaches the positive saturation voltage. The time required to complete the ramp is determined by the values of R_1 and of the feedback capacitor C.

The slope of the ramp describes the time period, because the x-axis of the waveform is measured in time units. A steep slope represents a short timing period, while a more gradual slope defines a longer timing period.

Changing Capacitor Current: A Critical Idea

FLASHBACK

When resistors are used as the input resistance and feedback resistance, the voltage gain of the inverting input of an op-amp is defined by the equation

$$A_V = - R_f / R_{in}$$

When the input or feedback resistance is actually a more complex impedance, as it is in the feedback circuit in Figure 10-3, the voltage gain is defined as

$$A_V = - Z_f / Z_{in}$$

The impedance of the feedback loop in Figure 10-3 depends on the frequency, if an **ac** signal is applied to the input. At **dc,** when the capacitor is fully charged, the voltage gain is simply

$$A_V = - R_f / R_{in}$$

The effective impedance of the capacitor during charge or discharge varies as the capacitor current varies, and the capacitor current changes as its accumulated charge changes. This changing capacitor current is used as negative feedback to correct the nonlinearity in the R-C charge curve.

FIGURE 10-4 Op-amp Integrator

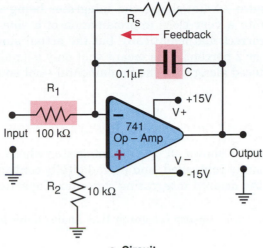

Input 100 kΩ

a. Circuit

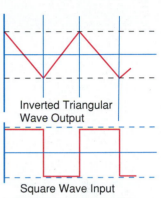

Inverted Triangular Wave Output

Square Wave Input

b. Inverted Triangular Wave Output Produced by a Square Wave Input to the Integrator

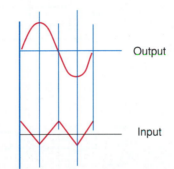

Output

Input

c. Approximation of a Sine Wave Produced as Output by a Good Triangular-wave Input to the Integrator

Resistor R_s is installed in nearly all practical op-amp integrator circuits to reduce the low-frequency (down to **dc**) voltage gain (see Figure 10-4(a)). Its purpose is to minimize the effect of input bias currents, which can add to the charging current and cause serious inaccuracies in the timing period. In fact, the input bias currents can cause a small timing capacitor to take on a full charge in a fairly short time, even when the input voltage is zero (grounded). If the input is at 0 volts, we don't want the timing to start yet. The feedback resistor (R_s) is typically about ten times the value of the input resistor (R_1), yielding a **dc** voltage gain of only 10.

A square or rectangular waveform is the electronic equivalent of the input switch in Figure 10-3. If the time constant of the timing circuit is long compared to the time period of the input waveform in Figure 10-4(b), the amplifier output will change directions before reaching saturation. The output waveform will then be a triangular waveform. The op-amp integrator converts rectangular or square waves into triangular waves with a linear slope, and it is frequently used for that application.

Figure 10-4(c) shows the integrator being used to convert a triangular wave into a very close approximation of a sine wave. The conversion is not quite correct mathematically, but the actual sine-wave distortion is typically only 1 or 2 percent. This means that only a small amount of harmonic energy is produced along with the fundamental—not enough to matter in many applications.

Electronic Control of the Integrator Ramp

Figure 10-5 shows an op-amp integrator whose ramp slope (timing rate) can be remotely controlled, and even digitally controlled, by using analog gates to select the desired integrating capacitor. Capacitors can be selected to be used

FIGURE 10-5 **Op-amp Integrator Uses Analog Gates to Select Capacitors**

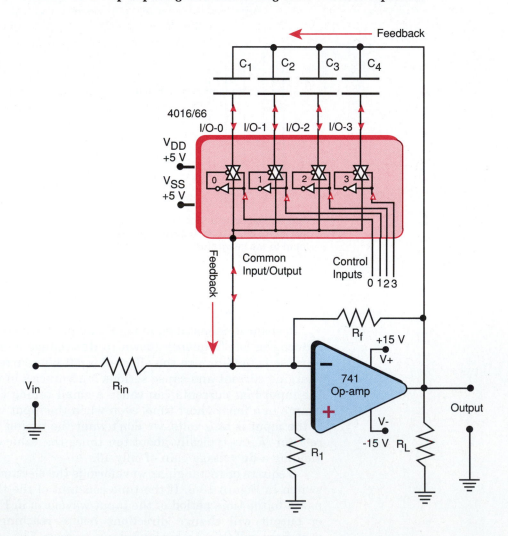

one-at-a-time, or they can be scaled to binary values that are to be connected in parallel combinations. Recall that the values of capacitors connected in parallel are additive. This means that selected values with ratios of 1, 2, 4, and 8 can be installed for binary timing selection, allowing for timing values from t to $16t$, in increments of t. Figure 10-6 illustrates the effect on the timing slope of selecting each of the four capacitors.

Practical Integrator Timer and Triangular Wave Generator

The op-amp integrator can be made into a practical time-delay circuit by adding a comparator to sense the end of the desired timing period. The circuit in Figure 10-7 consists of a comparator and a basic op-amp integrator with a fixed **dc** voltage applied to the inverting input. A push-button is placed across the timing capacitor to discharge it before a new timing cycle is begun. Depressing the push-button discharges the capacitor. When the push-button is released, the op-amp integrator output starts ramping down from its starting point at $+V_{sat}$. It will continue ramping to the negative saturation voltage, where it will sit until the reset button is pushed again.

The output of the integrator is connected to one input of a comparator. The other input of the comparator is connected to an adjustable reference voltage. Although the integrator will continue to ramp downward until it saturates, its output voltage will cross the comparator reference voltage at some time before it saturates. When the output voltage of the integrator crosses the reference

FIGURE 10-6   **Variations in Output Timing Slopes for the Circuit in Figure 10-5**

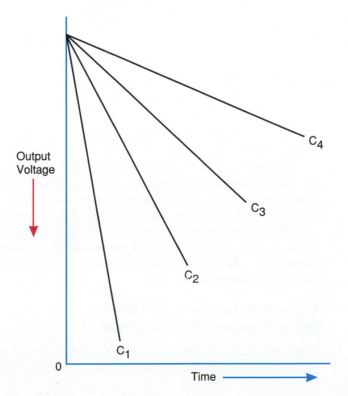

FIGURE 10-7   Time-delay Circuit Using Op-amp Ramp Generator (Integrator) and Comparator Voltage-level Detector

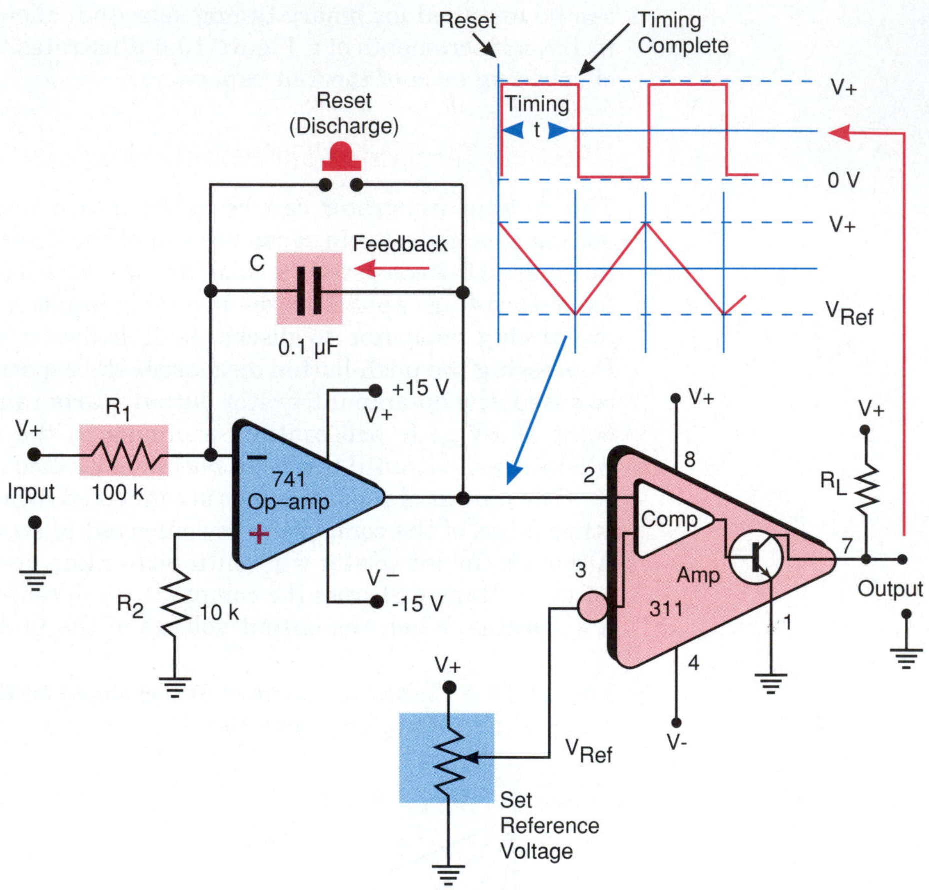

voltage setting, the comparator output suddenly switches, indicating the end of the timing cycle.

The comparator reference voltage can be adjusted to switch the comparator soon after the ramp starts or later in the output ramp voltage (later in the timing period). By adjusting the reference voltage, we can vary the time from the start of the ramp to the time the comparator switches.

The output of the comparator can be used to turn some electronic or electrical device on (or off) after a controlled time-delay period. Pushing the reset button resets both the integrator and the comparator, and the circuit is ready to start another timing cycle.

Converting the Time-delay Circuit into a Triangular Wave Generator

Since we can use the comparator's output voltage in Figure 10-7 to control some other electronic circuit, let's use it to control the integrator input voltage and form a closed-loop circuit like the one shown in Figure 10-8. In order to make the circuit work as a waveform generator, we must introduce a second trip point so that the ramp voltage can swing between two voltage levels. In Figure 10-8, some positive feedback has been added, introducing hysteresis and two trip points.

The comparator switches back and forth as the integrator ramp crosses the upper and lower trip points, causing the integrator to ramp down, then up, and then back down again endlessly.

When the integrator is ramping down from $+V$, the comparator output voltage is positive, providing a positive integrator input voltage and causing the integrator to continue ramping down. When the integrator's ramp voltage crosses the comparator's lower trip-point voltage, the voltage at the comparator output switches to its negative rail. The inverting op-amp input was at the comparator's positive rail voltage and was ramping downward.

FIGURE 10-8 **Triangular Wave Generator Using Op-amp Ramp Generator and Comparator Switch**

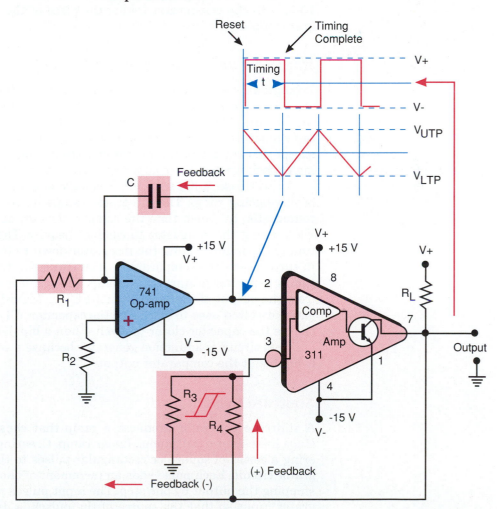

The comparator has now switched the integrator input to its negative rail, so the integrator output starts back up. When the ramp voltage crosses the comparator's upper trip point, the comparator switches back to the positive rail, and the ramp heads back down toward the comparator's lower trip point.

The output of the integrator is a continuous stream of triangles that form a triangular waveform. The circuit is a triangular-wave-signal generator. It is sometimes called an *oscillator*. The most common use of the term *oscillator* is to refer to sine-wave generators. If this circuit is to be called an *oscillator,* it should be called a *nonsinusoidal oscillator.* The integrator's output oscillates between the comparator's upper and lower trip points.

Technically, the circuit in Figure 10-8 is called a *function generator* because it produces two different useful output waveforms—one from the integrator output, and one from the comparator output. A function generator, by definition, produces two or more useful output waveforms. There may be some waveforms in a circuit that can't be used as outputs without upsetting the timing or some other circuit function, and they don't qualify.

The circuit in Figure 10-8 is an automatic version of the circuit in Figure 10-3, with the comparator taking the place of the manual switch at the integrator input.

Integrator and Comparator Sawtooth-wave Generator

Triangular waves ramp up and back down with the same slope. The sawtooth wave is a special case, in which each ramp is followed by a sudden return to the baseline. The sawtooth generator in Figure 10-9 is similar to the integrator/comparator-based triangular-wave generators we have just examined, except that the comparator turns on a transistor, abruptly discharging the timing capacitor when the comparator senses the end of the ramp time.

In this case, a negative input voltage is applied to the integrator to get a positive-going ramp. The integrator ramps up to the upper trip point of the comparator, at which time the comparator output switches to approximately $V+$, causing the transistor to conduct heavily. The transistor discharges the timing capacitor, driving the integrator down to the ramp-start voltage. This, in turn, resets the comparator, which turns the transistor off and allows the timing capacitor to start a new ramp. The result is a stream of sawtooth waves from the output of the integrator. J-FET or MOS-FET transistors and analog gates are often used to discharge the capacitor. All of these devices tend to discharge the capacitor closer to 0 volts than a bipolar transistor can.

This circuit is a function generator because a square-wave signal output is available at the comparator output.

Staircase Generators

A staircase generator produces a ramp that consists of a series of discrete steps instead of a continuous linear ramp. Creating a staircase involves delivering a series of square or rectangular pulses to the input of an op-amp integrator circuit. Each pulse adds an increment of charge to the timing capacitor, stepping the output by one step. The input pulses must be gated into the integrator input so that the source of the pulses is disconnected from the input

FIGURE 10-9   Sawtooth Generator Using Transistor for Fast Capacitor Discharge

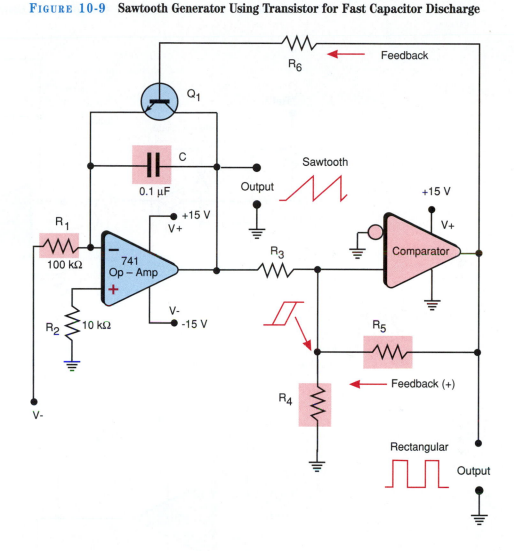

between pulses. If pulses are not gated in, the integrator will try to follow the input voltage back down during the time between input pulses. Figure 10-10 shows two versions of the staircase-generator circuit.

Analog-gate Version

The version shown in Figure 10-10(a) uses an analog gate to connect and disconnect a **dc** input voltage controlled by a square or rectangular input signal. This circuit allows the amplitude of the input signal to be varied independently of the amplitude of the input square (or rectangular) waveform. The analog gate provides excellent isolation of the input voltage from the integrator input during the time between pulses. The unity-gain-inverting op-amp is used to convert the negative-going staircase into a positive-going staircase. The gain of the inverting amplifier can be changed to alter the size of each

Figure 10-10 Staircase Generators (Stepped-ramp Generators)

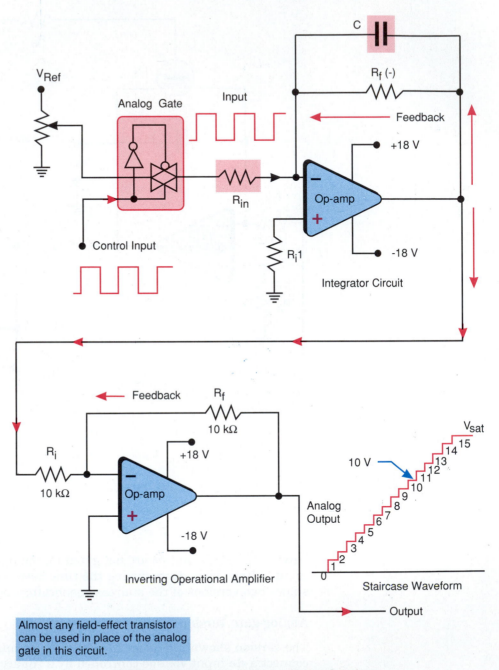

Almost any field-effect transistor can be used in place of the analog gate in this circuit.

a. Staircase Generator Using an Analog Gate Input

step, if that is desired. The circuit will continue the ramp until either the integrator or the inverting amplifier saturates.

Because the staircase voltage changes in discrete increments, the circuit is often used as an analog counter. For example, parts passing through a light beam on a conveyor belt can be used to produce the control pulse for the ana-

FIGURE 10-10 Continued

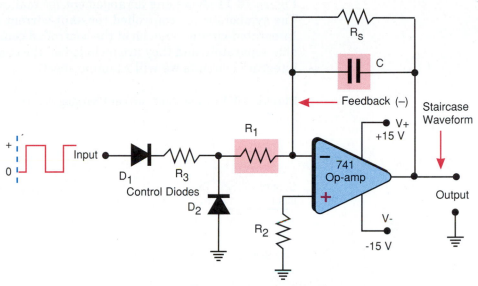

b. Diode Input Isolation Circuit is Simple, but Less Flexible

log gate. As each part breaks the beam, a standard-width pulse is delivered to the analog gate, stepping the staircase one step—representing one count. A comparator might be connected to the staircase output, with its trip point set to initiate some action at some predetermined count (voltage).

Diode Version

The version of the staircase circuit in Figure 10-10(b) uses a diode to provide isolation between the integrator and the source of stepping pulses between steps. The diode (D_1) is turned on only when the input pulse is positive. When the input pulse goes to zero, the diode turns off, disconnecting the integrator input from the input pulse source. Diode D_2 is optional; it is there to prevent negative pulses (if they exist) from affecting the integrator action. An inverting amplifier may be added to this circuit if a positive-going stairstep output is desired.

10.4 Waveform Generators That Use Constant-current Devices to Charge and Discharge Capacitors

A constant-current device can be used to charge and discharge timing capacitors, greatly improving the performance of R-C timing circuits. When those timing circuits are used as part of a waveform generator, the result is much-improved frequency accuracy and stability. We can use the familiar constant-current source—the current mirror—as a charging current source and discharging current sink. The current mirror replaces the resistor in the R-C timing circuit to eliminate the nonlinearity in the capacitor's charging curve.

To use the current mirror in conjunction with waveform generators, we need to have some way to turn the current mirror on and off. The circuit in Figure 10-11 shows one arrangement for making a controlled current mirror. The symbol for the controlled constant-current device is shown in the figure. Integrated-circuit versions of the controlled constant-current device are available separately, and they are included on the chip in some of the more complex integrated circuits we will examine shortly.

FIGURE 10-11   Constant-current Charging Source

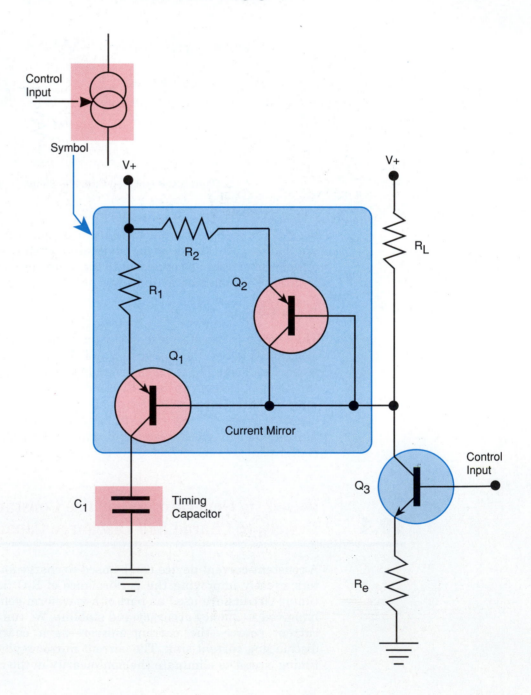

The circuit in Figure 10-11 is an ordinary current-mirror circuit, with an added control transistor (Q_3). When the base of Q_3 is at ground potential, its collector voltage rises to $V+$. This places $V+$ on both the base and the emitter of current-mirror transistors Q_1 and Q_2. Because there is no voltage difference between base and emitter, transistors Q_1 and Q_2 are turned off, and there is no collector current available to charge the timing capacitor.

If the base of the control transistor is made positive, it turns on, pulling its collector voltage to near ground level. There is now a base–emitter bias-voltage difference, so base–emitter bias current flows, and the collector current through Q_1 begins charging the timing capacitor. A similar circuit can be used to discharge the timing capacitor; when used in that application, it is called a *constant-current sink*.

Hysteresis Comparator with a Constant-current Charge/Discharge Circuit

Figure 10-12(a) is a flashback to the conventional hysteresis comparator square-wave generator we studied in Chapter 9. The circuit in Figure 10-12(b) is basically the same circuit, except that the timing resistor has been replaced by two constant-current circuits. One of the constant-current circuits is used to charge the timing capacitor, and the other is used to discharge it. Two constant-current circuits are required because the constant-current device can conduct in only one direction. The resistor it replaces can carry current in both directions. The inverting amplifier may be a transistor or an op-amp; but in most integrated circuit versions of Figure 10-12(b), a digital inverter is used. Since digital inverters are 5-volt devices, a 5-volt Zener is used to limit the comparator output voltage to 5 volts for the inverter.

How It Works

1   Assume that the capacitor is discharged. With 0 volts on the inverting input, the positive feedback voltage on the noninverting input causes the comparator output to go to $+V_{sat}$. The Zener diode (D_1) and R_3 limit the voltage input to the controlled current sources to $+5$ volts. This allows greater flexibility in selecting comparator supply voltages, whatever kind of inverting amplifier is used.

2   The positive voltage from the output of the comparator is fed back to the control input of the constant-current source (Z_1) and turns it on. At the same time, the inverting switching amplifier (inverter I_1) turns the constant-current sink (Z_2) off, so there is no discharge path for the timing capacitor. With the current source (Z_1) turned on, the timing capacitor starts charging.

3   The capacitor continues to charge until it reaches the comparator's upper trip point. When it slightly exceeds the trip-point voltage, the comparator output suddenly switches to ground. The comparator output voltage turns the current source (Z_1) off. The inverter inverts the 0-volt signal to a $+V$ signal, and turns the current sink on, starting the capacitor discharge part of the cycle.

4   When the capacitor has discharged to the lower trip-point voltage, the comparator output again goes high ($+V_{sat}$), and a new cycle begins.

FIGURE 10-12 **Improved Comparator Square-wave Generator**

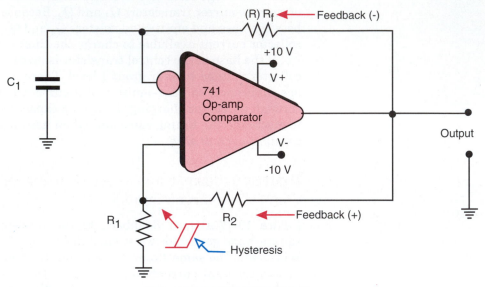

a. Square-wave Generator Circuit

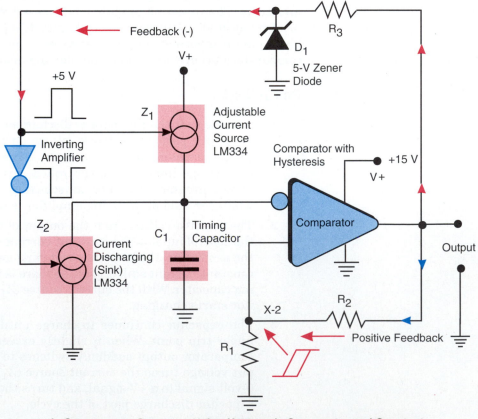

b. Square-wave Generator Using Hysteresis Comparator and Constant-current Sources to Charge and Discharge the Timing Capacitor

10.5   Introduction to Modulation

Because we will be discussing voltage-controlled oscillators and phase-locked loops in later sections of this chapter, we need to know something about modulation. Rather than trying to cover all the modulation bases here (there are a lot of them), we will focus only on those that are important to the circuits and systems in this chapter.

There are many instances in electronics where one signal is impressed on a second signal. This process is called **modulation.** For example, a radio wave with a frequency of several megahertz has music in the frequency range from 20 Hz to 20 kHz impressed on it. Music (audio) frequencies cannot be radiated into space as a radio wave, so the music information is used to modulate a high-frequency (radio) signal that can radiate into space. At the other end of the transmission, a receiver picks up a radio frequency signal whose power level varies at an audio rate. The radio frequency part of the signal is called the **carrier,** because it carries the information. The audio part of the signal is called the *modulating* (or *modulation*) *signal.*

At the receiver, we are only interested in recapturing the power variations in the radio-frequency signal. The process of disposing of the radio-frequency part of the signal and keeping only the power variations is called **demodulation.** The power variations can be applied to an audio amplifier and loudspeaker—and we have music. This kind of modulation is called **amplitude modulation,** or simply **AM;** it is illustrated in Figure 10-13(a).

Modulation is not confined to radio. A light beam might be varied in brightness by a music signal, and those variations detected by a photodetector at the receiving end. The light beam might also be used in an on/off mode to transmit digital ones and zeros. But amplitude modulation is not the only possibility. In addition to the familiar FM (frequency modulation), there are several pulse and digital modulation schemes.

In the case of **FM (frequency modulation)** the characteristic of the carrier that is varied at the modulating (possibly audio) frequency rate is the frequency of the carrier. The carrier has a rest or center frequency, and the modulating signal causes the frequency to deviate up and down from that center frequency at the modulating rate. Figure 10-13(c) illustrates frequency modulation.

A common example of nonelectronic frequency modulation is *vibrato* in music. A singer or violinist adds life to the music by varying the pitch of the note at about a 6-Hz rate. Instead of sounding a fixed-frequency 440-Hz A note, for example, the singer or musician alternately increases and decreases the 440-Hz frequency by a few hertz on either side of 440 Hz. The 440-Hz A is thus frequency-modulated at a 6-Hz rate.

There is also a music amplitude modulation known as *tremolo.* The modulating frequency for tremolo is also approximately 6 Hz. Adding tremolo to a guitar amplifier can result in some interesting effects, and it is easy to do. By the time you finish this text, you will know how to do it.

So far, we have been thinking in terms of sine-wave signals, but the voltage-controlled oscillators we will be studying involve modulating rectangular waveforms. Figure 10-13(d) illustrates a frequency-modulated rectangular waveform.

Figure 10-13 **Key Concepts of Modulation**

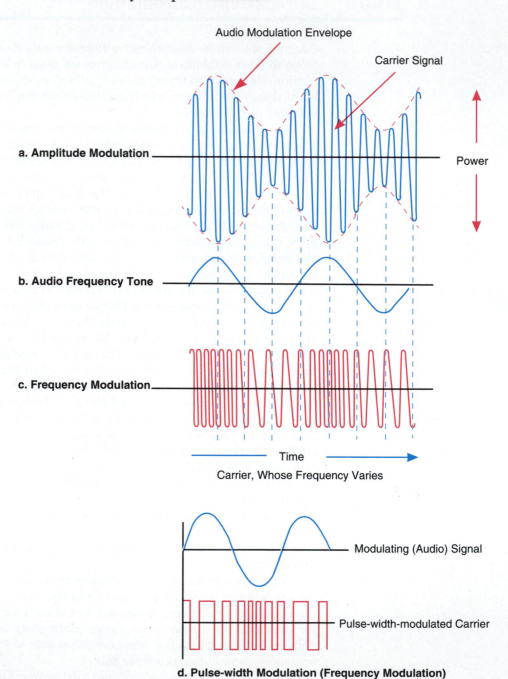

a. Amplitude Modulation

b. Audio Frequency Tone

c. Frequency Modulation

d. Pulse-width Modulation (Frequency Modulation)

10.6   Voltage-controlled Oscillators

The hysteresis comparator square-wave generator in Figure 10-12 can be converted into a square-wave oscillator whose frequency can be varied by an external control voltage. Voltage-controlled oscillators are sometimes called *voltage-to-frequency converters.* Voltage-controlled oscillators (VCOs) are available as a complete integrated-circuit subsystem, and they also form a part of the circuitry in other integrated-circuit devices.

Later, we will examine a commercial integrated-circuit voltage-controlled oscillator (VCO) and a device known as a phase-locked loop, which uses a VCO as a part of a larger subsystem.

The voltage-controlled oscillator circuit shown in Figure 10-14 is a modified version of the circuit we just studied (Figure 10-12), with two additional constant-current devices included. The additional current sources are biased for linear input voltage control, to allow an input voltage to control the amount of charging and discharging current. In the previous circuit, the current source and sink were either switched on (to some predetermined current level) or turned completely off. The two added constant-current devices in Figure 10-14 allow us to vary the amount of charging current linearly, using an external **dc** voltage or a varying modulating signal.

It is customary to refer to the resistor (R_t) in Figure 10-14 as the *timing resistor,* even though the timing capacitor's charging current does not flow through it. The timing resistor does control the timing indirectly, because it controls the bias on the linear current source and sink. The timing resistor sets the free-running frequency of the oscillator, which can then be varied by the control or modulating voltage. If the modulating signal is a sine (or other bipolar) waveform, the timing resistor is said to set the center frequency. In a properly designed VCO, the output frequency is a linear function of the modulating (or control) voltage.

10.7   555 Timer

The 555 timer integrated circuit and its immediate offspring form one of the most popular and useful IC classes. The 555, nicknamed the *triple nickel,* is a hybrid of analog and digital circuits and is equally at home with either analog or digital systems. The 555 is the first of a family of similar devices, most of which are modified or extended versions of the 555. The 555 is a bipolar transistor device, but there is also a pin-for-pin C-MOS replacement for the 555. The C-MOS version is ideal for low-power applications, and the high input impedance allows for long time delays using common capacitors. There are precision versions designed for precise timing applications; other versions feature a built-in digital counter that allows programmed time delays of many days. The key component of these devices is one or more comparators.

Some Typical Applications of the 555

The following 13 applications are a representative selection of the many possibilities for using the 555:

FIGURE 10-14   Voltage-controlled Oscillator

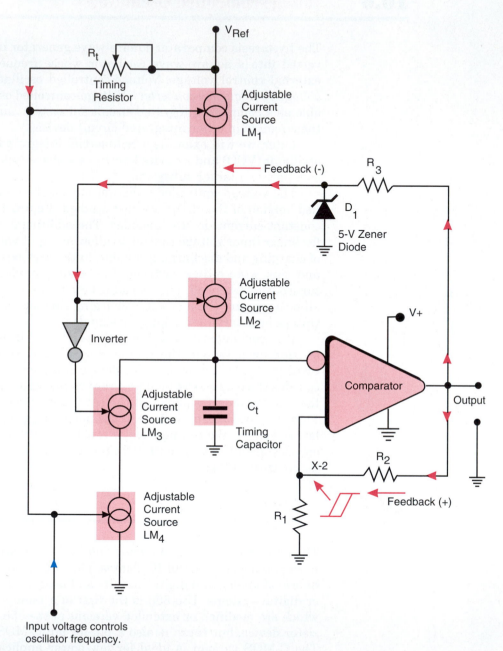

1 Turning some device on or off after a specified time delay.

2 Serving as a free-running square-wave generator.

3 Operating as a voltage-controlled oscillator.

4 Functioning as a pulse-width (FM) modulator.

5 Operating as a pulse-position modulator.

6 Working as a clock for digital circuits.

7 Serving as a window detector.

8 Stretching pulse widths.

9 Delaying pulses.

10 Serving as a function generator.

11 Generating linear ramps.

12 Converting analog waveforms into digitally compatible waveforms.

13 Performing common comparator functions.

There are many possible specific applications within each of these general headings. Indeed, at least one book is devoted entirely to this timer family. We will obviously not cover every timer function here, but we will examine some of the most important practical applications.

Monostable and Astable Operating Modes Defined

The two fundamental modes of operation for the 555 timer are **monostable** and **astable.** *Mono-* means single or one, and in this case implies a single stable operating condition. Although not included in the title, there is also a semi-stable or temporary state. The 555 output can have only one of two voltages: the positive rail voltage or ground. In the monostable mode, a trigger input pulse causes the output to leave the stable output voltage condition and switch to the semistable output state. The output voltage remains in the semistable state for a limited period of time, determined by the time constant of the external timing resistor-capacitor combination. At the end of the time constant period, the output switches back to the stable state. The output remains in the stable state indefinitely, until a new trigger pulse is applied.

The initial *a-* in *astable* means not or non-, indicating that in this mode the operating condition inherently lacks stability. In the astable mode, the 555 operates essentially as a monostable-mode device that automatically provides itself with a new trigger pulse at the end of each time-constant period. The result is a 555 output that swings back and forth, from rail to rail, generating a continuous rectangular output waveform.

Functional Parts of the 555 Timer

The 555 timer IC uses two comparators, a digital type flip-flop, and a single discharge transistor. The two comparators use an internal voltage divider to provide one reference voltage at about 33% and another at about 66% of the supply voltage. The flip-flop serves as a memory element to retain the most recent change in the condition of the two comparators. External connections to the comparators allow the circuit to be configured for various tasks.

Comparators

Figure 10-15 is a block diagram of the 555's internal circuits. The heart of the 555 is the pair of comparators and the voltage divider that provides their reference voltages. The voltage divider fixes the trip point of the upper comparator at two-thirds of the power supply voltage ($+V_{CC}$). The trip point of the lower

FIGURE 10-15   Functional Block Diagram of the 555 Timer

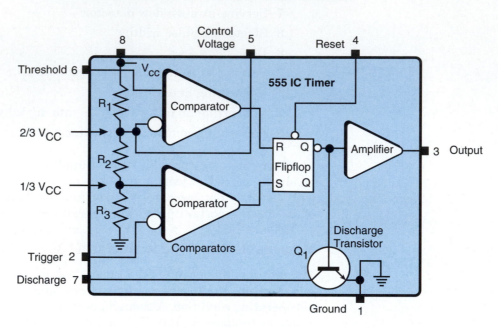

comparator is set at one-third of the power supply voltage. As the graph in Figure 10-16 indicates, the upper trip point is set at approximately the end of one time constant. Because of the voltage divider, this one-time-constant trip point applies for any pair of timing resistor-capacitor (R-C) values and any supply voltage. The timer can never operate beyond the first time constant, forcing it to operate always in the most linear part of the curve.

The lower comparator trip point is set at one-third of the supply voltage, which means that the timing is always between one-third and two-thirds of the first time constant. A small linear segment of the most linear part of the curve is all that is used for timing purposes.

Notice that the reference voltage divider is connected to the inverting input of the upper comparator and to the noninverting input of the lower comparator. The importance of this arrangement will become clear in a moment.

Flip-flop

Lower Comparator Action   Assume that the lower comparator trigger input voltage changes from any voltage above the $1/3$ of $+V_{CC}$ reference to a voltage slightly below the reference voltage. This causes the comparator output to switch, forcing the flip-flop to set by taking the 555 output pin to the positive $(+V_{CC})$ voltage.

Once the comparator is set, changes in the output state of the lower comparator have no effect on the state of the flip-flop or on the output voltage. Once the flip-flop is set, it will stay latched until the upper comparator forces it to reset. The 555 output pin stays at $+V_{CC}$.

FIGURE 10-16 Timing Diagram for the 555 Timer

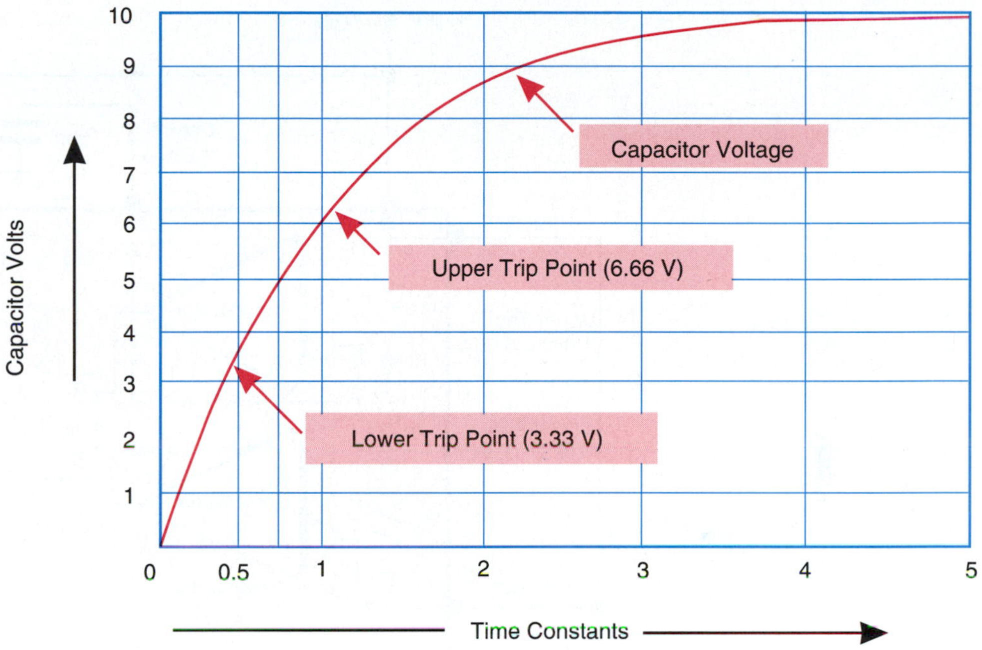

Upper Comparator Action Assume that the upper comparator's threshold input voltage changes from any voltage below the $^2/_3$ +V_{CC} reference voltage to a voltage slightly above the reference voltage. This will cause the comparator to switch, forcing the flip-flop to reset by taking the 555 output to ground.

Once the comparator is reset, changes in the output state of the upper comparator have no effect on the state of the flip-flop or on the output voltage. Once the flip-flop is reset, it will stay latched until the lower comparator forces it to set. The 555 output pin stays at ground.

Figure 10-17 illustrates the upper and lower trip points when both the threshold input and the trigger input are connected. The circuit in the figure is basically a window detector.

555 Timer in Monostable Mode

For the following discussion, examine the monostable timing circuit and its associated waveforms in parts (a) through (c) of Figure 10-18. The monostable timer configuration is sometimes called a *one-shot* because it executes a single output pulse upon receiving an input trigger pulse.

The input trigger pulse is a start-the-timer command. It can have any time duration, as long as it doesn't last longer than the desired timing period. The minimum timing period is 10 microseconds. The maximum timing period has no theoretical limit, but leakage currents in capacitors—particularly in those of larger values—provide a practical limit. If the capacitor's leakage current is about the same as the charging current, the capacitor never gets charged;

FIGURE 10-17 Threshold and Trigger Levels for the Window Detector

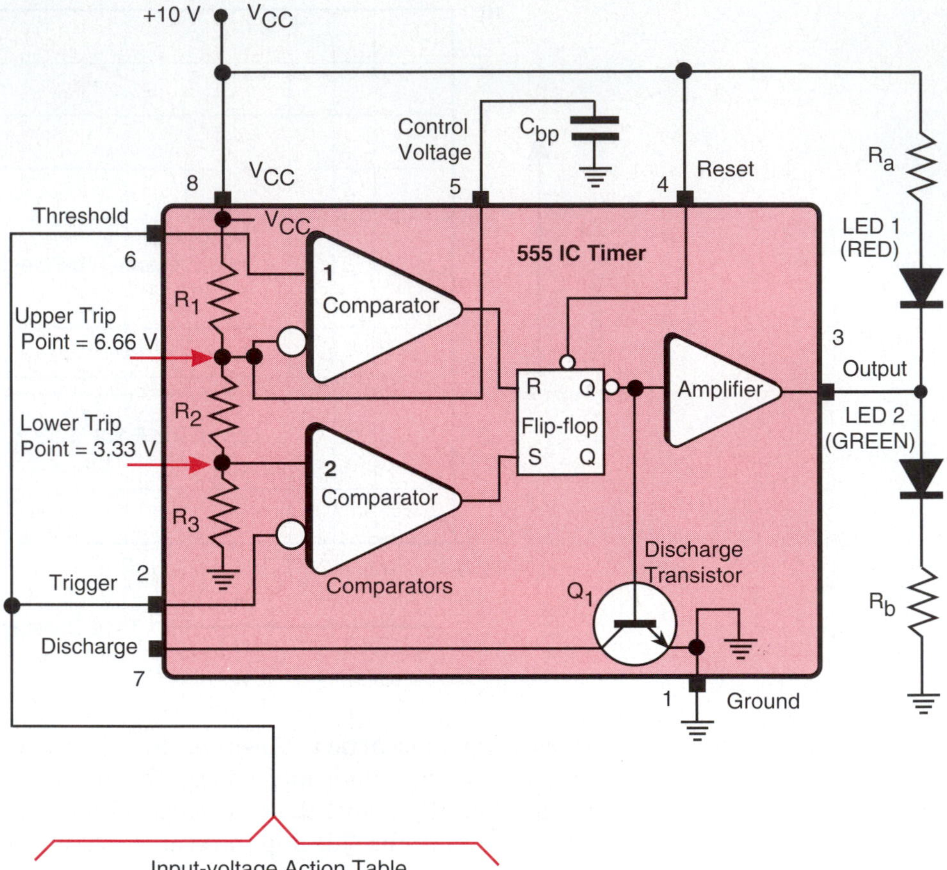

Input-voltage Action Table

Input voltage crosses UTP going down. Red LED stays on.	Input voltage crosses UTP going up. Red LED turns on.
Hysteresis	Dead band: no changes in output state when input voltage goes up or down.
Input voltage crosses LTP going up. Green LED stays on.	Input voltage crosses LTP going down. Green LED turns on.

— V_{CC}
— Upper Trip Point (UTP) 2/3 V_{CC}
— Lower Trip Point (LTP) 1/3 V_{CC}
— Ground

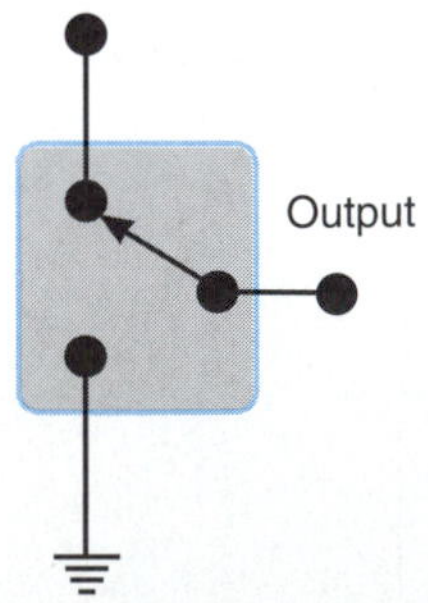

Equivalent Output Circuit

FIGURE 10-18 555 Timer in Monostable Mode

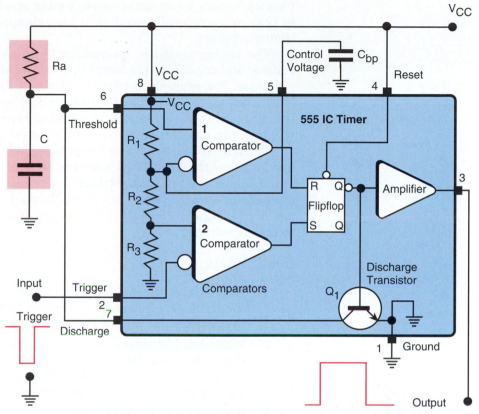

a. Monostable Circuit, Including the 555 Timer's Internal Circuitry

Notes:
1. The reset holds the output at ground when it is open or grounded. If the reset terminal is not used, it must be connected to V_{CC}.
2. Terminal #5 is the control voltage input. If a control voltage is not being used, terminal #5 may be connected to ground through a 0.01 µF capacitor. The capacitor serves as a decoupling capacitor, making the timing less responsive to power-supply voltage changes during the timing cycle.

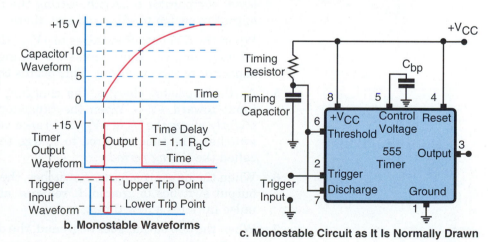

b. Monostable Waveforms

c. Monostable Circuit as It Is Normally Drawn

trying to charge it is like pouring water into a bucket that has a large hole in the bottom. Even if the leakage current is not large enough to keep the circuit from operating, it can cause severe timing errors. A timing period of up to 10 to 12 minutes generally requires no special capacitors and provides reasonable timing accuracy.

The timing period is determined by the values selected for the timing resistor and capacitor; it is completely independent of the duration of the trigger pulse. For example, a trigger pulse of a few milliseconds' duration may start a timing period lasting for 10 seconds. If we don't change the value of the timing resistor or the value of the timing capacitor, a 1-second trigger pulse will result in a repeat of the output timing period of 10 seconds.

The trigger pulse must be a negative-going pulse, and it must start above and go below the $1/3$ of $+V_{CC}$ reference voltage to start the timing cycle. Prior to the trigger start pulse, the trigger input must be held at a voltage above the comparator circuit's reference voltage. To start a timing cycle, the trigger voltage need not go all the way to 0 volts, but it must drop below the $1/3$ of $+V_{CC}$ reference voltage.

How It Works

1   Assume that the input trigger voltage is resting at some voltage above the $1/3$ of $+V_{CC}$ reference voltage, and that the flip-flop has been reset to make the 555 output switch to ground.

2   When the 555 output is at ground level, so is the collector of the discharge transistor (Q_1). The discharge transistor is turned on, shorting the timing capacitor, holding it in a discharged condition, and preventing it from charging through the timing resistor.

3   Assume that the trigger voltage is momentarily changed from above the comparator reference voltage and then below it. This causes the lower comparator to switch, setting the flip-flop and causing the 555 output to switch to $+V_{CC}$ and stay there.

4   When the 555 output switches to $+V_{CC}$, the discharge transistor turns off, removing the short across the timing capacitor and allowing the capacitor to start charging. This begins the timing cycle.

5   As the capacitor charges, the charging voltage pulls the threshold input toward $+V_{CC}$. When the capacitor's charging voltage rises to slightly above the $2/3$ of $+V_{CC}$ reference voltage, the upper comparator switches. The time required to charge the capacitor to $2/3$ of $+V_{CC}$ is called the *timing period*.

6   When the upper comparator switches, the flip-flop is reset and the 555 output switches to ground. It remains at ground until a new trigger pulse initiates a new timing cycle.

7   When the 555 output goes to ground, the discharge transistor is turned on, shorting across the timing capacitor and discharging it. The discharged capacitor is then ready to be charged for the next timing cycle.

8   The 555 output stays at ground and the capacitor remains shorted until the trigger input again drops below the $1/3$ of $+V_{CC}$ reference voltage, initiating a new timing cycle.

The Power-on One-shot Glitch

In our analysis of how the 555 timer works in monostable mode, we assumed that an initial flip-flop master-reset operation initially set up the timer. In cases where the power is simply turned on, there will be a single output pulse if the flip-flop takes on the wrong state. A flip-flop can take either of the two states when the power in turned on. As when we flip a coin, we can't be certain which state it will adopt until after it happens. A 555 output of $+V_{CC}$ allows the capacitor to begin charging without a proper trigger signal.

Figure 10-19(a) shows a modified monostable-mode 555 designed to produce a single output pulse every time the power is initially turned on. The configuration is often used to provide a master-reset pulse to a microprocessor or other system. The circuit produces one output pulse, and then does nothing until the power is turned off and then back on. The waveform in Figure 10-19(b) describes the operation of the circuit and illustrates how an unwanted power-on pulse can happen.

555 Timer in Astable (Oscillating) Mode

The 555 timer can be connected as a free-running rectangular-waveform generator. With a little extra timing circuitry, it can become a square-wave generator. And if a modulating voltage is applied to the control input, the 555 can be used as a voltage-controlled oscillator (VCO). Figure 10-20 shows the 555 connected in astable mode and provides examples of the circuit's associated waveforms.

FIGURE 10-19 Power-on Time-delay Circuit

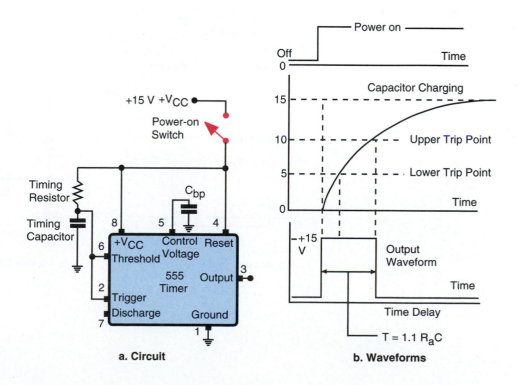

a. Circuit   b. Waveforms

FIGURE 10-20   555 Timer in Astable (Free-running Oscillator) Mode

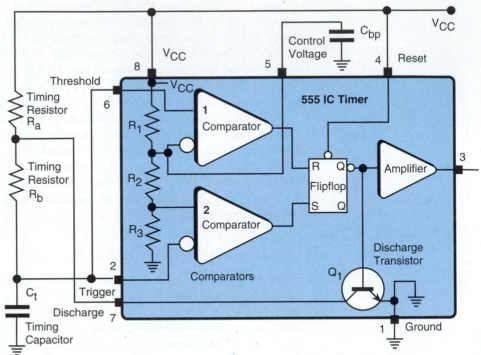

a. Astable Circuit, Including the 555 Timer's Internal Circuitry

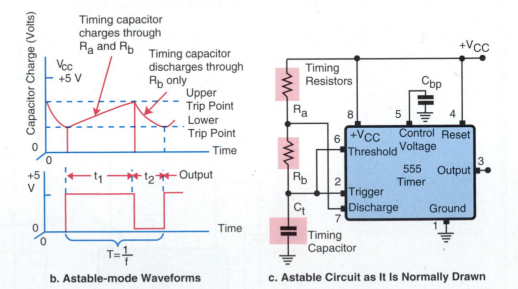

b. Astable-mode Waveforms

c. Astable Circuit as It Is Normally Drawn

FIGURE 10-20 Continued

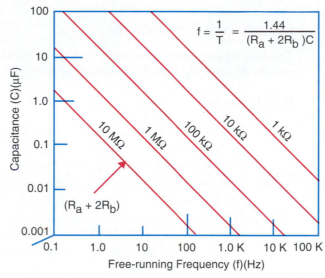

d. Timing Nomogram for the 555 Timer Astable Mode

Notice that there are two timing resistors in Figure 10-20(a), and that the threshold and trigger comparator inputs are connected. The circuit is essentially a self-triggering timing circuit. Refer to the circuit and waveforms in parts (a) through (d) of Figure 10-20 during the following discussion.

How It Works

1 Assume that the capacitor has been discharged, and that the 555 timer output is high.

2 Since the discharge transistor is off, the timing capacitor starts charging through timing resistors R_a and R_b, toward $+V_{CC}$.

3 When the capacitor's charging voltage crosses the $2/3$ of $+V_{CC}$ reference voltage, the upper comparator switches, resetting the flip-flop, and the 555 output goes low.

4 When the timer output goes low, the discharge transistor is turned on.

5 The discharge transistor starts discharging the timing capacitor through the timing resistor R_b. When the timing capacitor voltage discharges to a voltage slightly below the $1/3$ of $+V_{CC}$ reference voltage, the lower comparator switches, setting the flip-flop. The timer output goes high.

6 When 555 output is high, the discharge transistor turns off, and the timing capacitor again starts charging up toward the upper comparator's $2/3$ of $+V_{CC}$ reference voltage.

The timing capacitor charges and discharges, oscillating between the two comparators. The timer output switches from high to low, and then back to high, generating an endless train of rectangular output pulses. The output pulses are rectangular (but not square), because the timing capacitor charges through two timing resistors (R_a and R_b) and discharges through R_b alone.

Square-wave Output Modification

The 555 square-wave generator in Figure 10-21 requires equal resistance values for R_a and R_b, and the addition of a couple of steering diodes.

How It Works

1 During the charging part of the cycle, D_1 is forward-biased and the timing capacitor charges through R_a, bypassing R_b. Diode D_2 is reverse-biased, and R_b is disconnected.

2 During the discharge part of the cycle, D_2 is forward-biased and the timing capacitor discharges through R_b. Diode D_1 is reverse-biased because its anode is grounded through the discharge transistor and its cathode is positive. If the two timing-resistor values are equal, the charge and discharge times are equal, and the timer produces a square-wave output.

Adding a Constant-current Source for More Precise Timing

The timing-frequency accuracy of the astable 555 oscillator can be improved significantly by replacing the timing resistors with a controlled constant-current source. The LM-134, -234, and -334 controlled-current devices are true floating two-terminal constant-current sources, so they require no additional power-supply voltage. They feature a current accuracy of ±3%, and a voltage

FIGURE 10-21   **555 Timer as a Square-wave Generator**

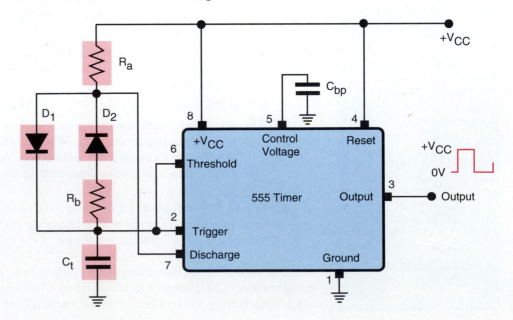

FIGURE 10-22 **555 Timer in Astable Mode Using an Adjustable Current Source for More Precise Timing**

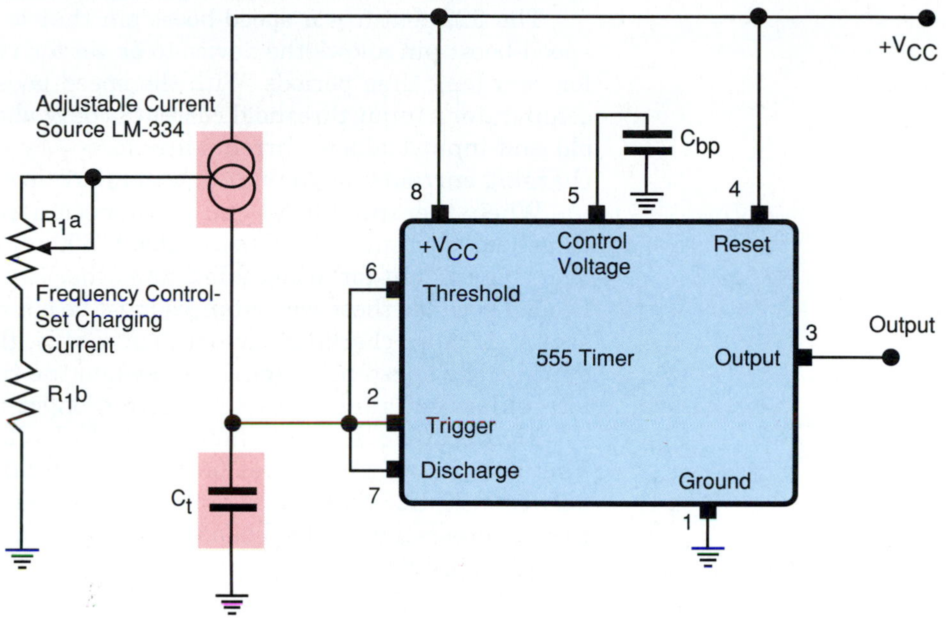

operating range of 1 to 40 volts. The current range is adjustable from 1 microampere to 10 milliamperes, using a single resistor or variable resistor.

Figure 10-22 shows the circuit of a 555 astable circuit with an LM-334 adjustable-current source replacing the timing resistors.

10.8   556 Dual Timer

The LM-566 dual timer contains two 555 timers in a single 14-pin dual-inline-package DIP. The two timers are independent circuits and share only the power-supply pins.

10.9   LM-322 and -3905 Precision Monostable Timers

The 322/3905 precision timer features more precise timing and longer time delays. The 322 is available in a 14-pin DIP or a 10-pin TO-5-style metal can. The 3905 is available only in an 8-pin MINIDIP package.

Timing periods of minutes or hours (for outdoor lights, sprinklers, and so on) are monostable timing operations. This kind of timing cycle is not repeated at regular intervals; instead, it is initiated randomly. Both monostable and astable timing periods can range from nanoseconds to days (or weeks), depending on the application. The 332/3905 timers are capable of very short (5 microseconds) to very long (2 hours) timing periods. Even longer—theoretically, nearly unlimited—timing periods are possible with the 322/3905, but capacitor leakage problems, circuit-board leakage paths, stray capacitances, and so on impose practical limits.

The 322 features a speed-boost pin that is unavailable on the 3905. The speed-boost pin allows the device to be set for very fast (short) time periods or for very long time periods. With the speed-boost terminal not connected, the comparator's input threshold current is only about 300 picoamps. Low threshold and input leakage currents are necessary for long timing cycles, because charging currents for timing capacitors are so small.

When the speed-boost pin is connected to $V+$, the comparator's input threshold current increases to about 300 nanoamps, providing additional charging current for Miller effect and other capacitances. The extra current is needed because these capacitances must be charged before the comparator can switch. If more charging current is available, the capacitors can charge more quickly. The speed-boost pin is used when the timing is in the microsecond-to-low-millisecond range; it is not used for longer time periods.

Figure 10-23 shows a functional block diagram of the 322/3905 timer. There are many similarities between the architectures of the 555 and of the 322/3905. Notice that the 322/3905 has only one comparator. This means that the hysteresis and its dead band have been eliminated, rendering astable operation impossible.

The Comparator and the Timing Values

The remaining comparator uses a voltage divider to obtain its reference voltage, but its divider provides a reference voltage of almost exactly 63.2% of the applied voltage. This places the trip point at exactly one time constant, making the timing equation exactly $T = RC$. In addition, the voltage to the voltage divider is provided by a built-in 3.15-volt voltage regulator.

The external timing (R-C) circuit can also use the internal regulator as a source of charging current, as shown in Figure 10-23(a); or some other source can be used. The waveforms in Figure 10-23(c) indicate that the timing capacitor is always allowed to charge to 2.0 volts before the comparator switches. This 2-volt value is derived by multiplying the regulator voltage by 0.632 (0.632×3.15 V $= 2.0$ V).

The Trigger Amplifier

In the 555, a comparator provides a lot of voltage gain for a trigger signal; but here, with the comparator gone, we need an amplifier. The flip-flop requires several volts on the set input to make it go. The trigger amplifier is an inverting transistor amplifier. The trigger input requires a voltage of at least +1.6 volts to set the flip-flop, as illustrated by the waveforms in Figure 10-23(c).

The Discharge Transistor

The discharge transistor is activated by the flip-flop and discharges the timing capacitor at the end of the timing cycle, just as it does in the 555 timer. In the 322/3905, however, the discharge transistor is internally connected to the timing R-C input on the comparator.

The Flip-flop and Normal/Invert Logic

The flip-flop is latched into the reset state by the upper comparator and remains reset until a voltage of 1.6 volts on the trigger input causes it to set. The Q output of the flip-flop is high (approximately $+V$) during the timing period (following the trigger pulse) and falls to ground (low) at the end of the timing period.

The normal/invert logic block is a switching amplifier that can be either an inverting amplifier or a noninverting amplifier, depending on the voltage level on the logic-sense control pin. If the control pin is connected to the +3.15-volt regulator, a high ($V+$) at the flip-flop's Q output is inverted to a low (ground) at the base of the output transistor. If the logic-sense control pin is grounded (low), the Q output of the flip-flop appears at the base of the output transistor without being inverted (normal).

The actual circuit that performs the normal/invert function is a digital logic gate called an **exclusive-Or gate.** Its operation is diagrammed in Figure 10-23(b).

How It Works

Assume that the logic-sense control pin is grounded, and that the circuit is connected as shown in Figure 10-23(a). Then we can trace the following action:

1 The trigger pulse starts the timing cycle.

2 The trigger pulse sets the flip-flop. The flip-flop's Q output is high ($+V$).

3 The normal/invert logic block is in the noninverting mode. So the high on the flip-flop's Q output is a high on the base of the output transistor.

4 Because the output transistor is connected as a common emitter inverting amplifier, a high on the base drives its collector low (to ground), and the LED goes dark.

5 The output stays low and the LED stays dark during the timing cycle.

6 When the timing capacitor charges to slightly above 2 volts, the comparator switches, resetting the flip-flop and causing its Q output to go low. This is the end of the timing cycle.

7 The low Q output is applied to the base of the output transistor (noninverted). The low base voltage turns the transistor off, the collector voltage rises to $+V$, and the LED turns on, indicating the end of the timing cycle.

8 When the flip-flop Q output (Not Q) goes high, it turns the discharge transistor on, discharging the timing capacitor and readying the circuit for the next timing cycle.

9 The flip-flop stays in this state until the next trigger pulse initiates a new timing cycle.

Output Circuit Variations

The output transistor can be connected as either a common emitter circuit or a common collector circuit. The common emitter circuit is an inverting ampli-

FIGURE 10-23   322/3905 Precision Monostable Timer

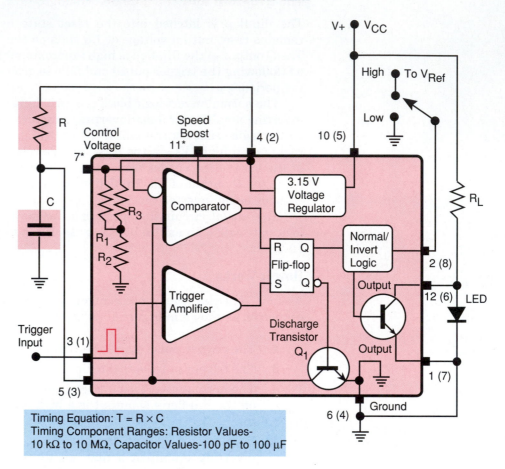

Timing Equation: T = R × C
Timing Component Ranges: Resistor Values-
10 kΩ to 10 MΩ, Capacitor Values-100 pF to 100 μF

Notes:
1. Pin numbers in parentheses are for the 3905 timer.
2. * Function is not available on the 3905 timer.

a. Complete 322/3905 Monostable Circuit

b. Detail of the Normal/Invert Block

FIGURE 10-23 Continued

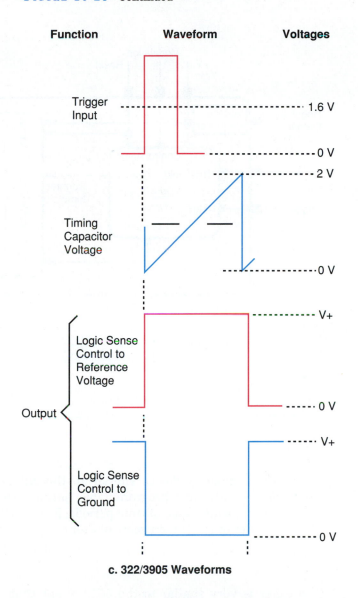

c. 322/3905 Waveforms

fier, while the common collector configuration is noninverting. Still, you can have the load turned either on or off during the timing period, with either output transistor configuration, by using the logic-sense control pin.

10.10   Adding a Counter for Very Long Time Delays

Any of several 555-like timers may be connected to a digital counter for very long time delays (ranging from hours to two or three weeks). Exar Corporation makes two timer/counter devices, the XR-2240 and XR-2250; and Intel

FIGURE 10-24   Timer Counter: XR-2240, XR-2250, and Intel 8260

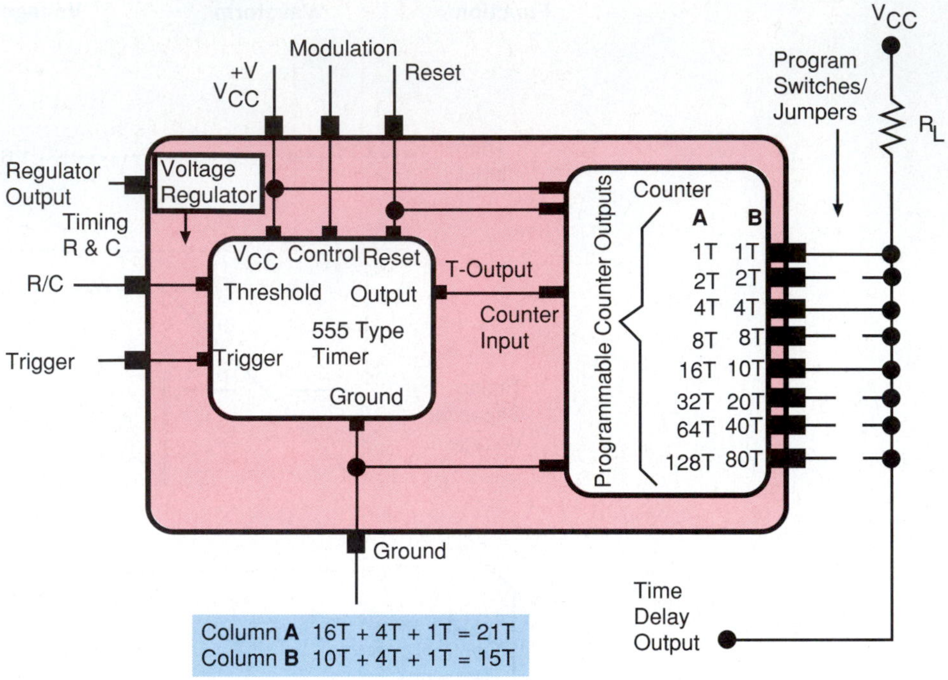

Column A   16T + 4T + 1T = 21T
Column B   10T + 4T + 1T = 15T

Notes:
1. Column A output values are for the XR2240 model.
2. Column B output values are for the XR2250 and Intel 8260 models.
3. The Intel 8260 model does not have an 80T output.

Corporation makes a device similar to the XR-2250, designated the 8260. These devices all have programmable outputs that allow them to be programmed to a wide range of time periods. Figure 10-24 is a generic functional block diagram of the three timer/counters.

Timer

The timer is very similar to the 555, except that its discharge transistor is connected internally, there is an on-board regulator, and some pins have been renamed. The timer can be configured as either monostable or astable. Each time the timer completes a timing cycle, it is registered as 1T, or one **time-base period.**

Counter

The counter counts the number of time-base timing cycles. Each output on the counter represents a specific count. The particular output remains at ground (low) until the time base has completed that specific number of timing cycles. Each output value is specified as the number of time base cycles that must be completed before that output switches off.

For example, look at the *B* column of output values in Figure 10-24. Assume that the timer illustrated in the figure is operating in astable mode and endlessly performs repeat timing cycles. The final time-delay output is held low by the three counter outputs with jumpers: 16T, 4T, and 1T. The counting is actually done in binary, which makes no difference to us, except that higher count values may happen before lower values as the count progresses. Other than as it relates to the following explanation, you need not even know that the numbers are processed in binary.

After the timer has completed sixteen time-base cycles, the 16T output turns off. The 4T and 1T outputs are still grounded, so the time-delay output remains low. After the next four time-base cycles, the 4T output turns off. The 1T output is still low, holding the final time-delay output at ground. After one more timing cycle, the last connected output (1T) turns off. The time-delay output no longer has any counting output pulling it to ground, so it rises to +V, indicating that the timing cycle is over.

If the timer had been set for a time-base period of 1 hour, for example, the final time-delay output in Figure 10-24 would go high only after

$$(1 \text{ hour} \times 16) + (1 \text{ hour} \times 4) + (1 \text{ hour} \times 1) = 21 \text{ hours}$$

The outputs for the 2250 and 8260 are defined in column *B*. With all jumpers connected, the final time delay is 165 times the time-base period. With a time-base period of 1 hour, the time delay is 165 hours, or nearly one week.

With all jumpers connected, the 2240 defined in column *A* provides a delay of 256 times the time-base period, or about ten days in this example. The 2240 has the longest delay capability, but the 2250 and 8260 are preferred when time must be specified in days–hours–minutes–seconds.

10.11   Voltage-timer Counter-controlled Oscillators

Using the 555 Timer as a Voltage-controlled Oscillator

The control input (pin #5) is connected to the top ($^2/_3$ V_{CC}) reference voltage divider. This access to the comparator reference voltage divider allows you to modify the normally fixed comparator reference voltages externally. Modifying the comparator reference voltages is accomplished by applying a voltage to the control input (pin #5). A **dc** voltage can be used to trim the timing period accurately, perhaps with a voltage-divider potentiometer.

If a **dc** voltage is used, the control-voltage source can be located at a considerable distance from the timer IC. Because the control voltage varies the reference voltage divider, the control-voltage range is directly related to the power-supply voltage. Increasing the power-supply voltage yields a wider range of timing or frequency control. The useful voltage on the control input pin (pin #5) ranges from 2 volts to (+V − 1 V). If the 555 timer is operated in the astable mode, with a power-supply voltage of 15 volts, varying the voltage on the control input can vary the frequency over a range of up to 3 decades.

The control pin is nearly always connected to ground with a small capacitor to filter out power-supply voltage transients. The capacitor is a good idea,

FIGURE 10-25 **dc Control Voltage for Adjusting the Frequency of a 555 Oscillator**

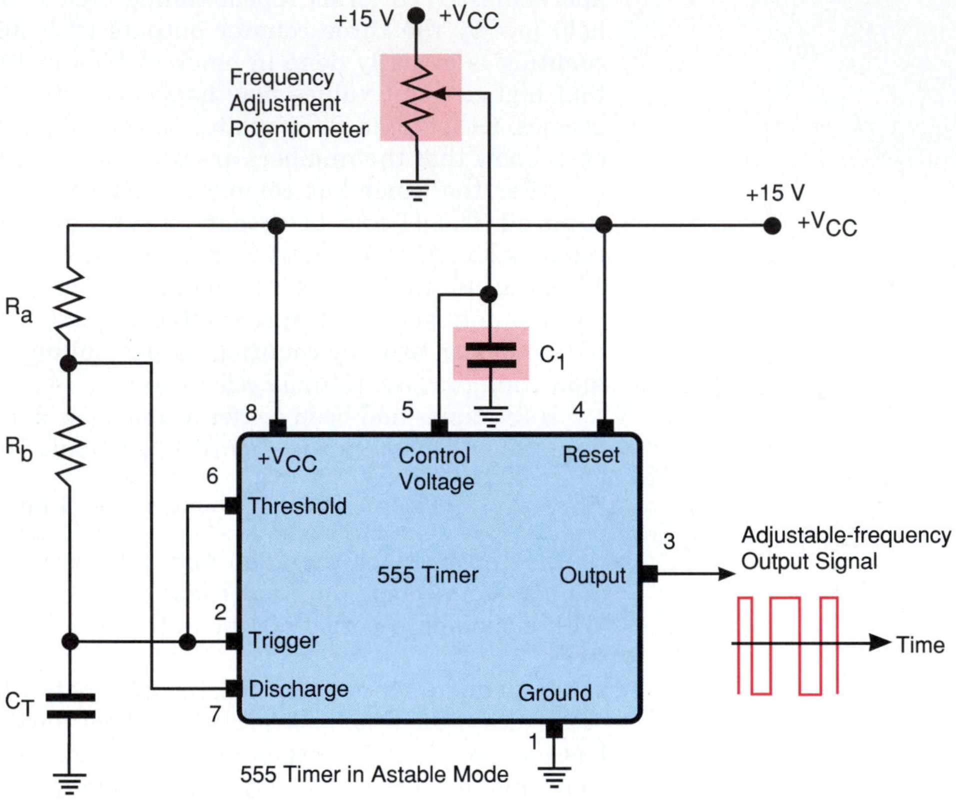

555 Timer in Astable Mode

even when **dc** is applied to the control pin. Figure 10-25 shows a typical **dc** control-voltage connection.

This vastly underused control input can also be used as a frequency-modulating input. The 555 can be modulated by an audio signal or in a variety of pulse-modulation formats. Figure 10-26 shows a 555 oscillator modulated by a sine wave. We will look at some practical modulated 555 timer circuits in Chapter 12, which deals with modulation and demodulation.

566 Integrated-circuit Voltage-controlled Oscillator

The 566 integrated-circuit VCO can use a control or modulating voltage to vary the output frequency by a 10:1 ratio. The oscillator can oscillate at frequencies of up to 1 MHz. The 566 can operate with a single-power supply voltage of up to 25 volts, or with a split +12.5-volt/−12.5-volt supply.

The circuit in Figure 10-27 shows the internal functional block diagram and external components for a center (no modulating signal) frequency of 10 kHz. The 566 presents a very sophisticated arrangement of circuits—one that we have examined in previous sections. Refer to Figure 10-27 during the following discussion.

FIGURE 10-26 Sine-wave Signal Applied to the 555 Control Input to Produce a Frequency-modulated Rectangular Wave Output

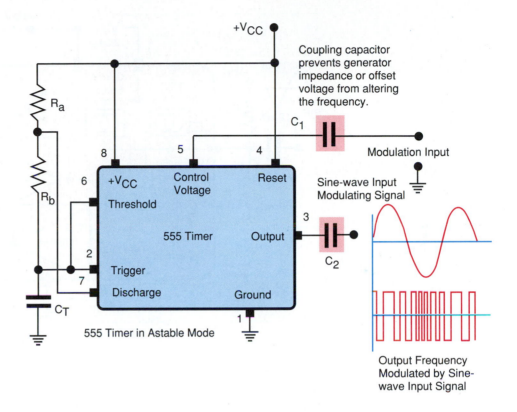

Output Frequency Modulated by Sine-wave Input Signal

How It Works

The Oscillator The oscillator is a basic comparator (with hysteresis) square-wave generator circuit, like the one in Figure 10-12. Positive feedback for hysteresis is provided by R_f and R_{in}. The comparator oscillator's output is buffered by an inverting output amplifier to eliminate any output loading effects on the comparator oscillator's operation.

The Capacitor Charging Circuit The external timing capacitor is charged and discharged by means of controlled constant-current sources built into the IC. The controlled constant-current sources in the 566 are more complex than the ones we used in previous examples, because of the 566's ability to work with a split-voltage power supply and to meet the desire for very good linearity.

Timing Capacitor Buffer Amplifier The timing capacitor is connected to a unity-gain high-input-impedance buffer amplifier to prevent any capacitor loading. The buffer amplifier delivers the capacitor charging voltage to the inverting comparator input and, at the same time, provides an external triangular-wave output that follows the timing capacitor's charge and discharge voltages.

FIGURE 10-27   **566 Integrated Circuit Voltage-controlled Oscillator (VCO)**

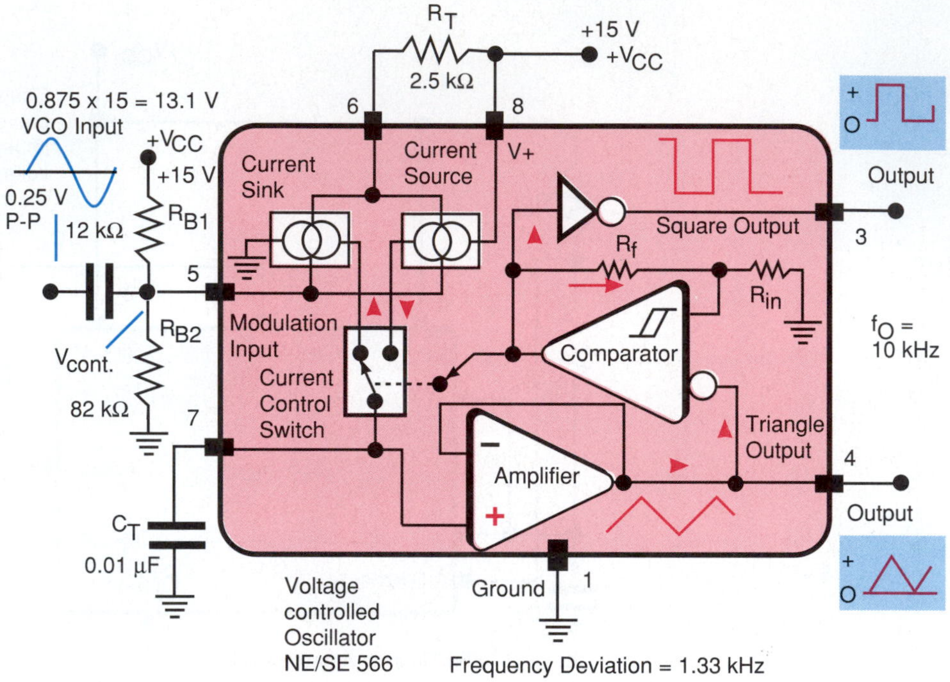

Voltage Frequency Control   Voltage control of frequency is accomplished by imposing a control voltage on the controlled constant-current sources, which in turn control the charging/discharging rate of the timing capacitor.

Some Special Requirements

1   The timing resistor (R_T) must be within the range of 2 to 20 kilohms.

2   The useful modulating input voltage ranges from $0.75 \times V_{CC}$ to V_{CC}. The modulation input should be biased (see R_{B_1} and R_{B_2}) about midway between $0.75\,V_{CC}$ and V_{CC}, if a bipolar modulating signal is to be used. Midway between 75% and 100% is 87.5% (of V_{CC}), so the normal pin #5 bias voltage is $0.875 \times V_{CC}$.

Calculating the Timing Component Values

1   The free-running frequency (f_0) is given by the following equation:

$$f_0 = 2(V_{CC} - V_{cont})/(C_T \times R_T \times V_{CC})$$

2   Assume that V_{cont} is biased to $0.875 \times V_{CC}$. Then

$$0.875 \times 15 = 13.1 \text{ V}$$

Example 10.1 | **Timing Component Calculations**

Calculate the timing component values for a center frequency of 10 kHz.

Solution The procedure is to make an educated guess at a capacitor value. If the timing resistor calculates out to a value between 2 and 20 kilohms, all is well. If the timing resistor is not in that range, you must try another capacitor value. Let's try a value of 0.01 microfarad for the timing capacitor.

Modifying the frequency equation above to solve for R_T, we get

$$R_T = 2(V_{CC} - V_{cont})/(f_0 \times C_T \times V_{CC})$$
$$= 2(15 \text{ V} - 13.1 \text{ V})/[10 \text{ kHz} \times (1 \times 10^{-8}) \text{ F} \times 15 \text{ V}]$$
$$= 3.8/(1.5 \times 10^{-3})$$
$$= 2.533 \text{ kilohms}$$

A value of 2.5 kilohms is within the range, so we have useful values.

Calculating the Frequency Deviation The amount of frequency deviation caused by a modulating signal voltage of ΔV_{in} (in volts P–P) is given by the following equation:

$$\Delta f = 2\Delta V_{in}/(C_T \times R_T \times V_{CC})$$

Example 10.2 | **Calculating Frequency Deviation**

Find the frequency deviation when the modulation input voltage is 0.25 V P–P.

$$\Delta f = 2\Delta V_{in}/(C_T \times R_T \times V_{CC})$$
$$= (2 \times 0.25 \text{ V})/[10^{-8} \text{ F} \times (2.5 \times 10^3) \times 15 \text{ V}]$$
$$= 0.5 \text{ V}/(37.5 \times 10^{-5})$$
$$= 1.33 \text{ kHz}$$

The total frequency deviation is 1.33 kHz, or 10 KHz + 700 Hz and 10 kHz – 700 Hz. The oscillator's output frequency will swing between 9.3 kHz and 10.7 kHz.

A **dc** voltage of 0.25 volts would also shift the oscillator frequency by 700 Hz. We will encounter voltage-controlled oscillators in specific applications in subsequent chapters.

Troubleshooting Waveform Generators

Waveform generators make troubleshooting easy. An oscilloscope will quickly tell you if a waveform generator is working properly, because it generates a distinctive output signal.

IC Waveform Generators

1 Because waveform generators output a signal—generally of a known waveshape—the first thing to check is the output signal. The oscillo-

FIGURE 10-28   Using a Norton Op-amp to Make a Pocket Function Generator

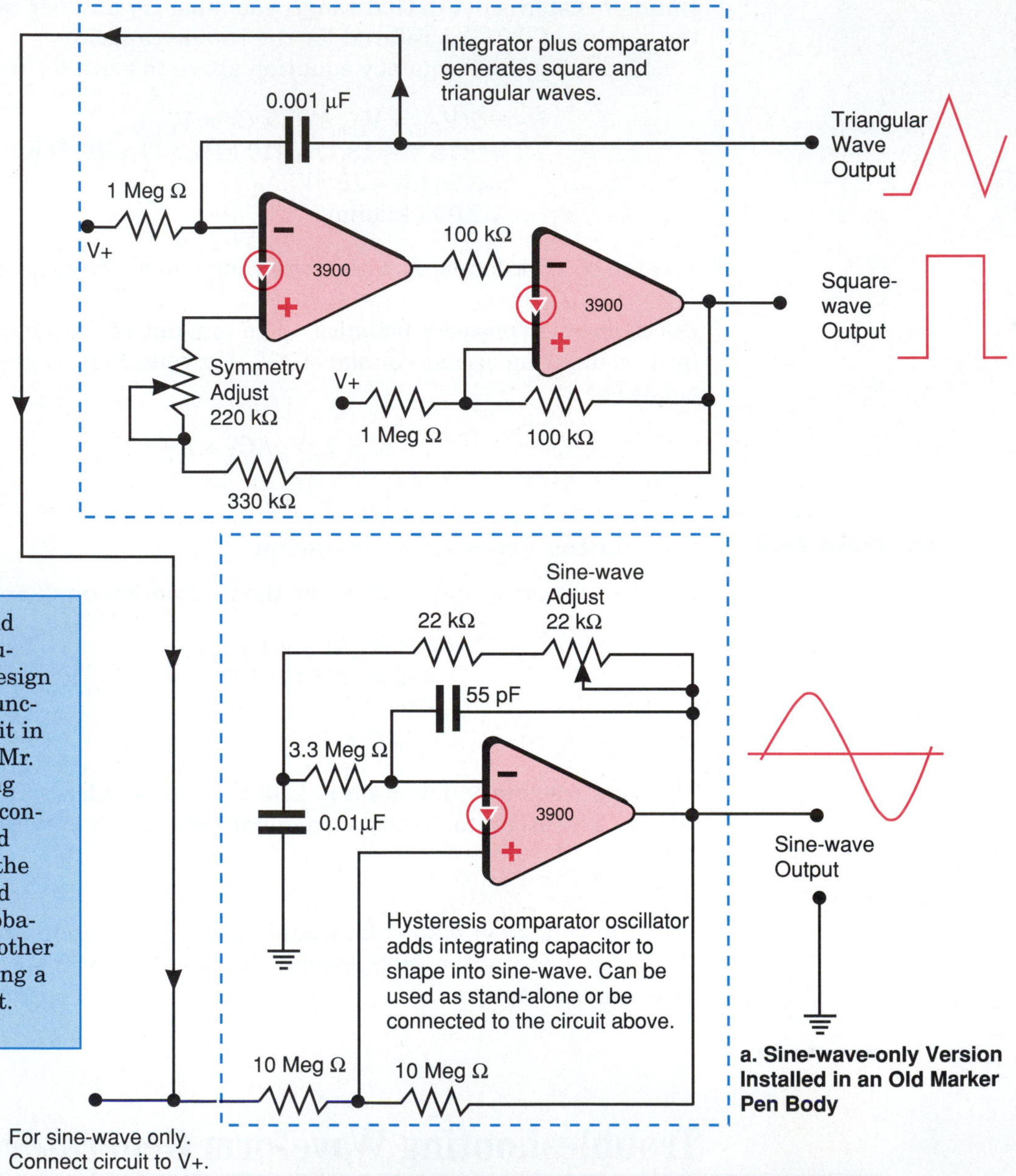

Integrator plus comparator generates square and triangular waves.

0.001 μF

1 Meg Ω

V+

Triangular Wave Output

100 kΩ

3900

3900

Square-wave Output

Symmetry Adjust 220 kΩ

V+

1 Meg Ω

100 kΩ

330 kΩ

My thanks to Donald Lillard, a former student, for the final design and testing of the function generator circuit in Figure 10-28. After Mr. Lillard got a working design, it became a contest to see who could build the circuit in the smallest case. An old marker pen was probably the winner, but other shapes made selecting a winner very difficult.

Sine-wave Adjust

22 kΩ   22 kΩ

55 pF

3.3 Meg Ω

0.01μF

3900

Sine-wave Output

Hysteresis comparator oscillator adds integrating capacitor to shape into sine-wave. Can be used as stand-alone or be connected to the circuit above.

10 Meg Ω   10 Meg Ω

a. Sine-wave-only Version Installed in an Old Marker Pen Body

For sine-wave only. Connect circuit to V+.

b. Schematic Diagram of the Complete Function Generator

scope is the proper tool for this task. In some cases, a digital logic probe may be a suitable output waveform indicator.

2   If the waveform is absent or is of incorrect frequency or amplitude, verify the **dc** power-supply voltages at the IC pins. If the voltages are missing or incorrect, you have probably found the problem. A common

NOTE

Reversing the power-supply connections on a 555 timer causes it to get very hot—often to the point of exploding. It is a little chip and a little explosion. The overheated 555 chip may or may not recover after the power is connected properly and cooled down. There are no available data about whether such overheating shortens the life of the chip.

troubleshooting problem with 555 timer circuits involves failing to test the voltage on the reset pin (pin #4) for V_{CC}. If the reset pin is grounded or open, the circuit cannot perform its timing function.

3   If you are working with a 555 timer circuit, refer to the action table in Figure 10-17, and try to force the comparators as described in section 9.12. This will either prove that the chip is bad or eliminate it as a suspect.

4   If electrolytic-type capacitors are used for long time delays, they can eventually become leaky or shorted. Electrolytic capacitors are much more prone to failure than are other capacitor types.

5   As usual, most problems result from poor mechanical connections, solder joints, connectors, and so on.

Op-amp-based Waveform Generators

1   Because waveform generators output a signal—generally of a known waveshape—the first thing to check is the output signal. The oscilloscope is the proper tool for this task. In some cases, a digital logic probe may be a suitable output waveform indicator.

2   If the waveform is absent, or is of incorrect frequency or amplitude, verify the **dc** power-supply voltages at the IC pins. If the voltages are missing or incorrect, you have probably found the problem.

3   If you are working with an op-amp circuit, refer to the op-amp troubleshooting section, and try to force the op-amp output voltage to change. This will either prove that the op-amp is bad or eliminate it as a suspect.

4   If electrolytic-type capacitors are used for long time delays, they can eventually become leaky or shorted. Electrolytic capacitors are much more prone to failure than are other capacitor types.

5   As usual, most problems result from poor mechanical connections, solder joints, connectors, and so on. Never remove a part from a circuit board unless you are very sure it is bad. Any time you remove a part, you run the risk of damaging the PCB.

The Digital Connection: A Digital/Analog Staircase Generator and Digital-to-Analog Converter

The circuit in Figure 10-29 derives its digital data from the output of a binary (digital) counter. The first op-amp is connected as a summing amplifier, with the resistors scaled for a 1–2–4–8 binary number system. The second op-amp is simply an inverting amplifier to cause an upward-moving staircase.

FIGURE 10-29   Digital Staircase Generator and Digital-to-Analog Converter

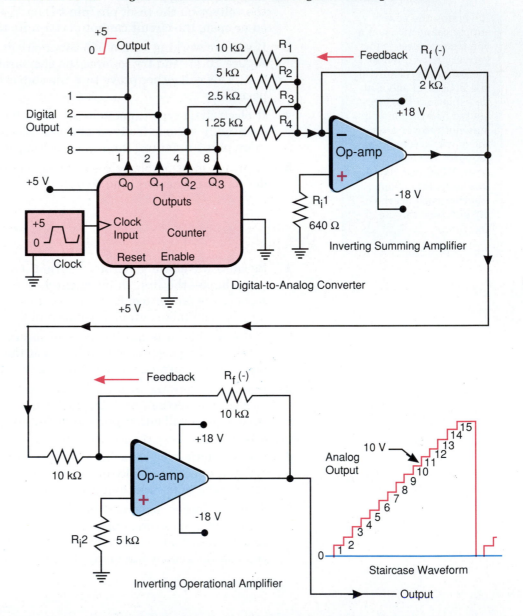

Summary of Q Voltage Values
(Assume that each input voltage from Q_0 to Q_3 = +5 V.)

Q_3 Output:

$$\text{Voltage Gain} = R_f/R_i = 2 \text{ k}\Omega/1.25 \text{ k}\Omega = 1.6$$
$$\text{Amplifier Output Voltage} = A_V \times +5 \text{ V} = 1.6 \times +5 \text{ V} = 8.0 \text{ V}$$

Q_2 Output:

$$\text{Voltage Gain} = R_f/R_i = 2 \text{ k}\Omega/2.5 \text{ k}\Omega = 0.8$$
$$\text{Amplifier Output Voltage} = A_V \times 5 \text{ V} = 0.8 \times +5 \text{ V} = 4.0 \text{ V}$$

Q_1 Output:

$$\text{Voltage Gain} = R_f/R_i = 2 \text{ k}\Omega/5 \text{ k}\Omega = 0.4$$
$$\text{Amplifier Output Voltage} = A_V \times 5 \text{ V} = 0.4 \times +5 \text{ V} = 2.0 \text{ V}$$

Q_0 Output:

$$\text{Voltage Gain} = R_f/R_i = 2 \text{ k}\Omega/10 \text{ k}\Omega = 0.2$$
$$\text{Amplifier Output Voltage} = A_V \times 5 \text{ V} = 0.2 \times +5 \text{ V} = 1.0 \text{ V}$$

How It Works

The counter counts from binary 0000 (decimal 0) to binary 1111 (decimal 15). The output voltage of the summing amplifier rises from 0 to 15 volts, in 1-volt steps. These steps represent the decimal numbers 0 through 15. The rising voltage generates a step waveform, called a *staircase waveform*.

Analog Timer with Digital Output   The comparator-based analog timer (see Figure 10-30) is a simple and conventional circuit with an adjustable reference voltage and a dial calibrated in seconds. The only trick is that the capacitor's charge voltage must increase at a constant rate of 1 volt per second.

The counter must also increase its count at one count per second. The comparator actually does all of the timing; and when it switches at the end of the preset delay time, it stops the binary counter. The counter retains the count it has reached at the end of the timing period. The reset button resets the counter to 0000 and discharges the capacitor to 0 volts at the same time.

Example 10.3 | ### Conversion Example

1   Set the reference voltage to 5 seconds (5 volts).
2   The capacitor charges at a rate of 1 volt per second for 5 seconds.
3   At the end of 5 seconds, the capacitor charge voltage is 5 volts. The comparator then switches to high, turning LED *B* on. This is the end of the timing cycle.
4   The enable pin of the counter was low at the start of the comparator timing cycle, enabling the counter to count the one-per-second clock pulses.
5   At the end of the comparator's 5-second timing period, the comparator output went high and disabled the counter. The counter was stopped at a count of binary 0101, or decimal 5.

Because the counter provides a binary time value equal to the analog time value, the circuit is a form of analog-to-digital converter.

Using Digital Pulses to Synthesize a Sine Wave   The circuit in Figure 10-31 can be used to convert a digital clock pulse (square wave) into a fairly respectable sine

FIGURE 10-30   Analog Timer for Controlling a Digital Counter

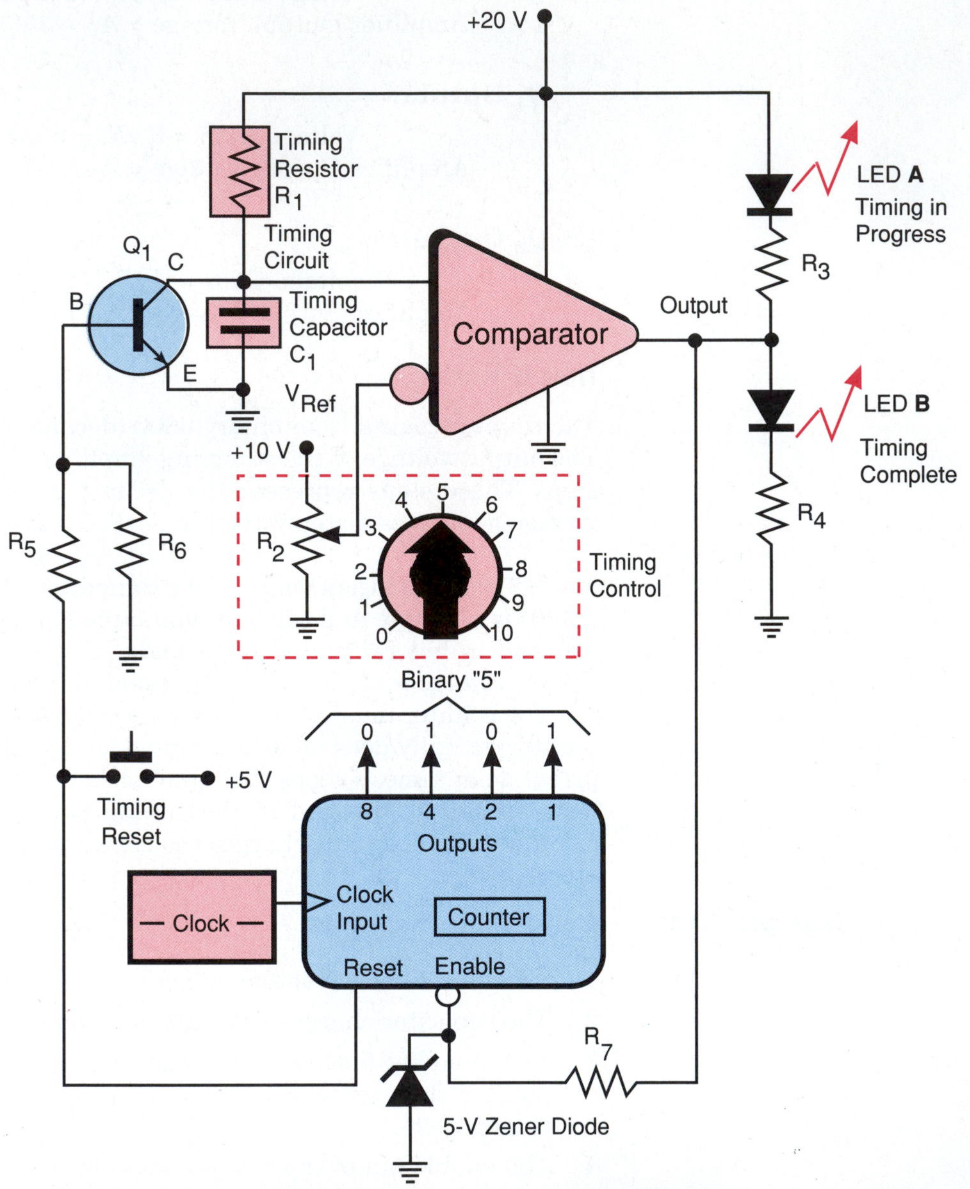

Note: The counter output can be connected to other digital circuits.

wave. This particular version uses a tricky exclusive-Or gate network to take care of both upward and downward voltage movement, to permit the use of a single counter and a single analog-gate package. As the counter counts, it causes the 1-of-8 binary decoder to turn analog gates D_0 through D_7 on in sequence (one at a time).

Four of the analog gates are connected to a positive voltage for the positive-going half of the wave. The other four analog gates are connected to a negative voltage to produce the negative-going half of the waveform.

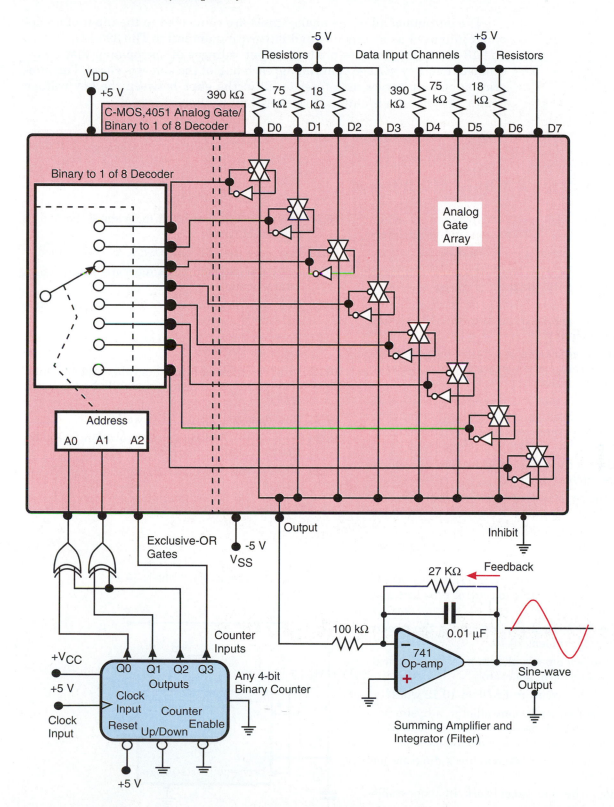

FIGURE 10-31 How Digital Pulses are Converted into a Sine-wave Signal by an Analog Gate Data Selector (Multiplexer)

The outputs of all of the analog gates are connected to the input of an op-amp configured as a summing and integrating amplifier. The 390-kΩ, 75-kΩ, and 18-kΩ input resistors produce output voltages at the op-amp that correspond to four key voltage points along each half of the sine waveform. The integrating op-amp serves as a filter to fill in the gaps between the key voltage points. The sine-wave output is cleaner than you might expect.

Questions and Problems

10.1 What is the basic problem involved in using resistor-capacitor (R-C) analog timing circuits.

10.2 Describe two ways of compensating for or correcting this problem.

10.3 Draw a schematic diagram of an op-amp integrator.

10.4 Why can the op-amp integrator be used as a linear timing circuit?

10.5 A resistor is connected in parallel with the feedback capacitor in an op-amp integrator. What is the resistor's function?

10.6 Draw the schematic diagram of a hysteresis comparator-based square-wave generator, with a constant-current charging and discharging source and sink.

10.7 Define the terms *source* and *sink,* as applied to constant-current devices.

10.8 Why is the constant-current version of the comparator-based square-wave generator superior to the resistor-capacitor version?

10.9 How does the 555 timer deal with the capacitor nonlinear charging problem?

10.10 Draw a schematic diagram of a ramp generator that uses a comparator to control the start and stop of the ramp.

10.11 Describe a practical application for the circuit you drew in Problem 10.10.

10.12 Draw the schematic diagram of a practical staircase waveform generator.

10.13 How does the circuit in Problem 10.12 work?

10.14 How is the frequency controlled in a hysteresis comparator-based oscillator that uses a constant-current source/sink charging circuit?

10.15 Draw the internal functional block diagram for a 555 timer.

10.16 How does the 555 timer work in monostable mode?

10.17 Draw the external block diagram of the 555 in monostable mode.

10.18 How does the 555 timer work in astable mode?

10.19 Draw the external block diagram of the 555 in astable mode.

10.20 How can very long time delays be had with a 555 timer?

10.21 How can the 555 be used as a voltage-controlled oscillator (VCO)?

10.22 Draw an internal block diagram of the 566 integrated-circuit voltage-controlled oscillator (VCO).

10.23 How does the 566 IC VCO work?

10.24 What is the 555 timer configuration in Figure 10-32?

10.25 What is the 555 timer configuration in Figure 10-33?

Figure 10-32 Circuit for Problem 10.24

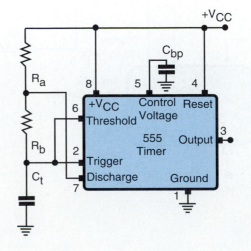

10.26 What the 555 timer configuration in Figure 10-34?

10.27 Why are the diodes D_1 and D_2 used in the 555 timer circuit in Figure 10-35?

10.28 What is the frequency of the astable 555 timer circuit in Figure 10-36?

10.29 What is the time-delay period of the monostable 555 timer circuit in Figure 10-37?

FIGURE 10-33 **Circuit for Problem 10.25**

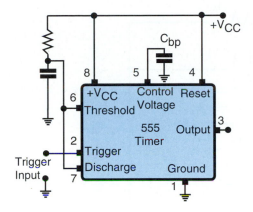

FIGURE 10-34
Circuit for
Problem 10.26

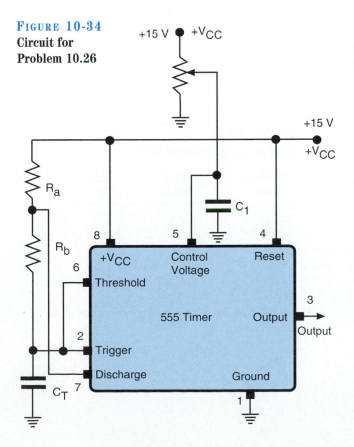

FIGURE 10-35 **Circuit for Problem 10.27**

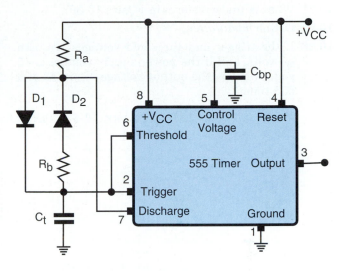

FIGURE 10-36 **Circuit for Problem 10.28**

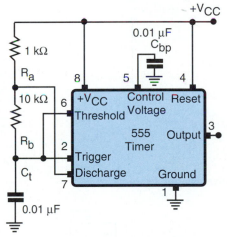

FIGURE 10-37 **Circuit for Problem 10.29**

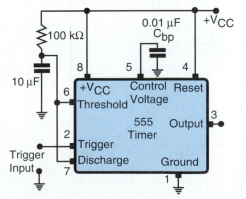

10.30 What is the circuit in Figure 10-38?

10.31 What is the waveform in Figure 10-39?

10.32 Define *modulation*.

10.33 If the trigger and threshold voltages are both +5 volts, and if the power-supply voltage is +5 volts, what is the output voltage of pin #3 of a 555 timer?

10.34 What is normally done with the control pin (pin #5) of a 555 timer when it is not being used? Why?

10.35 Complete Table 10-1 by filling in the trip-voltage levels for the 555 timer's comparators.

FIGURE 10-38 **Circuit for Problem 10.30**

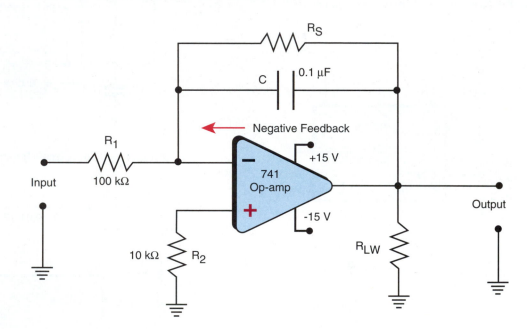

FIGURE 10-39 **Waveform for Problem 10.31**

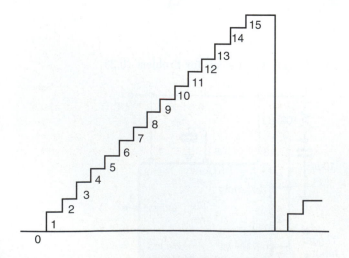

TABLE 10-1

Work Table for Problem 10.35

Threshold Voltage	Trigger Voltage	Output Voltage
+3 V	+5 V	
+5 V	+5 V	
+0.2 V	+0.2 V	
0 V	0 V	
+1.8 V	+1.8 V	
+3.5 V	+3.5 V	

Note: Assume a power supply voltage of +5 V.

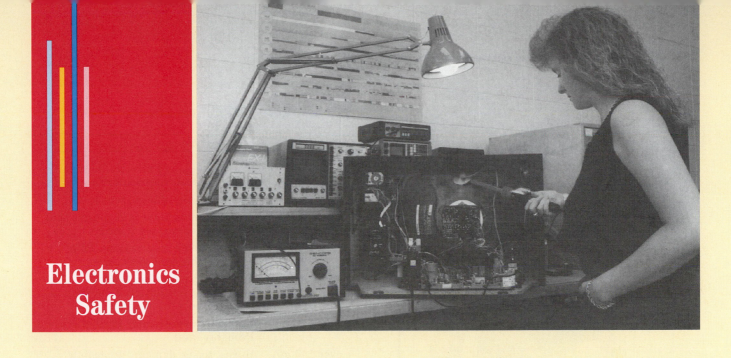

Electronics Safety

The risks of working in electronics may not seem great when compared to those encountered in some other technical and industrial fields. But there are hazards in this area, especially if you find yourself working with some of the higher voltages or frequencies. Service work on televisions, monitors, or other types of equipment that use voltages in the kilovolt range can be deadly if you don't take the time and precautions necessary to be safe. **ac** power-line voltages are dangerous, too, and electronics technicians work on line-powered equipment quite often. The level of the voltage isn't the only hazard, though; injuries can be caused by voltages as low as 12 volts. And electrical shock isn't the only danger an electronics technician has to contend with. Chemical hazards, injuries due to lifting and tripping, and accidents involving tools are other occupational hazards of the electronics technician.

You have heard it before, but it bears repeating: You are responsible for your own safety and for the safety of your co-workers. And safety should be an attitude, a habit, and a way of life. You should think about safety before you perform any measurement, adjustment, or other action on electrical equipment. Ask yourself, "Is this safe? For me? For my co-workers?" It only takes a moment to be sure.

Some Facts About Electrical Hazards

Did you know that only 100 microamperes of current through your heart can kill you? *Ventricular* *fibrillation* is the medical term for one response of the heart muscle to electrical shock. In response, the heart muscle stops beating regularly and instead just quivers randomly.

The human body is mostly water, which in itself is not a good conductor—but ionic solutions are. The water in human tissue contains sodium, potassium, and chlorine ions, which facilitate the movement of electrochemical stimuli throughout the body. Electrical current produced by an external potential overrides the normal low-level signals and causes a phenomenon called *polarization*. Depending on the level of current, the reaction can range from a mild tingling sensation to fibrillation and eventually to burns and tissue damage.

Although the internal resistance of the body is very low, skin resistance is relatively high, typically ranging from about 2.5 kΩ to 25 kΩ. If your hands are wet, this resistance may drop to as little as 300 Ω. Dry hands can have a resistance in the 100 kΩ range. Of course, for current to flow, an electrical potential and two contact points are required. The location of those contact points on the human body has an important bearing on the results of an electric shock.

Since current flow spreads out as it travels through the body, the current density at a given point will determine the effect. Thus, 100 μA through the heart can kill, but—because of the current density—a current of 100 mA from arm to arm

A 100 mA Current Through the Heart Can Kill

TABLE A
Effects of Various Levels of Shock Current

Current Level	Effects
100 µA	Current required directly through the heart to cause ventricular fibrillation.
1 mA	Lowest perceptible current level.
5 mA	Muscle contractions and spasms that can cause secondary injuries.
20 mA	Strong muscle contractions. Pain. Most victims cannot let go.
50 mA	Significant pain. Can cause ventricular fibrillation if duration is longer than a few seconds.
100 mA	Immediate ventricular fibrillation. Probably fatal.
over 100 mA	Burns, tissue damage, immediate death.
1A to 6A	Current levels used to restart the heart.

may be necessary for the same result. Table A shows several current levels and their effects on the body.

An interesting phenomenon occurs as the frequency of shock current increases: The severity of the shock decreases. While 100 mA at 60 Hz can kill you, the same current at 1 kHz, while noticeable, is neither painful nor life-threatening. The current at even higher frequencies is not even noticeable.

This is not to say that very high-frequency electromagnetic radiation is safe. Exposure to radiation in the ultraviolet, infrared, microwave, and nuclear ranges can cause direct damage to human tissue. Electronics technicians who work on microwave sources, radar, and other equipment that radiates energy must observe proper safety practices.

What You Can Do to Be Safe

The first and foremost rule is to know the risks. Obviously, you can't see the presence of high volt-age in a circuit, so you must treat every circuit with respect. As in firearm safety, treat every piece of equipment as if it were loaded. When in doubt, check for dangerous voltages with a voltmeter. But be careful. Some equipment (such as televisions) may have voltages high enough to damage the meter itself.

Each area of electronics has its own hazards. One danger that exists in television servicing involves the live chassis. Most televisions don't use a power-supply transformer (which would normally isolate the circuitry from the line voltage and ground). From the factory, the television chassis should be connected to the neutral side of the **ac,** but sometimes the **ac** plug is replaced incorrectly. This connects the 110-V **ac** line power to the chassis of the television. If you come in contact with the chassis and ground you can receive a serious shock.

Another shock hazard can occur in many kinds of electronic equipment if the insulation resistance of a transformer or capacitor decreases because of heat or for other reasons. The leakage current that is produced permits the chassis or other parts of the device to become live. Since very little current is required for a serious shock, leakage current can be dangerous. Figure B shows a simple test apparatus that enables the user to check for **ac** or **dc** potential due to leakage current. For safe operation, the voltage measured should never be more than 0.75 volts.

Anytime you have to perform measurements on a live piece of equipment, you should consider using an isolation transformer. An isolation transformer is just a 1:1 power transformer that isolates the equipment from the **ac** power line. Any voltages inside the equipment then have no return path through ground. This doesn't mean you are completely safe, however. If you make contact with two points inside the circuit that have a potential across them, you can still receive a shock.

Certainly you should protect yourself by working safely, but you also have a responsibility to your co-workers and customers. In consumer electronics servicing, using the correct parts to replace defective components has become a big issue. If you replace a critical component with a nonapproved part, and a fire, shock, or other accident occurs that threatens life or property, you and/or your employer may be legally liable for damages. Most critical components are designated on the schematic; but if you are in doubt, use the direct replacement.

Many other hazards lurk in an electronics shop. Besides the more common electrical ones just described, there are various mechanical hazards to consider. If you work with televisions or monitors, beware of the cathode-ray tubes. The implosion from a broken tube can send shards of glass flying. Never leave tools, spare parts, or packaging laying around. Numerous accidents result from tripping, falling, or improper lifting.

Nobody does everything right all the time, but good work habits and a safety-conscious attitude will go a long way toward preventing injury due to electrical hazards.

Leakage Current Test Setup

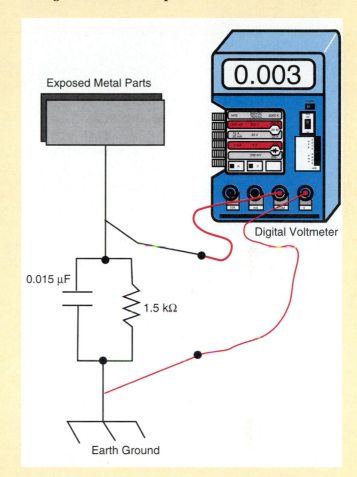

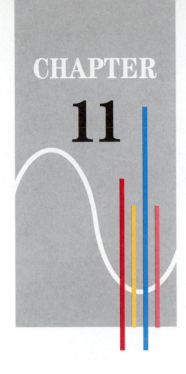

CHAPTER 11

Active Filters and Sine-wave Oscillators

OBJECTIVES

Upon completion of this chapter, you should be able to:

1 Define a first-order filter, and state its roll-off in dB.

2 Define a second-order filter, and state its roll-off in dB.

3 Define a third-order filter, and state its roll-off in dB.

4 Explain the difference between a second-order and a two-pole filter.

5 Define bandwidth in terms of Q.

6 Name the point on a band-pass curve where the bandwidth is measured.

7 Identify high-pass and low-pass filter circuits.

8 Explain the difference between an active filter and a passive filter.

9 Identify a Salen and Key active filter circuit.

10 Identify unity-gain high-pass and low-pass active filters.

11 Explain the operation of an op-amp integrator.

12 List three applications for an op-amp integrator.

13 Identify a twin-T notch filter circuit.

14 Identify a twin-T oscillator.

15 List the Barkhausen criteria.

16 Explain the operating principles of sine-wave oscillators in general.

17 Identify the phase-shift oscillator circuit.

18 Identify the Colpitts oscillator, and explain how positive feedback is obtained.

19 Identify the Hartley oscillator, and explain how positive feedback is obtained.

20 Identify the crystal-controlled oscillator, and explain how positive feedback is obtained.

21 Define *damping factor*.

22 Define *Q*.

23 Define *bandwidth*.

24 Define *ringing*.

25 Explain how a ringing L-C circuit operates.

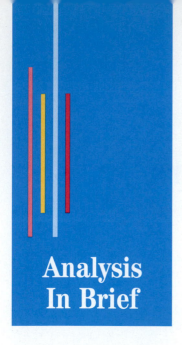

Active Filters in Brief

Passive Filter Review

The following equation defines the cut-off frequency for all first-order R-C filters:

$$F_c = 1/(2\pi RC)$$

THE FIRST-ORDER PASSIVE LOW-PASS FILTER

In the first-order low-pass filter, the resistor and the capacitor form a voltage divider. At low frequencies the reactance of the capacitor is very high, and most of the signal voltage is dropped across the capacitor. As the frequency increases, the reactance of the capacitor falls, and the capacitor voltage (output voltage) decreases at a rate of −20 dB per decade, while the frequency continues to increase. Low frequencies pass from the input of the filter to its output with little attenuation.

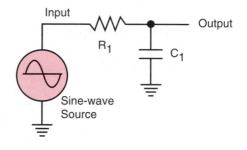

Low-pass Filter Circuit

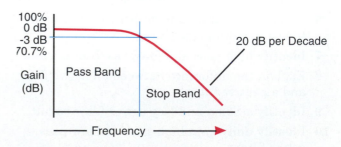

Low-pass Filter Response Curve

First-order Active Filters

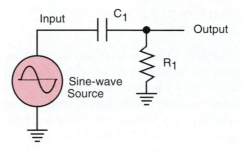

THE FIRST-ORDER PASSIVE HIGH-PASS FILTER

In the first-order high-pass filter, the resistor and the capacitor form a voltage divider. At high frequencies, the reactance of the capacitor is very low, and most of the signal voltage appears across the output resistor. As the signal frequency falls below the cut-off frequency, the reactance of the capacitor increases. As the voltage drop across the capacitor increases, the output voltage across the resistor decreases at a rate of –20 dB per decade. The cut-off frequency is defined as the point where the output voltage has fallen to 3 dB below its maximum value.

First-order Passive High-pass Filter Circuit

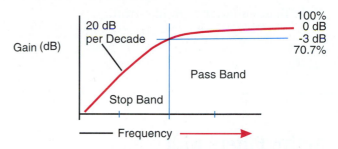

First-order Passive High-pass Filter Response Curve

THE FIRST-ORDER ACTIVE LOW-PASS FILTER

The simplest active low-pass filter is nothing more than a passive low-pass filter with an added voltage-follower buffer op-amp. The high op-amp input impedance prevents the filter from being loaded, while the op-amp output can drive a significant load. The –3-dB cut-off frequency is

$$F_c = 1/(2\pi RC)$$

where F_c is the 3-dB cut-off frequency in Hz, R is the resistance in ohms, and C is the capacitance in farads.

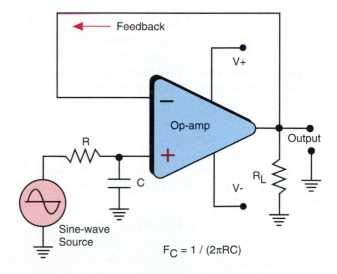

First-order Low-pass Active Filter

THE FIRST-ORDER ACTIVE HIGH-PASS FILTER

The simplest active high-pass filter is nothing more than a passive high-pass filter with an added voltage-follower buffer op-amp. The high op-amp input impedance prevents the filter from being loaded, while the op-amp output can drive a significant load. The –3-dB cut-off frequency is

$$F_c = 1/(2\pi RC)$$

Note: Both equations are identical.

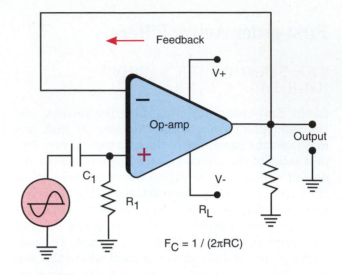

First-order High-pass Active Filter

Active Filters with Voltage Gain

Adding a voltage follower to a passive filter isolates the filter from loading, but it does nothing about the passive filter's insertion loss. A voltage gain that is at least large enough to make up for the passive filter losses may be required.

Adding a feedback network to control the voltage gain is often essential in more complex filters. Adding voltage gain is also very useful in these simple active filters.

CUT-OFF FREQUENCY

Adding the feedback network does not have an effect on the cut-off frequency. The –3-dB cut-off frequency is still

$$F_c = 1/(2\pi RC)$$

VOLTAGE GAIN

The voltage gain is that of any noninverting op-amp:

$$A_V = 1 + (R_f/R_{in})$$

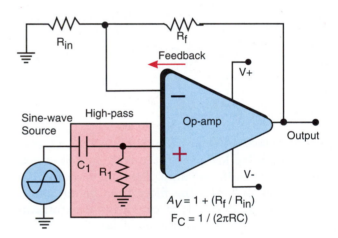

First-order High-pass Filter

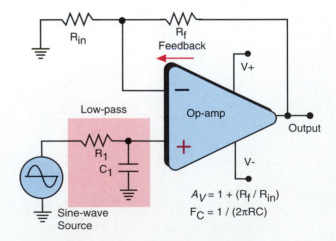

First-order Low-pass Filter with Voltage Gain

Second-order Active Filters

SECOND-ORDER LOW-PASS FILTER

The simplest form of second-order active low-pass filter uses two simple first-order active op-amp buffered filters connected in cascade. Each first-order circuit provides one of the two poles (breakpoints). The op-amps may provide voltage gain or may be unity-gain voltage followers.

Each filter section roll-off has 20 dB per decade, or 40 dB per decade total (6 dB per octave, or 12 dB per octave total).

The –6dB frequency is calculated as

$$F_c = 1/(2\pi RC)$$

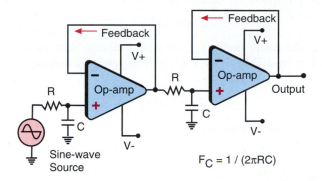

Two-stage (40 dB per Decade) Low-pass Filter

SECOND-ORDER HIGH-PASS FILTER

The simplest form of second-order active high-pass filter uses two simple first-order active op-amp buffered filters connected in cascade. Each first-order circuit provides one of the two poles (breakpoints). The op-amps may provide voltage gain or may be unity-gain voltage followers.

Each filter section roll-off has 20 dB per decade, or 40 dB per decade total (6 dB per octave, or 12 dB per octave total).

The –6dB frequency is calculated as

$$F_c = 1/(2\pi RC)$$

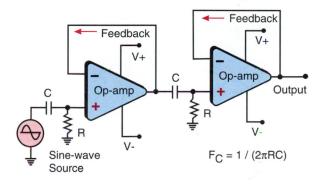

Two-stage (40 dB per Decade) High-pass Filter

SALEN AND KEY (VCVS) SECOND-ORDER ACTIVE FILTERS WITH POSITIVE FEEDBACK

The Salen and Key second-order filter uses positive feedback to simulate (partially) an inductor. This allows the circuit to act as a two-pole filter, with one capacitive and one inductive pole. The circuit opposite is a two-pole (second-order) L-C filter, which the Salen and Key partially simulates. The simulated L-C filter takes on several characteristics of the real L-C circuit. The circuit may be underdamped (Bessel function), overdamped (Chebyshev function), or critically damped (Butterworth response), depending on the damping factor. The damping factor is the reciprocal of *Q*, the figure of merit in an inductor or L-C circuit.

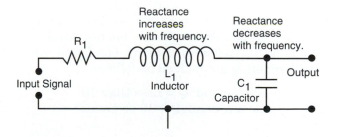

The Second-order LC Filter

Equations:

The equation for damping factor is

$$D = 1/Q$$

The damping factor for the equal-component-value Salen and Key filter with a Butterworth response is

$$D = 1.414$$

where the voltage gain (A_V) is

$$A_V = 1 + (R_f/R_{in}) = 1.586$$

Note: VCVS stands for *voltage-controlled voltage source.*

$$F_c = 1/(2\pi R_1 C_2)$$

Conditions for equal-component-value, second-order Butterworth Salen and Key filter:
Damping factor:

$$D = 1.414$$

Voltage gain:

$$A_V = 1.586$$

Component values:

$$R_1 = R_2, C_1 = C_2$$

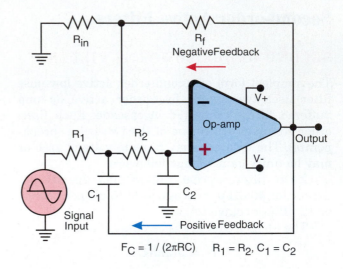

Salen and Key VCVS, Second-order, Low-pass Active Filter

BAND-PASS FILTERS

Band-pass filters are composed of a pair of filters: one high-pass and one low-pass. The frequencies that lie between the cut-off frequencies represent the band of frequencies that can pass through the filter, the pass band. There are two general classifications of band-pass filters: narrow-band and wide-band (broadband). Narrow-band and wide-band band-pass are defined as follows:

> Wide-band: Q = Less than 10
> Narrow-band: Q = 10 or greater

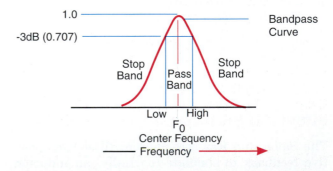

The Band-pass Curve

MULTIPLE NEGATIVE FEEDBACK SECOND-ORDER INVERTING BAND-PASS FILTERS

Unlike the Salen and Key filter, this circuit is an inverting amplifier and uses only negative feedback. The amplifier gain is generally set to unity by making $R_f = R_1$. This filter is often used as a band-pass filter. It can be used as a low-pass or as a high-pass filter, but it seldom is. Because there is so much component interaction, the circuit is difficult to design and nearly impossible to modify.

Equations

Band-pass center frequency equation:

$$F_O = \frac{1}{2\pi C}\sqrt{\frac{R_1 + R_2}{R_1 R_2 R_3}}$$

Voltage gain within the bandwidth:

$$A_V = R_3/2R_1$$

Quality (Q):

$$Q = \pi F_O C R_3$$

−3-dB bandwidth:

$$BW = F_0/Q$$

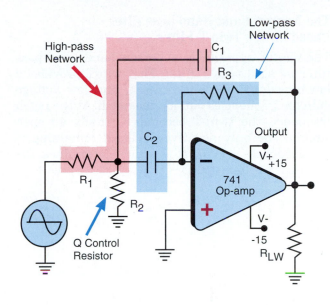

The Multiple-feedback Band-pass Filter

NARROW-BAND BAND-STOP AND BAND-PASS FILTERS

The Notch Filter

The twin-T filter network is basically a high-pass circuit in parallel with a low-pass circuit. This particular arrangement results in the equivalent of a nearly open circuit at the center frequency, called the *notch* frequency. The skirts of the notch curve are also steeper than those of the typical band-pass curve. The circuit can produce a notch depth of −60 dB with resistors and capacitors matched to within 1%. The circuit Q depends on relative component accuracy. The circuit can behave as an L-C tuned circuit; and when used with an amplifier, it is capable of ringing and oscillation.

The notch center frequency equation is

$$F_c = 1/(2\pi RC)$$

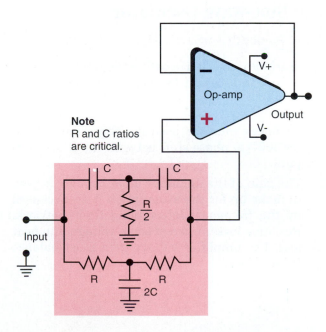

The Twin-T Notch Filter

The Narrow-band Band-pass filter
Based on the Twin-T Filter

The circuit opposite is a narrow-band band-pass filter using a twin-T filter. The twin-T provides a low-impedance feedback circuit, and low voltage gain for frequencies outside the notch. At the notch frequency, the twin-T becomes virtually an open circuit. The only negative feedback remaining is the feedback loop R_f/R_{in}, which determines the band-pass voltage gain.

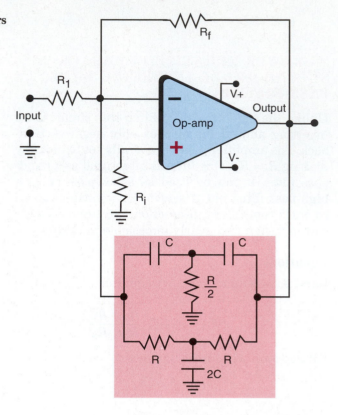

The Narrow-band Band-pass Filter

Sine-wave Oscillators in Brief
The Sine-wave Oscillator

THE BARKHAUSEN CRITERIA

The Barkhausen criteria define the conditions that must exist in a circuit for sustained oscillation to take place:

1   Feedback must be positive. There must be 0° (or 360°) of phase shift between input and output.

2   The gain of the amplifier must be high enough to make up for the attenuation (losses) caused by the frequency-determining filter network. Resistive losses convert some voltage into heat loss. The amplifier voltage gain must balance those losses.

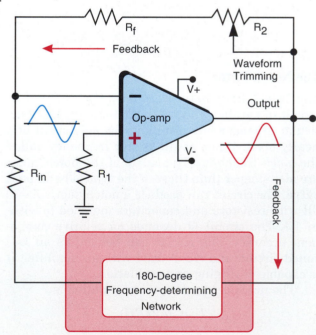

General Sine-wave Oscillator (with 180° Network)

Initially, transient noise occurs at power-on. Noise contains all frequencies; but after a few cycles, the filter removes all but the desired frequency.

Feedback frequency-determining networks are usually common passive-filter types, or resonant tank circuits.

Feedback networks can provide the necessary 0° or 360° phase shift for positive feedback by using either 0° or 180° phase shifts. 180° networks are much more common.

THE PHASE-SHIFT OSCILLATOR

The phase-shift oscillator uses three resistor-capacitor (R-C) networks, each designed to have a 60° phase shift ($3 \times 60° = 180$) at the frequency at which oscillation will occur. The requirement for positive feedback is satisfied only at the frequency at which the network causes a total phase shift of 180°.

Equations

Frequency of oscillation:

$$F_0 = 1/(2\pi RC\sqrt{6})$$

Voltage gain:

$$A_V = R_f/R_{in} = 29$$

Phase shift:

$$\phi = \tan^{-1} \times [1/(2\pi fRC)]$$

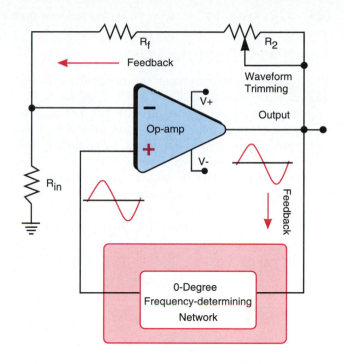

General Sine-wave Oscillator (with 0° Network)

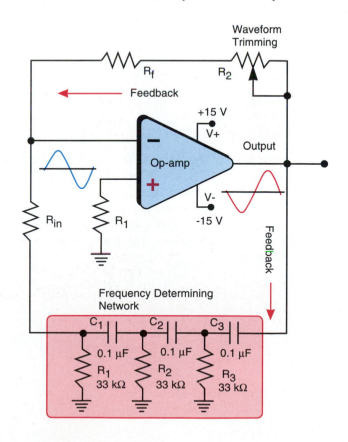

The Phase-shift Oscillator

THE TWIN-T OSCILLATOR

The twin-T oscillator places the notch filter in the negative-feedback loop. The twin-T filter has a low impedance at frequencies outside the notch, providing lots of negative feedback and almost no voltage gain. The positive feedback is separate and independent of frequency. However, the circuit can only use the positive feedback at the notch frequency. At the notch frequency, the negative feedback nearly vanishes, the voltage gain suddenly becomes very high, and the positive feedback starts the oscillation.

The equation for oscillation frequency is

$$F_0 = 1/(2\pi RC)$$

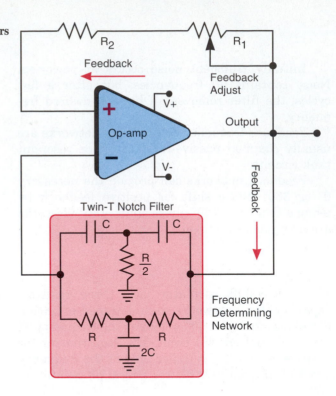

The Twin-T Oscillator

Inductor-Capacitor (L-C) Oscillators

THE ARMSTRONG OSCILLATOR

The Armstrong was the first of the tuned resonant circuit feedback oscillators. The modern Hartley and Colpitts are direct descendants of the Armstrong. The oscillating frequency is determined by the value of inductance and capacitance. Positive feedback is obtained by observing transformer secondary phasing.

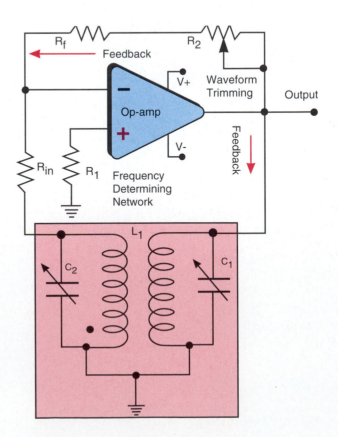

The Armstrong Oscillator

THE HARTLEY OSCILLATOR

The Hartley oscillator combines the transformer primary and secondary into a single-winding auto-transformer configuration. A single capacitor is used to tune the circuit to the desired frequency.

The oscillator frequency is given by the equation

$$F_0 = 1/(2\pi\sqrt{LC})$$

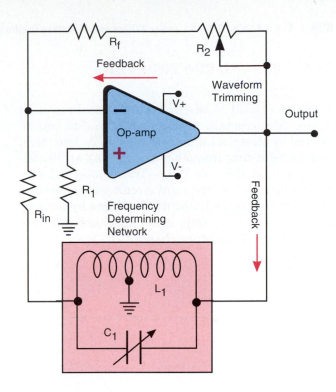

The Hartley Oscillator

THE COLPITTS OSCILLATOR

The Colpitts oscillator is essentially the same circuit as the Hartley, except that a tapped capacitor is used in place of a tapped coil as a phase reference. Which of the two circuits is preferable depends on the application. The Colpitts might be used in a television tuner, for example, because varactor diodes (voltage-variable capacitors) are normally used for tuning. There is no practical voltage-variable inductor available. In other instances, you might wish to tune the oscillator by moving a core (called a *slug*) in and out of the coil; in such a case, you might consider the Hartley version.

The oscillating frequency equation for both oscillators is

$$F_0 = 1/(2\pi\sqrt{LC})$$

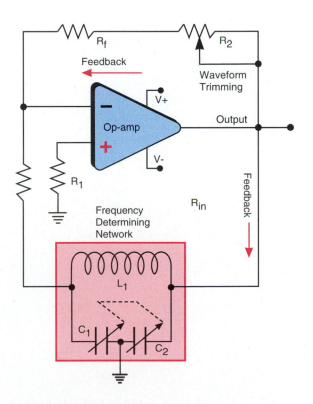

The Colpitts Oscillator

THE CRYSTAL-CONTROLLED OSCILLATOR

A quartz crystal is an electromechanical resonant device. The crystal's mechanical vibration can be excited by electrical signals applied to its two faces, and its vibrating frequency appears as an electrical signal at the two faces. The crystal behaves exactly like an L-C resonant circuit, except that its resonant frequency is not much altered by temperature. In contrast, coils and capacitors are very sensitive to temperature changes.

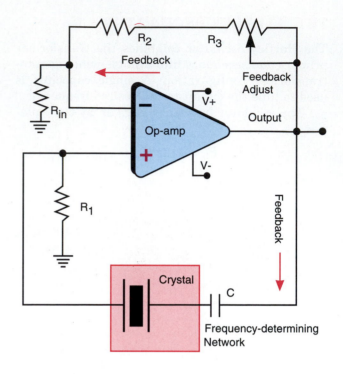

The Crystal Oscillator

11.1   Introduction to Filters

In this chapter, we will look at filters that are designed to pass certain frequencies and reject others. Simple resistance capacitance filters, called *passive filters,* are fine in theory, but they have some serious practical limitations. The primary problem with passive filters is that the circuits they are connected to tend to load them, altering their response characteristics.

If we add an op-amp to serve as a buffer between a passive filter and the circuit it drives, we have the simplest kind of active filter. The term *active* implies that an operational amplifier serves as part of the filter. Once we have included an operational amplifier, new possibilities arise, because we can introduce multiple feedback paths. The addition of multiple feedback paths allows us to assemble a complex filter network with fewer components than would otherwise be required.

The second part of this chapter deals with sine-wave oscillators. These oscillators generate a continuous stream of sine waves at some desired frequency. Oscillators require positive feedback in just the right amount to balance the signal voltage drop across the positive feedback network. The positive feedback network is generally some kind of filter that rejects all frequencies except the desired oscillating frequency.

If the voltage gain of the amplifier just equals the signal voltage loss in the positive feedback network—a loop gain of 1—the oscillator produces a clean sine wave. If the loop gain is unity (1), the circuit meets the Barkhausen criteria and is a sine-wave oscillator. Too much amplifier gain produces a distorted waveform, and too little amplifier gain fails to sustain oscillation.

11.2   Review of Passive Filters

Both passive and active filters are classified according to which part of a band of frequencies they allow to pass through with little or no attenuation. The rest of the frequencies in the band of frequencies are greatly attenuated.

A **high-pass filter** passes only frequencies at the high-frequency end of the band. A **low-pass filter** passes only frequencies at the low-frequency end of the band. A **band-pass filter** passes a selected narrower band of frequencies cut out of the band somewhere between the upper and lower frequency limits. A **notch filter,** also called a *band-stop* or *band-reject* filter, passes frequencies on either side of a narrower notch or band of frequencies. The frequencies in the notch-band are greatly attenuated, while frequencies on either side of the notch band pass freely.

Low-pass Filter

The low-pass filter in Figure 11-1(a) uses a resistor and a capacitor as a frequency-selective voltage divider. At higher frequencies, the capacitor has a low reactance, so most of the signal voltage is dropped across the resistor, leaving only a small voltage drop across the capacitor. Since the output voltage from the filter is taken across the capacitor, the filter output voltage is small at higher frequencies. At lower frequencies, the reactance of the capacitor is high, so most of the signal voltage passes through the filter network without being shunted to ground by the capacitor.

Frequency Response Curves

The frequency response curve shown in Figure 11-1(b) is standard for single-section low-pass filters. Notice that the response curve does not fall off a cliff; instead, it falls gradually at higher frequencies, as the capacitive reactance decreases. The term **stop band** is often used to describe the band of frequencies outside the pass band.

The plateau in the curve is not as flat as it appears, because frequency response curves are normally plotted on a semilogarithmic scale to get the whole curve on a sheet of paper. The vertical scale is usually a decibel scale, which also flattens smaller variations in amplitude.

The typical audio-frequency response curve, for example, covers a range (plotted on the horizontal axis) of from about 10 Hz to more than 10 kHz. This means that the highest frequency on the curve is more than 1000 times the lowest frequency. If we used a linear scale to plot the frequency, we would need over 1000 divisions on the horizontal axis of the graph. Obviously we would have to use either very small divisions or a very wide sheet of paper to plot the frequency response curve on a linear scale.

The plateau is not really flat if we expand it. The curve has no perfectly linear segment and no definite sudden change in the direction. We need some standard reference point to serve as the official end of the pass frequencies. The point 3 dB down (–3 dB) from the top of the curve has been adopted as the standard reference point; –3 dB corresponds to 70.7% up the curve from the bottom, or $0.707 \times$ Peak Value. A difference of 3 dB is also the minimum power level difference the human ear can detect.

Because of the decibel response curve of the human ear, that 200-watt audio amplifier you just purchased does not sound twice as loud as the 100-watt amplifier you traded in. In fact, if you hear any significant increase in power at all, it is mostly a product of your imagination. Doubling the power represents a 3-dB increase in power, and this 3-dB power increase is barely detectable, under good conditions, by a healthy human ear. It's true! The power output specification race is mostly a con game, intended to sell higher-power and more expensive stereo equipment. There are a multitude of factors beyond the power curve of the human ear, so the question of how loud it really is has no easy answer.

High-pass Filter

The high-pass filter in Figure 11-1(c) exchanges the resistor and capacitor locations in the voltage divider. In this case, higher frequencies see low capacitive reactance, so most of the signal voltage drop is developed across the resistor part of the voltage divider. The signal output voltage is taken across the resistor, and the filter output signal voltage is nearly equal to the input signal voltage. The high-frequency signal, in effect, passes through.

At lower frequencies, the reactance of the capacitor rises and most of the signal voltage is dropped across the capacitor, leaving only a small signal voltage across the output resistor. Figure 11-1(d) shows the typical response curve

FIGURE 11-1
Review of Passive Filters

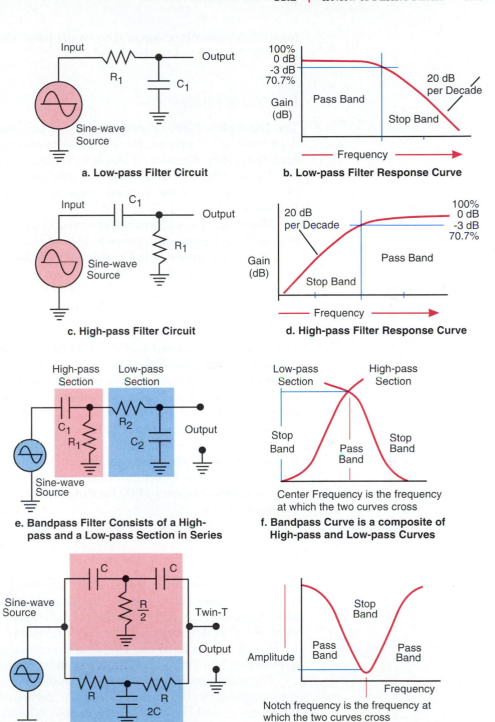

a. Low-pass Filter Circuit

b. Low-pass Filter Response Curve

c. High-pass Filter Circuit

d. High-pass Filter Response Curve

e. Bandpass Filter Consists of a High-pass and a Low-pass Section in Series

f. Bandpass Curve is a composite of High-pass and Low-pass Curves

g. Notch or Band-stop Filter Consists of a High-pass and a Low-pass Filter in Parallel

h. Band-stop Curve is a composite of High-pass and Low-pass Curves

for a high-pass filter. Again, the −3-dB point on the curve defines the transition from pass band to stop band.

Band-pass Filter

The band-pass filter (Figure 11-1(e)) is composed of a low-pass filter and a high-pass filter in series. By selecting the component values, you can cause the high-frequency signal and the low-frequency signal to roll off into the stop band at any desired frequency. The overlapping pass bands in the middle create a pass band of the desired range, with a center frequency located where you want it.

Most band-pass filters are symmetrical about the center frequency, but an asymmetrical band-pass curve is also possible. The response curve in Figure 11-1(f) is a composite of the high-pass and the low-pass filter response curves, as you might expect.

Notch Filter

The Notch filter shown in Figure 11-1(g) is a parallel combination of a low-pass and a high-pass filter. The two filter networks tend to load each other, and an extra capacitor and an extra resistor have been added to ensure a deep narrow notch. We will look at multisection filters shortly. This network is called a **twin-T** or *parallel-T* **filter** because each section resembles a letter *T*. The response curve for the notch filter is given in Figure 11-1(h).

This circuit is often used to eliminate undesirable frequency components within the spectrum. Many analog circuits are beset by stray 60-Hz signals from the **ac** power line that sneak into the signal information. A notch filter with a center frequency of 60 Hz can greatly attenuate stray power-line signal voltages.

Decades and Octaves

Two common semilogarithmic scales are used for the horizontal axis of frequency response curve graphs: decade and octave. A decade scale means that each major division on the horizontal axis is ten times the frequency of the previous division. For example, the major divisions might be 1 Hz, 10 Hz, 100 Hz, 1 kHz, 10 kHz, and so on. The decade scale allows a very large range of frequencies to be plotted on a standard sheet of paper.

The octave scale is often used for audio frequency response curves, because it corresponds to the musical scale. On keyboard instruments, middle A has a frequency of 440 Hz; the A above middle A has a frequency of 880 Hz; and the next-higher-frequency A has a frequency of 1760 Hz. The frequency doubles with each octave. The word *octave* is derived from *octal,* meaning eight, and is used because there are eight whole-note tones from A to A inclusive, from B to B inclusive, from C to C inclusive, and so on.

For example, if you start with a 440-Hz A on a keyboard and count the white keys to the right, you will find that the eighth key is an 880-Hz A. Figure 11-2 shows a keyboard and illustrates the frequency relationships.

FIGURE 11-2   Octaves and the Musical Keyboard

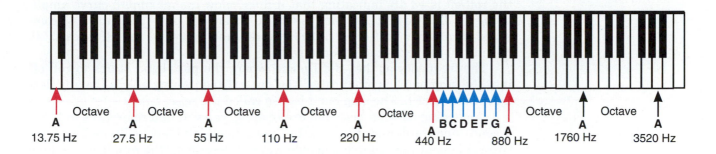

FIGURE 11-3   First-, Second-, and Third-order Low-pass Filters

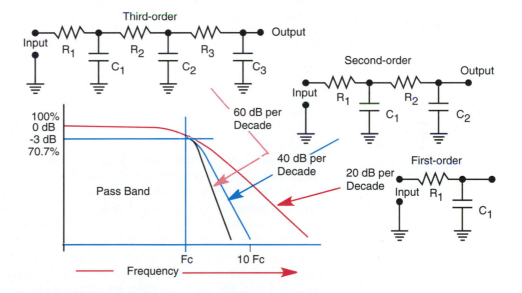

Order of a Filter

In the case of passive filters, the term **order** describes both the number of cascaded resistor-capacitor (R-C) filter networks and how fast the pass band rolls off into the stop band. Figure 11-3 shows first-order, second-order, and third-order low-pass filters, along with the roll-off curve for each. The single-stage filter always has a roll-off rate of 20 dB per decade. This is true for low-pass filters as well as for high-pass filters.

If we cascade (series-connect) two first-order filter networks, we get a second-order filter, with a roll-off rate of 40 dB per decade. Similarly, cascading three first-order filter networks yields a third-order filter with a roll-off rate of

60 dB per decade. The start of the roll-off is still measured at 3 dB down from the top of the curve. If we designate the top of the curve as 0 dB, the start of the roll-off is –3 dB, indicating that it falls below the maximum curve amplitude by 3 dB. The first-, second-, and third-order high-pass filters are also composed of cascaded first-order filter networks, and they have roll-off rates of 20, 40, and 60 dB per decade, respectively.

The frequency at the –3-dB point, where the pass band starts to roll off into the stop band, is often called the **cut-off frequency** and is usually designated as f_0 or f_c. The stop band is also called the *reject band*.

Depending on what book you are reading, you will find references to first-, second-, and third-order filters, or to 1-pole, 2-pole, and 3-pole filters. These are two equivalent and interchangeable ways of describing the number of simple filter sections and their respective roll-off rates.

11.3   Introduction to Active Filters

The simplest kind of active filter circuit consists of a passive filter network buffered by an op-amp, as illustrated in Figure 11-4. Some basic active filter circuits are connected as voltage followers to maximize the input impedance. Figure 11-4(a) illustrates this case for a low-pass active filter, and Figure 11-4(b) illustrates it for a high-pass active filter.

The addition of the voltage-follower op-amp does not alter the filter characteristics of the passive network in any way, but this is exactly the importance of the added op-amp. Nearly any load we might want to hang on the output of the passive filter has the potential to change the frequency at which roll-off occurs. The load may also alter the roll-off rate. The added op-amp isolates the filter network from a load, allowing the designer the freedom to connect any kind of load without altering the ideal filter characteristics.

An even more difficult problem occurs when first-order passive filters are cascaded to get a second-order (or higher-order) filter. The problem is that the second filter network loads the first section, altering its characteristics. Similarly, a third filter section loads sections one and two; and so on for each filter section added. Higher-order passive filters are difficult to design and often demand nonstandard component values, so they could be more expensive in the long run than an inexpensive op-amp. Theoretically, we could design the filter network to compensate for any damage the load and added filter sections might inflict, but in practice this is a difficult and often impractical approach.

Calculating the Roll-off or Cut-off Frequency (One Size Fits All)

Assuming that all capacitor values in the filter network are the same, and that all resistor values are the same, the following equation predicts the cut-off frequency:

$$F_c = 1/(2\pi RC)$$

FIGURE 11-4   Passive Filters with Voltage-follower Op-amp Buffers

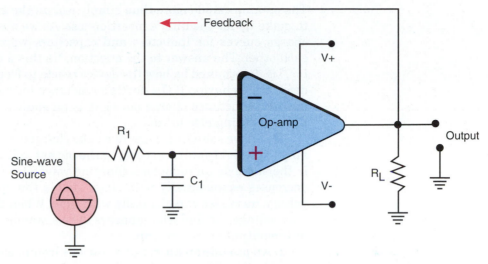

a. First-order Low-pass Active Filter

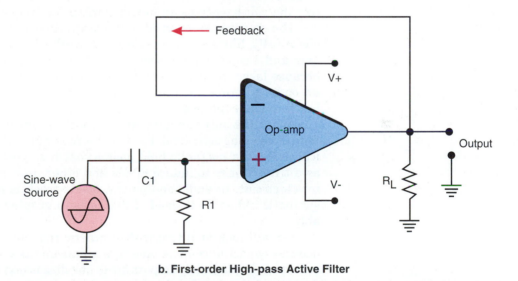

b. First-order High-pass Active Filter

The equation applies to filters of all orders—both passive filters and (most) active filters. Although it does not account for the loading of one passive filter section by the following section, it usually is valid for active filters because we use an op-amp to eliminate the loading problem. Some active filters add feedback to the filter network, which forces us to modify the cut-off frequency equation slightly.

Occasionally, one section of a second- or third-order passive filter has different values, to compensate for loading effects; but this is almost never necessary in active filter circuits, because we use an op-amp to eliminate the need for compensation.

Synthetic Inductance

The op-amp can do more than simply isolate the load and provide voltage gain to make up for the filter's insertion loss. As we observed earlier, reactance and charge curves for inductors and capacitors were inverted mirror images of each other. The answer to the question, "Is this a capacitor or is this an inductor?" is determined by how the device reacts to frequency and by its charge and discharge behavior. If the device's reactance increases with frequency, and if it has an inductance charge curve, it is an inductor—even though it may look like something else to you.

When we connect a capacitor to be charged at the inverting input of an op-amp (Figure 11-5(b)) and observe the output voltage waveform, we find that it is the same as an inductor's charging waveform (Figure 11-5(a)). The voltage decreases exponentially with time. This is the opposite of a capacitor, whose voltage increases exponentially with time. When the capacitor is charged with a **dc** voltage, we find what appears to be a standard inductive "charging" curve at the output of the op-amp.

If we provide an **ac** signal at various frequencies and plot a graph of amplifier output voltage, the plot will show that the output voltage of the amplifier increases as the frequency increases. The amplifier output voltage behaves like the voltage across an inductor when the frequency varies.

The op-amp can be used to convert a capacitor into an op-amp that looks, electrically, like an inductor. We can synthesize an inductor by using an op-amp and a capacitor. Synthetic inductors are very popular at low frequencies, because large inductors tend to be very expensive, and heavy. A capacitor and an op-amp represent a much less expensive and less bulky substitute for a low-frequency inductor.

About the only thing we can't do with a synthetic inductor is take advantage of the magnetic field, because there is none. However, the lack of a physical coil and a magnetic field means that the synthetic inductor can't behave as a transformer to induce 60-Hz and other unwanted electromagnetic fields into a circuit. In audio equipment, 60-Hz trash is a real headache; but using synthetic inductors instead of real ones in graphic equalizers helps considerably.

We will look at filters with synthetic inductors later in this chapter; but because capacitance in an op-amp's input can take on inductive properties, the concept is immediately important to our discussion of the damping factor. The damping factor is related to inductor/capacitor resonant circuits and is the reciprocal of the resonant circuit figure of merit Q:

$$D = 1/Q$$

Resonant Circuits, Ringing, and Q

FLASHBACK

You have studied resonant circuits, ringing, and Q in previous courses. Here is where you can put that knowledge to work. Let's take a little time to review important resonant circuit concepts as they apply to filters and oscillators.

FIGURE 11-5 Inductive and Capacitive Charge Voltages

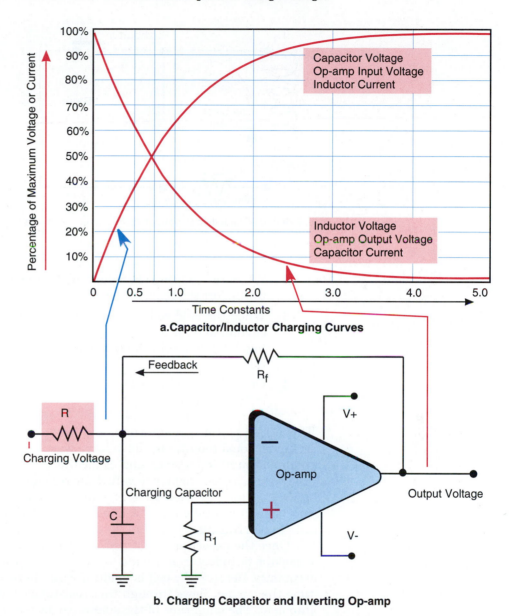

a. Capacitor/Inductor Charging Curves

b. Charging Capacitor and Inverting Op-amp

Resonant Circuit

If we push the button in Figure 11-6(a) momentarily and then release it, we will have charged the capacitor and set the following action in motion:

1 The capacitor discharges through the inductor, storing energy in its magnetic field.

FIGURE 11-6 **Design and Action of the Resonant Circuit**

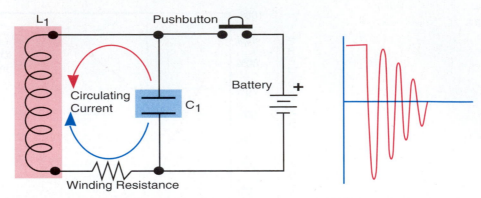

a. Inductive-Capacitive Resonant Tank Circuit   **b. Damped Sine-wave-ringing Waveform**

2 When the capacitor has discharged and can no longer provide current to support the magnetic field, the field collapses.

3 When the field collapses, it cuts through the turns of wire, inducing a current and returning its stored energy to the circuit. The current that had been stored in the magnetic field is now used to charge the capacitor.

4 Once the capacitor is charged, it again begins to discharge through the coil, starting a new cycle.

This action is called **resonance;** and in theory, the action should go on forever. Under ideal conditions, the voltage waveform is a perfect sine wave. Every combination of inductance and capacitance produces some specific-frequency sine wave. At that frequency, called the *resonant frequency,* the inductive reactance (X_L) and the capacitive reactance (X_C) are equal. When a resonant circuit (sometimes called a *tank circuit*) is started into oscillation, the circuit is said to be **ringing.**

Once the capacitor is charged, the circulating current begins to flow from capacitor to inductor and back to capacitor—back and forth, at the resonant frequency. Because the coil is made of wire, there is some resistance; and each time the current flows through that resistance, it converts some of the energy into heat. The amplitude of the sine wave decreases with each cycle until all of the energy has been dissipated. The sine wave produced is called a **damped waveform** (see Figures 11-6(b) and 11-7(a)–(c)).

Mechanical Resonant Circuit Parts (a) and (b) of Figure 11-8 illustrate the concept of a mechanical resonant circuit. One proof of nature's consistency is provided by the fact that the resonant circuit mathematics (with some normalization of units and terms) is the same in electronics, mechanics, acoustics, and thermal physical systems. The study of cross-system physics is called the study of *dynamical analogies.*

You may not be aware of it, but Ohm's law was the result of Georg Ohm seeing an analogy between the already well-understood physics of heat con-

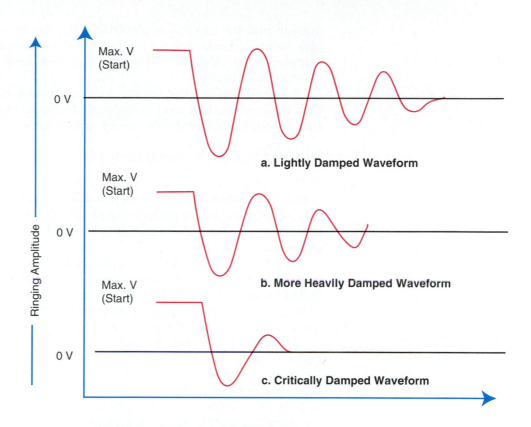

Max. V
(Start)

0 V

a. Lightly Damped Waveform

Max. V
(Start)

0 V

b. More Heavily Damped Waveform

Max. V
(Start)

0 V

c. Critically Damped Waveform

Ringing Amplitude

FIGURE 11-8
**Action of the Resonant
Circuit**

Support Beam

Spring
(Capacitor)

Weight →
(Inductor)

1   2   3   4   5

Time

a. Mechanical Resonant

If a pen were connected to the weight and the support beam
moved at a constant rate along the time line, the pen would
trace a sine wave.

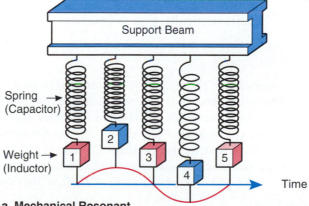

Amplitude

Time

Time

b. Mechanical Resonant Circuit Waveform

duction and the physics of electricity. His university colleagues thought his idea was so ridiculous that they had his office moved to a remote corner of the campus in an effort to widen the depletion zone between themselves and Ohm's crazy theory.

Later in this chapter, we will look at a device made from a quartz crystal that is a combination mechanical-and-electronic resonant device. We will use this crystal to control the frequency of an oscillator.

Damping Factor (*D*) and Quality (*Q*)

If the energy is dissipated within a very few cycles, the circuit is highly damped and has a high damping factor. If the wave takes many cycles to damp, it is lightly damped (or underdamped) and has a low damping factor. The actual numerical value for the damping factor *D* is found by taking the reciprocal of the quality *Q*.

Q is the other side of the damping coin and is a figure of merit for resonant circuits in general. A resonant circuit has a high *Q* value if it rings for many cycles before the waveform is fully dissipated. The coil and capacitor must be able to store a lot of energy—compared to the amount of energy lost to the coil's resistance at each cycle—for the circuit to have a high *Q* number.

If we are interested in having an efficient ringing circuit with which to build an oscillator, we want a high-*Q* resonant tank circuit. At other times, however, the resonant circuit just exists: we don't really want it at all, but we are stuck with it. In such cases, we usually want the tank to quit ringing as quickly as possible, so we may add some extra resistance deliberately to damp the wave quickly. The tank circuit will then be heavily damped, giving it a high damping factor and a low *Q*.

Mathematically, *Q* is defined as

$$Q = X_L/R$$

where X_L is the inductive reactance (in ohms) of the inductor at the resonant frequency, and *R* is the resistance (in ohms) in the circuit (mostly due to the coil resistance). We could have used the X_C value instead of X_L, because $X_C = X_L$ at the resonant frequency. Either the capacitor or the inductor is storing the peak energy at one time. We use X_L in the equation because the inductor contributes the resistance (and not the capacitor).

Since *Q* is the reciprocal of *D*, *D* is also the reciprocal of *Q*:

$$Q = 1/D$$
$$D = 1/Q$$

When we use an op-amp with voltage gain in an active filter, we don't have a free hand in the amount of voltage gain used. The gain of the op-amp controls the damping factor of the filter, which alters the characteristic response of the filter. There are three standard response curves, each associated with a specific damping factor and a specific op-amp (or loop) voltage gain. Each of the three standard response curves has its own particular kind of application. Figure 11-9 shows the response curves for various values of *D* and *Q*.

FIGURE 11-9 Response Curves for Various Damping
Factors and *Q* Values

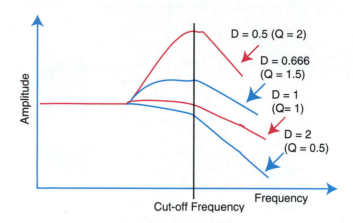

Standard Filter Response Curves: Butterworth, Chebyshev, and Bessel

These three special response curves—called Butterworth, Chebyshev, and Bessel functions—have been adopted as standard response curves. Each has been adopted because it is optimum for some specific class of filter functions. An infinite number of in-between response curves exist, but they don't represent optimum conditions for anything in particular.

The **Butterworth function,** with its 1.414 damping factor, is the flat (sometimes called flat–flat) response curve that is almost universally preferred for ordinary filters. It has a flat response and is exactly –3 dB below peak at the cut-off frequency. **The Bessel function,** with a damping factor of about 2, provides a very droopy roll-off and is often called an *all-pass filter* because it can be used as a phase shifter over a wide range of frequencies. The **Chebyshev function** produces a curve with an inductive peak, results in shallow damped ringing along the top of the pass band, but provides a sharper cut-off than the Butterworth. A bit more gain or a damping factor nearer to zero could lead to sustained oscillation, which we will cover later in this chapter. From here on, we will assume that we are dealing with a Butterworth response curve.

11.4 Active Filters with Voltage Gain

Although we can use an op-amp with voltage gain in a first-order active filter, we aren't free to specify the amount of voltage gain used in an active-filter op-amp. The gain of the op-amp controls the damping factor of the filter, which alters the characteristic response of the filter. Nearly all op-amp first-order filters have their voltage gain set to provide a Butterworth response curve. Figure 11-10 shows typical first-order high-pass (a) and low-pass (b) active filters with gain.

FIGURE 11-10 Simple Filters with Voltage Gain

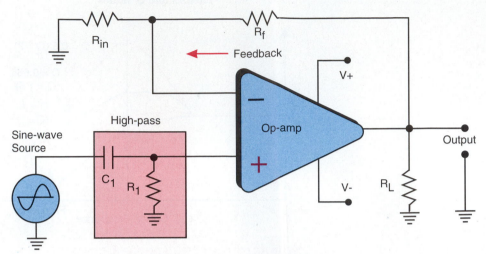

a. First-order High-pass Filter

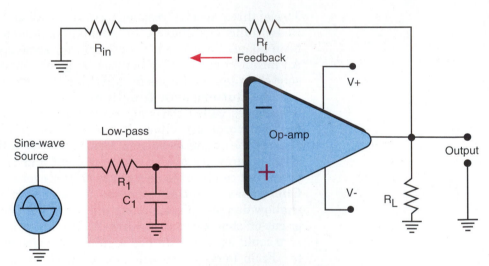

b. First-order Low-pass Filter

Active Filters and the Miller Effect

An often overlooked consideration at higher frequencies is the Miller effect at the input of an op-amp configured for a high voltage gain. The Miller-effect-influenced input capacitance of such an amplifier can alter the response of the passive filter network. Because the voltage follower has a unity (1) voltage gain, Miller-effect problems are eliminated.

At frequencies low enough for the Miller effect to be insignificant, we can use an op-amp with some voltage gain, and thereby (in some cases) reduce the number of op-amps required in the system. In high-pass and low-pass filter circuits like the ones shown in Figure 11-10, the filter network is normally connected to the noninverting input to take advantage of that configuration's high input impedance. Since the high input impedance does not load the filter network, it does not affect the filter response curve. Any Miller-effect problems are minimized by keeping the voltage gain low.

11.5 Filters with Feedback

Adding feedback—sometimes negative and sometimes positive—can significantly improve the effectiveness of filter circuits. Multiple feedback paths can be used to reduce the number of op-amps needed for higher-order and band-pass filters.

Introduction to the Integrator Circuit

The **integrator circuit** in Figure 11-11(b) can be classed as a low-pass filter, but it is much more than that. It is one of the most common and useful op-amp circuits, and it seems to crop up everywhere to do one job or another. The integrator is a linear version of the R-C time-constant circuit, which is a passive low-pass filter. Like the simple R-C time-constant circuit, the integrator accumulates charge on a capacitor over time.

The integrator can be used as a low-pass filter, a ramp generator, a timing circuit, an averaging circuit, a 90° phase shifter, or a waveshape-altering circuit. The integrator gets its name from the fact that it performs an electronic analog version of the mathematical operation called integration.

Figure 11-11(a) shows a passive R-C time-constant circuit or low-pass filter. The circuit in Figure 11-11(a) is sometimes called an *integrator,* but its performance is a poor simulation of mathematical integration, because the simple R-C time constant curve is not linear. The accumulated charge (in Figure 11-11(a)) during the first second is 0.632 volts. During the next second, the capacitor accumulates an additional charge of only 0.233 volts. Thus, the charge voltage increases at a rate of 0.632 volts per second during the first time constant and only at a rate of 0.2332 volts per second between the first and second time constants. It gets worse for each successive time constant. Only during the first time-constant period is the curve reasonably linear.

The op-amp integrator in Figure 11-11(b) provides a constant charging current to produce a 1-volt-per-second increase in capacitor charge voltage for each second of elapsed time until the amplifier runs out of negative power-supply voltage. This linear action is an accurate analog representation of mathematical integration.

The circuit shown in Figure 11-11(b) is an active integrator, with the timing capacitor installed in the feedback loop. Because of the negative feedback, the op-amp acts as a constant-current charging source for the capacitor, causing it to charge at a constant rate of x volts per second (or millisecond, or microsecond).

FIGURE 11-11   **Integrator**

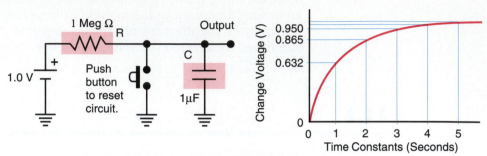

a. Passive R-C Circuit and its Charging Curve

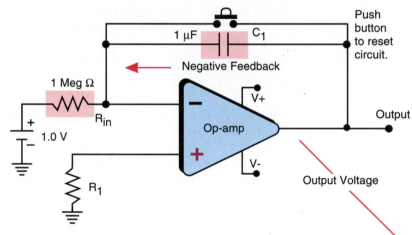

b. Op-amp Integrator Circuit with Values for Example 11-1

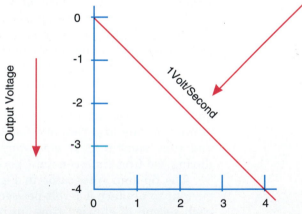

c. Op-amp Integrator Charging Curve

Output Voltage Equation

The integrator output voltage (see Figure 11-11(b)) is given by the equation

$$V_O = \{V_{in} \times [1/(R_{in} \times C_1)]\} \times \text{Time}$$

Example 11.1

Integrator Calculations

Given the circuit in Figure 11-11(b), find the integrator output voltage at the end of 4 seconds, and calculate the input current.

Solution

First let's make the usual assumption that the inverting input virtual ground is an actual ground, and assume that the push-button has just been released to start the capacitor charging from zero. Then the integrator output is

$$
\begin{aligned}
V_O &= -\{V_{in} \times [1/(R_{in} \times C_1]\} \times \text{Time} \\
&= -\{1 \times [1/(1 \times 10^6 \times 1 \times 10^{-6})]\} \times 4 \\
&= -(1 \times 1) \times 4 \\
&= -4 \text{ V}
\end{aligned}
$$

The output voltage of the integrator has changed at a rate of 1 volt per second, for a total accumulated voltage of negative 4 volts at the end of 4 seconds. Figure 11-11(c) shows the changing curve. Remember, this is an inverting amplifier, so the output voltage is a negative-going voltage when the input voltage is positive.

Now, using our initial assumption, let's find the input current through R_{in}:

$$
\begin{aligned}
I_{Rin} &= V_{in}/R_{in} \qquad \text{(Ohm's law)} \\
&= 1 \text{ V}/1 \text{ meg } \Omega \\
&= 1 \text{ } \mu A
\end{aligned}
$$

Example 11.2

More Integrator Calculations

Given the circuit in Figure 11-12, find the integrator output voltage at the end of 4 seconds, and calculate the input current.

Solution

First let's make the usual assumption that the inverting input virtual ground is an actual ground, and assume that the push-button has just been released to start the capacitor charging from zero. Then the integrator output is

$$
\begin{aligned}
V_O &= -\{V_{in} \times [1/(R_{in} \times C_f]\} \times \text{Time} \\
&= -1 \times [1/(5 \times 10^5 \times 5 \times 10^{-6})] \times 4 \\
&= -1 \times (1/2.5) \times 4 \\
&= -1 \times 0.4 \text{ V} \times 4 \text{ seconds} \\
&= -1.6 \text{ V}
\end{aligned}
$$

The output voltage of the integrator has changed at a rate of 0.4 volt per second for a total accumulated voltage of negative 1.6 volts at the end of 4 seconds.

Now, using our initial assumption, let's find the input current through R_{in}:

$$
\begin{aligned}
I_{Rin} &= V_{in}/R_{in} \qquad \text{(Ohm's law)} \\
&= 1 \text{ V}/0.5 \text{ meg } \Omega \\
&= 2 \text{ } \mu A
\end{aligned}
$$

FIGURE 11-12 Integrator Circuit for Example 11-2

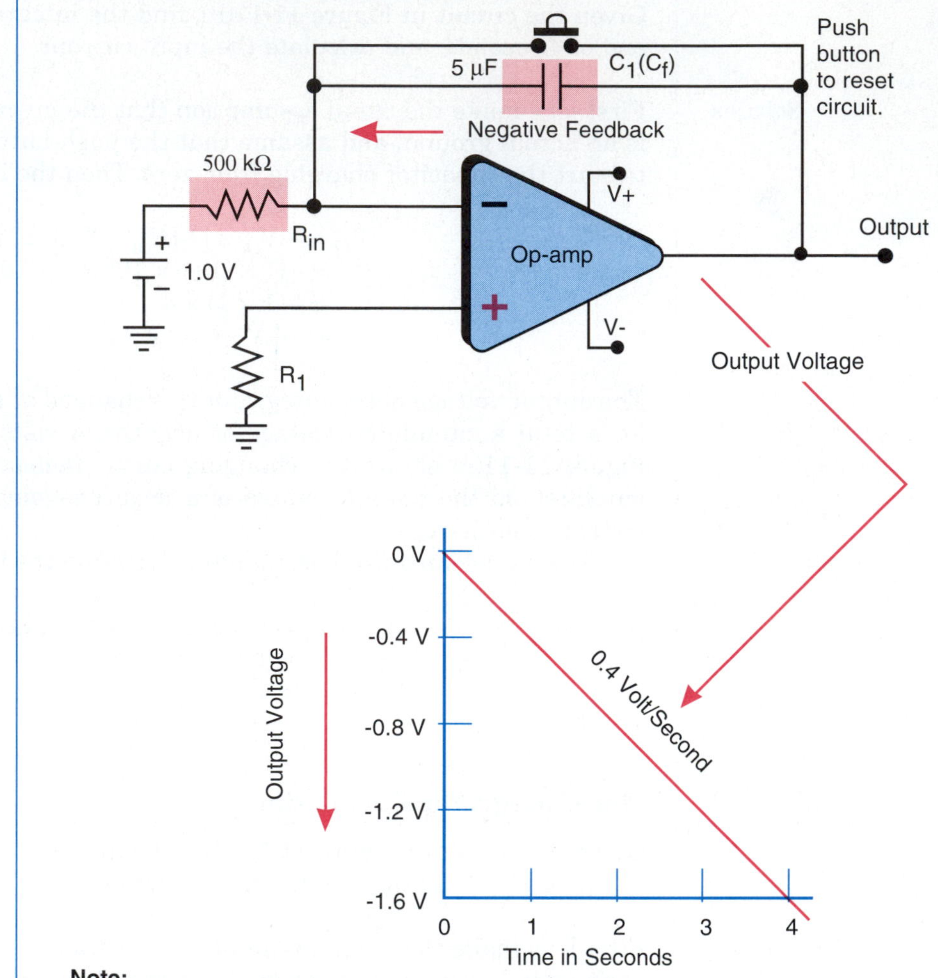

Note:
Because the inverting input is at a virtual ground, the output voltage and the charge voltage on the capacitor are the same.

How the Integrator Responds to Changing Input Voltages

Figure 11-13 illustrates the integrator action with several different kinds of changing voltages. The drawings all reflect the fact that the integrator is a time-delay circuit and tends to respond slowly to fast voltage changes. Abrupt changes in input voltage are converted into more gradual changes in the output voltages.

Recall that the sine wave is the only pure waveform with no harmonics. All other waveforms contain many harmonics. The higher-frequency harmonics show up in a waveform as sharp or abrupt changes in waveform slope or direction. Lower-frequency harmonics show up as more gentle changes in slope along the waveform.

It follows that, if we use the integrator as a low-pass filter on a waveform with sharp edges, those edges will be rounded off. Afterward, the edges of the waveform will be softened because we will have eliminated the high-frequency harmonics that caused the sharp edges.

As a matter of convenience, we tend to think of the integrator as a time-delay or slow-response circuit when input voltage changes are slow (or fixed **dc**). We tend to think of the integrator as a low-pass filter when voltage changes are happening at a high frequency. The integrator operates the same way in both cases. For example, a low-pass filter is defined as a circuit where lower frequencies (those below a certain value) applied to the input also appear in the output. Frequencies above that cut-off value are applied to the input but fail to show up as changes in the circuit's output voltage. Since most electrical noise consists of quick, sharp voltage changes, low-pass filters are often used to filter out noise.

Two Ways of Looking at the Same Thing

Version #1 You can say that the capacitor has a low reactance value at higher frequencies and, therefore, bypasses high frequencies to ground. The capacitor has a high reactance at low frequencies and does not bypass them to ground, so low frequencies appear in the circuit's output.

Version #2 You can say the circuit has a long time constant compared to how fast the input voltage changes at higher frequencies. If the input voltage changes so fast that the slower integrator can't keep up, the output voltage won't be able to follow the input waveform's voltage changes. Signal voltages at the higher frequencies fail to appear as output voltage variations and are lost. They have been filtered out because the integrator's (frequency) response is too slow. If the frequencies at the input are low enough, the integrator can keep up with their voltage variations, and they are faithfully reproduced in the circuit's output.

Both explanations are equally true, but we tend to select one or the other—whichever makes more sense—depending on the specific kind of circuit application at hand.

Figure 11-13(a) repeats the integrator circuit for your convenience. Figure 11-13(b) shows the integrator's staircase output waveform for the case in which a non-zero-crossing **dc** waveform is used as the integrator input signal.

How It Works

1   At time t_1, the input voltage rises suddenly to $+V_1$. Because the amplifier is an inverting amplifier, the output voltage goes toward the negative supply voltage.

2   As soon as the input voltage rises to $+V$, input current starts to flow through the input (timing) resistor. The amplifier output starts to deliver a constant charging current to the timing capacitor.

3   As the capacitor charges, the output voltage V_O ramps toward the negative supply voltage. The slope of the ramp down is determined by the R-C time constant.

4   At time t_2, the input voltage drops to zero. There is no input current to cause an op-amp output charging current, so the op-amp doesn't

FIGURE 11-13   How the Integrator Responds to Changing Input Voltages

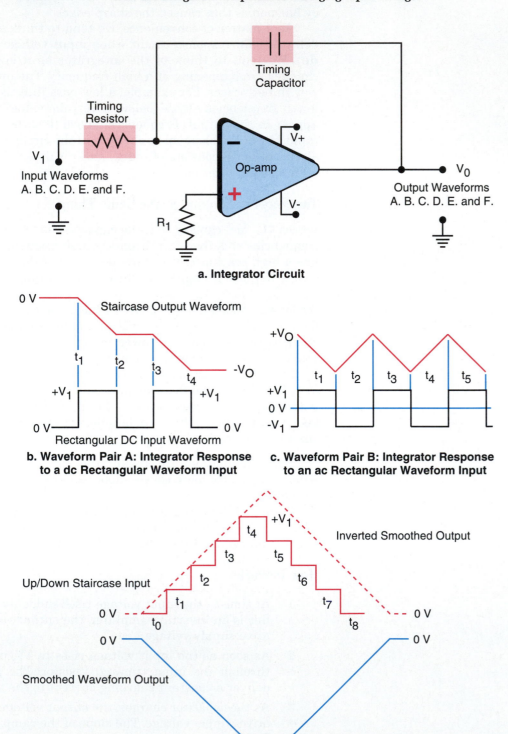

a. Integrator Circuit

b. Waveform Pair A: Integrator Response to a dc Rectangular Waveform Input

c. Waveform Pair B: Integrator Response to an ac Rectangular Waveform Input

d. Waveform pair C: Integrator Response to an Up-Then-Down Staircase Waveform Input

FIGURE 11-13   Continued

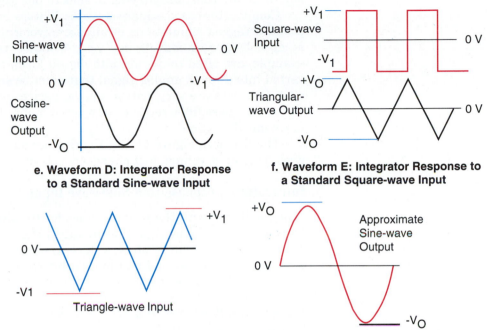

e. Waveform D: Integrator Response
to a Standard Sine-wave Input

f. Waveform E: Integrator Response to
a Standard Square-wave Input

g. Waveform F: Integrator Response to a Standard Triangular Waveform Input

change the charge voltage on the capacitor. The output voltage
remains constant until time t_3.

5   At time t_3, the input voltage suddenly rises to $+V_1$, starting input current flow and output charging current. The output starts another ramp
down.

Integrator's Response to an ac Rectangular Waveform

The difference between this input waveform (Figure 11-13(c)) and the previous
one is that this is an **ac** waveform. The waveform rises to a positive $V+$ and
stays there for a period of time. This starts the ramp down, as before. At time
t_2, however, this waveform doesn't stop and rest at zero; it switches all the way
to $-V_1$. The $-V_1$ input voltage causes the amplifier output V_O to ramp up
toward the positive supply voltage. The change in output voltage polarity
reverses the current flow to the capacitor, causing it to charge to $+V_O$. At time
t_3, the input voltage again goes positive, starting a new ramp down. The waveform, for obvious reasons, is called a triangular wave.

Integrators Response to an Up-Then-Down
Staircase Input Waveform

This example (Figure 11-13(d)) illustrates an important kind of smoothing
operation that integrators often perform. Any low-pass filter can be used to
knock off the corners of sharp waveforms, but the integrator is especially good

at it. This smoothing function is particularly important in modern technology, where we are constantly trying to smooth out ragged waveforms.

Comparator-based square-wave oscillators and other digital circuits can produce ragged simulations of analog waveforms that we must smooth into acceptable analog waveforms. Long-distance telephone transmissions, for example, are often converted into digital form, transmitted, and then reconverted into a rough analog signal that must be smoothed to be suitable to our ears. Our ears are very critical analog devices and do not tolerate distortion well. A low-pass filter removes unwanted higher harmonics, which we hear as harmonic distortion.

The circuit in Figure 11-13(d) works the same as the one in Figure 11-13(b) does. The input voltage just changes levels in a more complex pattern.

Integrator's Response to a Sine-wave Input

When matched with the proper values for the timing resistor and the capacitor, a sine-wave input signal to the integrator will output a cosine wave. The cosine wave is 90° out of phase with a sine wave and is said to be in **quadrature** with the sine wave. Later in this chapter, we will examine the quadrature oscillator, which takes advantage of this 90° shift.

Integrator's Response to a Standard Square-wave Input

This circuit (Figure 11-13(f)) operates exactly the way the **ac** rectangular wave in Figure 11-13(c) does. This is the special case of a rectangular wave called a square-wave.

Integrator's Response to a Triangular Wave Input

Many textbooks don't mention this possibility, because technically it is not quite true. The mathematics is wrong, and the sine wave produced is not perfect. For a single frequency and proper circuit adjustment, however, a sine wave with less than 2% harmonic distortion can be produced (see Figure 11-13(g)). Engineers have used this trick for many years as a simple, practical way to convert triangular waves into sine waves.

Differentiator Circuit

The **differentiator** is basically a high-pass filter, whose output voltage is proportional to how fast its input voltage is changing. The differentiator is not used very often, because it has fewer practical uses and tends to be noisy and unstable. The differentiator's two primary characteristics are:

1   No output voltage when there is no change in the input voltage.
2   An output voltage proportional to how fast the input voltage is changing (rate of change).

If you plot a graph of input-voltage versus time, you will see that the differentiator output voltage is proportional to the slope (Voltage/Time) in volts per second. The slope, or rate of change, is often called dv/dt—the differential function in calculus.

Because of the differentiator's inherent instability and the added input resistor and feedback capacitor needed to make the circuit practical, it does not provide a very accurate representation of mathematical differentiation.

Figure 11-14(a) shows a practical differentiator circuit that can be used to detect the edge of a square or rectangular pulse for purposes of releasing a precise timing pulse. Capacitor C_1 is the differentiating capacitor and resistor R_f is its associated timing resistor. Resistor R_1 is used to cause a high-frequency

FIGURE 11-14   Differentiator Circuit and Waveforms

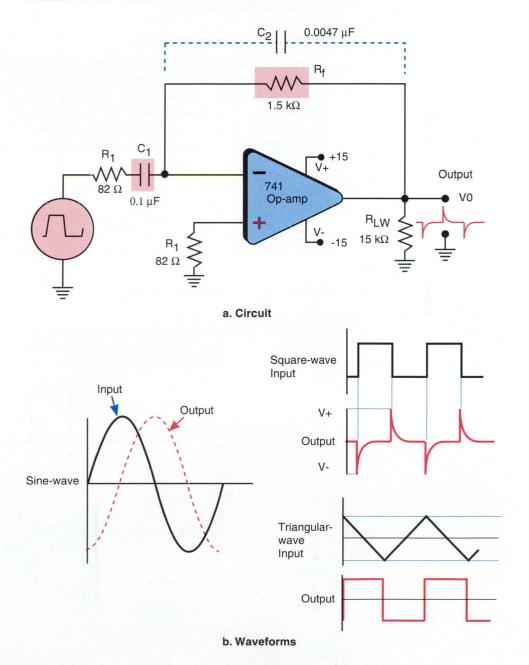

a. Circuit

b. Waveforms

roll-off to help prevent the amplifier from breaking into oscillation. Capacitor C_2, in the feedback loop, must be included to serve as an additional frequency-compensation capacitance, to limit the op-amp's high-frequency response. The resulting waveforms are shown in Figure 11-14(b).

What Is This Circuit?

The circuit in Figure 11-15 could be an op-amp with external feed-forward frequency compensation, an integrator circuit, or a differentiator circuit. Here is how to tell the difference.

Feed-Forward Compensation

Putting a small speed-up capacitor across the normal feedback components is a common technique called **feed-forward frequency compensation.** If done properly, this trick can be used to improve the speed of a wide variety of op-amp circuits, so it is often used. The clue is that the value of the capacitor in the feedback loop is very small. A typical range is from 1 to 30 picofarads. A capacitor this small is unlikely to be part of an integrator or differentiator circuit.

Integrator

The integrator circuit may or may not have a large resistor across the timing capacitor (in the feedback loop). The large resistance is there to help prevent input offset currents from charging the capacitor when the input voltage is

FIGURE 11-15 **Identifying Circuits**

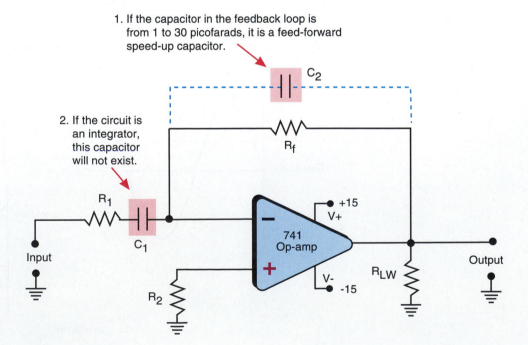

1. If the capacitor in the feedback loop is from 1 to 30 picofarads, it is a feed-forward speed-up capacitor.

2. If the circuit is an integrator, this capacitor will not exist.

zero. The extra resistor usually has a resistance about 10 times greater than the input resistance. The best guideline is to multiply R (in megohms) and C (in microfarads) to get the timing period. Then see if that value fits with the kind of circuit you have at hand.

Differentiator

Try to eliminate the other two possibilities first. Look for the resistor in series with the input capacitor. If it is there, you almost certainly have a differentiator circuit. In addition, the extra capacitor across the feedback resistor is typically only 10% (or less) of the value of the input capacitor.

11.6   Filters with Complex Feedback Networks

A number of filters use more than one feedback path. In the VCVS filter, the normal negative-feedback path is supplemented by a carefully controlled positive feedback. So-called multiple-feedback filters use two or more different feedback paths. Perhaps the most powerful of all filters with complex feedback networks are integrated circuits designed as universal filters.

Earlier in this chapter, we briefly investigated how a capacitor can be made to simulate an inductor. In the following sections, we will examine some circuits that behave as simulated inductances. As far as electronic circuits are concerned, if a device or circuit has the reactive and phase behavior of an inductance, it is an inductance. The previous example was rather simple and obvious, but most practical circuits have very complex transfer functions, and the mathematics they apply is well beyond the scope of this book (was that a sigh of relief?). Nonetheless, we can still get a working understanding of these circuits by examining what they do, without being precise about how they do it.

In most cases, we are interested in creating some kind of second-order filter when we resort to more complex feedback networks. The best way to get an effective two-pole filter is to use a capacitance and an inductance. Because synthetic inductance is easy, compact, available, and cheap, it is the logical choice. As you will see, it is easy to combine a synthetic inductor and a real capacitor to get an L-C (inductance–capacitance) circuit. These L-C circuits are so real that they can be made to ring, and they may be designed to have a high Q if we so wish. We can, in fact, use them to replace the normal coil and capacitor resonant tank circuit in some oscillators. In the case of filters, we deliberately keep the Q low (and the damping factor high) to prevent any ringing from altering the response curve.

Figure 11-16 shows a typical L-C low-pass filter circuit. The circuit is effective and produces two poles because the inductor and the capacitor exhibit opposite reactive behaviors. When the input frequency increases, the reactance of L_1 also increases, causing a larger and larger voltage drop across the inductor. This accounts for one of the poles. Meanwhile, as the frequency rises, the capacitive reactance decreases, shorting out what is left of the input voltage. This accounts for the second pole.

FIGURE 11-16 R-L-C Second-order Filter

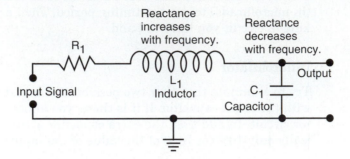

Salen and Key (VCVS) Filter with Positive Feedback

The **Salen and Key** second-order filter uses positive feedback to simulate the inductive pole of an L-C filter. Figure 11-17 shows the high-pass Salen and Key filter, and Figure 11-18 shows the low-pass version. The Salen and Key is also called the **VCVS** (for voltage-controlled voltage source) **filter.**

There are several variations on this theme, including a unity-gain voltage-follower version. However, only the equal component value version with a Butterworth response curve is very common. The cut-off frequency is defined by

$$F_C = 1/\left[2\pi(R_2R_1C_2C_1)\right]$$

The equation is the same for both low-pass and high-pass filters.

The damping factor is controlled independently of the frequency-determining components, by using the negative-feedback loop. The damping factor required for a Butterworth response curve is 1.414 ($D = 1.414$). The pass-band voltage gain of the op-amp must be

$$
\begin{aligned}
D &= 3 - A_V \\
A_V &= 3 - D \\
&= 3 - 1.414 \\
&= 1.586 \\
&= 1 + (R_f/R_{in})
\end{aligned}
$$

For the circuits in Figures 11-17 and 11-18, we need a ratio of R_f to R_{in} of 0.586. The standard values of 27 kΩ and 47 kΩ should do, yielding a voltage gain of

$$A_V = 1 + (27\ \text{k}\Omega/47\ \text{k}\Omega) = 1.575$$

The slightly lower gain of 1.575 may cause a very slight droop in the response curve, but that is usually preferable to a peak caused by too much gain. As usual, the absolute values of R_f and R_{in} are not very important; the ratio is the critical thing.

FIGURE 11-17 Salen and Key (VCVS) Second-order High-pass Filter with Positive Feedback

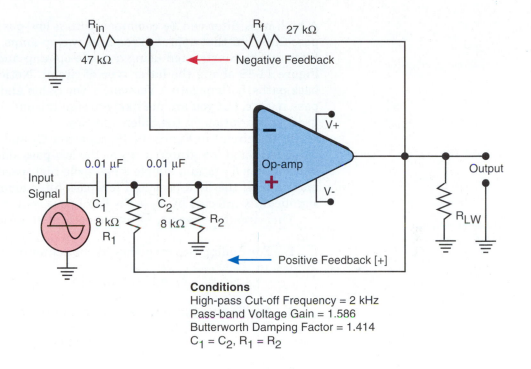

Conditions
High-pass Cut-off Frequency = 2 kHz
Pass-band Voltage Gain = 1.586
Butterworth Damping Factor = 1.414
$C_1 = C_2$, $R_1 = R_2$

FIGURE 11-18 Salen and Key (VCVS) Second-order Low-pass Filter with Feedback

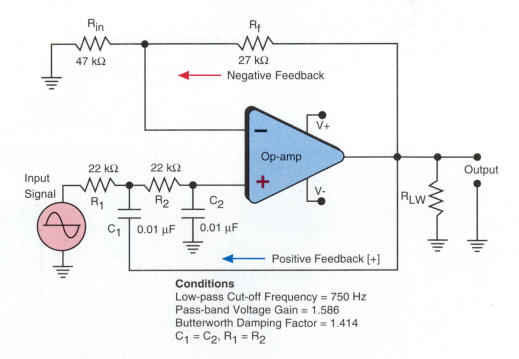

Conditions
Low-pass Cut-off Frequency = 750 Hz
Pass-band Voltage Gain = 1.586
Butterworth Damping Factor = 1.414
$C_1 = C_2$, $R_1 = R_2$

11.7 Multiple-feedback Band-pass Filters

A high-pass filter can be combined with a low-pass filter to produce a band-pass filter, but that approach requires two op-amps. Alternatively, a band-pass filter can be constructed using a single op-amp and multiple feedback paths. Figure 11-19 shows the latter type of circuit. Notice that there are two feedback paths, both negative. Versions of the Salen and Key can be used for band-pass service, but you are not likely to run into one. It is a possible but not very practical variation on the Salen and Key.

In the circuit in Figure 11-19, capacitor C_1 and resistor R_1 form a virtually conventional integrator, to act as the low-pass side of the filter. Capacitor C_1 and resistor R_3 form a simple single-pole high-pass filter. Resistor R_2 is common to both the high-pass and low-pass filter elements. The only parameter significantly affected by R_2 is the band-pass center frequency.

This circuit makes use of five important equations:

1 For finding the frequency at the center of the band-pass curve:

$$F_C = [1/(2\pi C)] \times \sqrt{R_1 + R_2 /(R_1 R_2 R_3)}$$

2 For finding the voltage gain within the pass band:

$$A_V = R_3/2R_1$$

3 For finding the circuit Q:

$$Q = F_C R_3$$

FIGURE 11-19 **Multiple-feedback-path Band-pass Amplifier**

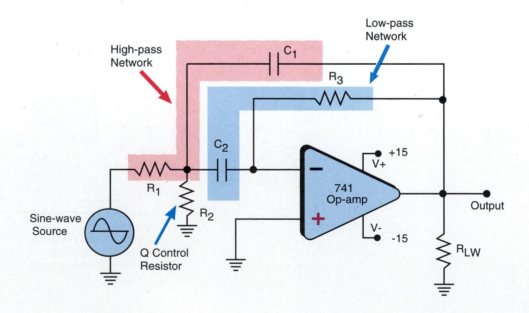

4 For finding the circuit damping factor:

$$D = 1/Q$$

5 For finding the –3-dB bandwidth:

$$BW = F_C/Q$$

The maximum practical value for Q in the multiple-feedback band-pass filter is about 10. This yields a fairly wide band-pass. The two basic classifications of band-pass filters are wide band and narrow band. This circuit fits the wide-band classification.

Twin-T Filters

The two circuits that follow (shown in Figures 11-20 and 11-21) are common circuits that take advantage of the narrow steep-sided notch characteristics of the **twin-T filter network.**

Narrow-band Band-pass Filter

The twin-T network in Figure 11-20 is installed in the negative-feedback loop of an op-amp. Frequencies outside the twin-T's notch pass freely through the

FIGURE 11-20
Twin-T Narrow-band Band-pass Filter

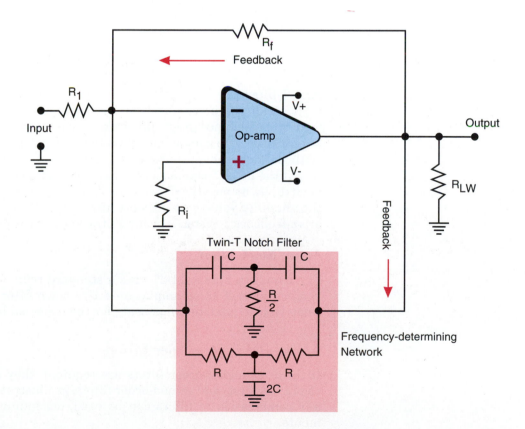

FIGURE 11-21 Twin-T Notch Filter

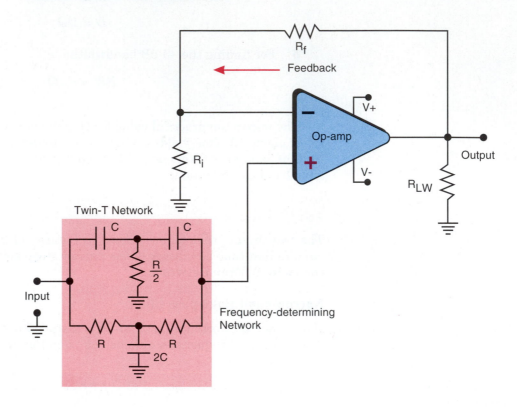

twin-T filter, providing a very large amount of negative feedback. At frequencies outside the notch, the filter provides the equivalent of a very low-value feedback resistance (a low impedance). For frequencies outside the notch, the amplifier voltage gain is much less than 1. The signal is attenuated, and there is almost no signal output voltage.

At the notch frequency, the filter acquires a very high impedance, so there is virtually no negative-feedback voltage. In the absence of negative feedback, the amplifier voltage gain shoots up to a value limited only by the added negative-feedback resistor, which sets the voltage gain at the notch frequency.

Active Notch Filter

The circuit in Figure 11-21 uses a standard twin-T notch filter in the noninverting input of an op-amp as an active notch filter. The circuit as shown has some voltage gain, to compensate for the insertion loss of the twin-T network.

Third- and Higher-order Filters

If third- or higher-order filters are required, they can easily be obtained by combining first- and second-order filters, as illustrated in Figure 11-22. Almost any combination of filters can be used, depending on the system's requirements.

FIGURE 11-22   Assembling Higher-order Filters: Third-order Active Filter Example

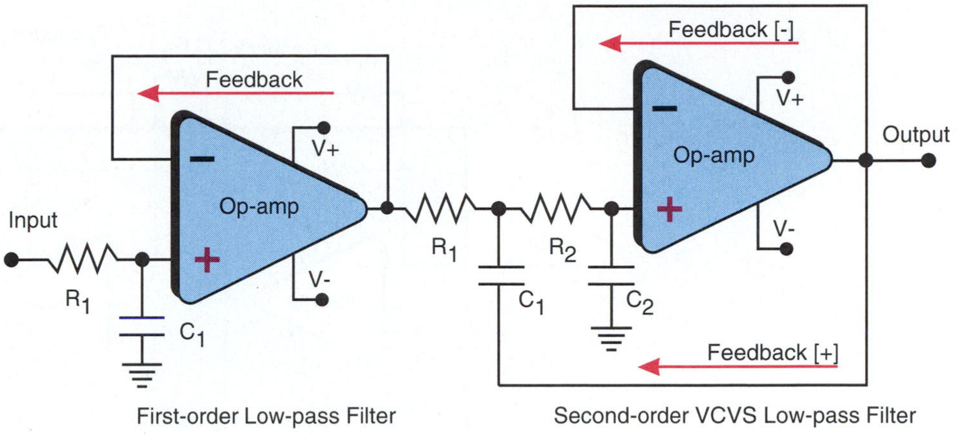

First-order Low-pass Filter   Second-order VCVS Low-pass Filter

11.8   State-variable Universal Filter

Figure 11-23 shows the **state-variable universal filter.** This type of integrated circuit (such as the LF-347) provides a single-package solution to most common filter problems. The state-variable circuit configurations were developed to serve as electronic analogs of physical systems in the days of analog computers. The mathematics that describes the operation of state-variable filters involves the use of calculus, but we can use the state-variable filter without doing a formal mathematical analysis.

The state-variable filter consists of two op-amps connected as integrators and a summing op-amp to add signals from the complex multiple-feedback paths. Several commercial integrated-circuit versions of this circuit offer a one-chip solution to most common filter problems. Because there are several possible configurations, you should consult the manufacturer's data manual in each case for specific applications.

The state-variable filter has outputs for high-pass, low-pass, and band-pass filters—all available at the same time. The high-pass and low-pass cut-off frequencies and the band-pass center frequency are always the same. Even though all three functions are available, it is common practice to optimize the circuit for only one of them at a time. For example, the damping factor selected for a low-pass filter probably won't yield the desired bandwidth for the band-pass output. If the band-pass filter is set for a high Q, the high- and low-pass filters might well have too peaked a response curve, and may even be unstable. You usually can't get every parameter the way you want it at the same time.

The Details

This information applies only to the configuration shown in Figure 11-23. The process is easy if we start by making some basic assumptions:

FIGURE 11-23   State-variable Universal Filter

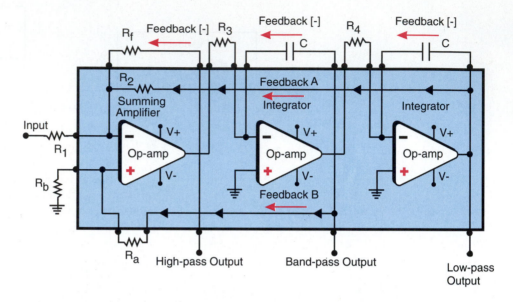

1   Set $C_1 = C_2$, and call it C. Select a reasonable value for C. It is a good idea to pick the capacitor value first, because capacitors come in fewer values and usually aren't adjustable.

2   Assume that $R_1 = R_3 = R_4$, and call it R_C.

3   Assume that all other resistor values are equal, and call their value R_d. Don't include R_a and R_b—which control the Q (or damping factor)—in this group.

The equations for the cut-off frequency for low-pass and high-pass filters, and for the center frequency for band-pass filters, is

$$F_C = 1/(2\pi\, R_C C)$$

The value of Q is given by the following equation, as well as by $D = 1/Q$:

$$Q = (R_a + R_b)/R_C$$

11.9   Introduction to Sine-wave Oscillators

An oscillator uses positive feedback to generate a sine-wave signal. There is always some kind of filter in the feedback loop to filter out all but the desired oscillation frequency.

Barkhausen Criteria

1   The feedback must be positive (in phase with the input signal).

2 The amplifier must provide enough voltage gain to make up for the insertion loss of the filter in the feedback loop. The total loop gain must be unity:

$$\text{Voltage Gain/Filter Loss} = 1$$

Positive feedback can be obtained in two ways:

1 The output signal can be fed back to the noninverting input through a filter that has 0° phase shift at the oscillating frequency.

2 The output signal can be fed back to the inverting input through a filter that has 180° phase shift at the oscillating frequency.

Figure 11-24 shows the general sine-wave oscillator implemented using an op-amp (part (a)) and a transistor amplifier (part (b)). Any circuit in this chapter that uses the op-amp inverting input and a frequency-determining filter with a 180° phase shift can be converted into a transistor amplifier circuit.

The purpose of an oscillator is to generate a continuous sine wave at some very specific frequency. The oscillator frequency normally must have some degree of frequency stability, but how stable the circuit must be depends on the application. Changes in part values within the frequency-determining network (filter) occur as a result of temperature variations is the biggest obstacle to maintaining frequency stability.

Most frequency-determining networks fall into one of two categories: resistor-capacitor (R-C) filters and narrow-band band-pass filters (of which tuned L-C resonant circuits are the most popular type).

11.10   Sine-wave Oscillators with Resistor-Capacitor (R-C) Frequency-control Networks

Capacitors are frequency-sensitive because their reactance is a function of the prevailing frequency. When capacitors and resistors are combined, the amount of phase shift that occurs is a function of capacitor and resistor values. We can use a combination of resistors and capacitors to convert negative feedback into positive feedback by making use of phase shift. We can also take advantage of other filter properties to provide positive feedback at the desired frequency.

Phase-shift R-C Oscillator

Figure 11-25 shows a simple R-C oscillator circuit called a **phase-shift oscillator.** This circuit uses a three-section R-C phase-shift network, with a 60° phase shift per section, to produce a total phase shift of 180° at the desired oscillating frequency. Any other frequency will produce a phase shift greater than or less than the required 180°, and the circuit will not oscillate.

Most oscillators are self-starting. There is always noise in the system, and noise contains energy at all frequencies. Typically, a large noise signal is induced when the circuit is powered up, which makes the start-up easy. When the power is first applied, some energy at the desired frequency gets through

FIGURE 11-24 General Sine-wave Oscillator

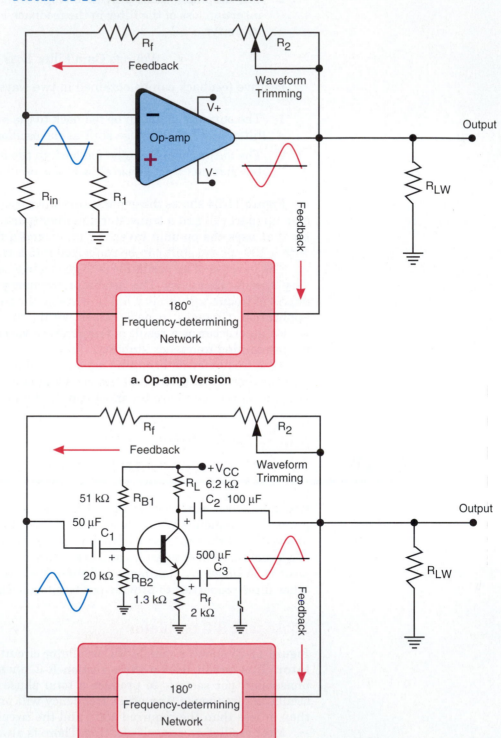

a. Op-amp Version

b. Transistor Version

FIGURE 11-25   R-C Phase-shift Oscillator

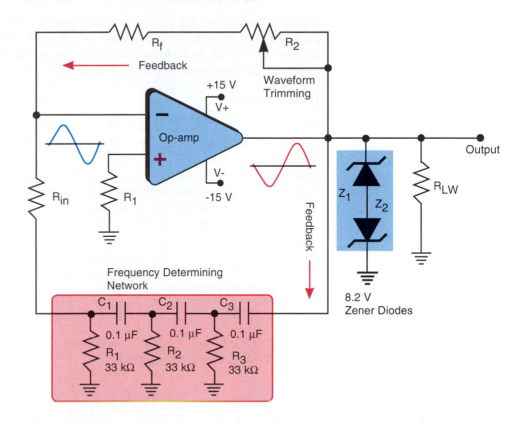

the frequency-determining filter and is amplified. The filter prevents the other frequencies from reaching the input, so they don't get amplified. In a few cycles, the signal output voltage builds up to its final peak-to-peak value.

It is common practice to control the voltage gain of the amplifier with negative feedback. If there is not enough gain, the circuit will not oscillate. If there is too much gain, the amplifier will become overdriven and produce a distorted output signal. The phase-shift oscillator requires a voltage gain of 29 to produce a perfect sine wave. Part of the negative-feedback resistance can be made variable to trim the waveform to a perfect sine wave, if necessary.

A common problem with R-C oscillators is their tendency toward amplitude instability, which may cause overdrive distortion even with the proper amplifier gain. Zener diodes are often used to impose an upper limit on the output voltage, to prevent overdrive distortion.

Given the conditions that $C = C_1 = C_2 = C_3$ and $R = R_1 = R_2 = R_3 = R_{in}$, the frequency of the phase-shift oscillator is defined by:

$$F_O = 1/(2\pi RC\sqrt{6})$$
$$= 1/(2\pi RC \times 2.45)$$

For this oscillator, the voltage gain must be

$$A_V = R_f/R_{in} = 29$$

Twin-T R-C Oscillator

Don't be too surprised that the twin-T notch filter network is back again in Figure 11-26. Sine-wave oscillators use a number of common filter circuits, and the narrow notch and steep slope into the notch of the twin-T support ideal oscillator frequency-determining behavior.

How It Works

The twin-T filter is installed in the negative-feedback loop. Frequencies outside the notch pass freely through the filter. The filter has a low impedance, which produces a very low amplifier voltage gain. For frequencies inside the notch, the filter impedance rises drastically, and there is almost no negative feedback at the desired oscillation frequency. Consequently, the amplifier gain is very high and oscillation takes place.

Some fixed negative feedback (R_1, R_2) is essential in this circuit, to prevent serious overdrive and waveform distortion. The twin-T circuit's frequency stability is significantly better than that of the phase-shift oscillator.

Given the conditions identified in Figure 11-26, the frequency of the twin-T oscillator is defined by:

$$F_O = 1/(2\pi RC)$$

FIGURE 11-26 **Twin-T R-C Oscillator**

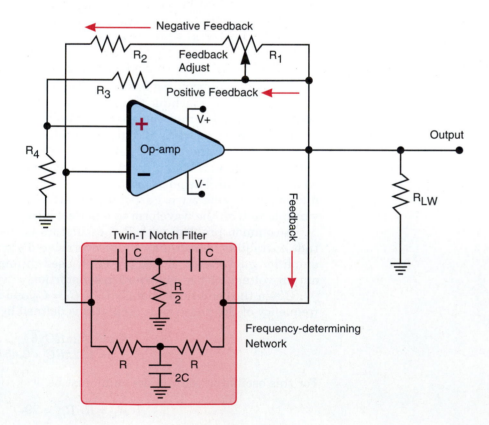

You have seen this equation before. The voltage gain for this circuit is

$$A_V = R_f/R_{in} = 10$$

Wien-bridge R-C Oscillator Circuit

The **Wien-bridge R-C oscillator,** depicted in Figure 11-27, is another popular sine-wave oscillator. Figure 11-27(a) shows the circuit drawn for the explanation. Figure 11-27(b) shows the circuit as it is usually drawn, in diamond-shaped bridge form.

Basically, the Wien-bridge oscillator is a version of the phase-shift oscillator, with a slightly more complex combination of series and parallel R-C circuits to provide the necessary 180° phase shift. The combination of parallel and series R-C circuits provides some natural compensation for temperature-caused part-value variation, which makes the Wien-bridge circuit more stable than the phase-shift oscillator.

The Wien-bridge oscillator nearly always has some form of output voltage limiting to prevent it from overdriving itself, and producing a distorted output signal. Whenever limiting of this sort is used, some distortion is introduced by the limiting action itself.

Remember once again that a sine wave is the only pure wave form, with no harmonics. If a waveform is not a true sine wave, it must contain some harmonic frequencies. A distorted sine wave has harmonics included, but the frequency-determining network blocks these harmonics, effectively removing them. The waveform has been cleaned up and the distortion has been removed. This works well for a moderate amount of distortion, but it can't deal with extreme distortion of the kind produced by a badly overdriven amplifier.

Given the conditions that $R = R_1 = R_2$ and $C = C_1 = C_2$, the frequency of the Wien-bridge oscillator is defined by:

$$F_O = 1/(2\pi RC)$$

The voltage gain for this oscillator is

$$A_V = 1 + (R_f/R_{in}) = 3$$

Quadrature Oscillator

The **quadrature oscillator** consists of two integrator circuits in series (see Figure 11-28). Recall that an integrator produces exactly 90° of phase shift at its cut-off frequency. If we use two integrators in series, each with 90°, we get the required 180° for oscillation and two outputs that are 90° out of phase. Sine waves that differ by 90° are said to be in *quadrature*—hence the name *quadrature oscillator*. One output is called the *sine-wave output* and the other is called the *cosine-wave output*.

A simple low-pass filter is placed in the positive-feedback loop to help reduce high-frequency noise and to remove distortion produced by the output-limiting Zener diodes.

The quadrature oscillator tends to have a little trouble starting if the value of R_2 is not at least 10% larger than the value of R_1.

FIGURE 11-27 Wien-bridge R-C Oscillator

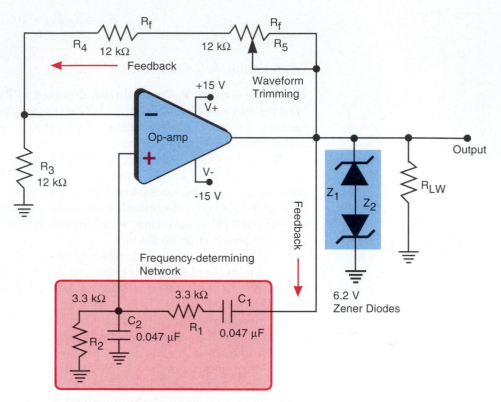

a. Drawn for Explanation

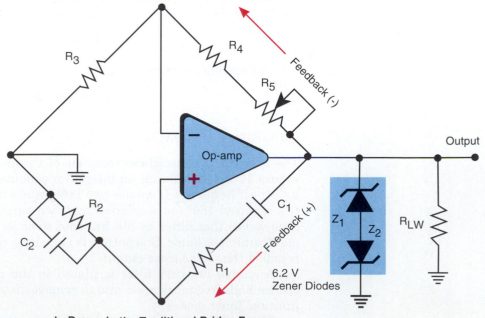

b. Drawn in the Traditional Bridge Form

FIGURE 11-28   Sine/Cosine Quadrature Oscillator

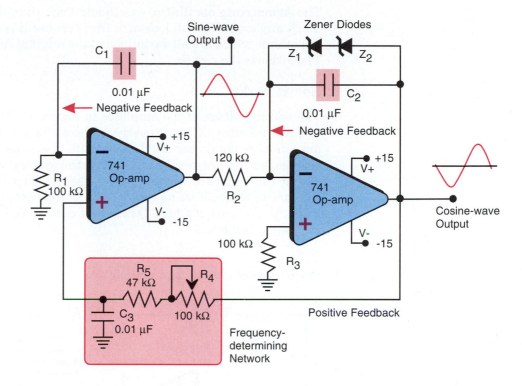

Given the conditions that $R_2 > R_1$, $R = R_2 = R_3$, and $C = C_1 = C_2 = C_3$, the frequency of the quadrature oscillator is defined by:

$$F_O = 1/(2\pi RC)$$

As a matter of convenience, this circuit is often constructed using a state-variable filter integrated-circuit package.

11.11   Sine-wave Oscillators with Inductor-Capacitor (L-C) Frequency-control Networks

Inductors tend to be too large, heavy, and expensive to use for audio and other low-frequency applications, so R-C oscillators generally are used. At radio frequencies, inductor-capacitor tuned resonant circuits are usually the simplest, cheapest, and most efficient frequency-determining network. The L-C tank circuit can have a very high Q, which means that the band-pass of the frequency-control filter is very narrow; and the narrower the bandpass (within reason), the better the frequency stability. Thus, at radio frequencies, L-C oscillators are overwhelmingly preferred.

Armstrong Oscillator

The **Armstrong oscillator** was the first circuit in this class, but it is not used much any more. We will look at it first because it is easier to understand than the others, which are all variations of the original Armstrong oscillator. Figure 11-29 shows the circuit.

How It Works

The output voltage of the amplifier in Figure 11-29 is applied to the primary of a transformer. If we properly connect the secondary to the input of the amplifier, we obtain our 180° phase shift. If we connect the secondary wires in reverse, we get negative feedback and the system won't work. The dots at the end of the transformer windings in the drawing indicate transformer leads that are in phase. If we now add a tuning capacitor to make the secondary and primary coils part of a high-Q resonant circuit, we have an efficient and stable sine-wave oscillator.

FIGURE 11-29 **Armstrong L–C Oscillator**

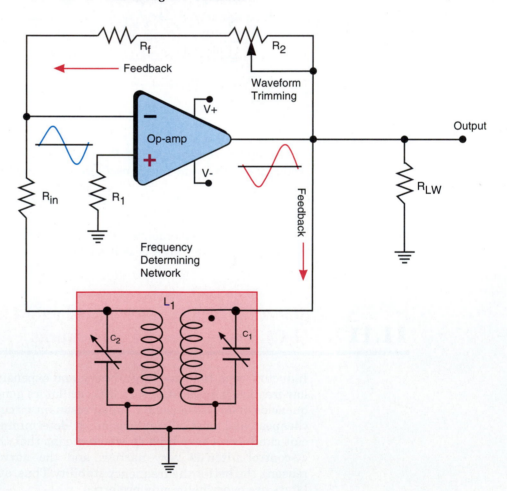

We can use different numbers of turns in the primary and secondary to step up or step down the feedback voltage for still greater flexibility.

Hartley Oscillator

The **Hartley oscillator,** shown in Figure 11-30, is a modern version of the Armstrong. It differs only in that the transformer consists of a tapped winding instead of two separate windings, and both coils are resonated by the same capacitor. The circuit shown in Figure 11-31 is a transistor version of the Hartley oscillator. A single transistor is easier to use at higher frequencies than an op-amp, and it is often much cheaper.

The frequency of the Hartley oscillator is defined by:

$$F_O = 1/(2\pi \sqrt{L_{EQ}C})$$

with

$$L_{EQ} = L_a + L_b + 2L_M$$

where L_M is the mutual inductance. The equation for L_M, in turn, is

$$L_M = K \sqrt{L_a \times L_b}$$

where K is the coefficient of coupling.

FIGURE 11-30 Hartley *L–C* Oscillator: Op-amp Version

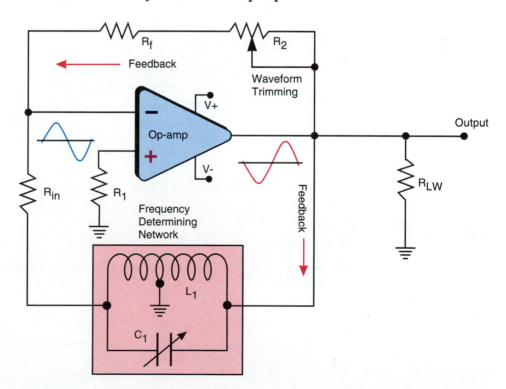

FIGURE 11-31 Hartley L–C Oscillator: Transistor Amplifier Version

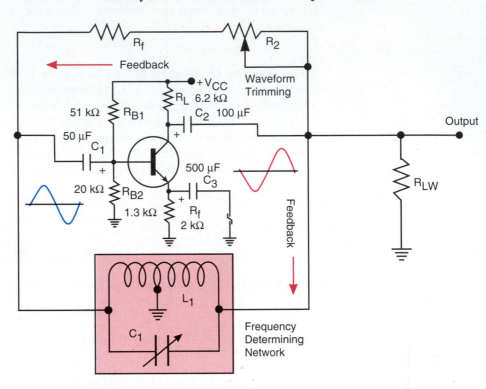

Colpitts Oscillator

The **Colpitts oscillator** (see Figure 11-32) is a modification of the Hartley that uses a tapped capacitor pair to provide a phase reference. Its characteristics are virtually identical to those of the Hartley, except that the Colpitts' tuning range can be much greater than the Hartley's. Both capacitors in the Colpitts oscillator must be adjusted together, if they are to be adjusted. Consequently, the capacitors are often fixed and the resonant frequency is adjusted by moving a core, called a **slug,** in and out of the coil. The frequency of the Colpitts oscillator is defined by:

$$F_O = 1/(2\pi\sqrt{LC_{EQ}})$$

where

$$C_{EQ} = (C_1 \times C_2)/(C_1 + C_2) \qquad \text{(capacitors in series)}$$

Crystal-controlled Oscillators

A quartz crystal is a piezo-electric crystal—a crystal that responds to external pressure or force by developing a proportional voltage between two of its faces. In addition, an electrical voltage applied to a piezo-electric crystal's faces causes the crystal to shrink or expand slightly.

FIGURE 11-32 Colpitts *L–C* Oscillator

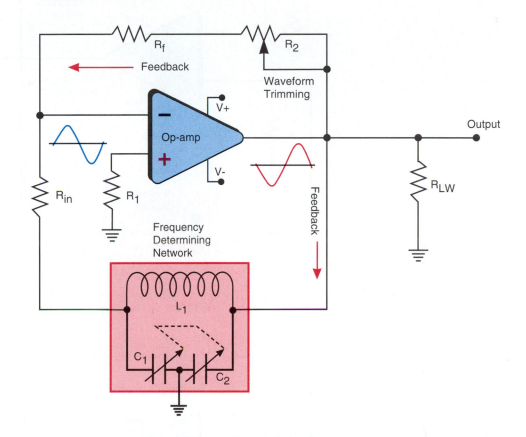

If a high-frequency voltage is applied across the crystal faces, the signal voltage causes the crystal to vibrate. If the frequency of the applied voltage is at the crystal's natural mechanical resonant frequency, it will generate a large aiding voltage across its faces at that frequency. The mechanical resonant frequency of the crystal depends on the crystal's physical dimensions.

When the crystal is resonating, it behaves exactly the way an L-C resonant circuit does and it can be used as a frequency-determining network in an electronic oscillator. The advantage of the crystal is that its resonant frequency is little influenced by temperature and other outside factors, so it can be used where extreme frequency stability is required.

Figure 11-33(a) presents the schematic diagram of a simple crystal-controlled oscillator, which acts as a series resonant circuit. The crystal's impedance is very high at all frequencies except its resonant frequency. At its resonant frequency, the impedance drops to a low value; and enough positive feedback is developed to sustain oscillation.

Crystals operate as either series (Figure 11-33(b)) or parallel (Figure 11-33(c)) resonant circuits. There are many variations on the crystal-controlled oscillator. Adding a series resonant crystal to a Colpitts oscillator (as in Figure 11-3(b)) makes the Colpitts very stable. The circuit is then classed as a **crystal-controlled oscillator.** The conventional tank allows the crystal frequency to be altered very slightly for precise tuning. It also allows for frequency

FIGURE 11-33
**Typical Crystal-controlled
Oscillator Circuits**

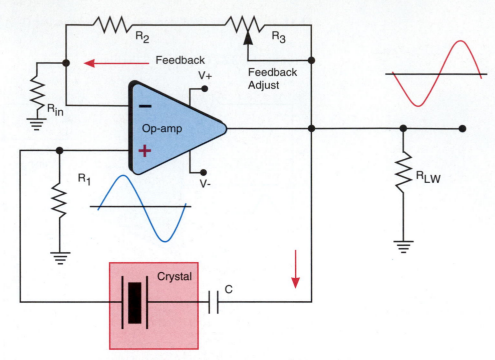

**a. Crystal Acts as a Series Resonant Circuit. The Crystal
Impedance is Low Only at Resonance.**

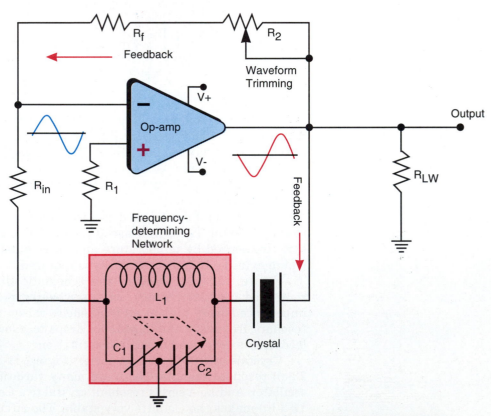

b. Adding a Crystal to a Colpitts Oscillator Makes the Colpitts Very Stable

FIGURE 11-33   Continued

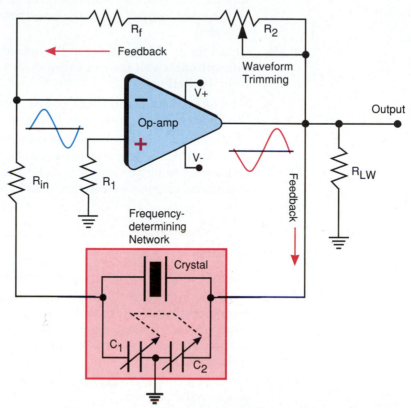

c. Crystal Operating in a Parallel Resonant Mode, Where the Impedance is High at the Resonant Frequency

modulation over a limited range. In a circuit that uses a parallel resonant crystal (as in Figure 11-33(c)), the crystal oscillates in a parallel resonant mode; the impedance is high at resonance.

The resonant frequency of a crystal is determined by the manufacturer and cannot be altered. The frequency is printed on the crystal's case, and the crystal is good only for that frequency and for harmonics of that frequency.

11.12   Troubleshooting Filters and Sine-wave Oscillators

Oscillators are easy to troubleshoot because they provide a sine-wave signal for you to view on an oscilloscope. Filters are more difficult to troubleshoot because you must use both a scope and a signal or function generator.

Troubleshooting Oscillators

Procedure

1   Connect an oscilloscope to the output of the oscillator and view the waveform.

2   The waveform should be a sine wave. If the waveform is distorted or missing, verify the **dc** voltages at the IC pins. If the waveform is slightly distorted, adjust the waveform trimming potentiometer, if one is provided.

3   Check for an offset voltage at the output of an op-amp oscillator. A significant offset voltage can cause distortion or can prevent the circuit from oscillating.

4   If an offset voltage is present and **dc** voltages at the IC pins are ok, use the op-amp forcing tests from Chapter 8 to test the op-amp.

5   The remaining possibilities are bad solder joints (or other mechanical problems) and a defective capacitor. Unless you are using electrolytic capacitors, capacitor failure should be the last thing you consider. All capacitors, except electrolytics, seldom fail.

Troubleshooting Filters

Procedure

1   Identify the kind of filter you are dealing with.

2   Find the approximate cut-off frequency, if the circuit is a low-pass or high-pass filter. Find the center frequency if the circuit is a band-pass filter. The equipment manufacturer's manual may provide the information. If not, you can calculate the appropriate cut-off or center frequency yourself.

3   Connect a sine-wave generator to the input of the filter and an oscilloscope to the output.

4   Set the sine-wave generator to a frequency near the cut-off or center frequency.

5   Vary the sine-wave generator frequency from a frequency below the cut-off (or center) frequency to a frequency well above the cut-off or center frequency, and observe the oscilloscope.

6   You should see a significant change in amplitude at (or near) the cut-off or center frequency. Specifically,

   a   If the circuit is a low-pass filter, the amplitude should be high for frequencies below cut-off and low for frequencies above cut-off.

   b   If the circuit is a high-pass filter, the amplitude should be low for frequencies below cut-off and high for frequencies above cut-off.

   c   If the circuit is a band-pass filter, the amplitude should be highest at the center frequency and should fall off for frequencies higher or lower than the center frequency.

   d   If the circuit is a notch filter, the amplitude should be lowest at the center frequency and should increase for frequencies higher or lower than the center frequency.

7   The waveform should be a sine wave. If the waveform is distorted or missing, verify the **dc** voltages at the IC pins. Check for an offset voltage at the output of an op-amp oscillator. A significant offset voltage can cause distortion or prevent the circuit from amplifying the signal.

8 If an offset voltage is present and **dc** voltages at the IC pins are ok, use the op-amp forcing tests from Chapter 8 to test the op-amp.

9 The remaining possibilities are bad solder joints (or other mechanical problems) and a defective capacitor. Unless you are using electrolytic capacitors, capacitor failure should be the last thing you consider. All capacitors, except electrolytics, seldom fail. A leaky or shorted capacitor may cause an offset voltage.

The Digital Connection: The Switched Capacitor Filter

The **switched capacitor filter (SCF)** is the latest of the integrated-circuit filter devices to appear on the market. The MF1O is a typical example of the universal SCF filter device. The SCF is designed to filter analog signals, but internally it is a mix of analog and digital circuitry. The main advantage of the SCF is that it requires no precision capacitors. The SCF's cut-off frequency is determined by the frequency of a digital clock instead of by resistor and capacitor values.

Figure 11-34 provides a simple illustration of how the SCF can be used as a digitally controlled integrator—which, as you know, can be used as a low-pass filter. Using additional integrators and circuitry permits the other filter functions to be performed. The MF10 IC contains all the required integrators and other control circuitry to perform almost any filter function. Let's take a closer look at Figure 11-34.

FIGURE 11-34
Digitally Controlled Switched Capacitor Filter

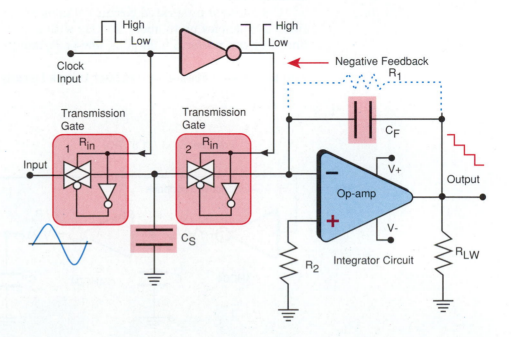

How It Works

The Integrator is the heart of the system. CS is a charge transfer storage element. The two analog gates are controlled by a digital clock. The inverter ensures two things:

1 When analog gate 1 is a closed circuit, analog gate 2 is an open circuit.

2 When analog gate 1 is an open circuit, analog gate 2 is a closed circuit.

The Action with the Clock

1 When the digital clock signal is high (+5 volts), capacitor CS is connected to the input signal voltage but not to the input of the integrator (op-amp).

2 When the digital clock pulse goes low (0 volts), capacitor CS is disconnected from the input signal source and connected to the input of the integrator (op-amp). The charge on CS is transferred to the integrator input.

The circuit periodically transfers a sample of the signal input voltage to the integrator. If we only clock-in a voltage sample once every few minutes, it will take a long time to stair-step the integrator to its limit. On the other hand, if we clock the signal into the integrator once every second, we will complete the staircase ramp much more quickly.

Since the slope of the integrator output represents a plot of voltage versus time, changing the digital clock's frequency effectively changes the time constant of the integrator. If we change the time constant of the integrator, we also change the cut-off frequency, if we use the integrator as a low-pass filter. Thus, we can vary the digital clock's frequency to tune the filter to the desired cut-off frequency.

Commercial integrated circuit versions of the SCF, such as the MF10, can filter frequencies up to about 20 kHz with a digital-clock frequency of 1 MHz, and they can perform high-pass, low-pass, band-pass, and notch functions.

FIGURE 11-35 **Phase-shift Oscillator Using a Linearly Biased C-MOS Digital IC**

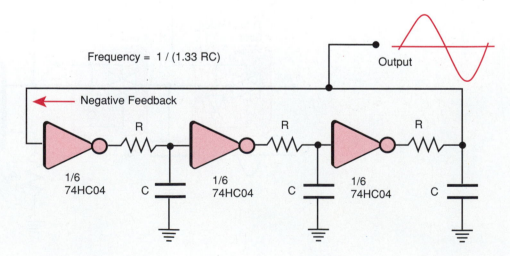

Frequency = 1 / (1.33 RC)

Phase-shift Oscillator Based on a C-MOS Digital IC

Figure 11-35 shows a simple phase-shift sine-wave oscillator constructed from a 74HC00-series C-MOS digital integrated circuit. Any of the inverting gate packages can be used for this purpose. The circuit looks very much like the common TTL digital-clock circuit; but this circuit produces a sine wave, which TTL devices cannot do. The oscillator can be used at frequencies from a few hertz to several megahertz and can be run on a 5-volt power supply.

QUESTIONS AND PROBLEMS

11.1 A first-order filter has a roll-off of how many dB per decade?

11.2 A second-order filter has a roll-off of how many dB per decade?

11.3 A third-order filter has a roll-off of how many dB per decade?

11.4 How does a two-pole filter differ from a second-order filter?

11.5 Identify the circuit and write the cut-off or band-pass center frequency equation for the circuit in Figure 11-37.

11.6 Identify the circuit and write the cut-off or band-pass center frequency equation for the circuit in Figure 11-38.

FIGURE 11-36   **Multiple Feedback Filter**

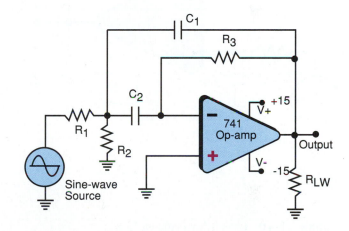

FIGURE 11-37   **Circuit for Problem 11.5**

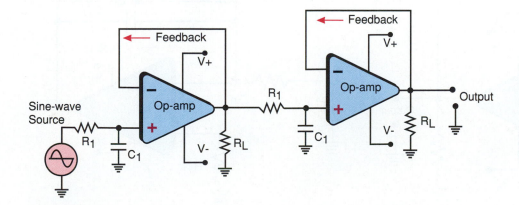

11.7 Identify the circuit and write the cut-off or band-pass center frequency equation for the circuit in Figure 11-39.

11.8 Identify the circuit and write the cut-off or band-pass center frequency equation for the circuit in Figure 11-40.

11.9 Identify the circuit and write the cut-off or band-pass center frequency equation for the circuit in Figure 11-41.

11.10 Identify the circuit and write the cut-off or band-pass center frequency equation for the circuit in Figure 11-42.

FIGURE 11-38 Circuit for Problem 11.6

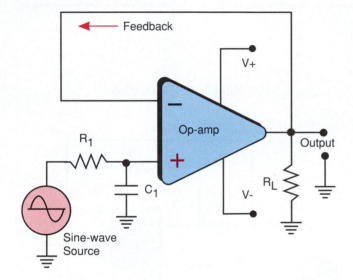

FIGURE 11-39 Circuit for Problem 11.7

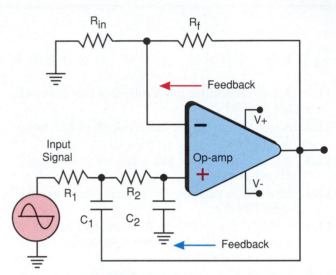

FIGURE 11-40 Circuit for Problem 11.8

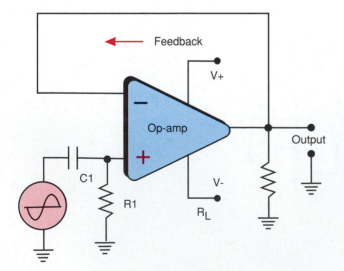

FIGURE 11-41 Circuit for Problem 11.9

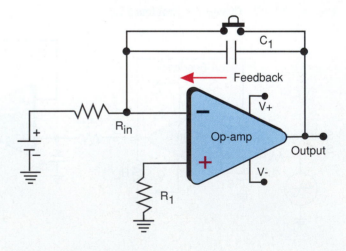

11.11 Identify the circuit and write the cut-off or band-pass center frequency equation for the circuit in Figure 11-43.

FIGURE 11-42 Circuit for Problem 11.10

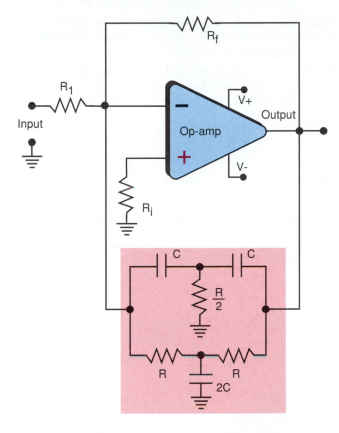

11.12 Identify the circuit and write the cut-off or band-pass center frequency equation for the circuit in Figure 11-44.

11.13 Identify the circuit and write the cut-off or band-pass center frequency equation for the circuit in Figure 11-45.

11.14 Identify the circuit and write the cut-off or band-pass center frequency equation for the circuit in Figure 11-46.

FIGURE 11-44 Circuit for Problem 11.12

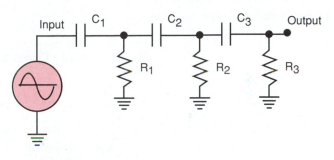

FIGURE 11-45 Circuit for Problem 11.13

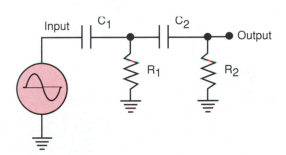

FIGURE 11-43
Circuit for Problem 11.11

11.15 What kind of response curve is most common in filters?

11.16 Define the term *damping factor*.

11.17 Define *Q*.

11.18 What is another name for a narrow band-stop filter?

11.19 Which of the following would a twin-T notch filter most likely be a part of?
  a. An active band-stop filter.
  b. An active band-pass filter.
  c. An oscillator.
  d. All of the above.

11.20 Define the term *oscillator*.

11.21 In what way might an oscillator be related to a signal generator?

11.22 Identify and explain the Barkhausen criteria.

11.23 In what two ways can the required positive feedback for oscillation be obtained?

11.24 Why might negative feedback be added to an oscillator circuit when positive feedback is required for oscillation?

11.25 If you were to use an L-C resonant circuit in an oscillator, would you want one with a high *Q* or one with a high damping factor? Why?

FIGURE 11-46   Circuit for Problem 11.14

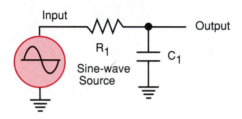

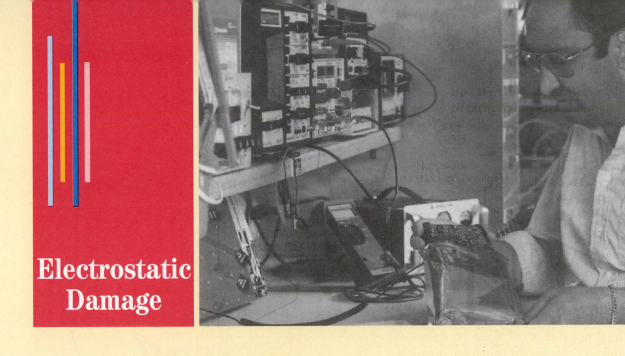

Electrostatic Damage

An interesting aspect of the development of electronics involves the phenomenon of electrostatic damage. In the "old days" of electronics, circuits used vacuum tubes, most of which required relatively high voltages to operate. Obviously, a technician working on a tube amplifier with a 400-volt power supply would be exceedingly unwise to reach carelessly into the circuitry.

These days, most circuitry operates at much safer voltages; but unfortunately, a lot of that circuitry is sensitive to high voltages, such as those that can build up through static electricity. A technician, having shuffled across a carpet and acquired a static charge, can damage or destroy components on or off a printed circuit board just by touching them.

When this phenomenon was first noticed, it was thought that the problem only existed in relation to sensitive components such as MOS-FETs and MOS integrated circuits. The extremely high input impedance of these devices would not allow static charges to bleed off slowly; and once the charge reached the breakdown potential of the device, current would flow, damaging the device internally. For this reason MOS devices were—and still are today—shipped in antistatic packaging, and special handling precautions are necessary.

In the years since then, more testing has been done; and results have shown that the problem, to varying degrees, exists for almost all semiconductor devices. Another important fact discovered was that, although some devices are destroyed outright, others are only damaged. This damage may not show up in conventional testing, but it will shorten the life of the component, so that failure occurs inexplicably some time after the product is in service.

What level of voltage is considered dangerous to electronic components? The answer depends on the particular component. The charge that builds up on someone who walks across a nylon carpet in dry air can reach 35,000 volts. Humidity tends to reduce the electrical potential, but a person could still create a charge of something over 2000 volts. Innocuous activities such as picking up a styrofoam coffee cup or sitting in a padded chair can create several thousand volts. Table A shows the approximate electrostatic voltage required to degrade or destroy various types of electronic components. Notice that a voltage as low as 30 volts can degrade a VMOS device. Even a standard bipolar transistor can be degraded by only 380 volts. MOS-FETs can be destroyed outright if subjected to 200 volts, and even a silicon-controlled rectifier (SCR), which is used in power applications, is not immune: 1000 volts can destroy an SCR.

Electrostatic charges can damage components in any of three ways: by discharge, by induction, or by polarization. As mentioned previously, discharge can cause damage by creating instantaneous large currents in a device. This may occur when the device's breakdown voltage is exceeded

TABLE A

Device Type	Electrostatic Voltage	
	To Degrade	To Destroy
V-MOS	30	1800
MOS-FET	100	200
EPROM	100	—
J-FET	140	7000
Op-amp	190	2500
Bipolar Transistor	380	7000
SCR	680	1000
TTL Logic	1000	2500

or when the breakdown voltage of the air is exceeded and a spark jumps to the device's terminal, thereby creating damaging currents.

Induction-related damage is caused by the electrostatic field set up by the static charge on an object near the device. If the device moves within the field, currents are induced in the device, again causing damage. The magnitude of the electric field and the speed at which the device moves through it determine the level of current induced.

Polarization occurs when the electric field remains stationary and polarizes the device. For instance, a styrofoam cup, placed next to the device, carries a positive charge due to handling. The electrostatic field of the cup draws electrons inside the device toward the cup, thereby polarizing the device. If you pick up the device by the positive end, it draws electrons from your hand into the device, charging it. If you then set the device down on a grounded surface it can discharge the electrons, causing current flow and damage.

What Can Be Done to Prevent Electrostatic Damage?

In recent years, a whole industry has sprung up around the manufacture of tools, accessories, chemicals, and other products designed to minimize the risk of damage due to electrostatic discharge (ESD). While these may be useful to some extent, the most important factor in preventing damage is to understand the problems that exist and to take precautions. If that necessitates the use of these products, then they should be considered.

ESD Protection Grounding Strap

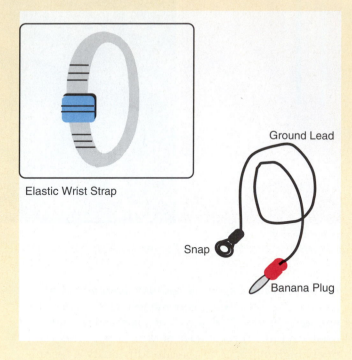

Elastic Wrist Strap

Ground Lead

Snap

Banana Plug

There are two main ways to reduce the risk of ESD:

- Create a static-safe work area.
- Handle sensitive components correctly.

A static-safe work area is one that discourages the build-up of static electricity for any reason. This can be accomplished by three expedients:

- Working on a bench with a conductive surface that is grounded.
- Covering the floor with a grounded conductive surface.
- Removing nonconductive materials such as plastics, nylon, and styrofoam from the workplace.

The first rule of proper component handling is to wear a proper grounding wrist strap. It is very important that these wrist straps be properly constructed. The strap should have a 1-MΩ resistance in series with it. Simply grounding your wrist via a wire could be unsafe if it ever became disconnected from ground and came into contact with a voltage

Electrostatic Damage Modes

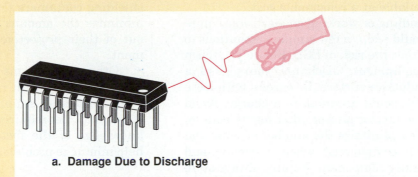

a. Damage Due to Discharge

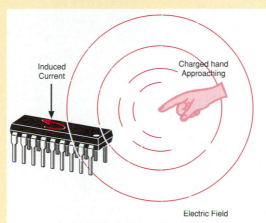

b. Damage Due to Induction

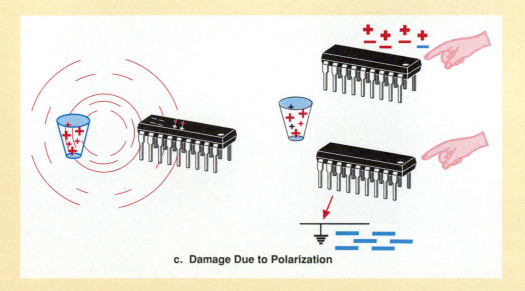

c. Damage Due to Polarization

in the equipment under test. Wrist straps are commercially available at a reasonable cost.

When handling or working on electronic equipment, you should adopt a few simple precautions to minimize further the risk of ESD. Before you touch a piece of equipment, discharge yourself to a grounded conductive surface. Be careful with loose clothing, which could accumulate a charge. Avoid touching the connector pins or other metal leads on circuit modules. Minimize the number of times you have to touch components when removing and replacing a faulty component. Finally, always store and carry components and modules in static-shielding containers that guard against ESD; and minimize the amount of time these modules are out of their protective container during replacement.

Other suggestions might also be mentioned here, especially with regard to special protective equipment and supplies. In specific situations these may be required; but in general, a basic knowledge of the factors explained here should help minimize problems with electrostatic damage.

CHAPTER 12

Advanced Analog Signal-processing Devices, Circuits, and Techniques

OBJECTIVES

Upon completion of this chapter, you should be able to:

1 Draw the internal block diagram of a typical sample-and-hold amplifier.

2 Explain how the sample-and-hold amplifier works.

3 Give three examples of situations where a sample-and-hold amplifier could be used.

4 Define each of the following sample-and-hold terms: **a)** *acquisition time;* **b)** *aperture time;* **c)** *dynamic sampling error;* **d)** *gain error;* **e)** *hold settling time;* **f)** *hold set-up voltage;* **g)** *capacitor droop*

5 Explain the purpose of a multiplexer when it is used with a sample-and-hold amplifier, and give a typical example of such an application.

6 Draw a block diagram of a typical analog multiplexer.

7 Explain the difference between an analog multiplexer and a multiplexer intended for digital service.

8 Draw a sketch showing how a CCD analog shift register is constructed, and briefly explain how it works.

9 Draw a block diagram of the ISD-1016 voice message system.

10 Draw a partial diagram of a CCD videocamera chip image matrix.

11 Explain the difference between linear mixing and nonlinear mixing of two sine-wave signals. (Assume that one sine wave is 1 MHz or greater and that the other is 1 kHz.)

12 Define *carrier wave.*

13 Explain amplitude modulation (AM), make sketches of the waveforms, and show the spectrum diagram.

14 Define the terms *side frequency* and *sideband.*

15 Describe the new frequencies created by amplitude modulation.

16 Define *balanced modulation.*

17 Explain why the carrier frequency in an AM transmitter can be suppressed.

18 Describe single-sideband AM, and explain how it differs from standard AM.

19 Draw spectrum diagrams for standard AM and for single-sideband AM.

Signal Processing in Brief

The purpose of analog circuits is to process analog signals. We will review some ways in which signals can be processed, and then we will take a first look at some new ones to be discussed in this chapter.

HIGH-FIDELITY AMPLIFICATION

We can amplify a signal without altering its waveform (high fidelity). The amplifying circuit for high-fidelity signal processing is shown at right.

FILTERING

We can amplify selected frequencies, while rejecting others (filters).

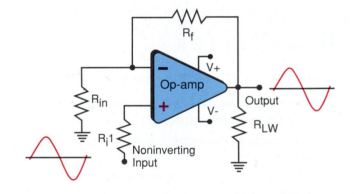

Linear High-fidelity Amplifier

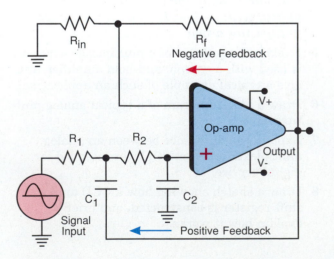

Frequency Filtering

WAVESHAPE ALTERATION ➤

We can deliberately alter the waveshape of a signal. The integrator at right is an example. Guitarists sometimes use a fuzzbox, which deliberately distorts the signal by overdriving an amplifier, to create a special sound.

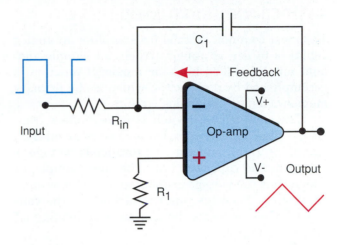

Waveshape-altering Amplifier

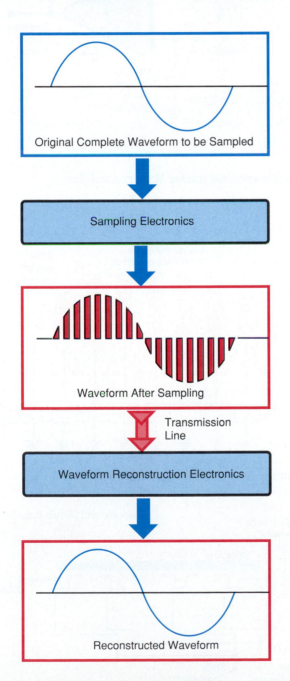

Waveform Sampling (Time Multiplexing)

◀ **TIME MULTIPLEXING**

We can take a series of samples from a waveform, send the samples to a remote destination, replace the missing parts of the signal and re-create the original signal. This is called *time multiplexing*.

Notice that the waveform (at left) is composed of a sample of the waveform, followed by a period of time during which there is no signal, followed by another sample.

We can put a second, alternately sampled signal into the empty time slots. This technique is used in dual-trace oscilloscopes to display two waveforms with a single electron beam.

Time multiplexing can be used to send two or more telephone conversations at the same time over the same pair of wires.

Time multiplexing is also popular in analog-to-digital converter systems, and for digital data handling in general.

SAMPLE-AND-HOLD AMPLIFIER

The most common method for sampling an analog signal is to use an integrated-circuit sample-and-hold amplifier. The hold (or memory) function is accomplished by a capacitor, which is the primary memory element in analog circuits.

The sample/hold switch is a field-effect transistor. The signal is sampled at a regular interval (sampling rate), and the instantaneous voltage is stored in the capacitor until the next sample, at which time the voltage is updated to that new sample instant. The high input impedance of the voltage follower prevents the charge from leaking off between samples.

The sampling process is always required before analog sound or picture signals can be converted into digital signals (digitizing) for computer signal processing.

The main memory (dynamic memory) in modern digital personal computers (PCs) also uses tiny IC capacitors as memory storage for +5-V and 0-V binary signals.

STORING ANALOG SIGNALS—MEMORY

The samples derived from a sample-and-hold amplifier can be stored in a bank of capacitors, one for each sample voltage. In the drawing at right, each sample is used to charge the first capacitor. The charge voltage previously stored in the first capacitor is shifted to the right and stored in the second capacitor. Each new sample causes the previous samples to shift one capacitor to the right, until the entire waveform is stored. The capacitor memory bank can remember the waveform until the charges leak off.

The samples can be shifted out of the capacitor memory bank; and with a little filtering, the waveform originally sampled by the sample-and-hold amplifier can be reconstructed.

The practical version of the analog memory system is an integrated-circuit device called a *charge-coupled device (CCD)*. The CCD was first used as an audio delay for echo effects. Advanced versions like the (ISD-1016) C-MOS analog voice message system can store up to 16 seconds of speech, and remember it for 10 years.

Special light-sensitive CCDs capture and store video images while they are scanned and processed in video cameras.

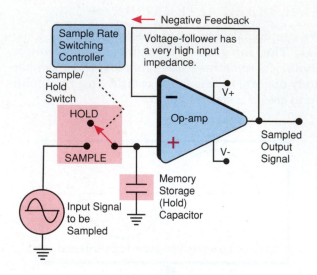

Sample-and-Hold Analog Memory Amplifier

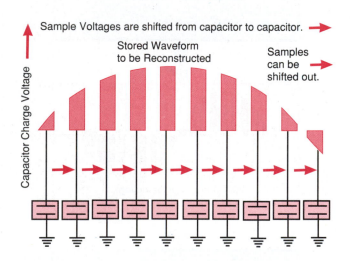

How the Analog Shift Register Memory Works

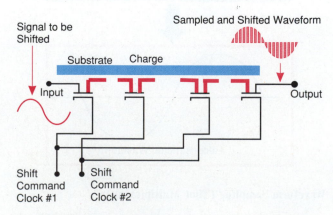

The CCD Charge-coupled Device Implementation of the Analog Shift Register Memory

Phase-locked Loops and Analog Multipliers in Brief

INTRODUCTION

The analog multiplier and the phase-locked loop are two of the most versatile analog ICs available for controlling frequency, frequency shifting, and frequency synthesis. They also handle chores involved in both frequency modulation (FM) and amplitude modulation (AM). Neither device is really new, but integrated-circuit fabrication has recently made these very versatile and powerful techniques common and inexpensive. Both devices are also used as part of larger-scale systems.

FREQUENCY MODULATION

Frequency modulation causes the frequency of a high-frequency sine wave or square wave called a *carrier* to vary in response to the voltage variations in a low-frequency (often audio) signal. Once the frequency-modulated (FM) carrier has been generated, it can be transmitted from place to place by way of:

a Radio transmission

b Wire or coaxial cable

c **ac** power lines or railroad tracks

d LASER or LED light beam sent direct (TV remote controls, for example)

e LASER or LED light beam sent over fiber-optic cables

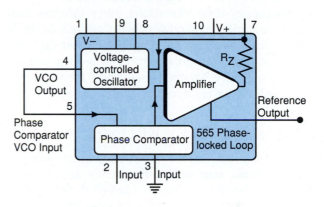

Phase-locked Loop

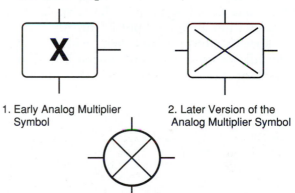

1. Early Analog Multiplier Symbol

2. Later Version of the Analog Multiplier Symbol

3. Round Version – Currently the Most Commonly Used Symbol

The Analog Multiplier Symbol

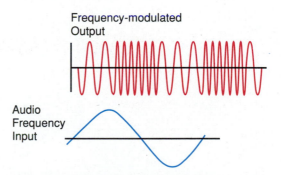

Frequency Modulation (FM) Sine-wave Carrier

SQUARE-WAVE CARRIER

When frequency modulation (FM) is used, the carrier frequency can be square (rectangular) because only frequency-time variations carry information; amplitude (voltage or power level) variations don't matter. Sine-wave carriers must be used for normal radio transmissions. Square-wave harmonics would interfere with assigned frequencies. Rectangular carriers are okay for cable, etc.

FREQUENCY-SHIFT KEYING

In its simplest form, frequency-shift keying switches from one audio frequency to another, when modulated by a binary (on/off) digital waveform. It is often used with computer modems to produce a 1200-Hz tone for a binary zero and a 2400-Hz tone for a binary one. The telephone network cannot handle digital pulses directly.

A special phase-locked loop called a *tone decoder* allows us to use multi-toned FM signals for a variety of remote control and other functions.

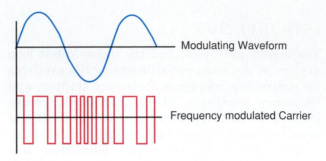

Square-wave Carrier

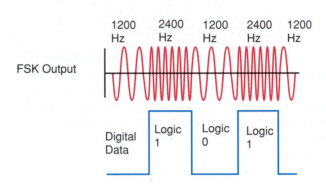

Frequency-shift Keying (FSK)

12.1   Introduction

This chapter is devoted to some advanced analog signal-processing circuits, devices, and techniques. The purpose of most electronics—analog or digital—is to process some kind of signal. So far, we have used linear amplifiers to increase the signal amplitude, and we have added reactive components to process certain frequencies while eliminating others. We have used op-amps as voltage comparators and switches. We have used op-amps to process one waveshape into a different waveshape with a new harmonic content.

Most of the waveshapes we have encountered so far have been simple basic waveforms, and we have not been very concerned about their harmonic content and so on. In this chapter, we will do some sophisticated (and sometimes tricky) things with those waveforms. Recall that when we change the shape of a waveform, we create new sine-wave frequencies. That concept will be important in this chapter.

Analog Memory Devices

In this chapter, we will look at totally electronic analog memory devices whose functions range from storing a single **dc** voltage as a capacitor charge voltage in a sample-and-hold amplifier to operating as a purely electronic voice recorder. In between, we will look at the analog shift register and some of its applications.

Signal Sampling

We will also examine sampling techniques that involve extracting pieces from the waveform and then storing just the pieces or transmitting them to another location. Later, an electronic circuit puts the pieces back together and fills in the missing parts to construct an exact replica of the original waveshape.

A common example of waveform sampling and time-multiplexing is observable in your school laboratory. Most dual-trace oscilloscopes have only one electron beam. But if each of the two waveforms (traces) is alternately sampled and displayed at different locations on the screen, you can see two waveforms at the same time.

In reality, half of each waveform is missing; and the traces would appear as waveforms made up of dashed lines if the sampling rate were slower. While the beam is tracing a section of waveform #1, it can't be tracing waveform #2, so there is a gap in #2. Then it is waveform #2's turn to be traced, leaving a gap in #1 during that time.

Waveform sampling and time-multiplexing allow the phone company to send several phone conversations simultaneously over the same pair of wires. The sampled waveform (signal) reconstruction on your end of the phone is so good that you have no trouble recognizing the voice on the other end.

Waveform sampling is also essential before an audio or video signal can be stored in an analog memory. Analog audio or video signals must be sampled before they can be converted into digital signals.

Modulation and Demodulation

The amplitude modulation technique was originally devised to make transmission of voice and music on radio waves possible. The first radio communications system required the transmitter's power to be turned on and off in a code (Morse) to send a message.

High-frequency (radio-frequency) sine-wave voltages radiate into space if connected to a piece of wire called an antenna. In contrast, when electrical signals in the audio-frequency range are connected to an antenna, nothing happens. If the antenna were long enough, audio frequencies would produce an electromagnetic field in space, too, but the wavelengths involved are completely prohibitive.

Modulation is a process of combining an audio (or video) signal with a radio-frequency signal in such a way that the radio frequency carries the audio signals with it when it radiates into space. We then call the radio frequency a **carrier wave.** The music (or whatever) can be recovered at a distant receiver by a process called *demodulation*.

Introduction to Modulation

There are two basic kinds of modulation: amplitude modulation and frequency modulation. Each form has its own advantages and disadvantages, and there are several variations on each process. We will examine both AM and FM and their important variations. We normally associate AM and FM with our favorite radio stations, but there are many other applications. In radio or television, we select a station or channel (using a filter) by selecting a particular carrier frequency. The receiver demodulates whatever information was originally used to modulate the carrier for that station or channel.

We can just as well send a number of modulated carrier frequencies at the same time along a coaxial cable or fiber-optic cable. At the other end of the cable, we can tune to the desired carrier frequency to get different channels.

Modulation allows us to send many different channels along the same cable at the same time, and we can even combine modulation with sampling and time multiplexing to send even more information on the same cable.

Modulation and sampling techniques are also used with digital data or with both analog and digital information on the same cable. In fact, sampling itself is a form of amplitude modulation. Any time one signal alters another signal, it is performing some kind of modulation.

FM and AM Devices and Circuits

In the past, a wide variety (hundreds) of circuits and devices performed the operations of modulation and demodulation. Although those circuits and many of the devices still exist, they have all been replaced by a handful of analog integrated circuits. These new IC devices are used to modulate signals for transmission over any kind of medium: wires, cables, light beams, fiber-optic cables, and of course radio waves and microwaves.

The **analog multiplier** integrated circuit can perform all of the operations involved in amplitude modulation and demodulation, including all of the variations, with hardly any external components. Complete AM transmitter and receiver integrated circuits are available, which include on-board analog multipliers. We will study the analog multiplier, and the circuits in which it is used.

The **phase-locked loop** integrated circuit is the mainstay of FM systems and is capable of performing nearly any FM-related task. The varactor diode is also important to FM systems. We will study the PLL and varactor diode, as well as the circuits in which they are used.

12.2   Analog Signal Sampling

Sampling a signal involves taking enough sample segments out of a waveform to be able to reconstruct it later. Sampling is a necessary prerequisite to storing the waveform in analog memory, or to converting the analog signal into a digital signal. Figure 12-1 shows a sampled sine wave.

If the waveform is sampled frequently enough, the gaps can be filled in by a simple filter, in much the same way that gaps in a full-wave power-supply pulse are filled in by a simple capacitor filter. More elaborate filter circuits may be required, depending on how much distortion is acceptable. Because compact disk players must reconstruct the original analog music signals from samples, you can appreciate how nearly perfectly a sampled waveform can be reconstructed.

FIGURE 12-1   **Sampling a Waveform**

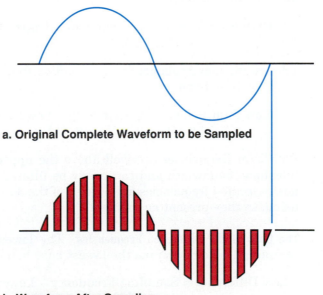

a. Original Complete Waveform to be Sampled

b. Waveform After Sampling
Note:
The sampling rate (frequency) must be at least twice the frequency of the waveform being sampled.

Nyquist Sampling Rate

The higher the sampling rate, the easier it is to reconstruct the sampled waveform and the less distortion we can expect. There is, however, an absolute minimum sampling rate called the **Nyquist criterion.** According to Nyquist, the sampling frequency (rate) must be at least twice the highest frequency to be sampled.

The process of sampling a waveform is a form of amplitude modulation, which we will examine in detail later in the chapter. One of the results of any amplitude modulation process is the creation of two new frequencies (called **side frequencies**). One of the new frequencies is the arithmetic sum of the two frequencies involved, and the other is the arithmetic difference between the two frequencies involved. In this case, one frequency involved is the frequency of the waveform to be sampled, and the other is the sampling frequency (sampling rate).

Figure 12-2 illustrates why the sampling rate (frequency) must be at least twice the highest frequency to be sampled. The problem is called **foldover distortion,** or more commonly, **aliasing distortion.** Let's examine Figure 12-2 to see what happens.

Suppose that we are working with a standard telephone-system bandwidth that extends from 300 Hz to 3 kHz. We will call this the **base band.**

When the Sampling Rate Is High Enough

For Figure 12-2(a), we will assume a sampling rate of 6 kHz—exactly twice the highest base-band frequency of 3000 Hz.

The Case of the Sum Frequencies   The highest sum frequency is the sampling frequency plus the highest frequency in the base band:

High Sum = Sampling Frequency + Uppermost Base-band Frequency
= 6 kHz + 3 kHz = 9 kHz

The lowest sum frequency is the sampling frequency plus the lowest frequency in the base band:

Low Sum = Sampling Frequency + Lowermost Base-band Frequency
= 6 kHz + 300 Hz = 6.3 kHz

Both sum frequencies are well above the upper limit (3 kHz) of the normal telephone bandwidth and can easily be filtered out, if necessary. Both of the newly created frequencies are outside of the 300- to 3000-Hz pass band we are using, so they present no problem.

The Case of the Difference Frequencies   The lowest difference frequency is the sampling frequency minus the lowest base-band frequency:

Low Difference = Sampling Frequency – Lowermost Base-band Frequency
= 6 Hz – 300 Hz = 5.7 kHz

This 5.7-kHz frequency is well above the upper base-band frequency of 3 kHz, so it lies outside the pass band we are using and causes no problem.

FIGURE 12-2   Aliasing (Fold-over) Distortion Example

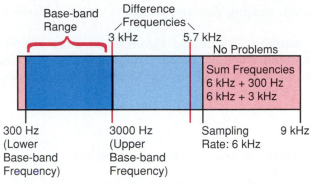

a. What Happens with a 6 kHz Sampling Rate (Frequency)

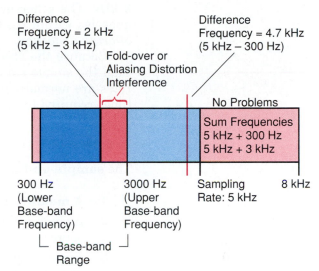

b. What Happens with a 5 kHz Sampling Rate (Frequency)

The upper difference frequency is the sampling frequency minus the highest base-band frequency:

High Difference = Sampling Frequency – Uppermost Base-band Frequency
= 6 kHz – 3 kHz = 3 kHz

This frequency is the same as the upper base-band frequency of 3 kHz, so it is at the upper edge of the 300- to 3000-Hz telephone pass band. In a perfect world, there would be no problem here. In the real world, a somewhat higher sampling rate is a good idea. Using a sampling rate of four times the highest frequency to be sampled has become a common practice. Because real audio does not consist of simple sine waves, harmonics are always present. A higher-than-minimum sampling rate greatly reduces the likelihood of harmonics' folding over into the normal pass band.

When the Sampling Rate Is Less Than Twice the Upper Base-band Frequency

If you look at Figure 12-2(a) and (b), you can see that the sum frequencies are always outside the 300-Hz to 3000-Hz telephone pass band, so they do not interfere with speech information within the band.

Figure 12-2(b) illustrates a problem that can arise with the difference frequency when the sampling rate is too low. In this example, the sampling frequency is 5 kHz, which is less than twice the highest base-band frequency.

The highest base-band frequency is 3 kHz. The sampling frequency is 5 kHz. The difference frequency is the sampling frequency minus the upper base-band frequency:

Difference Frequency = Sampling Frequency
– Uppermost Base-band Frequency
= 5 kHz – 3 kHz = 2 kHz

The 2-kHz difference frequency falls within the base-band range of 300 Hz to 3 kHz. The difference frequency thus interferes with legitimate speech frequencies within the base band, causing distortion. This is called *foldover* or *aliasing distortion.*

Because the speech (or music) in a sampled signal is far more complex than the simple sine waves in our example, higher-than-minimum sampling rates are normally used. In addition, special anti-aliasing circuits are sometimes required to take care of harmonics problems.

Sample-and-Hold Amplifiers

The **sample-and-hold amplifier** consists of five basic elements:

1 A **memory capacitor** (hold capacitor) for storing the sample.
2 An input amplifier to isolate the memory capacitor from the signal source.
3 A very high-impedance field-effect device switch to allow the sample voltage into the capacitor or to disconnect the input amplifier from the capacitor (during the hold time).
4 A very high-input impedance amplifier to prevent the circuit load from draining off the memory capacitor charge.
5 A **sample/hold control signal** to control the sample/hold (FET) switch.

Figure 12-3 shows a simplified schematic diagram of the sample-and-hold amplifier, showing all of its essential components. Two voltage-follower circuits are used because of the high input impedance required. FET switches come in numerous circuit variations, including the analog transmission gate. Two FET switches are shown in Figure 12-3, but the second one (FET-2) is required only when the op-amp output must return to zero between samples. This is sometimes used in systems where a second sampled signal fills the time slot between the first signal's samples.

It is uncommon to build a sample-and-hold amplifier from individual parts, since some excellent integrated-circuit versions are available for nearly any sample-and-hold task. The LF-398, shown in Figure 12-4, is a good example.

LF-398 Sample-and-Hold Integrated-circuit Amplifier

The LF-398 has one very useful feature: a comparator that can be used to control the sample/hold switch. The comparator can be set up to work with any appropriate reference voltage. This means that nearly any voltage or any waveform input signal can be used to control the sample/hold switch.

A standard 5-volt digital pulse is often used to control the sample/hold switch and the sampling rate. The LF-398 can accommodate almost any analog voltage as well. By reversing the logic-input and the logic-reference connections, we can give the input control voltage either polarity or digital phase. In short, almost any kind of signal can be used to control the sample/hold rate.

FIGURE 12-3 Simplified Sample-and-Hold Amplifier

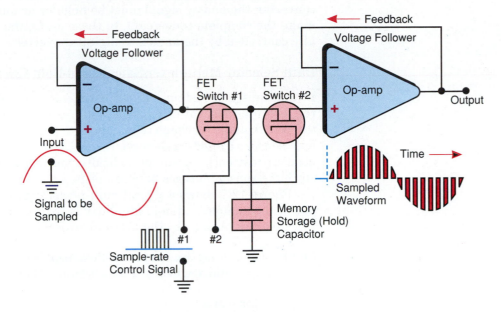

FIGURE 12-4 LF-398 Sample-and-Hold Amplifier

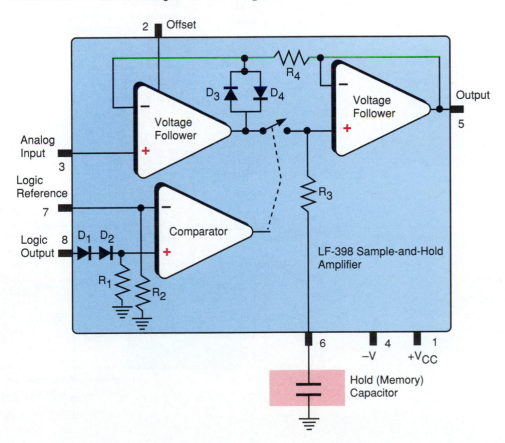

When a sample-and-hold amplifier is used as part of an analog-to-digital converter, the analog signal must be held for as long as the converter takes to make the complete conversion. In these cases, the sample/hold switch is usually controlled by the analog-to-digital converter.

Data Selector Multiplexer/Sample-and-Hold Circuit

Data acquisition systems often must monitor a number of different sensors at regular intervals. For example, a system may be required to sound an alarm if the peak voltage (perhaps representing a processing temperature) from any of four sensors exceeds a dangerous level. A data selector (multiplexer) can be used to test each sensor periodically, and a sample-and-hold amplifier can store the sensor voltages between tests.

Figure 12-5 shows a typical multiplexer/sample-and-hold system. The output of the sample-and-hold amplifier is fed to a block labeled "Analog Signal Processor" or "Analog-to-Digital Converter." In the preceding example, a single

FIGURE 12-5 **Using One Sample-and-Hold Amplifier to Sample Several Analog Signals, with the Help of a Data Selector (Multiplexer)**

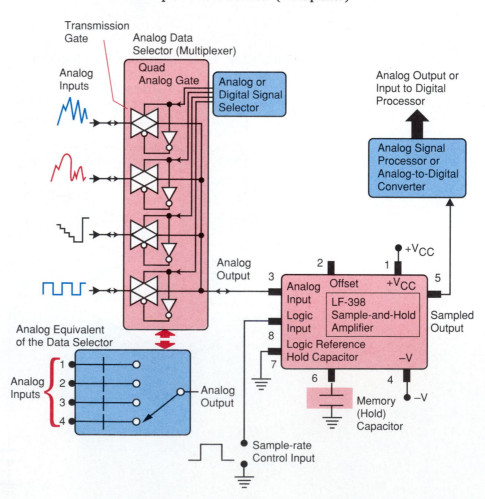

comparator can be used to sound an alarm if the output voltage of the sample-and-hold amplifier ever exceeds the comparator reference voltage (indicating an over-temperature condition at one of the stations). In many modern data acquisition systems, the output of the sample-and-hold amplifier is converted into digital data, and further processing is accomplished digitally.

12.3 Large-scale Analog Memory Systems

The **bucket-brigade** charge transfer device illustrated in Figure 12-6 represents the first and simplest kind of analog memory system that can remember an entire waveform. There are now some incredibly complex (for analog ICs) and powerful **charge-coupled devices (CCDs).** Commercial devices range from the early SAD-1024 to a single-chip speech record-and-playback device to the CCD chips that convert light-image values into the electrical signals used in videocameras.

Bucket-brigade Serial Analog Delay (SAD)

The primary use for the **serial analog delay (SAD)** is to impose audio signal delays, for reverberation and other special audio effects. The rate of the two-phase clock controls how fast the waveform is shifted through and thus determines the amount of signal delay.

How It Works

Figure 12-6 offers a simplified explanation of the operating mechanism of the bucket-brigade charge transfer device. The process works this way:

1 At the instant switch SW_1 closes, switches SW_3 and SW_5 also close; this is called **phase 1.** During phase 1, the charge from memory capacitor C_4 is transferred to C_5. Simultaneously, the charge from C_2 is transferred to C_3, and a new input sample charges C_1.

FIGURE 12-6 **How the Analog Shift Register Works**

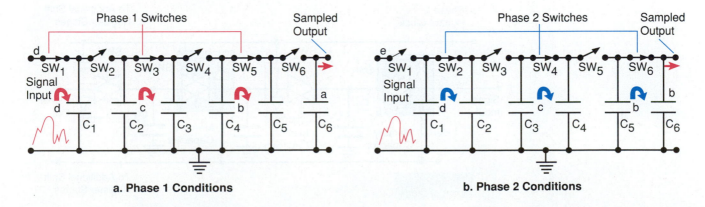

a. Phase 1 Conditions

b. Phase 2 Conditions

2 The phase 1 switches all open, and the phase 2 switches (SW$_2$, SW$_4$, and SW$_6$) closed. The signal voltage sample stored in capacitor C_1 during phase 1 is now transferred to the right, and loaded into C_2.

3 Each previously stored sample voltage is transferred one capacitor to the right. This operation can continue until the entire waveform resides in the device as sample voltages occupying all of the capacitors.

4 Once stored as samples in the analog shift register, the waveform can hold for a second or so, until capacitor charges begin to leak off. Because of the limited storage time possible, the devices are called *analog delay devices* instead of *analog memory devices*.

Figure 12-7 shows a possible electronic implementation of the SAD, using transmission-gate switches. The SAD-1024 and similar devices use simple N-MOS field-effect switches, but the trend is toward the kind of high-speed C-MOS technology originally developed for digital systems. Video applications have become very important, and these require a very high density of memory capacitors and a much faster response than do audio applications. Remember, any reverse-biased silicon junction diode is a capacitor, with the *n*-doped and *p*-doped spots representing the plates, and the depletion zone serving as a dielectric. Millions of these tiny silicon capacitors can be placed on a single chip.

SAD-1024A Analog Delay

The SAD-1024A analog-delay integrated circuit is so named because it has 1024 memory capacitors and support circuits. The memory is divided into two 512-cell sections, and the two sections can be connected in series to take advantage of all 1024 cells for better resolution. Each section has its own two-phase clock, so the shift rate can be different in the two sections. Some strange audio effects can be obtained when the two sections are connected in series with different clock rates. Normally, though, the same two-phase clock signals are used for both, if they are connected in series.

FIGURE 12-7 **Simplified Transmission-gate Version of the Analog Shift Register**

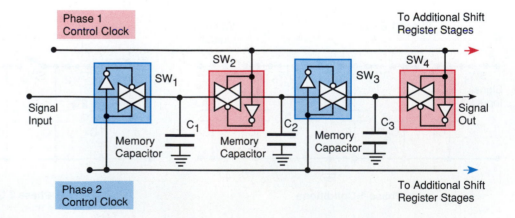

FIGURE 12-8 **SAD-1024A Analog Shift Register Memory**

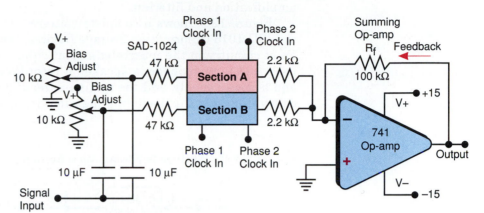

As shown in Figure 12-8, the parallel connection often uses different clock rates to provide two sounds with different delays, to simulate reverberation in a hall. The two bias adjustments are necessary to center the input waveform on the operating curve. There must be such a bias network on each section, even if they are connected in series. A coupling capacitor is used when the two sections are connected in series. The summing op-amp adds the two output signals to combine the two delays.

Clock frequencies of 3 kHz to 100 kHz can be used. Aliasing (foldover) distortion is sometimes induced on purpose, to obtain unusual vocal effects. But because the SAD-1024A is a bit critical, it does not make a good toy for the impatient novice.

ISD-1016 C-MOS Voice-messaging System Integrated Circuit

The ISD-1016 is essentially a "tape recorder" on a chip, but it uses no tape and has no moving parts. The ISD-1016 can record up to 16 seconds of speech, and play it back as many times as you like. The recorded message will last for up to ten years without having any power applied. Rewind is instantaneous, and you can replace an old message with a new one if you like. The chips may be cascaded into a string for longer messages. The voice quality is very good.

Again, research into digital memory has helped analog-systems designers find a required kind of electrically controlled read/write memory with long storage time and analog capabilities. The memory cell is a spin-off of the EEP-ROM read-only memory (ROM) used in computers to store permanent data. This kind of memory cell is too slow to be used in the lightning-fast main computer memory, but improved versions are now fast enough to be used as an audio analog storage cell.

The ISD-1016 has a dynamic automatic gain control (AGC) circuit that accounts for various sound levels and microphone distances, and it has a built-in audio amplifier.

The ISD-1016 is a very efficient approach when measured against the common digital method of audio storage. A similar digital system would require an 8- to 12-bit analog-to-digital converter, about 8 kilobytes per second of memory, an 8- to 12-bit digital-to-analog converter, and some additional amplification and filtering.

Figure 12-9 shows a complete voice-recording/playback system based on the ISD-1016 (Information Storage Devices, Inc., of Austin, Texas). There is a lot of circuitry in the integrated-circuit chip, but not much is required of the user. Add a microphone, a speaker, a record/playback switch, a start button, and a power supply, and the system is ready to go. Most of the required circuitry is on the chip, including the sampling clock. The heart of the system is

FIGURE 12-9 **ISD-1016 Solid-state Voice Recorder System**

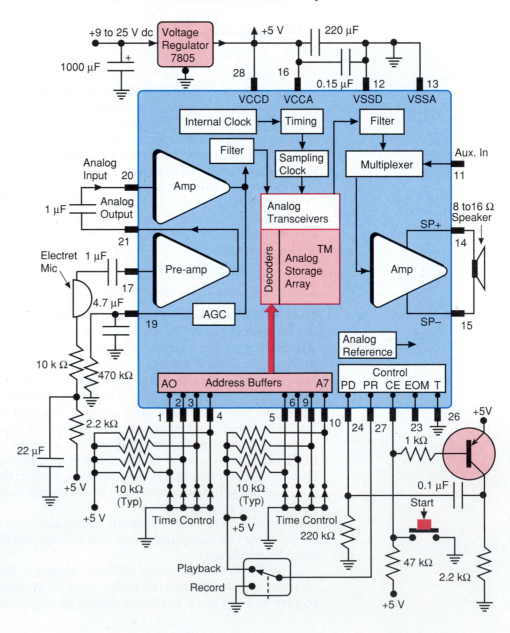

the memory array, called the *Analog Storage Array* (a registered trademark of Information Storage Devices, Inc.). The basic cell is still a capacitor—but a rather sophisticated, no-leak version.

This device is easy to use, and it is an experimenter's dream. Eight time-control switches are shown in Figure 12-9. These switches and their associated resistors allow you to arrange the time segments in any pattern consisting of 160 100-millisecond segments. Ordinarily, these switches wouldn't be there; instead, the time segments would be hard-wired on the PCB. If you ground all eight address buffer inputs, you will get one 16-second record/playback segment. The transistor acts as an automatic reset circuit. When more than one chip is used, the second chip provides the reset pulse.

Charge-coupled Device (CCD) Video Camera Element

Figure 12-10 shows a simplified diagram to illustrate how a modern CCD image element works. A number of different cell types are available, but all are based on some form of charge storage in an integrated-circuit capacitor. The imaging cell in Figure 12-10(a) is the most common kind of cell used in video cameras.

How It Works

The FET device gate is biased slightly positive to attract any electrons freed by heat or incident light. The electron–hole pairs generated by heat constitute an unwanted background noise and can sometimes be a problem.

Electron–hole pairs are generated by light energy that penetrates the substrate. Freed electrons are pulled up to the underside of the silicon dioxide insulator below the gate. Electrons accumulate in the charge well between the

FIGURE 12-10   CCD Video Camera Imaging Element

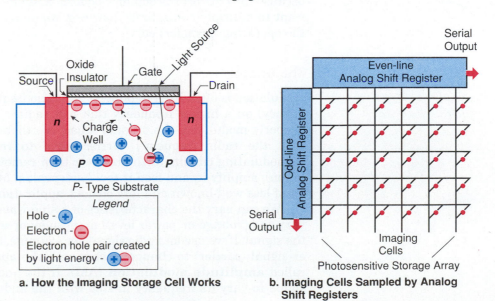

a. How the Imaging Storage Cell Works

b. Imaging Cells Sampled by Analog Shift Registers

source and drain wells. The number of electrons that accumulate in the charge well is a function of light intensity and exposure time.

The source–drain circuit is periodically sampled by a pair of analog shift-register scanning circuits and is shifted out of the registers one cell-charge value after another (see Figure 12-10(b)). The cells are integrated onto the chip in a matrix connected to appropriate shift registers. In the standard video camera, the odd and even lines are scanned alternately to produce interlaced scanning, which minimizes video flicker.

12.4   Introduction to Analog Multipliers

A device called an **analog multiplier,** once used exclusively in analog computers, now fills many communications needs—particularly modulation and demodulation. Standard multiplier integrated circuits are still available, but most of the ones used in communications systems are scaled-down, modified, or less accurate versions called *balanced modulators mixers,* and other names to indicate the intended IC application.

Modulation is the process of impressing information, in the form of an electrical signal, on a second signal called the **carrier**. Stated another way, when one signal waveform alters the shape of another, some form of modulation is taking place. New frequencies are always created by the modulation process.

Linear mixing is, mathematically, an additive operation, producing a voltage sum. Nonlinear mixing is multiplicative and results in a mathematical product of the two signals.

Actually, all electronic devices are nonlinear along part of their operating (transfer) curve. Nature doesn't seem to have much interest in linear things. We usually go to considerable effort to avoid those parts of the operating curve, because they distort the signal waveform and alter the harmonic content. If we want to multiply two signals, however, we do not avoid the nonlinear part of the operating (transfer) curve.

Modulation

Modulation is often used to impress an audio frequency (or some other low frequency) on a higher frequency that can radiate into space as a radio wave. Properly modulated, the original music or other information is carried along with the radio frequency carrier wave to receivers many miles away. Demodulating circuits at the receiver then remove the original music (modulation), amplify it, and feed it to a loudspeaker. Modulation is also useful in a lot of less well-known and less obvious applications.

We can vary the characteristics of a sine-wave signal in two ways: We can vary the (voltage or power-level) amplitude; or we can vary the frequency of the signal. If we use an audio signal, for example, to cause the voltage of another signal (carrier) to change in step with the audio signal, this procedure is called **amplitude modulation (AM).** If the audio signal causes the carrier signal to vary its frequency in step with the audio variations, it is called **fre-**

quency modulation (FM). Figure 12-11 compares amplitude modulation, frequency modulation, and the process of linear mixing (which is not a form of modulation). We will look at amplitude modulation and multiplication circuits first.

Analog Multiplication and Amplitude Modulation

Amplitude modulation, heterodyne frequency shifting, and heterodyne mixing are all multiplication processes. A modified multiplier circuit is also used to form a true waveform-to-RMS converter. *RMS* stands for *root mean square,* which involves the mathematical process of squaring a waveform, finding the mean (average) value of the squared waveform, then taking the square root of the mean. A true RMS converter can use analog multiplier circuits to convert any common waveform into a correct RMS value. In most cases, however, the multiplication operation is of greater interest for its final effect on frequencies and waveforms than for its numerical results.

Until recently, it was not common to think of amplitude modulation as a multiplication process, because various modulator circuits existed that had little in common from a hardware viewpoint. In reality, they were all multiplier circuits, in the sense that they all performed the same multiplication process in approximately the same way. After we examine the mathematics of analog multiplication, which applies to all multiplier devices, we will concentrate on the specialized communications integrated-circuit versions. The original four-quadrant industrial multiplier is still used, but it is a precision device and is not as common as the modified and scaled-down versions.

Multiplication by Adding Logarithms

As it happens, the full operating curve of any diode or bipolar transistor traces an almost perfect logarithmic curve, if we don't leave out the curved part near the junction voltage. We usually avoid using that part of the curve, however; remember crossover distortion?

If we allow a signal voltage to pass through a diode or a transistor, using a big enough range of the logarithmic curve, the signal output will be the log of the input signal. We know from basic math that adding logarithms is a way of performing multiplication. An archaic calculating machine, called a *slide rule,* multiplied by adding logs; and we add decibels and other such units similarly. A multiplier converts the multiplier and multiplicand signals into their electrical logarithmic signals and then adds them.

In linear amplifier circuits, we often go to considerable trouble to avoid the new frequencies and the waveform distortion that result from accidental (and unwanted) waveform multiplication. Analog circuits can be either linear or nonlinear, and we are sometimes careless about our use of the two terms. Analog circuits are not necessarily linear.

There are several practical reasons for multiplying two sine waves. The first application involves shifting a frequency. For example, a signal received by a satellite dish is at such a very high frequency that it would be very expensive to amplify and demodulate. The satellite frequency is also much too high

FIGURE 12-11 Linear Mixing, Amplitude Modulation, and Frequency Modulation

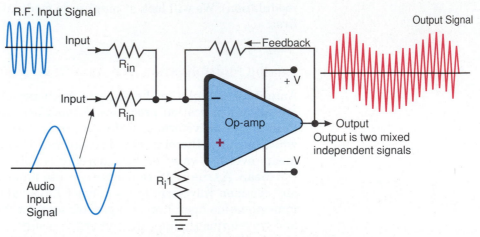

a. Linear Mixing of Two Sine Waves

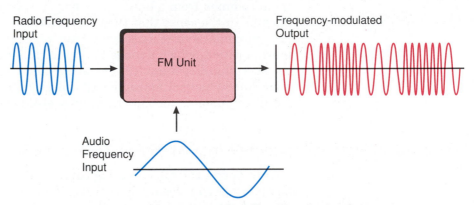

b. Frequency Modulation of One Sine Wave by Another

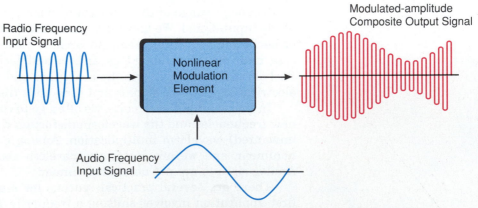

c. Amplitude Modulation of One Sine Wave by Another

a frequency for an ordinary television receiver to handle. As soon as the dish antenna has captured the signal, the frequency is down-shifted to a frequency compatible with standard TV receivers. This process is called **frequency shifting** or **heterodyning.**

Multiplying a fixed carrier frequency by a varying audio frequency produces a kind of amplitude modulation. If both inputs of the multiplier are driven by the same sine-wave frequency, the value of that frequency in the output appears to be twice the value of the input frequency. Most frequency doublers require tuned resonant circuits and work at only one frequency. The multiplier doubler requires no tuned circuits and works at any frequency.

An analog multiplier can be used to do the frequency shifting. Over the years, a variety of different circuits, called **mixers,** have been used to perform the frequency shift or heterodyne operation. The modern version of the mixer is a special integrated-circuit analog multiplier called the **double-balanced mixer.** The analog multiplier has become the universal frequency-management device and is often a critical part of radio circuits-on-a-chip.

When two sine-wave frequencies are multiplied, two products are produced in the multiplier output. The first product is the sum of the two input frequencies $(f_1 + f_2)$, and the second is the difference between the two input frequencies $(f_1 - f_2)$. These two frequencies are called the *upper side frequency* (sum) and the *lower side frequency* (difference).

If you remember the discussion of sampling and the problems with side frequencies and aliasing distortion, you should recognize this as a case of analog multiplication. In fact, there are some significant similarities between the sample-and-hold amplifier and some multiplier circuits.

The Mathematics

You might think that generating the two side frequencies is an addition or subtraction process, but it is not. The proof is based on a fairly complex trigonometric identity that will be presented in a box at the end of this section. If your trigonometry is out of date, you will have to take it on faith that the two sidebands are generated by multiplying two sine-wave frequencies. The equation is

$$V_{out} = 2.5 \text{ V } [\sin 2\pi \,(10{,}000 \, t)] \, [(\sin 2\pi \,(1{,}000t)]$$

Carrier Signal or Frequency to Shift (f_C)	Shifting or Modulating Signal (f_M)

When frequency shifting is required, one sine-wave signal is the signal whose frequency is to be shifted. The second frequency is the shifting frequency, called the *heterodyne* or *beat frequency,* and is generated by an oscillator called a *local oscillator* or **beat-frequency oscillator (BFO).** The term *beat frequency* is often used in tuning musical instruments. When two notes have very nearly the same frequency, the difference side frequency becomes an audible thurum-thurum sound, called a beat. Because only the lower side frequency is normally used for this purpose, the upper side frequency has no special name.

Multiplying Two Sine-wave Frequencies

Since modulation is an expanded case of frequency shifting, we will combine the two concepts in this section.

Example 12.1

Multiplier Calculations

Given the following data and the drawings in Figure 12-12, calculate the multiplier output voltages, and determine the output frequencies.

Solution

Our equation for output voltage is

$$V_{out} = [(V_{CP} \times V_{MP})/S_F] \times [\sin 2\pi(f_C t) \times [\sin 2\pi(f_M t)]$$

where:

V_{out} = Multiplier Output Voltage
S_F = Multiplier Scale Factor
V_{CP} = Peak Voltage of the Carrier (Frequency to Be Shifted)
V_{MP} = Peak Voltage of the Local Oscillator (or Modulating Sine Wave)
f_C = Carrier or Frequency to Be Shifted (in Hertz)
f_M = Frequency of the Shifting or Modulating Signal (in Hertz)

We also have the following values:

Modulating Sine Wave = 1 kHz,   at 10 V *P–P*
Carrier Sine Wave = 10 kHz,   at 10 V *P–P*

Figure 12-12   **Balanced Modulator/Demodulator (Multiplier) System**

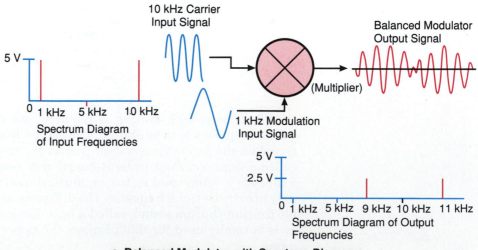

a. Balanced Modulator with Spectrum Diagrams

FIGURE 12-12 Continued

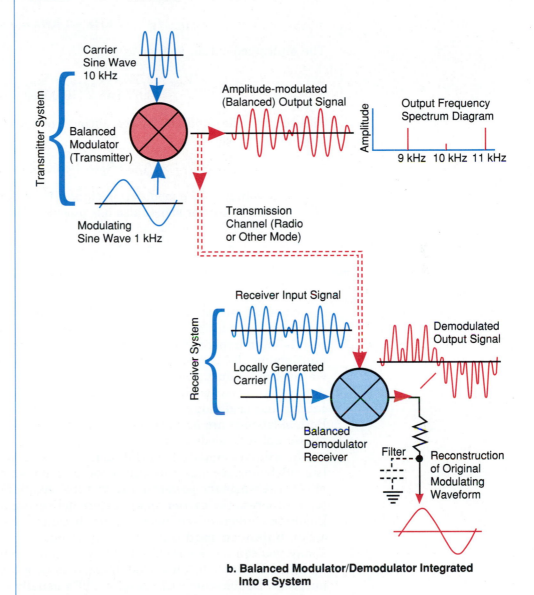

b. Balanced Modulator/Demodulator Integrated
Into a System

where 10 volts peak-to-peak is equivalent to 5 volts peak. The sum frequency is

$$f_C + f_M = 10 \text{ kHz} + 1 \text{ kHz} = 11 \text{ kHz}$$

The difference frequency is

$$f_C - f_M = 10 \text{ kHz} - 1 \text{ kHz} = 9 \text{ kHz}$$

The multiplied output voltage is

$$V_{out} = 2.5\text{V} \times (V_{carrier} \times V_{(modulation)})$$
$$= 2.5\text{V} [\sin 2 (10,000t)] \times [\sin 2 (1000t)]$$

Substituting the equation for the output voltages when two sine waves are multiplied, we get

$$V_{out} = [(V_{CP} \times V_{MP}) / S_F] \times [\sin 2 (f_C t)] \times [\sin 2 (f_M t)]$$

The peak voltage of the modulating signal (V_{MP}) is 10 volts peak-to-peak or 5 volts peak. The carrier signal also has a peak voltage (V_{CP}) of 5 volts peak.
   If the scale factor is 20, we have

$$V_{out} = (V_{CP} \times V_{MP})/S_F$$
$$= (5\text{V} \times 5\text{V})/20 = 1.25\text{V}$$

   When an analog multiplier is used as an amplitude modulator, the carrier frequency is multiplied by the modulating frequency. The multiplication results in an upper sideband and a lower sideband, which are the products of multiplying the two sine waves. When we multiply two numbers together, we get the product. The multiplier and the multiplicand are not part of the product, and that is the way it works in the analog multiplier. The original carrier and modulation signals do not appear in the output of the multiplier. Both sidebands are product terms and do appear in the output of the multiplier/amplitude modulator.

   In ordinary amplitude modulation, the carrier signal exists along with the two sidebands. Because the carrier contains no information, it is often preferable not to waste the power to transmit the energy. Many communications systems suppress the carrier signal before delivering the power to the antenna. Before the integrated-circuit multiplier became available, a number of circuits, called **balanced modulators,** balanced out or canceled the carrier signal. Today, you can buy a special integrated-circuit version of the analog multiplier, called a double-balance modulator, that is optimized for communications purposes. Double-balanced modulator ICs usually have some added circuitry to allow the carrier signal to be added back into the output signal if ordinary AM is desired.

Balanced Modulator System

The original balanced modulator, which consisted of a pair of transformers and a modified diode bridge, was called a **diode ring balanced modulator.** The ring modulator used the diodes as the nonlinear elements to mix the carrier and modulation frequencies. The purpose of the bridgelike diode connection and the two transformers was to balance out the carrier frequency. The circuit added two versions of the carrier that were 180° out of phase with each other, and the carrier frequency vanished from the output of the modulator. Only the two sideband frequencies remained to be amplified and delivered to the transmitting antenna. The output of a balanced modulator contains only the two sidebands; the carrier has been eliminated.

Figure 12-12(a) shows the amplitude-modulated output waveform and a spectrum diagram indicating that there is no carrier frequency. Some AM systems—including the commercial standard broadcast band—do transmit the carrier along with both sidebands, and all three are picked up at the receiving antenna. Ordinary AM is not a very efficient system because the carrier frequency itself does not contain any information (music or what have you). The modulation information is contained in the sidebands. Whatever power we produce to radiate the carrier frequency into space is wasted.

In addition, both sidebands contain exactly the same modulation information, so one of them is redundant and can be eliminated. Radio communications systems can be found that use both sidebands plus the carrier, both sidebands but no carrier, one sideband plus the carrier, or one sideband only.

The system shown in Figure 12-12(b) uses two sidebands and no carrier, which is the normal output from any balanced modulator. Once the modulator has produced the modulated signal, it can be transmitted to some distant receiver. At the receiver, the same balanced modulator circuit serves as a demodulator, with the help of a little filtering to recover the original music or other modulation.

Notice that one of the inputs on the balanced modulator receiver calls for a locally generated carrier. A signal of the same frequency as the original carrier is generated by a beat-frequency oscillator (BFO) in the receiver. This local oscillator is necessary to provide a reference frequency for the demodulator (multiplier). The system shown in Figure 12-12 is a double-sideband, suppressed-carrier system. Sometimes a low-power sample of the carrier is transmitted to synchronize the local BFO, since its frequency is critical to proper demodulation. The sample is called a *vestigial (partial) carrier.* In some receivers, the two sidebands (or one full and one vestigial) can be used to synchronize the BFO. Integrated circuits are available to handle the local carrier-generation chores.

Modern balanced modulator circuits, being multipliers, naturally eliminate the carrier; but most of them have provisions to reinject the carrier or a vestige of it into the output, so that any of the several forms of amplitude modulation can be used.

The transmission channel for modulated signals is often radio, but it may be wire, cable, direct light, laser beam, or a light beam on a fiber-optic cable. Each carrier frequency can be used as a separate transmission channel, just as you use different carrier frequencies to select different TV channels to view different programs. With modulated carriers, a single cable or other medium can carry a number of simultaneous transmission channels. Carrier frequencies can be selected by employing simple filter circuits.

Integrated-Circuit Balanced Modulators

Figure 12-13 shows a simplified schematic diagram of a typical balanced modulator IC. Here is how it works:

1   Transistors Q_5 and Q_6 form a conventional differential amplifier circuit, with dual current sources for emitter (current-mode) feedback.

FIGURE 12-13   **Schematic Diagram of a Typical Balanced Modulator/Demodulator**

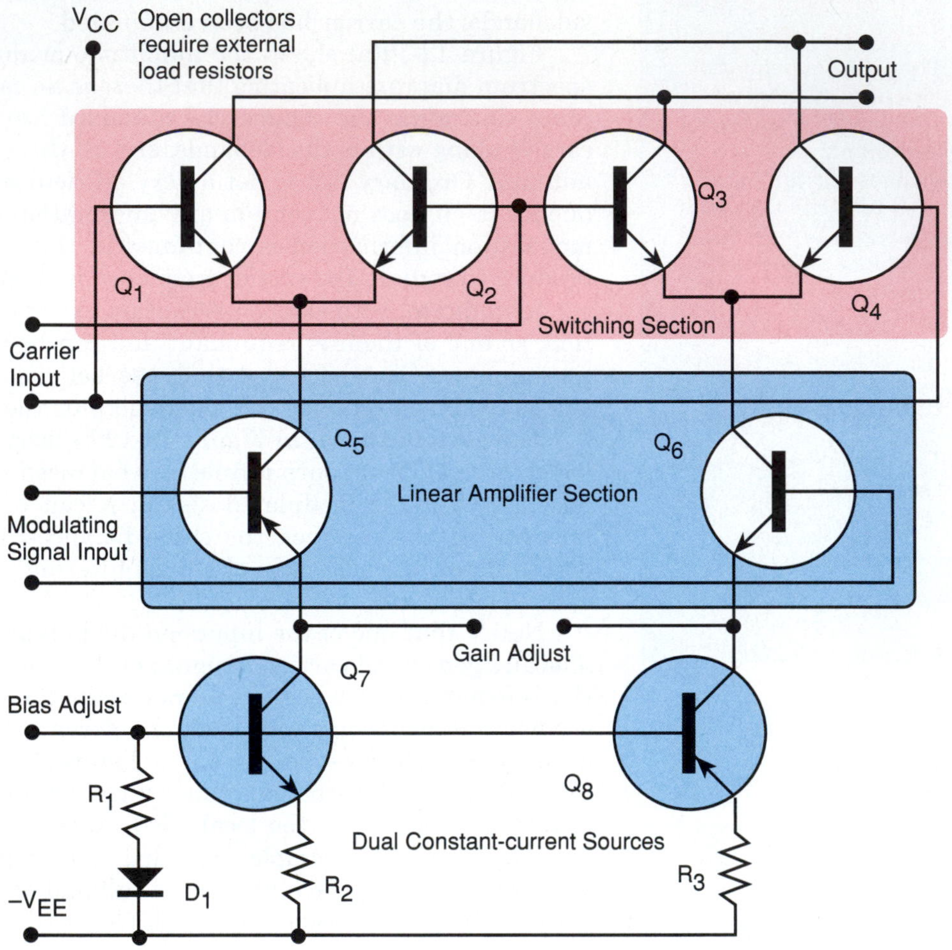

2   A resistor across the gain-adjust terminals provides feedback to set the voltage gain of the differential amplifier, and an emitter-current bias adjustment is provided.

3   The Q_5–Q_6 differential amplifier is operated as a conventional linear amplifier.

4   Transistors Q_1, Q_2, Q_3, and Q_4 form two differential amplifier circuits that have no negative feedback in their emitter circuits. This means they have no feedback compensation for temperature or other variations.

5   Feedback is not required, however, because the transistors are normally operated in a switching mode. The base–emitter junctions are either reverse-biased or fully turned on. The linear range is not used.

6   About 25 mV is needed to switch the base–emitter junctions, so carrier-voltage levels must be between 50 and 200 mV for proper operation.

Usually, the carrier-voltage level must be carefully controlled to ensure proper operation. It may be a little fussy.

7   With a modulating signal applied to the linear amplifier section, the carrier voltage switches the modulating signal on-and-off at the carrier frequency. This is a classical waveform-sampling process.

8   The two switching (sampling) differential amplifiers are cross-coupled to cancel the carrier in the output.

9   The balanced modulator circuit is essentially a waveform-sampling circuit (no hold) with provisions for eliminating the original carrier. A small capacitor can be connected from the carrier input to the output and used to reinject the carrier if a carrier is required.

The Analog Devices AD-630 Balanced Modulator

The Analog Devices company has taken a slightly different approach to the switching (sampling) part of the circuit. The AD-630 in Figure 12-14(a) uses a high-speed comparator to perform the switching. The carrier is applied to one input of the comparator. The advantage of this scheme is the flexibility of the comparator. The reference voltage can be adjusted to suit available carrier levels, which is generally a lot easier than maintaining careful control over the carrier. This is particularly true when the device is used as a demodulator. The AD-630 can be implemented easily for various modulation and demodulation operations—some of them too specialized to cover here.

The two linear amplifiers can be cross-coupled to cancel the carrier, if this is required. The comparator status output is used whenever a carrier signal is required—in a synchronous detector, for example. Circuits connected to the comparator status output do not load the original carrier signal source.

Figure 12-14(b) shows the AD-630 connected as a balanced modulator.

This comparator-switching approach is also used in contemporary transmitters and receivers on-a-chip integrated circuits. Parts (a) and (b) of Figure 12-15 provide block diagrams for typical integrated-circuit transmitter and receiver chips, respectively.

The Transmitter

The transmitter generates a carrier frequency, using a crystal-controlled oscillator. An audio (modulation) amplifier is provided for a low-output voltage microphone or other source. Amplitude modulation takes place in the balanced modulator. The modulated carrier is coupled to a power-amplifier transistor that is capable of 100 to 200 milliwatts of radio frequency (RF) power. An external high-power transistor amplifier can be added if larger power levels are required. The IC low-power transmitter is often called an *exciter* when it is used as a modulator to drive higher-power radio-frequency amplifiers.

A tuned resonant circuit connected to the output transistor eliminates unwanted frequencies and ensures the most efficient use of the RF energy. The output tuned circuit is constructed as a transformer to serve as an impedance-matching transformer between the transmitter and the antenna's transmission-line impedance.

FIGURE 12-14   Analog Devices' AD630 Balanced Modulator

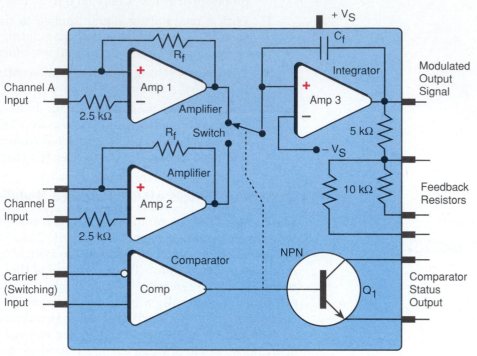

a. AD 630 Functional Block Diagram

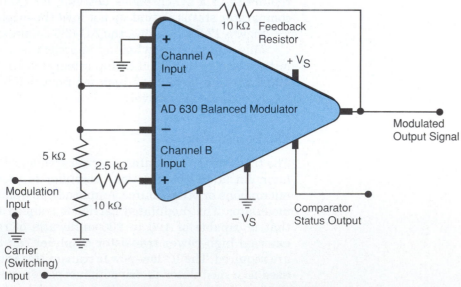

b. AD 630 Balanced Modulator Block Diagram

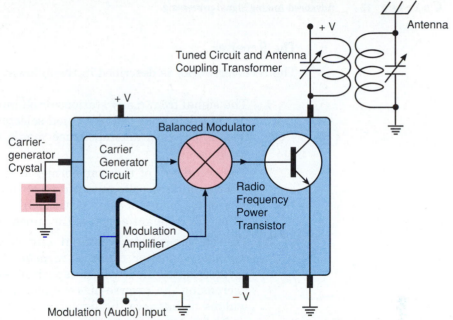

a. Balanced Modulator-based Amplitude-modulated Transmitter

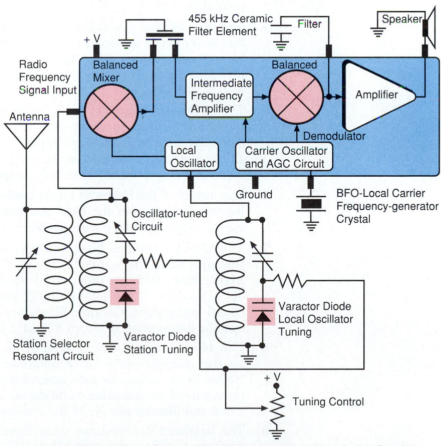

b. Balanced Mixer / Demodulator-based Amplitude Modulation Receiver

The Receiver

The receiver works as described in the following steps:

1   The signal from many stations or channels is picked up by the antenna and delivered to the station-selector resonant circuit. The transformer provides an impedance match to the transmission line and antenna.

2   The secondary of the transformer is a parallel resonant circuit that delivers a large voltage to the receiver input only if it is tuned to the carrier frequency of some station. Other carrier frequencies are rejected by the resonant action of the tuned circuit.

3   The station-selector resonant circuit is tuned by using a voltage-variable capacitor called a *varactor* or *varicap*. The varactor is a reverse-biased diode whose capacitance can be varied by increasing or decreasing the reverse-bias voltage. Varying the reverse-bias voltage widens or narrows the depletion zone. Changing the depletion zone width is equivalent to changing the thickness of the dielectric in an ordinary capacitor. Varactor tuning is the preferred tuning method for higher frequencies, for which suitable varactor capacitance values are available.

4   The local oscillator, which is also tuned by a varactor diode, tracks the frequency of the incoming station-selector frequency. The local oscillator is tuned in such a way that the difference frequency between the incoming radio-signal frequency and the local oscillator frequency is always 455 kHz.

5   This difference frequency is produced in the balanced mixer. The local oscillator is used to down-shift the frequency, from whatever the incoming selected frequency happens to be, to 455 kHz. The frequency-shifting process is called *heterodyning,* so it is only reasonable that a receiver using it should be called a **superheterodyne receiver.**

6   The output of the balanced mixer contains the sum and the difference frequencies between the local oscillator frequency and the incoming signal frequency, along with assorted unwanted frequencies.

7   The 455-kHz ceramic filter is a very high-Q electromechanical filter that acts much like a quartz crystal. The ceramic filter eliminates all frequencies except the difference frequency of 455 kHz. All that comes out of the filter is the 455-kHz down-shifted radio carrier and its associated modulation. Ceramic filters are only available in a handful of specific frequencies, of which 455 kHz is by far the most common.

8   The 455-kHz signal is amplified by an amplifier called an *intermediate-frequency amplifier* or *IF amplifier*. The IF amplifier gain is controlled by an automatic gain control to account for the different signal strengths of the incoming radio signal. The AGC voltage is often a rectified and filtered sample of the audio-signal output voltage.

9   The balanced demodulator must have a carrier frequency signal to perform the demodulation. Most IC receivers have provisions for either using a received carrier (if it is transmitted) or generating a car-

FIGURE 12-16   Some Balanced Modulator/Multiplier Symbols

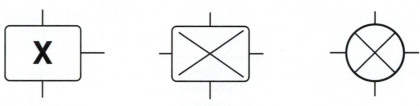

a. Early Analog Multiplier Symbol

b. Later Version of the Analog Multiplier Symbol

c. Round Version – Currently the Most Commonly Used Symbol

rier locally for single-sideband reception. The BFO local carrier oscillator may or may not be used.

10 The balanced demodulator output high-frequency ripple is filtered out, leaving a clean audio signal to deliver to the audio amplifier. Many of the IC receivers available have enough power output to drive a small speaker.

Figure 12-16 attempts to clarify some of the symbols used for multipliers, balanced modulators, and balanced mixers.

12.5   Frequency Modulation Systems

Frequency modulation requires an oscillator to vary its frequency in step with some modulating signal. There is often a conflict between the stability of the oscillator frequency and the ease with which the modulating signal can change the oscillator frequency. We will examine several approaches to frequency modulation.

Frequency-modulating a Sine-wave Carrier

The process of frequency modulation (FM) involves altering the carrier frequency in step with a modulating signal. This implies that the oscillator that generates the carrier must be a voltage-controlled, variable-frequency oscillator. We have already encountered voltage-controlled oscillators in the form of comparator-based square-wave oscillators. We can and do use these VCO square-wave oscillators as modulated carrier generators in FM systems.

At times, however, the carrier oscillator must be a sine-wave oscillator. Sine-wave oscillators are positive-feedback oscillators, whose frequency depends on the value of capacitors, inductors, crystals, and sometimes resistances. In most cases, at higher frequencies, we are restricted to using inductor-capacitor (L-C) or crystal-controlled resonant circuits to determine the oscillator frequency.

We must have either a voltage-variable inductor or a voltage-variable capacitor to modulate the frequency of an L-C oscillator. Voltage-variable inductors exist, but barely. Voltage-variable capacitors, or varactors, are common and inexpensive.

Figure 12-17 Frequency Modulation

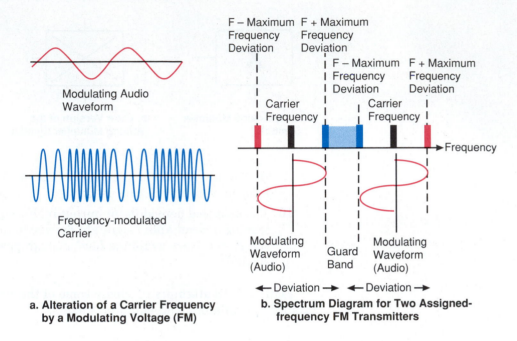

a. Alteration of a Carrier Frequency by a Modulating Voltage (FM)

b. Spectrum Diagram for Two Assigned-frequency FM Transmitters

A second problem with FM systems, if they are radio-wave systems, involves the Federal Communications Commission (FCC). Each commercial transmitter is assigned a resting carrier frequency and an allowable amount of deviation within which to modulate the carrier. The signal must not stray beyond those assigned boundaries. Figure 12-17 illustrates the situation.

You must swing the carrier frequency up and down to modulate it, but you cannot let it vary too far. The guard band is provided in case of accident, to avoid actual interference with the adjacent station. What all this means is that you must have a precise center (rest) frequency for the carrier and good control over the amplitude of the modulating signal.

A stable carrier center frequency nearly always demands a crystal-controlled oscillator, since crystal-controlled oscillators fight to stay on frequency. We must therefore try to vary a frequency that wants to be stable. The conflict is most often resolved by using a crystal-controlled Colpitts oscillator with one or more varactor diodes to modulate the frequency.

This set-up requires a fairly high-frequency carrier, because the crystal-controlled Colpitts can only be forced to change frequency by a very small percentage of the resonant frequency of the crystal. Figure 12-18 shows a crystal-controlled Colpitts oscillator with a varactor diode for frequency modulation.

Voltage-controlled Oscillator as a Frequency Modulator

In systems that have a less critical need for a precise center frequency than FCC-controlled commercial radio does, voltage-controlled oscillators make excellent frequency-modulation devices. The fact that these comparator-based

FIGURE 12-18 **Varactor Diode–modulated Colpitts Oscillator**

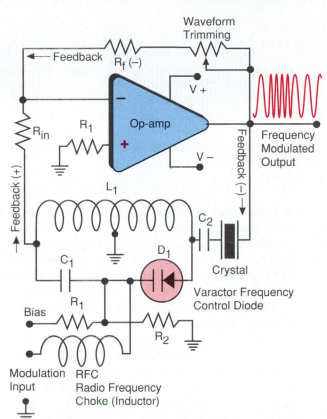

oscillators produce a square or rectangular output waveform is not much of a problem, since the square-wave frequency can vary in step with a modulating signal and since waveshape is not very important in FM systems.

A square wave does carry many odd harmonics, and all the extra frequencies can be a problem in some situations. A square wave should not be radiated into space as a radio signal, for example, because all of the harmonics represent real signals with real energy content. Computers use square-wave signals and must be shielded to prevent those many harmonic frequencies from radiating and causing interference with various communications frequencies. Those open slots in the back of the computer must be filled with an add-on card or a metal blank to prevent leakage of the higher harmonic frequencies.

Square-wave carriers can be used in closed cables, light-beam transmitters, fiber optics, and other places where the harmonics cause no problems. If necessary, an FM square-wave carrier can easily be converted into an FM sine-wave carrier. All that is required is a tuned resonant circuit, tuned to the fundamental frequency of the square-wave carrier.

Most FM receivers use a clipping amplifier, called a *limiter,* to convert the received sine-wave carrier into a nearly square wave. The result is that noise (most of which is a form of amplitude modulation) gets clipped off the top of the waveform. Figure 12-19 illustrates why it is much easier to have noise-free

FIGURE 12-19   **Electrical Noise in FM and AM Systems Compared**

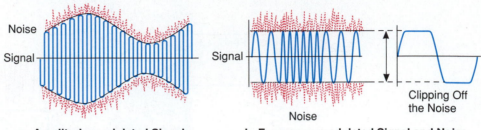

a. Amplitude-modulated Signal and Noise

b. Frequency-modulated Signal and Noise

reception in FM systems than in AM systems. Better noise immunity is FM's main claim to fame.

We have already studied comparator-based voltage-controlled oscillators (VCOs), and we need only apply a modulating signal to the control pin of a VCO to have an FM oscillator. Even the 555 timer can serve as a respectable (if limited) FM oscillator if we apply a modulating signal to its control pin (pin 5). The VCO is a natural FM modulator and can be the primary element in a number of specialty FM transmitters.

Many wireless intercoms or remote stereo speaker systems use the **ac** power line as a transmission cable for an FM transmitter/receiver system. It is possible to establish several communications channels on the **ac** power line by frequency-modulating several different carrier frequencies. Communications over the **ac** power lines is called *carrier current communications* and has been used for more than 50 years by power companies over quite long distances. The railroads use the rails as a radio frequency transmission line. Such transmission lines radiate very little energy into space when properly terminated in their characteristic impedance. The LED transmitting–receiving system used in TV remote controls is strictly digital; but infrared FM transmitters are used to transmit analog TV sound to headphone receivers. The FM transmitters and receivers use the same basic circuits, whatever the channel medium:

a   Radiation into space (radio).

b   Transmission cables—coaxial cable, twin lead, twisted pair, etc.

c   Exotic transmission lines—power lines, railroad tracks, fencing, etc.

d   Light beam—direct-beam LED or laser, fiber-optic cable, etc.

Characteristic Impedance

FLASHBACK

When we talk about the 300-ohm twin lead for a TV antenna or a 72-ohm coaxial cable, we are talking about the **characteristic impedance** of a transmission line. If your **ac** electronics text covered this topic, the present subsection's coverage will be a flashback for you. Some **ac** texts don't cover it, so it may be new to you.

The concept of characteristic impedance is important in communications systems and even in digital systems, where a disk-drive cable acts as a transmission line. Every transmission line has a characteristic impedance at the frequency at which it is being used. Impedance mismatches between the cable and the equipment to which it is connected cause a power loss. More important than the simple power loss is the problem of signals being reflected back down the transmission line if the line is not terminated in a matching impedance. Signals that are reflected back can occur in a variety of phases and can cause an assortment of problems, such as lost power, TV ghosts, and lost data in digital systems.

We won't do a real mathematical analysis here, just some simple arithmetic to demonstrate the concept. Figure 12-20(a) shows the equivalent circuit for a length of cable. Let's call the length 1 inch.

Inductance and Capacitance

Any time current flows, a small magnetic field is created, and energy is stored there. If the current changes, the wire exhibits some inductance, no matter

FIGURE 12-20 Characteristic Impedance

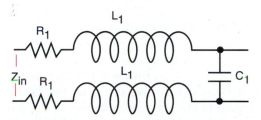

a. Equivalent Circuit for 1 Inch of Example Cable

$Z_L = 1\,\Omega$

$X_C = Z_1$ through $Z_n = 10\,k\Omega$

b. Simplified Equivalent Circuit for Each Cable Inch

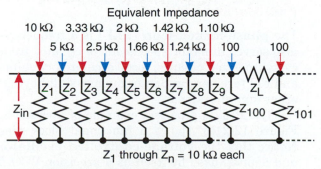

Z_1 through $Z_n = 10\,k\Omega$ each

c. Effects of Adding Additional Lengths of Cable (Simplified Model)

how short it is. In high-frequency circuits, the inductance in even a short piece of wire can be significant.

Because a transmission cable consists of two wires with insulation between them, the cable has capacitance. So the equivalent circuit in Figure 12-20(a) is a close approximation of the actual cable.

Because the inductance and the capacitance are spread out along the cable, they are called *distributed inductance* and *distributed capacitance*. The circuit, as we have drawn it in Figure 12-20(a), is called a *lumped constant* equivalent of distributed impedances. This description will be extremely simplified, so for further study you may want to find a book that covers it in greater detail.

In Figure 12-20(b), the cable circuit has been simplified into two equivalent resistances, representing the inductive and capacitive reactances at some particular frequency. Z_1 is the parallel impedance (capacitive reactance) of the first inch of cable; Z_2 is the parallel impedance of the second inch; and so on, through Z_{101}. Each inch of cable has a parallel impedance of 10 kilohms. The series resistance/inductance value is not shown in the first few inches, because it doesn't become a factor until the cable gets a little longer.

Starting with inch number 1, we have a Z_1 of 10 kΩ. Adding the second inch (Z_2) places a 10 kΩ impedance in parallel with Z_1—or 10 kΩ in parallel with 10 kΩ—for an equivalent 5 kΩ. Thus, adding the second inch reduces the total impedance by 50%. Adding the third inch reduces the total impedance to 3.33 kΩ, a 33% reduction.

Each added inch decreases the total impedance, but each successive added inch has a smaller effect on the total than the preceding inch. By the time we get to inch 100, adding another inch decreases the total impedance by only 1%.

Somewhere along the line, the series impedance Z_L becomes significant relative to the total impedance. In Figure 12-20(c), the impedance Z_{101} would reduce the total impedance to 99 ohms, but the series impedance Z_L of 1 ohm brings that total back up to 100 ohms.

In any cable, at a particular frequency, a point is reached where adding more inches (or feet) of cable causes no significant change in the total impedance of the cable. This is called the *characteristic impedance* of the cable. The load for a cable should have the same impedance or resistance value as the characteristic impedance of the cable.

Phase-locked Loop

The **phase-locked loop (PLL)** is a frequency-modulation workhorse, but it also serves in motor speed-control circuits, frequency-multiplier and frequency-synthesizer circuits, local oscillator circuits for frequency shifting, in TV and FM receivers, and in frequency-shift keying (FSK) circuits in modems and other digital systems.

The PLL is a perfect example of a closed-loop negative-feedback system. Figure 12-21(b) shows a functional block diagram of a phase-locked loop. Notice that the **voltage-controlled oscillator** (VCO) is very sophisticated and is often used by itself as a precision VCO. The PLL can be used for both FM modulation and demodulation, as well as in a number of other useful applications. Because the phase-locked loop is such a fine example of a closed-loop negative-feedback system, let's relabel the blocks using feedback terminology (see Figure 12-22(b)).

The function of the phase-locked loop is to hold the voltage-controlled oscillator exactly in phase with the input signal (pins 2 and 3).

How It Works

1 The VCO is set to free-run at the same frequency as the input signal.

2 If the phase of the input signal frequency changes slightly, the phase comparator senses the difference between the input and VCO phases

FIGURE 12-21 565 Phase-locked Loop IC

FIGURE 12-22 Phase-locked Loop in Standard and Negative Feedback Terminology

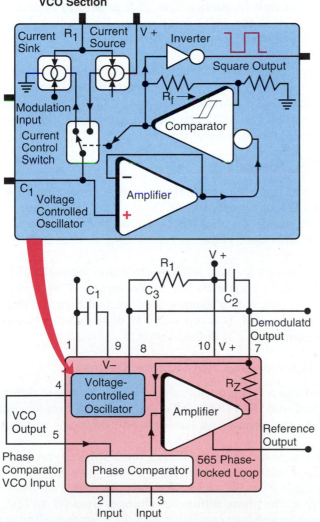

a. Approximate Functional Block Diagram of the VCO Section

b. Functional Block Diagram of the 565 Phase-locked Loop

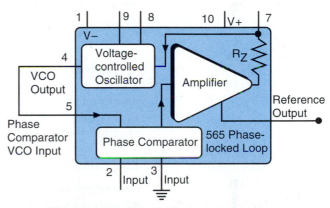

a. PLL Functional Block Diagram in Standard Terminology

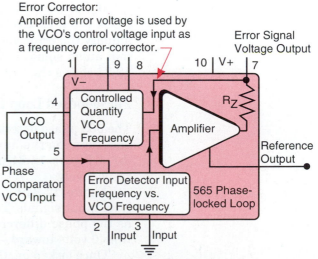

b. PLL Functional Block Diagram in Negative Feedback Terminology

and produces an error voltage. A frequency change must begin with a change in phase, since a phase change can be viewed as a frequency change of less than one cycle. An error signal results even when a small change in phase occurs.

3   The error voltage from the phase comparator is fed to an amplifier. The amplified error voltage is filtered by a low-pass filter (C_2 and R_Z) to produce a **dc** error voltage. The **dc** error voltage is then used to adjust the VCO in the proper direction to bring it back into phase with the input signal.

4   Once the VCO is back in phase with the input signal, the error voltage vanishes and the VCO's phase is not adjusted further.

Like all negative-feedback loops, the PLL corrects for any error (phase difference) between the reference quantity (input signal) and the controlled quantity (VCO frequency). If the input signal frequency changes, the VCO follows the input signal to that new frequency. The PLL is said to be locked on the input signal frequency, and the VCO frequency tracks (follows) and matches the input signal frequency changes (within its locking frequency range).

All this may seem a bit pointless. Why go to the trouble to generate one frequency, in phase, exactly like another frequency we already have? There is no point to it, unless the signal frequency changes. Then, the VCO must follow the signal frequency to its new value. The VCO can change frequency only if an amplified error voltage is applied to its control input.

Phase-locked Loop as an FM Demodulator

Now suppose that we have an input frequency that changes continuously, as we normally would have in a frequency-modulated (FM) signal. A continuously changing input frequency causes the phase detector to produce new error voltage values continuously, with each value corresponding to each new frequency.

If the input frequency changes in step with some audio frequency, the PLL's error voltage will also be in step with the audio signal. If we simply amplify the error voltage and apply it to a loudspeaker, we will hear the original audio signal reproduced. The PLL is a natural FM demodulator. Figure 12-23 shows the PLL connected as an FM demodulator.

Phase-locked Loop Phase Detector Lock and Capture Ranges

The 565 phase detector produces a 0-volt output signal when the phase difference between the input signal and the VCO is 90° (quadrature). Parts (a) and (b) of Figure 12-24 illustrate the circuit and its **lock** and **capture ranges,** respectively.

Figure 12-24(b) shows the phase detector action for low-level and high-level input signals. The output voltage increases from 0 volts toward +0.7 volt as the phase difference changes from 90° to 0°. The output voltage increases from 0 volts toward −0.7 volt as the phase difference changes from 90° to 180°.

Once locked on, the phase-locked loop can follow (track) frequency changes over a range that is much wider than the capture range, as illustrated in Figure 12-24(c). If the input and VCO frequencies differ by more than 180° or

FIGURE 12-23 **Phase-locked Loop Connected as an FM Demodulator**

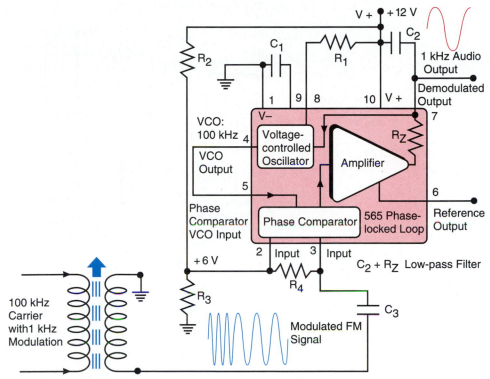

0°, as shown in Figure 12-24(b), the circuit will fall out of lock, and a new signal capture will have to be made. The lock and capture ranges and the VCO frequency can be calculated by using the following three equations (refer to Figure 12-23 for parts identification).

First, the equation for the VCO's free-running frequency f_{out} (in Hz) is

$$f_{out} = 1.2/(4R_1C_1)$$

Note: R_1 and C_1 are the VCO's timing components.

Second, the equation for the lock range f_L (in Hz) is

$$f_L = +(8\,f_{out})/V$$

where f_{out} is the VCO's free-running frequency, and V is the total power supply voltage:

$$V = (+V) - (-V) \text{ volts}$$

Third, the equation for the capture range f_C (in Hz) is

$$f_C = +f_L/[(2)(3.6)(10^3)(C_2)]$$

Note: C_2 is the low-pass filter capacitor.

FIGURE 12-24   **Phase-locked Loop: Lock and Capture Characteristics**

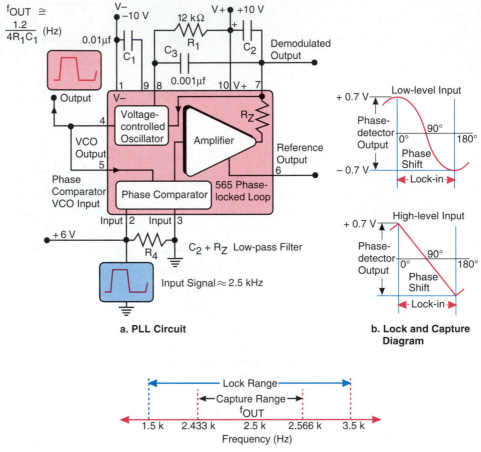

a. PLL Circuit

b. Lock and Capture Diagram

c. Phase Detector Characteristics

Specifications for the 565 Phase-locked Loop

1   Frequency range: 0.001 Hz to 500 kHz.

2   V+ voltage: +6 V to +12 V.

3   V− voltage: ground to −12 V.

4   Phase comparator input impedance (pins 2 and 3): 10 kΩ

5   Triangular-wave amplitude (pin 9): typically 2.4 V *P–P*, with a +6-V, −6-V power supply.

6   Square-wave amplitude (pin 4): typically 5.4 V *P–P*, with a +6-V, −6-V power supply.

7   Bandwidth adjustment range: 1% to slightly less than 60% of the VCO's free-running frequency.

Introduction to a Practical but Simple FM Transmitter System

The simple FM transmitter receiver shown in Figures 12-25 and 12-26 uses a 555 timer as a VCO for the FM transmitter and employs a 565 phase-locked loop as the FM demodulator.

The VCO in a second 565 PLL or a dedicated VCO IC could have been used as the FM modulator; but you are already familiar with the 555 timer, and it is capable of a little more output power than the 565 VCO. Students have built several variations of the circuit in the laboratory, using visible and infrared LEDs and photoreceptors. Minor circuit changes may be necessary to accommodate different photoreceptors, and the op-amp gain may have to be changed to accommodate different microphones or other input signal-source levels.

One student adapted the system as a baby monitor by coupling the transmitter and receiver to the **ac** power line to make a carrier-current system. The transmitter was plugged into the baby's room, while the receiver could be plugged-in in any room where the parents happened to be. Another student used a coax cable and two transmitter/receiver pairs to transmit two selectable audio channels to backyard speakers.

A third student built a fiber-optic demonstrator, but the cost of the cable and the student's limited budget kept the distance down to only a few feet.

Ordinary radio transmission is also possible; but at 35 to 40 kHz, the wavelength is so long that the required antennas are too long to be practical. It is difficult get a range greater than a few feet. Student-created input circuit and output circuit modifications are shown in Figure 12-27.

The point of mentioning these variations is to emphasize that the same basic modulation and demodulation devices can be used to send data—either analog or digital—over a variety of media. Radio is but the most familiar medium to most of us.

FIGURE 12-25   **Light-beam Version of a Simple FM Transmitter**

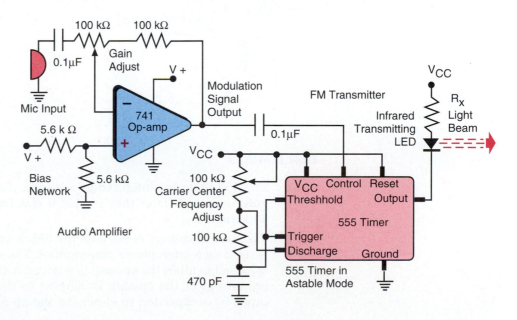

FIGURE 12-26   Light-Beam Version of a Simple FM Receiver

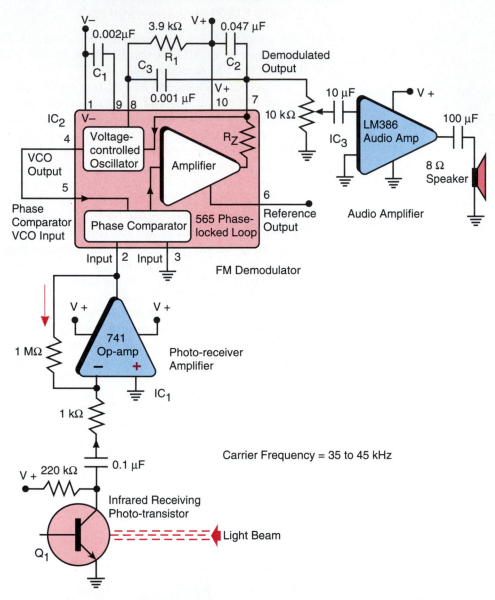

How It Works

1   The 555 timer is configured as a standard stable free-running, rectangular-wave oscillator that runs at a rest frequency of between 35 and 40 kHz.

2   The control voltage in pin 5 on the 555 is connected to an op-amp that is used as a microphone preamplifier. The op-amp noninverting input is biased to allow the op-amp to work on a single power supply. The signal output of the op-amp is coupled to the 555 modulation (control) input by a capacitor, to eliminate the op-amp's output offset voltage.

The offset voltage is about $\frac{1}{2}V+$ because of the single-power-supply op-amp biasing. The audio voltage from the op-amp output varies the reference voltage on the 555's upper comparator, causing the 555's frequency to vary accordingly.

The output circuit in Figure 12-25 consists of a light-emitting diode (visible or infrared) and a current-limiting resistor. Because the LED is not operated continuously, a current of 80 mA or so can be used for a high light-output value. A 35- to 40-kHz tuned circuit can be added to the output circuit to allow two transmitters to share the same light beam while using two different carrier frequencies. Multiple carrier frequencies are commonly used with commercial fiber-optic systems.

Figure 12-27 shows how the transmitter can be modified to use a coaxial cable or the **ac** power lines as the transmission medium. If the transmitting frequency were higher, it would be practical to connect output A on the output transformer to an antenna. The tuned output transformer converts the 555's rectangular waveform into a frequency-modulated sine wave and provides an impedance match to the characteristic impedance of the coaxial cable or power line. A power transistor can easily be added at the 555 output if more power output is required.

Once the FM signal is converted into a sine wave, another tuned circuit at the receiver can select a particular carrier out of several. The two high-voltage capacitors isolate the system from the power lines to prevent the possibility of electrical shock and to keep the **ac** power from damaging the electronic components.

Ordinary TV twin leads, or even speaker wire, can be used instead of the coaxial cable, but with far greater energy loss and noise pick-up.

564 Phase-locked Loop

The intermediate frequency for standard FM broadcast (88 to 108 MHz) is 10.7 MHz. The 565 can't handle that high a frequency. The 564 phase-locked loop, shown in Figure 12-28, is a more sophisticated version that can work at frequencies up to 50 MHz. If you look at the diagram in Figure 12-28, you will notice that the phase-detector block is shown as a balanced modulator and that there is a special digital output. The balanced modulator (multiplier) has become the most common phase-detector circuit in integrated-circuit phase-locked loops.

Balanced Modulator as a Phase Detector

If two sine-wave signals of the same frequency are applied to the inputs of a balanced modulator, the modulator output will consist of a sine wave (of twice the input frequency) and a **dc** component. The voltage value of the **dc** component is proportional to the phase angle between the two input signals. The output voltage is directly proportional to the cosine of the angle between the two frequencies. The circuit in Figure 12-29 uses a balanced modulator as an FM demodulator. The circuit is sometimes called a *product detector,* because the circuit is a multiplier and because the output of a multiplier is a product. The

FIGURE 12-27   **Converting the Receiver to Carrier Current or Coax Transmission**

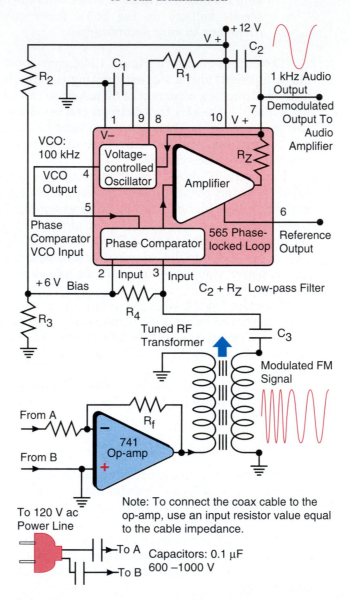

circuit has also been called a *quadrature detector*, but several other circuits that do not use balanced modulators are also called *quadrature detectors*. **Balanced-modulator phase detector** and **product detector** are the preferred terms.

How It Works

1   The quadrature circuit is tuned to produce a 90° phase shift at the carrier rest frequency. A slug-tuned coil or variable capacitance is used to adjust the quadrature circuit. A parallel resistance is often used to

FIGURE 12-28   Functional Block Diagram for the 50-MHz
564 Advanced Phase-locked Loop

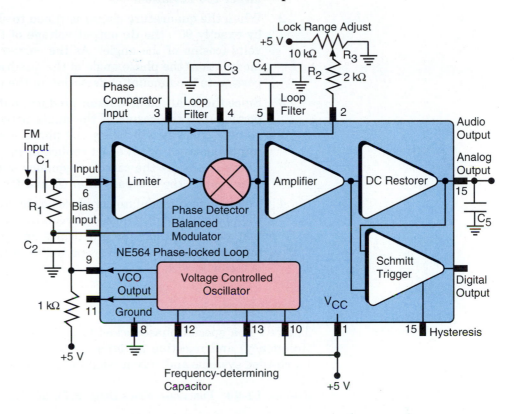

FIGURE 12-29   Balanced Modulator as a Phase
Detector/FM Demodulator

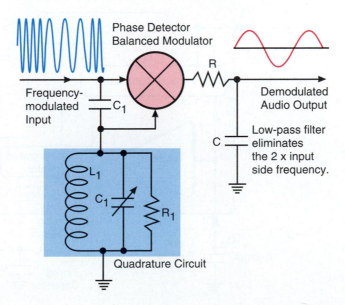

lower the Q of the tuned circuit to widen the band pass, allowing for a larger FM deviation.

2   When the quadrature circuit is tuned to shift the carrier signal phase by exactly 90°, the **dc** output voltage of the balanced modulator is 0 volts (cosine of the angle). As the carrier frequency changes (due to modulation) the phase angle of the quadrature circuit changes in proportion to the amount of deviation in the carrier frequency.

3   Since the balanced modulator produces a **dc** output voltage that is proportional to the cosine of the angle between the two inputs, the **dc** voltage changes will follow the phase angle changes. The **dc** output component will be an exact replica of the original modulating signal. There is also 2× the carrier frequency output, but that voltage is filtered out by the low-pass filter (R and C).

4   The balanced modulator produces a **dc** voltage that is proportional to the phase angle difference between the two inputs. It makes an excellent phase detector for practically any purpose. Consequently, most modern IC phase-locked loops use a balanced modulator for phase-detector service.

Special-purpose 567 PLL Tone Detector

The 567 is a special phase-locked loop IC for detecting a specific audio tone (frequency) and rejecting all others. The tone detector is used in telephone touch-tone systems and various signaling applications. Figure 12-30 shows the

FIGURE 12-30   Functional Block Diagram for 567 Tone Decoder

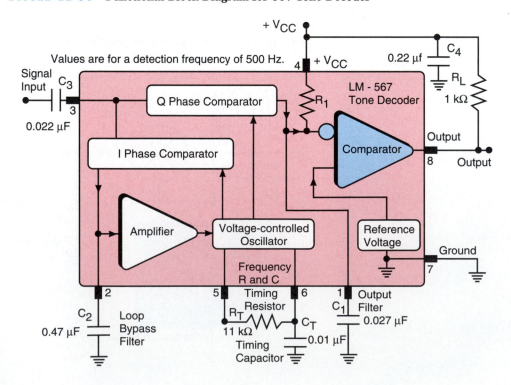

functional block diagram of the 567 tone-decoder circuit. The 567 tone decoder covers a range of frequencies from 0.01 Hz to 500 kHz and a bandwidth that can be adjusted up to 14% of the detection frequency.

How It Works

The 567 has two balanced-modulator phase detectors. The second phase detector eliminates the need for a quadrature tuned circuit (which is not simple when audio and subaudio frequencies are involved). It operates as follows:

1   If the VCO and input frequencies are the same, the I (in-phase) detector produces a zero output voltage and the VCO receives no error voltage. If the phases are different, a proportional error voltage is produced. So far, this is just an ordinary phase-locked loop.

2   The Q (quadrature) phase detector compares the tone input signal with the output of the I phase detector. This yields an output voltage that is the sine of the phase difference between the VCO and tone-signal input phases. Since the sine of 90° is 1, we get an output voltage when the two signals are in phase. The tone frequency bandwidth is controlled by the **loop band-pass filter.**

Figure 12-31 shows a typical multiple-tone-detector signaling receiver system. Several possible tone frequencies arrive at the bank of tone decoders. When a tone decoder recognizes its frequency, it produces a +5-volt output. This is a standard digital logic voltage, so digital AND logic gates can be used. Two tones (frequencies) must be present at the same time to activate an AND gate and produce a final output. Using two simultaneous tones allows a reasonably large tolerance in the frequency of each tone, but it makes false triggering unlikely. The use of two not very precise frequencies permits the incoming signal to be used even if the transmitter's tone frequency has drifted badly.

Phase-locked Loop as a Frequency Multiplier and Frequency Synthesizer

Although the following frequency-multiplier and frequency-synthesizer circuits depend on a digital counter for their operation, they are strictly mainline analog functions.

FIGURE 12-31
Multiple-tone Decoder Circuit

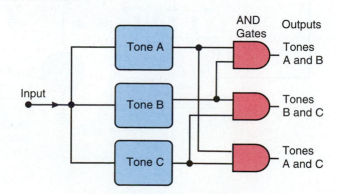

Frequency Multiplier

The circuit in Figure 12-32 uses a digital counter to trick the VCO into phase-locking, on a frequency that is four times higher than the input signal. Because the counter outputs only one pulse for every four produced by the VCO, the phase comparator's input frequencies are actually the same. The VCO stays in lock, even though it is running at a higher (multiple) frequency than the incoming (input) signal.

The input frequency has been multiplied by four, but any other multiple could be used by changing the counter value. The rectangular waveform from the frequency-multiplied VCO output can be converted into a sine-wave signal by using a resonant tank circuit, as shown.

FIGURE 12-32 **Phase-locked Loop as a Frequency Multiplier**

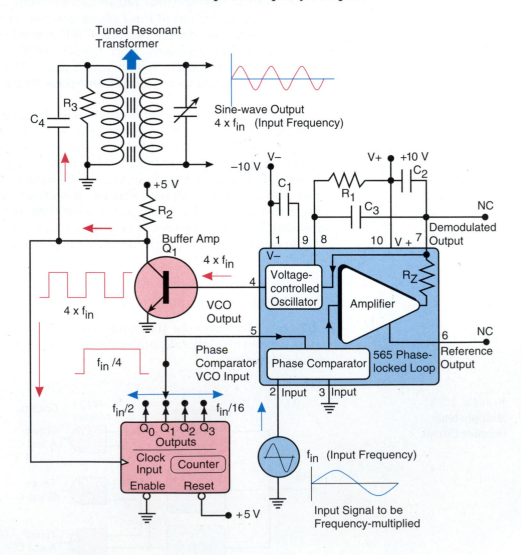

Frequency Synthesizer

The frequency synthesizer is a phase-locked loop crystal-controlled oscillator circuit that can produce a multitude of crystal-accurate (and -stable) frequencies, using only two crystals. CB radios with 40 crystals, for example, would be expensive. The PLL frequency synthesizer in Figure 12-33 can produce all 40 carrier frequencies inexpensively. The circuit is actually an extension of the frequency multiplier circuit, but it adds a balanced mixer that uses a crystal oscillator as a heterodyne frequency shifter. The output frequency of the VCO is first frequency-down-shifted by heterodyning the VCO output with the crystal oscillator, and then further lowered by counting down the difference frequency. The combination of an appropriate crystal frequency and count selection can generate a number of different desired frequencies. Again, a tuned resonant circuit can be used to convert the rectangular-wave VCO output into a sine wave.

FIGURE 12-33 **Phase-locked Loop as a Frequency Synthesizer**

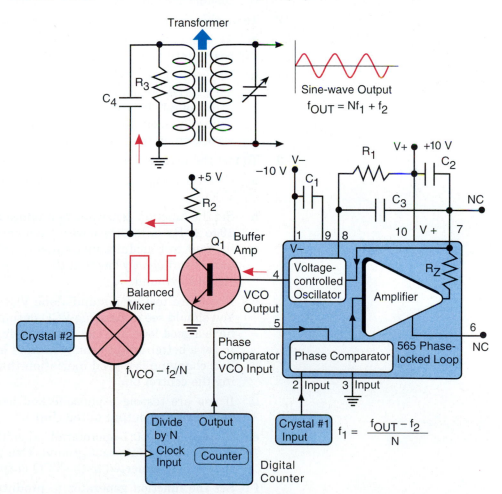

Troubleshooting Voltage-controlled Oscillators and Phase-locked Loops

Phase-locked loops contain a voltage-controlled oscillator as a subsystem. You can test the voltage-controlled oscillator in a phase-locked loop or a stand-alone voltage-controlled oscillator using the following procedures.

1 Connect an oscilloscope to the output of the VCO.

    a If you can't lock the square or rectangular output waveform from the VCO on the scope, it probably means that a modulating signal is connected to the control voltage input. The control voltage signal is changing the VCO frequency so rapidly that the scope can't get a lock on the output signal.

    b Disable the modulating signal to get a good steady waveform.

2 To disable the modulating signal, proceed as follows:

    a As usual, you don't want to disconnect anything from the circuit board until you have identified the defective component.

    b Use a clip lead or clip-on digital test clip to connect a 0.1- to 0.22-microfarad capacitor between the modulation control pin on the IC and ground. This will bypass audio and higher frequencies, disabling the modulating signal. You should see a steady rectangular waveform on the scope.

3 To test the modulation input function, proceed as follows:

    a Connect a signal or function generator to the control pin of the VCO.

    b Set the function generator to produce a sine wave at a frequency of 10 to 20 Hz. Select the lowest frequency your scope will lock on easily. Your test-modulating frequency, injected by the function generator, is sufficiently low that the test capacitor won't bypass the test signal to ground.

    c If you are testing a stand-alone VCO, observe the scope pattern. You should see a changing VCO output frequency, even if you don't have a good lock. You can lower the function generator's frequency to get a better lock. Even if you can't lock the pattern, the fact that it is changing is a good indication that your test signal is activating the control pin.

    d If you are testing a phase-locked loop, connect the scope to the demodulating output of the chip.

    e Connect a test 0.1-microfarad capacitor between one of the phase comparator inputs and ground. Use the phase comparator input that is not connected to the VCO output.

    f Set the function generator to produce a 10- to 20-Hz sine wave. Vary the function generator's frequency, and look for changes on the scope. The demodulator pin output may appear chopped because the low-pass filter capacitor was selected for a higher frequency.

g Any changes in the function generator's frequency that seem to cause changes in the demodulated output indicate that the PLL is operating. This test is a little broad, but it is usually adequate. A total lack of response to changes in the function generator's frequency generally indicates a PLL problem.

h If the PLL seems to respond, check the signal that is normally applied to the phase comparator input. Remove the test capacitor (which was installed to disable the phase comparator input signal) first. Now, you will want to examine the phase comparator input signal on the scope. If there is no signal here, you can be fairly certain the problem is not in the PLL; consequently, you will have to find out where the signal got lost and then verify that part of the circuit. Move the scope test probe from each stage back to the preceding stage until you find the signal again. Isolate and test the stage where the signal doesn't get through.

Testing Multipliers, Balanced Modulators, and Mixers

1 Install a 0.1- to 0.22-microfarad capacitor between each of the two inputs on the multiplier, balanced modulator, or mixer IC. This disables the existing input signals. Then connect a scope to the chip's output pin.

2 Connect a 0.22-microfarad capacitor from each of the two inputs of the multiplier, balanced modulator, or mixer to the output of a function generator. The same signal will be fed to both inputs. Set the function generator to produce a 20-Hz sine wave.

3 Verify that the signal displayed on the scope is twice the frequency produced by the function generator. The sum of the frequencies should be the only signal output. The original function-generator frequency and the difference frequency should have been eliminated by the chip.

QUESTIONS AND PROBLEMS

2.1 Draw the internal block diagram of a typical sample-and-hold amplifier.

How does the sample-and-hold amplifier work?

ive three examples where a sample-and-hold plifier could be used.

each of the following sample-and-hold

ition time

time

ampling error

e. *hold settling time*
f. *hold set-up voltage*
g. *capacitor droop*

12.5 What is the purpose of a multiplexer when it is used with a sample-and-hold amplifier? Give a typical example.

12.6 Draw a block diagram of a typical analog multiplexer.

12.7 What is the difference between an analog multiplexer and a multiplexer intended for digital service?

12.8 Draw a sketch showing how a CCD analog shift register is constructed. How does it work?

12.9 Draw a block diagram of the ISD-1016 voice message system.

12.10 Draw a partial diagram of a CCD videocamera chip image matrix.

12.11 What is the difference between linear mixing and nonlinear mixing of two sine-wave signals? (Assume that one sine wave is 1 MHz or greater and that the other is 1 kHz.)

12.12 Define *carrier wave*.

12.13 What is amplitude modulation? Make sketches of the waveforms, and show the spectrum diagram.

12.14 Define the terms *side frequency* and *sideband*.

12.15 Describe the new frequencies created by amplitude modulation.

12.16 Define *balanced modulation*.

12.17 Why is it possible to suppress the carrier frequency in an AM transmitter?

12.18 Describe single-sideband AM. How does it differ from standard AM?

12.19 Draw spectrum diagrams for standard AM and for single-sideband AM.

12.20 Draw the block diagram of a balanced modulator using an analog multiplier or commercial mixer IC.

12.21 What is needed to detect or demodulate an amplitude-modulated signal?

12.22 Draw the block diagram of a frequency doubler using an analog multiplier. Draw a spectrum diagram of the output.

12.23 How does frequency modulation work? How does it differ from amplitude modulation?

12.24 Draw the circuit for a frequency-modulated oscillator, using a varactor diode.

12.25 Draw the functional block diagram of a frequency-modulated transmitter using a voltage-controlled oscillator (VCO) and an infrared light-emitting diode as the transmission element (light-beam medium).

12.26 Draw the functional block diagram of a receiver for the transmitter in Problem 12.25, using a phase-locked loop as the FM demodulator.

12.27 How does the phase-locked loop recover the audio from the FM signal?

12.28 How does the phase-locked loop work?

12.29 Draw the functional block diagram of a frequency multiplier using a phase-locked loop and a digital counter.

12.30 How does a phase-locked loop frequency synthesizer operate?

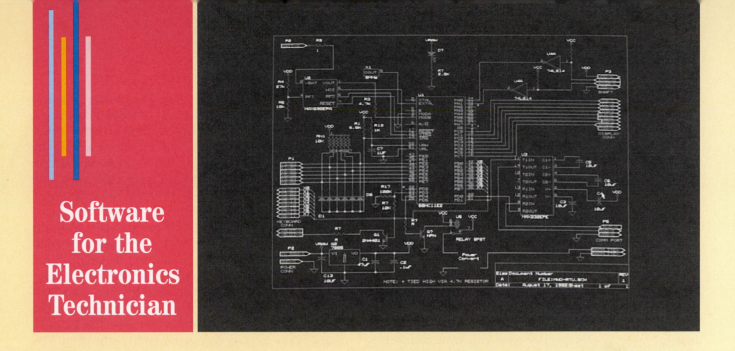

Software for the Electronics Technician

Whether we like it or not, computers are creeping into almost every aspect of our lives. In our homes and work places, they have infiltrated everything from the thermostat to the typewriter to the machine tools on the manufacturing floor. For electronics technicians, the computer's presence is felt on several levels. Since the computer is an electronic device, our role may be to sell, install, service, design, or test them as part of our jobs. But just as the computer has proved to be an effective tool in many other disciplines, technicians are now using computers to make their jobs easier and to improve speed, efficiency, and productivity.

Just take a look in a few of the current trade magazines, and you will find a whole range of software written for use in electronics. From printed circuit design and mechanical drafting to expert systems, and from circuit simulation to data acquisition, software is available to enhance almost any aspect of the technician's job. Because the number of applications and the specific products available are constantly changing, the following discussion merely offers an overview of some of the types of software available.

Design Software

Design software can roughly be divided into two categories: applications designed specifically for electronics use, and general design software that can be applied to uses in electronics.

In the second category fall computer-assisted design (CAD) programs such as AutoCAD and many others that have been created to do general drafting. Programs such as AutoCAD permit the technician to make almost any kind of drawing related to electronics design. Printed circuit-board layouts, schematics, chassis and mechanical drawings in two or three dimensions, isometrics, orthographics, and others can be done on CAD. One advantage of using a general CAD program is its flexibility. Drafting personnel are already familiar with these programs; and the file formats are common and transferable between many applications. Although programs like AutoCAD are expensive and require significant time to learn, they are very powerful and allow the user to configure many functions to obtain the best results for a particular application.

Design programs created specifically for electronics applications typically offer more of the features that are important to the electronics design process. Programs like OrCAD, for example, provide comprehensive libraries of schematic symbols. Another feature takes component labels, numbers, and values off the schematic and provides lists for later documentation. Many of these programs use the schematic when designing the printed circuit board and automatically route and reroute traces from components pin to pin.

For drawings that include more pictorial elements, programs such as MicroGraphics Designer can be used. Whereas older drawing programs such as PC Paintbrush were pixel-based (which caused "jaggy" drawings when expanded), Designer is element-based (and therefore produces smooth lines and angles). Although it lacks many of the features of a CAD program, Designer makes up for it in flexibility, ease of use, and relatively low cost.

A number of programs are now available that allow you to "breadboard" a circuit on screen and test it by using an emulation feature. Electronics Workbench is one such program. It has two separate types of design and emulation: analog and digital. In this program, specific electronic components can be selected from a menu bar on the right side of the screen and positioned on the "workbench." With a mouse, the user can interconnect component inputs, outputs, and power-supply connections. When the circuit is complete, the user presses a "switch," and the program starts an emulation. Test equipment such as meters, oscilloscopes, and signal generators can be placed on screen and connected to do fairly realistic testing of the circuit. Modifications based on the results of the emulation can easily be introduced to the circuit, which can then be retested.

Along the same lines, but not as graphically based, programs like PSPICE allow circuit descriptions to be programmed in, using simple commands. The program is then run, and the result of the circuit action is placed in an output file. It can be examined in this form or sent to another program to be converted into displayable waveforms.

Other, simpler programs have been written to simplify various aspects of design, such as active filter calculations, capacitor charge curves, and transistor amplifier design. Many of these programs are available in shareware form through bulletin boards or other sources.

Service-oriented Software

Servicing computers themselves differs from conventional circuit troubleshooting. Most signals are transitory: they just don't stay around long enough to permit you to look at them on an oscilloscope. Because of this, troubleshooting procedures must be different. One advantage of troubleshooting computers is that they can do some of the work for you. If the computer is at least partially functioning, it can run diagnostic tests on itself to locate the area where the fault lies. Diagnostic software enables you to run these kinds of tests. Frequently, the computer will identify a disk drive or other peripheral problem that can be cured simply by replacing the module. On other occasions, it can narrow the problem down to a specific area, and you can conduct further tests of the identified area thereafter.

In several areas of electronics, expert systems software can be very helpful. One case is in complex systems, where an individual or small group of experts have established how the system performs under specific conditions. Rather than forcing each user to relearn all that experience, the data are formatted in software so that it asks questions of the person doing the troubleshooting. The answers guide the program toward the correct solution to the problem.

A simpler version of this exists in the consumer electronics servicing business. In this case, the software manufacturer gathers data from experienced service personnel and structures these in software. The software provides solutions to problems by relating specific symptoms to specific models of equipment. In most cases, the recommended repair procedure cures the problem.

Data acquisition boards use software packages that turn your computer into one or more pieces of bench test equipment. The advantage here is that, as long as the hardware can accommodate the signals coming in, the software can transform the computer screen into practically any kind of instrument you need. Meters, oscilloscopes, signal generators, frequency counters, waveform analyzers, spectrum analyzers, and more can be simulated on the computer screen. Data can be stored for future examination, for record keeping, and for other purposes.

One other type of software that can be very useful to you as an electronics technician is educational software. Electronics is a field where we can never afford to stop learning. Computers can help us in the process of continuing education. Whether we review the basis of transistor circuits or work our way through tutorials on the latest developments in semiconductor technology, computers at home or at work allow us to pursue our interests thoroughly and efficiently.

In a well-structured electronics shop, we would never question the need for tools, a multimeter, an oscilloscope, a power supply, and various other pieces of test equipment. The computer is a multi-purpose tool—powerful because it can be used in so many ways to extend our capabilities. Much of the power of the computer, however, is found in software. Knowing what the computer can do and how it can be made to do it is the key to unlocking that power to enhance our roles as electronics technicians.

Power Control and Interfacing

OBJECTIVES

Upon completion of this chapter, you should be able to:

1 Explain the operation of the basic linear voltage regulator.

2 List the differences between the linear regulator and a class-A power amplifier.

3 Describe the feedback controller components in a linear voltage regulator circuit:
   a Controlled quantity
   b Reference quantity
   c Error detector
   d Amplifier
   e Error corrector

4 Explain how the voltage is stepped down in a switching regulator.

5 Explain how the voltage is stepped up in a switching regulator.

6 Explain how the voltage is inverted in a switching regulator.

7 Describe the feedback controller components in a switching regulator circuit:
   a Controlled quantity
   b Reference quantity

   c Error detector
   d Amplifier
   e Error corrector

8 List the advantages and disadvantages of a switching regulator.

9 Describe the condition of the three junctions in a forward-biased SCR.

10 Describe the condition of the three junctions in a reverse-biased SCR.

11 Describe the action of the gate in an SCR.

12 Explain why the gate cannot turn the SCR off after it has turned it on.

13 Identify the two common trigger devices used with SCRs.

14 Describe the TRIAC.

15 Name the most common device used to trigger TRIACs, and describe its construction.

16 Explain how duty-cycle control can simulate analog variable control.

17 Describe the major difference between an SCR and a TRIAC duty-cycle controller.

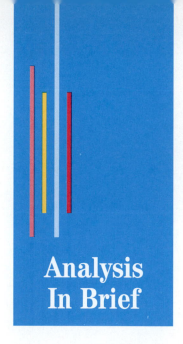

Analysis In Brief

Linear Voltage Regulators in Brief

THE CLASS-A LINEAR VOLTAGE REGULATOR

The common linear voltage regulator is a modification of the class-A power amplifier. The voltage regulator modifies the class-A amplifier by removing provisions for an input signal. The input coupling capacitor is removed, the two negative feedback resistors are combined into a single feedback resistor, and the capacitor at the midpoint of the feedback resistors is removed. The split feedback resistor and bypass capacitor were included in the audio power amplifier to prevent the feedback loop from correcting for audio signal changes and thereby eliminating the sound.

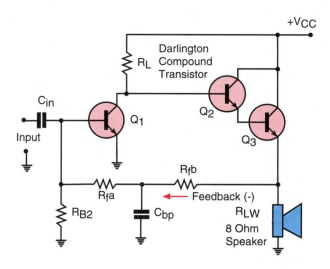

The Class-A Amplifier Circuit

THE VOLTAGE REGULATOR ADDS A REFERENCE VOLTAGE

The class-A power amplifier already has all of the necessary feedback elements for a feedback-control circuit except the reference quantity.

1   The controlled quantity is the load (output) voltage.
2   The reference voltage is the added Zener diode.
3   Q_1 and Q_2 serve as the error corrector.
4   Q_1 is the error detector and amplifier.

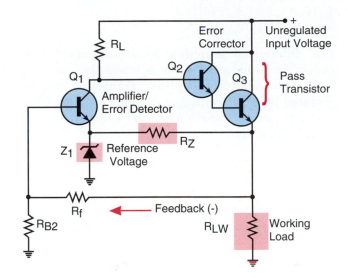

The Linear Voltage Regulator Circuit

HOW THE LINEAR REGULATOR CIRCUIT IS USUALLY DRAWN

The linear voltage regulator is usually drawn in a way that makes it look quite different from the class-A power amplifier. The circuit at right shows how it is usually drawn. The only difference in this drawing and the previous class-A amplifier circuit, however, is the source of the Zener diode voltage. Both connections are used.

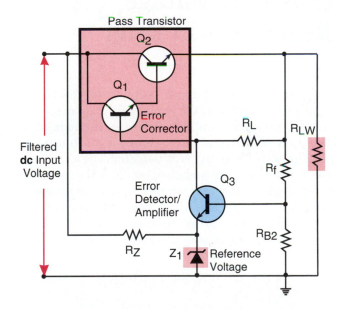

How the Voltage Regulator Circuit is Normally Drawn

THE INTEGRATED CIRCUIT VARIATION

The differential amplifier is the standard low-power amplifier used in integrated circuits. The single error-detector/amplifier transistor can be replaced by a differential amplifier, as shown. This is the preferred variation of the circuit when it is used for ICs.

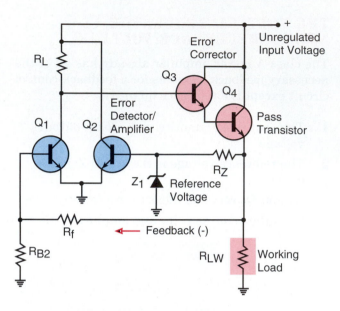

The Integrated-circuit Version of the Linear Voltage Regulator

Switching Voltage Regulators in Brief

SWITCHING REGULATORS

The switching regulator is similar to the linear regulator, except that the error-corrector transistor is switched on and off by a pulse-width modulator. The pulse-width modulator serves as both a reference quantity and an error detector. The error-corrector transistor is switched at 20 kHz or more, which makes filter components small and inexpensive. The energy stored in the inductor can be arranged to add to or subtract from the filter capacitor charge, to step the voltage up or down.

THE STEP-DOWN SWITCHING REGULATOR

The step-down switching regulator uses the reactance of the inductor to drop some of the voltage, resulting in a step-down action. The reactance produces a voltage drop without a heat loss.

THE STEP-UP SWITCHING REGULATOR

The step-up switching regulator adds the inductor's collapsing field voltage to the capacitor charge voltage to yield a voltage step-up.

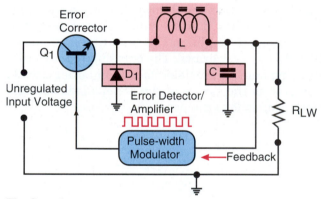

The Step-down Switching Regulator

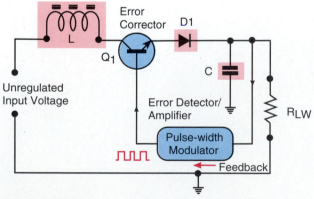

The Step-up Switching Regulator

THE NEGATIVE OR INVERTING SWITCHING REGULATOR

The inverting regulator uses only the opposite-polarity collapsing-field inductor voltage.

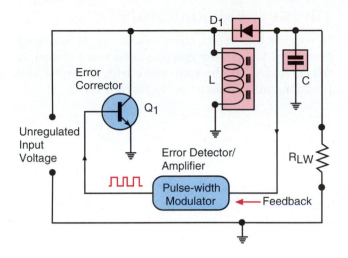

The Negative or Inverting Switching Regulator

The Thyristor Family in Brief

THE THYRISTOR FAMILY

The thyristor family is a group of four-layer diodes, each with three junctions. The SCR and its bidirectional equivalent, the TRIAC, can control power levels of hundreds of amps and hundreds of volts. Other common members of the family—the Shockley four-layer diode and the DIAC—are used to trigger SCRs and TRIACs. All thyristors conduct in an avalanche mode.

THE FORWARD-BIAS MODE

In the forward-bias mode, the SCR has two junctions forward-biased and one junction reverse-biased. A small trigger voltage applied to the gate initiates an avalanche in the middle (reverse-biased) junction. Once avalanche starts, the SCR allows a current flow, limited only by the load resistance. Once started, the avalanche can be stopped only by effectively opening the anode/cathode circuit. The gate no longer has any influence.

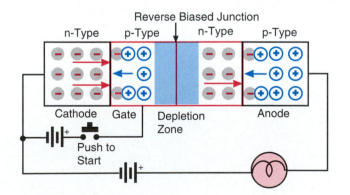

The Silicon-controlled Rectifier (SCR) in Forward-bias Mode

THE REVERSE-BIAS MODE

In reverse-bias mode, there are two reverse-biased junctions. The center junction is forward-biased. A trigger pulse on the gate can't trigger any of the junctions into avalanche. No current flows.

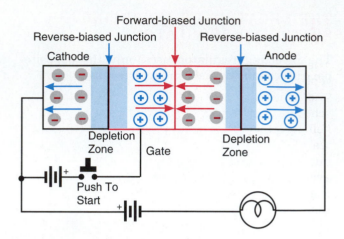

The Silicon-controlled Rectifier (SCR) in Reverse-bias Mode

SYMBOLS

The symbols at right all apply to members of the thyristor family except for the unijunction transistor (UJT), which is used to trigger SCRs.

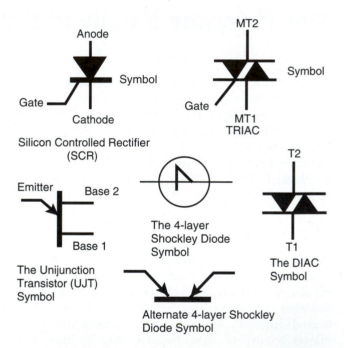

Symbols

THE SILICON-CONTROLLED RECTIFIER (SCR)

The SCR (and the TRIAC) is used as a modern replacement for relays in industrial applications. Ladder notation is commonly used in industrial schematics. The circuit to the right shows SCRs being used to turn on high-power loads. Pressing a start button triggers the SCR into avalanche conduction, turning on the motor or lamp. Once the avalanche has started, the **dc** power line must be opened (by means of the Master Reset button) to turn it off.

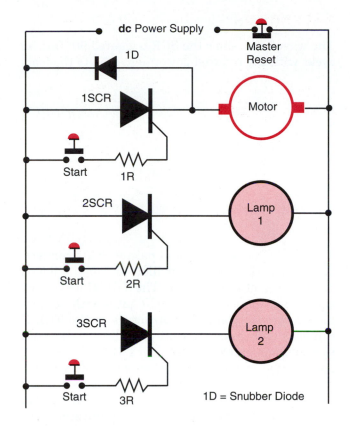

Industrial Schematic for an SCR Ladder Circuit

DUTY-CYCLE CONTROL USING AN SCR AND A FOUR-LAYER DIODE TRIGGER

Duty-cycle control is used to simulate continuous analog-like control at high power levels. The household lamp dimmer is a common example. The SCR turns the load on for variable percentages of the power-line **ac** cycle. The R_2-C_1 time-delay circuit starts charging the capacitor at the start of each half-cycle. The capacitor must charge to the four-layer diode's breakover voltage before it fires and turns on the SCR. By delaying SCR firing by varying amounts, we can vary the average power to the load.

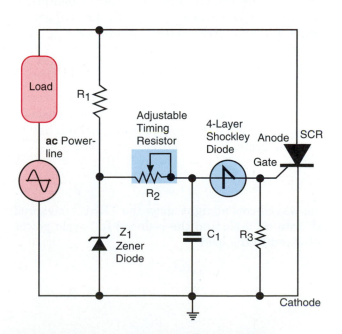

Duty-cycle Power-control Circuit Using a Silicon-controlled Rectifier and Four-layer Diode Trigger

WAVEFORMS

The waveforms show the SCR triggered 90° into **ac** cycle, with the reduced power delivered to the load.

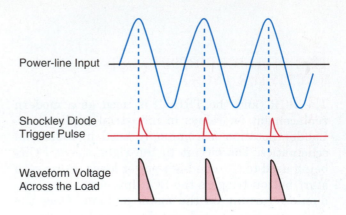

Power-line Input

Shockley Diode Trigger Pulse

Waveform Voltage Across the Load

Waveforms

THE TRIAC

The TRIAC is an integrated pair of SCRs connected in opposite directions to allow for conduction on both halves of the **ac** cycle. The TRIAC solves the problem of connecting two SCR gates with no common reference point. The structure of the TRIAC connects both gates internally and brings them out to a common gate terminal. The TRIAC is designed specifically for 60-Hz operation, and it may be too slow for some applications. Two SCRs, connected back-to-back, may be triggered with a special transformer for higher-frequency application, but at higher cost. The DIAC is two four-layer diodes connected in opposite directions for **ac** operation. The DIAC is designed specifically for triggering TRIACs in duty-cycle control circuits.

THE DUTY-CYCLE CONTROL CIRCUIT

The TRIAC/DIAC duty-cycle control circuit works exactly like the SCR/four-layer diode circuit, except that the TRIAC circuit works on both halves of the **ac** cycle.

WAVEFORMS

The waveforms at right show the TRIAC triggered 90° into **ac** cycle and the reduced full-cycle power delivered to the load.

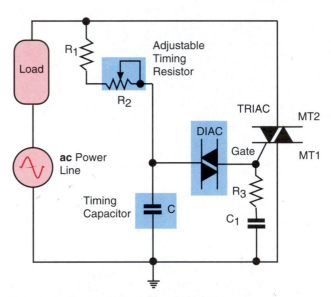

Load

R_1

Adjustable Timing Resistor

R_2

ac Power Line

DIAC

TRIAC

MT2

Gate

MT1

R_3

Timing Capacitor

C

C_1

Duty-cycle Power-control Circuit Using a TRIAC Trigger and a DIAC Trigger

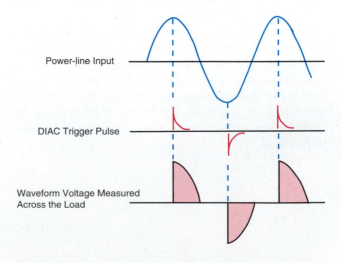

Power-line Input

DIAC Trigger Pulse

Waveform Voltage Measured Across the Load

Waveforms

Power-level Interfacing in Brief

TRANSLATING FROM ANALOG VOLTAGES TO STANDARD DIGITAL VOLTAGE LEVELS, USING A 311 COMPARATOR

The 311 comparator was designed to translate voltage levels. It has independent power-supply connections for the comparator and output stages. It is most commonly used to translate analog voltages into standard digital logic levels, but it can be used for other voltage-level translations as well.

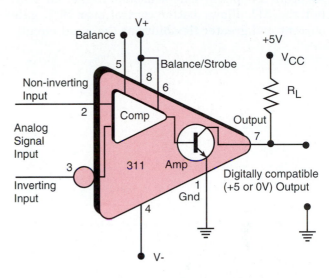

Using a 311 Comparator as an Analog-Digital Interface

USING THE 311 COMPARATOR AS AN OP-AMP FOR ANALOG-TO-ANALOG VOLTAGE TRANSLATION

The 311 can be converted into an op-amp by adding a negative feedback loop. The voltage gain should be kept near unity for best results. The circuit to the right shows how the 311 can be used to translate one analog voltage into another.

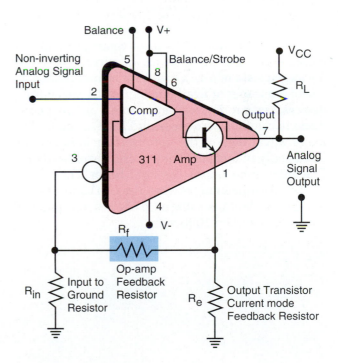

Voltage-level Translation Using a 311 Comparator as an Op-amp

USING THE 311 TO TRIGGER AN SCR

An ordinary op-amp can be used to trigger an SCR, but the 311 allows better control over SCR gate current and greater flexibility at the input circuit.

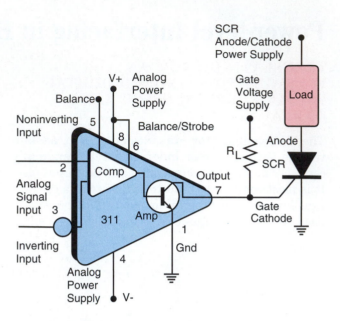

Interfacing a Comparator to an SCR

OPTICAL ISOLATORS (OPTICAL COUPLERS)

The optical isolator is the most common power- or voltage-level interface device in current use. The device consists of a LED with its beam directed on one of a number of available light-activated output devices. Output devices range from a simple photodiode to complex zero-crossing TRIAC circuits. The major advantage of the opto-isolator is its totally electrically independent input and output circuits. The only connection is the light beam. There is no electrical connection between the LED and the output circuit. Voltage differences between input and output can exceed 1000 volts.

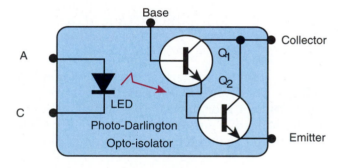

Opto-Isolator with Photo-Darlington Output

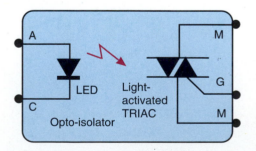

Opto-Isolator with Photo-Trac Output

POWER MOS-FETS

Recent manufacturing developments have produced MOS-FETS with several hundred volts of isolation between the gate and the source/drain circuit. Power MOS-FETs are available for hundreds of volts and 100 amps or so. They can easily be paralleled for higher currents. Power MOS-FETs can be used to control **dc** power, as well as **ac** power. They create less noise than do most other high-power switching devices, and do well in harsh environments such as automotive applications (like the one shown at right).

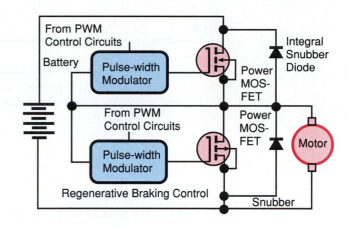

Power FETs Used in a Possible Controller for Electric Automobiles

Analog/Digital Converters in Brief

ANALOG-TO-DIGITAL AND DIGITAL-TO-ANALOG CONVERTERS

There are three basic types of analog-to-digital converters.

1 Ladder comparators coupled to a digital encoder to convert analog outputs into binary. These devices are called *flash converters* and are used for audio and video signal conversion. Flash converters are available in 8-, 16- and 32-bit integrated-circuit varieties.

2 Successive approximation converters that are very digital and beyond the scope of this book. They, too, are available in integrated-circuit packages. They are slower but cheaper than flash converters.

3 Slope or ramp converters, like the circuit illustrated to the right. These devices find applications in digital multimeters and other applications where low cost is critical and speed is less important.

Digital-to-analog converters are generally some form of op-amp summing amplifier. These, too, are available as precision integrated-circuit packages. Serious AD or DA converters are far too complex to build from discrete circuits. You will not be repairing any real-world converters.

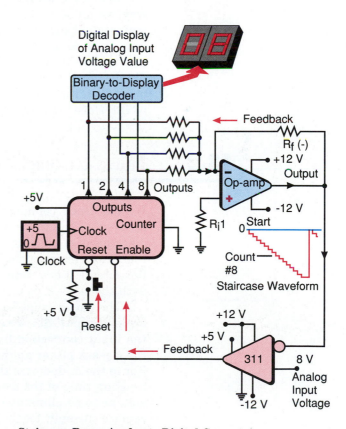

Staircase Ramp Analog-to-Digital Converter

13.1   Introduction

Power control—of motor speeds, lamp brightness, and a variety of other power-demanding devices—is one of the most common tasks performed by electronic circuits. Most of our electronic signal-processing circuits operate at levels from milliwatts to a few watts; no matter what kind of electronic processing is being done, it is done at low power levels. When our low-power circuits must control high-power devices, we must add some kind of power control device to the low-power processing circuitry.

In most cases, the relatively high voltages and current required for power application differ significantly from those used by the low-power control circuits. And because the higher voltages and current required by power devices are often hazardous to the low-power circuits they are connected to, we often must provide some kind of protective isolation between high- and low-power circuits. In other cases we must convert analog signals into digital signals for a computer to use. In other cases, we must convert digital data into analog information.

Whenever we want to connect two circuits that are not compatible in some way, we must add circuitry called **interface circuits.** An interface acts as a transitional link between the two different circuits. Digital interface circuits tend to be complex, but most analog interface circuits are comparatively simple—unless they must interface with digital circuits. Technically, a power supply designed to convert a 120-volt **ac** power line voltage into low-voltage **dc** power is an interface circuit. We don't normally refer to power-supply circuits as interface circuits, but they do meet the definition.

13.2   Linear-feedback Voltage-regulator Circuits

We discussed commercial voltage regulators in Chapter 3 but deferred our scrutiny of the internal circuitry until now. Figure 13-1 shows the schematic diagram of a common class-A power amplifier. The simplest linear voltage-regulator circuit is nothing more than a variation of the class-A power amplifier circuit. Let's compare the basic voltage-regulator circuit in Figure 13-2 with the class-A power amplifier in Figure 13-1.

The power amplifier has an input connection to the base of Q_1 that acts as a signal input. Since we have no interest in inputting a signal, we can remove the input connection from the voltage-regulator circuit, as in Figure 13-2. In the class-A power amplifier, we split the feedback resistor and added a capacitor to the midpoint of the feedback resistor. We added the capacitor to slow the reaction time of the feedback loop, so it would not view the signal as an error voltage to be eliminated. The voltage regulator has no signal input, so we can replace the split feedback resistor and the capacitor with a single resistor.

Now we can view the former class-A amplifier as just a negative-feedback control circuit. The Darlington pair (Q_2 and Q_3) serves as the error corrector. The error-corrector power transistor is often called the *pass transistor* in voltage regulators, but its true function is to work as an error corrector. Transistor

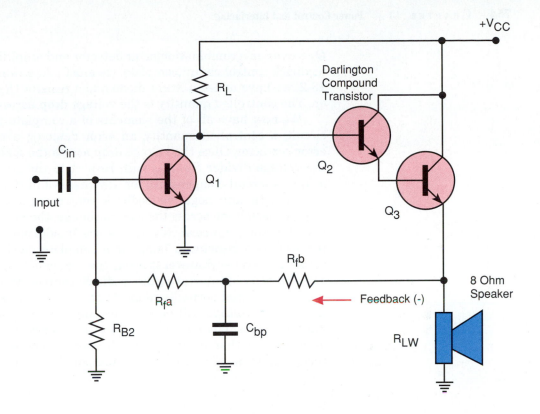

FIGURE 13-1
Class A Audio Power Amplifier

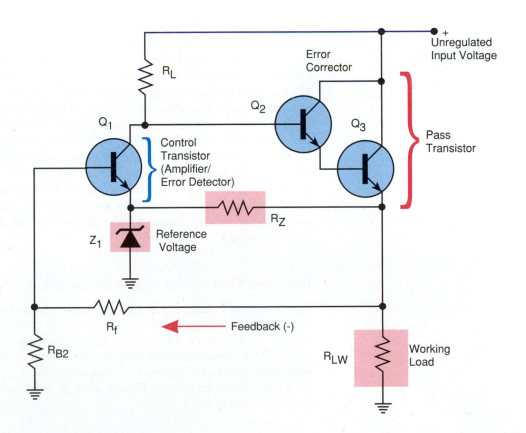

FIGURE 13-2
Linear Voltage Regulator

Q_1 serves as a combination error detector and amplifier. To make the negative-feedback control circuit complete, we must add a reference quantity. In Figure 13-2, we have added a Zener diode and a resistor (R_Z) to turn the Zener diode on. The controlled quantity is the voltage drop across the working load.

We now have all of the elements of a complete negative-feedback control circuit: a controlled quantity, an error detector, a reference voltage, and an error corrector. Once the voltage drop across the load has been established by the voltage divider R_{B1}/R_{B2} with respect to the Zener reference voltage, the feedback circuit compensates for any changes in the load voltage.

As with any negative-feedback control circuit, it doesn't matter what causes the voltage across the load to change; the voltage difference will be corrected by the error corrector transistors. In addition, it doesn't matter whether the load is a resistor, a lamp, or a complete electronic circuit. The voltage across the load—whatever it is—is the controlled quantity and (therefore) the only quantity of interest to the feedback control circuit.

The terminal normally connected to the load is called the *regulator output.* The output voltage will be same for any load we connect. Some kind of load must be there, but resistors R_{B1} and R_{B2} serve as a permanent low-current dummy load that ensures that the output voltage is always correct, even though the real working load has temporarily been disconnected.

Ripple Reduction

When the circuit was a class-A power amplifier, we installed some time delay in the feedback loop to prevent the circuit from correcting errors quickly enough to cancel signal variations. When we converted the class-A power amplifier into a regulator, we removed the R-C circuit, resulting in lightning-fast error correction. We removed the provision for an input signal, but we still have a signal of sorts, called **ripple.** Now that we allow the feedback circuit to respond almost instantaneously, the feedback circuit treats the ripple voltage changes as just another error voltage and corrects for them. Consequently ripple voltages are virtually eliminated—from 1 volt or more to a few millivolts.

Linear or class-A regulators are not very efficient because they are basically class-A power amplifiers. They are, however, very effective regulators, and the amount of power they waste as heat is tolerable. Because some voltage drop must occur across the error corrector (pass transistor), the unregulated input voltage must always be somewhat higher than the regulator's output voltage.

Different View of the Linear Regulator

The schematic diagram of the voltage regulator we have been discussing is usually drawn in a way that makes it look quite different from the circuit in Figure 13-2. Figure 13-3 shows how a voltage regulator circuit is normally drawn. The two schematics (Figures 13-2 and 13-3) are identical, with one very minor exception. In Figure 13-2, the Zener turn-on resistor is connected to the load side of the circuit; in Figure 13-3, it is connected to the input side. Both connections are used.

FIGURE 13-3 · **How the Regulator Circuit is Normally Drawn**

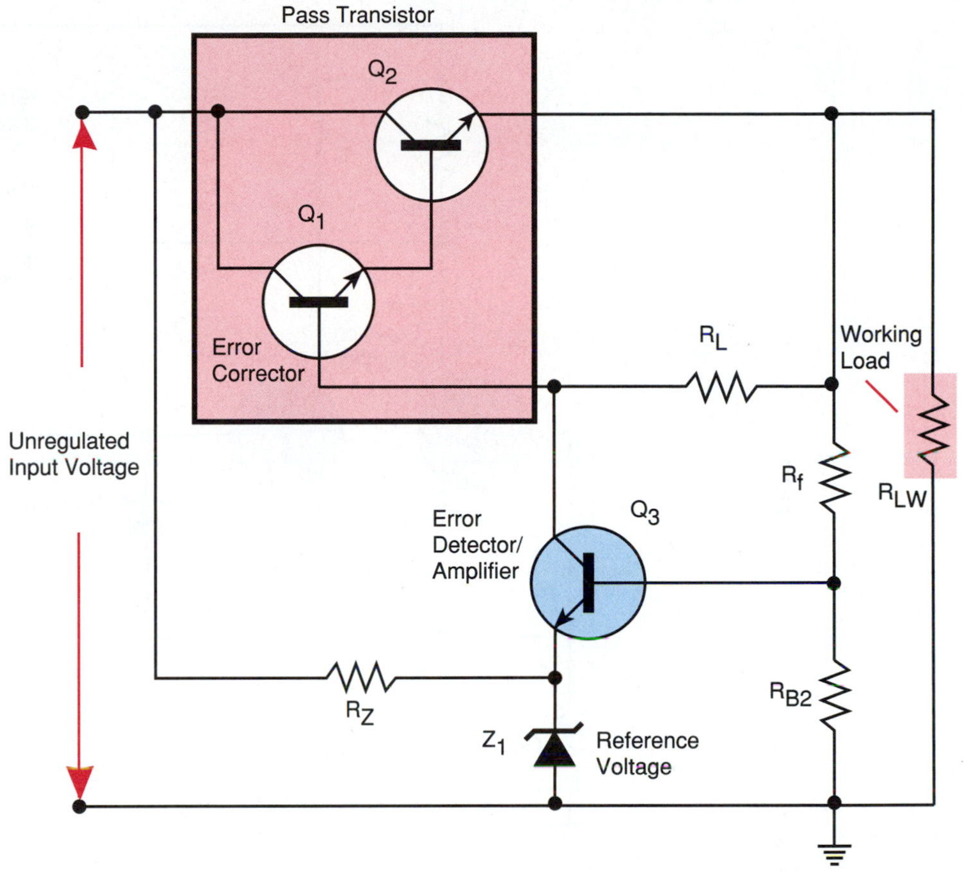

Integrated-circuit Modifications

The differential amplifier is almost universally used in integrated circuits rather than single transistors. Figure 13-4 illustrates how the circuit can be modified to use differential amplifier. In this case, the reference voltage is established at the noninverting input, and the negative feedback is applied to the inverting input. The base of Q_2 is anchored at the reference voltage, and any difference in voltage between the two inputs becomes an error voltage for controlling the error-corrector transistors Q_3 and Q_4.

13.3 Switching Regulators

Switching regulators can also function as negative-feedback regulators; and although they are more complex, they are less expensive, because they operate at a higher frequency than the 60-Hz (or 120-Hz full-wave) line frequency.

FIGURE 13-4   **Integrated Circuit Version of the Linear Regulator**

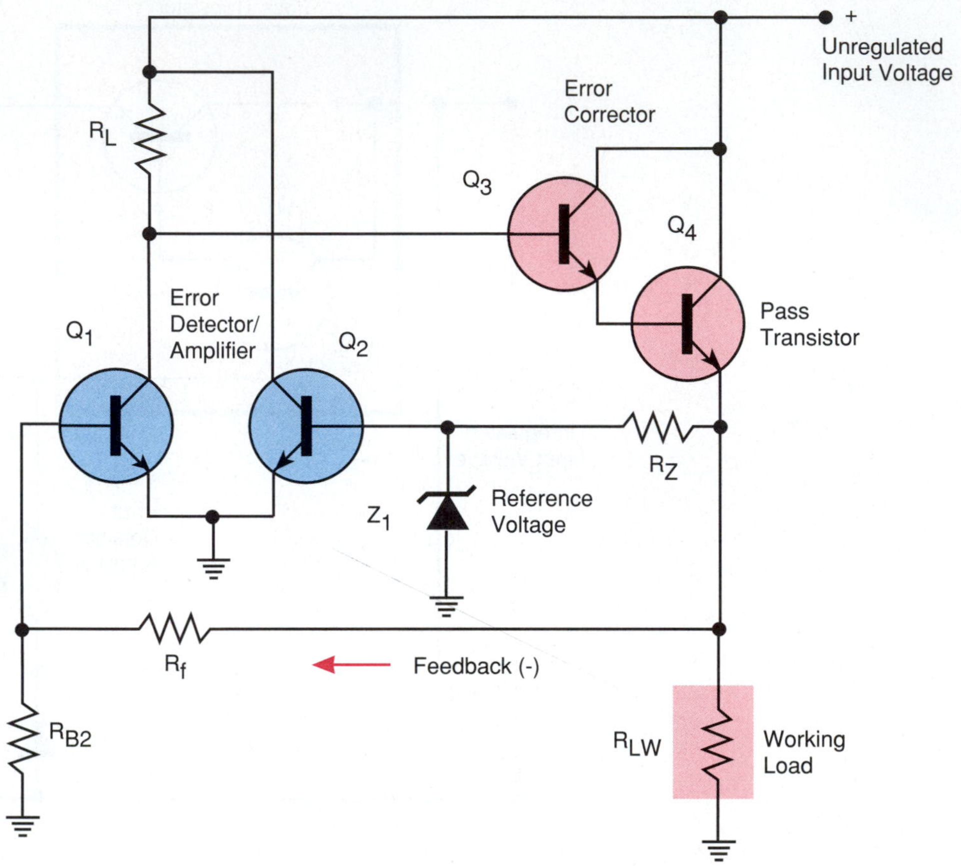

Voltage step-up or step-down is accomplished by inexpensive inductors (or transformers, if isolation is required).

The input to the regulator is often a full-wave bridge connected directly to the **ac** line, with some filtering. This poorly filtered full-wave voltage is sampled at a rate of 20 kHz or higher.

The error-corrector transistor shown in Figure 13-5 is operated in a switching mode. The transistor is switched on and off at 20 kHz or more; inductors and transformers designed for 20-kHz operation are smaller and much less expensive than transformers designed for 60-Hz operation. The power transformer in conventional power supplies accounts for most of the cost, weight, and size of the power supply.

Switching power supplies are efficient and inexpensive, in spite of their greater complexity. The comparator-based pulse-width modulator, which is the heart of the feedback circuit, is available in several integrated-circuit forms. The ICs are also inexpensive, so the increased complexity doesn't increase the cost significantly. Because of the high sampling rate (frequency), final filtering can be accomplished by relatively small capacitors and inductors. The output voltage can be stepped up or down by adding or subtracting the collapsing field voltage stored in an inductor.

FIGURE 13-5   **Step-down Switching Regulator**

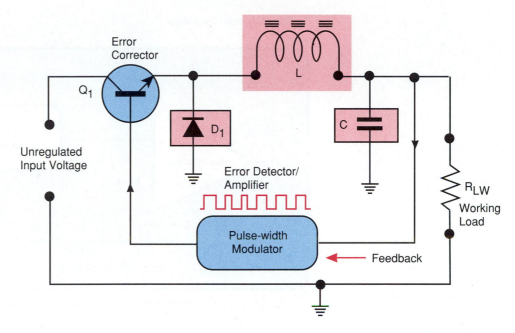

Switching regulators are difficult to troubleshoot because of the complex nature of the pulse-width modulator's feedback loop, and because a component failure anywhere in the feedback loop often destroys other components in the circuit. On the other hand, switching power supplies are so inexpensive that they are generally replaced anyway, because it is not economical to repair them.

Many modern computers use switching power supplies, and they seem to have a higher failure rate than supplies based on linear regulators. They can also develop subtle noise problems that are often very hard to pin down. Switching regulators are used less often to power analog circuits than to power digital systems. Digital systems have a greater immunity to noise signals from a power supply than do analog systems, as a result of analog circuits' infinite resolution.

Pulse-width Modulator Error Detector/Amplifier

Figure 13-6(a) presents a simplified version of the pulse-width modulator IC shown as a block in Figure 13-5. We examined the pulse-width modulator in Chapter 9, and here we have a chance to apply it to a practical circuit. In this case, the variable voltage is the **dc** output voltage of the regulator (the controlled quantity). The on-board triangular-wave generator serves as the reference quantity. The waveforms shown in Figure 13-6(b) provide examples of how the pulse width changes as the regulator's output voltage changes. The output pulses of the pulse-width modulator turn the error-corrector transistor on and off. If the output voltage drops, the error-corrector transistor stays on for longer periods, and its off time is shortened. If the output voltage rises, the error corrector remains on for shorter periods, and the off time is lengthened.

FIGURE 13-6 **Pulse-width Modulator**

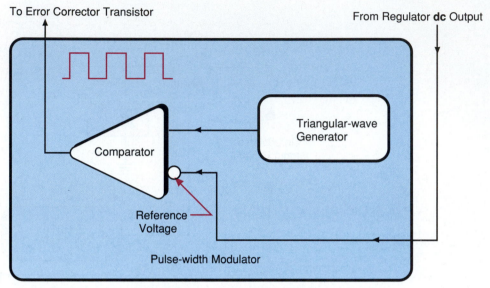

To Error Corrector Transistor

From Regulator **dc** Output

Comparator

Triangular-wave Generator

Reference Voltage

Pulse-width Modulator

a. Simplified Functional Block Diagram

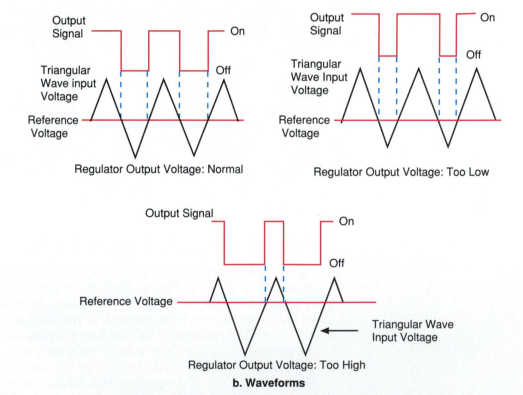

Output Signal

On

Triangular Wave input Voltage

Off

Reference Voltage

Regulator Output Voltage: Normal

Output Signal

On

Triangular Wave Input Voltage

Off

Reference Voltage

Regulator Output Voltage: Too Low

Output Signal

On

Off

Reference Voltage

Triangular Wave Input Voltage

Regulator Output Voltage: Too High

b. Waveforms

Inductor-capacitor (L-C) Filter Circuit

The filter circuit consisting of L and C in Figure 13-5 averages the pulses and smooths them into a **dc** value proportional to the width of the (on) pulses provided by the error-corrector transistor. Because the pulses are rectangular, the inductor interprets the steep rise and fall of the pulses as being equivalent to a high frequency; consequently, it exhibits a high reactance, with a significant voltage drop.

The inductor value can be selected to drop the desired amount of voltage, leaving the rest to provide the intended **dc** output voltage. The configuration in Figure 13-5 is a step-down version for that reason. Diode D_1 acts as a snubber to eliminate the inductor's collapsing field voltage.

Step-up Switching Regulator

If we rearrange the filter network as shown in Figure 13-7, the same basic circuit becomes a step-up regulator. The difference here is that the inductor is in series, but on the input side of the error-corrector transistor. In addition, the snubber diode delivers the collapsing field voltage to the capacitor instead of bypassing it to ground. Diode D_1 adds the collapsing field voltage to the existing capacitor charge voltage. The output voltage is greater than the input voltage.

Negative or Inverting Switching Regulator

If we want the circuit to produce a negative output voltage when the input voltage is positive, we can rearrange the filter network as shown in Figure 13-8. In this case, the positive pulses delivered by the error-corrector transistor

FIGURE 13-7
Step-up Switching Regulator

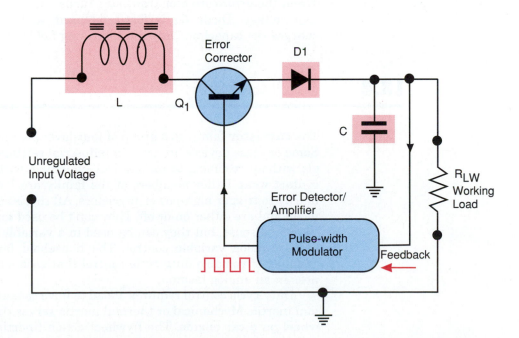

FIGURE **13-8**   **Negative or Inverted Switching Regulator**

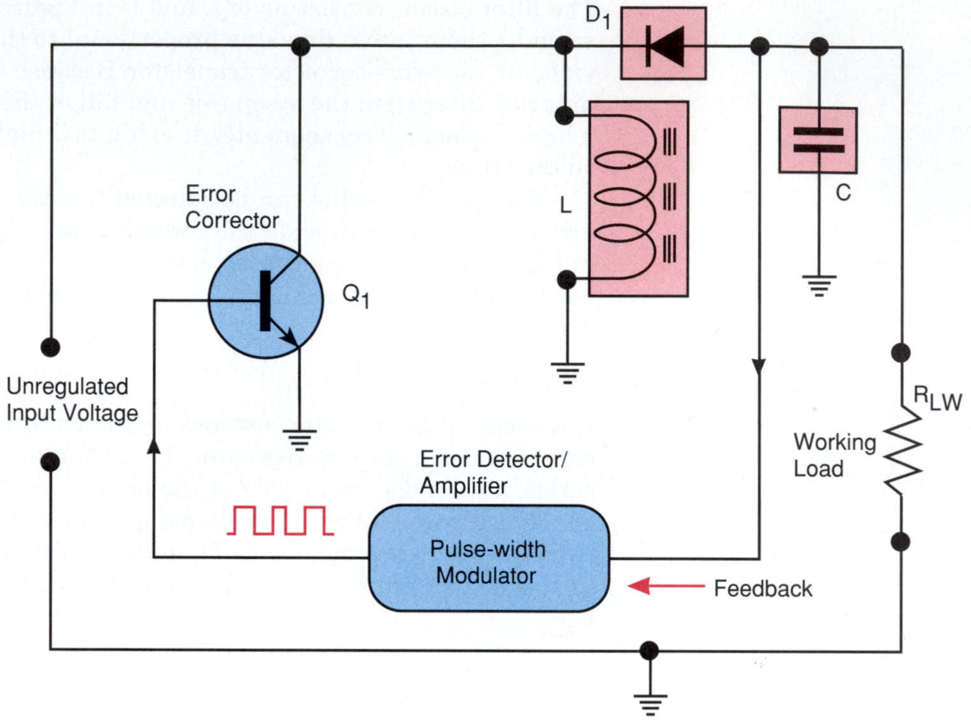

are delivered to the inductor (L) when the error-corrector transistor is turned on, but diode D_1 prevents the positive pulses from charging the capacitor (C). When the error-corrector transistor turns off, the collapsing field has a negative voltage. Diode D_1 is forward-biased, and the collapsing field voltage charges the capacitor. Only the collapsing field voltage is used in this case.

13.4   Thyristor Family

The **thyristor** family is a group of four-layer p–n–p–n silicon junction devices. Some of these devices are used in industrial settings to switch high power levels without resorting to mechanical switching, with its attendant arcing and contact wear. Other members of the family are low-power devices, used primarily to trigger high-power thyristors. All thyristors operate in an avalanche mode and are either on or off. They can't be used for linear variable control in the usual sense, but they can be used in a variable duty-cycle mode that simulates analog variable control. The household lamp dimmer is a common example of variable duty-cycle control that seems to control the lamp brightness in an analog fashion.

Duty-cycle control requires a load to have considerable mechanical or thermal inertia. Mechanical or thermal inertia serves the same function as the flywheel on a car engine. The flywheel action (inertia) averages the individual cylinder explosions into what seems to be a smooth, continuous rotation. An incandescent lamp filament doesn't cool down immediately when the power is shut off, so it can average a string of pulses into an average value. The ratio of on time to off time is called the **duty cycle.** A high duty cycle delivers power to the load most of the time. A low duty cycle delivers power to the load only part of the time.

Duty-cycle control simulates the continuously variable analog power control we might get from a variable resistor, but without the latter's heat dissipation. At high power levels, resistive control is not very practical, and it certainly is not an efficient way to get the job done. Duty-cycle control of high power levels is not new. Members of a family of gas-filled devices called *thyratrons* and *ignitrons* were once used for high-power duty-cycle control. The gas-filled devices have since been replaced by the thyristor family because thyristors are smaller, cheaper, and more efficient than their gas-filled equivalents. Reactance-control devices have also been used for efficient control of high power levels, because a reactance can provide an **ac** voltage drop without the heat and power waste of resistive control. Reactive controllers are heavy, large, and very expensive. During World War II, the United States used gas-filled devices to control the positioning of large guns on navy ships. A near miss by an enemy shell often sufficed to shatter the glass and disable the control system. German ships used reactance-control devices that looked like (and were built like) large transformers. They were almost impossible to knock out. United States' ships were soon equipped with gas-filled devices housed in a steel container instead of a glass bottle.

Silicon-controlled Rectifier

All members of the thyristor family operate in the avalanche mode. The avalanche mechanism is illustrated in Figure 13-9. The avalanche mode drives a reverse-biased junction beyond the Zener knee, deep into avalanche and a

FIGURE 13-9   The Avalanche Mechanism

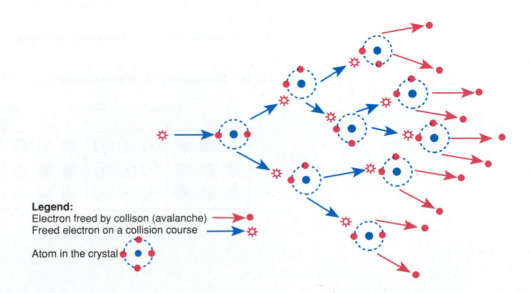

Legend:
Electron freed by collison (avalanche)
Freed electron on a collision course
Atom in the crystal

potentially self-destructive condition. The current must be limited to a safe value, and the heat must be removed from the junction effectively if the device is to survive. Thyristors are specifically designed to survive avalanche-mode operation, but the user must provide current limiting and proper heat sinks where these are needed. Normally, the load itself provides the necessary current limiting, as long as the load current does not exceed the manufacturer's specified limit.

Figure 13-10 shows the structure of the **silicon-controlled rectifier (SCR).** It is a four-layer device, with an anode, a cathode, and a control element called the *gate*. The first four-layer device was called the **Shockley four-layer diode,** named after one of the co-inventors of the transistor. The Shockley diode is a low-power four-layer diode that has no gate. Shockley's four-layer diode is still used to trigger the gate to turn on SCRs in duty-cycle control circuits.

Like ordinary diodes, four-layer diodes have a reverse-bias mode and a forward-bias mode and are designed to allow current to flow in only one direction. Figure 13-11(a) shows the silicon-controlled rectifier in its forward-bias condition. Two of the junctions are forward-biased, but the center junction is reverse-biased.

If we ignore the gate for now, the device operates as a simple four-layer diode, and no current flows in either direction until the anode-to-cathode voltage is increased beyond the Zener knee. Once the critical Zener voltage is exceeded, the middle junction begins the regenerative action that quickly builds to full avalanche. The center depletion zone becomes saturated with carriers, and a heavy current flows. The lamp turns on, and its resistance is the only thing that keeps the device from destroying itself. The two outside junctions were already forward-biased before avalanche started, so current flows freely through them.

Suppose that each of the junctions has a Zener knee voltage of 200 volts. The battery voltage in Figure 13-11(a) would have to be raised to 200 volts or more to start conduction. Once the avalanche has started, it is not easy to stop. Conduction continues until the battery voltage is reduced to a few volts, at which point the current drops below a value sufficient to sustain the avalanche, and the four-layer diode turns off. The minimum current required to sustain avalanche is called the **holding current.**

FIGURE 13-10 **Structure of the Silicon-Controlled Rectifier (SCR)**

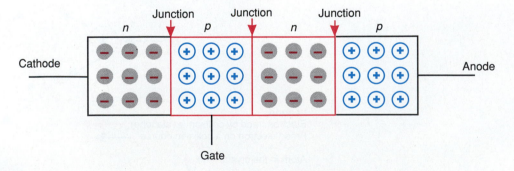

Gate as a Control Element

Now let's look at the gate operation in Figure 13-11(a). Assume that the Zener knee voltage of each of the three junctions is still 200 volts, but that the anode-cathode battery voltage is only 100 volts. Since the circuit still needs 200 volts to start conduction, no current is flowing through the center reverse-biased junction or through the lamp. The avalanche is an unstable mechanism and is easily triggered (like a real snow avalanche).

If we press the push-button to connect a small momentary voltage to the gate in Figure 13-11(a), the small gate current is enough to trigger the avalanche, even though the anode-to-cathode voltage is well below the middle

FIGURE 13-11   The Silicon-controlled Rectifier

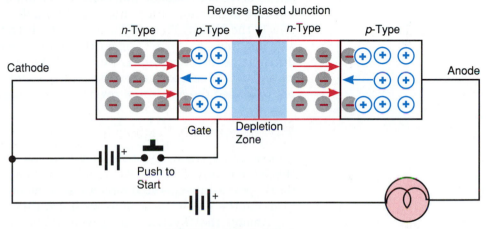

a. Forward-biased Mode

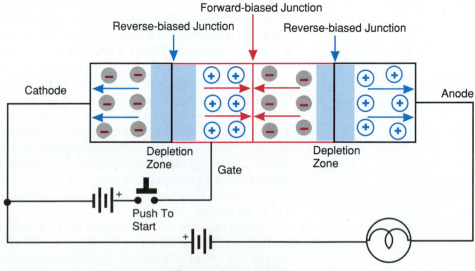

b. Reverse-biased Mode

junction's Zener voltage. Once the gate current has initiated the avalanche process, the avalanche continues to build up quickly to the current limit determined by the load resistance. The lamp turns on, and it stays on after we release the push-button. The gate no longer has any influence on the conduction, nor can we use it to turn off the SCR. We can only turn the SCR off by opening the anode-to-cathode circuit, or by reducing the anode/cathode current below the holding current value.

Reverse-Bias Mode

Figure 13-11(b) shows the anode/cathode supply voltage connected to reverse-bias the SCR. Now, we have two reverse-biased outside junctions and a forward-biased center junction. If we assume that each junction has a Zener knee voltage of 200 volts, it would take 400 volts between the anode and the cathode to drive both outside junctions into avalanche. The reverse-bias **blocking voltage** is therefore 400 volts. In addition, the gate is connected so that it only affects the already forward-biased junction. The gate can't trigger the two outside junctions, so it can't trigger the avalanche process. The SCR does not conduct in this direction unless we exceed the blocking voltage. Normally, we are careful not to exceed the reverse-bias blocking voltage, just as we are careful not to exceed the maximum reverse-bias voltage on an ordinary diode.

Thyristor Family Symbols and Package Styles

Figure 13-12 shows symbols and common case styles for members of the thyristor family. The **TRIAC** is a bidirectional SCR designed specifically for 60-Hz power-line-operated circuits, such as household lamp dimmers. We will examine the TRIAC as soon as we finish this section on SCRs.

Among the thyristor low-power trigger devices in Figure 13-12 is one device that is not a member of the Thyristor family. The **unijunction transistor (UJT)** is not an avalanche-mode or a four-layer device. It is included here because it is commonly used to trigger SCRs when they are used as duty-cycle controllers. The Shockley diode is a low-power SCR without a gate. The **DIAC** is a bidirectional version of the Shockley diode that is designed to act as a trigger device for TRIACs when they are used as duty-cycle controllers.

Figure 13-13 shows a typical conduction curve for a silicon-controlled rectifier. Although it takes a little study, the curve indicates that a higher voltage is needed to turn the device on, and a lower voltage is needed to turn it off. The difference between turn-on anode voltage and the much lower voltage at which the SCR turns off is another case of hysteresis. All thyristors have a large hysteresis voltage. The hysteresis is implied by the holding current section of the curve. Large hysteresis voltages may sometimes present a challenge when we must turn off avalanche devices, but hysteresis is an essential property for R-C oscillators. Ordinary diodes and Zener diodes have no hysteresis, or too little to be useful.

The trigger devices in Figure 13-12 are usually used as the active element in R-C oscillators called **relaxation oscillators.** The term *relaxation* relates to the fact that a charged capacitor represents an electrical stress on the dielectric, which relaxes when the capacitor is discharged. A relaxation oscillator continuously charges and discharges a capacitor in alternate cycles of

FIGURE 13-12 Thyristors and Their Trigger Devices

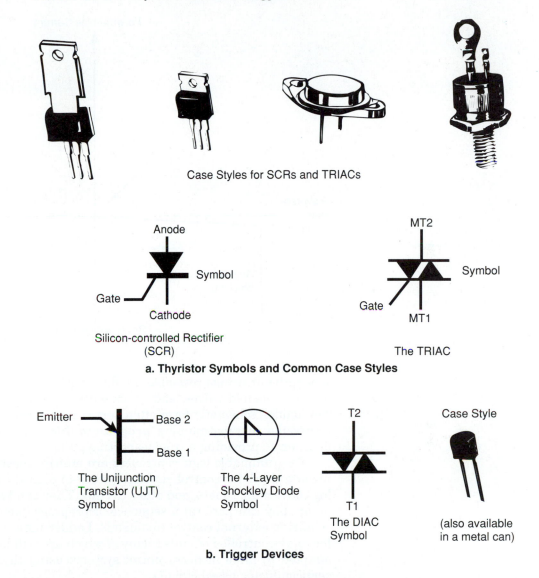

Case Styles for SCRs and TRIACs

a. Thyristor Symbols and Common Case Styles

Anode

Gate

Symbol

Cathode

Silicon-controlled Rectifier
(SCR)

MT2

Gate

Symbol

MT1

The TRIAC

b. Trigger Devices

Emitter

Base 2

Base 1

The Unijunction
Transistor (UJT)
Symbol

The 4-Layer
Shockley Diode
Symbol

T2

T1

The DIAC
Symbol

Case Style

(also available
in a metal can)

dielectric stress and relaxation. Relaxation oscillators constitute a class of oscillators in which the control device must have some hysteresis.

13.5 Introduction to Industrial Ladder Schematic Diagrams

Because members of the thyristor family of devices are specifically intended to control high voltages, currents, and power levels, they find many applications in industrial settings. Industrial electronics often use a different set of symbols and a schematic form called **ladder notation.** In ladder notation, the

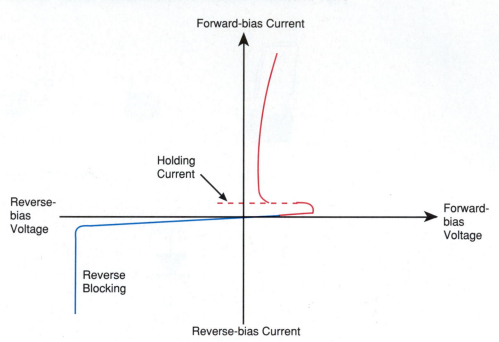

FIGURE 13-13   Silicon-controlled Rectifier Conduction Curve

schematic drawings resemble a ladder with rungs composed of relays, SCRs, motors, solenoid valves, and various devices used in industrial applications. The ladder schematic form often serves as a kind of flow chart to guide the technician in programming **programmable logic controllers (PLCs),** in addition to providing a troubleshooting guide.

Programmable logic controllers are multipurpose "smart boxes" that may include thyristor control circuits, analog-to-digital converters, digital-to-analog converters, timers, and other circuits. They can be controlled by a computer, or they can perform a sequence of preprogrammed control tasks without requiring external control commands. Ladder notation was originally devised for relay-controlled circuits (many of which are still in service), but it has been adapted to more modern control systems using thyristors, PLCs, and other semiconductor based circuits.

The following discussion represents a brief introduction. If you find yourself working in an industrial environment where industrial notation is used, you will have a multitude of new symbols and methods to learn. In some industrial environments, you may find a mix of conventional and industrial schematic diagrams.

Industrial Ladder Schematics Using Relays

Relays are electromechanical switches that use a small amount of power to actuate electrical contacts that control devices that demand larger power. Relays are also amplifiers, in the sense that they control a large power level with a smaller power level. They often have very large power gains. Figure 13-

FIGURE 13-14 Industrial Schematic Notation Using Relays

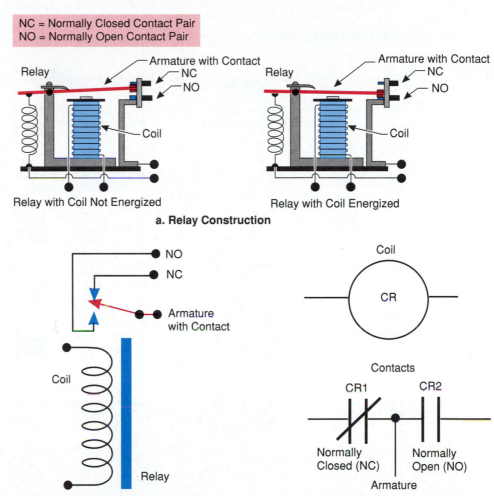

NC = Normally Closed Contact Pair
NO = Normally Open Contact Pair

a. Relay Construction

Relay with Coil Not Energized   Relay with Coil Energized

b. Conventional Schematic Symbol

c. Industrial Schematic Symbols

14(a) shows how simple relays are constructed. Many relays activate several sets of contacts.

Figure 13-14(b) shows the conventional relay symbol, while Figure 13-14(c) shows relay symbols that are used in industrial ladder schematics.

The ladder circuit shown in Figure 13-15 shows the push-button connected to the relay coil in the open condition. The open circuit has the following conditions:

Push-button Open

1 The coil is not energized.

2 Contact pair CR1 is open, so the motor is not running.

3 Contact pair CR2 is closed, so lamp #1 is on.

FIGURE 13-15 Industrial Ladder Schematic Notation Using Relays

FIGURE 13-16 Industrial Schematic for an SCR Ladder Circuit

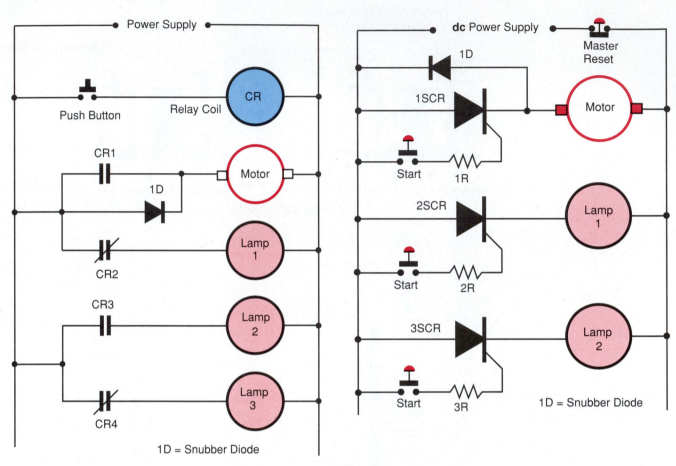

4   Contact pair CR3 is open, so lamp #2 is off.

5   Contact pair CR4 is closed, so lamp #3 is on.

Notice that a snubber diode called 1D extends across relay contact pair CR_1. A motor is an inductive device and can cause contact arcing when the magnetic field collapses. Notice also that the diode is called 1D instead of D_1. Industrial schematics often use 1R or 1C instead of R_1 or C_1, and so on.

Now let's press the push-button. With the circuit closed, we observe the following conditions:

Push-button Closed

1   The coil is energized.

2   Contact pair CR1 is closed, so the motor is running.

3   Contact pair CR2 is open, so lamp #1 is off.

4   Contact pair CR3 is closed, so lamp #2 is on.

5   Contact pair CR4 is open, so lamp #3 is off.

SCR Ladder Circuit

A ladder control circuit is shown in Figure 13-16. If we press the push-button for 1SCR, the motor starts. The SCR stays on after we release the push-button and continues to run until we open the power-supply circuit by pressing the master reset button. When the reset is pressed, the SCR turns off; it stays off until we press the start button again. The two lamp circuits work in the same way.

Resistors 1R, 2R, and 3R limit the SCR gate current to a safe value. Diode 1D is a snubber that serves to protect the SCR from the motor's collapsing field.

Thyristors can tolerate huge overcurrents for short periods of time. The amount of peak current the SCR can survive depends on how long the overcurrent exists. Up to 100 times the normal current is safe if the overcurrent lasts for only part of an **ac** cycle. The manufacturer provides overcurrent ratings for different time periods. A thyristor will survive any overcurrent that does not cause excessive junction temperatures.

This overcurrent capability is important because motors draw starting currents that are much higher than their normal running current. Even incandescent lamps draw about 10 times more current when the filament is cold than they do when the filament is hot. The filament resistance is very temperature-dependent, and the temperature required to produce light output is very high.

13.6   Relaxation Oscillators

Unijunction Transistor (UJT)

The unijunction transistor is not a member of the thyristor family, but its primary use is to provide trigger pulses to SCRs in duty-cycle control circuits. Figure 13-17 shows the construction of a unijunction transistor and its approximate equivalent circuit. The UJT consists of a bar of n-type silicon with an embedded p-type element to form a p–n junction. The p-type area is positioned part way down the block, and the resistance of the n-type block forms a voltage divider. This voltage divider sets the voltage required to forward-bias the junction at a fixed percentage of the base-1-to-base-2 voltage. When the emitter voltage reaches a value high enough to forward-bias the junction, there is a sudden increase in emitter–base 1 current. At the same time, the n-type block is filled with emitter-injected carriers, causing a sudden drop in the resistance between base 1 and base 2. The drop in interbase resistance causes a large current flow of up to 2 amps. The emitter current is typically from 2 to 10 microamps when the UJT is turned on. The UJT has a large hysteresis voltage, which makes it suitable for use as a relaxation oscillator.

UJT Relaxation Oscillator

Figure 13-18 presents the schematic diagram of a UJT relaxation oscillator. The simplicity of the circuit and its relative independence of base 1/base 2 sup-

FIGURE 13-17   The Unijunction Transistor

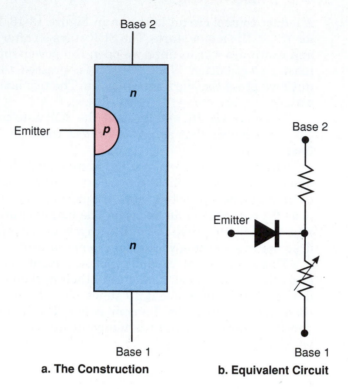

a. The Construction   b. Equivalent Circuit

ply voltage make it a popular trigger device for SCR duty-cycle control circuits. There are not a lot of UJTs to choose from, because they are dedicated to SCR trigger circuits.

How the Unijunction Relaxation Oscillator Works (see Figure 13-18)

1   Capacitor C_1 charges through the timing resistor until the charge voltage is high enough to drive the emitter junction into forward bias. The voltage required to forward-bias the junction is determined by the ratio of the voltage divider formed by the n-type block. The voltage divider ratio is called the **intrinsic standoff ratio** in UJT jargon. The actual voltage required to forward-bias the junction depends on the supply voltage and is always a fixed percentage of the supply voltage.

2   When the capacitor charge voltage drives the junction into forward bias, the interbase resistance drops suddenly, resulting in large current flow through R_{B1} and a large voltage drop across it.

3   As soon as the junction goes into forward bias, the timing capacitor discharges quickly through the emitter–base 1 path. The output is a short-duration pulse. When the capacitor voltage falls, the junction recovers its reverse-bias condition, and the capacitor starts charging for the next cycle.

FIGURE 13-18   The Unijunction Transistor Relaxation
(R-C) Oscillator

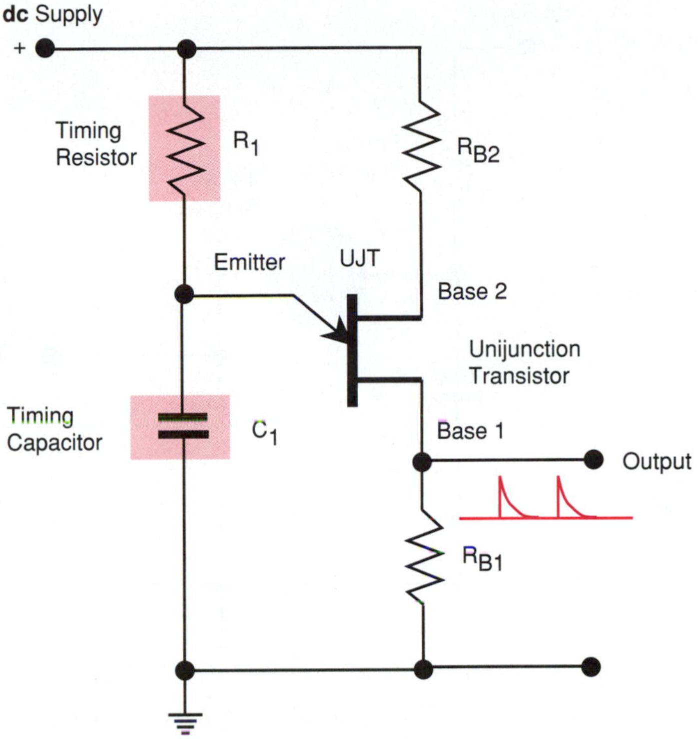

The circuit outputs a string of pulses that are well suited to the job of trig-gering an SCR. The waveshape of the pulses is not very useful for anything but trigger-pulse service. The UJT requires a very small emitter current, so small capacitors can be used for timing. High-quality capacitors are inexpensive in small values.

Shockley Four-Layer Diode Relaxation Oscillator

The Shockley four-layer diode is basically a small SCR without a gate (see Figure 13-19). It has built-in hysteresis that makes it suitable for relaxation oscillator service for SCR triggering. The UJT circuit is generally superior to the Shockley diode circuit, because the timing capacitor in the Shockley circuit must store enough energy to trigger the SCR gate, including some overdrive current. This results in larger, more costly timing capacitors and somewhat less reliable triggering.

How It Works

The timing capacitor (C) charges through the timing resistor (R) until the mid-dle junction of the Shockley diode is driven into avalanche-mode conduction. When the diode turns on, it discharges the capacitor through R_1, which simu-

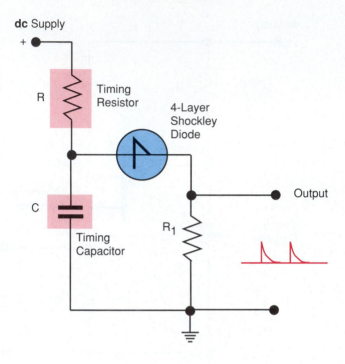

FIGURE 13-19   The Four-layer Shockley Diode
Relaxation Oscillator

lates an SCR gate-to-cathode path. When the capacitor discharge current falls
below the diode's holding current, the diode turns off, and a new charging cycle
begins.

13.7   Duty-cycle Variable-power Control Circuits

Duty-cycle control circuits provide a way to simulate continuous analog control of lamp brightness, motor speed, oven temperature, and various other
power applications. Duty-cycle control takes over when power levels become
higher than those at which power-transistor linear control becomes impractical because of excessive heat production and power loss.

SCR Half-wave Duty-cycle Controller

The SCR duty-cycle variable controller shown in Figure 13-20 uses a Shockley
four-layer diode relaxation oscillator to trigger an SCR. The alternate load
position (Figure 13-20) is often used with motor-speed controllers to introduce
current-mode feedback that helps to adjust for changing loads on the motor
shaft.

The power control circuit in Figure 13-20 is a half-wave circuit at its maximum. The timing circuit and trigger diode allow us to reduce the portion of
the positive half-cycle during which the SCR conducts and powers the load. We
can adjust the SCR firing angle from 0° of the sine wave (full on) to near 180°

FIGURE 13-20 Duty Cycle Power Control Circuit Using a Silicon-controlled Rectifier and a Four–layer Diode Trigger

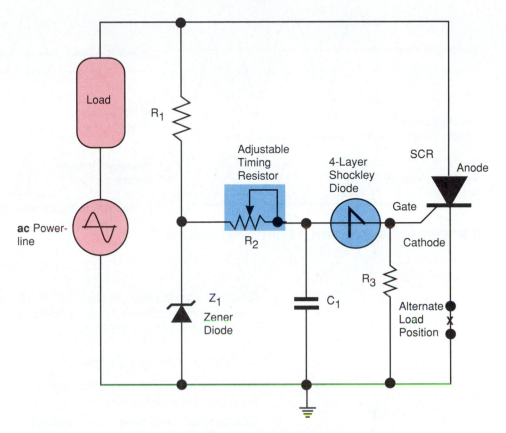

(off). By varying the firing start angle with respect to the **ac** line voltage, we can control the length of the load's on time for each cycle. The load averages the on and off times into a lamp brightness or motor speed.

How It Works

Figure 13-20 shows the circuit, and Figure 13-21 shows the waveforms involved. Examine both drawings as you read the following explanation.

Complete Half-wave Conduction

1   Assume that the sine-wave input voltage is in its anode-negative half-cycle, but rising toward zero crossing.

2   As soon as the sine wave crosses the zero reference line, the SCR anode becomes positive and is in a condition to turn on (forward-biased).

3   If the trigger diode fires and turns on the SCR gate at the very start of the half-cycle, the SCR turns on at the start of the half-cycle (see Figure 13-21(a)). Power is delivered to the load when the SCR turns on. Once it has turned on, the SCR stays on as long as its anode is positive. The SCR anode remains positive until the power-line voltage

FIGURE 13-21 **SCR Waveforms**

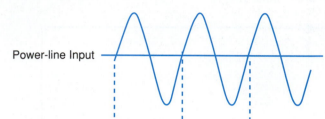

Power-line Input

Shockley Diode
Trigger Pulse

Waveform Voltage
Across the Load

a. SCR is Triggered at the Start of Each Positive Half-cycle

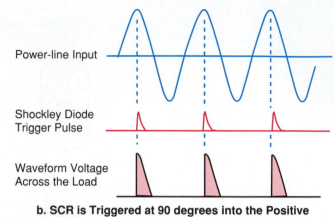

Power-line Input

Shockley Diode
Trigger Pulse

Waveform Voltage
Across the Load

**b. SCR is Triggered at 90 degrees into the Positive
Half-cycle**

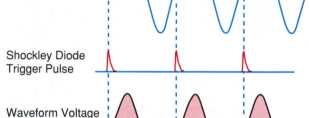

falls to near 0 volts. It then turns off the load power, and the SCR remains off until the trigger diode turns it on again at the start of the next positive-going half-cycle.

Delaying the Trigger Pulse

1   When we set R_2 to a significant resistance value, resistors R_2 and C_1 become a time-delay circuit.

2   The timing starts when the input sine wave becomes slightly positive on the SCR anode. The R-C time constant delays the firing of the trigger diode until the capacitor has charged to the breakover voltage. The trigger diode then fires and turns the SCR on.

3   Figure 13-21(b) shows the power delivered to the load when the trigger diode firing is delayed until 90° into the power-line waveform. The SCR only conducts for one-fourth of the **ac** cycle. The load averages the shorter conduction time into a lower average load power than we had when we started SCR conduction at the start of the half-cycle. We can adjust R_2 for any firing angle between 0° and 180°, from full-load power to zero-load power. When the SCR is on, it has a very low voltage drop and discharges the capacitor.

Zener diode Z_1 ensures against damage to the trigger diode. The Zener may or may not be included. The trigger diode can normally block the negative half of the **ac** waveform. Resistor R_3 protects the SCR from false triggering due to noise sources. Sometimes a capacitor or a combination of a resistor and a capacitor is used to filter out noise transients, depending on how much noise is likely to be present. Multiple trigger pulses are often present, but only the first one counts. Once the SCR is triggered on, it stays on; additional trigger pulses have no effect.

TRIAC Duty-cycle Control Circuit

The TRIAC is a bilateral version of the SCR that can be used to control both halves of the **ac** cycle. It can be used simply to turn high-power **ac** loads on or off; but we will examine the duty-cycle TRIAC control circuit first, because it is so similar to the SCR circuit we just examined.

It is possible to connect two SCRs in opposite directions for full-cycle **ac** control, although the two gates will have different reference points in the circuit. SCRs are sometimes used back-to-back, but the trigger circuit is fairly complex and requires a trigger transformer to control the two gates. Two back-to-back SCRs may still be used if switching frequencies are beyond the TRIAC's limit.

The TRIAC is designed to combine the two gates into a single gate input. The gating conditions vary somewhat with different line and trigger phases but not enough to matter. Because of its more complex internal structure, the TRIAC is limited to 60-Hz operation. 400-Hz devices are available, but typically as expensive military parts.

The DIAC consists of two back-to-back four-layer diodes integrated into a single device, and is the preferred trigger device for a TRIAC. The DIAC presents no special problems because it includes no gates. Figure 13-22 shows the DIAC conduction curve, illustrating its symmetrical, bidirectional operating characteristics.

The structure of the TRIAC and its four possible conduction and gating modes are shown in Figure 13-23. The terminals are called *main terminal 1*

FIGURE 13-22   DIAC Conduction Curve

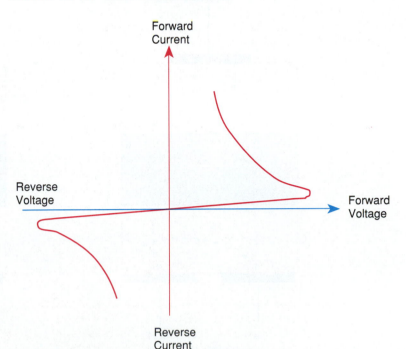

Figure 13-23 The Four Conduction Modes of the TRIAC

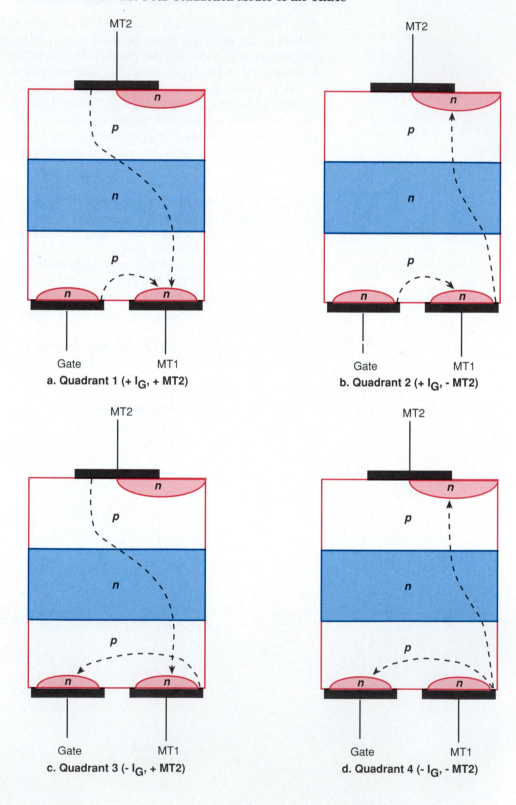

a. Quadrant 1 (+ I_G, + MT2)

b. Quadrant 2 (+ I_G, - MT2)

c. Quadrant 3 (- I_G, + MT2)

d. Quadrant 4 (- I_G, - MT2)

(MT1), *main terminal 2* (MT2), and *Gate*. MT_1 and MT_2 change places from anode to cathode, depending on which half of the **ac** cycle is being applied.

Full-cycle TRIAC Duty-cycle Control Circuit

The full-cycle TRIAC duty-cycle controller shown in Figure 13-24 works exactly like the SCR controller presented in Figure 13-20, except that conduction occurs on both halves of the **ac** waveform and the capacitor charges in both negative and positive directions to provide firing angle delay for both halves of the cycle.

The waveforms given in Figure 13-25 show the relationships among the **ac** power-line voltage, the gate trigger pulses, and the power delivered to the load. In this case, we can vary the time constant of the trigger delay to vary the power that goes to the load. However, the TRIAC/DIAC combination allows load power variations from full off to full cycle on. The SCR is limited to half-cycle operation, from full off to half-power to the load.

On/Off TRIAC Power Controller (Solid-state Relay)

Electromechanical relays were the mainstay of high-power switching for many years, but they required maintenance and frequent replacement. TRIAC (and SCR) circuits have been supplanting most relays in new industrial equipment.

FIGURE 13-24   **Duty-cycle Power Control Using a TRIAC and a DIAC Trigger**

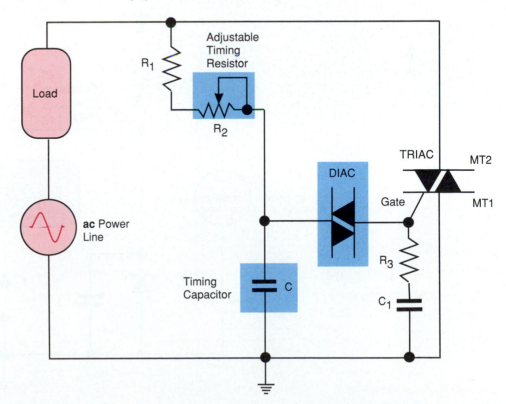

FIGURE 13-25   TRIAC Full-cycle Duty Cycle Control
Circuit Triggered at 90° and 270°

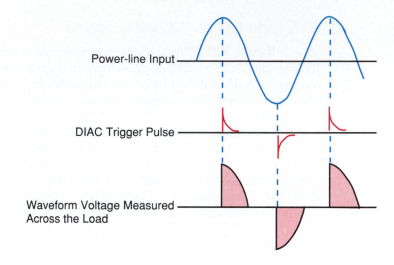

FIGURE 13-26   Power Controller Using a Zero–crossing Detector

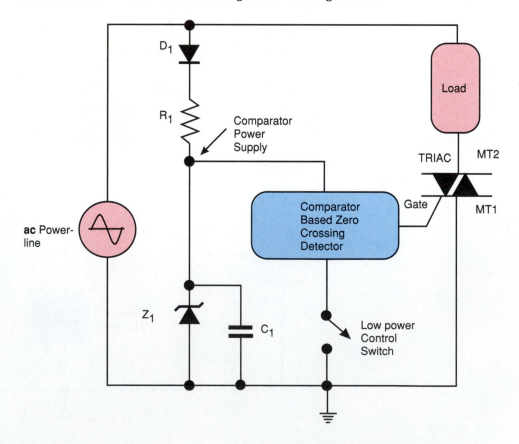

Figure 13-26 shows the circuit most commonly used for **ac** power switching. The solid-state relay generally uses a comparator-based zero-crossing detector to turn the TRIAC (or SCR) on. Turning the load on near 0 volts virtually eliminates problems associated with high motor starting and other start-up surge currents that tend to be deleterious to control devices and loads.

13.8   Interfacing Techniques

Interface problems are very common, and many solutions are available for each problem, but some approaches stand out as superior to most others. In this section, we will look at some of the most straightforward and useful approaches. In particular, we will study two kinds of interface problems in this section. The first involves interfacing two circuits that operate at different voltage, current, or power levels and can't be connected without some intervening interface device. The second involves interconnecting analog and digital circuits.

Using a 311 Comparator to Interface Circuits with Different Operating Voltages

The 311 comparator was particularly designed as an analog-to-digital voltage-level interface. Analog input voltages may be any voltage or any waveform with a variety of voltage levels. Digital circuits use +5 volts and ground as their two binary voltage levels. The 311 is designed as a comparator to convert analog waveforms into binary two-state voltage levels. The 311 is designed to operate with independent input and output power-supply voltages. This design

FIGURE 13-27   Using the 311 Comparator as Analog/Digital Interface

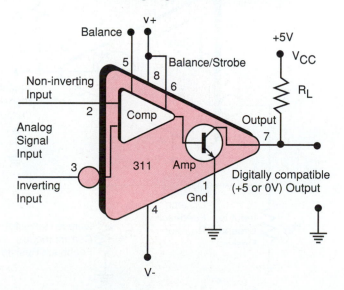

allows simple voltage-level translations for analog-to-digital interfacing. Figure 13-27 shows the most common configuration of the 311.

A comparator is basically an op-amp with no negative feedback. A 311 comparator can be connected as an op-amp simply by adding a feedback loop. The circuit in Figure 13-28 shows how a 311 comparator can be used as an analog voltage-level interface to link two analog circuits that use different supply voltages. The voltage gain of the circuit in Figure 13-28 should be kept low (near unity), because the 311 is a better comparator than it is an op-amp. In most applications, unity gain is what you need anyway. There are several possible variations, including single-supply operation of the op-amp portion of the 311. Bias resistors can be added to the noninverting input in the same way that other op-amps are biased for single power-supply use. Alternatively, a noninverting signal could be taken from the emitter of the output amplifier.

The 311 can be used to interface either analog or digital voltages to the gate of a thyristor, as illustrated in Figure 13-29. Ordinary op-amps can also be used to drive a thyristor gate, but the 311 offers far more flexibility than an ordinary op-amp does.

Optical Isolators

Optical isolators or optical couplers are the safest, most common, and most varied of all available interface devices. Opto-isolators use a light-emitting diode (LED) as the input circuit. The LED may be visible red or infrared. From

FIGURE 13-28   **Voltage Level Translation Using a 311 Comparator as an Op-amp**

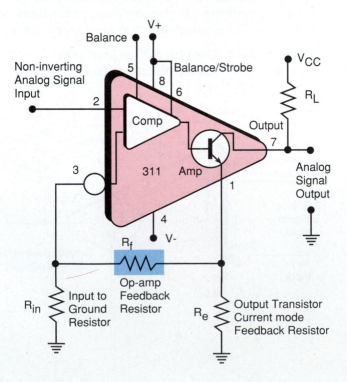

FIGURE 13-29 Interfacing a Comparator to an SCR

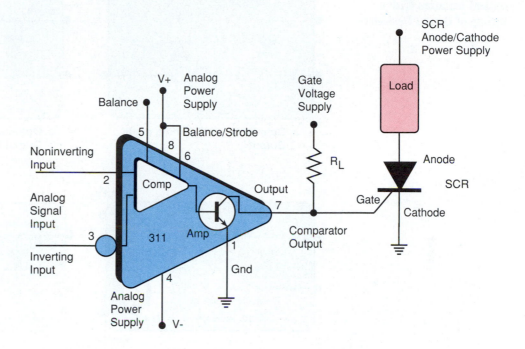

a troubleshooting standpoint, it doesn't matter which kind of LED is used, because we can't see into the sealed opto-isolator case, anyway. There is no visible clue.

Output devices range from a simple photodiode to complex zero-crossing TRIAC circuits. Figure 13-30 shows the most common opto-isolator configurations. Package styles also come in a wide variety, from 6- or 8-pin DIPs to slot detectors to fiber-optic cables between the input LED and the output detector.

The output devices in Figure 13-30 show the control element, the base of a transistor, the gate of a thyristor, and so on, of the output devices brought out to the outside as a connection. Some devices permit external connection to the control element, while others do not.

Power MOS-FETs as Combined Power-control and Interface Elements

The insulated-gate **MOS-FET** power transistor provides power control for voltage levels of several hundred volts and virtually any desired current, because power MOS-FETs can easily be connected in parallel. International Rectifier Corp., for example, manufactures a wide variety of high-performance power MOS-FETs with the trade named HEXFET.®

The HEXFET® line has proved to be very resistant to hostile environments, such as heavy industry and automotive service. Isolation is achieved between input and output by insulation of the MOS-FET gate from the sub-

FIGURE 13-30
Optical Isolators With a Variety of Output Devices

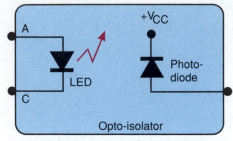

a. Opto-isolator with Photo-diode Output

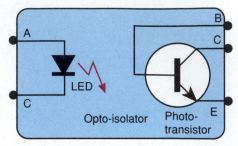

b. Opto-isolator with Photo-transistor Output (4N28)

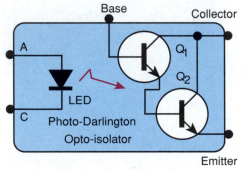

c. Opto-isolator with Photo-Darlington Output (4N31)

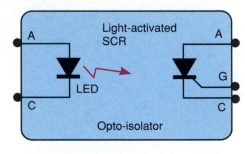

d. Opto-isolator with Photo-SCR Output (PS3000)

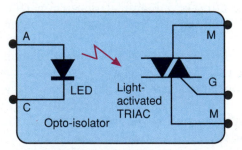

e. Opto-isolator with Photo-TRIAC Output (MCP3022)

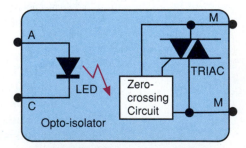

f. Opto-isolator with Photo-TRIAC Output with Zero Crossing Detector

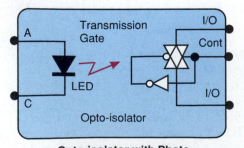

g. Opto-isolator with Photo-transmission Gate Output

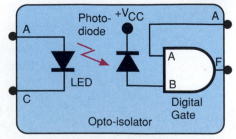

h. Opto-isolator with Photo-diode/ Logic Gate Output (MCL2601)

strate, with current devices providing several hundred volts of isolation. The HEXFET® line also includes devices with built-in insulation between the case and heat sink, but with excellent thermal conductivity. Isolation between device and heat sink is up to 2000 volts.

Power MOS-FETs are also quieter when switching than are most other power devices available, and they are capable of fairly high switching rates, making them suitable for duty-cycle control in environments where electrical switching noise can be a problem. Power MOS-FETs are probably the best device for controlling **dc** in high power applications.

Figure 13-31 shows how a complementary pair of power MOS-FETs might be used in an electric car. A motor can become a generator when it is turning and is connected to a load. Because a generator converts the inertial energy of the moving car through the wheels, we could use the energy thus generated to help recharge the battery, and slow down the car at the same time. This is not perpetual motion—only a way of recovering some of the energy that would normally be wasted in friction braking.

This kind of loaded generator braking, called *dynamic* or *regenerative braking,* has long been used on diesel-electric locomotives. In the locomotive case, however, the dynamic braking energy is customarily wasted as heat in large load resistors.

FIGURE 13-31   Power FETs Used in a Possible Controller for Electric Automobiles

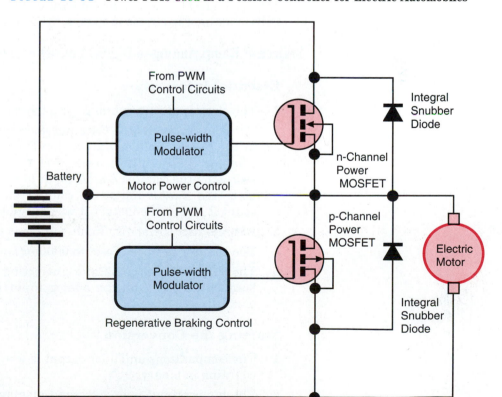

Analog/Digital Conversion

You are unlikely ever to have to troubleshoot an analog-to-digital converter. They tend to be complex and must be precision devices, so they are typically designed as inexpensive integrated-circuit productions. It is not practical to construct analog-to-digital converters from individual parts, like the one shown in Figure 13-32.

There are three common types of analog-to-digital converters: the flash converter, consisting of a comparator ladder and a digital binary encoder; the successive approximation converter; and the slope or ramp converter, like the simplified version in Figure 13-32. There are many variations of each form.

Ladder comparators, which we have already investigated, yield the fastest converter type because they process all analog levels at the same time. The only conversion time is comparator switching time, so they have earned the name of **flash converters.** A very large number of comparators are needed for practical applications, and they are expensive. The flash converter is well-suited to high-quality audio and video analog-to-digital conversion.

Successive approximation converters use digital components that are beyond the scope of this text, but you will find them covered in your digital course.

The **slope** or **ramp converters** are the kind used in digital multimeters and similar systems where low cost is critical and high-speed conversion is not. There are several variations of the slope converter. The one we will examine generates a staircase ramp instead of a smooth ramp of the kind that might be produced by an op-amp integrator.

Let's examine the circuit in Figure 13-32 to see how it works.

Staircase Ramp Analog-to-Digital Converter, Step-by-Step

Essential Information

1   The clock is always running, producing a string of digital pulses.

2   The counter counts the clock pulses in binary, as long as it is enabled by the *Enable* input on the counter. The comparator output determines whether the counter is permitted to count clock pulses or not, by switching the counter *Enable* input low or high. If the comparator disables the *Enable* input, the counter stops counting but retains the count it had at the instant it was disabled.

3   Assume that the analog input voltage is 8 volts **dc.**

4   The counter has been reset to 0000 (Q) outputs.

5   The output of the comparator's inverting amplifier is low at the start, because the +8 V on the analog input is dominant. The counter is enabled.

Starting the Conversion

1   The comparator's amplifier output is low, allowing the counter to start counting in binary.

2   The binary count is decoded and displayed on the digital display (or sent to some other digital circuit).

FIGURE 13-32 Staircase Ramp Analog/Digital Converter

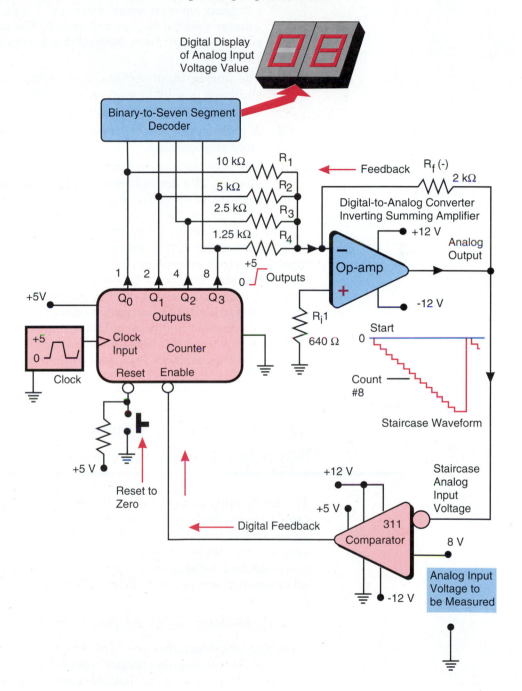

3   At the same time, the counter outputs are fed into a weighted summing op-amp. The summing op-amp is called a *digital-to-analog converter* in this case.

4   Each time the binary counter increases by 1 count, the output of the summing op-amp steps down by 1 volt. On the eighth count, the counter's digital display will read 8. The staircase input voltage to the comparator will have stepped down from 0 V to –8 V (+ a millivolt or two).

5   The analog input voltage is +8 volts, and the output voltage of the staircase is a little over –8 volts. The comparator switches low. The comparator output stage inverts the low to a high.

6   A high on the *Enable* input of the counter disables the counter and stops the counting.

7   The counter stops counting, but it retains the 8 count it had when it stopped. The display remains at 8, and the summing amplifier output remains at –8 volts. Press the reset button to start a new conversion.

The important concept in this circuit is that no real conversion from analog 8 volts has actually happened. Instead, two circuits—one analog and one digital—are both processing data in their own forms. They are synchronized so that the analog voltage of the staircase reaches 8 volts at exactly the same time as the digital counter reaches a digital count of 8. No actual conversion from analog to digital has actually taken place, but the outcome is the same as if a real conversion had occurred. It is important to understand that there is not really any direct relationship between analog and digital data that we can use for direct translation. Still, we can find clever ways to fake it.

Troubleshooting Power Supplies and Interfaces

Troubleshooting power circuits presents some hazards not normally encountered in most electronic circuits. High voltages and currents can be dangerous, so you should troubleshoot and repair power circuits with the power off whenever possible. Fortunately, you can generally remove SCRs, TRIACs, and other power devices without worrying about circuit board damage. They are generally mounted on a separate heat sink and can easily be removed and tested.

Troubleshooting Switching Power Supplies

Switching power supplies are hard to troubleshoot, but you can check **dc** voltages and the rectangular output waveform at the output of the pulse-width modulator chip. Make sure that there are power-supply voltages on the PWM chip.

Testing SCRs and TRIACs

Calculate a resistor value in Figure 13-33 for R_1, based on the following rule of thumb:

FIGURE 13-33   Testing SCRs and TRIACs

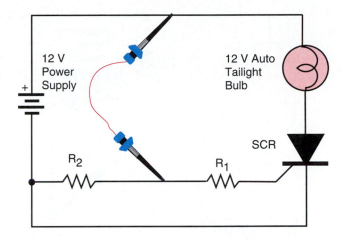

Note: Reverse the battery to test TRIACs in the reverse direction.

$$\text{Gate current} = 1 \text{ mA}$$

for each continuous anode (or MT_2) current amp. Use 6 volts for the calculation. Make R_1 equal to R_2.

Connect the test lead as shown. The lamp should light. Remove the test lead, and the lamp should stay on. Reverse the battery to test the opposite direction in a TRIAC.

Troubleshooting SCR and TRIAC Duty-cycle Controllers

Three test points are shown in Figure 11-34. The typical waveforms for each test point are shown on Figure 13-35. TP1 may be an inverted version of the one shown in Figure 13-35; it reflects the load variations. You can't test directly across the load unless you use a battery-operated scope. The waveform should change as you vary R_2.

Test point 2 tells you whether the trigger oscillator is working. The trigger pulse will change rates if the oscillator is working properly. Test point 3 is mostly a backup oscillator test. The actual waveform varies a great deal, depending on several factors.

The Digital Connection: What in the World Is Fuzzy Logic?

Fuzzy logic lies somewhere between the analog and the digital worlds. Fuzzy logic was invented by Lotfi Zedah, a professor at the University of California, in 1964. By 1973, he had defined practical applications for it. The Japanese

FIGURE 13-34   Troubleshooting SCR and TRIAC Duty Cycle Controllers

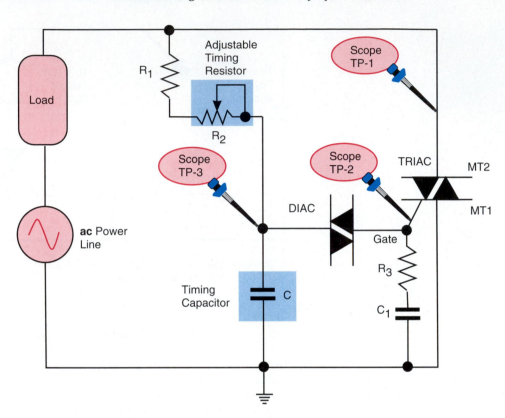

jumped on the fuzzy logic bandwagon early, and the controllers now used for the Sendai trains are based on fuzzy logic; these trains are without question the smoothest operating in the world. A multitude of products in Japan and Europe are based on fuzzy logic.

In the United States, almost no products are based on fuzzy logic. The nation's leading engineering experts called it imprecise and mathematically unsound, and ignored it. Since these experts control research grants and advise major companies, fuzzy logic found no place in the United States. We are just now chasing the bandwagon down the street, trying to catch up with it. We again exported our technological creativity to Japan and Europe at no charge.

The video camcorder from Japan that corrects for jiggling of the camcorder and makes the final results as steady as if the instrument had been on a tripod is based on fuzzy logic. In Japan the term *fuzzy* is synonymous with "good technology" and is widely hyped. American advertisements for Japanese and European products don't mention it, because here *fuzzy* has a negative connotation. NASA has used some fuzzy logic in its control systems and is working on other applications; and a few American companies have used fuzzy logic controllers in industrial control; but in general, fuzzy logic has not been welcome in the United States.

Fuzzy logic has found a place in automatic control systems because control systems generally interact with the analog world of natural phenomena,

FIGURE 13-35   **Troubleshooting TRIAC Duty Cycle Control Circuits**

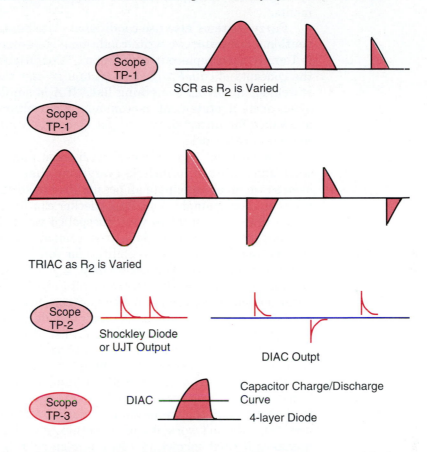

Note: The capacitor charge/discharge waveform may vary a great deal.

where it is usually impossible to build a sufficiently complex mathematical model to deal with nature's infinite complexity and resolution. Fuzzy logic gets results because it can function without precise values for every variable in the mathematical model on which the system is based. Fuzzy logic can deal with the imprecise values that we usually have to work with.

The binary logic used to design conventional control systems and computers is called *crisp logic*. The truth table for deriving a Boolean equation (the standard binary logic in modern digital systems) requires that the results of every possible condition be defined. We must assign a binary value of true or false (0 or 1) for every possible condition. If you are not sure, you must still fill in a value, yes or no. There is no provision for uncertainty. In fuzzy logic, allowances for uncertainty are part of the system.

The power of modern digital computers and controllers lies primarily in the "If . . . , then . . ." conditional statements that allow the computer to behave differently depending on the outcome of a particular operation. Without the conditional statement, the computer would be nothing more than a programmable calculator. The computer could not do a search and replace operation, for example, without being able to implement conditional statements.

Fuzzy systems also use conditional statements, in a fuzzy form. In a conventional computer, we write conditional statements in the form, "If the word is Tammy, then replace it with Tammi." Conditional statements are limited to the concepts of *equal to, less than,* and *greater than.* There is no conditional statement involving something like, "If A is similar to B, then ... " In fuzzy logic, such a statement is common. Conventional logic is crisp, with no allowance for uncertainty, and yet electronic controllers must deal with the uncertain real world.

Conventional logic can deal with some uncertainty, given enough hardware and software to include everything that might possibly go wrong; but even so we must anticipate all possibilities in advance and include them in the program. For example, a spelling checker can find a group of correctly spelled words that are similar to your misspelled word. The spelling checker uses a complex set of rules to make a crisp comparison with words stored in its dictionary and displays the possibilities for you to select from. But if your word is not in the dictionary, you get no help. If you transpose two letters in a word, you may get some pretty bizarre results, or none at all. If you violate the preprogrammed rules, you get nothing or nonsense.

Logic controllers must make their own decisions on the fly. They can't wait for a human operator to select from a list of possible actions when they run into a condition that was not taken into account in the programming model. A crisp system may crash or do nothing.

Fuzzy logic behaves more like a human operator. It takes its best guess and does something reasonable. In an emergency, taking no action or crashing is almost always the wrong thing to do; but a fuzzy system takes some action, and if that doesn't work, it tries something else. The fuzzy controller may make mistakes it can't correct, just as a human operator may, but more often than not it copes with the unexpected quite well.

Before fuzzy logic, uncertainty was the province of statistical or probability models; and probability is the tool used in conventional controllers to account for uncertainty. But probability is based on faith to a large extent. When you flip a coin, probability tells you that it should come up heads 50% of the time. You have faith that flipping a coin is a perfectly fair way to decide who gets the last piece of cake, but is it really perfectly fair?

Any gambler can tell you that there is often some order hidden in seeming randomness. The gambler's lucky streak often involves the same numbers coming up several times in defiance of the odds. Chaos theory and fractal geometry have uncovered order in things we have always assumed were random. The gambler's lucky streak is probably a case of order within randomness that we don't yet understand.

The odds against your winning the lottery may be 20 million to 1. But the odds are also 20 million to 1 against the person who actually wins it. For the winner, probability theory failed to be good predictor. Similar things happen in controllers. A part fails. Even though the odds against failure were 20 million to 1, it failed on the second try, possibly disastrously. Taking polls is a probability exercise that attempts to predict the outcome of an election, or the future success of a new product, by polling a sample of the population. Ironically, pollsters have unwittingly introduced a fuzzy set into their pie charts.

The pie wedge vaguely labeled *"undecided"* or *"not sure"* is a fuzzy set. The pollsters tend to discount that segment, but if it includes enough people who have not yet decided but will decide to vote *for* the ballot proposition at the polling booth, the fuzzy set may determine the final outcome. The set is fuzzy, because it lumps careful voters who will later make an informed decision together with people who probably won't vote and with people who always vote against any proposition that might increase taxes, but don't yet know how this proposition might effect taxes.

Nearly all human decisions are fuzzy in the sense that we hardly ever know everything there is to know about a subject. Traditional logic requires an accurate mathematical model of the system that a controller must control. Fuzzy logic is based on rules, approximations, and some "try it, and if it doesn't work try the next most reasonable rule."

One of the best examples of a fuzzy logic device is a camera designed for news photographers. In the middle of a riot, the photographer has no time to make camera adjustments. The fuzzy logic data base in the camera's memory contains a number of carefully selected archetype photos. The camera looks at the scene and compares it to the photos in its memory. It finds the stored photo most like the real scene and uses the settings that were used to make the original archetype photo to set the camera for the real scene.

Conventional logic would have to find an exact match between the real scene and the photo stored in memory, which of course would never happen. Alternatively, a conventional logic system would have to execute a complex mathematical algorithm to determine the best match. By then, the riot might be over.

Fuzzy logic is not the answer to all of our control and computer problems, and many existing fuzzy logic controllers are a mix of conventional digital logic and fuzzy logic. However, fuzzy logic is better (and far less expensive) than crisp binary logic at dealing with the complexities of the analog world. The next generation of computers will likely be a combination of conventional digital and fuzzy circuits, along with some neural networks, which attempt to emulate human brain operations.

Multimedia offers a fertile ground for fuzzy logic because it involves trying to duplicate the enormous complexity and infinite resolution of sights and sounds in the real analog world. The sheer digital computer power required in current computers is pushing both the technology and the cost-limit envelopes. We have reached a point where sheer speed and more hardware are probably not a viable approach for the future.

Fuzzy logic has been reduced to silicon chips, selling in Japan for as little as $5.00. You won't find them in your electronics catalog at this writing, but your next new car may have fuzzy logic controllers. It may be a while before the term *fuzzy* becomes acceptable. In the meantime, you will probably use some products that use fuzzy logic chips, without your knowing it.

Several major American companies are currently working on fuzzy logic research, but they tend to be secretive in a highly competitive field. Most of the information available about fuzzy logic is written in Japanese or German. Books on the subject were not available in this country, in English, until 1982, and they are still hard to find. Expect to find fuzzy logic becoming a standard part of the electronic engineering curriculum very soon.

QUESTIONS AND PROBLEMS

13.1 How does the basic linear voltage regulator operate?

13.2 What are the differences between the linear regulator and a class-A power amplifier?

13.3 What are the feedback-controller components in a linear voltage-regulator circuit?

13.4 How is the voltage stepped down in a switching regulator?

13.5 How is the voltage stepped up in a switching regulator?

13.6 How is the voltage inverted in a switching regulator?

13.7 What are the feedback controller components in a switching regulator circuit?

13.8 What are the advantages and disadvantages of a switching regulator?

13.9 What is the condition of each of the three junctions in a forward-biased SCR?

13.10 What is the condition of each of the three junctions in a reverse-biased SCR?

13.11 How does the gate in an SCR work?

13.12 Why can't the gate turn the SCR off after it has turned it on?

13.13 What are the two common trigger devices used with SCRs?

13.14 What is a TRIAC, and how does it work?

13.15 What is the most common device used to trigger TRIACs, and how is it constructed?

13.16 How can duty-cycle control simulate analog variable control?

13.17 What is the major difference between an SCR and a TRIAC duty-cycle controller?

13.18 How does a unijunction transistor (UJT) operate?

13.19 What requirement must a load satisfy in order to be used for duty-cycle control?

13.20 How do the Shockley four-layer diode and the DIAC differ?

13.21 Explain the operating mode used in thyristors.

13.22 How does the operating mode in thyristors differ from that used in ordinary diodes?

13.23 In what environment would you be likely to find ladder schematics?

13.24 How is a relay constructed, and what is its function?

13.25 List three ways the 311 comparator can be used to interface devices that have different power-supply voltage requirements.

13.26 How is an optical isolator constructed, and how does it operate?

13.27 List several common output devices used in opto-isolators.

13.28 What main feature of the opto-isolator makes it so popular?

13.29 List three reasons for using power MOS-FETS in power-control systems.

13.30 List three conditions under which power MOS-FETS are preferable to SCRs or TRIACs.

13.31 Define the term *snubber diode*.

13.32 List three kinds of analog-to-digital converters.

13.33 What is a *flash* analog-to-digital converter?

13.34 How does the staircase slope type of analog-to-digital converter work?

13.35 What common analog circuit is used for digital-to-analog conversion?

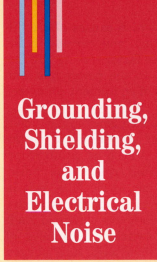

Grounding, Shielding, and Electrical Noise

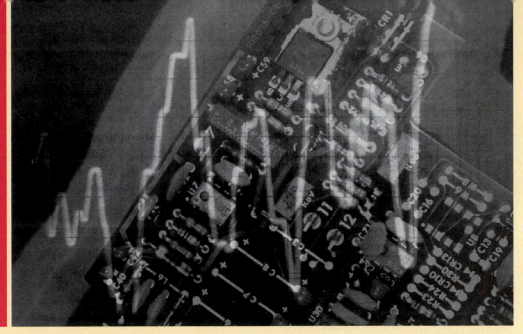

Imagine for a moment that you are the lone electronics technician working with a crew of industrial maintenance electricians on the campus of a university. On Monday morning, the foreman calls you into his office and tells you that the variable-frequency drive that controls an exhaust fan motor in one of the research buildings has been shutting itself down randomly over the weekend. Experiments with toxic chemicals are conducted under the fume hood, and this fan (when it operates properly) ensures the safety of the lab techs who perform the experiments.

"You have to get this problem fixed. If the fan quits at the wrong time somebody could inhale toxic fumes and get sick, or worse . . ."

After performing every test you can think of, calling the manufacturer for advice, and finally replacing the whole variable-frequency drive unit, you observe that the problem still persists.

What do you do?

With increasing regularity, computerized (and other) electronic equipment exhibits behavior that cannot be explained or observed with ordinary methods. Often the problem is rooted in the phenomenon of electrical noise, which finds its way into the equipment's circuitry. The task of determining that noise really is the source of the problem, and then effectively eliminating it, is a big challenge for the electronics technician.

Detecting and eliminating electrical noise constitute an entire field of specialization. More and more equipment falls victim to this kind of problem, largely because of the many types of equipment being used that produce noise in one form or another. Laboratories, industrial plants, and office complexes are just a few of the environments affected by these problems.

Electrical noise reaches electronic equipment in two different ways. The first is by electromagnetic induction. Current flowing through an electrical conductor creates an electromagnetic field around the conductor. At higher frequencies, these fields radiate easily through space (as in the case of a radio transmitter) and induce currents in other conductors within that field. Although manufacturing standards require that electrical and electronic equipment limit their radio frequency (RF) noise, many sources contribute to the mishmash of signals all around us. To minimize the problem of noise pickup, equipment is built with electromagnetic shielding at the source, to limit the amount of noise radiated. Sensitive electronic equipment must also be grounded properly and shielded to prevent current from being coupled into the equipment. This is also why properly grounded shields are necessary on signal wiring in most situations.

The second way noise may be coupled from a source to a receiver is directly through common wiring such as the **ac** power line. Although most

Noise Induction and Grounding

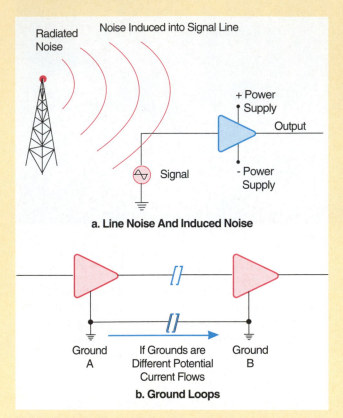

a. Line Noise And Induced Noise

b. Ground Loops

electronic power supplies are designed to shunt noise signals to ground, they are not always perfectly efficient, especially when high levels of noise are present on the line.

Even though manufacturing standards limit noise, it is often present due to faulty equipment. In addition, a poorly designed or improperly installed wiring system, as well as other sources, can allow noise to get through.

Probably one of the biggest reasons that noise problems are increasing is the nature of the loads connected to the commercial power system. Many types of equipment use microcomputer and other new technologies. In some cases, this equipment controls the current draw of heavy industrial loads. An example is the variable-frequency drive mentioned earlier. These units control the speed of **ac** motors by controlling the amplitude, frequency, and phase of the **ac** waveform supplied to the load. The frequency of the current drawn by the load often has no relationship to the line frequency. Add to this the typically inductive nature of motor and

other industrial systems, which causes phase differences between the voltage and current in the system. Cumulatively, these factors create odd frequency noise components on the power line, and this can affect many types of equipment connected to the power source.

Another source of line-frequency-independent noise is the switching power supplies found in computers and other equipment such as uninterruptible power supplies. Besides continuous noise, short-duration transients can cause false switching, firing, and resetting of electronic systems. Transients are caused when heavy electrical loads are started, stopped, or switched. Examples of transients include power-line sags and surges, blackouts, frequency inaccuracies, and other short-term voltage changes. Lightning is another major cause of transients. Characterized by extremely high voltages of short duration and very fast voltage-rise times, lightning is difficult to protect against.

Although they are not a source of noise themselves, ground loops are a common cause of noise entering electronic systems. Ground loops are caused when two interconnected pieces of equipment, located some distance apart, are electrically referenced to the earth ground at each of their locations. If a potential difference exists between those two earth grounds, current will flow from one location to the other. This current may fluctuate, depending on the potential difference, causing a noise signal to be superimposed on the interconnecting wiring.

Many techniques are used to reduce noise, but for brevity, only a few categories will be outlined here.

Grounding

The most important consideration in grounding is that, for any system, there should be only one electrical reference point. This eliminates the noise created by ground loops and allows shielding and filtering to shunt noise signals to that common point. All signal processing is then handled with reference to that point. One of the most important conventions dealing with signal wiring is called *source grounding*. For best results, unbalanced signals should be referenced to ground at the end that is the source of the signal. This is also true for shielding. One exception to this rule involves signals from low-level sensors, where the shield should be grounded at the receiving end only.

Filtering

Much of the objectionable noise carried into electronic equipment is high-frequency-related. Using low-pass filters in many systems is a simple way of eliminating noise. Sometimes band-pass or notch filters may be required to remove specific noise frequencies. In some situations, analyzing the noise components present may require using some kind of spectrum analyzer. Once the problem frequencies are identified, they may be eliminated with appropriate filters or other methods.

Transient Devices

As mentioned earlier, power-line-based transients come in a variety of forms. Spikes and other overvoltage transients can be clamped or limited by using devices such as capacitors, Zener diodes, varistors, gas discharge tubes, and silicon avalanche diodes. The peak voltage, rate of rise, and duration of the spike dictate what type of suppression is required. Power-line sags can sometimes be handled by constant-voltage transformers. Almost all of these problems can be eliminated by using an uninterruptible power supply, which charges a battery from the line and then resynthesizes the line power from the battery voltage.

As previously mentioned, solving noise-related problems in electronic systems is becoming a large and important field. The information given here only touches the surface of this area. Solutions are diverse and applications are specific. The technician who becomes knowledgeable about the underlying conditions that cause noise and the practical techniques that can be used to solve these problems can expect to be in demand in the future.

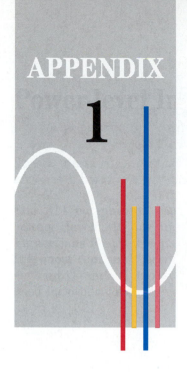

APPENDIX 1

Answers to Odd-numbered Questions and Problems

1.1 The sine wave is the only pure waveform because it consists of a single frequency and has no harmonic frequencies.

1.3 All waveforms consist of at least one sine wave. Other waveforms add additional sine waves of simple multiples (1, 2, 3, 4, etc.) of the fundamental sine-wave frequency. The additional sine waves are called *harmonics*.

1.5 See Figure A-1.

FIGURE A-1

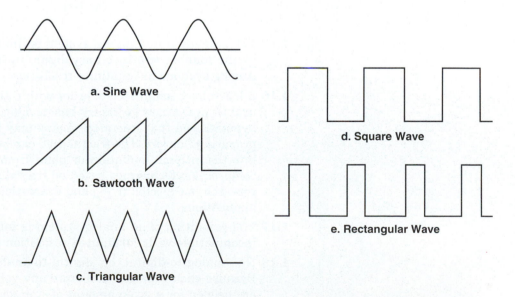

a. Sine Wave

b. Sawtooth Wave

c. Triangular Wave

d. Square Wave

e. Rectangular Wave

1.7 The attack time is the time from the initiation of a sound (as in plucking a string) to the time when it reaches its full volume. The decay time is the time it takes a sound to fall from full volume to inaudibility.

1.9 A square wave contains harmonics, and harmonics contain energy. The 500th harmonic of 200 kHz falls within the FM band. Receivers are sensitive to a few microvolts.

1.11 Please see Table A-1.

TABLE A-1   Binary Count Table

Binary Count Table

	Binary Column Heads			
	8	4	2	1
0	0	0	0	0
1	0	0	0	1
2	0	0	1	0
3	0	0	1	1
4	0	1	0	0
5	0	1	0	1
6	0	1	1	0
7	0	1	1	1
8	1	0	0	0
9	1	0	0	1
10	1	0	1	0
11	1	0	1	1
12	1	1	0	0
13	1	1	0	1
14	1	1	1	0
15	1	1	1	1

Binary Numbers

1.13 The resolution in a digital system is limited by the system—usually by the number of bits (binary digits) in the system's standard word. Analog systems have infinite resolution.

1.15 A fractal is a small seed of order within the seeming chaos in all natural things. A new geometry has developed around fractals, and it is hypothesized that fractal geometry may be based on a fundamental property of nature. The fractal seed can be mathematically expanded into the original mountain or plant from which it was taken. Data-compression techniques based on fractal geometry already exist, but research is expected to expand dramatically the number of practical applications.

1.17 The resolution of an analog system is infinite and is technologically incompatible with our limited-resolution monetary system.

1.19 No analog-to-digital or digital-to-analog converter is required. Because each capacitor can store any voltage level, fewer capacitors are needed for a given amount of storage.

1.21 a An analog signal carrying digital data.

1.23 a Always analog.

2.1 The transition (depletion) zone is an area on both sides of the junction that has almost no holes or electrons.

2.3 c The element or compound from which it is made. Most diodes are made of silicon and have a 0.6- to 0.7-volt junction voltage. Gallium arsenide light-emitting diodes have junction voltages near 1.2 volts.

2.5 d A zone on either side of the junction that has very few available carriers.

2.7 c Electrons move toward the p-block, and holes move toward the n-block.

2.9 e The current (in silicon devices) either triples or doubles for each 10-Celsius-degree increase in temperature. There seems to be credible authority for both values. See the note in the main text.

2.11 a The voltage has virtually no effect on the reverse-bias current.

2.13 A reverse-biased junction consists of a p-type block (conductor) and an n-type block (conductor) separated by the depletion zone (an insulator). Increasing the reverse-bias voltage makes the depletion zone wider (thicker), decreasing the capacitance.

2.15 *Recovery time* is another name for storage time.

2.17 $R_J = 25$ mV/I_D: junction resistance equals the constant 25 mV divided by the current through the junction (in mA). The 25-mV-value is an empirical constant.

2.19 Avalanche conduction is a case of carrier multiplication by collision and is an extension of the Zener part of the operating curve.

2.21 Peak reverse voltage, maximum continuous forward current, and maximum peak forward current.

2.23 The chief failure mechanism is excessive junction temperature.

2.25 a As dynamic resistance.

2.27 Peak inverse voltage (P_{IV}) is the Zener voltage threshold.

2.29 The maximum continuous forward (bias) current.

2.31 V_F is the forward-bias voltage drop. It mostly consists of the 0.6- to 0.7-V junction voltage, but it may include some ordinary resistance voltage drop as well.

2.33 Approximately 0.6 to 0.7 volt.

2.35 a 2.8 volts.

    b 2400 volts.

    c 2.8 volts.

    d The leakage current will increase by an amount controlled by the diode that has the smallest increase.

2.37 A resistor must be added in series with each diode.

2.39 Yes.

2.41 It is faster because it does not have a storage-time or junction capacitance problem.

2.43 The Schottky diode's forward resistance is higher, and its reverse resistance is lower.

2.45 Coherent waves have a single frequency and are in phase. Incoherent waves are a band of frequencies with a variety of phases.

2.47 LEDs are made of gallium arsenide or gallium arsenide phosphide compounds.

2.49 b Reverse-biased mode.

2.51 Isolating two circuits with different power-supply voltages. Coupling a digital circuit to an analog circuit.

2.53 The current flowing through the circuit is approximately 1 amp. If you assumed a junction voltage of 0.6 volt, the current is 0.94 amp. If you used 0.7 volt, the current is 0.93 amp.

2.55 a The voltage drop across the resistor is 9.3 to 9.4 volts (depending on your answer to Problem 2.53).

2.57 a The diode is reverse-biased.

2.59 The voltage drop across the diode is approximately 10 volts. The meter used to measure the voltage acts like a 10-megohm (or so) resistor to complete the circuit. This is a practical concept worth some discussion.

2.61 The diode junction resistance is an open circuit (infinite).

3.1 60 Hz.

3.3 120 Hz.

3.5 The half-wave rectifier is approximately 32% efficient. The full-wave circuit is approximately 64% efficient.

3.7 The rectifier converts **ac** into pulsating **dc.**

3.9 The power available from the plugs in homes, offices, and so on is alternating current (**ac**), at a higher voltage than most modern electronic devices can tolerate. Most semiconductor devices, transistors, integrated circuits, and so on cannot tolerate alternating current. Batteries are fine, but you know the problems with batteries! A good power supply provides a battery-like **dc** power that is always ready.

3.11 The output of a rectifier is a pulsating voltage, which is not suitable for most electronic devices. The filter eliminates enough of the pulsations to make the output **dc** voltage useful.

3.13 a $1.4 \times 40 \times 0.316 = 17.696$

   b $1.4 \times 40 \times 0.636 = 35.616$

3.15 See Figure A-2.

3.17 A clipper circuit clips off a portion of a waveform.

3.19 A clamper adds a **dc** reference voltage to a waveform.

3.21 Half-wave rectifier circuit.

3.23 17.1 V or 17.8 V.

FIGURE A-2

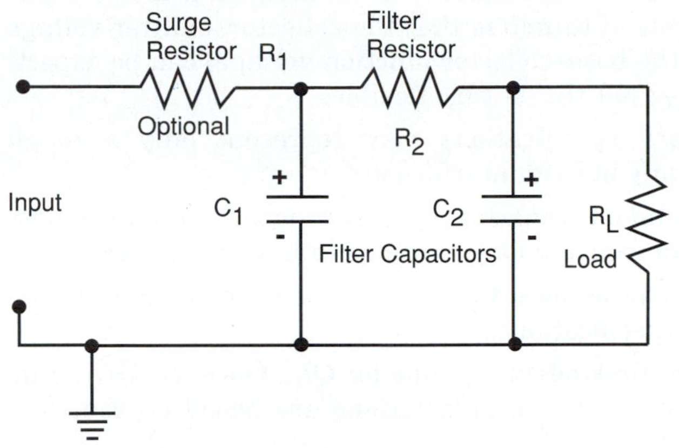

Surge Resistor R_1 Filter Resistor

Optional

Input

C_1 R_2 C_2 R_L Load

Filter Capacitors

2-stage Pi (π) Filter

FIGURE A-3

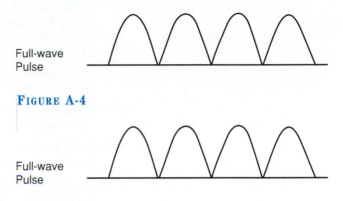

Full-wave Pulse

FIGURE A-4

Full-wave Pulse

3.25 See Figure A-3.

3.27 The capacitor will depolarize, get hot, and possibly explode.

3.29 The ripple voltage will drastically increase, because you will have a half-wave pulse instead of a full-wave pulse. The filter capacitor will not be damaged, however.

3.31 See Figure A-4.

3.33 The output waveform is a half-wave pulse.

3.35 400 V

3.37 Replace it. At best, the capacitor is leaky and on its way to total failure.

3.39 A full-wave pulse is much easier to filter and requires much smaller and less expensive filter capacitors. The extra diodes are very inexpensive compared to the cost of electrolytic capacitors.

3.41 Full-wave voltage doubler.

3.43 1 Provides a precise output voltage in spite of changing line voltages and load currents.

    2 Over 10,000 times more effective at filtering out ripple than equivalent capacitors.

    3 Very inexpensive compared to filter capacitors.

    4 Takes up very little space.

    5 Provides both overcurrent and overtemperature protection.

    6 Provides automatic shutdown for short-circuit protection.

4.1 The bipolar transistor can be connected in three basic configurations: common emitter, common collector, and common base.

4.3 The BJT's three elements are called the base, the emitter, and the collector.

4.5 In normal operation, the collector–base junction is reverse-biased, and the base–emitter junction is forward-biased.

4.7 The voltage across the collector–base junction is normally several volts, but it is different with different circuit designs; it is not a constant. A common rule of thumb is that the collector–emitter voltage (within 0.7 volt of the base–collector junction voltage) can be expected to be about $\frac{1}{2} V_{CC}$, but this is very flexible.

4.9 Most manufacturers' specifications may represent only a rough approximation for any individual transistor.

4.11 Since we can't depend on transistor specifications, we must use negative feedback to compensate for variations in key parameters.

4.13 Negative feedback can be used to correct for a wide range of variations in transistor specifications.

4.15 The circuit designer first selects a value for QI_C. Once the QI_C value is established, all other circuit calculations are based on that QI_C value.

4.17 Beta is the current gain of the transistor and has a more or less typical value of 100. Beta is the factor by which any input current is amplified to appear in the output (collector) as a larger current.

4.19 For a transistor circuit to amplify a signal properly, a specific resting (idling) collector current must be established. This resting current is called the *quiescent collector current*. The output signal current swings above and below the QI_C idle current.

4.21 The quiescent collector current is set by providing a fixed base current. If beta is 100, a 1-mA fixed base current produces a quiescent collector current of 100 mA.

4.23 Nothing can be done to keep beta from changing, but negative feedback can be used to detect and correct any error in the quiescent collector current that results from changes in beta.

4.25 Biasing.

4.27 HFE or hfe.

4.29 Beta = 100.

4.31 $V_{R_f} = QI_C \times R_f = 20 \text{ mA} \times 200 \ \Omega = 4 \text{ V}$.

4.33 $R'_e = 25/QI_C = 25/20 = 1.25 \ \Omega$.

4.35 I_C doubles.

4.37 Zero mA.

4.39 There is no change.

4.41 The emitter resistor.

4.43 $I_B = 700 \ \mu\text{A}$, $I_C = 60 \text{ mA}$.

4.45 $I_B = 500 \ \mu\text{A}$.

4.47 The quiescent collector current, QI_C.

4.49 All functional analog transistor circuits must include negative feedback. Negative feedback corrects for any differences in transistor parameters.

4.51 Negative feedback is a self-adjusting opposing voltage that maintains a constant quiescent collector current in spite of any circuit changes.

4.53 The dummy load forms a fixed half of a voltage divider. The transistor serves as the variable half of the voltage divider.

4.55 The working load is some device—a lamp, for example—that does some useful work. The dummy load produces only waste heat and a desired voltage drop. The dummy load does no useful work.

5.1 Worksheet #1:

Step 1.	2 V.
Step 2.	1.3 V.
Step 3.	1.0 mA.
Step 4.	25 Ω.
Step 5.	4.7 V.
Step 6.	3.75 mW.
Step 7.	4 V.

Worksheet #2:

Step 1.	10.4 kΩ.
Step 2.	2.5 kΩ.
Step 3.	2 kΩ.
Step 4.	3 V P–P.
Step 5.	96.
Extras	
Step 6.	3.75 W.
Step 7.	1.25 mA P–P.

5.3 Worksheet #1:

Step 1.	Not applicable.
Step 2.	10 V.
Step 3.	21.3 kΩ.
Step 4.	25 Ω.
Step 5.	5 V.
Step 6.	4 kΩ.
Step 7.	5 V.

Worksheet #2:

Step 1.	R_{B_2} [only] = 22 kΩ.
Step 2.	690 kΩ.
Step 3.	21 kΩ.
Step 4.	4 V P–P.
Step 5.	Unity.
Step 6.	4 mW.
Step 7.	1 mA P–P.

5.5 Worksheet #1:

Step 2.	10 V.
Step 3.	0.5 mA (per transistor).
Step 4.	50 Ω (each transistor).
Step 5.	3.1 V.
Step 6.	6.2 kΩ. (We don't know what the working load value is.)
Step 7.	V_{CE} = 6.9 V.

Worksheet #2:

Step 1.	R_B = 20 kΩ.
Step 2b.	R_{in} = 10 kΩ.

Step 3. $Z_{in} = 6.7$ kΩ.
Step 4. $V_{op\text{-}p\,(max)} = 3.1$ V P–P
Step 5b. $A_v = 50$
Step 6. $P_{o\,(max)} =$
Step 7. $I_{o\,(max)} =$ mA P–P

6.1 The input impedance of the driven (bypassed) transistor stage represents a working load (R_{LW}) of about 2 kΩ, which yields an optimum voltage gain of 66. If the working load is not the input impedance of another transistor stage, higher voltage gains are possible.

6.3 Because of the reflected impedance in a transformer, a fairly low impedance connected to the secondary can appear as a very high impedance to the transistor's collector circuit in the primary. The transistor sees a high-impedance working load, and produces a high voltage gain.

6.5 The tuned resonant circuit (at the resonant frequency) yields a much higher collector-load impedance than does an untuned transformer.

6.7 The dynamic resistance of a transistor can simulate a 1- to 2-megohm resistor without the excessive voltage drop a real resistor would develop. Thus, we can have very high effective collector load values for high voltage gains, and very high effective values of emitter feedback resistance for outstanding circuit stability.

6.9 Bootstrapping is a limited form of positive signal feedback that can be used to increase the effective value of input circuit resistances.

6.11 a Class A uses a single transistor in a linear circuit. It amplifies the full sine wave.
b A class B amplifier uses two transistors, one for each half of the sine wave.
c Class AB is identical to class B in operation, except that a small bias is added to prevent crossover distortion.

6.13 Crossover distortion happens in class-B operation because one transistor turns off 0.6 to 0.7 volt above the sine-wave zero value. The second transistor doesn't turn on until the sine wave goes 0.6 to 0.7 volt negative. There is a 1.2- (or 1.4-) volt gap where the amplifier does not respond to the input signal. It is a very irritating form of distortion. A small constant bias can cure the problem.

6.15 Two-stage untuned transformer-coupled, common emitter transistor amplifier.

6.17 See Figure A-5.

6.19 a The input impedance of the inverting input is drastically reduced by the Miller effect.
b The input impedance of the noninverting input is drastically increased by bootstrapping (limited positive feedback).

6.21 Class-A operation uses a single transistor, biased to some midpoint quiescent-current value. The signal causes the collector current to swing above and below the Q-point value.

FIGURE A-5

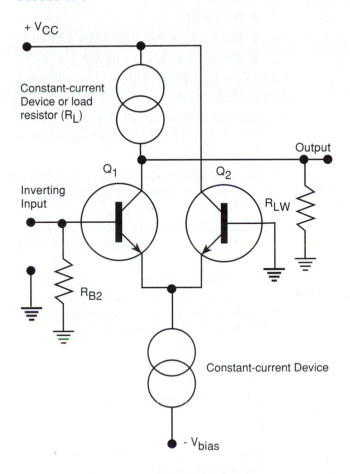

6.23 Class-C operation requires the transistor to work for only a small portion of the sine wave. The rest of the sine-wave energy must be completed by an L-C resonant circuit. Class C is only useful for single-frequency amplifiers.

6.25 Integrated circuits can't dissipate the heat developed in high-power applications.

6.27 A basic differential amplifier using Darlington-pair transistors.

6.29 A multistage differential amplifier with a constant-current device in place of the emitter feedback resistor. A common collector output stage is included (Simplified op-AMP circuit).

6.31 The circuit is a two-stage class-A power amplifier with two-stage voltage-mode negative feedback.

6.33 2.2 kΩ:

   1 Beta = 100.

   2 Beta $\times R'_e$ = 100 $\times$ 25 Ω = 2.5 kΩ.

   3 22 kΩ//2.5 kΩ = 2.2 kΩ.

6.35 5 kΩ:

    1 Beta = 100

    2 Beta × R_e' = 100 × 55.6 Ω = 5.56 kΩ

    3 47 kΩ//5.56 kΩ = 4.97 kΩ.

6.37 $QI_C = E_{bias}/R_f$ = 10 V/10 kΩ = 1.0 mA.

6.39 A_V (total) = $A_{V(1)} \times A_{V(2)}$ = 33.4 × 170.8 = 5,705.

6.41 5.3 V P–P: The lesser of V_OP–P (max) = $2QI_C \times R_{Lac}$ = 2(0.001 A) × (4270 Ω) = 8.54 V P–P and V_OP–P (max) = V_{CE} − 1; $V_{CE} = V_{CC} - V_{R_L}$ = 10 − 4.7 = 5.3 V P–P.

7.1 **a** An enhancement-mode transistor is normally off (no channel current). The transistor must be biased on by a **dc** bias, a signal, or both.

    b A depletion-mode transistor is normally on (channel current is flowing). The transistor must be biased off by a **dc** bias, a signal, or both.

7.3 Most devices conduct some quiescent channel current, so a signal can swing the channel current up or down, without added bias. An enhancement-only is normally off without bias, and a minimum threshold voltage must be applied to the gate to start channel current flowing.

7.5 Fairly low at zero bias.

7.7 The device would be below the turn-on value. The gate would have no control over channel current. Transconductance would have no meaning.

7.9 The reverse-bias voltage on the two junctions must be sufficiently high to produce depletion zones wide enough to meet in the channel to cause pinch-off.

7.11 I_{DSS} is the source-to-drain current when the gate is shorted to the source to ensure a true zero-bias condition.

7.13 gm is the amplification figure of merit for FETs. gm is called *transconductance* and is defined, mathematically, as: $gm = I_{DS}/V_{GS}$.

7.15 The figure of merit of an FET. See Problem 7.13.

7.17 $gm = gm_0 \times [1 - V_{GS}/V_{GS(off)}]$.

7.19 You could rearrange the gm equation, but consulting the manufacturer's data is a much better option.

7.21 **a** Without the bypass capacitor: $A_V = R_{Lac}/R_f$. (This should look familiar.)

    b With the bypass capacitor: $A_V = gm \times R_{Lac}$.

7.23 The input impedance is approximately equal to the value of the source-to-ground resistor. It usually has a low value compared to the dynamic channel resistance and is dominant.

7.25 $gm = 2 K \times (V_{GS} - V_T)$.

7.27 The source resistor is both a bias resistor and a feedback resistor. Any value you use for it will be a compromise between the optimum feedback and the optimum bias.

FIGURE A-6

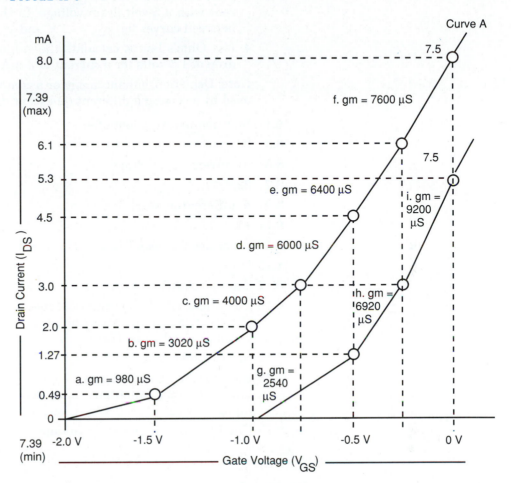

7.29 The channel resistance can be varied by changing the gate voltage, but there is no amplification. The channel behaves very much like an ordinary resistor. There is no gain, and the transconductance (*gm*) does apply.

7.31 5.7 mA (approximate).

7.33 2.5 mA (approximate).

7.35 *gm* = 1800 microsiemens (approximate).

7.37 −3.4 V.

7.39 8 mA, 5.3 mA.

7.41 V_{GS} = −0.25 to −0.5 V.

7.43 For depletion-mode operation, assume a QI_C of about 3 mA.

   1 Using the curves in Figure 7-25, find the V_{GS} (gate voltage) required for a source-to-drain current (I_{DS}) of 3 mA.

   2 Assume some drain-source voltage—ideally, somewhere near the middle of the V_{DS} (drain voltage) range. We will select 5 volts.

3 Find the curve with a V_{GS} value that will yield a 3-mA drain current with a 5-volt drain voltage. In this case, we must interpolate between curves for $V_{GS} = -1$ V and $V_{GS} = -2$ V; call it -1.5 V.

4 Use Ohm's law to calculate the value of resistor R_S for a voltage drop of 1.5 volts for a current of 3 mA. $R_S = 500\ \Omega$

Note: Depletion, enhancement, or a combination of both modes can be used by selecting a different quiescent drain-current value.

8.1 **b** A noninverting amplifier.

8.3 23.

8.5 An inverting amplifier.

8.7 22.

8.9 **d** differential amplifier.

8.11 15.

8.13 **c** voltage follower.

8.15 Unity (1).

8.17 See Figure A-7.

8.19 The Miller effect is negative feedback, since it is fed from the output of the op-amp to the inverting input. The inverting input resistance becomes nearly 0 Ω. This is often called a *virtual ground*.

8.21 See the answer to Problem 8.19.

8.23 Because the resistor values are equal, the voltage gain is unity for all inputs, yielding an output voltage that is a simple sum of the three input voltages: $V_O = 0.25$ V $+ 0.62$ V $+ 0.41$ V $= 1.28$ V.

8.25 One-half the power-supply voltage.

FIGURE A-7

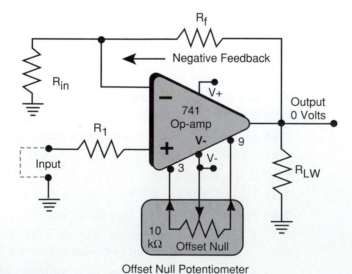

8.27 a The input circuit is not a differential amplifier.

b Because the input circuit demands a high-value series resistance, the input impedance of the Norton is as high as or higher than the input impedance of the noninverting input of the conventional op-amp.

c The output impedance of the Norton is several thousand ohms, compared to 75 ohms or so for a conventional op-amp.

8.29 All circuits have some phase shift. The amount of phase shift increases as the frequency increases. At some frequency, negative feedback is phase-shifted into positive feedback. Positive feedback causes the amplifier to become unstable and (often) to break into oscillation.

8.31 Unity gain bandwidth is the upper frequency at which the op-amp voltage gain falls to 1 (unity).

9.1 The amplifier is operated without negative feedback—open loop.

9.3 By connecting the negative power terminal to ground, making the negative rail ground (0 volts).

9.5 Only one.

9.7 10 volts.

9.9 Excellent immunity to unwanted signal noise or signal distortion.

9.11 The extra drive voltage ensures that the comparator will switch in spite of small amounts of noise, power-supply voltage variations, and other variable factors.

9.13 The time (in microseconds or nanoseconds) required for a comparator output to switch from a logic low (ground) to a logic high (+ 5 volts), or from a logic high to a logic low. Response time is used only when standard digital voltages are used as the rail voltages.

9.15 Comparator with hysteresis, window detector, and comparator ladder circuit.

9.17 The 311.

9.19 To allow some circuit to enable or disable the comparator output.

9.21 to **9.25** See Table A-2.

TABLE A-2
Answers to Problems 9.21 Through 9.25

Problem Number	Reference Voltage	Input Voltage	Output Voltage	LED On/Off
9.21	+1.5 V	−3.1 V	+10 V	Off
9.23	−3.1 V	−2.2 V	Ground	On
9.25	+6.4 V	−4.4 V	+10 V	Off

9.27 to **9.31** See Table A-3.

TABLE A-3
Answers to Problems 9.27 Through 9.31

Problem Number	Reference Voltage	Input Voltage	Output Voltage	LED On/Off
9.27	+1.5 V	−3.1 V	−10 V	On
9.29	−3.1 V	−2.2 V	+10 V	Off
9.31	+6.4 V	−4.4 V	−10 V	On

9.33 The inverting input.

9.35 +10 volts and ground.

9.37 A comparator ladder.

9.39 0.3 V   (1.2 V/4 resistors).

9.41 Current limiting resistors or the LEDs.

9.43 7.5 volts.

9.45 15 V.

9.47 3.48 volts.

9.49 9.96 volts.

FIGURE A-9

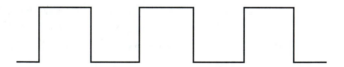

9.51 See Figure A-8.

9.53 See Figure A-9.

9.55 Window detector; 11 V, 9 V.

FIGURE A-8

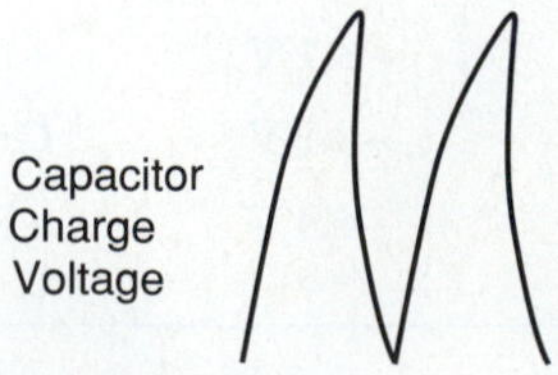

Capacitor
Charge
Voltage

9.57 Hysteresis provides a dead band that is immune to noise and prevents an instability called *jitter*. It yields a much more positive action than does a comparator without hysteresis.

9.59 Hysteresis.

10.1 Resistor-capacitor timing circuits are nonlinear. Only the first time constant is linear enough to be useful. As the capacitor charges, the charge voltage opposes the power-supply voltage, causing the charging current to decrease.

10.3 See Figure A-10.

10.5 The resistor bleeds off the unwanted capacitor-charging current that results from any op-amp offset voltage.

10.7 A constant-current source generally provides current derived from the power supply. A constant-current sink usually carries the current to ground. There is some confusion because some people use electron flow, and others use conventional current flow.

10.9 The 555 timer has two fixed trip points: one at about $\frac{1}{3}$ of the supply voltage, and one at about $\frac{2}{3}$ of the supply voltage. This ensures operation within the first time constant. The 555 always operates on the most linear part of the R-C timing curve.

10.11 A ramp generator.

10.13 The reference voltage sets the step voltage. A clock runs the analog gate on and off to determine how fast the staircase rises. Each opening and closing of the analog gate places a specific charge on the integrator's timing capacitor, causing the output to change by one step per clock pulse.

10.15 See Figure A-11.

FIGURE A-10

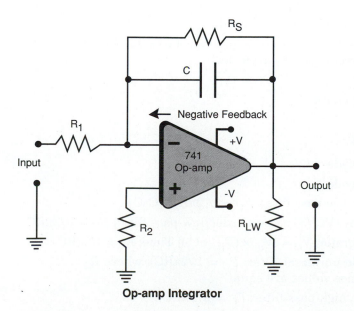

Op-amp Integrator

FIGURE A-11

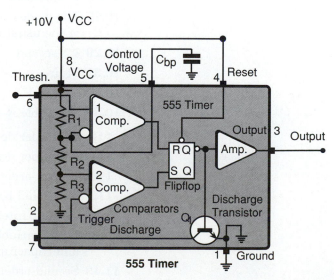

555 Timer

FIGURE A-12

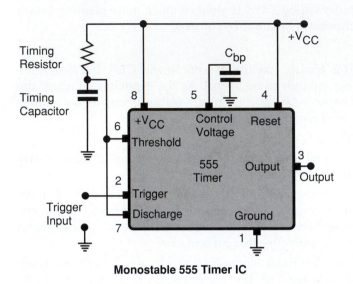

Monostable 555 Timer IC

FIGURE A-13

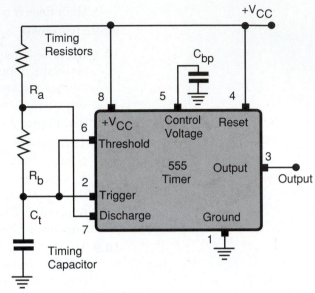

555 Astable Circuit

10.17 See Figure A-12.

10.19 See Figure A-13.

10.21 The control pin (pin 5) can be connected to a variable **dc** voltage. Changing the voltage on pin 5 changes the comparator trip points and changes the astable circuit's frequency.

10.23 The 566 is based on a hysteresis comparator square-wave generator with constant-current source and sink to charge and discharge the timing capacitor. The frequency is altered by adjusting the constant-current devices.

10.25 Monostable (one-shot) mode circuit.

10.27 The diodes enable the output to produce a true square wave instead of the usual rectangular waveform.

10.29 1.1 second

10.31 Staircase waveform.

10.33 Near 0 V.

10.35 See Table A-4.

11.1 20 dB per decade

11.3 60 dB per decade

11.5 Second-order, low-pass filter: $F_C = 1/(2\pi R_2 R_1 C_2 C_1)$.

11.7 Salen and Key (VCVS) second-order, low-pass filter: $F_C = 1(/2\pi RC)$.

11.9 Op-amp integrator: $V_O = [V_{in} \times (R_{in} \times C_f)]$ Time; $F_C = 1/2\pi RC$.

11.11 State variable universal filter: $F_C = 1(/2\pi RC)$, where $R_1 = R_3 = R_4$ and capacitance values are equal.

11.13 Second-order, high-pass filter: $F_C = 1(/2\pi RC)$.

FIGURE A-14

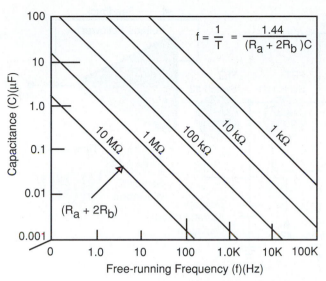

$$f = \frac{1}{T} = \frac{1.44}{(R_a + 2R_b)C}$$

Timing Nomogram for the 555 Timer Astable Mode

TABLE A-4
Table for Problem 10.35 (Assuming a Power-supply Voltage of +5 V)

Threshold Voltage	Trigger Voltage	Output Voltage
+3 V	+5 V	dead band
+5 V	+5 V	0 V
+0.2 V	+0.2 V	+5 V
0 V	0 V	+5 V
+1.8 V	+1.8 V	+5 V
+3.5 V	+3.5 V	0 V

11.15 Butterworth

11.17 Resonant circuit figure of merit: $Q = X_L/R$

11.19 d All of the above

11.21 A signal generator is an oscillator. The term signal generator is usually used to describe an oscillator designed to be used as a laboratory instrument.

11.23 a Applying the feedback signal to the amplifier's inverting input through a feedback network with a 180 degree phase shift.

b Applying the feedback signal to the amplifier's noninverting input through a feedback network with zero degree phase shift.

11.25 High Q yields better oscillator frequency stability because a high Q resonant circuit has a narrow bandwidth.

12.1 See Figure A-15.

12.3 1 To capture an instantaneous analog voltage at some specific time.

2 All analog-to-digital signal (waveform) converters must use sample-and-hold circuits.

3 When two or more signals must be time-multiplexed to share a single transmission channel.

12.5 A multiplexer switches from one input sample source to another (time multiplexing). An example of multiplexing is when two (or more) sampled telephone messages share the same phone line, seemingly carrying both messages at the same time.

12.7 Digital multiplexers are unidirectional and usually designed to operate only with +5-V (or +3-V) standard digital signals. Analog multiplexers can pass analog voltages within a wide range and are often bidirectional.

FIGURE A-15

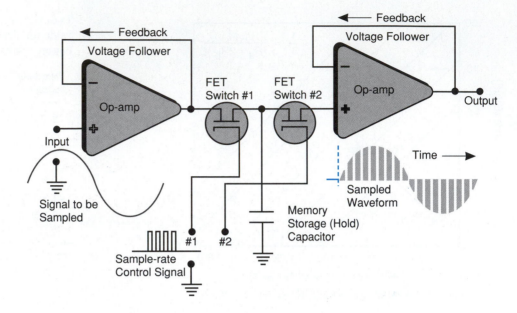

12.9 See Figure A-16.

12.11 a When the waveforms are mixed linearly, no new harmonics are produced. You will still have only a 1-kHz signal and a 1-MHz signal.

FIGURE A-16

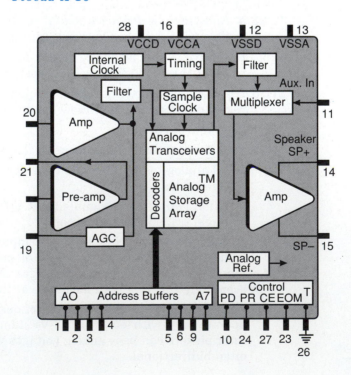

b If the signals are mixed in a nonlinear circuit, two new frequencies are produced: (1 MHz + 1 kHz) and (1 MHz − 1 kHz). The two original frequencies are still present as well.

12.13 See Figure A-17.

12.15 See the answer to Problem 12.13.

12.17 All of the modulation is carried in the two sidebands.

12.19 See Figure A-18.

12.21 The carrier frequency is far above our hearing range, and the sound signal is encoded on the carrier as carrier power variations. To hear the sound, we must extract the carrier power variations and get rid of the carrier.

FIGURE A-17

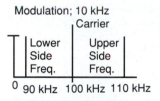

a. Spectrum Diagram

b. Waveform

FIGURE A-18

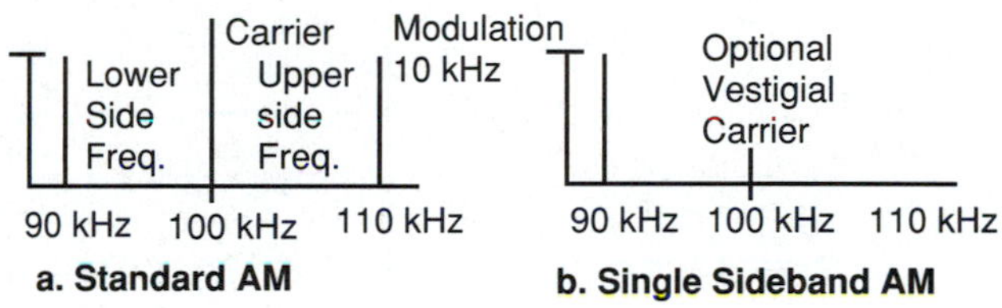

a. Standard AM

b. Single Sideband AM

12.23 FM shifts the carrier frequency in step with the modulating signal. AM varies the carrier power (amplitude) in step with the modulating signal.

12.25 See Figure A-19.

FIGURE A-19

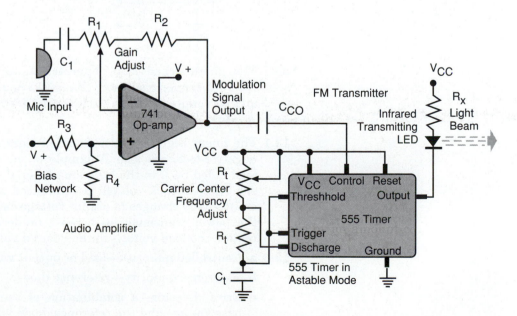

FIGURE A-20

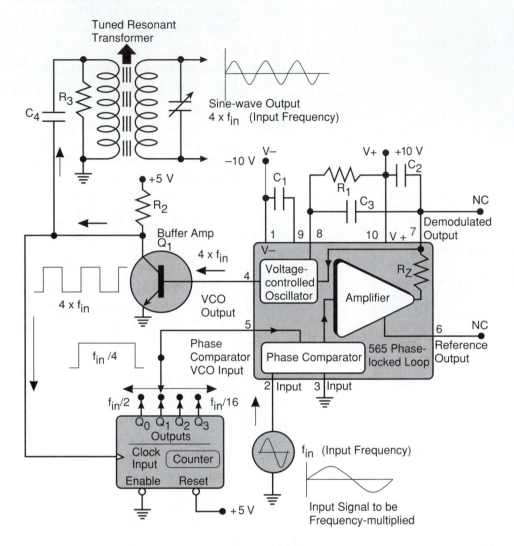

12.27 The phase-locked loop generates an error signal whenever the received carrier frequency changes. The error signal variations are an exact representation of the original modulating signal.

12.29 See Figure A-20.

13.1 The linear regulator is a modified class-A power amplifier with a negative-feedback loop. A sample of the output voltage across the load is fed back to the base small-signal driver transistor (amplifier). The feedback voltage is referenced to a reference diode in its emitter. Any changes in output voltage cause the amplifier to adjust the power transistor to increase or decrease its voltage drop to restore the load voltage to the desired value.

13.3 a Controlled quantity—load or output voltage.

  b Reference quantity—reference diode.

  c Error detector—a combination of the small-signal transistor's base voltage and the reference diode voltage in the emitter.

    d Amplifier—a small-signal driver transistor.

    e Error corrector—power transistor or power Darlington pair.

13.5 Energy is stored in the inductor's magnetic field and then is returned to the circuit when the field collapses. The voltage from the collapsing field is added to the capacitor charge voltage.

13.7 **a** Controlled quantity—the regulator output voltage.

    b Reference quantity—the triangular waveform applied to the pulse-width modulator.

    c Error detector—the pulse-width modulator's comparison of the output voltage (using a comparator) with the reference triangular waveform.

    d Amplifier—the pulse width modulator is also the amplifier.

    e Error corrector—the (power) pass transistor.

13.9 In the forward-biased SCR, two junctions are forward-biased and the middle junction is reverse-biased.

13.11 In forward-bias mode, the small gate current initiates avalanche in the middle reverse-biased junction, starting current flow through the SCR anode/cathode circuit. Once avalanche has started, the gate no longer has any influence. In the reverse-biased SCR, the middle junction is already forward-biased, and the gate has no effect on the two reverse-biased junctions. The gate is therefore inactive.

13.13 The Shockley four-layer diode and the unijunction transistor.

13.15 The DIAC—an integrated pair of back-to-back Shockley four-layer diodes.

13.17 The SCR is unidirectional and yields a half-wave pulse (maximum) in an **ac** circuit. The SCR is faster than the TRIAC and is more useful for **dc** power control. The TRIAC is designed for full-cycle **ac** at 60 Hz, and it is cheaper and easier to use in 60-Hz **ac** circuits than the SCR.

13.19 The load must have significant mechanical, thermal, or electrical inertia.

13.21 The avalanche mode is used by devices in the thyristor family. Avalanche operation involves driving a reverse-biased junction over the Zener knee into avalanche. The avalanche current must be limited (usually by the load) to a safe value for the thyristor.

13.23 Ladder schematics are often used in an industrial control environment. Originally intended for relay circuits, they have been adapted for use with thyristors and as a guide for programming programmable logic controllers.

13.25 **1** Interfacing analog voltages to standard digital voltages.

    2 Interfacing one analog circuit to another that has a different power-supply voltage.

    3 Interfacing an analog or digital circuit to a thyristor.

13.27 Light activated; diodes, transistors, Darlington transistor pairs, thyristors, and logic gates.

13.29 1. Input-to-output circuit isolation

2. Duty cycle **dc** power control

3. Where fast switching with low noise is required

13.31 All inductive devices, relay coils, motors and so on, produce a collapsing field voltage that is potentially destructive to semiconductor devices. A diode can be connected to be forward biased when the field collapses to absorb the collapsing field energy. A semiconductor diode can tolerate very high peak forward currents for a short time.

13.33 The Flash converters use a comparator ladder circuit combined with a digital binary encoder.

13.35 A summing op-amp circuit.

Blank Transistor Circuit Analysis Worksheets

You may photocopy these flowcharts (worksheets) for use in working the problems in the text. You will also find these worksheet blanks in the student *Troubleshooter's Guide.*

Current Mode Transistor Circuit Analysis Worksheet

Sheet #1: dc Voltages

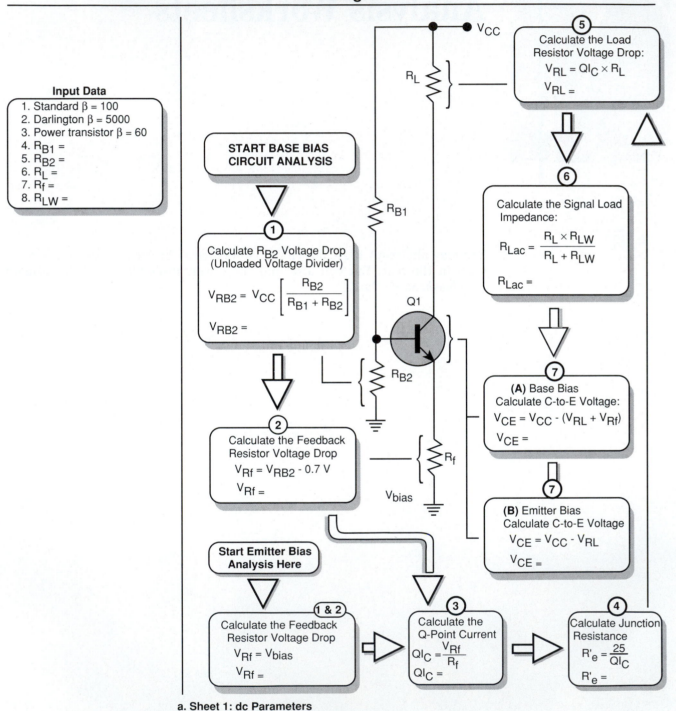

Input Data
1. Standard $\beta = 100$
2. Darlington $\beta = 5000$
3. Power transistor $\beta = 60$
4. $R_{B1} =$
5. $R_{B2} =$
6. $R_L =$
7. $R_f =$
8. $R_{LW} =$

START BASE BIAS CIRCUIT ANALYSIS

① Calculate R_{B2} Voltage Drop (Unloaded Voltage Divider)

$$V_{RB2} = V_{CC}\left[\frac{R_{B2}}{R_{B1} + R_{B2}}\right]$$

$V_{RB2} =$

② Calculate the Feedback Resistor Voltage Drop

$V_{Rf} = V_{RB2} - 0.7\ V$

$V_{Rf} =$

Start Emitter Bias Analysis Here

1 & 2 Calculate the Feedback Resistor Voltage Drop

$V_{Rf} = V_{bias}$

$V_{Rf} =$

③ Calculate the Q-Point Current

$$QI_C = \frac{V_{Rf}}{R_f}$$

$QI_C =$

④ Calculate Junction Resistance

$$R'_e = \frac{25}{QI_C}$$

$R'_e =$

V_{CC}

R_L

R_{B1}

Q1

R_{B2}

R_f

V_{bias}

⑤ Calculate the Load Resistor Voltage Drop:

$V_{RL} = QI_C \times R_L$

$V_{RL} =$

⑥ Calculate the Signal Load Impedance:

$$R_{Lac} = \frac{R_L \times R_{LW}}{R_L + R_{LW}}$$

$R_{Lac} =$

⑦ **(A)** Base Bias Calculate C-to-E Voltage:

$V_{CE} = V_{CC} - (V_{RL} + V_{Rf})$

$V_{CE} =$

⑦ **(B)** Emitter Bias Calculate C-to-E Voltage

$V_{CE} = V_{CC} - V_{RL}$

$V_{CE} =$

a. Sheet 1: dc Parameters

Current Mode Transistor Circuit Analysis Worksheet

Sheet #2: Signal Parameters

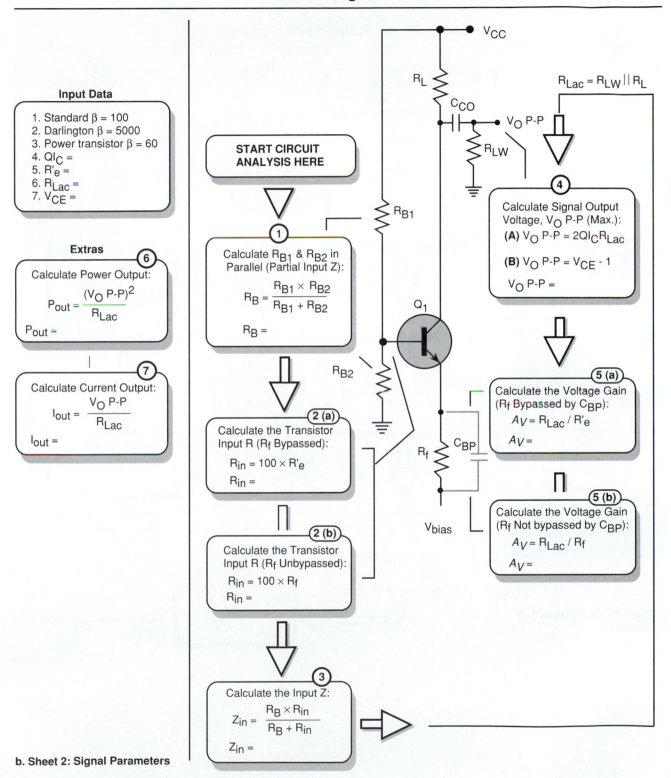

Input Data

1. Standard $\beta = 100$
2. Darlington $\beta = 5000$
3. Power transistor $\beta = 60$
4. $QI_C =$
5. $R'_e =$
6. $R_{Lac} =$
7. $V_{CE} =$

Extras

6 Calculate Power Output:

$$P_{out} = \frac{(V_O \text{ P-P})^2}{R_{Lac}}$$

$P_{out} =$

7 Calculate Current Output:

$$I_{out} = \frac{V_O \text{ P-P}}{R_{Lac}}$$

$I_{out} =$

START CIRCUIT ANALYSIS HERE

1 Calculate R_{B1} & R_{B2} in Parallel (Partial Input Z):

$$R_B = \frac{R_{B1} \times R_{B2}}{R_{B1} + R_{B2}}$$

$R_B =$

2 (a) Calculate the Transistor Input R (R_f Bypassed):

$R_{in} = 100 \times R'_e$

$R_{in} =$

2 (b) Calculate the Transistor Input R (R_f Unbypassed):

$R_{in} = 100 \times R_f$

$R_{in} =$

3 Calculate the Input Z:

$$Z_{in} = \frac{R_B \times R_{in}}{R_B + R_{in}}$$

$Z_{in} =$

$R_{Lac} = R_{LW} \| R_L$

4 Calculate Signal Output Voltage, V_O P-P (Max.):

(A) V_O P-P $= 2QI_C R_{Lac}$

(B) V_O P-P $= V_{CE} - 1$

V_O P-P $=$

5 (a) Calculate the Voltage Gain (R_f Bypassed by C_{BP}):

$A_V = R_{Lac} / R'_e$

$A_V =$

5 (b) Calculate the Voltage Gain (R_f Not bypassed by C_{BP}):

$A_V = R_{Lac} / R_f$

$A_V =$

b. Sheet 2: Signal Parameters

Voltage Mode Transistor Circuit Analysis Worksheet

Sheet #1: dc Voltage and Currents

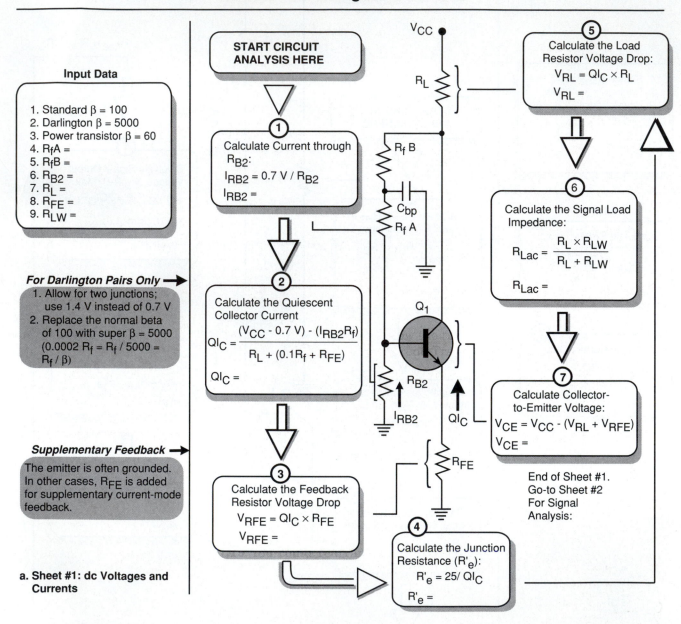

Input Data

1. Standard $\beta = 100$
2. Darlington $\beta = 5000$
3. Power transistor $\beta = 60$
4. $R_fA =$
5. $R_fB =$
6. $R_{B2} =$
7. $R_L =$
8. $R_{FE} =$
9. $R_{LW} =$

For Darlington Pairs Only →

1. Allow for two junctions; use 1.4 V instead of 0.7 V
2. Replace the normal beta of 100 with super $\beta = 5000$ ($0.0002 R_f = R_f / 5000 = R_f / \beta$)

Supplementary Feedback →

The emitter is often grounded. In other cases, R_{FE} is added for supplementary current-mode feedback.

a. Sheet #1: dc Voltages and Currents

START CIRCUIT ANALYSIS HERE

① Calculate Current through R_{B2}:
$$I_{RB2} = 0.7 \text{ V} / R_{B2}$$
$$I_{RB2} =$$

② Calculate the Quiescent Collector Current
$$QI_C = \frac{(V_{CC} - 0.7 \text{ V}) - (I_{RB2}R_f)}{R_L + (0.1R_f + R_{FE})}$$
$$QI_C =$$

③ Calculate the Feedback Resistor Voltage Drop
$$V_{RFE} = QI_C \times R_{FE}$$
$$V_{RFE} =$$

④ Calculate the Junction Resistance (R'_e):
$$R'_e = 25 / QI_C$$
$$R'_e =$$

V_{CC}

R_L

$R_f B$

C_{bp}

$R_f A$

Q_1

R_{B2}

I_{RB2} QI_C

R_{FE}

⑤ Calculate the Load Resistor Voltage Drop:
$$V_{RL} = QI_C \times R_L$$
$$V_{RL} =$$

⑥ Calculate the Signal Load Impedance:
$$R_{Lac} = \frac{R_L \times R_{LW}}{R_L + R_{LW}}$$
$$R_{Lac} =$$

⑦ Calculate Collector-to-Emitter Voltage:
$$V_{CE} = V_{CC} - (V_{RL} + V_{RFE})$$
$$V_{CE} =$$

End of Sheet #1.
Go-to Sheet #2
For Signal
Analysis:

Voltage Mode Transistor circuit Analysis Worksheet
Sheet #2: Signal Parameters

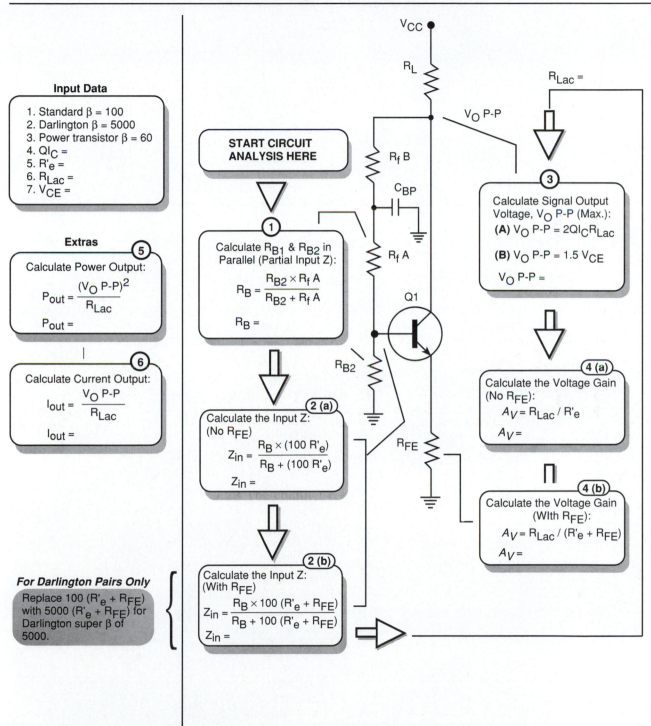

Input Data

1. Standard β = 100
2. Darlington β = 5000
3. Power transistor β = 60
4. QI_C =
5. R'_e =
6. R_{Lac} =
7. V_{CE} =

Extras

5 Calculate Power Output:
$$P_{out} = \frac{(V_O \text{ P-P})^2}{R_{Lac}}$$
P_{out} =

6 Calculate Current Output:
$$I_{out} = \frac{V_O \text{ P-P}}{R_{Lac}}$$
I_{out} =

For Darlington Pairs Only

Replace 100 (R'_e + R_{FE}) with 5000 (R'_e + R_{FE}) for Darlington super β of 5000.

START CIRCUIT ANALYSIS HERE

1 Calculate R_{B1} & R_{B2} in Parallel (Partial Input Z):
$$R_B = \frac{R_{B2} \times R_f A}{R_{B2} + R_f A}$$
R_B =

2 (a) Calculate the Input Z: (No R_{FE})
$$Z_{in} = \frac{R_B \times (100\ R'_e)}{R_B + (100\ R'_e)}$$
Z_{in} =

2 (b) Calculate the Input Z: (With R_{FE})
$$Z_{in} = \frac{R_B \times 100\ (R'_e + R_{FE})}{R_B + 100\ (R'_e + R_{FE})}$$
Z_{in} =

V_{CC}

R_L

R_{Lac} =

V_O P-P

$R_f B$

C_{BP}

$R_f A$

Q1

R_{B2}

R_{FE}

3 Calculate Signal Output Voltage, V_O P-P (Max.):
(A) V_O P-P = $2QI_C R_{Lac}$
(B) V_O P-P = $1.5\ V_{CE}$
V_O P-P =

4 (a) Calculate the Voltage Gain (No R_{FE}):
$$A_V = R_{Lac} / R'_e$$
A_V =

4 (b) Calculate the Voltage Gain (WIth R_{FE}):
$$A_V = R_{Lac} / (R'_e + R_{FE})$$
A_V =

b. Sheet #2: Signal Parameters

Glossary

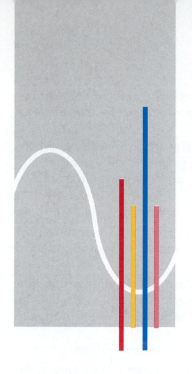

acceptors: trivalent atoms used to dope *p*-type silicon

acquisition time: the time required for a circuit to sample a signal and reach a stable condition

active: using semiconductor devices and/or vacuum tubes or other amplifying devices

adder: in an analog circuit, a summing amplifier

address buffer: the inputs of a voice message IC that control message time and analog memory location (address)

aliasing distortion: distortion caused by sampling a signal at too low a sampling rate; also called foldover distortion

alternating current (ac): electricity where the polarity constantly reverses; in a waveform: any waveform that crosses the zero reference line

amplification: the process of controlling a larger power, voltage, or current using a smaller power, voltage, or current signal

amplifier: an electronic circuit having the capability of amplification and designed specifically for that purpose

amplitude: the voltage, positive or negative, seen as the height of a waveform on an oscilloscope

amplitude modulation (AM): radio waves where the amplitude changes but the frequency is constant. Information is carried as variations in carrier wave power

analog: an electronic simulation of natural phenomenon; features infinite resolution and any waveform

analog delay: temporary storage of analog waveforms as charges in analog memory capacitors

analog gate: a gate compatible with analog voltages. Can be used to control analog or digital signals; bidirectional

analog multiplier: a circuit that can modulate and demodulate data, both analog and digital; originally used in analog computers

analog processor: any circuit able to process analog signals

analog storage array: the capacitor memory section of an electronic voice-message system

analog storage devices: capacitors used to store analog voltages

analog technology: applying to any electronic circuit that is not digital

analog-to-digital converter: converts analog voltages into the binary voltages used by digital electronics

analog transmission gates: an electronic circuit used to pass, block, or attenuate analog signals; bidirectional

AND gate: a logic gate that outputs a high only when both inputs are high (logic 1)

anode: the positive electrode

aperture time: time window to allow signal samples to enter a sample-and-hold amplifier

archetype waveforms: sine, square, rectangular, triangle, and sawtooth waveforms

Armstrong oscillator: an oscillator that uses a transformer in its feedback network to achieve the required 180-degree voltage-phase shift

astable mode: having no stable output mode

attenuation: reducing the amplitude of a signal

audio-taper potentiometer: pertaining to potentiometers with a nonlinear resistance element corresponding to the logarithmic curve of the human ear

automatic gain control: circuit used in receivers to adjust the output sound level to compensate for antenna signal strength variations

avalanche mode: when a semiconductor device is in reverse bias and breaks over and it begins to conduct electricity in the reverse direction; some devices such as SCRs and Zener diodes normally operate in this mode; for most diodes and transistors this condition is destructive

averaging amplifier: an amplifier whose output voltage is the mathematical average of two or more input voltages

balanced modulation: amplitude modulation in which the carrier frequency is cancelled-eliminated

band-pass filter: a filter that consists of a high-pass and a low-pass filter that only allows through a certain narrow band of frequencies

Barkhausen criteria: oscillation will take place if the feedback is positive, and if the amplifier gain makes up for all losses in the feedback loop

barrier zone: depletion zone

base: the control or input terminal of a bipolar junction transistor

base band: the bandwidth of the telephone lines or other system with a defined bandwidth

base spreading resistance: transistor base resistance owing to a thin base element and few available carriers

base-to-emitter junction resistance: current dependent resistance defined by the Shockley relationship

beat-frequency oscillator: an oscillator whose signal is nonlinearly mixed with another signal to produce sum and difference frequencies; heterodyning

Bessel function: all pass filter function

beta: the current gain of a transistor-measures independent of any circuit-statistical value in practice

beta-independent: where the Q-point value in a circuit is independent and not controlled by beta; negative feedback must be used to make a circuit independent of beta

bias current: base circuit idling current

biased diode clipper: a clipper that uses a **dc** voltage source to bias a diode

biasing: establishing an idling-(no signal) current or voltage value

bias voltage: the voltage applied to a transistor or other device to set a specific idling current; also forward or reverse bias of a diode

BI-FET op-amp: op-amp composed of both bipolar and field effect transistors

bistable latch: circuit that has two stable outputs; also known as a flip-flop

binary system: the number system in base-2 that is used in digital electronics; when counting in binary only two conditions are possible: 0 or 1, high or low

bipolar junction transistor: a three layer p-n-p or n-p-n, two junction transistor

blocking voltage: threshold voltage before conduction can begin

bootstrapping: using positive feedback to increase the effective value of a resistance or impedance

break point: the point on a frequency response curve where the voltage drops to 0.707 of the maximum voltage

bucket brigade device: analog delay using capacitors to store waveform samples; also called analog shift register

bulk resistance (Rb): the ohmic resistance of a forward-biased diode

Butterworth function: also called flat-flat response filter function, the most common filter frequency response curve

bypassed: using a capacitor to shunt signal frequencies around a high-value resistor

capacitor droop: capacitor charge voltage decrease resulting from leakage paths

capture range: range of frequencies or phase in which a phase-locked loop can lock on a signal

carrier: radio frequency carrier wave that is generated by the transmitter; usually carries information impressed as amplitude or frequency modulation; also electrons and holes as carriers of electricity

carrier wave: see **carrier**

cathode: the negative electrode

CB radio: radio transmitter/receiver that operates within an FCC-assigned CB frequency band

ccd charge-coupled device: analog capacitor storage medium

center-tapped transformer: a transformer with an output lead connected to the center of the secondary winding

channel pinch-off: field effect transistor in which the channel is closed and current can't flow through the channel

characteristic impedance: beyond a certain length, adding more cable has almost no effect on cable impedance

charge-coupled devices: analog storage devices based on integrated circuit silicon capacitors

charge well: an electronic "well" that allows a silicon capacitor to retain a charge for years

Chebyshev function: filter with an inductive peak; allows ringing

clamper: a circuit that adds a **dc** level to an **ac** signal; also called a **dc** restorer

class-A amplifier: an amplifier that conducts 100% of the input cycle; also called a linear amplifier

class-AB amplifier: modified class-B biased to eliminate amplifier crossover distortion

class-B amplifier: an amplifier that conducts about 50% of the input cycle

class-C amplifier: an amplifier that conducts less than 50% of the input cycle

class-D amplifier: a digital type pulse-width modulated amplifer

clipper: circuits used to eliminate a portion of an **ac** signal; also called a limiter

closed-loop voltage gain: the gain of an op-amp with a feedback loop

CMOS: a logic family made up of complementary MOS-FET's one n-channel, one p-channel

coherent light: light having only one wavelength, phase, and frequency

collector: bipolar transistor connection, generally the output terminal

collector-emitter saturation voltage VCE (Sat): collector to emitter voltage when the collector-base junction is driven into forward bias

Colpitts oscillator: sine-wave oscillator that uses a pair of tapped capacitors and an inductor to produce the 180-degree phase-shift in the feedback network

common-base circuit: BJT circuit in which the base is common to the input and output

common-collector circuit: BJT circuit in which the collector is common to the input and output

common-drain circuit: FET circuit in which the drain is common to input and output

common-emitter circuit: BJT circuit in which the emitter is common to input and output

common-gate circuit: FET circuit in which the gate is common to the input and output

common-mode voltage gain: the voltage gain of a differential amplifier, connected in the differential mode

common-source circuit: FET circuit in which the source is common to the input and output

comparator: an op-amp circuit used to compare voltages

comparator ladder: group of comparators that work together to create an LED readout display such as the digital sound output meter on a stereo

complementary: generally refers to transistor pairs, one n-p-n and one p-n-p, or one n-channel and one p-channel FET

complementary symmetry: circuit using a matched pair of *n-p-n/p-n-p* transistors

composite waveform: two or more waveforms combined to form a new waveform

constant-current device: using a collector-base junction, in reverse bias, to simulate a high value resistor with a small voltage drop

constant-current sink: constant-current device between the circuit and ground

constant-current source: constant-current device between power supply voltage and the circuit

controlled quantity: quantity to be controlled by a negative feedback circuit; voltage, current, frequency, etc.

covalent bond: an atomic bond held together by shared electrons

crossover distortion: distortion in a class-B amplifier; it occurs when neither of the transistors in the circuit is conducting near the zero part of the input waveform

current carriers: electrons and holes

current-differencing amplifier: special circuit used in the Norton op-amp

current hogging: a condition that happens when diodes are connected in parallel; only one diode gets enough voltage to turn on the junction and conduct all of the current

current mirror: specific circuit version of a constant-current device

current-mode feedback: feedback derived from current-induced voltage drop; generally the voltage drop across an emitter or source resistor

cut-off: an amplifier operating state where the output current is nearly zero

damped waveform: a sine wave whose amplitude decays with time

dangling bond: in semiconductor materials, where an electron is missing; also called a hole

Darlington pair: a pair of directly connected transistors in which the beta value is the product of the beta values of the two transistors

dB: short for decibel: the unit for measuring sound volume level; dB scale also used in frequency response curves

dc offset: the **dc** component of a sine wave above or below its normal zero reference line; also offset of an op-amp output voltage to other than zero with no signal

dc restorer: see **clamper**

decade: times ten; used in frequency response curves (as in, each decade is 10 times some previous frequency)

demodulation: extracting the original information signal from a modulated and transmitted carrier wave

depletion zone: the transition zone in semiconductor devices where there are no holes or free electrons; it occurs where *p*-type and *n*-type material come together; in silicon devices it creates a 0.6–0.7 V threshold. No conduction occurs at less than 0.6–0.7 V. Can be widened by increased reverse voltage

deviation: in frequency modulation, the deviation above and below the rest frequency

DIAC: a four-layer trigger diode for TRIAC triggering

dielectric: the insulating material in a capacitor

differential amplifier: amplifies the difference between two input voltages; the input circuit of an op-amp driven by both the inverting and noninverting inputs

differential voltage gain: voltage gain of a differential amplifier when both inputs are driven

differentiator: a circuit whose output is proportional to the rate of change of its input signal; not often used

digital amplifier: switching amplifier with standard +5 V and 0 V output levels

digital technology: having to do with digital circuits and systems

digital-to-analog converter: a circuit that converts binary voltages into analog or natural voltage values

diode: a semiconductor device that allows current flow in only one direction

diode ring balanced modulator: the original balanced modulator using diodes and two transformers; used for amplitude modulation with a suppressed carrier

DIP: short for dual-inline package

discharge transistor: an internal timer IC transistor used for discharging the external capacitor in the timing circuit

distortion: any signal information in the output of an amplifier that was not present in the input signal

distributed capacitance and inductance: capacitance and inductance spread along two or more conductor-wire lines; also present in printed circuit boards, traces, and internal integrated circuit "wiring"

donor atom: impurity atom in n-type semiconductor that supplies free electrons

doping: adding impurities to a semiconductor to make it n-type or p-type

double balanced mixer: the modern device used for frequency shifting

double side-band, suppressed carrier: amplitude modulation system in which the carrier is eliminated but both side bands are transmitted

down shifted: the process of heterodyning two frequencies to produce sum and difference frequencies. Difference frequency is said to be down shifted

drain: the FET's equivalent of a BJT's collector

drop-out voltage: the minimum voltage required by a voltage regulator to make it function, usually just a few volts higher than the output. When the voltage drops below the drop-out voltage, the regulator ceases to function

DSP: short for digital signal processor; a new technology that allows talking, animated watches and picture phones. Data compression is a key feature

dummy load: a usually resistive load that does no useful work

duty-cycle: the ratio of pulse width to cycle time, measured as a percentage of the on-cycle time to the off time

duty-cycle power control: controlling power loads by delivering power during varying portions of the **ac** cycle; simulates analog control

dynamical analogies: cross-system physics

dynamic memory: digital memory using silicon capacitors as memory elements

dynamic output resistance: the effective resistance measured as a change in current corresponding to a change in voltage

dynamic resistance: see **dynamic output resistance**

dynamic sampling error: waveform error due to waveform sampling

efficiency: ratio of useful power to the power wasted as heat

emitter: one of the terminals on a bipolar transistor, usually common to the input and output of a circuit

emitter bias: a bias circuit using a second power supply connected to the emitter

emitter-follower circuit: common collector circuit; a unity gain transistor amplifier circuit

enhancement: a normally off device; it must be turned on by bias and/or a signal

error corrector: any device used to correct an error in a negative feedback circuit

error detector: some form of comparator used to detect an error in a negative feedback system

exciter: low-power radio frequency amplifier (often used as a modulator) to drive high-power radio frequency amplifiers

Exclusive-Or GATE: a digital comparator; if only one input is high and the other low, it puts out a high; if inputs are both high, or both low, it outputs a low digital value

feedback path or loop: a signal path from the output of an amplifier back to the input

feedback voltage: voltage fed back from output to input of an amplifier; usually a percentage of the output voltage

feed-forward frequency compensation: using a capacitor across the feedback components to prevent oscillation or instability

field-effect transistor (FET): a three-terminal, voltage-controlled device used in amplification and switching applications

field emission: when a voltage is high enough to tear electrons from their shells, freeing them for conduction

figure of merit: a quality rating; details vary with particular devices

filter capacitor: a capacitor that is used to reduce the ripple in pulse-wave **dc;** usually at the output of a rectifier

filtered dc waveform: a pulsating half or full-wave pulse that has been filtered by a capacitor or more complex filter circuit

filtering: reducing the pulsating **dc** variations (ripple) of the output of a rectifier

flash converter: an analog to digital converter based on a comparator ladder

flip-flop: a bistable latch whose output depends on a pulse delivered to its set (S) or reset (R) inputs

floating device (floating circuit): a circuit not referenced to the system ground

FM sine wave: frequency modulated sine-wave carrier

FM square wave: a frequency modulated square-wave carrier

foldover distortion: see **aliasing distortion**

forward-bias condition: when a semiconductor is in the forward conducting condition; for example: when positive is applied to the anode and negative to the cathode of a diode; also when the base-emitter junction of a transistor is biased into the conducting state

forward-current transfer ratio: transistor current gain; see also beta

forward voltage: the voltage across a forward-biased *p-n* junction

fractal: a tiny pattern in natural things that carries the information to reconstruct the geometry of an entire tree, mountain range, etc.

frequency: the number of cycles per second of a waveform measured in hertz (Hz)

frequency compensation: a high-pass filter to prevent instability and oscillation in an amplifier

frequency modulated oscillator: an oscillator whose frequency is varied in step with an audio or other modulating signal

frequency modulation (FM): a system in which the carrier frequency is varied in step with an audio or other modulating signal

frequency shifting: lowering a frequency to make it compatible with another circuit; heterodyning two frequencies and using the difference frequency; also called down shifting

frequency synthesizer: a phase-locked loop circuit capable of generating many different frequencies

full-wave bridge: a four-diode full-wave rectifer, generally drawn in a diamond shape

full-wave dc pulse: the output of a full-wave **dc** rectifier

full-wave voltage doubler: a circuit that produces 2X the input voltage with a full-wave pulse output

fundamental waveform: the original sine waveform, and not a harmonic, determines the frequency of a complex waveform

fuzzy logic: a multivalued digital-integrated circuit logic that allows true, false, approximately true, and approximately false conditions; standard digital logic allows only true or false conditions

gain-bandwidth product: a constant, equal to the unity gain frequency of an op-amp, voltage gain times bandwidth

gate: the FET equivalent of a BJT base, usually the control element

glitch: an unwanted power-on pulse or other hard-to-define transient voltage

ground: the common connection of a circuit; reference point for power supply voltages

half-wave dc pulse: one half of an **ac** sine wave

half-wave rectifier circuit: outputs half-wave **dc** pulses with an **ac** sine-wave input

half-wave voltage doubler: a circuit that uses a clamper and a half-wave rectifier and filter to produce a **dc** output equal to the peak-to-peak voltage from a sine-wave input

harmonics: waveforms that are 1, 2, 3, etc. times the frequency of the original sine wave; harmonics are added to produce waveforms other than a sine wave

Hartley oscillator: an oscillator that uses a tapped inductor and a parallel tuning capacitor in its feedback network to produce the 180-degree phase-shift required for oscillation

heterodyne or beat frequency: nonlinear mixing of two frequencies. The resulting difference frequency is called the beat frequency

heterodyning: see **heterodyne** and **frequency shifting**

HFE: hybrid parameter of equivalent beta in bipolar transistors

high/low logic level: +5 volts and ground or +3 volts and ground are standard. Represents binary 0 and 1.

high-pass filter: one that passes higher frequencies, and rejects lower frequencies

holding current: the minimum value of forward current required to maintain avalanche conduction

hold settling time: time required for a sample and hold amplifier to become stable after sampling a voltage

hole: see **dangling bond**

hybrid parameters: transistor specifications that describe the operation of transistors without regard to any circuitry

hysteresis: the voltage difference between the two trip points of a comparator; the difference between turn-on and turn-off voltages in other devices

Hz: short for Hertz, the measurement of frequency; one cycle per second is one Hertz

IC: short for integrated circuit, a semiconductor device that contains a complete electronic circuit on a single semiconductor chip

incoherent beam: ordinary light beam with many frequencies, wavelengths, and phases

input impedance: the input impedance of an amplifier or related circuit or device measured while in operation; includes resistance, reactance, and opposing voltages

integration: a mathematical process involving the accumulation of a quantity over time

integrator circuit: a circuit whose output is an accumulation of input voltage over time

interface: the place where two distinctly different elements come together

interface circuit: a circuit that bridges two otherwise incompatible circuits or systems

intrinsic: a natural quality of something

intrinsic standoff ratio: in unijunction transistors, a threshold voltage established when the device is fabricated

inverter: a one input digital logic gate; if a logic high is input, the output is a logic low; if a low is input, the output is high

inverting summing amplifier: an amplifier that adds two or more voltages and outputs the inverted sum

junction: where p-type and n-type semiconductors meet; where the depletion or transition zone is formed

junction capacitance: when a diode is reverse biased, it has the properties of a capacitor and will hold a charge; junction capacitance can be varied by changing the bias voltage

junction field-effect transistor (J-FET): voltage-controlled junction field-effect transistor

junction resistance: the current dependent resistance in a semiconductor p-n junction; to calculate it use the Shockley relationship; $R'e = 25/QIc$

Kirchhoff's loop equation: using Kirchhoff's law, all series voltage drops must equal the source voltage; used to analyze series circuits

ladder notation: a form of schematic diagram used in industrial electronics and programmable logic controllers

LASER: short for Light Amplification by Stimulated Emission of Radiation; LASERS produce a coherent beam of light, with all waves at the same frequency (color) and phase

leakage current: the current that leaks through a diode when it is in the reverse-bias condition—resulting from heat-ruptured bonds

light-emitting diodes (LEDs): a type of diode that produces light when it is in the forward-biased condition; usually made of a gallium arsenide compound instead of silicon

light intensity: the amount of light per unit area that is received by a given photo detector

light-sensitive (photo) diodes: diode with an exposed junction light striking the junction increases leakage current; transistors and other devices may also be constructed to be sensitive to light

limiter: a circuit that cuts off part of a waveform above and/or below a specific voltage

linear: a straight-line graph relating two values; also faithful reproduction of a signal with no new harmonics added

linear amplifier: an amplifier that produces an output with little or no distortion of the input waveform; no new harmonics are added by the amplifier

linear taper potentiometer: one that changes resistance at a constant rate

local oscillator: an oscillator used in heterodyning as a beat-frequency oscillator

lock and capture ranges: ranges of phase and frequency within which a phase-locked loop can lock onto a frequency

log: abbreviation for logarithmic

log-tapered potentiometer: an audio-taper potentiometer with a nonlinear resistance corresponding to the sound power-level response of the human ear

lower-side frequency: the difference frequency when two frequencies are heterodyned

low-pass filter: one that accepts lower frequencies and rejects higher ones

lumped constant: equivalent amount of distributed capacitance, inductance, or impedance in a wire line, cable, or circuit board

maximum collector voltage: absolute maximum collector voltage

maximum forward dc current: maximum sustained diode current when it is conducting

maximum power dissipation: maximum power a device can dissipate safely as heat

maximum reverse dc voltage: maximum voltage without driving a junction into Zener conduction

mechanical resonant circuit: a mechanical system that behaves like an electronic resonant circuit; spring action is equal to capacitance and mass behaves like an inductor

memory capacitor: analog memory storage device

Miller effect: negative voltage-mode feedback that drastically lowers an amplifier's input impedance; also capacitance multiplier

mixers: frequency shifting ICs; nonlinear frequency mixers

modem: short for modulator/demodulator; a device that allows computers to send and receive digital data through the telephone lines as audio tones

modulation: using one signal to alter the frequency or amplitude of another signal

monostable mode: having a single stable output voltage and a temporary output voltage state

MOS-FET: Metal Oxide Semiconductor-Field Effect Transistor

multiplexer: an electronic selector switch

multistage amplifier: two or more amplifiers connected in series, cascaded

NAND gate: a logic gate that outputs a high logic voltage except when both inputs are high

narrow-band: a small range of frequencies

negative feedback: the return of a portion of the output signal to the input such that it is 180 degrees out-of-phase with the input signal

noninverting summing amplifier: an amplifier that adds two or more input voltages; the output sum voltage is not inverted

nonlinear amplifier: any amplifier that produces an output waveform that is different from the input waveform

nonlinear switching amplifier: an amplifier whose output is either the supply voltage or ground regardless of the input waveform

NOR gate: a logic gate that outputs a high when all inputs are low

Norton operational amplifier: a special low-power op-amp with a unique current differencing input circuit

notch-filter: a filter that passes all but a narrow band of frequencies

n–p–n transistor: transistor with an n-type collector and emitter and a p-type base

n-type semiconductor: semiconductor material that is doped with impurity atoms that have 5 electrons in their valence shell to add free electrons

Nyquist criterion: the minimum sampling rate of a signal without aliasing (foldover) distortion

octave: a twofold change in frequency; the frequency doubles with each octave

offset voltage: dc component that shifts the zero reference of a waveform; in op-amps, a small **dc** output voltage when both inputs are grounded

open-feedback loop: open-feedback circuit that provides no feedback

open-loop voltage gain: the gain of an amplifier with no negative feedback

operational amplifier (op-amp): a universal integrated circuit amplifier that can be used for any amplification purpose such as a pre-amp, post-amp, or a comparator

optical isolator (opto-isolator): a component that uses light instead of electricity to connect two circuits so that they are electrically isolated; they can be used to connect a high-power to a low-power circuit

order: in filters, the amount of attenuation is proportional to the number (order) of filter sections; corresponds to the number of high-pass or low-pass filter sections

OR gate: a logic gate that requires only one input to be a logic high to output a logic high

oscillation: positive feedback supplies an amplifier's input signal from its own output; if conditions are right, an oscillator produces a continuous output signal; see also **relaxation oscillator**

output impedance: consisting of amplifier internal output impedance, dummy load, and working load in real circuits

overdrive: providing excess input voltage or current to a switching amplifier

parametric: a transistor circuit design method based on the hybrid parameters

paraphase: having two outputs 180 degrees out of phase

pass-band: the band of frequencies that a band-pass filter or amplifier lets through

passive filter: a type of filter containing only passive elements such as resistors, capacitors, and inductors

peaking coil: an inductor used as a collector load to improve high frequency response

peak inverse voltage (PIV): it is the maximum momentary voltage that can be applied to a semiconductor diode before it goes into avalanche or Zener conduction; destructive

peak reverse voltage (PRV): see **peak inverse voltage (PIV)**

peak-to-peak ripple voltage: measurement of remaining voltage after half-wave or full-wave pulses have been filtered

phase-locked loop: a negative feedback control circuit in which frequency of an internal oscillator is the controlled quantity. Used for FM demodulation, frequency synthesis, and other tasks

phase-shift oscillator: sine-wave oscillator that uses a three section R-C phase-shift network to produce a 180-degree phase shift at the desired frequency

Pi: in filters; the Pi filter designation is used because the circuit as it is usually drawn resembles the Greek letter Pi; low-pass filter used in power supplies; in math, 3.14

piezo electric effect: when a voltage is applied to a crystal, it vibrates; pressure applied to a crystal produces a small voltage between the faces

Pi filter R-C network: power-supply filter network consisting of two capacitors and one resistor or inductor

pinch-off voltage: the gate-to-source voltage in an FET large enough and of the right polarity to close the source-to-drain channel

p-n junction: where n-type and p-type semiconductors join, creating a 0.6–0.7 V depletion zone and barrier voltage

p-n-p transistor: a transistor with a p-type collector and emitter and an n-type base

positive feedback: returning an in-phase signal from the output to the input, used in oscillators and bootstrapping; otherwise usually undesirable

power amplifier: an amplifier designed to produce large output powers

power dissipation: conversion of electrical power into heat

power gain: power amplification, power gain = output power/input

power product detector: circuit using a balanced modulator to demodulate an FM signal

programmable logic controllers (PLCs): used in industrial control systems; often use ladder schematics

p-type semiconductor: semiconductor material that is doped with impurity atoms that have 3 electrons in their valence shell to create holes (dangling bonds) that can accept free electrons

pulse-width modulation: a type of rectangular wave frequency modulation, it is used in certain communication applications such as fiber-optic telephone lines, TV remotes, and in switching power supplies

pure wave: a sine wave

push-pull amplifier: a type of class-B amplifier with two transistors, one for each half-cycle of the sine-wave signal

Q-point: the **dc,** (no signal), operating (idle) collector current of a transistor

Q: figure of merit for resonant circuits. $Q = XL/R$. Q is the reciprocal of damping factor (D)

quadrature: sine waves that differ by 90 degrees

quadrature oscillator: an oscillator with two outputs, each 90 degrees out-of-phase with respect to the other

quiescent collector current: the **dc** (no signal) operating (idle) collector current of a transistor amplifier (Q-point)

rail voltage: the power-supply input voltages of a comparator; special term, applies only to comparators

RAM: short for random access memory; the main read/write memory in a computer

ramp: a voltage that slopes up or down at a constant rate

R-C timing circuit: timing circuit using a resistor and capacitor to measure time

R'e: transistor junction resistance; see **Shockley relationship**

real time: the time that actually exists; natural time, as it is happening

recovery time: the time required for a device to change from an on-state to an off-state

rectangular waveform: a waveform similar to a square wave except that on and off times are not equal

rectifier: a device composed of diodes that converts **ac** into pulse-wave **dc;** usually the wave-form derived from rectifiers with a sine-wave input

reference quantity: in negative feedback control systems, a voltage used as a standard against which variations in the controlled quantity are compared

regulation: the act of maintaining a constant power-supply voltage in spite of changes in line voltage and load currents

relaxation oscillator: an oscillator based on an R-C timing circuit and a switching device with hysteresis that does not produce a sine-wave output; waveform is generally one of the archetype waveforms

relays: electromechanical devices in which an electromagnet is used to open or close electrical contacts

resistive region: in FETs, an operating mode in which the source-drain channel acts as a voltage-controlled resistance

resolution: the smallest increment measurable

reverse-bias condition: when the polarity is such that no current flows through the *p-n* junction of a semiconductor device; for example, connecting positive to the cathode and negative to the anode of a diode

reverse-bias leakage current: see **leakage current**

ringing: results from deliberate or incidental resonant circuit; the resonant circuit circulates current until it is damped out by resistive losses

ripple: the fluctuation in the **dc** output voltage of a rectifier circuit

RMS: short for root mean square; the effective voltage 0.707 X peak

roll-off: the decrease in gain of an amplifier or loss in a filter above or below certain critical frequencies

Salen & Key filter: a filter that uses positive feedback to partially simulate an inductor

sample-and-hold amplifier: an amplifier with an input storage capacitor to sample and hold an input voltage

sampling: extracting parts of a signal at regular intervals; the entire waveform can be reconstructed from the samples

saturation: when the collector-base junction becomes forward-biased

sawtooth waveform: waveform with a gradual slope up (or down) followed by an abrupt fall (or rise)

Schmitt trigger: a comparator with hysteresis provided by positive feedback

Schottky barrier diode: metal-silicon junction which has a very low junction capacitance and is used in circuits that require a fast change between one bias state and another

self-bias: in FETs, bias voltage produced by current flow through a source resistor

semiconductors: material that conducts electricity too well to be an insulator but not well enough to be a conductor; examples are: silicon, germanium, and carbon. Conduction is heat dependent

serial analog delay (SAD): also called analog shift register; uses capacitors to store analog waveform samples

Shockley four-layer diode: an avalanche device used to trigger thyristors; member of the thyristor family

Shockley relationship: an equation used to calculate the base-to-emitter junction resistance of a transistor (or diode) using a constant 25 millivolts and the current in milliamperes; Ohm's law with a constant voltage of 25 mv

side band: a range of side frequencies composed of a carrier frequency and various sum and difference frequencies when a modulating frequency varies

side frequency: the sum or difference frequency when a carrier is modulated by a constant frequency

signal: a varying voltage, containing information to be processed; see also **waveform**

signal input impedance: the total opposition to input circuit current flow, including resistance, reactance, and opposing voltages

signal parameters: signal characteristics such as signal voltage gain, signal input, and output impedances

silicon: the primary semiconductor that is used today for transistors, diodes, and other devices (also found on beaches; sand is mostly silicon dioxide)

silicon-controlled rectifier (SCR): a four-layer diode (thyristor) with a control gate to turn it on; operates in avalanche mode

sine wave: the only pure electrical wave; all electrical waves are derived from combinations of sine waves of harmonic frequencies

single-sideband AM: a form of amplitude modulation in which only one side band is radiated

slew-rate: the rate of change of the output voltage of an amplifier or comparator in response to a step voltage input

slope (ramp) converter: analog to digital converter that uses an R-C timing circuit for the analog function and a counter for the digital function

snubber diode: a diode used to short-circuit the energy from the collapsing field of an inductor; used to protect semiconductor devices

source: the FET equivalent of a BJT's emitter; also used as the source of a signal delivered to an amplifier

speed-boost: a timer IC input terminal used for very high frequencies

square wave: an electrical wave that contains an infinite number of odd harmonics; the equivalent of a switch being turned on and off at regular intervals; on time equals off time

square-wave generator: a circuit that produces a continuous stream of square waves

square-wave oscillator: same as a square-wave generator

stability factor: a stability figure of merit for bipolar transistor circuits; values between 1 and 10 result in adequate Q-point stability

staircase waveform: waveform that goes up or down in steps

standard AM: amplitude modulation in which the carrier wave and both side bands are transmitted

state-variable universal filter: a filter that performs high-pass, low-pass, and band-pass filtration all at the same time

static resistance: simple Ohm's law **dc** value

stop band: frequencies outside the pass-band

storage time: in switching transistors; the time required to clear carriers out of the depletion zone, when a transistor is switched from on to off

sum frequency: one of the two new frequencies resulting from the nonlinear mixing (heterodyning) of two frequencies; the other new frequency is the difference frequency

summing amplifier: a circuit that produces an output voltage that is proportional to the sum of the input voltages

super-beta: the product of the two beta values in a Darlington pair

superheterodyne receiver: receiver that carries out frequency shifting to produce a constant intermediate frequency regardless of station or channel frequency

swamping: using a large constant voltage to minimize the effect of small voltage changes

switched capacitor filter (SCF): a digital integrator used as an analog filter

tank circuit: resonant circuit consisting of a capacitor and an inductor

thermal energy: heat energy

thermal runaway: a regenerative more heat more leakage, more heat destructive event

threshold voltage: a minimum voltage required to initiate some action

thyristor: a family of four-layer semiconductor devices used for high-power duty cycle power control; thyristors are avalanche mode devices

time multiplexing: taking a series of samples of a waveform sending it, then replacing the missing segments at the receiver; several conversations or other data combinations can be sent along a single cable or phone line

transconductance: FET equivalent of BJT beta; the control of source-drain current by the gate-source voltage; the ratio of output current to input voltage

transistor: silicon-based amplifying devices; two types; bipolar junction transistors (BJTs) and field effect transistors (FETs)

transistor input resistance: beta times base-emitter resistance, includes $R'e$ and external emitter resistance

transition zone: see **depletion zone**

transmitter: a device that sends out an electrical signal via antenna, cable or other media; usually involves a high-frequency carrier modulated by analog or digital information

TRIAC: a four-layer bidirectional controlled rectifier used for 60-Hz power control; a member of the thyristor family

triangle waveform: a waveform that slopes up gradually, then gradually slopes downward; may be **ac** or **dc**

trigger: the activating input for some electronic device; usually a pulse

trip-point voltage: the threshold voltage needed to cause the output of a comparator to change states

TTL: transistor-transistor logic; a family of logic gates and other digital devices based on bipolar transistors

twin-T filter: a filter that consists of a high-pass and a low-pass filter with an extra capacitor and resistor; also known as a notch filter

unbypassed: without a bypass capacitor for signal frequency; see also **bypassed**

unijunction transistor: device designed to trigger thyristors; it is not a member of the thyristor family

unity-gain: a voltage gain of one

unity-gain bandwidth: the bandwidth of an amplifier with negative feedback set for a voltage gain of unity (1)

upper-side frequency: the sum frequency when two signals are mixed; nonlinearly heterodyned; see also **lower-side frequency**

vacuum tube: the first electronic amplifier devices; they have been replaced by transistors in most applications

Varactor diode: special diode used as a voltage variable capacitor; also known as a varicap

VCC: the main power terminal of a circuit or IC

VCO: voltage-controlled oscillator; an oscillator whose frequency can be varied by appying a control voltage

VCVS filter: a filter that has negative feedback supplemented with a carefully controlled positive feedback

vestigial carrier: a reduced power carrier signal sent though space to help a balanced modulator demodulate a radio carrier signal

virtual ground: the Miller effect reduces the input impedance of an op-amp's inverting input to near zero ohms—called a virtual (almost) ground

voltage divider base bias: using a voltage divider to provide the proper bias voltage for a transistor

voltage follower: a noninverting op-amp with negative feedback and a voltage gain of (1) unity

voltage gain: the ratio of output voltage to input voltage; voltage gain = output voltage/input voltage

voltage-mode feedback: negative feedback in which the feedback path is from output to inverting input

voltage multiplier: a power-supply circuit that charges capacitors in parallel and discharges them through the load in series to add the charge voltages

voltage regulator: an integrated circuit that is used with a filter capacitor to change pulse-wave **dc** into pure **dc** and maintain a constant voltage

waveform: a graphical representation of voltage changes in a circuit, often displayed on an oscilloscope

Wien-bridge R-C oscillator: often drawn showing frequency determining network in a diamond shape

window detector: circuit using two comparators, each within its own trip-point voltage

working load: load device such as a loudspeaker that is driven by an electronic circuit

Zener diode: a diode that is operated in the reverse-bias mode and is usually used as a voltage regulator

Zener knee: the part of the voltage vs. current curve of a diode when the reverse current jumps from microamperes to milliamperes or amperes

zero-crossing detector: a comparator with one input tied to ground. The comparator changes output states each time the input waveform passes through zero volts

Index